ENVIRONMENTAL SCIENCE AND ENGINEERING

J. Glynn Henry
and
Gary W. Heinke

*with contributions by other staff members
of the University of Toronto:*

Ian Burton

F. Kenneth Hare

Thomas C. Hutchinson

William J. Moroz

R. Ted Munn

O. J. C. Runnalls

Donald Mackay

Prentice Hall, Englewood Cliffs, N.J. 07632

Library of Congress Cataloging-in-Publication Data

HENRY, J. GLYNN
    Environmental science and engineering / by J. Glynn Henry and Gary
W. Heinke ; with contributions by other staff members of the
University of Toronto, Ian Burton . . . [et al.].
       p.   cm.

    Includes bibliographies and index.
  ISBN 0–13–283177–5
    1. Environmental engineering.   2. Environmental protection.
3. Human ecology.   I. Heinke, Gary W.   II. Burton, Ian.
III. Title.
TD146.H45   1989
620.8—dc19   88–15543
                                      CIP

Editorial/production supervision and
    interior design: Carolyn Fellows and Joseph Scordato
Cover design: Ben Santora
Cover Photograph: Caroline Tesiorowski
Manufacturing buyer: Mary Noonan

 © 1989 by Prentice-Hall, Inc.
A Divison of Simon & Schuster
Englewood Cliffs, New Jersey 07632

Printed in the United States of America
10 9 8 7 6 5 4 3 2 1

ISBN 0-13-283177-5

Prentice-Hall International (UK) Limited, *London*
Prentice-Hall of Australia Pty. Limited, *Sydney*
Prentice-Hall Canada Inc., *Toronto*
Prentice-Hall Hispanoamericana, S.A., *Mexico*
Prentice-Hall of India Private Limited, *New Delhi*
Prentice-Hall of Japan, Inc., *Tokyo*
Simon & Schuster Asia Pte. Ltd., *Singapore*
Editora Prentice-Hall do Brasil, Ltda., *Rio de Janerio*

Contents

not important (handwritten annotation)

11 WATER SUPPLY, *Gary W. Heinke*   372

12 WATER POLLUTION, *J. Glynn Henry*   409

do this

do this

Preface

This book is intended for a university-level introductory course in environmental sciences and engineering. Such a course is now part of the core of all civil and most chemical engineering programs and, in some cases, is followed with one or more advanced environmental courses as part of the core or elective component of the undergraduate curriculum. However, too few disciplines expose their students to environmental problems. In the opinion of the authors, it is timely that *all* engineering students take one course which helps them to understand the basic nature of the natural systems—the atmospheric, aquatic, and terrestrial systems—and how technology affects these and can be used to minimize damaging impacts. Even science students specializing in environmental matters now lack exposure to technology and would benefit by an introduction to the technology of environmental control. Thus, we believe that there is a need for the type of book we have written.

Many good textbooks on the technical aspects of environmental engineering are available for upper and graduate-level courses, but few technical books are available for an introductory environmental engineering course. A major shortcoming of these introductory books is the lack of a quantitative approach: few, if any, numerical examples and practice problems are included. We believe that it is essential for an engineering student not only to qualitatively understand a problem, but to have the ability to express the problem and the solution quantitatively.

Our book is designed for a one-semester course covering about 40 lecture hours. It will normally be taught in the second or third year of a four-year program. Students may be assigned certain sections, which need not then be covered in a lecture. A tutorial period is helpful, but not essential, since students can obtain practice in problem solving through home assignments.

After its introductory chapter, Part I deals with the basic causes of environmental problems: population and economic growth, including industrialization and urbanization (Chapter 2) and energy growth (Chapter 3). A chapter on natural environmental hazards, such as floods, droughts, earthquakes, and volcanic eruptions (Chapter 4) is included to provide a comparison with human environmental disturbances. Chapter 5 completes Part I by highlighting the interaction of many environmental problems through two case studies: the global buildup of carbon dioxide, and the regional problem of acid rain.

Part II introduces the student to the scientific background necessary for the understanding and treatment of environmental problems. It is assumed that most students will have taken first-year mathematics, physics, and chemistry courses. Some may also have had, or be taking, introductory courses in fluid mechanics and thermodynamics, which are helpful, but not essential. Chapter 6 is a review of the essentials of physics and chemistry and an introduction to material balances and reaction kinetics. It is followed by the fundamentals of climatology and meterology (Chapter 7), microbiology and epidemiology (Chapter 8) and ecology (Chapter 9). In Part II the instructor should select, from the material presented, whatever is appropriate for a particular class. Our selection of the disciplines covered in Part II was designed to complement the knowledge which most engineering students would have before taking an introductory environmental course.

Part III, on technology and control, contains six chapters on traditional engineering topics: water resources (Chapter 10), water supply (Chapter 11), water pollution (Chapter 12), air pollution (Chapter 13), solid wastes (Chapter 14), and hazardous wastes including nuclear wastes (Chapter 15). Each chapter provides information on the current state of the art and, where applicable, on the lack of knowledge and the research required to improve the situation. In the final chapter (Chapter 16), the student is introduced to the methodology of environmental impact assessment, the strategies available for improvement of the environment and the role of ethics in environmental management. Depending on the lecture time available, instructors may select only certain chapters, or parts of chapters, for study.

Illustrative examples of practical problems are used to clarify topics under discussion. Problems are provided at the end of each chapter, so that students can practice what has been preached. Part of the requirement in many of these is for the student to make reasonable assumptions for missing data, just as a practicing engineer must do. Thus, for many problems there will be several "right" answers.

The student will have to become familiar with the symbols and units of measurement in the environmental field. Consistent symbols are used throughout the book and represent the most frequent usage in the field. The most common symbols are listed in Appendix C. Others, of importance only to Chapter 7, are listed at the end of that chapter. The *Système International* (SI) *d'Unites* is used in this book. However, because North America will continue to use the American Engineering System (AES) for years to come, it is also used in the text. Appendices A-1 to A-4 provide information on the SI system, including conversion factors, examples, and a list of terms used in environmental engineering. Data throughout the book are provided in SI and AES units where appropriate, and either of these two systems may be used in particular examples and problems. We are fully aware that there is a price to be paid for using both systems in this book. But the student must obtain a feel for realistic values for many quantities. For example, how much water do people require for their daily use? To know the amount in liters and in gallons is more taxing than to know it in any one of those units alone and leads to a greater likelihood of error. However, we believe that our approach is a sensible one. If you, the instructor, disagree, it is easy to use only one system in your course and still use the book effectively.

A comment on what we omitted from the book is appropriate. At one time, potential topics for chapters, or parts thereof, included the fundamentals of fluid mechanics and thermodynamics, noise pollution, food and food protection, mineral resources, forest resources, and occupational health. The reasons for ultimately not including these topics were (a) our conviction that they are less important for achieving the objectives of the book than the material included, (b) our belief that only a limited number of topics could be presented, even in an intensive one-semester course, and (c) because we wanted to produce a textbook which could be reasonably priced.

About 60 lecture hours will be needed to cover the complete text, so instructors in a one-semester course must be selective in their choice of material. Indicated in the following table is the allotment of lecture hours for Environmental Engineering I as taught at the University of Toronto. Other suggested time allocations, depending upon the emphasis of the course, are also shown. A two-hour-per-week laboratory/tutorial was added to the University of Toronto course in 1986–87 which incorporates six laboratory assignments, six problem sets, and two field trips.

		CHAPTER	COURSE EMPHASIS & SUGGESTED LECTURE HOURS				
	No.	Brief Title	Env. Eng I	Air Pollution	Water & Wastewater	Solid & Haz. Waste	Complete Text
I	1	Problems	1	1	1	1	1
	2	Growth	2	2	2	2	3
	3	Energy	3	4	3	3	4
	4	Natural Hazards	1	1	1	—	2
	5	Disturbances	2 (9)	3 (11)	2 (9)	2 (8)	3 (13)
II	6	Chemistry	5	6	5	5	6
	7	Climatology	2	4	—	—	4
	8	Microbiology	2	2	3	2	4
	9	Ecology	2 (11)	2 (14)	3 (11)	2 (9)	3 (17)
III	10	Water Resources	2	—	3	2	3
	11	Water Supply	3	2	4	3	4
	12	Water Pollution	4	2	6	4	6
	13	Air Pollution	4	6	—	3	6
	14	Solid Waste	2	2	2	4	4
	15	Hazardous Waste	3	2	3	4	4
	16	Management	2 (20)	2 (15)	2 (20)	3 (23)	3 (30)
	TOTAL HOURS		40	40	40	40	60

Blending of ideas and writing styles in a book with 16 chapters by nine different people can be difficult. However, we chose this approach since probably none of us alone, because of lack of time and lack of expertise in all areas, would have written the book. All of the contributors have given parts of the introductory course in Environmental Engineering I taught at the University of Toronto since 1975. Most of the text material has been used since 1984 at the University of Toronto and the University of Alberta. Comments from students and colleagues were helpful and appreciated. Particular thanks are due to a number of graduate students and associates for their assistance: Messrs. Abhay Tadwalkar, Durga Prasad, Ted Bowering, Jim Sato, John Bontje, Bob Mayberry, Perrichiyappan Senthilnathan and Da-hong Li. We also wish to acknowledge the helpful comments and ideas received from our colleagues: Professors Barry Adams, Phil Byer, George Ganczarczyk, Phil Jones, Pat Seyfried, and Steve Hrudey. The typing of the final manuscript and of the several earlier drafts was the responsibility of Ms. Ampy Pural and Ms. Catherine Riendeau. Their patience and thoroughness are much appreciated. Others who assisted at times included Kathy De Lory, Ingrid Smith, Rosario Henriquez, Diana Alli, Saroj Pathak and Diane McCartney.

We were especially pleased that Dr. Albert E. Berry, Canada's most outstanding environmental engineer, was able to review the entire manuscript at an early stage. The book benefited greatly from his forthright comments and criticism. In addition to his long service in the Ontario Department of Health and as General Manager of the Ontario Water Resources Commission, he taught in both the School of Hygiene and the Department of Civil Engineering at the University of Toronto for a total of 37 years. He was also the only person ever to serve as president of both the American Waterworks Association and the Water Pollution Control Federation. His death on October 26, 1984, in his 90th year marks the end of an important era in Ontario's and Canada's efforts at environmental improvement.

Toronto, Ontario   J. Glynn Henry
Canada   Gary W. Heinke

J. GLYNN HENRY, Professor of Civil Engineering at the University of Toronto, has been on the full-time staff since 1973. From 1977 to 1986, he was Chairman of the Environmental Engineering Program, a collaborative undertaking by four graduate engineering departments. He is a graduate in civil engineering of Queen's University, Princeton University, and the University of Toronto. He spent over twenty years in the consulting engineering field as Principal and Director of R. V. Anderson Associates, Toronto, before joining the university. His responsibilities included all environmental and research activities of the firm, including the design and construction of over twenty major wastewater treatment projects. He lectures on water supply and pollution control and on advanced environmental design. His major research interests include the biological treatment of municipal wastewater and hazardous wastes. His work has resulted in over 70 technical publications and reports. He is a registered Professional Engineer in Ontario, and has been a consultant to various agencies of the Canadian Government.

GARY W. HEINKE, Professor of Civil Engineering at the University of Toronto and Dean of the Faculty of Applied Science and Engineering, has been on the full-time staff since 1968. He was Chairman of the Department from 1974–1984. He is a graduate of the University of Toronto in civil engineering and of McMaster University in chemical engineering, and spent ten years in consulting engineering in the municipal and environmental field before joining the university. He has taught an introductory environmental engineering course at Toronto for the past ten years and graduate courses in water and wastewater treatment processes. His major research interests include cold-climate environmental engineering, physical/chemical treatment, and public health engineering. His work has resulted in about 50 technical articles and reports. He is a registered Professional Engineer in Ontario and the Northwest Territories and a Fellow of the Canadian Society for Civil Engineering. He consults regularly to federal, provincial, and municipal governments, and to industry and consulting firms, with special emphasis on northern environmental work.

IAN BURTON, Professor of Geography at the University of Toronto, was Director of the Institute for Environmental Studies from 1979 to 1984. He is a graduate in geography and in water resources and resources management of the University of Birmingham and the University of Chicago. Before joining Toronto, he taught environmental courses at Indiana, Queen's, Clark, and East Anglia Universities. His major research interests include natural environmental hazards and their risk assessments. He has edited or contributed to nine books and written over 100 scholarly papers, reports, and reviews. Appointments by Canadian governments, by several universities, and by the Ford Foundation on resource management, on development of environmental programs, and on water resources planning have given him a broad perspective on environmental problems in North America, Africa, and India. Many consulting firms have employed his expertise on flood control matters.

F. KENNETH HARE, Professor of Geography, was Provost of Trinity College from 1980 to 1986 at the University of Toronto, and Director of the Institute for Environmental Studies from 1975 to 1979. He is a University Professor, the University of Toronto's highest academic honor. He was educated at the University of London (Kings and Imperial College; the London School of Economics) and the University of Montreal. He holds six honorary doctorates, the Patterson and Massey Medals, and the Patron's Medal of the Royal Geographical Society. He is a Fellow of the Royal Society of Canada and an Officer of the Order of Canada. He has served on the National Research Council (Canada), the National Environment Research Council (U.K.), and as a Director of Resources for the Future, Inc. (Washington, D.C.). A lasting interest in his wide range of atmospheric research has been the bioclimatology of the boreal forest, as well as northern climatic variation. He has studied surface energy and water balances in North America, the climatology of the desert margin, and the circulation of the north polar stratosphere. He has published about 150 papers, books, and monographs. He is Chairman of the Climate Planning Board of Canada, is active in the World Climate Programme, and was responsible for the convening and editing of the overview papers for the 1979 U.N. World Climate Conference. He himself wrote the paper on climate variability and variation. In 1976, he was the senior author of the background paper on climate prepared for the U.N. Conference on Desertification.

THOMAS C. HUTCHINSON, Professor of Botany at the University of Toronto, was Chairman of the Department from 1976 to 1982. He is also cross-appointed as Professor of Forestry and has a long standing association with the Institute for Environmental Studies. He was educated at the University of Manchester and the University of Sheffield. His major research interests include studies on effects of acid rain and heavy metals on terrestrial and aquatic ecosystems, impacts of oil spills in the arctic and physiological mechanisms by which plants have adapted to pollution stress. He has edited two books and has authored over 130 scientific articles, reports and book chapters. He organized and chaired the 1st International Conference on Heavy Metals in the Environment held at Toronto in 1979, and acted as Chief Editor for the proceedings. His knowledge of ecological stress has been sought by WHO in Europe and by the Canadian Government for studies in the Arctic.

DONALD MACKAY, Professor of Chemical Engineering and Applied Chemistry, joined the University of Toronto in 1967 after working in the petrochemical industry. All his degrees were obtained at the University of Glagow. He lectures on unit operations in Chemical Engineering and on energy and environmental issues. His major research interests include the behavior of toxic substances in the environment, the modeling of toxic organic substances in the environment including quantification of partitioning reactivity, persistence, transport and accumulation. His studies on oil spills on land and water have taken him to the Canadian Arctic and East Coast offshore regions. He has authored over 200 scientific articles and reports, contributed chapters to several books and co-edited a text on hydrocarbons in the environment. He is a registered Professional Engineer in Ontario and a Fellow of the Chemical Institute of Canada. His expertise has been sought by Canadian governments, the U.S. Environmental Protection Agency, the National Bureau of Standards and by many industrial organizations.

WILLIAM J. MOROZ, Director of the Department of Environmental Studies and Assessments at Ontario Hydro from 1980 to 1985, is Adjunct Professor in the Department of Mechanical Engineering at the University of Toronto. He was a director of an environmental consulting firm in Toronto for ten years, and was a professor at the University of Toronto and Director of the Center for Air Environment Studies at the Pennsylvania State University. He is a graduate in mechanical engineering of the University of Toronto and the University of Michigan. His main research interests are in air pollution. He has published about 40 technical papers, is a registered Professional Engineer in Ontario and Pennsylvania and a Fellow of the Royal Meteorological Society. As an advisor to the Ontario Ministry of Health, he supervised the Canadian team for an International Joint Commission study on transboundary pollution.

R. TED MUNN, Head of Environmental Programs, International Institute of Applied Systems Analysis (Laxenburg, Austria), is also an Associate of the Institute for Environmental Studies at the University of Toronto. He was educated at McMaster University (physics), the University of Toronto (meteorology), and the University of Michigan (civil engineering). He was previously Chief Scientist of the Air Quality Branch of Environment Canada, and Professor of Physics and Geography at the University of Toronto. His major accomplishments have included the design of a global environmental monitoring system, later adopted as the basis for the present world system, preparation of a WHO manual on designing urban air pollution systems for epidemiological studies, and assisting in the preparation of a Clean Air Act for São Paulo, Brazil. He is Editor-in-Chief of the annual, 1,000-page International Journal of Boundary Layer Meteorology, has edited or written six books, and has authored about 140 publications and reports.

O. JOHN C. RUNNALLS, Professor of Energy Studies in the Faculty of Applied Science and Engineering, University of Toronto was appointed to this new Chair in 1979. He was selected for this important post because of his wide experience in energy matters gained in holding senior positions with Atomic Energy of Canada Limited, Energy, Mines and Resources of the government of Canada, and Uranium Canada Limited. Since 1983 he has served also as Chairman of a new Centre for Nuclear Engineering at the University. He obtained all his degrees in engineering from the University of Toronto. His current research interests include energy systems studies, uranium supply and demand, nuclear fuel development, nuclear materials technology, and radioactive-waste management. He has published over 100 scientific and technical papers and reports. He is a registered Professional Engineer in Ontario, and a Fellow of the Royal Society of Canada.

PART 1
CAUSES OF ENVIRONMENTAL PROBLEMS

1

The Nature and Scope
of Environmental Problems

Gary W. Heinke

1.1 WHAT IS THIS BOOK ABOUT?

The objective of this book is to introduce engineering and science students to the inter-disciplinary study of environmental problems: their causes, why they are of concern, and how we can control them. The book

- Provides a description of what is meant by environment and by environmental systems
- Gives information on the basic causes of environmental disturbances
- Reviews or introduces basic scientific knowledge necessary to understand the nature of environmental problems and to be able to quantify them
- Covers the current state of the technology of environmental control in its application to water, air, and land pollution problems
- Exposes the considerable gaps in our current scientific knowledge of understanding and controlling many of the complex interactions between human activities and nature
- Points out that there are many environmental problems which could be eliminated or reduced by the application of current technology, but which are not dealt with because of society's lack of will to do so, or in many instances because of a lack of resources to do so.

1.2 SOME IMPORTANT DEFINITIONS

Where they are first used in the book, definitions are introduced in block form, as shown here, or printed in bold type.

Environment is the physical and biotic habitat which surrounds us; that which we can see, hear, touch, smell, and taste.

System, according to Webster's dictionary, is defined as "a set or arrangement of things so related or connected as to form a unit or organic whole; as, a solar system, irrigation system, supply system, the world or universe."

Pollution can be defined as an undesirable change in the physical, chemical, or biological characteristics of the air, water, or land that can harmfully affect the health, survival, or activities of humans or other living organisms.

When the goal of improving environmental quality is taken to be improving human well-being, the word "environment" broadens to include all kinds of social, economic, and cultural aspects. Such broadness is unworkable in many real situations and impractical in a textbook designed for a one-semester course. Our examination of environmental problems is therefore limited by our definition of "environment."

1.3 INTERACTION OF SYSTEMS

In later chapters of Part III, we shall deal with a number of different environmental problems associated with water, air, or land systems. Many of these problems will apply only within one of these systems, justifying the breakdown into these categories. Such a classification is also useful for easier comprehension of related problems within one system. Moreover, it is sensible because, for managerial and administrative reasons, such subfields as air pollution, water supply, wastewater disposal, and solid waste disposal are often dealt with separately by governmental agencies.

Unfortunately, many important environmental problems are not confined to an air, water, or land system, but involve interactions between systems. A current example is the acid rain problem stemming from the emission of sulfur dioxide and nitrogen oxide gases into the atmosphere from the stacks of generating stations, smelters, and automobile exhausts. These gases are then transported by air currents over wide regions. Rainfall "washes them out," creating acid rain which is harmful to aquatic life, forests, and agricultural crops. In Chapter 5, two examples of interaction between systems that cause major environmental disturbances are presented—the buildup of atmospheric carbon dioxide, a global problem, and the acid rain problem, normally of a regional nature.

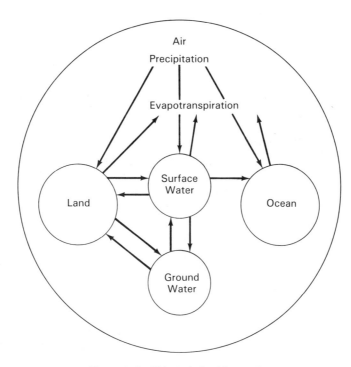

Figure 1–1 Water–air–land interactions

While many environmental problems discussed in later chapters are local or regional, and may be dealt with effectively at these levels, others must be viewed from an overall water–air–land interaction standpoint on a national, continental, or global basis. A simple illustration of this interaction is shown in Figure 1–1 and helps to explain how an insecticide like DDT is now ubiquitous.

1.4 ENVIRONMENTAL DISTURBANCES

Many major improvements to our standard of living can be attributed to the application of science and technology. A few examples are noted here. Can you think of others?

- The production of more and better quality food
- The creation of housing as protection from extremes of climate and as living space
- The building of fast and reliable means of transportation
- The invention of various systems of communication
- The invention of machines to replace human or animal power
- The supply of safe water and the disposal of wastes
- The elimination of many infectious diseases

- The elimination of most water-borne diseases in the developed world through improved water technology
- The availability of leisure time through greater productivity, providing the opportunity for cultural and recreational activities
- The protection from the worst effects of natural disasters such as floods, droughts, earthquakes, and volcanic eruptions.

With these improvements, however, have come disturbing side effects, such as lost arable land, disappearing forests, environmental pollution, and new organisms resistant to controls. Many effects originally considered to be just nuisances are now recognized as potential threats to nature and to humans. In an agrarian society, people lived essentially in harmony with nature, raising food, gathering firewood, and making clothing and tools from the land. The wastes from animals and humans were returned to the soil as fertilizer. Few, if any, problems of water, land, or air pollution occurred (Figure 1–2). For the small settlements which grew up, the supply of food, water, and other essentials and the disposal of wastes had to be kept in balance with the changing community, but no serious environmental problems were created.

The cities of ancient times, particularly those of the Roman Empire, had systems to supply water and to dispose of wastes. The aqueducts supplying the ancient city of Rome (population about 1 million) with safe water from the Apennine Mountains, and the Cloaca Maxima, the best known and one of the earliest sewers to be built, are examples of such systems. The municipal technology of ancient cities seems to have been forgotten for many centuries by those who built cities throughout Europe. Water supply and waste disposal were neglected, resulting in many outbreaks of dysentery, cholera, typhoid, and other waterborne diseases. Until the middle of the nineteenth century, it was not realized that improper waste disposal polluted water supplies with disease-carrying organisms. The industrial revolution in nineteenth-century Britain, Europe, and North America aggravated the environmental problems since it brought increased urbanization with the industrialization. Both phenomena, urbanization and industrialization, were and are fundamental causes of water and air pollution which the cities of that time were unable to handle.

Rapid advances in technology for the treatment of water and the partial treatment of wastewater took place in the developed countries over the next few decades. This led to a dramatic decrease in the incidence of waterborne diseases. Figure 1–3 illustrates the waste disposal cycle for an industrialized society. Note that all wastes discharge into the environment, and thus pollute our water, air, and land systems.

Following the Second World War the industrialized countries experienced an economic boom fueled by a burgeoning population, advanced technology, and a rapid rise in energy consumption. During the 1950s and 1960s this activity significantly increased the quantity of wastes discharged to the environment. New chemicals, including insecticides and pesticides, used without sufficient testing for their environmental and health effects, caused, and continue to cause, enormous problems not anticipated when they were introduced. Unfortunately, the problem is worsening as the variety and amounts of pollutants discharged to the environment increase inexorably while the capacity of our air, water, and land systems to assimilate wastes is limited.

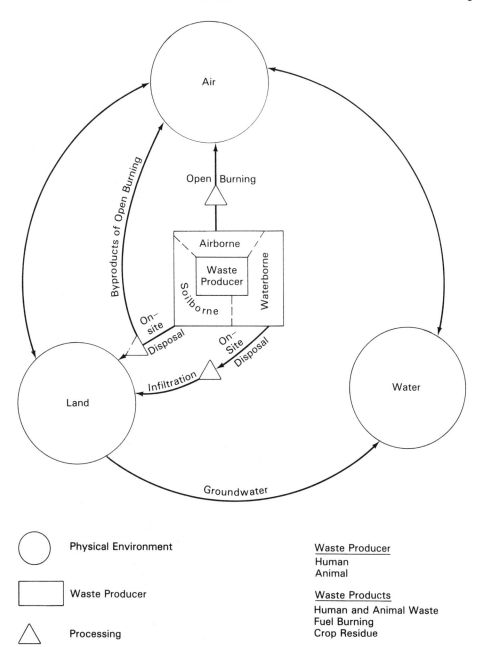

Figure 1–2   Waste cycle—agrarian society

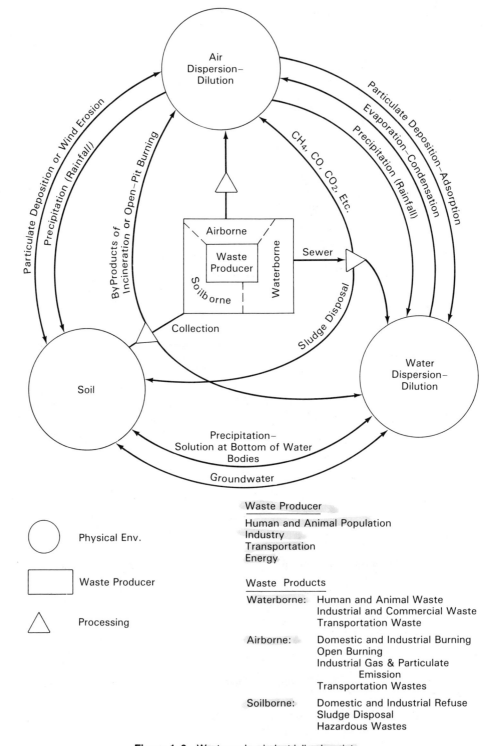

Figure 1-3 Waste cycle—industrialized society

1.5 PUBLIC AWARENESS AND ACTION

A few voices began to speak out about these new problems. Among the most effective crusaders to heighten public awareness were Rachel Carson in *Silent Spring* (1962); G. Hardin in his famous essay *The Tragedy of the Commons* (1968); Paul Ehrlich in *The Population Bomb* (1968); D. H. Meadows, *et al.*, in *The Limits to Growth* (1972); Barry Commoner in *The Closing Circle* (1971); Paul and Ann Ehrlich in *The End of Affluence* (1974); Barbara Ward and Rene Dubos in *Only One Earth, The Care and Maintenance of a Small Planet* (1972); and Erik R. Eckholm in *Losing Ground* (1976), *The Picture of Health* (1977), and *Down to Earth* (1982). These are fascinating books, available in convenient paperback form, that provide extremely important and stimulating reference reading.

Another reason why pollution came to the forefront in the United States was expressed by Goldman (1967):

> Finally public attention was directed to pollution for unusual reasons. By the mid-1960's our government had almost run out of domestic crusades to conduct. After years of battle, legislation had been adopted to contend with most of the major challenges; medicare had been approved, as well as programs in highway expansion, poverty control, urban renewal and education improvement. Pollution control was something that could evoke a similar missionary spirit among voters and politicians alike. There could be no Great Society if the water, air and dumps were dirty!

Goldman also stated that "there is reason to believe that it is only the very wealthy countries, able to afford the luxury of clean water and air, that can make a fuss about it." One may argue with this statement, but until recently, it was often cited by the delegates of developing countries at the United Nations and its agencies. They generally did not wish to heed the advice from the developed countries urging them "not to make our mistakes over again by omitting pollution controls for new industrial developments!"

In most countries of the Western world, legislation to control many aspects of pollution was introduced from the late 1960s to the late 1970s. In the United States, the agency created in 1970 to administer environmental programs is called the Environmental Protection Agency (EPA). All of the states followed by establishing environmental boards or agencies. Similar developments occurred in other western countries and to some extent in other parts of the world. An encouraging start was made, but much remains to be done. The United Nations focused on the problem by organizing a Conference on the Human Environment in 1972 in Stockholm. Later United Nations conferences dealing with population, food, women's rights, desertification, human settlements, science and technology, and the Third World continued the emphasis on environmental problems.

Eckholm (1982) describes the enormous task faced by the Third World countries:

> Reasonably clean and plentiful water, clean excreta-disposal facilities, and the practice of sanitary principles are together essential to better health; . . . yet more than half the people in the Third World (excluding China) do not have reasonable access to safe water supplies; three out of four have no adequate waste disposal facilities, not even a bucket latrine.

During the decades of the 1960's and 1970's the percentage of the Third World residents with ready access to clean water and sanitary facilities rose significantly. But as population soared, the absolute numbers lacking these necessities still climbed. Against this backdrop, the United Nations declared the 1980's the International Drinking Water Supply and Sanitation Decade. The hope—known to be hollow even as it was announced—was that Third World governments and international aid donors would drastically step up their investments in water and sanitation, providing these goods to all by 1990. Achieving this goal would require a three-fold to five-fold increase in expenditures over the 1979 investment level of $6–$7 billion, one-third of which was provided as international aid. It would also require ending the urban bias in water and sanitation spending, wider use of simple technologies, and pursuit of new forms of community involvements and education to ensure that new wells and latrines are better maintained than they have often been.

The needed funds sound large until they are compared to other global expenditures. Meeting the financial needs of the Decade would require global spending of some $80 million a day— this in a world that lays out more than $250 million a day on cigarettes and $1.4 billion a day on arms. No genuine political commitment to providing universal access to water and sanitation has emerged among aid givers or most Third World governments.

The difficult economic times of the late 1970s and the early 1980s forced changes in the priorities of the public and its governments. Inflation, unemployment, and energy became the major concerns, and understandably so. Disarmament (or better, the lack thereof), the threat of nuclear war, crime, education, medical care, family breakdown, and racial and sex-related discrimination compete for the politician's attention. The galloping increases in energy costs of the past and the enormous increases in the social costs of welfare and unemployment have caused huge financial deficits for governments of the developed world and have brought many underdeveloped countries to the brink of financial disaster. It will take extraordinary statesmanship and wisdom at national and international levels to steer us through the next decade. How high the priority for environmental improvement will be in these difficult times remains to be seen. However, it seems clear that public concerns about the health and safety aspects of toxic and hazardous wastes will continue to increase for a long time.

1.6 QUANTIFICATION OF ENVIRONMENTAL PROBLEMS

As a future engineer or scientist, it is not sufficient for you to understand the causes and effects of environmental problems in qualitative terms only. You must also be able to express the perceived problem and its potential solution in quantitative terms.

Many environmental problems are very complex. Often the problem can be divided into several components, which can be analyzed by making material or mass balances on each component, which then leads to a solution for the total system. Material balances are a very effective tool in this regard and are introduced in later chapters. Problems are provided at the end of each chapter, and illustrative examples are used where appropriate throughout the book. Both SI (Système International) and AES (American Engineering

System) units are used in the text. Conversion factors and practice problems are given in Appendices A1 through A3.

PROBLEMS

1.1. List three pollution problems with which you are personally familiar. Explain briefly why each is a problem, and describe how more than one system (air, water, land) is involved in each case.

1.2. Study the table of contents of this book, and then list those environmental problems that appear to be missing.

1.3. Find out which agencies are responsible for environmental management of (a) water supply, (b) water pollution, (c) air pollution, (d) solid wastes, and (e) hazardous wastes at (*i*) the local community level, (*ii*) the state or provincial level, and (*iii*) the federal level.

1.4. "Poor nations cannot afford the luxury of environmental control." Discuss.

1.5. You are taking this course because someone has decided it should be part of your core curriculum, or because you have decided to take it as one of your electives. What do you hope to get out of the course? Make a list of a number of points you can think of now. At the end of the course, go back to this problem and review your list. Have your expectations changed? Have they been satisfied?

REFERENCES

ASHBY, LORD ERIC. "What Price the Furbish Lousewort." *Environmental Science and Technology* 14 (1980): 1176. Also presented at the Fourth Conference on Environmental Engineering Education, June 19–20, 1981, University of Toronto.

CARSON, R. *Silent Spring*, Boston: Houghton Mifflin, 1962.

COMMONER, B. *The Closing Circle*. New York: Knopf, 1971; Bantam, 1972.

ECKHOLM, E. P. *Down to Earth: Environment and Human Needs*. New York and London: Norton, 1982.

ECKHOLM, E. P. *Losing Ground: Environmental Stress and World Food Prospects*. New York and London: Norton, 1976.

ECKHOLM, E. P. *The Picture of Health: Environmental Sources of Disease*. New York and London: Norton, 1977.

EHRLICH, P. R. *The Population Bomb*. New York: Ballantine, 1968.

EHRLICH, P. R., and EHRLICH, A. H. *The End of Affluence*. New York: Ballantine, 1974.

GOLDMAN, M. I., ed. *Controlling Pollution, The Economics of a Cleaner America*. Englewood Cliffs, N.J.: Prentice Hall, 1967.

HARDIN, G. "The Tragedy of the Commons." *Science* 162 (1968), 1243.

HOLDGATE, M. W., KASSAS, M., and WHITE, G. F. *The World Environment 1972–1982*. Dublin: Tycooly, 1983.

MEADOWS, D. H., MEADOWS, D. L., RANDERS, J., and BEHRENS, W. W. *The Limits to Growth.* New York: Universe Books, 1972. Also New York: Signet, 1972.

WARD, B. *Human Settlements, Crisis and Opportunity.* Ottawa: Information Canada, 1975.

WARD, B., and DUBOS, R. *Only One Earth, The Care and Maintenance of a Small Planet.* New York: Norton, 1972.

2

Population
and Economic Growth

Gary W. Heinke

2.1 INTRODUCTION

Until about 250 years ago, humanity existed in relatively small numbers with limited technology. Any environmental disturbances caused by people were local, and usually well within the environment's capacity to absorb them. However, in the last two centuries, four developments have occurred which have created environmental problems beyond nature's assimilative capacity.

First, there has been an explosive growth of population, creating enormous environmental pressures because of the sheer numbers of people on earth. Second, this growth, particularly in the developed countries, has been accompanied by new industrial processes, whose wastes have caused environmental damage. Third, population growth and industrialization have given rise to urbanization—the movement of people from small settlements to cities and towns—thus increasing local environmental problems, because of the high density of people and industry. Finally, the explosive growth of energy use and the introduction of many new products, particularly since World War II, have added more environmental stress.

These developments have generally had a negative, and in some areas a disastrous, impact on the physical environment. The economic successes and high standards of living of people in the urban centers of the more developed nations have been accompanied by the consumption of natural resources such as water, timber, mineral deposits, energy supplies, and land. The growing domestic and industrial demands for more products and the corresponding depletion of natural resources cannot be sustained indefinitely without severe environmental disorder.

2.2 POPULATION GROWTH

2.2.1 The Nature of Population Growth

Population growth is often characterized as exponential; that is, it increases (or decreases) by a fixed percentage of the existing total number over a unit period of time. Mathematically, this can be expressed as

$$P = P_0 e^{rt} \tag{2.1}$$

where P = the future size of the population
P_0 = the current size of the population
t = the number of years for the extrapolation
r = the assumed constant growth rate for each of the t years (as a fraction)
e = the base of natural logarithms

The growth rate r is usually expressed as a percent increase per year, or as the increase in the number of people per 1,000 population per year. Currently, the world population growth rate is approximately 1.8 percent per year, or 18 people per year per 1,000 population. For any country, the growth rate of a population is determined by four principal components: births, deaths, immigration, and emigration. Growth rate can be defined by the equation

$$r = (b - d) + (i - e) \tag{2.2}$$

where b, d, i, and e are the birth rate, death rate, immigration rate, and emigration rate, respectively, expressed as either numbers per 1,000 population per year or percent per year. The excess of births over deaths is referred to as the natural increase of population, and the difference between the number of immigrants and emigrants is called **net migration.** Another useful basis for expressing exponential growth is that of **doubling time.** Simply, the doubling time refers to the length of time necessary for the quantity being considered to double in size when growing at a constant growth rate r. An approximation used to estimate doubling time is

$$T_{db} \approx \frac{70}{r} \tag{2.3}$$

where T_{db} = the doubling time in years
r = the growth rate as a percentage per year

Figure 2–1 is a graphical presentation of the population statistics for Canada from 1851 to 1981. The illustration shows that the annual growth rate underwent tremendous

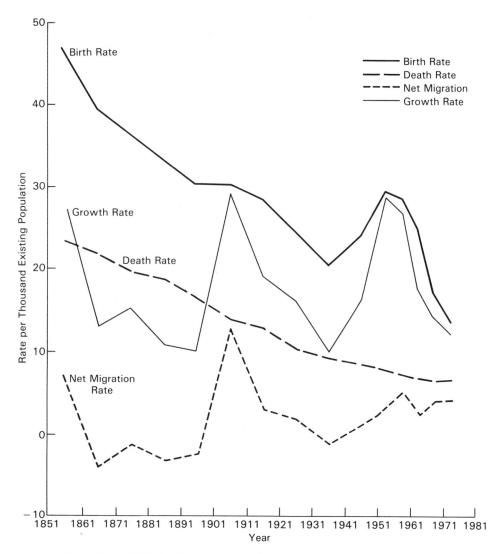

Figure 2–1 Birth rate, death rate, net migration, and growth rate for Canada: 1851–1981 per 1,000 population. Source: *Canada Yearbook*, 1978–79 (updated to 1981)

increases in the decade after 1896 and for 20 years after 1936. Since the natural increase in population prior to 1906 was declining, it is evident that a high influx of immigrants caused the jump in the growth rate between 1896 and 1906. On the other hand, the soaring growth rate between 1936 and 1956 was due to a high birth rate (the postwar baby boom), as well as another period of high immigration.

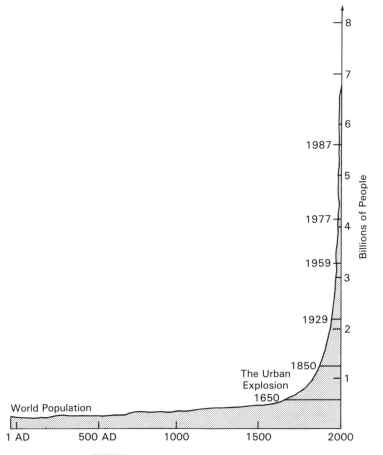

Figure 2–2   World population. Source: *Towns and Cities* (Gage Studies Series), by R. D. Bramwell. Copyright 1977, Gage Educational Publishing Limited. Reproduced by permission of the publisher.

Worldwide population trends have been outlined by Bramwell (1977):

Figure 2–2 gives a graphic picture of the growth of world population over the last two millennia. The population of the world during the Christian Era was some 300 million people. By about 1650—often referred to as the beginning of the modern era of science and technology—that population had increased to 500 million, but since then growth has been so explosive that the population in 1983 stood at about 4.7 *billion*. In other words, it took most of humanity's time on earth—perhaps 500,000 years—for the population to reach one billion, which is the estimated population in 1800. But it took only 130 years (to 1930) to add a second billion, only a further 30 years (to 1960) to add a third billion, and a mere 15 years more (to 1975) to add a fourth billion, giving an estimated world population of 4 billion in 1975. The world population has therefore grown at rates which have increased from about two percent *per thousand years* during the Paleolithic Era to *two percent per annum* in the mid-1950s—a thousandfold increase.

David Suzuki, the renowned geneticist and environmental activist, used bacteria to illustrate the impossibility of continued exponential growth of population, GNP, energy use, pollution, or in fact anything else that grows steadily in proportion to its size. He suggested (1986) imagining a test tube with a bacterial medium in it:

> At 11:00 we introduce one bacterial cell with a doubling time of one minute. So at 11:00 there is one cell, at 11:01 there are two, at 11:02 there are four, and so on until at 12:00, the tube is full. The question is when is the tube half full? The answer, of course, is at 11:59. If you were a bacterium, when would you become aware that there was a space (or population) problem? At 11:58, the tube would be $\frac{1}{4}$ full, at 11:57, $\frac{1}{8}$ full, etc. If a bacterium were to say to its mates at 11:55, "I think we've got a space problem" he'd be laughed out of the tube— any sensible bacterium could see it was 97 percent *empty*! Yet they'd only be 5 minutes away from being full.

> Suppose at 11:58, some enterprising cells got out, scoured the planet for new resources, and came back with three test tubes of food.

> That is a phenomenal find, three times the known supply! (Can you imagine how reassured we'd be if we made such an oil find?) How much time would that buy? At 12:00, the first tube would be full, at 12:01, the second would be filled and at 12:02, all four would be packed! Quadrupling the amount of food would only buy two more minutes of time if growth continued at the same rate.

2.2.2 Population Growth in More Developed and Less Developed Regions

It is instructive to examine the rapid growth period of the last 250 years in more detail, as presented in Figure 2–3. Because of the very significant differences in population growth between the more developed regions (MDR) of the world and the less developed regions (LDR), they are presented separately in the two portions of the graph. On a large-scale division of the world into these two regions, net migration between them is small compared to overall growth and can be neglected. Therefore, the overall growth rate in each of the two regions is determined as the difference between the birth rate and the death rate. It is clear from the figure that the following have occurred:

- Birth rates have dropped dramatically in the MDR from about 40 per annum per 1,000 population in about 1800 to less than 15 now. In the less developed regions, the high birth rates of about 40 per annum per 1,000 population continued until the middle of this century but have dropped sharply in the last 25 years to about 25 per annum per 1,000 population.
- Death rates have declined sharply in the MDR from about 35 in 1800 to less than 10 per annum per 1,000 population now. There are suggestions that about 10 is a minimum, and that it may increase slightly for the rest of this century because of aging population. The same decline has occurred in the LDR, but it started much later, near the beginning of this century, to reach levels approximately equal to those of the MDR now.

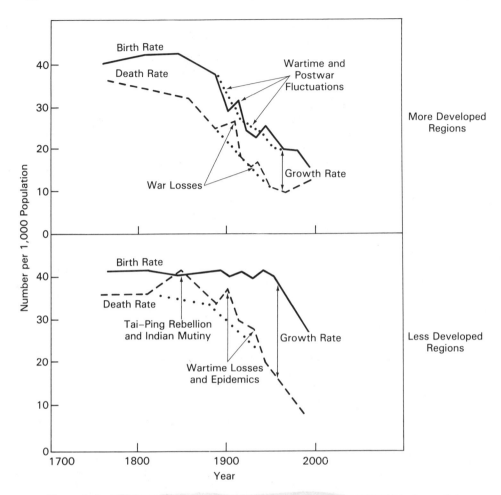

Figure 2–3 Growth rate for the more developed and less developed regions. Source: Data to 1950 from United Nations (Population Studies No. 48, 1971); data after 1950 from United Nations (Population Studies No. 78, 1981)

More Developed Regions (MDR)	Less Developed Regions (LDR)
Europe, U.S.S.R., U.S.A., Canada, Japan, Temperate South America, Australia, New Zealand.	All other areas

- Growth rates, shown graphically in Figure 2–3 as the vertical distance between the birth rate and death rate lines, have stayed about the same in the MDR for the past 200 years at about 10 per annum per 1,000 population, but with important variations, such as the effects of war and postwar baby booms. There are, of course, also important variations within the many nations making up the MDR. For the LDR, the graph shows very clearly that growth rates have increased greatly from about 5 prior to 1900 to over 20 per annum per 1,000 population for most of this century. The graph shows that it is decreasing death rates, and not increasing birth rates, which are responsible for the population growth. Improved public health measures and improved agricultural food production in the less developed regions have dramatically lowered the death rate. Consequently, historically high birth rates are no longer offset by high death rates, and the result is sharply increased growth rates. In more developed regions, declining death rates have been experienced with the advent of improvements in sanitation and medicine in the 1800s. Subsequently, there has been a lowering of birth rates brought about in part by urbanization. In the less developed regions, the decrease in birth rates did not start until after World War II. The efforts of the past 30–40 years in birth control campaigns have been successful in decreasing the birth rate. The rate of decrease in the birth rate is now approximately equal to the rate of decrease in the death rate, graphically shown as being about parallel in the figure.

It is important to understand the difference in the situation with regard to population growth in the MDR and LDR countries in order to appreciate the trends in world population growth and the related socioeconomic implications. One simple fact illustrates the point: of every 10 people living today, four live in one or the other of the two less developed countries, China and India.

Tables 2–1 and 2–2 indicate that about 70 percent of the world's population lives in the less developed regions. Declines in the more developed regions' growth rate have been outweighed by growth rates in the less developed regions, and are likely to be balanced at a constant growth rate of about 1.7 percent in the period 1975–2000. In "absolute terms," the socioeconomic implications of a constant 2.5 percent growth rate for less developed countries, such as many in South and Central America, are that in order to just maintain their present standard of living, which is much lower than in developed countries, every school, hospital, house, road, water treatment facility, market, power plant, etc., would have to be duplicated within the next 28 years. The phrase "absolute terms" is used because these projections do not account for such things as the replacement of services that have worn out or the increasing per capita consumption of goods and services. In many of the

TABLE 2–1   POPULATION DATA FOR MORE DEVELOPED AND LESS DEVELOPED REGIONS

	Population (millions) (1980)	Annual births per 1,000 population (1975–80)	Annual deaths per 1,000 population (1975–80)	Estimated growth (%) (1975–2000)	Doubling time (years) (1975)	Estimated population (millions) (2000)
MDR	1,131	15.8	9.4	0.61	115	1,272
LDR	3,301	33.0	12.1	1.98	35	4,878

Source: United Nations, *Population Studies, No. 78*, 1981.

More Developed Regions (MDR) Europe, U.S.S.R., U.S.A., Canada, Japan, Temperate South America, Australia, New Zealand

Less Developed Regions (LDR) All other areas

TABLE 2–2   WORLD POPULATION DATA

	Population (millions) (1980)	Annual births per 1,000 population (1975–80)	Growth rate* (%) (1950–1975)	Growth rate* (%) (1975–2000)	Doubling time (years) (1975)	Doubling time (years) (1973)	Estimated population (millions) (2000)
WORLD	4,432	28.5	11.4	1.6	44	35	6,119
NORTH AMERICA	249	16.3	9.1	0.94	75	87	299
Canada	24	16.2	7.7	1.31	53	61	35
United States	223	16.3	9.3	.85	82	63	260
EUROPE	484	14.5	10.5	.31	226	99	512
U.S.S.R.	265	18.3	9.0	.81	86	70	310
AFRICA	470	46	17.2	2.96	24	28	853
ASIA	2,486	32–37	12–14	1.9–2.2	32–37	30	3,583
LATIN AMERICA	364	33.6	8.9	2.25	31	25	566
OCEANIA	24	21.8	9.0	1.35	52	35	30

Source: For all data except statistics for Asia and 1973 doubling time: United Nations, *Population Studies, No. 78*, 1981.

Statistics for Asia from: *World Population*, 1977.

1973 doubling time estimates from: Population Reference Bureau (1973), *1973 World Population Data Sheet*, Washington, D.C.

* U.N. median variant.

less developed regions the scale of industrialization and technology necessary to just maintain the present standards for the rapidly increasing population simply does not exist. Slowing the growth rate in such countries is of paramount importance to the future well-being of the population.

In the most recent population trends there are indications of a slight decrease in the growth rate in LDRs and, consequently, in the world population growth rate. However, it will require some years to determine whether this decrease will be a lasting one. The average world growth rate in 2000 is estimated at 1.6 percent.

2.2.3 Population Parameters

A few of the more frequently used population parameters are as follows:

Age structure refers to the distribution of ages in the population.

Population pyramid is a graphical representation of age and sex distribution where male population is plotted on the negative (left) side of the horizontal axis and female population is shown on the other side. Figures are indicated as percent of total population occurring within each age group.

Fertility is a measure of the number of annual live births in a population.

> **Total fertility rate** is the number of children the average woman has during her lifetime.
>
> **Replacement growth** occurs when the total fertility rate is about 2.1.
>
> **Zero population growth** occurs when the death rate equals the birth rate and the net migration is zero.

Figure 2–4 shows three **population pyramids** for the 1974 populations of Mexico, the United States, and Sweden. As can be seen, a typical pyramid has the ages marked off on the vertical axis, with age zero at the origin. The male population is plotted on the negative (left) side of the horizontal axis, while the female population is shown on the positive (right) side. The scaling on the horizontal axis is in percent of total population occurring within each age classification. This percentage type of scaling, in contrast to showing total numbers, facilitates comparison between two or more countries or regions. The pyramid itself is composed of 18 horizontal bars on the male and female side, with each bar representing five-year intervals.

The leftmost pyramid in the figure shows the age structure for a rapidly growing population, that of Mexico. This pyramid, with sharply tapering sides, is typical of a population with a large base of young people. Rapid growth will occur as each large,

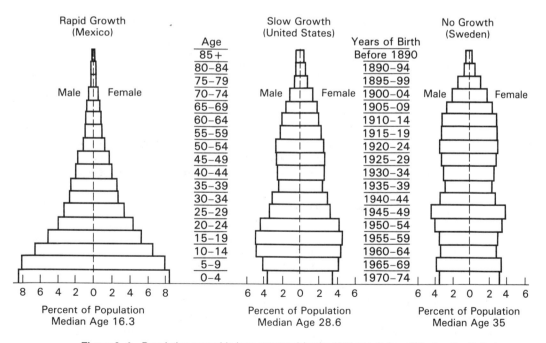

Figure 2–4   Population pyramids (age structure) for the 1974 population of Mexico, the United States, and Sweden. Source: Zito, 1979

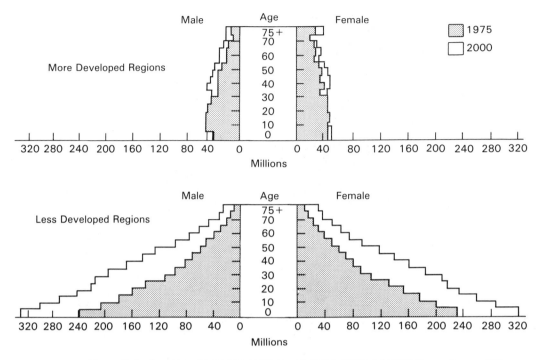

Figure 2–5 Age-sex composition of world population, 1975 and 2000. Source: Barney, 1980

young age group moves up the pyramid into its reproductive years. In sharp contrast is Sweden's pyramid, on the right, which is almost straight-sided until the oldest age groups are reached. Sweden's base age groups are no larger than those above it, implying that each reproductive generation has simply been replacing itself for so long that we do not witness a swelling in the numbers of Sweden's young people. Sweden's pyramid is characteristic of a no-growth population. The United States' pyramid in the middle represents a population whose stage of development is between that of Mexico and Sweden. Many of the United States' people are in age classes which are in their early reproductive years as a result of the growing up of the children of the World War II baby boom. Note, however, that the base age classifications are actually shrinking in size, thus making the United States' pyramid typical of a slow-growth population.

Figure 2–5 shows an illustration of the population pyramid for the world's population in 1975 and 2000, shown separately for the MDRs and LDRs. The disproportionate growth between these two regions and the implications for further social, economic, and environmental pressures in the LDRs are emphasized by this graph.

Fertility is a measure of the number of annual live births in a population, whereas the **total fertility rate** is the number of children the average woman has during her lifetime. The total fertility rate in the United States was about seven 200 years ago and has steadily fallen to about two now. Total fertility rates for selected countries and regions of the world are listed in Table 2–3.

TABLE 2–3   TOTAL FERTILITY RATES FOR
SELECTED COUNTRIES AND REGIONS

Country/Region	Total Fertility Rate 1979	
World Total	4.02	
More Developed Regions		2.03
Less Developed Regions		4.78
Africa	6.40	
East Asia	3.01	
China		3.13
Japan		1.80
South Asia	5.49	
Indonesia		4.92
Philippines		5.64
Latin America	4.92	
Haiti		5.92
Dominican Republic		5.00
North America	1.82	
Canada		1.87
United States		1.82
Europe	1.97	
Sweden		1.68
Germany		1.44
USSR	2.38	
Oceania	2.83	

Source: United Nations, *Population Studies, No. 72*, 1980.

Replacement Growth occurs at a fertility rate of about 2.1. One might expect that if each couple has two children, then the couples are just replacing themselves and no growth will occur in the population. In fact, however, this is normally not so, for two reasons. First, for most countries the population age distribution is such that more people are entering their reproductive years than are leaving them. Therefore, even if the total fertility rate—that is, the average number of children for each woman—were exactly two, there would still be more women in their fertile years than before, and therefore the population would continue to grow for several generations. Second, replacement growth requires the birth of just over two children per woman, because some females will die before they reach their reproductive years, some will be unable to have children, some will choose not to have children, and, in general, there are slightly smaller percentages of female children than male children. The exact number required for replacement growth will vary from country to country depending on the age distribution of the children and the percentage of male and female children.

From the table, one can see that such countries as the United States, Canada, Germany, and Japan are below replacement growth, the U.S.S.R. is near replacement growth, and such countries as China, Indonesia, and most African countries are above replacement growth.

Zero population growth is a term which is frequently misunderstood. Neglecting any immigration or emigration from a country, zero population growth occurs only when the death rate equals the birth rate. Replacement growth does not mean zero growth in the population, for the reasons discussed earlier, unless replacement growth rates have occurred for a long time.

2.2.4 Population Projections and Methods

Population projections are needed by an engineer or scientist for the design of facilities for a community, region, or nation and to control future environmental impacts in which population growth will play a major role. The length of time in the future for which estimates need to be made may be short term (up to 10 years) for the kinds of facilities which can be extended relatively easily after the end of the period, or long term (up to 50 years) for those facilities which would be very costly to duplicate or extend in the near future.

Both short-term and long-term population projections depend heavily on past records. They also depend heavily on the accuracy of predictions for future growth of the commercial and industrial activity of a region, which can have an important effect on the net migration to the region. Many other factors are difficult to assess, but can have important influences on population growth and long-term projections. Some of these are people's attitudes toward having children, economic conditions, wars, natural disasters, technological developments, and government attitudes. The "science" of population projection still requires a lot of crystal-ball gazing! Nevertheless, population planners have learned much in recent years, and engineers can draw on their help to obtain the best possible population forecasts for projects under consideration. In most communities and regions the local planning commissioner's office will provide this information. However, it is important for engineers to know something about the data and the assumptions behind the forecasts in order to appreciate the possibility of fluctuations in the predictions. Often the projections are given as two or more alternative estimates, such as high, medium, or low projections. It is wise to treat estimates with caution and to build into the design of facilities flexibility for changes in case actual growth is substantially higher or lower than past projections.

Graphical Methods.   Graphical projections of past population data are made to estimate future population. Projections can be made by simple graphical extension, by arithmetic or geometric extension, or by least square regression lines. Sometimes it is helpful to use growth curves of similar but larger communities for comparison in making the graphical extension. Graphical methods are simple to use and easy to apply; however, the results from different estimators may vary, depending on the experience of the persons making the projection.

Mathematical Methods.   For mathematical methods, population growth is assumed to follow a mathematical relationship. Growth may be arithmetic or exponential or follow one of the many mathematical formulations that have been proposed for various situations. A common method is to use an expression which describes an "S" curve, as shown in Figure 2–6.

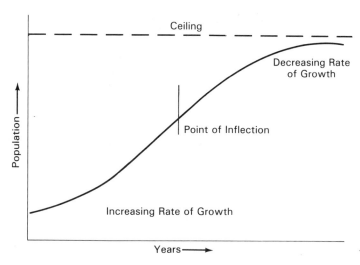

Figure 2–6 S-shaped population growth curve

Component Method.   The component method of projection is the most complex of the three, but it usually yields a very detailed picture of a population's future characteristics and numbers. Fertility and mortality trends are taken into account in this projection by breaking down the base population into age and sex components (usually five-year groupings) and then applying age- and sex-specific fertility and mortality rates to each group. The effects of immigration and emigration can be calculated as well, on the assumption of a future number of migrants. These migrants should then be subdivided into age and sex components so that appropriate fertility and mortality assumptions can be applied to them also.

The choice of method of projecting population often depends on the completeness or lack thereof of past population data. When only total numbers are available, and no information on birth and death rates or on immigration and emigration rates exists, a simple graphical or mathematical projection is the only choice. By contrast, when sufficient data are available, the component method can be employed, including the use of a sophisticated computer. A more detailed discussion of population projections is beyond the scope of this book. However, one good text on the subject is A. H. Pellard, F. Yosuf, and G. W. Pollard's *Demographic Techniques* (Sydney, Australia: Pergamon Press, 1974.)

Example 2.1

Estimate the population of the town of Waterville in the year 2000 based on the following past population data.

Year	1900	1910	1920	1930	1940	1950	1960	1970	1980	1983
Mid-year population	10,240	12,150	18,430	26,210	22,480	32,410	45,050	51,200	54,030	54,800

The town has been actively trying to increase its industrial growth, but has had only limited success over the past 10 years.

Solution   A graph of the past population data is made. Since actual populations between the 10-year census intervals are not available, one can connect the data points by an assumed linear change between censuses. One can speculate on the reasons for some of the changes, based on experience elsewhere: the drop in population during the depression of the 1930's, most likely caused by people moving away because of a lack of jobs; the rapid growth in the war and postwar boom periods of the 1940s and 1950s; and a slowdown in growth during the difficult 1970s. What about the period to the year 2000? The statement of the problem specifically mentions recent attempts, with only limited success, to attract industry. Will the 1980s be a repeat of the 1930s? Will the drive to achieve more jobs through new industries succeed? Could the rate of growth in the coming years be as high as during the 1940–60 rapid growth period?

Without further information on the particular town and region, it is not possible to specifically answer these questions. In fact no one, even with extensive information, can make more than an informed, educated guess. The best we can do are the following projections.

High Projection: Assume current growth rates will continue for five years, followed by an increasing rate of about two-thirds of the maximum previously experienced growth rate to the year 2000.

$$(1980\text{--}83) = \frac{54,800 - 54,030}{3} = 257/\text{year}$$

$$(1950\text{--}60) = \frac{45,050 - 32,410}{10} = 1,264/\text{year}$$

High population in 2000 = 54,800 + 5 × 257 + 12 × 2/3 x 1,264
= 54,800 + 1285 + 10,112 = 66,187 = 66,200

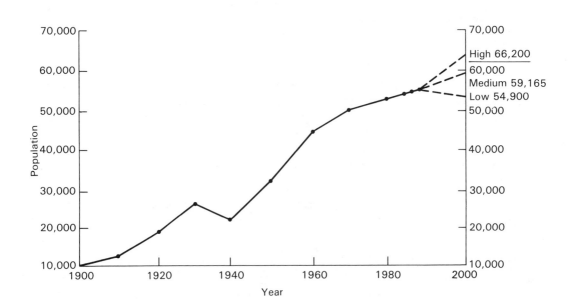

Medium Projection: Current growth rates will continue for the next 17 years.

Medium Population in 2000 = 54,800 + 17 × 257 = 59,165

Low Projection: There will be a population drop of 100 per year over the next five years, followed by a slow growth at 50 per year. The large drop of the 1930s will not be repeated because of the existence of social assistance programs keeping unemployed people in the community.

Low Population in 2000 = 54,800 − 5 × 100 + 12 × 50 = 54,900

Thus, the estimated range of population growth is from

Present population of 54,800 to   High of 66,200
                                   Medium of 59,165
                                   Low of 54,900

or a range of 11,300, or about 20% of the size of the present population. In terms of works which would fall under the purview of the town engineer, this range of population translates into about the following:

3,300 housing units

20 miles (32 km) of new roads

20 miles (32 km) of sewers and water mains

Extensions to the water treatment plant

Extensions to the waste treatment plant

Extensions to the hospital

A new high school

Several new public schools

A regional shopping center

Several neighborhood shopping centers

3 new fire halls

2 new libraries

Most importantly, the growth or lack of growth will have important financial impacts on the revenue of the town, for capital expenditures and for operating and maintenance expenditures. Therefore, what appears to be a small extension of the upper right-hand corner of the population graph has very important consequences for the future of the town.

2.2.5 Momentum of World Population Growth

A presentation of some very thought-provoking population projections for the world made by T. Frejka in 1973 is a fitting end to the discussion of material on population growth. Frejka's projections correspond to various assumptions regarding the year when a worldwide level of replacement fertility would be reached. As shown in Figure 2–7, the two lowest projections are based on replacement growth being reached by the early 1970s or early 1980s, with a corresponding stabilization of world population at 5.7 billion or 6.5 billion, respectively, about 100 years afterwards. It is now apparent that these two lowest projections

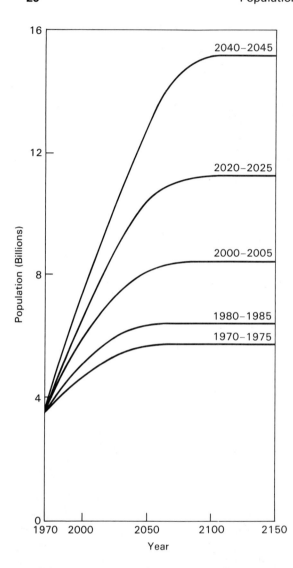

Figure 2–7 The momentum in world population growth. Source: T. Frejka's "The Prospects for a Stationary World Population," *Scientific American, Inc.*, March, 1973.

will be exceeded. This is because, while the need to restrict the birth rate seems obvious to the social planner, population growth has been fostered by social attitudes which had validity in the past and, in some parts of the world, are still valid today. Two hundred years ago, a woman was expected to have four to seven children to ensure the community's survival: half of these children would die before reaching fertility, and many women died during childbirth. But these childrearing patterns did not immediately alter with the falling death rate. In fact, only recently have social attitudes begun to change. Furthermore, the lack of social welfare legislation in developing countries influences social attitudes towards having children: in a country where few people can look forward to a pension, people expect their children to support and shelter them in their old age.

Frejka's highest projection has world population at about 15.1 billion after replacement fertility is achieved, by the year 2040. However, with world growth appearing to be slowing down (in 1976 the growth rate in 91 countries, or 70 percent of the world's population, decreased over the previous decade), a worldwide replacement fertility level is conceivable by the year 2000 or shortly thereafter. Thus, Frejka's middle projection seems most likely, and yet it would result in a stabilized world population of about 9 billion! This information has staggering implications. It means that in less than 100 years from now there will be about twice as many people in the world as there are today. To feed, house, clothe, and employ them will be a monumental task.

2.3 INDUSTRIALIZATION

For most people, the word "industrialization" is connected historically to the Industrial Revolution of the eighteenth and nineteenth centuries, with its origin in Great Britain. It is, however, a continuing phenomenon, which is still spreading globally, now affecting many of the less developed regions of the world. Because industrialization is perceived as bringing higher standards of living, it is a goal pursued by all nations.

The Industrial Revolution began in Great Britain during the eighteenth century. It was distinguished by numerous technological inventions, including, most notably, the spinning jenny and the steam engine. Such traditional trades as sewing, flour milling, brewing, and shoemaking were transformed into mechanized ventures. The real advance was in the thermodynamic conversion of heat energy into kinetic energy. This enabled mines to be pumped dry, coal and ore to be moved long distances, and machinery for many tasks to operate almost free of human or animal motive power.

The ultimate benefit of the Industrial Revolution for the individual was a higher standard of living through higher wages. The higher wages generated increased purchasing power. This meant that individuals were making more demands for products, resulting in increased resource consumption and the output of more airborne and waterborne effluents from factories. Industrialization was also accelerated in the more developed nations by exploitation of cheap labor, land, and the resources of less developed regions of the world.

2.3.1 Measures of Economic Growth and Industrialization

The most widely used economic indicator of a country's standard of living is the Gross National Product (GNP).

Gross National Product (GNP) is the sum of all personal and governmental expenditures on goods and services within a country, including the value of net exports.

Gross Environmental Improvement (GEI) is a component of the GNP that includes the costs of environmental improvements, such as money spent on reforestation or pollution control measures.

With the preceding definition of GNP in mind, it is important to recognize not only what the GNP is, but also what it is not. The GNP does not by itself reflect the country's economic health and well-being, nor does it reveal the distribution of wealth in a nation. Also, whether the environmental impacts of the goods and services consumed are beneficial or harmful is not disclosed by the GNP, nor does it indicate the extent of the depletion of natural resources. The dollar value of all goods and services is merely totaled to yield this one economic statistic, very widely used and useful, but providing an incomplete picture. Several people have argued for inclusion of resource depletion and environmental damages in an economic indicator. Economist John Hardesty* claims that in modern technological societies components of the GNP can in some way be linked to environmental destruction, and that a high GNP most likely also reflects a high rate of resource depletion. An example Hardesty cites is the North American auto industry, which accounts for possibly 10 percent of the GNP. A greater production of automobiles will help to maintain a growing GNP, but will result in more wastes from steel production, more toxic gases released from gasoline engines, and further degradation to our landscape from highways, parking lots, and scrap yards. In response to the need to quantitatively relate the GNP to environmental damage and resource depletion, a number of alternatives to the GNP have been suggested. Notable among these is a statistic called the Gross Environmental Improvement (GEI). Other proposed alternatives that include environmental effects have not been refined to the extent that they can be used in place of or in conjunction with the GNP.

The wide gap in economic growth between the rich nations of the more developed regions and the poor nations of the less developed regions becomes evident if we express the GNP on a per capita basis. Fig. 2–8 illustrates the wide range in per capita GNP in various regions of the world and relates this GNP to population growth rate. A high per capita GNP in a country means that in that country there will likely be a large number of cars, high steel production, an abundance of food, good housing, etc. It also means that the fuel for this industrialized society, namely its natural resources and those of other countries supplying them to the consuming nation, are being depleted and that the waste products of industrial production—air, water, and land pollutants—will create environmental problems. Thus, from an environmental point of view, one person living in a high-GNP country can create pressure on the environment and resources equivalent to that caused by perhaps hundreds of people in a low-GNP country. In this sense, statistics on GNP are useful comparisons between countries in determining relative pressures on the environment and on resources: the benefits and conveniences of a high standard of living in the developed countries can be translated into potential environmental damage and resource depletion through the use of per capita GNP statistics.

A word of warning, however, about making direct comparisons of GNP per capita data between countries and regions which have very different characteristics: to conclude that a person in an urbanized, developed country with a $10,000 GNP per capita figure is ten times better off than a person in a rural, underdeveloped country with a $1,000 GNP per capita figure is, of course, erroneous. Many of the needs of the urban dweller are not required by the rural inhabitants, who may also provide most of their food, clothing, and

---

* In Treshow, M., *The Human Environment* (New York: McGraw-Hill, 1976), p. 288.

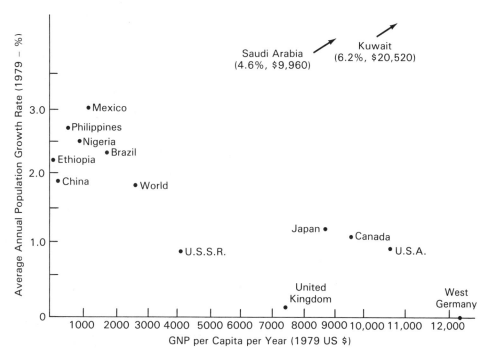

Figure 2–8 Average annual population growth rate vs. Per Capita GNP for selected Countries. Source: Adapted from 1981 World Bank Atlas: *Gross National Product,*
Population and Growth Rates

fuel by their own hands, in which case these products would not be included in the GNP.

Steel production has been acknowledged as another major index of industrial development. Henry Bessemer's discovery of the process of converting pig iron to steel by
burning off impurities by introducing air to the charge was critical to the Industrial Revolution. Prior to that, iron had been smelted by roasting ore over coke. With the advent
of steel, the construction of railways, ships, bridges, and heavy machinery, which had
previously been very limited due to the structural inadequacies of iron, was possible.

Global steel production increased dramatically from 0.5 million tons in 1870 to 28
million tons in 1900. In 1981, world steel production was approximately 700 million (metric)
tonnes. Table 2–4 is a ranking of the top 16 steel-producing countries in 1981. Notice that
all but one country, China, are considered to be more developed nations. It is not surprising
to see that the two acknowledged economic superpowers, the Soviet Union and the United
States, head the list together with Japan. However, in recent years the rate of growth of
steel production in developed countries has declined, while it has increased sharply in some
less developed countries, an indication of the spread of industrialization to the less developed
regions.

The fact that industrialization is spreading is highlighted by another industrial statistic,
the annual growth rate of industry. This statistic encompasses the mining, manufacturing,

TABLE 2–4   SIXTEEN LARGEST STEEL-PRODUCING COUNTRIES, 1981

Rank	Country	Steel production (millions of metric tons)	Share of world production, %
1 (1)	U.S.S.R.	150.0 (119.9)	21.0 (20.1)
2 (2)	U.S.	109.6 (109.2)	15.5 (19.0)
3 (3)	Japan	101.7 (88.5)	14.4 (15.4)
4 (4)	W. Germany	41.5 (40.4)	5.9 (7.0)
5 (7)	China	35.6 (18.1)	5.0 (3.2)
6 (9)	Italy	24.8 (17.4)	3.5 (3.0)
7 (6)	France	21.2 (22.8)	3.0 (4.0)
8 (8)	Belgium/Luxembourg	16.1 (17.7)	2.3 (3.1)
9 (10)	Poland	15.6 (11.5)	2.2 (2.0)
10 (5)	U.K.	15.6 (24.1)	2.2 (4.2)
11 (11)	Czechoslovakia	15.1 (11.4)	2.1 (2.0)
12 (12)	Canada	14.6 (11.1)	2.0 (1.9)
13 (15)	Romania	13.5 (6.7)	1.9 (1.2)
14 (13)	Spain	12.9 (7.7)	1.8 (1.3)
15 (16)	India	10.8 (6.7)	1.5 (1.1)
16 (38)	S. Korea	10.7 (6.45)	1.5 (0.08)

World Total: 708.3 (573.9)

(Figures for 1971 in brackets for comparison.)

Source: American Iron and Steel Institute, *Annual Statistical Report, 1981* and *Annual Statistical Report*, 1971 (Washington, D.C.: *American Iron and Steel Institute*, 1982 and 1972, respectively).

construction, electricity, and gas industries. The multiplier or ripple effect of industrialization is included in this index because such variables as the labor supply, the availability of raw materials, and the supporting infrastructure, such as road, rail, or air transport, are considered. Since these variables are related to population, consumption, and pollution, the index serves as an indirect measure of the annual increase in environmental degradation. In recent years this index has been highest in the less developed countries, ranging between 4.5 and 7.2 percent per annum. The more developed nations averaged about 3.2 percent per annum. The reason for the difference is that more developed nations already have large, well-established industrial bases and have reached a plateau in their industrial growth.

Statistics on the growth of steel production have been provided as but one important measure of industrialization. Many others could be cited, e.g., statistics regarding aluminum, cement, building materials, plastics, fertilizer, farm machinery, automobiles, and airplanes. Such statistics are provided by national and international organizations annually, or at least every few years. They generally confirm that increases in industrial production have occurred worldwide, being particularly rapid in the more developed countries in the two to three decades following World War II, and now in certain of the less developed countries.

The more developed countries have tried to pass on the lessons learned from ignoring environmental effects. However, the newly developing countries are usually under severe financial constraints, and not surprisingly, the advice is ignored. To the poorer, overpopulated countries, industrial growth with its resulting employment and higher GNP is more important than environmental protection. It is therefore to be expected that environmental damage will grow in the less developed regions as population increases and economic growth occurs.

2.3.2 Technology of Production

The post-World War II period has seen an unprecedented growth in the economy of most of the more developed countries, particularly the United States, Japan, and West Germany. In the United States this growth has occurred in many sectors of the economy—in agriculture and forestry, manufacturing, communications, transportation, the resource industries, and others. New products were mass produced—television sets, stereos, air conditioners, detergents, snowmobiles, plastics, computers, microcomputers, synthetic fiber goods, synthetic fertilizers, and insecticides. Some of these replaced goods that were less desirable. For example, detergents replaced soaps, synthetic fibers replaced wool and cotton, and synthetic fertilizers replaced compost and animal wastes as fertilizers. In some instances, the new products and the wastes from their production were later found to have been quite harmful to the environment. Numerous examples demonstrate this point. Barry Commoner, the eminent U.S. ecologist, discusses these in his book *The Closing Circle* (1972):

- The use of pesticides, particularly DDT, to control insects and thus increase agricultural production, has had very serious side effects on the wildlife and man. This led a few years ago to the banning of DDT in some countries.
- Through the concentration of livestock on small land areas where they are fed artificial foods to obtain high productivity, ''agricultural industries'' have emerged. Chicken hatcheries, broilers, and pig and beef feedlots may produce more organic wastes in a country than the domestic wastes from the people. Because of land shortage, the wastes which previously were put on the land are now often dumped untreated or poorly treated into rivers, contributing greatly to water pollution.
- The intensive use of synthetic fertilizers, particularly nitrogen, has led to high nitrate levels in surface waters and groundwaters. Nitrates cause methemoglobinemia, or ''blue babies'' disease, in infants. The problem of eutrophication of lakes is another side effect of the overuse of synthetic fertilizers.
- The production of synthetic organic chemicals as raw material for synthetic fibers, pesticides, detergents, plastics, and synthetic rubber has increased greatly. Since chlorine is frequently used in these processes, its production has increased sharply as well, requiring in turn greater production of mercury, since mercury produces chlorine. Increased release of mercury to surface waters, however, has resulted, via the food chain process, in high concentrations of mercury in fish and mercury poisoning of people who eat a steady diet of such fish. Synthetic products also require high energy for their production, and creation of this energy further contributes to environmental pressures.

- The vast increase in the number of automobiles (and until recently of high-powered automobiles), together with the shift in the transportation of goods from rail to trucks, has significantly increased air pollution problems.
- The introduction of the nonreturnable bottle and the throwaway can has greatly increased the solid waste disposal problem and the litter problem.
- The emergence of food packaging, as well as other kinds of packaging, has generated high quantities of solid wastes to be collected and disposed of.
- The enormous increase in the production of electric power has provided a growing source of pollution problems. Sulfur dioxide (SO_2) and various nitrogen oxide (NO_x) emissions from power plants are major contributors to acid rain. Potential radioactive emissions from the operation of nuclear power plants, together with the disposal of low- and high-level radioactive wastes as well as high-temperature cooling water discharges, are other environmental problems.

Many other cases could be cited. One of the questions at the end of this chapter will challenge you to prepare a particular case study. In sum, it is clear that the production and use of many new products since World War II are in large measure responsible for the much increased environmental disturbances evident in many of the highly industrialized countries. These disturbances go far beyond those that can be explained by increases in population and economic growth alone.

2.4 URBANIZATION

Urbanization refers to an increase in the ratio or urban population to rural population. Historically, the seeds of urbanization may have been sown as far back as 7000 to 5000 B.C., in what has been called the Agricultural Revolution. Gradually, the nomadic hunters and food gatherers of those early times settled down in increasing numbers to domesticate animals and grow food. The result of this transition was the development of a food surplus, which freed people from toiling on the land. There soon developed a division and specialization of labor amongst this newly emerging nonagrarian population group. From these primitive social developments, society developed complex, interrelated social structures, recognizable today as cities. The first cities arose along the Tigris and Euphrates rivers between 4000 and 3000 B.C., in what is now Iraq. Environmental factors played the largest role in the development of these early cities. Nearby flat land and rich soils were necessary for cultivation. In addition, there was a need for easy access to and from the site, and water had to be readily available. The floodplains of the Tigris and Euphrates, as well as those of the Nile and Indus Rivers, were ideally suited for these purposes. This early urbanization led to possibly the first, and certainly one of the most disastrous, environmental impacts in history, namely, the destruction of forests in the Middle East in order to provide lumber and fuel for the cities. The resulting soil instability, the consequent desertification, and ultimately the loss of productive land were the tragic consequences from which this region still suffers.

It was not until the eighteenth century, however, that modern urbanization really accelerated. The limited urbanization that had taken place before then was almost entirely due to migration from rural areas to towns by people who were no longer needed in agriculture. The spur to urban growth over the last 200 years has been technological development, which has stimulated industrialization and increased the demand for labor in the cities.

2.4.1 Definition of Urbanization

Statistics on urbanization trends and rates in various countries are difficult to compare. The problem arises because of the many different definitions of urbanization. Some countries distinguish between urban and rural areas by the size of the community. But at what point does a concentration of people become urban? Five thousand? Ten thousand? A hundred thousand? What may be defined as urban in one country may be rural in another. For example, areas with as few as 400 inhabitants are designated as urban in Albania, while in Japan the lower limit is 50,000 inhabitants. Urban status in some other countries is assigned on the basis of density. In Sweden, urban areas are those built-up areas with less than 200 meters between houses. In India, places having a density of not less than 1,000 persons per square kilometer where at least three-quarters of the male adult population is employed in nonagricultural work are called urban. Other countries define urban areas in terms of the extent of urban characteristics, such as the number of plazas or schools or the availability of sewers, electric, or water supply facilities. Still other countries classify urban areas by the type and extent of administrative control exercised over them.

The many definitions of urban areas which have resulted from historical, cultural, and administrative differences among nations make it difficult to discern a common pattern. Most often, a population of 20,000 is used as the size above which an area is called urban, and this is the criterion used in this text.

The current situation.   Accelerating urban growth in the last half of the twentieth century has been a global phenomenon. It has been most dramatic in the less developed regions of the world, proceeding at a rate of four percent or more during the post-World War II period. In the more developed regions during the same period, urbanization averaged about two percent, which is about double the population growth rate in these regions. Although the population growth rate in the MDR has been decreasing, the urban proportion has grown from 55 percent to 70 percent of the total population. Much of this increase is due to the decline in rural population through rural-to-urban migration rather than to the arrival of new immigrants to the cities.

The extremely rapid growth of the urban population in the LDR compared to that in the MDR has, in just 30 years, eliminated the difference in urban population between the two regions. In 1950, the MDR had almost twice as large an urban population as the LDR, but by 1980 the urban population in the LDR exceeded that in the MDR (800 million to 790 million) (Figure 2–9). United Nations projections suggest that the LDR will have over $\frac{3}{4}$ of a billion more urban population than the MDR by the end of the century. Note from

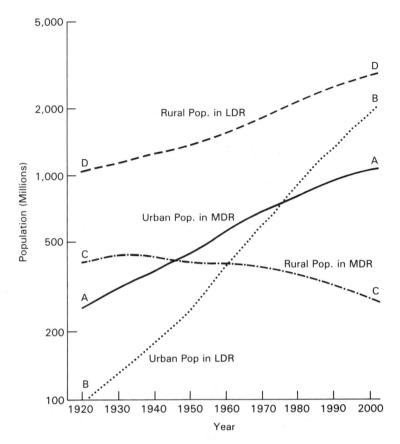

Figure 2–9 Urban and rural population in more developed and less developed nations. Source: World Housing Survey, UN, New York, 1976

the figure that although population growth rates are declining in the rural areas of the MDR, the same trend is not evident in the LDR.

The trend toward greater urbanization will have important ramifications for both the more developed and the less developed regions. If the economic growth rate in a country does not exceed the rate of urban population growth, then urban living conditions in that nation will not improve. On the other hand, if the economy does keep pace with urban growth, the implication may be that most of the resources will be consumed to support the urban population with little, if any, remaining to develop the rural economy. Agricultural output per farmer will have to increase to provide for the increasing numbers of urban inhabitants and the declining or slower growing rural population.

2.4.2 Growth of Cities

One of the current characteristics of urbanization is the trend toward "urban giantism." It has been estimated that in the period 1950 to 1975 cities with 5 million inhabitants or

TABLE 2–5   ESTIMATED PERCENTAGE OF URBAN POPULATION BY CITY-SIZE CLASSIFICATIONS IN 1950, 1975, AND PROJECTION TO 2000. (FOR WORLD, MDR, AND LDR)

	WORLD			MDR			LDR		
	1950	1975	2000	1950	1975	2000	1950	1975	2000
Total	100	100	100	100	100	100	100	100	100
Over 5 million	6.6	12.6	20.9	9.0	14.2	16.4	2.2	10.9	23.5
2 to 5 million	10.2	10.4	13.0	10.5	8.9	13.3	9.5	11.8	12.9
1 to 2 million	7.8	9.5	10.0	8.6	10.1	10.3	6.5	8.9	9.9
500,000 to 1 million	9.6	9.9	8.9	9.1	10.0	9.3	10.6	9.8	8.6
200,000 to 500,000	11.8	12.7	10.7	11.6	12.5	11.5	12.1	12.9	10.2
100,000 to 200,000	8.6	8.0	6.6	8.2	8.0	7.1	9.3	8.1	6.3
Other urban	45.4	36.9	29.9	43.0	36.3	32.1	49.8	37.6	28.5

Source: *United Nations Population Studies No. 63*, 1979.

more almost doubled their share of the total urban population, while cities with less than 100,000 inhabitants declined in relative importance. This trend is reflected in Table 2–5. Also evident in this table is the tendency toward "urban giantism" in many of the less developed regions. For example, the percentage of urban inhabitants of the LDR living in cities over 5 million has been projected to increase from 2.2 percent in 1950 to 23.5 percent in the year 2000. Similar statistics for the MDR show a jump from 9 percent in 1950 to only 16.4 percent in 2000. Therefore, cities such as Mexico City, São Paulo, and Buenos Aires in the LDR will soon have much larger populations than cities such as Tokyo, New York, and London in the MDR. Table 2–6 traces the recent population history with projections to the year 1990 for a few of the large cities to further illustrate this point.

TABLE 2–6   RANK AND POPULATION OF SELECTED CITIES FROM 1950 TO 1990 (POPULATION IN MILLIONS)

	1950		1960		1970		1980		1990*	
CITY	Rank	Pop'n.	Rank	Pop'n.	Rank	Pop'n.	Rank	Pop'n.	Rank	Pop'n.
TOKYO	4	6.7	3	10.7	2	14.9	1	20.2	1	24.9
NEW YORK	1	12.3	1	14.2	1	16.3	2	17.9	3	20.1
MEXICO CITY	16	2.9	13	4.9	6	8.6	3	13.8	2	21.6
SHANGHAI	5	5.8	5	7.4	4	10.0	4	13.2	5	17.1
SÃO PAULO	21	2.4	16	4.4	10	7.8	5	12.5	4	18.6
PEKING	26	2.2	14	4.5	13	7.0	6	11.2	6	16.4
LONDON	2	10.2	2	10.8	3	10.5	7	11.1	15	11.7
LOS ANGELES	11	4.0	8	6.5	8	8.4	8	10.7	11	13.0
BUENOS AIRES	9	4.5	7	6.7	9	8.3	9	10.4	12	12.3
RHEIN RUHR	3	6.8	4	8.7	5	9.3	10	10.0	18	10.6
OSAKA	12	3.8	11	5.7	11	7.6	11	10.0	14	12.0
PARIS	6	5.4	6	7.4	7	8.4	12	10.0	17	11.3

* Estimate of population.

Source: World Housing Survey; 1974, Dept. of Economic and Social Affairs, United Nations, 1976.

Global density in 1976 was 20 people per square kilometer. This does not seem to be very high, but it does not take into account the human tendency to gather in clusters—cities. Only 30 percent of the earth's land is potentially arable. The remainder, in mountains, frigid areas, deserts, etc., is of little use for agriculture. For all practical purposes, the limited arable land must support the world's growing population.

In contrast to the low global density, the density in urban areas may be greater by two orders of magnitude. For example, the density of Hong Kong is over 4,000 persons per square kilometer. The problems of water supply, waste disposal, housing, and transportation created by such high densities are staggering.

2.5 ENVIRONMENTAL IMPACT

It is important to recognize the impacts which urbanization and industrialization have on the environment. The environmental impact matrix provides a convenient inventory and display of these impacts. The pioneering work in this area was done by Leopold et al. (1971) and has been reviewed by Munn (1975). The matrices for the impacts of urbanization and industrialization are illustrated in Tables 2–7 and 2–8. The horizontal axis lists various aspects of urbanization or industrialization, while the vertical axis contains the components of the environment—the atmosphere, hydrosphere, lithosphere, human impacts, and others as appropriate. The elements of the matrix identify potential interactions between each activity and each environmental characteristic. Questions about each element in the matrix can then be considered. For example, does the mining industry affect air quality? (Answer: "Yes," from particulate matter released from open-pit operations, and the gaseous and particulate emissions from processing.) The matrix technique ensures that most questions are asked. If an impact is missed, it is because of ignorance of its existence rather than forgetfulness.

The impacts identified can then be classified as severe, moderate, slight, and zero, or a numerical scheme may be used. The classification is ultimately subjective and should preferably be done by several individuals, each influencing the opinion of others, in the hope that an informed, impartial consensus will emerge. Particular emphasis is often placed on environmental changes which are irreversible, for example, severe terrain disturbances, extinction of rare or endangered species, or widespread contamination.

The environmental impacts of *urbanization* are many and varied (Table 2–7). The predominant atmospheric effect of urbanization is the alteration of the atmosphere's chemistry through the release of massive quantities of CO_2, oxides of sulfur and nitrogen, dust, particulate matter, noxious and toxic chemicals, etc. The sources of these contaminants are diverse: industry, most forms of transportation, the heating of buildings, municipal incinerators, sewage treatment works, open fires, and landfill sites. In addition, significant heating of air masses over urban centers occurs as a result of reradiation from heat-absorbing surfaces such as roads, parking lots, and rooftops. This is in addition to the heat released from all types of combustion and industrial systems. The combustion of hydrocarbons, particularly those used in the transportation sector, also give rise to photochemical "smog"

TABLE 2-7 ENVIRONMENTAL IMPACTS OF URBANIZATION

Environmental component	Urban component			
	Population (Numbers and Density)	Land use	Transportation	Services
ATMOSPHERE	Increasing release of carbon dioxide, decreased oxygen production, as plant colonies are destroyed by spreading urban areas	Increased average temperatures for most urbanized areas	Air pollution from combustion of fuels; Creation of photochemical smog; Emission of lead from some engines	Particulates, noxious fumes from incinerators, landfills, sewage treatment works, etc.
HYDROSPHERE	Greater demand on water resources (both surface and subsurface)	More intense use of hydrologic resources causing increased pollution load	Rain, surface waters polluted with lead; Drainage patterns altered by infrastructure	Leaching of pollutants from landfills; Discharges from sewage outfalls; Pollution from boats
LITHOSPHERE	Increased transformation of uninhabited agricultural or unutilized land to urban users	Complete changes due to construction, landscaping, etc.	Disruption or disfigurement of landscape, etc.	Sanitary landfill of urban wastes and installation, repairs of services disturb landscape
HUMAN IMPACTS	Psychological impacts of high-density living	Psychological impacts	Increased noise levels; Health effects of noise, air pollution	

TABLE 2-8 ENVIRONMENTAL IMPACTS OF SELECTED GROUPS OF INDUSTRIES

Environmental consequences	Industrial component					
	Petrochemical	Metals	Food/beverage	Mining	Agriculture	Pulp and paper
ATMOSPHERE	Emissions to atmosphere from refining, processing plants (noxious, toxic)	Particulate, gas emissions during forging, working fabrication	Noxious fumes from food processing	Particulate matter from surface mining, transportation; Noxious, toxic fumes from smelting	Drift of agricultural sprays; Dust, pollen escape due to field operations	Release of noxious fumes during processing
HYDROSPHERE	Plant emissions to receiving water bodies	Discharge of mill pickling liquors; Other waste disposal to water bodies; Heavy metal releases (intentional, unintentional)	Wastes often have high organic content	Runoff from mine tailings; Processing wastes disposed of directly into water bodies	Runoff to surface and percolation to subsurface waters of pesticides, fertilizers; Silting of water bodies due to poor farming practices	Contaminated factory wastes (mercury, organics); Silt from deforested slopes; Loss of wildlife habitat
LITHOSPHERE	Disposal of waste solids, sludges to landfill; Accidental spills during transport, storage	Disposal of slag, waste products from processing		Dumping of mine tailings, processing wastes; Disruption of agriculture, forestry, recreation by open-pit mines	Erosion of land surface; Depletion of organic material, necessary soil microorganisms, etc.	Breakdown of ecosystem in clear-cut areas; Erosion of unprotected land
HUMAN IMPACTS	Some products, wastes toxic to many life forms; Disruption of life style from emissions to all 3 spheres	Health effects of released toxics in air, water		Health hazard to miners (mercury, asbestos, coal mining)	Health effects of biocides, polluted water, etc.	

as a result of the interaction of various by-products of the combustion process and energy from solar radiation.

The impact of urbanization upon the hydrosphere is severe because of the large volumes of pure water which must be provided and the correspondingly large volumes of used water requiring disposal. Storm water also has an impact. Although the total quantities of runoff from rainfall may not be altered significantly, the rate and characteristics of the runoff may be changed sufficiently to cause damage or inconvenience. The rate at which water runs off a paved road or parking lot, or off a smooth pitched roof, is considerably greater than the rate it runs off a rural or forested area (such as a golf course or park). As a result, water can accumulate rapidly in an urban drainage system, and if an overflow occurs, extensive flood damage is possible. Moreover, these stormwaters are often contaminated by chemicals or particulates adsorbed or absorbed during rainfall, or material such as oil being washed off streets and parking lots. Degradation of water resources by stormwater is a problem in most urban environments. Another potential contributor to the contamination of the hydrosphere is the drainage, called leachate, that comes from landfills of municipal solid wastes or toxic and hazardous wastes.

From a visual inspection of the urban environment, you would conclude that the lithosphere was the part of the environment most dramatically altered by urbanization. The original state of the environment appears to have changed irreparably. The elevations of the surface have been altered, rivers diverted, and lowlands either excavated for harbours or filled in for building. The "water edge" in many cities has been pushed farther into the lake to facilitate development and expansion of industry, transportation, and recreational facilities. In fact, the construction of buildings and roads has revamped the character of the region. Native ecosystems are replaced by urban patterns. Circulation of air is altered (on a local scale) by the presence of obstructions, such as tall buildings and smokestacks. Transportation, both public and private, is responsible for substantial alteration of the landscape because of the construction of roads, railroads, parking lots, airports, harbors, and warehousing and shipping facilities. The provision of municipal services accounts for some of the changes observed in the urban environment, viz., the existence of water towers, pumping stations, reservoirs, sanitary landfills, and other structures.

The human impacts of urbanization tend to be rather difficult to define and assess. The health effects of noise, air, and water pollution, and the psychological stresses caused by high density and a relatively "fast-paced" environment are not easily quantified. Many of the effects are not particularly harmful in isolated contacts, but continued exposure to inhalation of low-level concentrations of, for example, lead may be a much more serious problem. The psychological impacts are the least understood and, as a result, the most difficult to evaluate. However, there are few people who would deny that these stresses do exist.

The environmental impacts of *industrialization* tend to be a little easier to establish compared to those of urbanization, because the focus is on a smaller group of interests. Table 2–8, which presents the environmental impacts of selected groups of industries, is arranged in a fashion similar to Table 2–7 for display of the effects of urbanization. Although the table is reasonably self-explanatory, a brief review using the mining industry as an example may be helpful.

The impact of the mining industry upon the environment is substantial. Open-pit mining and the transportation of ores contribute particulate matter to the atmosphere. Processing of the metal ores (smelting, roasting, etc.) contributes oxides of sulfur and nitrogen to the atmosphere, depending on the material being processed. Various gaseous emissions may be noxious, toxic, or, in the case of the oxides, precursors of acid rain.

Runoff from mine tailings may wash hazardous materials into nearby surface or subsurface water resources. Occasionally, processing wastes are discharged directly to the receiving water body, where they impair water quality and affect aquatic life. The most obvious impacts of mining upon the lithosphere are (1) the residues from the dumping of tailings and processing wastes directly upon the landscape, and (2) the disruption of many activities, such as agriculture, forestry, and recreation—particularly from open-pit mining and quarrying.

The impact of the mining industry on human health and well-being is a subject of much debate. However, the adverse effects of the sustained exposure of miners to certain minerals like coal (causing black lung disease) and to asbestos (causing asbestosis) have been established beyond doubt. Noise pollution from mining or quarrying operations near inhabited areas may also have negative effects on the health and well-being of the local population.

2.6 THE DILEMMA OF INDUSTRIALIZATION AND URBANIZATION

Industrialization and urbanization are worldwide phenomena. More and more people will be living in larger and larger cities. These high-density communities pose a special challenge in the provision of potable water, clean air, waste disposal, transportation, and recreational space. Modern communication has made the world a global village, and has raised expectations in most of us for a better life. It will take enormous ingenuity, diplomacy, and determination for the world's leaders and those who help them—scientists, engineers, lawyers, economists, and managers—to guide development over the next century. In order to influence governmental policies on these matters, pressure groups have emerged which often put their case forward in a biased and exaggerated way. It is not surprising that on a particular environmental issue reports appear which are diametrically opposed to each other. We have all witnessed this in the popular press, radio, TV, and also in the scientific field. It becomes difficult at times to know whom and what to believe.

There are groups that claim that continued economic growth by nations is an impossible goal that will inevitably lead to the failure of world society and environmental disaster. They argue that a steady-state economy is a necessary and desirable future state of affairs, although the timing of this will vary considerably among nations depending on their current state of development. For example, Daly (1977) asserts that "A U.S. style high-mass consumption, growth dominated economy for a world of $4\frac{1}{2}$ billion people is impossible. Even more impossible is the prospect of an ever growing standard of per capita consumption for an ever growing world population." Besides Daly, authors like Meadows et al. (1972), Schumacher (1973), Ward (1976), and Ward and Dubos (1972) have dealt with the controversial topic of limited growth.

On the other hand, virtually every nation is attempting to increase its share of the global economy. Multinational companies compete vigorously for world markets. Underdeveloped countries, attempting to industrialize, find that their much lower wage scale gives them a competitive edge over the developed countries in certain fields. Economic recession and its resulting unemployment have a predictable effect on how governments view the apparent conflict between economic growth and environmental protection. Unfortunately, when the issue is presented simplistically as jobs versus a clean environment, the pressure on politicians to allow industry to ''defer'' pollution control measures is often irresistible.

Two U.S. reports highlight the controversy between the ''doom and gloom'' and the ''things are getting better'' environment philosophies. The *Global 2000 Report to the President of the United States* (Barney, 1980), commissioned by President Carter, was produced to a large extent by government and quasi-government agencies. Although it may be presumptious to classify such a massive report into one of the two categories mentioned, on balance it does belong in the ''doom and gloom'' category. An updated version, originally entitled *Global 2000 Revised* and produced by a group of independent scientists under the leadership of J. L. Simon and Herman Kahn, disagreed completely in many aspects with the earlier version. The report, later retitled *The Resourceful Earth*, (Simon and Kahn, 1985) stated that ''if present trends continue, the world in 2000 will be less crowded, less polluted, more stable ecologically and less vulnerable to resource-supply disruption.'' Two serious studies on the same topic thus have very different conclusions!

One of the most comprehensive reports on the effects of environmental abuse on the world's economy was carried out over a three-year period by the Norwegian Prime Minister, Gro Harlem Brundtland, and her 22-member UN commission (United Nations, 1987). The report warned that pollution and the overuse of resources threaten to radically alter the planet and the lives of many species upon it, including the human species. The Bhopal chemical accident (the worst industrial accident to date), the African famine, and the deaths of about 20 million children per year from diseases related to unsanitary drinking water and malnutrition were a few of the calamities noted. The prediction for the 1990s was that there would be even more disasters—particularly droughts and floods, which are most directly associated with environmental mismanagement.

A similar warning was expressed by Lester Brown (1987), president of the Worldwatch Institute, who noted that human use of the air, water, land, forests, and other systems that support life on earth was pushing those systems over ''thresholds'' beyond which they cannot absorb such use without permanent change and damage.

In recognition of the importance of environmental protection, the World Bank announced in 1987 that environmental factors would be one of the primary considerations in all of the bank's future lending and policy decisions.

The large gap in the quality of life between the world's richer and poorer nations is expected to widen because of the higher population growth rates in the less developed regions. The developed nations have recognized the uneven distribution of wealth, but have found it difficult to divert sufficiently large amounts of their resources to aid the poorer countries, since they themselves are faced with difficult problems of slow economic growth and inflation. The escalating costs of resources, particularly of energy, have caused very serious economic problems for all, but especially for the poor countries, who must pay for their energy imports with whatever exports they can manage. For the three-quarters of the world now residing in the less developed regions that aspire to reach the same standard of living as the one-quarter in the more developed regions, global energy and resource consumption would have to increase approximately tenfold for that to occur. Considering the present energy reserves and their value, this is clearly impossible. Furthermore, a tenfold increase in the consumption of energy and resources could also mean a tenfold increase in pollution, which would be difficult or (more likely) impossible for the environment to assimilate. Ultimately, we must face the question of whether the standard of living in the richer countries will have to come down to allow an increase in the standard of living in the poorer countries. Can this happen by peaceful means? We must hope that a way can be found.

PROBLEMS

2.1 Define or explain the following terms:
   (a) Population growth
   (b) Natural increase of population
   (c) Net migration
   (d) Zero population growth rate
   (e) Replacement growth rate
   (f) Total fertility rate
   (g) Age structure
   (h) Gross National Product
   (i) Urban vs. rural
   (j) Population density

2.2 The following are the average annual demographic statistics for Canada for the period 1970–75 (Source: *Canada Yearbook, 1978–79*):

Birth rate	15.8 per 1,000 population per year
Death rate	7.4
Immigration	7.6
Emigration	3.2

Calculate the average annual rate for the period and the corresponding doubling time.

2.3 In 1977 Ghana had a population of 11.3 million. There were 542,400 births and 192,100 deaths during that year. Assume that the net migration was zero.
   (a) What was the birth rate and death rate in that year?
   (b) What was the rate of natural increase in population?
   (c) What was the approximate doubling time in years?
   (d) What will the population be in 1987 if this growth rate persists for 10 years?

2.4 (a) Explain in words or by an equation what is meant by exponential growth or decay.
   (b) Give five examples which follow exponential growth or decay.

2.5 Obtain the average annual demographic statistics for your country (or state or province) for the most recent period available, i.e., the birth rate, death rate, immigration, and emigration rate. Calculate the average annual population growth rate from this information.

2.6 Distinguish between zero population growth and replacement growth. Which is likely to occur in the examples provided in Figure 2–4?

2.7 Sketch hypothetical population pyramids for:
   (a) A rapidly growing population
   (b) A declining population
   (c) A stable population
   (d) A stable population having recently experienced major war casualties
   (e) A stable population experiencing recent large immigration of young people

2.8 A distinction has been made between the more developed and less developed regions of the world in the material presented on population and economic growth. Why do you think this has been done?

2.9 State and discuss the reasons why world population is going to at least double over the next 100 years.

2.10 List the advantages and disadvantages for a country having
   (a) a high population growth rate
   (b) a high economic growth rate.

2.11 List several of the methods which have been used to control population growth. Which would you be prepared to support in your country, assuming you live in
   (a) a country with a stable population
   (b) a country with a rapidly growing population.

2.12 The book by Meadows et al. (1972) caused considerable controversy when it was published. It is now available in paperback (Signet Books, New American Library). Read it and prepare a short essay on your reaction to it.

2.13 A Western statesman was quoted as having posed the following dilemma in relation to Meadows's book: "I know what I need to do to solve the problems raised in this book, but if I institute them, I will be defeated at the polls and replaced by someone who knows less about it than I do." What is your reaction to his perceived dilemma?

2.14 The Industrial Revolution has produced social, economic, and environmental changes. State one example of each and discuss the consequences.

2.15 Explain what is meant by the term *urbanization*. List and explain as many environmental implications of urbanization as you can think of which have not been mentioned in Table 2–7.

2.16 Prepare a case study for a product which has been invented, produced, and marketed in recent years, and which has caused significant environmental problems. What, in your opinion, can and should be done about such a product? Use the matrix method to demonstrate the product's environmental impact.

2.17 Select a specific industry which is of major importance in your area, and prepare an environmental impact matrix for it. Discuss your findings.

REFERENCES

BARNEY, G. O. *The Global 2000 Report to the President of the United States*, Vol. 2. Washington, D.C.: Government Printing Office, 1980.

BLAIR, T. L. *The International Urban Crisis*. London: Hart-Davis MacGibbon Ltd., 1974.

BRAMWELL, R. D. *Towns and Cities: Yesterday, Today and Tomorrow*. Agincourt Ontario Canada: Gage Educational Publishing Ltd., 1977.

BREESE, G. W. *Urbanization in Developing Countries*. Englewood Cliffs, N.J.: Prentice Hall, 1969.

BROWN, LESTER, ed. *State of the World, 1987*. Washington, D.C.: Worldwatch Institute, 1987.

Canada Yearbook, 1978/79. Ottawa: Ministry of Industry, Trade and Commerce, 1979.

Commodity Research Bureau, Inc. *Commodity Year Book, 1981*. New York: Commodity Research Bureau, Inc., 1981.

DALY, H. E. *Steady-State Economics*. San Francisco: W. H. Freeman, 1977.

EHRLICH, P., EHRLICH, A., and HOLDREN, R. *Population, Resources, Environment*, San Francisco: W. H. Freeman, 1977.

FOIN, F. C. *Ecological Systems and the Environment*. Boston: Houghton Mifflin, 1976.

FREJKA, T. *The Future of Population Growth*. New York: Wiley, 1973.

HAUSER, P. M., ed. *World Population and Development*. Syracuse, N.Y.: Syracuse University Press, 1979.

HODGES, L. *Environmental Pollution*. 2d ed. New York: Holt, Rinehart and Winston, 1977.

KURIAN, G. T. *The Book of World Rankings*. New York: Facts on File, 1979.

LEOPOLD, L. B., CLARKE, F. E., HANSHAW, B. B., BALSEY, R., Jr. *Geological Survey Circular 645*. Washington, D.C.: U.S. Government Printing Office, 1971.

MACNEILL, J. W. *Environmental Management*. Ottawa: Information Canada, 1971.

MASTERS, G. M. *Introduction to Environmental Science and Technology*. New York: Wiley, 1974.

MEADOWS, D. H., MEADOWS, D. L., RANDERS, J., and BEHRENS, W. W. *The Limits to Growth*. New York: Universal Books, 1972. Also available as a Signet Book from the New American Library, 1972.

MUNN, R. E. *Environment Impact Assessment, Scope 5*. Toronto: Wiley, 1975.

SCHUMACHER, E. F. *Small Is Beautiful*. New York: Harper & Row, 1973.

SIMON, J. L., and KAHN, HERMAN, eds. *The Resourceful Earth*. New York: Oxford University Press, 1985.

SUZUKI, D. *"Exponential Growth Is Merely Another Case of False Worship."* *The Toronto Star*, 11 January, 1986.

TRESHOW, M. *The Human Environment*. New York: McGraw-Hill, 1976.

United Nations. *Concise Report on the World Population in 1977*. (Population Studies, No. 63.) New York: United Nations, 1979.

United Nations. *Concise Report on the World Population in 1979*. (Population Studies, No. 72.) New York: United Nations, 1980.

United Nations. *Prospects of Population: Methodology and Assumptions*. (Population Studies, No. 67.) New York: United Nations, 1979.

United Nations. World Commission on Environment and Development. *Our Common Future*. London: Oxford University Press, 1987.

United Nations. *World Housing Survey: 1974*. New York Dept. of Economic and Social Affairs, 1976.

United Nations. *World Population Trends and Policies*. Vol. 1. (Population Studies, No. 62.) New York: United Nations, 1979.

U.S. Department of Commerce, Bureau of Census. *World Population, 1977*. Washington, D.C.: U.S. Government Printing Office, 1979.

WARD, B. *Human Settlements, Crisis and Opportunity*. Report prepared for Information Canada, 1975. Vancouver, B.C.: UN Conference on Human Settlement, 1976.

WARD, B., and DUBOS, R. *Only One Earth*. New York: Norton, 1972.

WARREN, K. *World Steel*. New York: Crane, Russak, and Co., 1975.

World Bank. *1981 World Bank Atlas*. Washington, D.C., The World Bank, 1981.

World Bank. *World Development Report, 1981*. London: Oxford University Press, 1981.

Zito, G. V. *Population and Its Problems*. Syracuse, N.Y.: Syracuse University, 1979.

3

Energy Growth

O. J. C. Runnalls

Donald Mackay

As indicated in the preceding chapter, the world's population and its economic output will continue to grow for at least the next several decades. Much larger growth in both population and gross national product is projected for the less developed countries as compared to the more developed ones (Barney, 1980). Currently, three of every four inhabitants of the earth live in the less developed countries, and two-thirds of these—over two billion people— rely on the gathering of wood and crop and animal wastes to provide fuel for cooking and warmth (World Bank, 1981).

Clearly, the world faces substantial increases in energy consumption, particularly in those disadvantaged areas where population growth is still high but individual expectations for improvement are also understandably high. The production of energy brings with it the inevitable consequence of environmental disturbance. Whether we consider the denudation of forests to supply wood for the people of the developing world, or the atmospheric pollution that accompanies the generation of electricity in coal-burning power plants, environmental problems grow as energy requirements rise.

The purpose of this chapter, therefore, is to examine the availability of energy sources in the future and the environmental impacts from increased energy output.

3.1 SOURCES OF PRIMARY ENERGY

The sources of primary energy available to mankind have often been categorized as either renewable or nonrenewable. They might also be thought of in terms of the description adopted by Putnam (1953), who used the phrases ''energy income'' and ''energy capital.'' **Energy income,** or renewable energy resources, comprises those resources which are being continuously renewed because of the presence of tidal forces, wind, falling water, thermal

TABLE 3–1 AVAILABLE ENERGY SOURCES

Renewable (energy income)	Nonrenewable (energy capital)
Hydroelectric energy	Crude oil
Tidal forces	Natural gas
Geothermal heat	Coal
Biomass (wood, animal refuse, vegetable matter, etc.)	Nuclear fission
	Synthetic oil (from oil sands and oil shales)
Wind	
Solar input	
Ocean heat	

gradients in the ocean, geothermal heat, direct solar input, the generation of vegetable and animal matter, and so on. **Energy capital,** or nonrenewable energy resources, refers primarily to fossil fuels, which were deposited on earth hundreds of millions of years ago, or to radioactive minerals, which were present when the planet was formed. When such materials are mined, the quantity of energy capital is reduced. Actually, the fossil fuels are being replaced in nature, but at a rate which is so slow on the time scale of human development, as to be insignificant. Hence, oil, natural gas, and coal can be considered nonrenewable in the practical sense. The radioactive fuels uranium and thorium are not being replenished either. In fact, over a long time span, measured in billions of years, they are being transformed through radioactive decay processes to stable elements.

The currently available sources of energy are listed as renewable or nonrenewable in Table 3–1.

During the twentieth century, the annual consumption of primary energy provided commercially in the world has increased more than ten-fold, as shown in Figure 3–1. Part of the increase was required by a growth of about two and one-half-fold in the world's population during that period. Another important part of the rise in energy consumption, however, was a consequence of increasing mechanization, particularly in the industrialized world. This is illustrated in Figure 3–2, where the growing importance of machine energy in the twentieth century in one of the industrialized countries, the United States, is readily apparent.

Wood served as the predominant fuel in the world until about 1875, as illustrated in Figure 3–3. Then it began to be supplanted by coal. The percentage contribution of coal to the world's primary energy supply reached a peak some forty years later, at which time its use began to decline as oil and natural gas grew in importance. Now there are those who feel that oil may have passed its peak in the contribution to the world energy supply.

The growth of machine energy has been facilitated during the 1900s by the rapid development of internal combustion engines and the provision of liquid fuel for them by sophisticated transportation, refining, and distribution systems. The rising consumption of petroleum products has led to intensive worldwide programs in the search for crude oil and natural gas deposits. Many such deposits have been discovered during the twentieth century. Most of the nations which are principal users of these two commodities, however, do not have appreciable domestic supplies and must look to other countries to obtain them. This imbalance in the supply of and demand for liquid and gaseous fuels forms the basis for a serious energy supply problem, which was first recognized by the world in the 1970s.

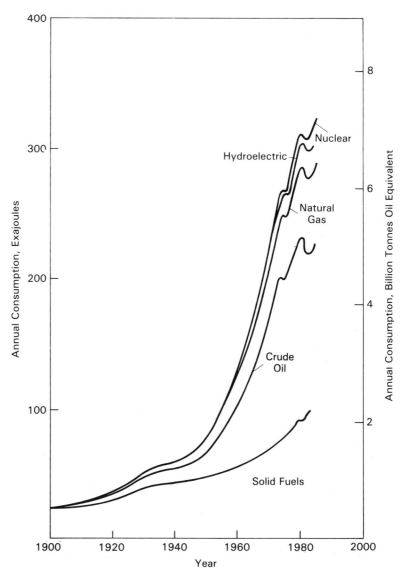

Figure 3–1 World's primary energy consumption during the twentieth century. Sources: World Energy Conference, *Survey of Energy Resources, 1986; BP Statistical Review of World Energy, 1986*

3.2 CURRENT CONSUMPTION OF ENERGY

The world's annual consumption of commercially provided energy in 1985 was about 330 EJ (exajoule = 10^{18} joules (J)) and was subdivided as shown in Figure 3–4 (see Table A-1.3 in Appendix A-1 for SI prefixes). Appreciable quantities of energy are provided internally in some industries by combustion of wastes, recycling of residues, etc., and are not accounted for in the normal commercial sense. A good example is the forest products

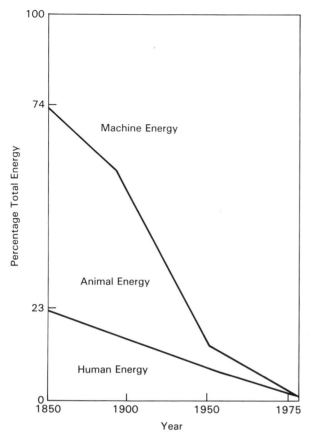

Figure 3–2 The growth of machine energy in the U.S.A. since 1850. Source: Wyatt, 1978

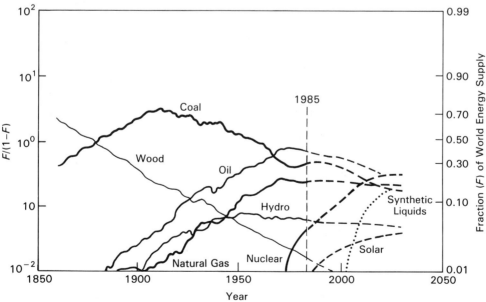

Figure 3–3 World Energy Sources 1860–2030. Sources: Hafele, 1981 [Copyright 1981 by the International Institute for Applied Systems Analysis. Reprinted by permission from Ballinger Publishing Company]; *BP Statistical Review of World Energy*, 1986; Runnalls, 1987

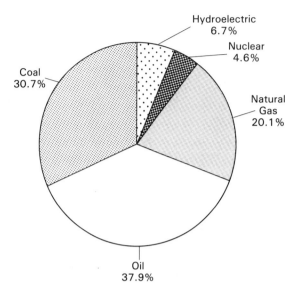

Coal
30.7%

Hydroelectric
6.7%

Nuclear
4.6%

Natural
Gas
20.1%

Oil
37.9%

Figure 3–4 World consumption of commercially provided energy, 1985. Source: *BP Statistical Review of World Energy*, 1986

industry. Noncommercial energy also plays a particularly important role in the developing countries, where wood and animal refuse still serve as essential sources of heat. In 1985, an estimated 30 EJ of such internally generated energy was produced in the Western world. Hence, total energy consumption in 1985 was about 360 EJ.

Using the percentages shown in Figure 3–4, it is possible to express the 1985 consumption of commercially provided energy in terms of exajoules as shown in Table 3–2.

The use of other units is also widespread in the literature. Hence, it is often desirable to describe energy outputs in alternative ways. The data in Table 3–3 are intended to aid in the conversion process.

Now, if the data in Table 3–2 are expressed in more usual units for each commodity, the results outlined in Table 3–4 are obtained.

Note from Figure 3–4 that more than one-third of the world's primary energy in 1985 was provided by oil. The enormous consumption of crude oil lies at the root of the world's energy problem. The growth of this appetite over the past 20 years is depicted graphically in Figure 3–5. What is immediately apparent is that there has been a doubling in demand during that time. Of necessity, production has risen to meet that demand, as

TABLE 3–2 WORLD CONSUMPTION OF COMMERCIALLY PROVIDED ENERGY, 1985

Commodity	Percentage	Quantity, EJ
Crude oil	37.9	125.1
Natural gas	20.1	66.3
Coal	30.7	101.3
Hydroelectric energy	6.7	22.1
Nuclear power	4.6	15.2
Total	100	330

Source: BP Statistical Review of World Energy, 1986

TABLE 3–3 ENERGY OUTPUTS AND CONVERSION FACTORS

Energy form	Heat value	
	SI units	AES units
(a) Energy Output		
Crude oil	38.512 TJ/10^3 m³	5.803 × 10^6 BTU/barrel
Natural gas	37.229 TJ/10^6 m³	1.000 × 10^6 BTU/10^3 ft³
Bituminous coal	29.993 TJ/10^3 tonne	25.800 × 10^6 BTU/short ton
Electricity	(i)• 10.5 TJ/GWh	10.000 × 10^6 BTU/MWh
	(ii)•• 3.6 TJ/GWh	3.412 × 10^6 BTU/MWh
(b) Conversion Factors		
Crude oil	1 m³ = 6.293 barrels	
	1 L = 0.264 gal	
	1 barrel = 42 gal	
Natural gas	1 m³ = 35.3 ft³	
Coal	1 tonne = 1.1023 short ton	
	= 2204.6 lb	
Energy	1 kJ = 0.948 BTU	

• For primary energy calculations, this value is adopted for hydraulic, nuclear, and purchased electricity; it is the equivalent thermal energy of a coal-burning plant assuming the conversion efficiency is similar.

•• For secondary energy calculations, such as conversion of electrical to thermal energy, as in resistance heating, this value is adopted.

TABLE 3–4 WORLD CONSUMPTION OF COMMERCIALLY PROVIDED ENERGY, 1985

Commodity	Quantity
Crude oil	8.9 × 10^6 cubic meters per day
Natural gas	62.9 trillion cubic feet
Coal	3.3 billion tons
Hydroelectric energy	2.2 × 10^{12} kilowatt hours thermal
	7.2 × 10^{11} kilowatt hours electrical
Nuclear power	14.5 × 10^{11} kilowatt hours thermal
	4.8 × 10^{11} kilowatt hours electrical

indicated in Figure 3–6. Many of the principal consumers, however, mainly in the industrialized Western world, do not possess substantial conventional crude oil deposits, as illustrated in Figure 3–7.

The centrally planned economies in the U.S.S.R., Eastern Europe, and China are collectively self-sufficient in crude oil supply at present. However, this is not the case for the Western world. The United States, for example, consumed nearly 50 percent more oil in 1985 than was produced within its own borders, even though its output was second only to the U.S.S.R. Production within the U.S. reached a peak in 1970 and has been declining slowly since then. As of the end of 1985, the proven reserves of U.S. crude oil* stood at

* Proven reserves are defined as that volume of oil remaining in the ground which geological and engineering information indicates to be recoverable from known reservoirs under existing economic and operating conditions.

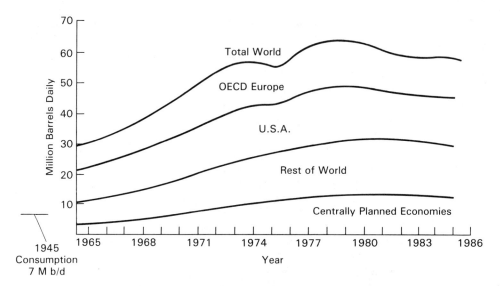

Figure 3–5   World's crude oil consumption, 1965–85. Sources: *BP Statistical Review of World Energy*, 1986; *Energy, Mines and Resources, Canada*, 1987

5.7×10^9 m³, only seven times larger than the 1985 consumption rate of 8.4×10^8 m³. The U.S. relies heavily on foreign supplies and imported nearly $50 billion worth of oil in 1985.

Western European nations and Japan are large consumers of oil products as well. Except for the United Kingdom and Norway, with their newly discovered deposits under

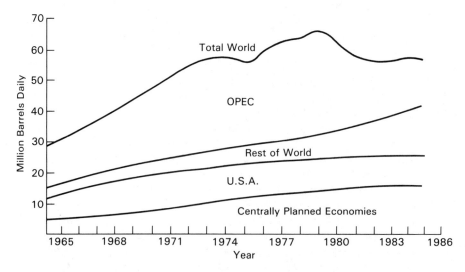

Figure 3–6   World's crude oil production, 1965–85. Sources: *BP Statistical Review of World Energy*, 1986; *Energy, Mines and Resources, Canada*, 1987

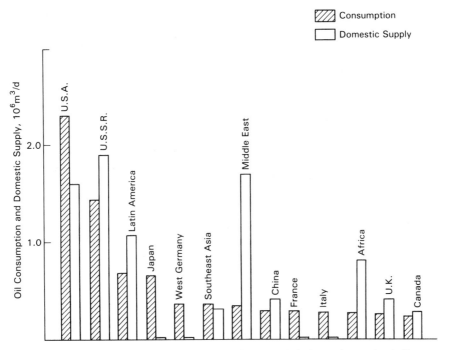

Figure 3–7 World's principal consumers of oil and their domestic supply of oil products in 1985. Source: *BP Statistical Review of World Energy*, 1986

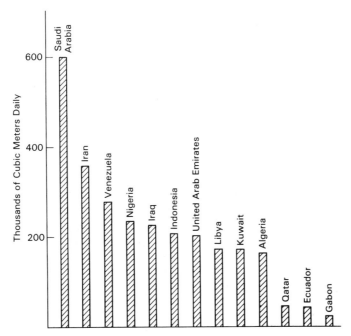

Figure 3–8 Production of crude oil in OPEC countries in 1985. Source: *BP Statistical Review of World Energy*, 1986

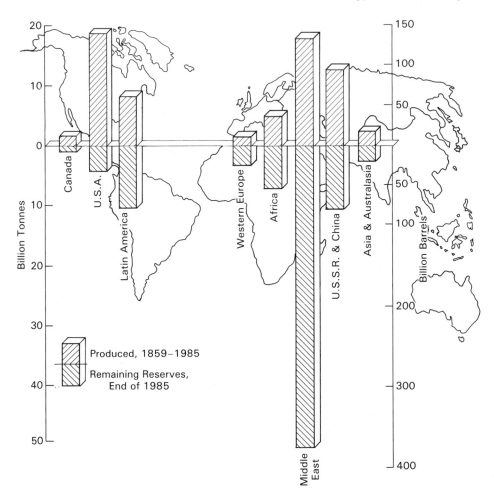

Figure 3–9 Total discovered oil. Source: *BP Statistical Review of World Energy*, 1986

the North Sea, most of these countries rely heavily on imported oil obtained from foreign producers, many of whom belong to the Organization of Petroleum Exporting Countries (OPEC). The volumes of crude oil that each of the 13 OPEC members produced in 1985 are shown in Figure 3–8. Comparing this with the previous figure, it will be noticed that the U.S.S.R., the United States, and Saudi Arabia are the three largest producers. Mexico is the largest non-OPEC oil-exporting country.

More than 40 percent of the crude oil requirements for the world (excluding the U.S.S.R., Eastern Europe, and China) was provided in 1985 by OPEC countries. Nearly 60 percent of that output came from the six members of OPEC located in the Middle East: Saudi Arabia, Iraq, the United Arab Emirates, Kuwait, Iran, and Qatar. Nature has favored them with large, easy-to-recover conventional crude oil reserves, ten times larger than those remaining in North America, for example, as illustrated in Figure 3–9. Consequently, their impact extends far beyond their numbers, which total at present less than 50 million people.

Saudi Arabia, with a population of less than ten million, has the greatest influence in the OPEC organization, because it is by far the largest producer among, and has the largest production potential of, all 13 members. Fortunately for Western countries, its policies on production and pricing appear to be the most moderate ones. In general, however, the Middle East may be unstable politically and subject to the ever-present threat of local conflicts, which have the potential of escalating into military confrontation between the superpowers.

Unquestionably, access to the oil resources in the Middle East is essential for the

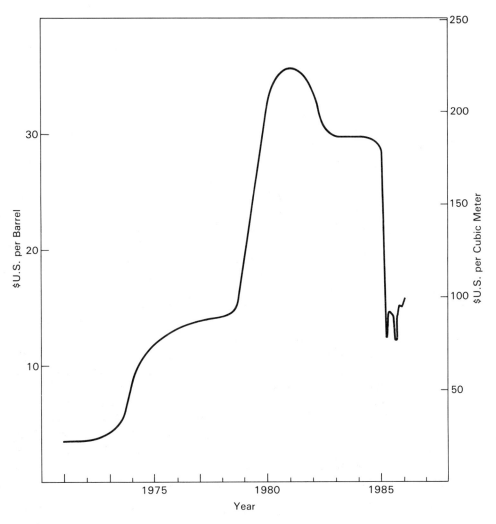

Figure 3–10 Average cost of crude oil imports at Montreal. Source: *Energy, Mines, and Resources, Canada, 1987*

TABLE 3–5 PERCENT CHANGES IN OIL CON-
SUMPTION IN SELECTED AREAS AND COUNTRIES
OF THE WORLD

Country/Area	1985 over 1979
United States	− 16.6
Canada	− 24.9
Latin America	+ 2.9
Western Europe	− 19.8
Japan	− 24.1
Middle East	+ 31.0
Africa	+ 25.6
South Asia	+ 41.6
Southeast Asia	− 1.1
U.S.S.R.	+ 4.8
China	− 3.8

Source: *BP Statistical Review of World Energy*, 1986

economic health of the Western world. Any attempt by the Soviet Union to restrict such access could lead to dire consequences. Meanwhile, the price shocks imposed by OPEC on oil consumers, first in 1973–74, and again in 1979–80, resulted by 1981 in more than a tenfold increase in current dollars in the world price for oil as shown in Figure 3–10. This heavy economic burden has led oil-dependent nations to search for energy alternatives on the one hand, while attempting to curb existing appetites on the other, even though prices have since fallen back to near $100 per cubic meter (see Figure 3–10).

Efforts to reduce oil consumption have been particularly apparent in countries of the Western world, as indicated in Table 3–5. This might have been expected, since they were, and still are, excessive users. Some would argue, however, that a significant portion of the reduction was due to decreased economic activity brought on by the huge increases in the price of oil in the 1970s. Other contributing factors have been the conversion to cheaper energy forms and measures to conserve energy. As a result, the demand for oil and its price per barrel have decreased significantly since 1980.

It is plain from Table 3–5 that the growth rates in consumption in many of the developing areas of the world are still at high levels. Such statistics are somewhat misleading, however, because populations in the developing areas are increasing annually at a rate about four times higher than those in the developed countries, and per capita energy consumption is much less in the underdeveloped areas. For example, Western Europe consumed more crude oil in 1983 than did all of Africa, Latin America, Middle East, South Asia,* and Southeast Asia** combined. The disproportionately high growth rate in the poorer countries and the excessive consumption by the richer nations will impose severe economic and environmental pressures on the world community during the coming decades.

* South Asia: Afghanistan, Bangladesh, Burma, India, Nepal, Pakistan, Sri Lanka.

** Southeast Asia: Brunei, Hong Kong, Indonesia, Malaysia, the Philippines, Singapore, South Korea, Taiwan, Thailand, Papua New Guinea, and the Southwest Pacific Islands.

3.3 FUTURE CONSUMPTION AND AVAILABILITY OF ENERGY SOURCES

After the end of World War II, the United States Atomic Energy Commission became concerned about public policy problems related to the possible development and use of nuclear generating stations. As a result, in 1949 Mr. Palmer Putnam, a consulting engineer, was asked to investigate the maximum plausible world demands for energy extending over the period 1950 through 2050. The results (Putnam, 1953) still make for fascinating reading today.

The author cast himself in the role of a hypothetical world trustee whose task was to identify and describe those contingencies which could affect energy consumption during the subsequent 100 years. One important component of the overall equation was to estimate the maximum world population prior to predicting the requirements for energy. According to Putnam's carefully projected results, the world's population could grow to 2.9 billion in 1975, 3.7 billion in 2000, and 6 billion in 2050.

At the time of the forecast in 1950, the world population was about 2.3 billion. Now, nearly 40 years later, the actual population has grown considerably beyond Putnam's forecasts. According to recent United Nations data, the population reached the 5 billion mark in 1987. Most of the unanticipated increase occurred in the developing countries, primarily because of a significant decline in infant and childhood mortality, as discussed in Chapter 2. It now seems likely that the world will be inhabited by at least 9 billion people by 2050 (see Figure 2–7). The demands of such a large populace on the world's energy resources will tax the ingenuity of those charged with resource recovery and the conversion of wastes to useful forms of heat and power.

Energy consumption has been predicted to more than double or even triple between 1985 and 2060, as shown in Figure 3–11. The uncertainty alone in energy needs by then will be larger than the world's total consumption in 1985. One estimate outlining how this huge future demand might be met is depicted in Figure 3–12, with coal playing the dominant role as an energy source in the twenty-first century. Serious environmental problems from gaseous emissions will have to be solved, however, before coal can be burned on the scale envisaged in the figure if a safe environment is to be preserved.

Future energy growth rates in the less developed regions, sometimes referred to as the South (Asia, Africa, Latin America), may be nearly double those in the more developed regions or North (North America, Europe, U.S.S.R., industrialized countries of the Pacific, South Africa), as shown in Figure 3–13. Nonetheless, levels of energy consumption in 2060 could still be seven times lower per capita in the South compared to the North (Frisch, 1986). The dominant sources of energy in the South in the early decades of the next century will be oil and hydroelectric power (see Figure 3–14), simply because nature has provided that part of the world with a greater endowment of those resources.

External demand from the North for oil supplies will be particularly severe as the South moves to use more and more of its indigenous deposits internally. Thus, pressures will develop in the North to find economic substitutes for ever-more-expensive and diminishing quantities of crude oil.

The available production data support the suggestion that world oil output may already

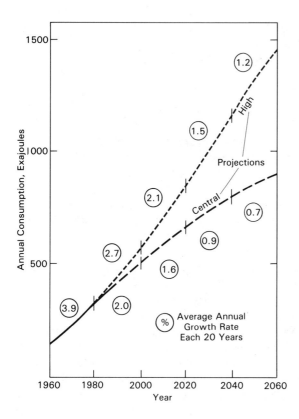

Figure 3–11 World primary energy production. Source: Frisch, by permission of the World Energy Conference, 1986

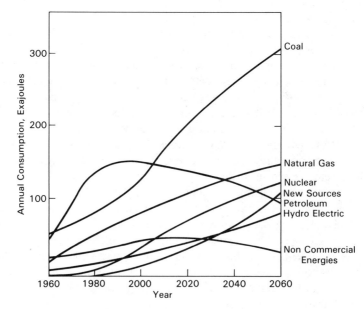

Figure 3–12 Evolution of world energy supplies. Source: Frisch, by permission of the World Energy Conference, 1986

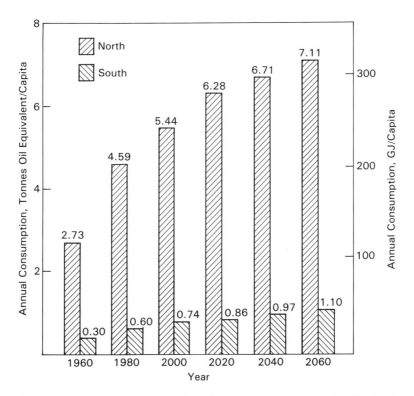

Figure 3–13   Levels of energy consumption. Source: Frisch, by permission of the World Energy Conference, 1986

have peaked. Oil output figures since 1970, tabulated in Table 3–6, indicate that the Western world was particularly affected by the OPEC-induced oil price rises in 1973–74 and 1979–80, whereas the countries of the Eastern Bloc were relatively unperturbed.

To illustrate the magnitude of the production rates outlined in the table, we need only to note that the proven reserves of oil for the whole of Western Europe, principally those under the North Sea, would be required to supply little more than a single year of world consumption. Furthermore, the North Sea discovery was one of only three major finds made in the world during the past 15 years, the others being in Mexico and at Prudhoe Bay in Alaska.

We face the clear prospect, therefore, of a decreasing contribution of oil to the world's primary energy needs. Natural gas could reach that position as well midway through the next century, after which time the quantities of oil and gas recovered are expected to drop as deposits are exhausted. Now, if oil and gas are to decline in importance as sources of energy in the twenty-first century and beyond, how will it be possible to meet the energy demands of a still-growing world population? Clearly, if major increases in energy production are to be achieved, they must come from coal and nuclear sources; the renewables, hydroelectric energy, and unconventional oil and gas, primarily from sands and shales, will be important also, but not nearly to the same extent. Similarly, potentially new sources,

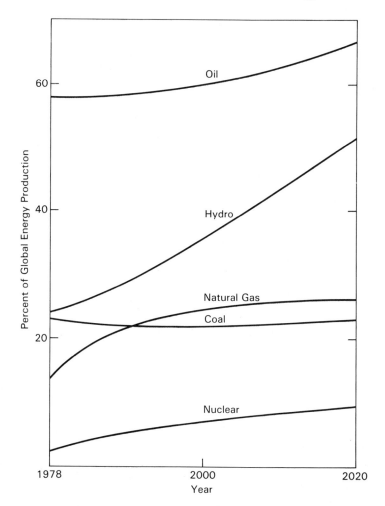

Figure 3–14 Energy production in less developed regions. Source: Frisch, by permission of the World Energy Conference, 1983

from thermonuclear fusion for example, are not expected to make an appreciable contribution to the world's energy needs before 2050.

More economic and efficient use of power—nuclear power in particular—is likely to be achieved during the next 40 years by the development of energy storage techniques which will permit reactors to operate at higher outputs. One such storage medium is hydrogen, produced by the electrolysis of water. Another may result from the development of materials which become superconducting at liquid nitrogen temperatures and above.

The greatest single fossil fuel resource available is coal, as indicated in Table 3–7. The resource-to-demand ratios for both coal and uranium tabulated in this table appear comfortably large. However, as indicated earlier in Figure 3–11, the utilization of these two resources could grow appreciably during the coming decades. Then, resource-to-demand

TABLE 3-6   WORLD OIL PRODUCTION, 1970–85

Year	NCW*	10^3 m³ per day CPE**	World total
1970	6,369	1,282	7,651
1971	6,686	1,388	8,074
1972	7,028	1,484	8,512
1973	7,685	1,617	9,302
1974	7,568	1,750	9,318
1975	6,972	1,880	8,852
1976	7,547	2,003	9,550
1977	7,824	2,122	9,946
1978	7,784	2,244	10,028
1979	8,165	2,290	10,455
1980	7,628	2,341	9,969
1981	7,049	2,347	9,396
1982	6,686	2,381	9,067
1983	6,601	2,407	9,008
1984	6,812	2,421	9,233
1985	6,708	2,404	9,112

* Non-communist world.

** Centrally planned economies: China, U.S.S.R., Albania, Bulgaria, Cuba, Czechoslovakia, East Germany, Hungary, Kampuchea, Laos, Mongolia, North Korea, Poland, Romania, Vietnam, Yugoslavia.

Source: *BP Statistical Review of World Energy*, 1986

ratios, particularly for uranium, might be measured in decades rather than centuries. However, once fast breeder reactors and advanced thermal reactors are substituted for current thermal reactor designs, known uranium and thorium resources will be more effectively utilized and the energy contribution from nuclear power should be increased as a consequence by fifty to sixtyfold. Nuclear fission would then provide considerably more energy in the future than that available from all remaining fossil fuel resources.

Another important factor in determining the market share that each energy source will capture is the economic competitiveness between them. In North America, for example,

TABLE 3-7   WORLD NONRENEWABLE ENERGY RESOURCES

Source	Proved resources, EJ	World demand 1985, EJ	Resource/demand ratio
Coal	40,123	102.1	393
Crude oil	4,344 ⎫	125.8	39
Oil sands and shales	582 ⎭		
Natural gas	3,314	69.0	48
Uranium*	1,164	15.2	77

* Current thermal reactors.

Sources: *BP Statistical Review of World energy*, 1986; Frisch, 1986

TABLE 3–8 MONTHLY RESIDENTIAL
ELECTRIC BILLS
1000 kW.h, January, 1985

City	Cost $ U.S.
New York	143
Detroit	98
Chicago	73
Atlanta	59
Halifax	49
Toronto	38
Montreal	29
Winnipeg	27

Source: Ontario Hydro, 1985

home heating is an important market, and the principal competition for it is between fuel oil, natural gas, and electricity. Regional variations can have a decisive influence on the results, as illustrated by the figures in Table 3–8 for the cost of electricity.

Example 3.1

If fuel oil sold for $1.00 U.S. per gallon in the United States and Canada in 1985, in which of the cities listed in Table 3–8 would electric heating have been competitive with fuel oil? Use a conversion efficiency of 60% for oil heating and 100% for electric heating.

Solution The heat content of fuel oil is 38.675 TJ/1,000 m³. Hence, the heat content of 1 gallon = $(38.675 \times 10^{12})/(6,293 \times 42) = 1.46 \times 10^8$ J.

But in most furnaces, no more than 60% of the heat evolved is captured in the house, the balance being lost up the chimney. In other words, the conversion efficiency is 60%.

So the heating cost with oil @ $1.00 for $0.6 \times 1.46 \times 10^8$ J = $11.42/GJ.

For electric heating at 100% conversion efficiency, the cost per gigajoule, utilizing the figures in Table 3–8, is as follows (1,000 kW.h = 3.6 GJ):

City	Cost/GJ, $U.S.
New York	39.72
Detroit	27.22
Chicago	20.28
Atlanta	16.39
Halifax	13.61
Toronto	10.56
Montreal	8.06
Winnipeg	7.50

Hence, electric heating was cheaper than oil heating in Toronto, Montreal, and Winnipeg in 1985.

Yet another factor influencing the future use of energy sources is the environmental and social impacts of each source. This applies particularly to the use of coal and nuclear power. The hazards of mining and transporting huge quantities of coal and the disabilities, such as black lung disease, generated in miners, are important considerations. The environmental damage incurred by exhausting into the atmosphere ever-growing quantities of residues from the burning of coal may place limits on the use of that fuel. Although the removal of sulfur and nitrogen-containing gases from the combustion process offers the greatest technical challenge, reduction in heavy metal effluents is also an important task. Then, looming over all such environmental concerns in its potential impact on the world's climate is the effect of significantly increasing the CO_2 content in the atmosphere (see Chapter 5).

Nuclear power is considered by many technically trained people to be the most benign of all energy sources as far as environmental impact is concerned. Others, however, are fervent campaigners against the nuclear option, claiming that the potential dangers from low-level radiation hazards, nuclear accidents, and the proliferation of nuclear weapons render it intolerable as a future source (see Chapter 15).

Other energy sources also have risks associated with their production. Many of these problems are environmental in nature, and are considered further in the next section as well as in other chapters.

3.4 ENVIRONMENTAL IMPACTS OF ENERGY DEVELOPMENT

Having examined the past, present, and future patterns of energy production and consumption, we turn to a discussion of the environmental impacts of these technologies. Many books and reports have been written on the environmental impacts of energy development. Most discussions of energy technology and politics include an account of environmental aspects, since it is generally recognized that environmental considerations can constrain energy production, especially when the environment is exposed to risk. Most texts on environmental issues contain sections on topics such as air pollution from automobiles and power plants, thermal pollution from power plants, and water pollution from oil spills. Some books and reports which explore the energy-environment issue in more detail include those by Fowler (1975), Tuve (1976), Chigier (1981), and Biswas (1973). The matrix format of listing the impacts of energy on environment developed in this last book is followed here.

It is useful to discuss first some recurring themes in the energy-environment issue. These are (*i*) a comparison of current energy usage in an industrial society with the "background" of solar radiation which we receive and with the minimum energy necessary as food for survival, (*ii*) the volumes of fuels which we generate, process, and use, (*iii*) the changing picture of energy availability, and (*iv*) the issue of toxic substance dispersion from energy development.

Background of solar radiation and food energy.   Commercial energy consumption by society can be compared with the "background" of energy received by solar radiation and the minimum food energy necessary for survival. An example is as follows, and it is suggested that the reader undertake a similar calculation for a specified country or region (see Problem 3.10).

Example 3.2

Compare and discuss the energy consumption or flow rates in units of J/year, gigawatts (GW), and watts per person as fuel, food, and solar radiation for a country of population density 20 persons/km^2, an area of 1 million km^2 and a fuel energy consumption rate of 250 GJ per person per year. Solar radiation reaching the ground is approximately 150 W/m^2. The average person consumes food containing 2,000 "calories" (actually kilocalories) per day. Note: 1 (kilo) calorie = 4,182 J.

Solution

FUEL ENERGY

$$\text{Total energy (J/year)} = 250 \times 10^9 \text{ (J)} \times 20 \text{ (persons/km}^2) \times 10^6 \text{ (km}^2)$$
$$= 5 \times 10^{18} \text{ J/year}$$
$$\text{(Energy is usually expressed in watts = J/s)}$$
$$\text{J/s = watt} = 5 \times 10^{18}/(365 \times 24 \times 3{,}600) = 1.59 \times 10^{11} \text{ W} = 159 \text{ GW}$$
$$\text{Watts/person} = 1.59 \times 10^{11}/20 \times 10^6 = 7{,}950$$

SOLAR ENERGY

$$\text{Total energy (watts)} = 150 \text{ (W/m}^2) \times 10^6 \text{ (m}^2/\text{km}^2) \times 10^6 \text{ (km}^2)$$
$$= 1.5 \times 10^{14} \text{ W}$$
$$= 150{,}000 \text{ GW}$$
$$\text{J/year} = 1.5 \times 10^{14} \text{ (J/s)} \times 3{,}600 \text{ (s/h)} \times 24 \text{ (h/day)} \times 365 \text{ (day/year)}$$
$$= 4.7 \times 10^{21} \text{ J/year}$$
$$\text{Watts/person} = 1.5 \times 10^{14}/20 \times 10^6 = 7.5 \times 10^6$$

FOOD ENERGY

$$\text{J/year} = 2{,}000 \text{ ("cal"/day} \cdot \text{person)} \times 4{,}182 \text{ (J/"cal")} \times 365 \text{ (days/year)}$$
$$\times 20 \times 10^6 \text{ (person)}$$
$$= 6.1 \times 10^{16} \text{ J/year}$$
$$= 1.94 \times 10^9 \text{ J/s} = 1.94 \text{ GW}$$
$$\text{Watts/person} = 1.94 \times 10^9 \text{ (w)}/20 \times 10^6 = 97$$

SUMMARIZING THE DATA, WE HAVE:

	J/year	GW	Watts/person	Ratio
Fuel	5×10^{18}	159	7,950	1
Solar	4.7×10^{21}	150,000	7.5×10^6	943
Food	6.1×10^{16}	1.94	97	0.012

COMMENTS

Note that the population density used in this example, 20 persons/km^2, is typical of the U.S., about one-tenth that of the U.K., and ten times that of Canada. Human energy needs are small compared to solar radiation, but it is the wide dispersal or "dilution" of solar energy which makes it difficult to exploit. Food energy needs are very small, although each person is generating nearly the same heat as a 100-W bulb.

The fuel energy consumption of 7,950 watts (7.95 kW) is 80 times the food consumption, thus leading to statements that "modern industrialized man has the equivalent of 80 energy slaves."

If the population density increases by 100 times or more (as may occur in urban areas) fuel and solar energy flows become more equal, particularly during low rates of solar insolation in winter. This leads to local climatic modifications. Hence, winters in large cities are often milder than those in suburbs.

Fuel volumes.   The potential for environmental disruption by energy-related activities can also be elucidated by calculating the mass or volume of fuel used by each person per year. Obtaining this mass of fuel, separating it from other undesired materials, transporting it, and eventually burning it all have environmental impacts. Table 3–9 indicates the energy densities of selected fuels in units of MJ/L (megajoules or 10^{12} J/liter). By "energy density," we mean the amount of energy contained in a unit volume of the fuel. Coupling this to the per capita energy usage shows that we are each responsible for the production and movement of a considerable volume of energy-related materials annually. For example, an individual in an industrialized country may "consume" 400,000 MJ per year, which corresponds to over 10,000 L of oil, 6,000 L of coal, 11 million L of natural gas, or 45,000 L (45 m³) of wood. These volumes are substantial, especially when it is noted that obtaining the fuel may require the removal of other material (e.g., rock in coal mining).

Although each stage from extraction to marketing may be disruptive in itself, particular problems arise from combustion when the fuel contains even small percentages of undesirable materials, such as sulfur. For example, even if only one percent of the volume of coal or oil is sulfur, considerable quantities of sulfur, usually in the form of sulfur dioxide, are emitted into the atmosphere following combustion. This creates major air and water pollution problems, as discussed in Chapters 5, 7, and 13.

Uranium and other nuclear fuels present a unique situation, in that the volumes required for energy production are exceedingly small. However, the hazards associated with radioactivity in these materials counterbalance this small amount, since radioactivity is an exceedingly potent "poison" (see Chapter 15).

TABLE 3–9 ENERGY DENSITIES OF
SELECTED FUELS

Fuel	Energy density, MJ/L
Natural gas	0.036
Oil or gasoline (petroleum)	35
Coal (solid)	65
Wood	9

Availability.   In the exploitation of energy resources, there has been in the past an understandable tendency for an industrialist to exploit first the cheapest, closest, richest, and least contaminated energy sources. These sources lead to maximum profit. Rightly or wrongly, less desirable sources tend to be left for future generations. Accordingly, rich coal seams and abundant oil supplies close to the surface and low in sulfur tend to be exploited first; uncomplicated and nearby hydroelectric schemes were the first to be developed; onshore petroleum has been preferred to offshore; where available, natural gas has been used as an energy source in preference to coal or coal gas; and cheap imported oil, when it is available, has been preferred to expensive domestic oil. A general consequence is that future energy supplies may (1) tend to come from more remote areas, incurring longer transportation distances, (2) be more contaminated with undesirable elements, and (3) be more "dilute" (e.g., coal seams may be thinner or oil production rates lower). Also, the fossil fuels exploited may lie deeper in the earth and thus be more difficult and expensive to find and produce. In total, it is likely that future energy developments will be more environmentally disruptive and will often occur in areas which have traditional competing uses such as fishing, agriculture, and tourism. These areas may be populated by individuals who have a life style which has relatively light energy demands compared to the industrial urban energy system and who understandably resent the intrusion of energy developments into their "backyard."

Toxic substances.   In 1962, Rachel Carson published her now classic book *Silent Spring*, an account of the adverse effects of pesticides, largely used for agricultural purposes, on nontarget organisms (or victims) such as birds. Since then, there has grown a widespread concern about the dissemination of toxic substances throughout the environment. These may be metals such as lead or mercury, organics such as DDT or PCBs (polychlorinated biphenyls), inorganics such as sulfur compounds or asbestos, or radioactive materials. The toxic effects are, fortunately, rarely lethal to humans, but there may be lethal or sublethal effects on other organisms, resulting in ecological changes. For example, there may be loss of reproductive capacity in fish or birds, or behavioral changes affecting predator-prey relationships. It is thus prudent for us to observe the natural environment closely for change, since a toxic effect on birds may act as a warning of a potential human effect. We are, after all, constructed from the same biochemical building blocks.

The energy industries handle considerable quantities of these hazardous substances, for example, uranium for nuclear fuels and sulfur compounds from the combustion of oil and coal. Of particular concern is the generation of substances which induce mutations or cancer. Examples are polynuclear aromatics, such as benzopyrenes, and various heterocyclic organic compounds, which include the elements nitrogen and sulfur. These compounds may be produced during combustion or in the synthesis of synthetic liquid fuels from coal. They are present in exhaust from diesel engines; thus, any change from gasoline to diesel-powered automobiles, which may be desirable from an energy viewpoint, can cause increased adverse health effects.

In the past, engineers and scientists have often failed to predict, and thus to control, the adverse effects of new energy developments on the environment and on human health. Hydroelectric dams have caused fish kills and siltation, sulfur dioxide has harmed forests and lakes, and uranium mine tailings have produced undesirably high levels of radioactivity. These mistakes are partly excusable in that there was no well-tried mechanism for assessing future environmental impacts. No longer do we have this excuse, however. Practical techniques and (in many countries) regulatory mechanisms now exist to ensure that these avoidable impacts do not occur, or at least that they are minimized (see Chapter 16). One approach is to compile environmental impact matrices in which we examine on the one hand the environment and on the other the activity, in this case, the energy development, and ask in a systematic, exhaustive manner what the impacts will be of each component of the energy development on each component of the environment. We use this approach in the next section.

3.5 ENVIRONMENTAL IMPACT MATRICES

The environmental impact matrix was introduced in Chapter 2 and again provides a convenient inventory and display of the impacts of energy production. As illustrated in Figure 3–15, the matrix is compiled with the horizontal axis listing the components of the development, such as exploration, mining, transportation, or utilization. On the vertical axis are components of the environment—the atmosphere, the hydrosphere, the lithosphere, and human impacts. The matrix indicates potential interactions between activities and the environment and provides answers to such questions as; Does oil exploration affect water quality? (Answer: yes, by oil spills.) Or does the use of hydroelectricity affect air quality? (Answer: probably no.) With the methodical matrix method, potential impacts are not likely to be overlooked.

In the matrices adopted in this section for the environmental impacts of energy, we use four columns and four rows.

Col. 1. Exploration
This includes the search for fuel sources prior to any production. Exploration is often unsuccessful and possibly carried out in remote areas in competition with other uses of the land or water.

Col. 2. Extraction, Production, and Processing
This includes the removal of the fuel from its present location by mining, drilling wells, constructing dams, refining the fuel (e.g., in an oil refinery), and, in the case of electricity generation, the production of thermal effluents.

Col. 3. Transmission
This is the transportation of fuel from its production site by pipeline, road, rail, tanker, or electricity transmission line to the site of use.

Col. 4. Use and Disposal
This includes the generation of products of combustion, spent fuels, and oxides of sulfur, nitrogen, and carbon.
    Note that the processes do not necessarily occur in exactly the order just given.

Environment	Exploration	Extraction, production, processing	Transmission	Use and disposal
		Energy activity		
Atmosphere	Emissions of H_2S and hydrocarbons as a result of a blowout	Refinery emissions of SO_2, H_2S, CO_2, NO_x, and hydrocarbons	—	Emissions of SO_2, CO_2, and hydrocarbons
Hydrosphere	BLOWOUTS AND SPILLS FROM EXPLORATORY WELLS AT SEA, LEADING TO OIL CONTAMINATION	BLOWOUTS AND SPILLS Brine and drilling chemicals disposal Refinery effluents	TANKER ACCIDENTS, LEADING TO OIL CONTAMINATION	Groundwater contamination by leaking tanks
Lithosphere	Blowouts and spills on land	Blowouts and spills Sludge disposal	Pipeline construction and spills Damage to permafrost	Used oil disposal
Human impacts	Disruption of life style	Interference with fisheries	Interference with fisheries or land use Disruptions of life style during construction	Hydrocarbons and polynuclear aromatic hydrocarbons from combustion

Figure 3-15 Environmental impacts of oil

For example, oil is usually extracted, transported, and then refined, whereas natural gas is extracted, processed to remove sulfur, and then transported.

Row 1. Atmosphere
This includes the immediate atmospheric environment around the development and the impact of the long-range transportation of pollutants, especially when tall stacks are used to disperse the pollutant more widely in an attempt to solve a local problem by dilution.

Row 2. Hydrosphere
This includes fresh water—rivers, lakes, and groundwater—and the oceans. We also include biota indigenous to the water, ranging from bacteria to fish and marine mammals.

Row 3. Lithosphere
This includes soil, rock, and bottom sediments of rivers, lakes, and oceans, including the attendant vegetation and animal life.

In the above three categories, we have included the resident biota as potential victims of the environmental impact. We give special consideration to human impacts in the final category:

Row 4. Human Impacts
This includes human welfare in its broadest sense, including effects on health, the economy, security, life style, social structure, and aesthetic considerations.

The impacts with the greatest severity are shown capitalized in Figures 3–15 through 3–19. Some impacts are "chronic," with continuous emission and a continuous effect, e.g., oil refinery wastewater discharge, while others are occasional and accidental, with a massive effect, which may occur once in five years and last for a few months, e.g., an oil spill. Comparing these fundamentally different impacts is challenging.

3.5.1 Environmental Impacts of Oil

Figure 3–15 gives the environmental impacts of oil.

During oil exploration, either onshore or offshore, there is a risk of blowout should control of the well be lost. This can lead to severe and prolonged oil spills, which are harmful to the marine environment. The 1979 Ixtoc blowout in the Gulf of Mexico is an example. Regaining control of a blowout can take considerable time and often requires drilling a relief well to interrupt the oil flow. This problem is particularly severe in northern climates, where ice formation and movement may seriously impede or even prevent drilling of relief wells.

Oil spills may result in mortality to birds and contamination of shorelines, resulting in severe biological effects on intertidal and near-shore organisms, including valuable shell fisheries. There may also be fouling of vessels, nets, and harbor facilities, requiring expensive cleanup. The impact of oil on the open ocean environment is more difficult to assess, but it is likely that the spill will have some effect on fisheries and, in general, on organisms present in the ocean surface waters. The issue of oil pollution has been reviewed in a number of texts, including those by Nelson-Smith (1973) and Malins (1977).

Exploration, production, and transmission activities may have a profound effect on life styles in remote areas. For example, in northern Canada and Alaska—especially in traditional Inuit communities—people are not well equipped to withstand the social pressures of modern industrial life.

There is also a risk of spillage during oil production and collection on land prior to its transmission to petroleum refineries. At these refineries there is a potential for emission of hydrocarbons, sulfur oxides (SO_x), which can cause lake acidification and human respiratory problems, hydrogen sulfide (H_2S), which is highly toxic and odorous, carbon dioxide (CO_2), which may lead to the "greenhouse effect" (see Chapter 5), oxides of nitrogen (NO_x), which cause photochemical smog and acidification, and some odorous substances. Refineries also generate liquid effluents which may contain hydrocarbons, phenols, ammonia, and other toxic substances. Dissolved organics in these effluents are normally treated by biological oxidation processes, and oil is removed by physical separation. Sludges are inevitably formed which consist of mixtures of hydrocarbons and organisms often contaminated with metals, particularly nickel and vanadium, which are usually present in crude oils.

During oil production, there is potential for damage due to contamination of the local environment by drilling chemicals, muds, and brine (a salt solution), which is often produced from the oil formation in association with the crude oil.

Oil is transported both by tanker and by pipeline in very large quantities, and neither of these is immune from causing environmental pollution. In fact, there have been several impressive oil tanker accidents resulting in widespread contamination of coastal regions, notably the Torrey Canyon and the Amoco Cadiz incidents in the English Channel. The effects are generally similar to those of blowouts discussed earlier. Oil releases from pipelines are usually less severe because the oil is more easily controlled, although there may be fouling of agricultural land leading to loss of productivity. A special problem is the construction of pipelines for transmission of oil or natural gas in northern climates where the ground is underlain by permanently frozen ground called permafrost. Normally, oil is pumped hot in order to diminish its viscosity and thereby reduce power requirements. Therefore, the pipeline must be well insulated to prevent the ground from thawing, which could result in land subsidence called thermokarst. A common engineering practice is to build the pipeline on piles, or in trenches with adequate insulation, thus separating it from the vulnerable ground. The first major pipeline of this type was the trans-Alaska pipeline from Prudhoe Bay to Valdez. Such pipelines often pass through sparsely populated wilderness areas where the construction activities can have severe impacts on local life styles, fisheries, trapping, and land use generally. A controversial, and as yet unresolved, issue is that of the possible interference with caribou migration by above-ground pipelines which may effectively present a barrier to these animals in northern Canada and Alaska.

The use and disposal of crude oil results in emissions of hydrocarbons, oxides of sulfur, nitrogen, and carbon dioxide, all of which may cause environmental and health problems. Some of the hydrocarbons produced during incomplete combustion are polynuclear aromatics such as benzopyrenes, which are potential carcinogens. A final impact is the disposal of used oil, particularly lubricating oil, which may be contaminated with lead. In principle, it is obviously desirable to reprocess this oil in order to recover the valuable hydrocarbons and eliminate this source of pollution, but often it is disposed of in landfill sites.

3.5.2 Environmental Impacts of Natural Gas

Figure 3–16 shows the environmental impacts of natural gas. It is readily apparent that these impacts are considerably less severe than those of oil. Generally, the impact of releases of natural gas (methane) into the environment is minor, provided that there is no accompanying fire or explosion. The methane tends to dissipate rapidly into the atmosphere, having little adverse environmental effect.

Again, there is a risk of emission of hydrocarbons during blowouts, but probably more severe is the risk of hydrogen sulfide (H_2S) emissions during exploration and production. Hydrogen sulfide, which normally accompanies natural gas in petroleum reservoirs, is highly toxic. The usual procedure is to remove the hydrogen sulfide from the methane at a "gas plant" close to the well. Repeatedly, gas plant malfunctions have resulted in severe H_2S emissions, requiring rapid evacuation of large areas downwind. As with oil, there is an impact from the disposal of brine and drilling chemicals. Also, construction of gas pipelines can be disruptive and may be particularly difficult in the permafrost regions.

One area of increasing concern is the possibility of severe accidents occurring as a result of transportation of liquefied natural gas (LNG). It is not economical to transport gaseous natural gas by tanker; liquefaction greatly increases the energy density of the cargo, permitting more to be carried per voyage. These cargoes must be maintained at very low temperatures, i.e., below the boiling point of natural gas, or at elevated pressures. If the containing vessel fails, there will be emission of the liquid methane, which then evaporates very rapidly, creating a highly hazardous condition in which fire and explosion are likely. There have been several LNG fires, but fortunately no major incidents involving, for example, loss of an entire cargo. Extreme care is necessary in designing the liquefaction, transportation, and gasification facilities, and there is a compelling incentive to locate these facilities far from human habitation.

3.5.3 Environmental Impacts of Coal

The environmental impacts of coal are presented in Figure 3–17.

Normally, coal is recovered by one of two processes. In strip mining, coal which lies close to the surface is removed by earth-moving equipment, the overburden of soil being removed and stored for later replacement. The area can then be filled in and revegetated after the removal of the coal. In mining for deeper deposits with substantial quantities of sulfur and metal compounds which have remained inert in their subsurface environment throughout geological times, exposure of the mined minerals to oxygen and water initiates a series of reactions, particularly oxidation and dissolution, which were previously not achievable. The result is the formation of oxides of sulfur (and hence sulfuric acid), metal solutions, phenols, and various other compounds, many of which are environmentally

Environment	Energy activity			
	Exploration	Extraction, production, processing	Transmission	Use and disposal
Atmosphere	Emissions of gas and H_2S during an accidental blowout	Gas plant emissions of H_2S, SO_2, and hydrocarbons	—	Emissions of CO_2, NO_x
Hydrosphere	Blowouts	Blowouts & drilling	—	—
Lithosphere	—	Disposal of chemicals	Construction of pipeline Damage to permafrost	—
Human impacts	—	LNG ACCIDENTS H_2S emissions	LNG ACCIDENTS Disruption of life style during construction	—

Figure 3–16 Environmental impacts of natural gas

	Energy activity			
Environment	Exploration	Extraction, production, processing	Transmission	Use and disposal
Atmosphere	—	Emissions of SO_2 and PNAs from processing to gas or liquid fuel Coal dust dispersal	—	EMISSIONS OF SO_2, NO_x, CO_2, and particulates
Hydrosphere	—	LEACHING OF ACIDS AND METALS Organic compounds formed with ''synfuels'' Siltation	—	Thermal effects
Lithosphere	—	DISRUPTION FROM STRIP MINING AND SUBSIDENCE Slag heaps	—	Fly ash disposal
Human impacts	—	LUNG DISEASE MINE SAFETY	—	Exposure to emissions from combustion and coke ovens

Figure 3–17 Environmental impacts of coal

harmful. Coal piles and slag heaps thus can generate considerable quantities of these substances, and the leachate must be controlled and prevented from entering surface waters or groundwaters.

An important human issue is that of the safety and health of those working in the coal mines. Problems of mine safety are well known, and there have been many tragic losses of life in underground accidents. Less dramatic, but as devastating, is the severe toll on human health by diseases such as black lung, which are caused by exposure of workers to particulates or dust in coal mines.

A related and growing concern is human exposure to the products of synthetic fuel production from coal. Coal is in many respects an inconvenient fuel when compared to crude oil and natural gas. Conversion of the coal into a liquid or gaseous fuel results in easier transportation, distribution, and use. Traditionally, coal has been used to produce "coal gas" (a mixture of carbon monoxide and hydrogen) in many countries, although in the U.K. it has been displaced recently by the more convenient natural gas produced from the North Sea. Considerable research is under way into coal gasification, a liquefaction process which usually involves high-temperature and often high-pressure conversion of the coal into liquid or gaseous products. During these processes, the synthetic organic compounds formed often have structures which do not normally occur in biological organic material, and they can be very toxic. Notable are polynuclear aromatic hydrocarbons (PNAs), organo-nitrogen and sulfur compounds, many of which have offensive smells. If coal conversion to gas or liquid fuels plays a major role in future energy economies, then there may be significant environmental impacts resulting from the formation and dispersion of these substances into the air and water environments.

As the demand for electricity increases and hydroelectric sites become fully exploited, and as oil becomes less available, there will be a tendency for coal to be again used as a source of electric power. This may result in the emission of large quantities of oxides of sulfur, nitrogen, and carbon, as well as particulate matter, into the atmospheric environment. It is well established that emissions of sulfur and nitrogen oxides into the atmosphere over a prolonged period of time can result in the acidification of downwind regions, often many hundreds or thousands of kilometers distant. Notable in this regard is the acidification of lakes in Scandinavia from emissions in Central Europe (especially the Ruhr Valley) and the U.K., and in Ontario and New York State from major emissions in the U.S. (from Ohio in particular). Especially vulnerable are lakes which have low buffering capacity, i.e., which contain only small amounts of natural alkaline material such as bicarbonate ion which can neutralize the acid that falls into the lake in rain or snow or by dry deposition. The problem of acidification, discussed in detail in Chapter 5, is probably one of the most severe energy-caused environmental problems. A considerable technical effort is being devoted to devising systems for removing sulfur from the coal or from the flue gas. Unfortunately, these processes are likely to be very expensive.

The generation of electricity from coal involves a loss of approximately two-thirds of the energy, in the form of waste heat, from the generating plant. The waste heat is dissipated through both the flue gases and the cooling water. Discharge of the cooling water can result in "thermal pollution" of the local aquatic environment; for example, a lake or river may be subjected to an unusual temperature rise. Most serious are situations in which

		Energy activity		
Environment	Exploration	Extraction, production, processing	Transmission	Use and disposal
Atmosphere	—	—	—	—
Hydrosphere	—	SILTATION, CHANGES IN FLOW CHARACTERIS- TICS OF SURFACE WA- TER & GROUNDWATER	—	—
Lithosphere	—	SUBMERGENCE OF LAND, LOSS OF ANIMAL HABI- TAT	Transmission lines	—
Human impacts	—	Disruption of life style from loss of land	—	—

Figure 3–18 Environmental impacts of hydroelectric development

a river is subjected to repeated heating by several power plants. Under conditions of low flow, and especially during the summer, the temperature of the water may rise beyond limits tolerable to aquatic life and fish mortality may result. In such cases, the use of air as a cooling medium is preferable, although it is generally more expensive.

3.5.4 Environmental Impacts of Hydroelectric Development

Figure 3–18 shows the environmental impacts of hydroelectric development. Of all the forms of energy production, hydroelectric power is probably the least environmentally disruptive, in the sense that there are rarely any chemical effects. The major effects result from changes to the local region arising from submergence of land, loss of animal habitat, and changes in local hydrology—for example, alteration of flow characteristics of rivers and groundwater. In some cases, there may be severe siltation problems. Increasingly, remote hydroelectric sites are being sought, often in wilderness areas with a sparse population of individuals who rely on trapping and a simple life style. To them, the loss of their land or the disruption of their life style can be devastating.

3.5.5 Environmental Impacts of Nuclear Power

Figure 3–19 lists the environmental impacts of nuclear power.

Paramount is the question of environmental contamination by radionuclides as a result of mining, accidents involving discharge of fuel, failure of the nuclear reactor system, or contamination from spent fuel. In the minds of many, the nuclear power issue is intimately, but irrationally, linked with the horrors of Hiroshima, Nagasaki, and nuclear warfare in general. No other environmental issue has developed such widespread concern.

Except for the 1986 explosion at the Chernobyl nuclear station in the Ukraine, the remaining few nuclear power plant accidents, like the Three Mile Island reactor breakdown in Pennsylvania in 1979, have all resulted in very minor releases of radioactive material

	Energy activity			
Environment	Exploration	Extraction, production, processing	Transmission	Use and disposal
Atmosphere	—	Accidents Radon emissions from mine tailings	—	—
Hydrosphere	—	Accidents Leachate from mine tailings	—	Thermal effects
Lithosphere	—	Accidents Tailings contamination	Transmission lines	DISPOSAL OF SPENT FUEL AND WASTE
Human impacts	—	ACCIDENTS AND MINE-PLANT EXPLOSIVE MINING HAZARDS	Accidents during fuel transport	Exposure to wastes TERRORISM

Figure 3–19 Environmental impacts of nuclear power

and no recorded deaths. Even if the Chernobyl victims are included (31 deaths at the time and an estimated 2,000 extra cases of cancer in Europe over the next 50 years),* a much greater loss of life and damage to health can still be attributed to the coal industry (see also Section 15.2.1.).

There have been cases of undesirably high exposure of miners to radioactive materials, both in the mine and in mineral processing. Also, mine tailings may contain residual but low-level radioactivity, and in some cases these tailings have not been disposed of in the proper manner. There is much concern about the issue of disposing of or reprocessing spent fuel. This fuel is highly radioactive and will remain so for centuries. Processing the fuel to ensure that this and future generations are secure from exposure is a considerable technological challenge, but possibly an even greater sociopolitical challenge. A worrisome aspect is the possibility of terrorism resulting from a small group of individuals obtaining quantities of radioactive material.

As with the generation of electricity from coal, there may be a thermal effect on local water used for cooling purposes.

The disposal of nuclear wastes is discussed more fully in Chapter 15.

Example 3.3

A 100-MW coal-burning electricity generating station operates at 33% efficiency burning coal containing 5% ash and 2% sulfur. Assume a coal heat content of 30 kJ/g. If 95% of the ash and 50% of the sulfur are trapped before emission from the stack, calculate (a) the emission rate to the atmosphere of ash and sulfur (in SO_2), in g/s, and (b) the volume of SO_2 generated at 20°C and atmospheric pressure. (c) If the sulfur is emitted into an urban area of stagnant air 500 m high and 5 km in diameter, how long it would take for the SO_2 concentration to reach the undesirable level of 0.3 ppm (part per million)?

Solution

(a)
$$\text{Coal use} = \frac{100 \times 10^6 \text{ J/s}}{30 \times 10^3 \text{ J/g} \times 0.33} = 1.0 \times 10^4 \text{ g/s}$$

	Removed	Emitted
Ash production $= 1 \times 10^4 \times 0.05 = 500$ g/s	475 g/s	25 g/s
Sulfur production $= 1 \times 10^4 \times 0.02 = 200$ g/s	100 g/s	100 g/s

(b) The molecular weight of SO_2 = 64. Therefore,

$$100 \text{ g/s } SO_2 = \frac{100}{64} = 1.56 \text{ moles}$$

From Equation 8.34,

$$\text{Volume } SO_2 = \frac{nRT}{P} = \frac{1.56 \text{ (mol)} \times 8.314 \left(\frac{\text{Pa m}^3}{\text{mol K}}\right) \times 293 \text{ (K)}}{101,325 \text{ (Pa)}}$$

$$= \underline{0.037 \text{ m}^3/\text{s}}$$

* Editorial, *London Free Press*, March 26, 1987, London, Ontario.

(c) $\text{Volume of air} = 500 \times \dfrac{\pi}{4}(5000)^2 = 0.98 \times 10^{10} \text{ m}^3$

$$\text{Concentration of SO}_2 = \frac{\text{Volume of SO}_2}{\text{Volume of air}}$$

$\begin{matrix}\text{Volume of SO}_2 \\ \text{for 0.3 ppm}\end{matrix} = \dfrac{0.98 \times 10^{10} \times 0.3}{10^6} = 2.94 \times 10^3 \text{ m}^3$

$\begin{matrix}\text{Time to reach} \\ \text{0.3 ppm SO}_2\end{matrix} = \dfrac{\text{Volume SO}_2 \text{ (m}^3)}{\text{Emission rate SO}_2 \text{ (m}^3\text{/s)}} = \dfrac{2.94 \times 10^3 \text{ (m}^3)}{0.037 \text{ (m}^3\text{/s)}}$

$= 7.95 \times 10^4 \text{ s} = \underline{22 \text{ hrs}}$

Example 3.4

A petroleum refinery processes 100,000 barrels of crude oil per day and operates with an energy efficiency of 90%. The oil and the petroleum products have energy contents of 35 MJ/L. (a) What are the energy and fuel volume flows to and from this refinery, in J/day? (b) If oil is supplied by a pipeline in which it flows at 0.8 m/s, what pipe diameter is required? (c) If the refinery stocks a five-day supply of oil (in case of interruption of supply), what will be the storage tank volumes required? Suggest dimensions, assuming that the tanks are cylinders of height half their diameter. 1 barrel of oil is 159 liters.

Solution

(a)   100,000 b/day = 159×10^5 L/day of oil = 159×10^2 m^3/day of oil
Product: 90,000 b/day (at 90% efficiency)
                                    = 143×10^5 L/day of product = 14,300 m^3 of product
Energy in: 159×10^5 L/day $\times$ 35 $\times$ 10^6 J/L = 5.57 10^{14} J/day
                    out: at 90% efficiency = 5.01×10^{14} J/day

The pipeline diameter is $Q = uA$, where the velocity $u = 0.8$ m/s and the flowrate $Q = 159 \times 10^2$ m^3/d. Therefore, the area of the pipe is

$$A = \frac{Q}{u} = \frac{159 \times 10^2 \text{ (m}^3\text{/d)}}{24 \times 3{,}600 \text{ (s/d) } 0.8 \text{ (m/s)}}$$
$$= 0.23 \text{ m}^2 = 2{,}300 \text{ cm}^2$$

and the diameter $D = 54$ cm.
(c) The volume of storage tanks required for a five-day supply of oil is

$$V = 159 \times 10^2 \times 5 = 7.95 \times 10^4 \text{ m}^3$$

Choose 10 tanks, each of volume 7,950 m^3 and height d/2. Then

$$\text{Volume of tank} = \left(\frac{\pi d^2}{4}\right)\left(\frac{d}{2}\right)$$

Therefore,

$$d = \sqrt[3]{\frac{7{,}950 \times 2 \times 4}{\pi}} = 27.3 \text{ m}$$

Thus, 10 tanks 27.3 m in diameter and 13.6 m high are required.

3.6 CASE STUDY: CANADA'S ENERGY SITUATION

For nations to balance their energy needs with the energy supplies available will become increasingly difficult in the future. That it can be accomplished on a regional basis has been demonstrated by the Florida Power and Light Corporation, which serves a large portion of the state. On a national scale, neither the United States nor Canada has developed a comprehensive long-term energy policy. Canada's energy situation will serve to illustrate the difficulty in implementing such a plan. Canada is one of the few industrialized countries in the world which has the potential to become self-sufficient in energy supplies for the long term. The fundamental reason is that the country, with its large land mass and relatively small population, has been well endowed with natural resources, including crude oil, natural gas, oil sands, coal, uranium, and hydroelectric potential. Unfortunately, however, most of the resource-rich areas are separated by large distances from the principal consuming regions in the nation. Not only does this add to transportation costs, but because inter-provincial trade occurs, the federal government becomes involved in resource management and taxation policy. As a result, federal-provincial confrontations have arisen in the past over such issues as pricing, revenue sharing, exploration incentives, and ownership of offshore resources, all to the detriment of the consuming public. Hence, it is not certain whether the country will ever achieve long-term energy self-sufficiency.

Canada's National Energy Board has reported on the possible supply of and demand for Canadian energy over the period 1980 through 2000 (NEB, 1981). The results illustrate the nature of the problems and the uncertainties that may be encountered in the provision of future energy supplies.

The primary energy demand in Canada in 1980 was 10.3 EJ. Assuming a population growth averaging 1.0 percent per year and an annual increase of 3.2 percent in real gross national expenditure during the period 1980–2000, energy requirements could grow to 16.2 EJ by 2000 (NEB, 1981). The demands would be met as shown in Figure 3–20. A large fraction of the coal produced, as well as some oil and natural gas, is burned to produce electricity, and the total role of electricity in the energy demand picture is shown in Figure 3–21.

Two important components on the supply side for the future are oil and natural gas. The locations of the major sources are shown in Figure 3–22. Comparing the NEB's base-supply case for crude oil and equivalent products with its middle-demand assumptions, we see that demand will exceed supply by 183,000 m^3/d in 2000, as shown in Figure 3–23. Of course, the level of required imports could be reduced significantly if oil prices, which collapsed sharply in 1986, were to recover to the level where profitable investment in new oil sands recovery plants could be made. However, as long as such recovery does not occur, the modified base supply curve in the figure is not likely to be achieved. Hence, the base supply curve seems the more likely one at present.

The domestic supply of natural gas is more encouraging. As of year-end 1985, established reserves were estimated to total 77.4 EJ (NEB, 1986). Hence, at 1985 production rates of about 2.7 EJ/yr, this equates to nearly 30 years of supply. If all of the available production capability were to be utilized, then frontier and offshore supplies from the Beaufort Sea, the Arctic Islands, and the Atlantic Coast would not be required before 2000.

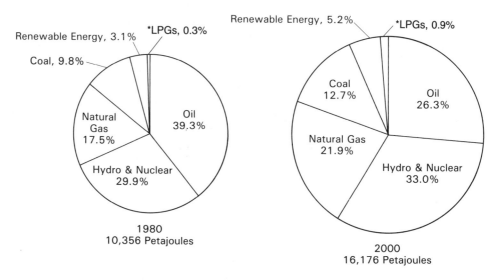

Figure 3–20   Primary energy demand in Canada. Source: NEB, 1981.
* Gas plant liquid petroleum gases only.

Undoubtedly, new gas discoveries will be made which will extend the period of Canada's heavy reliance on natural gas. The prospects for continuing to export large volumes of natural gas to the United States beyond the early 1990s, however, appear doubtful.

In 1985, Canada enjoyed a positive balance of trade in energy commodities, as

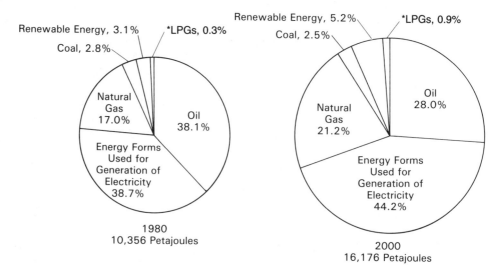

Figure 3–21   Primary energy demand in Canada showing total role of electricity.
Source: NEB, 1981
* Gas plant liquid petroleum gases only.

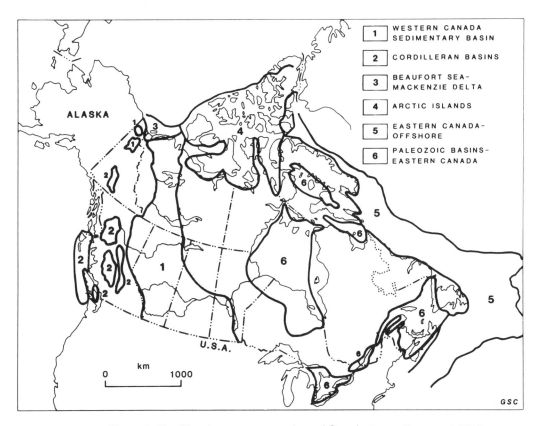

Figure 3–22 Oil and gas resource regions of Canada. Source: Procter et al. (1984)

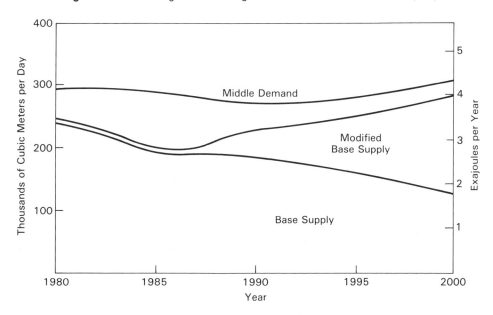

Figure 3–23 Supply of and demand for crude oil and equivalent products in Canada. Source: NEB, 1981

TABLE 3–10 CANADA'S ENERGY TRADE, 1985
($ MILLIONS)

	Exports ($ Can)	Imports ($ Can)	Balance ($ Can)
Crude oil	5,930	3,701	2,229
Oil products	2,318	1,419	899
Natural gas	3,912	—	3,912
Liquid petroleum gases	991	122	869
Coal	1,996	887	1,109
Coal products	34	136	− 102
Electricity	1,408	8	1,400
Uranium	822	28	794
Total	17,411	6,301	11,110

Source: *Energy, Mines and Resources, Canada,* 1986

indicated in Table 3.10. The net dollar inflow to the country from energy trade was $11 billion, (Canadian) approaching the $14.5 billion inflow from total merchandise trade. The energy trade picture should remain healthy into the 1990s, when significant decreases in exports of oil and natural gas are likely to occur. At that time, Canada may have to import large quantities of offshore crude oil.

Many other industrialized countries in the world will also be seeking oil supplies from sources that may not be able to supply the total required. Hence, Canadian politicians may be forced in the 1990s to answer a question that should have been resolved many years ago, namely, how to ensure that the country's vast potential from oil sands, coal, uranium, and hydroelectric sources can be developed and utilized with the objective of ultimately achieving energy self-sufficiency.

Other countries will face more critical decisions on energy needs than Canada or the U.S., and the problems on a global basis will be even greater. The need for cooperative world planning in energy use, population growth, and environmental protection cannot be ignored indefinitely if severe economic and social upheaval are to be avoided.

PROBLEMS

3.1. List the available energy sources for the world in the next 40 years in order of decreasing importance according to your perception.

3.2. If you were designing a cylindrical tank to contain one day's supply of the world's consumption of crude oil, what dimensions would you have to specify? Assume that the height of the tank is half its diameter.

3.3. What volume of uranium fuel would be needed to supply the same amount of thermal energy as in Problem 3.2 from natural-uranium-fueled power reactors operating with a fuel burnup of 7,300 MWd (thermal)/tonne U? (Assume a U density of 9.300 kg/m³.)

3.4. As a group project, evolve a case study similar to the one given in this chapter for Canada, using a country of your own choice.

3.5. The Syncrude oil sands plant in northern Alberta is designed to produce 129,000 barrels per day of synthetic crude oil. Assume that all of this output is delivered to an oil-fired power plant in Edmonton which operates with an efficiency of 35% and a capacity factor* of 80%. What would be the gross electrical output of the power plant?

3.6. How much has world energy consumption increased during the twentieth century, and why has it thus increased?

3.7. Name the three largest nation-producers of crude oil in the world in 1985, and tabulate their outputs. Did any of the three export crude oil? If so, how much?

3.8. Assume you are building a house which can be heated with either oil, natural gas, or electricity. An oil or gas furnace would operate with a conversion efficiency (i.e., heat into the house divided by heat content of fuel) of 60%. An electric furnace is assumed to have a conversion efficiency of 100%. It is estimated that 1,400 gal of light fuel oil would be required to provide heat for the house.

The prices for delivery of the three energy sources in your area for the first heating season will be as follows:

fuel oil	$1.25 per gallon
natural gas	0.75¢ per cubic foot
electricity	4.3¢ per kilowatt-hour

Prices are expected to increase over the next six years in the following way:

fuel oil	25%
natural gas	20%
electricity	10%

Calculate the heating costs for each of the three energy sources for the first heating season and for each of the five succeeding ones, and plot the results. Which heating source would you choose, and why?

3.9. An office building has 500 workers, each of whom spends eight hours in the building, five days per week. There are 150 office machines, each generating 500 W for five hours per working day. Calculate the annual (52-week) amount of energy generated by the workers and the machines assuming a worker "energy" of 100 W. If each worker were to be paid 3¢ per kWh for the heat provided, what would be the total annual payment and the payment per worker?

3.10. A country of area 200,000 km^2 has an annual energy consumption of 10^{15} BTU/year. Calculate the total wattage of solar radiation, fuel, and food for the country for an average density of one person per km^2. Assume a food consumption of 8 MJ/person-day and a solar radiation of 150 W/m^2 reaching the ground.

3.11. An individual living and working on a farm consumes 500,000 MJ of energy per year, of which 40% is oil, 10% is coal, 20% is natural gas, 5% is wood, and 25% is electricity. Calculate the fuel volumes used annually (except for electricity).

* Capacity factor = Actual energy production ÷ perfect production

3.12. An electricity generating plant is to produce 200 megawatts (MW) of electricity from coal. Assuming that it operates at 35% thermal efficiency, calculate the coal utilization in tonnes per year, given that the coal has a heat content of 30 kJ/g. If the cooling-water flow accepts the unconverted 65% of heat and rises in temperature by 5°C, what must be the flow of cooling water in cubic meters per second? The heat capacity of water is 4.18 J/g·°C, and its density is 1000 kg/m³.

3.13. (a) Select an energy development facility such as an oil refinery, hydroelectric dam, electricity generating station, or coal mine in an area with which you are familiar. Calculate the energy flows occurring in the facility from published performance data such as refinery throughput or wattage.

    (b) Compile an environmental impact matrix for the facility in part (a), listing all impacts and identifying environmental technologies which have been installed—for example, wastewater treatment.

REFERENCES

BARNEY, G. O. *The Global 2000 Report to the President of the U.S.* Washington, D.C.: Pergamon Press, 1980.

BISWAS, A. K. *Energy and Environment.* Ottawa: Environment Canada, 1973.

BP Statistical Review of World Energy. London: British Petroleum Company p.i.c., 1986.

CARSON, R. *Silent Spring,* New York: Fawcett Crest, 1962.

CHIGIER, N. *Energy Combustion and Environment.* New York: McGraw-Hill, 1981.

Energy, Mines and Resources, Canada. Current Energy Statistics. Ottawa: 1986.

Energy, Mines and Resources, Canada. Current Energy Statistics. Ottawa: 1987.

Environment Canada Planning and Finance Service. *Report No. 1.* Ottawa: Environment Canada, 1974.

FOWLER, J. M. *Energy and the Environment.* New York: McGraw-Hill, 1975.

FRISCH, J. R. "Energy Abundance: Myth or Reality?" Paper presented at the 13*th* Congress of the World Energy Conference, Cannes, France, October 1986.

FRISCH, J. R. "Energy 2000–2020: World Prospects and Regional Stresses." Paper presented at the World Energy Conference, London, 1983.

HAFELE, W. *Energy in a Finite World*: Global Systems Analysis, Vol. 2. Cambridge, Mass.: Ballinger Publishing Co., 1981.

LEOPOLD, L. B. CLARKE, F. E., HANSHAW, B. B. and BALSEY, J. R. "A Procedure for Evaluating Environmental Impact." *Geological Survey Circular 645*, Washington, D.C.: U.S. Government Printing Office, 1971.

MALINS, D. C. (ed.) *Effects of Petroleum on Arctic and Subarctic Marine Environments and Organisms*, Vols. 1 and 2. New York: Academic Press, 1977.

MUNN, R. E. *Environmental Impact Assessment, Scope 5.* Toronto: Wiley, 1975.

NEB (National Energy Board of Canada). *Annual report 1986*, Ottawa: National Energy Board of Canada, 1987.

NEB (National Energy Board of Canada). *Canadian Energy Supply and Demand, 1980–2000*, Ottawa: National Energy Board of Canada, 1981.

NELSON-SMITH, A. *Oil Pollution and Marine Ecology*, New York: Plenum Press, 1973.

Ontario Hydro. *Statistical Data*. Toronto: Ontario Hydro, 1985.

PROCTER, R. M., TAYLOR, G. C., and WADE, J. A. "Oil and Natural Gas Resources of Canada." Geological Survey Paper 83–31, Ottawa: Energy, Mines and Resources Canada, 1984.

PUTNAM, P. C. *Energy in the Future*. New York: Van Nostrand, 1953.

RUNNALLS, O. J. C. *Planning Energy Supplies for People*. Paper presented at the Canadian Nuclear Association Conference, Saint John, New Brunswick, June 15, 1987.

TUVE, G. L. *Energy, Environment, Populations and Food*. New York: Wiley, 1976.

World Bank. World Development Report. London: Oxford University Press, 1981.

World Energy Conference. *Survey of Energy Resources*, Oxford: Holywell Press Ltd., 1986.

WYATT, A. *The Nuclear Challenge*. Toronto: Book Press, 1978.

4

Natural
Environmental Hazards

Ian Burton

4.1 INTRODUCTION

Natural environmental hazards are those conditions or processes in the environment which give rise to economic damage or loss of life in human populations. Natural hazards are distinguished from human environmental disturbances by the fact that they owe their origin to the "God-given" environment rather than to human action. The most important natural hazards include floods, droughts, earthquakes, tornadoes, and fire. Examples of human environmental disturbances include air pollution, water pollution, improper disposal of toxic wastes, the hazards associated with the failure of the manufactured parts of our environment (e.g., a building or bridge collapse), and the accidental release of radiation from a nuclear generating station, or chlorine gas from a ruptured tank car in a train derailment.

This distinction between natural and human environmental disturbances is useful because it suggests where attention should be directed in seeking to alleviate or control the hazards. In examining flood hazards, for example, it is clearly necessary to focus upon the natural processes of precipitation, runoff, and stream behavior to mitigate their effects. In examining problems of pollution or technological hazards, it is the industrial processes and the design of engineered systems that demand attention, in addition to the natural processes which may be affected by pollution. Understanding the natural processes is a necessary part of natural hazards management, but, as is shown later, it is not sufficient by itself because natural hazards are not in fact entirely "natural." Nor for that matter are human environmental disturbances entirely due to the activities of people. This issue is examined in more depth after the problem of natural hazards has been described in more detail.

The topic of natural environmental hazards owes its importance to two facts. First, the damage and loss of life inflicted upon human society are often substantial catastrophic events, making the problem of natural hazards a salient one for the people at risk and for their governments. Second, in the field of environmental control, the largest history of experience is found in the record of how people have coped with natural hazards. Studies of that experience are a potential source of understanding and wisdom in finding effective ways to deal with the more recently identified environmental problems covered in other chapters.

4.2 CLASSIFICATION AND MEASUREMENT OF NATURAL HAZARDS

Natural hazards clearly include a wide range of different phenomena. They can be classified according to their principal causal process (Table 4–1). In this chapter, we focus primarily upon geophysical hazards rather than biological hazards; some of the biological hazards are treated in Chapter 8. Geophysical hazards can be separated into those that relate to the processes of the atmosphere (climatic and meteorological phenomena) and those that relate to the geological and geomorphological processes of the earth's crust and its surface.

The investigation of natural hazards is allocated to different scientific disciplines. Thus, meteorologists and hydrometeorologists study weather, storm formation and behavior, the intensity of rainfall, and other factors that give rise to floods. Hydrologists concern

TABLE 4–1   CLASSIFICATION OF NATURAL HAZARDS BY PRINCIPAL CAUSAL AGENT

Geophysical		Biological	
Climatic and meteorological	Geologic and geomorphic	Floral	Faunal
Snow and ice	Avalanches	Fungal diseases (for example, athlete's foot, Dutch elm disease, wheat stem disease, rust)	Bacterial and viral diseases (for example, influenza, malaria, smallpox, rabies)
Droughts	Earthquakes		
Floods	Erosion (including soil erosion and shore and beach erosion)		Infestations (for example, of rabbits, termites, locusts)
Fog		Infestations (for example, weeds, phreatophytes, water hyacinth)	
Frost			Venomous animal bites
Hail	Landslides		
Heat waves	Shifting sand	Hay fever	
Tropical cyclones	Tsunamis	Poisonous plants	
Lightning and fire	Volcanic eruptions		
Tornadoes			

Adapted from Burton and Kates, 1964

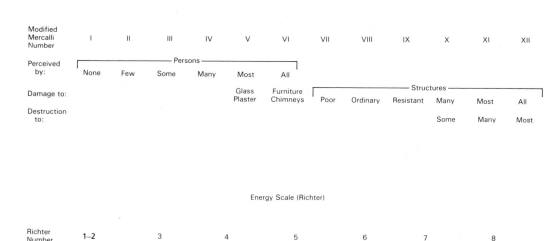

Figure 4–1 A comparison of the Richter and Modified Mercalli scales for earthquake magnitude. Source: THE ENVIRONMENT AS HAZARD by Ian Burton, Robert W. Kates, Gilbert F. White. Copyright © by Oxford University Press, Inc. Reprinted by permission.

themselves with flood magnitude and frequency and flood forecasting. The fields of geology and geophysics are subdivided into specializations such as seismology (dealing with earthquakes) vulcanology (having to do with volcanoes) and geomorphology (treating erosion and landslides). Usually, these specialists study the basic physical mechanisms and processes and are less concerned with how to control and manage the related hazards. This gives rise to some fundamental differences in approach, as exemplified in the attempts to develop scales of measurement for natural hazards. Different sorts of measurements are needed for different purposes.

How are natural hazards measured? There are two main approaches. The first is directed to the geophysical process and tries to measure its size by volume or energy. The second looks to the impacts and tries to measure those.

The difference between these approaches is clearly seen in the two scales that have been devised for the measurement of earthquakes (see Figure 4–1). The **Richter scale** measures earthquakes in terms of the energy released, in ergs.[1] This energy is measured by a seismograph, a very delicate instrument calibrated so that the amount of the displacement of the pen reflects the amount of energy released as transmitted by the seismic waves. The

---

[1] An erg, a unit of work or energy in the centimeter-gram-second system, is the work done by a force of one dyne acting in the direction of the force through a distance of one centimeter. One dyne is the force which, acting for one second on a mass of one gram, gives it an acceleration of one centimeter per second per second.

range of earthquake magnitude is extremely large, from the barest tremor which can be detected only by an instrument and is not directly perceived by human beings, to massive movements which shake down buildings. To accommodate this large range, the Richter scale is logarithmically constructed, which often causes confusion in its interpretation. News reports of earthquakes commonly make use of the Richter scale. Clearly, however, this scale conveys very little information except to experts, since, for example, the public is likely to assume that a Richter 6 earthquake is only twice as severe as a Richter 3 earthquake. Thus, an entirely misleading impression may be created.

The **Modified Mercalli scale,** on the other hand, tries to measure not the earthquake itself, but its impact upon people. An earthquake is designated Number V if it is strong enough to be perceived by most people and causes damage to glass and plaster. At level X, an earthquake causes damage to many structures and the destruction of some.

The fundamental difference between the two approaches stems from the different objectives of the geophysical scientist (Richter scale) and the hazard manager (Modified Mercalli scale). Whenever and wherever it may occur, a Richter 6.5 earthquake always involves the same energy release. The measurement is standard and universal, and exists independently of the presence of human settlement in the earthquake zone. The *impact* from two earthquakes may bear no relationship to their level on the Richter scale. A very strong earthquake (Richter 7) may cause less damage in a sparsely populated area than a much weaker earthquake in an area of concentrated human settlement. The advantage of the Richter scale, therefore, is its universal applicability on a constant basis anywhere in the world. Its big disadvantage is that it conveys no information about the actual amount of damage incurred.

The impact of an earthquake as measured by the Modified Mercalli number reflects both the character of human settlement in the earthquake zone and the strength of the earthquake. Poorly constructed buildings or those located on unstable slopes or soils are likely to suffer much more damage than those of proper construction and foundation. As measured in these terms, the earthquake records a higher or lower level on the Modified Mercalli scale according to the quality of construction. The Modified Mercalli scale therefore gives a measure of the impact at the expense of losing the universality of the Richter scale. Both scales serve useful purposes, but there is no means whereby one can be reliably converted to the other. The Richter scale is a geophysicist's scale; the Modified Mercalli scale is a hazard manager's scale.

Similar problems of measurement exist for all other natural hazards. In almost all cases there is some equivalent of the Richter scale. Thus, tropical cyclones can be measured in terms of their central pressure, the pressure gradient from the center to the periphery of the storm, the wind velocity, and the speed of movement of the whole weather system. Floods are usually measured in terms of the discharge of water at a given point on the river, and the rise and fall of water levels as reflected in a flood hydrograph. And blizzards can be measured according to the depth of snow accumulation and the associated wind speeds. There are few equivalents of the Modified Mercalli scale for other hazards, and reliance is often placed on monetary (dollar) estimates of the damages. This partly reflects the emphasis of scientific interest in the geophysical processes themselves. More important is the difficulty in producing scales of impact for some natural hazards. Consider, for

TABLE 4–2   FACTORS AFFECTING THE IMPACT OF SNOWSTORMS
IN URBAN AREAS

Frequency of major snow-storm events	The more frequent, the more prepared the city will be and the less impact there will be per unit of snowfall.
Slope of terrain, especially streets and highways	Snow accumulations on sloping streets have a much greater disruptive impact on traffic than similar amounts of snow on level ground.
Time of occurrence	Snowstorms occurring at night or in the middle of the day have less impact on traffic than do snowstorms at rush hour. Snowstorms occurring during the weekend have less impact on traffic than those occurring on weekdays.
Associated temperature level	At temperatures close to freezing, applications of salt to highways clears them quickly. At lower temperatures, the snow has to be ploughed and physically removed.
Availability of public transportation	A city with a well-developed public transit system, including a subway (underground) railway system, has an alternative means of transportation not available in cities that are more heavily dependent on private cars.

example, the impact of a heavy snowstorm on a major city. The same snowstorm as measured in depth of snow accumulation and wind speed will have a different level of impact according to a series of factors listed in Table 4–2. This variability of impact applies to other natural hazards. For example, the volume of discharge in a river may bear little or no relation to the amount of flood damage. And the moisture deficit as measured by the Thornthwaite water balance method (Thornthwaite and Mather, 1955) or the Palmer drought index (Palmer, 1965) does not measure the actual damage suffered by agriculture. Both these measures of moisture deficiency are climatic scales that do not take into consideration the drought resistance of various crops or the cultivation methods used.

Good scales of measurement for the impact of natural hazards are generally lacking. The best available yardstick is economic loss or damage. The difficulty with such measures is that the unit of measurement itself keeps changing (see Section 4.5).

4.3 WHAT IS A NATURAL HAZARD?

In the opening paragraph of this chapter, we gave a commonsense definition of natural hazards. Viewed in the light of our discussion about measurement, it is now clear that just as it takes two to make a quarrel, so it takes two to make a hazard—namely, nature and humans. Where humans and their works are absent, there can be no natural hazards.

This does not mean that on an uninhabited continent (for example, North America

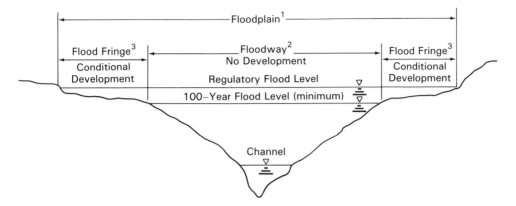

Figure 4–2 The hydrological definition of a flood plain and its use in land-use regulation.
Source: Ontario Ministry of Natural Resources

Notes
1. The *flood plain* is defined by a flood frequency, e.g., the 100-year flood or by a large flood of record. For example, in Ontario the floods caused by the passage of Hurricane Hazel in 1954 are used in some river basins.
2. The *floodway* is the lowest part of the flood plain where no development is permitted and which is reserved for the passage of flood flows.
3. The *flood fringe* is an area of flood plain land where filling and development *may* be permitted subject to land use and building code regulations designed to minimize damage.

before any settlers had arrived from across the Bering Strait) no floods or earthquakes occurred, but simply that when such events occurred they were not hazards to people. The word "flood" now has two meanings: in its commonsense meaning it is a natural hazard, but in another, stricter, sense a flood is merely an extreme geophysical event.

Thus, to understand natural hazards, we must understand extreme geophysical events. One important characteristic of extreme geophysical events is their probability. In the case of floods, for example, it is usual to describe the flood of a given magnitude in terms of its return period or recurrence interval, i.e., how often it may be expected to occur. A 100-year flood at any point on a river is that discharge of water which may be expected to occur on average once in a hundred years. As shown diagrammatically in Figure 4–2, a floodplain may be defined in terms of a specific flood frequency. A line can be drawn on the map showing the areas expected to be flooded by the 100-year flood. Beyond that line floods may still occur, but with a lower frequency. Similarly, within the line floods occur more frequently, until one reaches the river channel itself. In most humid and temperate environments, rivers reach the top of their banks almost once annually.

This introduces another element into the definition of a natural hazard. When an event becomes so frequent that it is part of the normal condition—as is water in the channel of a river—it is no longer a hazard. Similarly, at the other extreme, when an event is expected to occur very rarely on a human time scale, it ceases to be a natural hazard for all practical purposes.

So to restate our definition:

> A **natural hazard** is an extreme event in nature, potentially harmful to humans and occurring infrequently enough to be considered not part of the normal condition or state of the environment, but often enough to be of concern on a human time scale.

The distinction between hazards and normal conditions is important for an understanding of hazard management or adjustment as described later in Section 4.6. The Canadian Arctic is certainly a difficult place to live. So are the world's hot deserts. These are examples of harsh environments. But to cultural groups such as the Inuit in Canada and the pastoral nomads of the Sahel in Africa, their harshness does not constitute a hazard. It is only when unusual events occur that a hazard exists for a society that is well adjusted to its environment.

Not only can the environment change due to natural events; human actions can also result in drastic environmental change. A consideration of natural hazards therefore involves an appreciation of how changes in the environment through natural events compare with changes by human action.

4.4 EXTREME EVENTS AND ENVIRONMENTAL CHANGE

When extreme events occur in nature, they have a direct impact on humans by causing property damage, deaths, and injuries. They also have an indirect impact by changing the character of the environment.

There has been a long-standing controversy among students of the history of the earth about the relative importance of extreme events versus gradual change. Those who emphasize the importance of extreme events (sometimes called catastrophists) can point to the role of floods in erosion and deposition, the role of earthquakes in mountain building, and the role of sudden glaciation in shaping the landscape of mountains and lakes. On the other hand, the so-called uniformitarians emphasize the slow evolution of the earth under the long continuation of processes which can be observed every day.

Until very recently the forces of nature, including both extreme events and gradual processes, have far outweighed the effect of human impacts on the environment, except on a local scale. The modification of climate by volcanic eruptions is a well-known phenomenon (see Chapter 7). The dust particles in the atmosphere increase the albedo, resulting in colder temperatures over large regions at the surface of the earth for periods as long as two years following the eruption. For example, the great volcanic eruption of Mount Tomboro in Indonesia in 1815 led to two successive years of cold, wet growing seasons throughout the world beginning in 1816—"the year without a summer." The effects were aggravated in Britain, France, Germany, and the Netherlands by the economic consequences of the Napoleonic wars, and much suffering resulted (Post, 1977). Volcanic eruptions can also release poisonous gases, as happened in the Cameroon in August 1986, when 1,700 people were killed and 10,000 were otherwise affected by the toxic emissions.

By contrast, air pollution caused by human action has been heavily concentrated in urban areas. Approximately 4,000 "excess deaths" have been attributed to the London, England, great smog episode of December 5–9, 1952 (Larsen, 1970; Auliciems and Burton, 1973). Valid comparisons of these diverse kinds of events in terms of human consequences are not easily achieved.

The changes that have occurred in the hydrological cycle by the extraction of groundwater, the deforestation and urbanization of watersheds, cloud seeding, and reservoir construction all seem very small by comparison with the scale of natural events. They can, of course, have major impact on a small scale, but seem insignificant on a global basis when compared with the vast forces of nature.

This conventional view that the greatest environmental impacts are from natural hazards has been questioned in the past few years. It is now generally accepted that the burning of fossil fuels has substantially increased the carbon dioxide content of the atmosphere and that this may lead to significant climatic change—specifically, a global warming—by the middle of the next century (see Chapter 5). Recognition of the possibility of humans changing the environment on a global scale, either deliberately or inadvertently, has led to a major redirection of scientific effort toward the study of biogeochemical cycles (White and Tolba, 1979).

Most extreme geophysical events do not cause a permanent change to the environment. They may be regarded as a fluctuation or temporary disequilibrium from which environmental systems will return to a more "normal" or equilibrium state. Of course, these temporary environmental changes have severe impacts on human society, largely because they are extreme and short lived. They represent a departure from the normal conditions to which man is adjusted. By contrast, most changes from human activities take place slowly and thus give opportunity for adjustment. Where societies are adjusting to deteriorating environmental conditions, it is more difficult to measure impacts because too many variables are changing simultaneously.

The study of the impacts of natural hazards is therefore doubly important: it has value both in its own right and for what we can learn that is applicable to human environmental disturbances.

4.5 IMPACTS AND TRENDS

Everyone is affected by natural hazards, not only the obvious "victims." We tend to think of the term "victims" because the impact of hazards is dramatically reported when a disaster occurs. Major disasters are one end of a spectrum that extends to minor disturbances or fluctuations in the natural environment that cause small losses. The losses caused by many minor events, however, can, in aggregate, amount to more than the losses recorded from major disasters.

The full spectrum of impacts is suggested in Figure 4–3, which shows approximate estimates of impacts of various kinds (from death to taxes) that are expected to result from a recurrence of the famous 1906 earthquake in San Francisco. By recurrence is meant the same-size earthquake (8.3) as measured on the Richter scale as occurred in 1906. The impacts now would be very different since the city has changed so much since 1906. In that event, 450 people were killed and 514 city blocks containing 28,000 buildings were

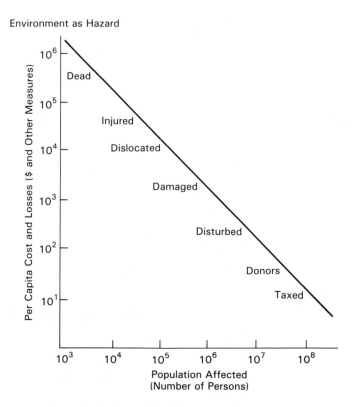

Figure 4–3 Loss sharing: future San Francisco earthquake. Source: THE ENVI-
RONMENT AS HAZARD by Ian Burton, Robert W. Kates, Gilbert F. White. Copyright © by Oxford
University Press, Inc. Reprinted by permission.

almost totally destroyed by the earthquake or the subsequent fire which continued for four
days.

Scientific simulation (Algermissen et al., 1972) and social scenarios (Cochrane et
al., 1974) permit some estimates of the impacts of a recurrence. Thus, it is estimated that
an earthquake of the magnitude of that of 1906 would today result in deaths in the Bay
area in the range of 2,000 to 10,000, and there could be as many as 40,000 injured,
depending upon the time of day the earthquake occurred. An additional 20,000 people
might be dislocated—uninjured but homeless. The scale of loss of residential buildings
would be greatly affected by the subsequent occurrence of fires and the ability to control
these in the aftermath of an earthquake.

Many more people would suffer financial loss. In some cases, this would be caused
by damage to buildings and other property or by indirect loss such as loss of earnings. Per
capita losses might well be in the thousands of dollars. In addition to those who suffer
damage in the area of physical impact many others would suffer loss as a result of the
disruption. The normal functioning of the economy in a wide area would be disturbed.

As in all major disasters, many more people would make voluntary contributions to
an earthquake disaster relief fund. These donors are people who voluntarily agree to share
the loss by making a financial sacrifice through the Red Cross or various private charitable

organizations. Beyond that, the government of the United States would undoubtedly provide disaster assistance extending perhaps to several billion dollars, depending on the scale of the disaster. The entire population of the United States as taxpayers would thus contribute to the costs of relief and rehabilitation, and the economic ripple effect would no doubt extend into Canada and beyond.

In major natural disasters, the number of deaths can be larger than in any human environmental disturbance except war. It is believed that 3.7 million people perished in the great Hwang Ho (Yellow River) floods in China in August 1931. Some 830,000 are thought to have died in the earthquake that struck Shensi Province, China, on January 23, 1956. Such estimates are notoriously unreliable, however. The tropical cyclone which brought a storm surge seven meters above normal high tide to the outer islands of the Ganges delta in Bangladesh (then East Pakistan) in November 1970 was initially reported to have killed more than a million people. More realistic estimates later put the death toll at around 225,000, but accurate figures are not available.

High loss of life as a result of floods, earthquakes, and droughts now occurs almost exclusively in developing countries. In developed, industrial societies, loss of life is usually very low by comparison, but property damages can be very high. The tropical storm, Agnes, that brought floods to the eastern United States (especially Virginia, Pennsylvania, and New York) in June 1972 was a very similar meteorological phenomenon to the Bangladesh cyclone of 1970. The damage caused was estimated at $3.5 billion, but only 118 people died. The damages were so high because much urban development had taken place on the narrow floodplains on which many eastern towns were located. Loss of life was comparatively so low because effective warning and evacuation plans were used to remove 250,000 people from their homes and out of danger. There was no effective warning and evacuation scheme that could have moved the 225,000 people who died in Bangladesh.

The pattern is a familiar one. In poorer countries, natural disasters tend to cause high loss of life and comparatively lower economic losses. In more developed industrial societies, loss of life is generally low but economic damages can be very high.

In the floods due to Hurricane Hazel in Toronto in 1954, the death toll was 81 and the damages were estimated at $3.5 million in the Don Valley alone. Similar patterns were recorded in the Fraser River (British Columbia) floods of 1948 and the recurrent floods on the Red River in southern Manitoba, including the City of Winnipeg (Burton, 1965).

In highly organized industrial societies, the costs of disruptions caused by natural hazards and other perturbations of the system can be higher than the direct damages. Few detailed estimates have been made, but a study of the impact of the Mississauga, Ontario, train derailment and evacuation of November 11–17, 1979[1] showed that while physical damages were very small, the cost of the disruption to the 225,000 people evacuated was $68.7 million, by a conservative estimate (Burton, Victor, and Whyte, 1981).

The health impacts of natural hazards of the geophysical type are now generally small in the developed countries, but continue to be a major concern of relief workers and disaster assistance teams in developing countries because epidemics of typhoid and cholera will likely break out due to the poor sanitary conditions common among the survivors.

---

[1] The derailment resulted in fires and explosions of tank cars carrying propane and toluene, and a tank car carrying 90 tons of chlorine developed a large hole. The evacuation was necessitated by fear of a sudden escape of chlorine gas.

The social and psychological impacts of the major hazard events are much more difficult to assess. In the management of emergencies, concern is often expressed that the threat of natural hazards and the associated social disruption, when an event occurs, will result in panic, social disorders such as looting and increased crimes of violence, and psychological distress. While all these do occur on occasion, the evidence from recent studies strongly suggests that well-established and normally healthy societies are slow to panic, do not resort to looting or violent crime, and are resistant to psychological harm. In fact, a healthy society commonly responds with a burst of constructive energy and community spirit. Volunteers man the dikes, fill sandbags, help search for the missing, take care of the injured, and provide shelter for the homeless. In many instances the official governmental emergency services could not cope without considerable help from volunteer organizations, which is often forthcoming in abundance.

In societies where there are already serious problems concerning interracial or class hostility, lack of trust in responsible government, and latent social or political unrest, the occurrence of a hazard event may be seized as an opportunity to manifest these social ills. For example, looting has been associated with natural hazard events in American cities. And again, the separation of East Pakistan (Bangladesh) from Pakistan occurred soon after the cyclone of 1970. The unpopularity of the national government in Islamabad (West Pakistan) was attributed partly to its lack of concern for the victims of the cyclone. Similarly, the overthrow of the Emperor Haile Selassie in Ethiopia in 1973 was attributed in part to the failure of his government to respond effectively to the drought that affected his country in the early 1970s.

Natural hazards probably facilitate, rather than cause, such political and social events. They do, nevertheless, appear to have a great deal to do with the timing of social or political disturbances by providing a pretext for the expression of discontents that already exist.

The impacts of major hazard events are well known and well documented. While there are difficulties in the precise measurement of impacts, at least some measurements are made. However, with a few exceptions, there is no systematic record of hazard losses that permits firm conclusions to be drawn about trends.

There are good reasons why this is so. In many developing countries, the government apparatus is not large or strong enough to devote time to the gathering of such data. In many of the smaller, developed, industrialized countries, the occurrence of extreme natural events is relatively infrequent, because of the limited extent of the national territory. Where statistics are collected, they tend to be for a specific hazard and not for all hazards.

The best data readily available come from the United States. As one example, Table 4–3 provides information on the impact of tornadoes. The table shows the frequency of tornadoes together with information on deaths and property losses. The number of tornadoes reported in recent decades is substantially higher than in the early decades of this century, presumably because of more complete reporting rather than an actual increase in frequency of tornadoes. The number of deaths varies considerably from year to year, and no clear trend can be identified. The size of property losses, on the other hand, does appear to be increasing, even with an adjustment for inflation. The change in the number of tornadoes reported, however, casts doubt on the comparability of later figures with those of earlier years.

Deaths from four natural hazards—lightning, tornadoes, floods, and hurricanes—in

TABLE 4–3   NUMBER OF TORNADOES, TORNADO DAYS, DEATHS, AND RESULTING
LOSSES BY YEARS, 1916–73

Year	Number tornadoes	Tornado days	Total deaths	Most deaths in single tornado	Total property losses†
1916	90	36	150	30	6
1917	121	39	509	101	7
1918	81	45	135	36	7
1919	64	35	206	59	7
1920	87	50	498	87	7
1921	105	55	202	61	7
1922	108	64	135	16	7
1923	102	50	109	23	6
1924	130	57	376	85	7
1925	119	65	791	689	7
1926	111	57	144	23	6
1927	163	62	540	92	7
1928	203	79	92	14	7
1929	197	74	274	40	7
1930	192	72	179	41	7
1931	94	57	36	6	6
1932	151	67	394	37	7
1933	253	96	362	34	7
1934	147	77	47	6	6
1935	180	77	70	11	6
1936	151	71	552	216	7
1937	147	75	29	5	6
1938	213	75	183	32	7
1939	152	75	87	27	7
1940	124	62	65	18	7
1941	118	57	53	25	6
1942	167	66	384	65	7
1943	152	61	58	5	7
1944	169	68	275	100	7
1945	127	66	210	69	7
1946	106	65	78	15	7
1947	165	78	313	169	7
1948	183	63	140	33	7
1949	249	80	212	58	7
1950	199	88	70	18	7
1951	264	113	34	6	7
1952	240	98	230	57	7
1953	422	136	515	116	8
1954	550	159	36	6	7
1955	595	153	126	80	7
1956	503	155	83	25	7
1957	856	154	191	44	8
1958	563	166	66	19	7
1959	604	156	58	21	7
1960	616	172	47	16	7

TABLE 4–3   *(Continued)*

Year	Number tornadoes	Tornado days	Total deaths	Most deaths in single tornado	Total property losses†
1961	698	169	51	16	7
1962	658	152	28	17	7
1963	464	141	31	5	7
1964	703	156	73	22	7
1965	901	181	295	44	8
1966	585	150	99	58	8
1967	929	173	114	33	8
1968	660	171	131	34	8
1969	608	155	66	32	8
1970	652	171	72	26	8
1971	989	192	156	58	8
1972	741	194	27	6	8
1973	1109	208	87	7	9
Means 1956–73	708	153	93	—	—

NOTE: Estimated losses are based on values at time of occurrence.

† Storm damages in categories:
5. $50,000 to $500,000
6. $500,000 to $5 million
7. $5 million to $50 million
8. $50 million to $500 million
9. $500 million and over

Source: U.S. National Oceanographic and Atmospheric Administration

the United States from 1940–75 are shown in Table 4–4. The long-term trend appears to be down for deaths from hurricanes and lightning and up, or at best unclear, for tornadoes and floods. The trends are hidden within a high range of variation on a year-to-year basis.

The most comprehensive survey of natural hazards in the United States to date concluded that aggregate (not per capita) damages from natural hazards are increasing in most cases (Table 4–5). In some instances there is evidence that the number of deaths is declining or staying about the same. While it is not possible to be precise about the trends in damages, it appears likely that in many developed, industrial societies the losses from natural hazards continue to rise. There is a somewhat better record in fatalities (see Table 4–6).

There have been few attempts to estimate losses from natural disasters on a worldwide scale. A statistical summary for the 35-year period 1947–81 (Thompson, 1982) indicated a total worldwide loss of life of 1,208,000 people, giving an average of 34,514 deaths per year. The number of natural disasters per year as recorded in the *New York Times Index* shows a generally downward trend from 1955 to 1975 and a strong upward trend since 1975 (see Figure 4–4). There are no world estimates for loss from natural hazards or disasters in monetary terms.

Flood losses in the United States have often been quoted as an example of rising damages (White et al., 1958). Even here, however, the unreliability of the data casts doubt on such conclusions. As shown in Figure 4–5, there is a great deal of year-to-year variability

TABLE 4–4   U.S. STORM FATALITIES FOR 1940–75

Year	Deaths			
	Lightning	Tornado	Flood	Hurricane
1940	340	65	60	51
1941	388	53	47	10
1942	372	384	68	8
1943	432	58	107	16
1944	419	275	33	64
1945	268	210	91	7
1946	231	78	28	0
1947	338	313	55	53
1948	256	139	82	3
1949	249	211	48	4
1950	219	70	93	19
1951	248	34	51	0
1952	212	229	54	3
1953	145	515	40	2
1954	220	36	55	193
1955	181	126	302	218
1956	149	83	42	21
1957	180	192	82	395
1958	104	66	47	2
1959	183	58	25	24
1960	129	46	32	65
1961	149	51	52	46
1962	153	28	19	4
1963	165	31	39	11
1964	129	73	100	49
1965	149	296	119	75
1966	110	98	31	54
1967	88	114	34	18
1968	129	131	31	9
1969	131	66	297	256
1970	122	72	135	11
1971	122	156	74	8
1972	94	27	554	121
1973	124	87	148	5
1974	104	361	89	1
1975	92	60	113	53
Total (36 years)	7,124	4,892	3,277	1,879
Annual average	198	136	91	52

Source: Michael Mogil and Herbert S. Groper, ''NWS Severe Local Storm Warning and Disaster Preparedness Programs,'' *Bulletin of the American Meteorological Society* 58 (4), April 1977, pp. 318–19.

TABLE 4–5   TRENDS IN DEATHS AND
DAMAGES 1954–78
From selected natural hazards in the
United States

Hazard	Damages	Deaths
Avalanche	+	+
Coastal erosion	+	N/A
Drought	?	N/A
Earthquake	+	+
Flood	+	+
Frost	+	N/A
Hail	+	N/A
Hurricane	+	−
Landslide	+	0
Lightning	?	−
Tornado	+	+
Tsunami	N/A	?
Urban snow	+	+
Volcano	N/A	0
Windstorm	+	+

Source: White and Haas, 1975

which tends to mask long-term trends. Major expenditures on flood control began in the United States with the passage of the Flood Control Act of 1936. The record of flood losses since that date, when reduced to constant-dollar terms, does not provide clear evidence of a trend. However, an increasing level of damage is a likely interpretation of the data in the figure.

4.6 ADJUSTMENTS AND THEIR CLASSIFICATION

In a simplified but generally valid way, we can view the history of our attempts to cope with natural hazards as being divided into three periods: preindustrial, industrial, and postindustrial.

4.6.1 Preindustrial Approach

In traditional preindustrial societies, the means were not generally available to control or to attempt serious modifications of the natural environment. People used their ingenuity to defend themselves from the perils of the environment, using the technology they possessed and their experienced judgment as to what nature might do. Small wood and paper houses in Japan were either resistant to earthquakes, or, if they did collapse, were unlikely to crush those trapped inside. Alpine villagers designed houses with steep pitched roofs to withstand heavy snowfall and located them in places where they knew, from experience, that avalanches were unlikely to occur. Peasant farmers in tropical savannah climates practiced intercropping of a variety of plants as a protection against drought. Peasant farmers in India timed the planting and harvesting of rice and other crops to harmonize with the arrival of the monsoon rains.

TABLE 4–6   DISASTER DEATHS, 1954–78 (UNITED STATES)

Fiscal year	Hurricanes	Tornadoes	Other windstorms	Floods	Flash floods	All other storms[1]	Other disasters[2]	Lightning	Total
1954–55	101	136	—	34	1	—	11	220	503
1955–56	8	86	2	83	211	4	9	181	584
1956–57	336	113	3	99	8	5	12	149	725
1957–58	2	102	2	17	3	—	3	180	309
1958–59	—	49	—	63	19	—	1	104	236
1959–60	4	53	1	15	2	61	12	183	331
1960–61	132	15	3	16	4	28	22	129	349
1961–62	45	23	2	27	28	29	3	149	306
1962–63	8	29	—	29	10	37	21	153	287
1963–64	2	44	3	34	—	15	82	165	345
1964–65	45	280	2	103	36	4	35	129	634
1965–66	72	92	33	16	6	9	2	149	379
1966–67	—	90	3	16	—	5	16	110	240
1967–68	19	139	4	32	6	8	94	88	390
1968–69	2	50	3	17	7	48	4	129	260
1969–70	272	78	—	50	1	3	19	131	554
1970–71	9	145	1	22	—	1	73	122	373
1971–72	2	22	1	500	19	13	12	122	691
1972–73	—	31	1	96	9	—	37	94	268
1973–74	—	412	3	60	11	5	7	124	622
1974–75	3	48	5	35	13	2	15	104	225
1975–76	32	40	6	48	7	38	9	92	272
1976–77	2	11	9	23	142	45	11	74	317
1977–78[3]	—	21	10	196	—	154	4	97	482
Total	1,096	2,109	97	1,631	543	514	514	3,178	9,682

[1] Snowstorm, hailstorm, cold weather, electrical storm, etc.

[2] Earthquake, landslide, drought, structural collapse, epidemic, etc.

[3] Ended June 30, 1978.

Source: American Red Cross disaster relief reports.

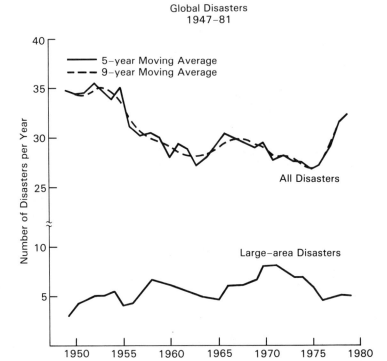

Figure 4–4   Global disasters, 1947–81 Source: Thompson, 1982

Everywhere, the rhythms and technologies of traditional societies were attempting to use the resources of the natural environment while trying to avoid the impacts of hazards. For much of the time this worked successfully, and it failed only when extreme events of high magnitude occurred.

In a few instances, preindustrial societies did organize to build major works for the control of water. The construction of polders in the Netherlands, the massive flood-control dikes along the Hwang Ho River in China, and the irrigation systems in the Tigris and Euphrates valleys are well-known examples. These systems were part of the development of advanced civilizations and certainly permitted a greater density of population to be supported on the land. Yet when major tidal surges, floods, or droughts occurred, disaster resulted on a scale that previously would not have occurred.

4.6.2 Industrial Approach

Early major water-control schemes in the industrial period were the precursors of the application of modern technology to the control of the environment. From the early nineteenth century to the present day, technological control systems have been designed and built on an increasing scale and at an accelerating rate. Many of these systems are designed to provide protection from natural hazards or from extreme fluctuations in natural systems. In the last few decades, many large dams have been built to store water for the purpose

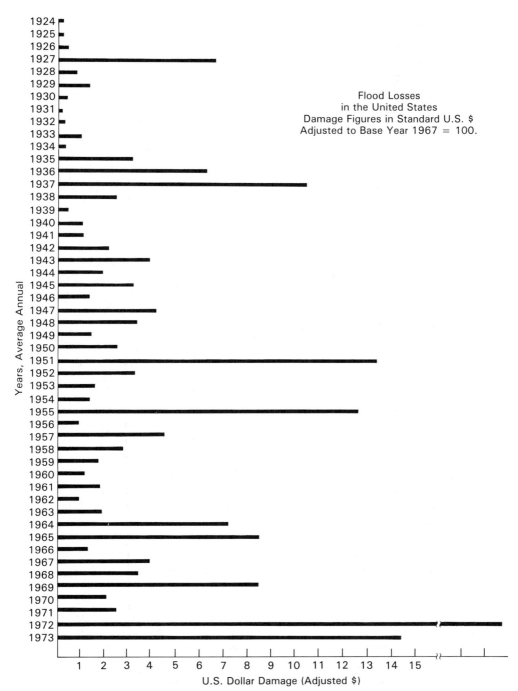

Figure 4–5   Flood losses in the United States (1924–73) (Damage figures in standard U.S. $ to base year 1967 = 100). Source: U.S. National Oceanic and Atmospheric Administration. Data compiled by David Fletcher and Bob Sargalis.

of flood control and to provide a reliable supply of water for irrigation in areas of low and uncertain rainfall.

Other examples of the application of science and technology to control nature include cloud seeding to produce rain in drought areas and modifications of the force and track of hurricanes. Various technologies have been developed for hail suppression, fog dispersal, and avalanche control. Sea walls and groynes are built to protect against coastal flooding and erosion. In the category of biological hazards, massive applications of chemical pesticides are made to control pests. Practically wherever one looks at natural hazards, there is evidence of attempts at control or of research and development to achieve such control. In fact, the discovery that the deep disposal of liquid wastes in Colorado triggered a series of very small earthquakes has even led to the suggestion that these strains in the earth's crust could be gradually alleviated in a controlled process.

Many of these technological achievements have brought great benefit to human society. Clearly, by the criterion of benefit-cost analysis, the benefits to society of environmental control have generally exceeded the costs. In flood control, for example, the costs of building, maintaining, and operating dams, dikes, and channels are commonly exceeded, sometimes by far, by the value of the flood damages that they prevent. Most project feasibility studies for dams have the built-in requirement that the flood-control and other benefits must exceed the costs.

Why, then, is there evidence of rising losses from natural hazards in general? Two possible explanations that are sometimes given can be quickly dismissed. First, the increase is not due to the declining value of the dollar through inflation: estimates of damage from natural hazards are made in constant-dollar terms by discounting present-day dollars back to a common base value. Second, the increase is not due, except perhaps in a small way, to changes in the environment: the perception that floods, earthquakes, or other extreme events occur more frequently now than before turns out to be largely false when the record is examined. Similarly, while it is true that climate changes, the trends are too long-term to be reflected in the relatively short period during which damage data have been collected. Also, short-term fluctuations have an impact, but average out over decades.

The better explanations for rising losses from some natural hazards are to be found in the limitations of technology and in changes in human society. Technologically, it is not practical to design and construct hazard-control systems to accommodate the very low-frequency, high-magnitude events. In flood control, for example, the design storm usually has a recurrence interval of 100 years. That is, the conditions producing the flow of water that a dam is designed to control are expected to occur on the average once every hundred years. The larger the dam, the higher the marginal cost of each additional increment of storage capacity. As the marginal costs of construction tend to go up with scale, so do the benefits decline. Whatever the economic losses likely to be caused by a 100-year flood may be, the average annual benefits of prevention, when reduced to present value, may amount to very little. In the economics of environmental control, it usually pays to control the more frequent rather than the rarer events. This means that when the design capacity of the system is exceeded, floods (or other hazards) will continue to occur. Thus, although a spillway is provided to carry the excess flow safely past a dam when its storage capacity is full, the residents of the floodplain below the dam may not be safe from flooding.

The second explanation for the growth in damages from natural hazards is that changes in human society are the cause. As populations grow and economies develop, and as people concentrate in cities, there is a greater accumulation of property and wealth to be damaged. Other things being equal, losses from natural hazards might be expected to grow along with population and gross national product. Insofar as environmental control systems are effective, they may be expected to reduce losses or, at any rate, to keep the increase in losses to less than the increase in population and GNP. The indications, however, are that in some instances at least, the reverse is true.

If expanding population and physical property were randomly distributed over the face of the earth, or over a national territory, then one might expect losses to rise in step with development. If environmental control systems were always effective, and if people sought to avoid areas known to be hazardous, then losses should decline. However, losses have increased because people have not avoided hazard zones, but seem almost deliberately to have chosen to put themselves in the path of danger. In the case of floods this may be because of a false sense of security generated by flood-control works: knowledge that the floodplain is now protected up to a 100-year design discharge appears to give people confidence and encourage them to build on floodplains. In other places, people flock to Alpine ski resorts and encourage the development of settlements in the paths of avalanches. And elsewhere, waterfront homes are built in hurricane zones only slightly above mean high-tide levels. In these last cases, the recreational and amenity value of the site is often what draws people regardless of the existence of coastal defenses or avalanche protection.

In some cases the actual physical existence of environmental control systems generates the confidence which leads to disaster. In other cases, where there is little or no protection against extreme events in nature, the confidence seems to be part of a widely shared faith in the power of technology and our ability to control nature. The benefits of this hubris are good while they last. However, disaster is almost certain to follow, sooner or later, for those that first ventured in or those that come later.

Field research in many hazardous locations (Burton, Kates, and White, 1978) suggests that few people occupy hazardous sites in total ignorance. The reasons that people move to and stay in places that they know to be hazardous are many. Overconfidence and miscalculation of risk is no doubt a strong factor in many cases. In others, the perceived recreational, aesthetic, or economic benefits exceed the perceived risk. In still other instances, each of the occupants is in the hazard zone only temporarily, as in new Alpine ski resorts, and thus is willing to take the risk, or does not consider the matter. In some cases—e.g., in the outer islands of the Ganges delta or on the fertile slopes of volcanoes, the lack of alternative economic opportunity may literally force the choice. In more affluent societies there is often a strong expectation that if the worst does happen, government disaster assistance will be provided.

4.6.3 Postindustrial Approach

For all its achievements, the application of technology to the control of environmental systems is increasingly seen to be deficient unless it also takes into account both its own inherent limitations and the likely future behavior of people in social and economic systems.

Failure to do so can result in the loss of many of the benefits of environmental control. The application of technology to the control of natural hazards can also pave the way for larger disasters and promote a sense of dependency upon government among the public and even among large private organizations.

What is now being consciously sought, therefore, is a more flexible response to hazards in which environmental control systems will be blended with a set of social and economic policies that seek to bring about a more harmonious relationship of human development with the natural environment, especially in its more extreme fluctuations.

For these purposes, it is now common to speak of adjustments to hazards, and to include within the set of adjustments all possible actions that might be taken to achieve a balance. There are five sets, or kinds, of adjustments:

1. Sharing and bearing losses, or *acceptance*
2. Hazard control, or *technological control*
3. Social adjustments, or *regulation*
4. Radical use change and migration, or *relocation*
5. Emergency planning, or *emergency measures*

These may be thought of as a sequence of changing responses or of learning behavior in the face of mounting experience with hazards of increasing severity.

Acceptance.   The most common response to natural hazards even today is to accept the losses. This is true both because many hazard events are quite minor and it is easier to suffer the loss than to spend the time and resources required on an active response. Many droughts, for example, do not become severe, but are recorded by farmers as a soil moisture deficiency which results in a lower yield rather than a total crop loss. Moderate and expected snowfalls cause delays and inconvenience which are accepted with complaints, but with little or no corrective action taken.

Where actual or expected losses have too high an impact upon individuals, the family, or the community, sharing mechanisms are developed. In traditional societies extended families bring help. Where this is insufficient, the circle of assistance widens to the tribe or larger social groups. In modern societies the same informal processes of sharing occur, especially in emergencies, and to these are added the more formal arrangements of insurance schemes, disaster relief, and governmental assistance including compensation.

Technological control.   A second set of adjustments consists of those aimed at controlling the natural events themselves. As we have seen, these adjustments have deep historical roots but have reached their full flowering in the present day. The wish of people to control nature is long standing. Anthropological studies reveal that propitiation of the gods, to prevent catastrophic floods, or rain dances to alleviate drought are expressions of this wish. Modern attempts to control nature, despite their limitations, are generally accepted as more effective.

Regulation.   There is a wide range of possible adjustments in the operation of human society which can reduce vulnerability to natural hazards. An obvious approach that has been alluded to before is to keep people and property away from hazardous areas. This can be done by means of land-use planning and regulations, a particularly effective adjustment to floods (see Figure 4–2). Where hazards are more widespread and not confined to definable locations, other planning devices such as building regulations (earthquake-resistant structures) or cropping patterns (adjustments to drought and hail) can be adopted.

Many social policies have an indirect and often unintentional effect on the vulnerability of a society to natural hazards. For example, urban renewal or redevelopment programs may increase or decrease future flood losses, transportation policy affects disruption from snowstorms, and building codes may change the extent of tornado-related damage.

Relocation.   An extreme form of social adjustment is use change and migration. Just as hazards are created by human use and occupation of hazardous lands, so too may they be reduced or eliminated by changing use or by (temporary or permanent) mass migration away from danger. For example, residential and other property on the floodplains of the Don and Humber rivers in Toronto, Canada, was compulsorily purchased by government after the Hurricane Hazel floods of 1954; the buildings were demolished, and the land use was converted to open space for recreational purposes (Burton, 1965). Similarly, the entire population of the South Atlantic island of Tristan da Cunha was evacuated by Britain following a volcanic eruption in 1961 (Blair, 1964). And many people were evacuated from the Mount St. Helen's region of Washington after a volcanic eruption in 1980. Migration away from areas hit by drought is also a well-known phenomenon in the drought polygon of southeastern Brazil, in the Sahel zone of North Africa, and elsewhere.

Emergency measures.   Many organizations have set up emergency units in recent years. Their purpose is to have emergency organizational plans in place to cope with unexpected events based on experience gained from previous natural disasters. The details of the plans will of course vary with the nature of the disaster.

4.6.4 Classification

For any natural hazard, there are so many theoretical possibilities for adjustment that it becomes helpful to classify them into types. Three main classifications are in common use. The first, already described, is the distinction between those adjustments directed at control of the environment or the natural processes themselves and those which involve changes in human society or in the pattern of social action and behavior. In Western industrial nations, after a period of heavy emphasis on the adjustments of the first kind, there has been a general broadening of response to include more social adjustments.

The second classification of adjustments is based on the criterion of timing, i.e., it specifies those actions to be taken before, during, and after the hazard event. In traditional societies the adjustments were largely confined to those emergency actions that could be taken during the event. In the absence of effective warning systems people were often caught totally unprepared, and emergency actions, including flight, proved to no avail, as is dramatically observed in the excavated ruins of Pompeii, Italy.

With the rise of modern science and technology, an emphasis on hazard prevention has developed in which the occurrence of a natural disaster is almost invariably followed by an inquiry and then programs of action directed at "never letting this happen again."

Because extreme natural events are the result of random fluctuations in natural processes, there will always be a future consequence which, given the passage of enough time, exceeds the magnitude of all previously experienced events. Therefore, all adjustments after the event are also adjustments prior to the next occurrence, whether greater or lesser than previously experienced.

The effectiveness of adjustments in reducing hazard-related losses has to be evaluated in terms of the relationship between the environment and the changing character of human settlements. Will society slowly achieve a less vulnerable state in relation to extreme natural events? In other words, do we learn from hazard experiences? If making adjustments to natural hazard events is part of a learning process, then hazard-related losses ought to decrease over time. The fact that they have not done so suggests that there is more to learn about how to manage natural hazards.

A third classification of adjustments is seen in the distinction between actions taken by private organizations, such as companies, and by government at all levels from local to federal. Here again, a clear trend may be discerned in the selection of adjustments: the rise of large-scale urbanized societies in the modern world has been accompanied by a decline in attention to adjustments at the individual and household level and a growth in the responsibility of organizations, especially governments, to protect people from natural hazards. That is, as has occurred elsewhere, in the realm of natural hazards there has been a decline in individual self-reliance and a growth in dependency upon the state.

4.7 A THEORETICAL PERSPECTIVE: FUTURE POSSIBLE RESPONSES

In the introduction to this chapter, it was suggested that natural hazards are not entirely natural: while the physical or environmental processes that give rise to extreme events are natural, the intensity of their consequences or impacts depends a great deal on what people have decided to do or not to do about them.

A traditional view of flood hazards is illustrated in Figure 4–6(a). Floods are seen as events in the natural environment that impinge upon human society and cause deaths and damages. An alternative widespread view (Figure 4–6(b)) is that the forces of nature can be controlled or modified to eliminate or lessen the impacts on society. This view can be represented as a positive feedback model as shown in Figure 4–6(c).

Systematic research on and observation of the effects of policies based on the model or theoretical approach represented in Figure 4–6(c) have revealed that the positive feedback efforts to control floods also produce negative feedback effects which provide human populations with incentives to expand floodplain activities, adding more population and property. The negative feedback effect can be further reinforced by government relief and

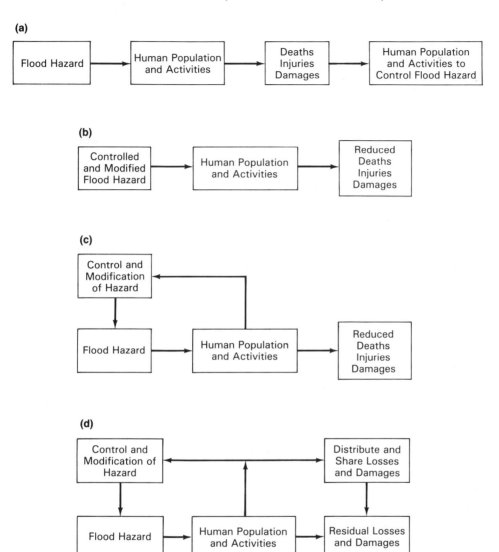

Figure 4–6 Hazard models—I

rehabilitation programs designed to distribute losses or share them with the wider community, as illustrated in Figure 4–6(d). The "residual" losses and damages can become higher over time than they would have been without these feedback effects.

Identification through research of the processes of reinforcement that increase damages has led to shifts in public policy. Chiefly, these changes have been intended to widen the options available to decision makers, specifically to include a range of social adjustments. These include improved policies for sharing losses, strengthened procedures for emergency planning, and, on occasion, even steps to change use in more radical ways to facilitate migration away from hazard zones.

For any given hazard, the range of social adjustments is large and can often be further increased by research or policy innovations. In the case of floods, for example, insurance has generally not been available from the private insurance industry except on a very limited scale. Coastal property on the Atlantic and Gulf coasts of Florida and adjacent states is sometimes insured with Lloyds of London. Premiums, however, are very high.

Among the reasons for the lack of private insurance against floods is the fact that many householders expect not to be flooded during their term of residence, even though they know that they live on a floodplain. A reason given by the insurance industry is that the narrowly defined extent of the risk (only those resident on floodplains) means that there is an insufficiently wide basis over which to spread the risk: everybody needs fire insurance, but only those on flood-prone areas need flood insurance.

To make this social adjustment available to floodplain residents in the United States, the federal government has passed legislation to create a federally sponsored flood insurance scheme (Kunreuther, 1977) which is marketed by the private insurance industry and underwritten by the government. A danger was perceived in this action, namely that by making flood insurance available, the process of development of floodplain lands might be accelerated, resulting in higher damages and big insurance claims. To offset this danger, the U.S. federal government requires that in order to qualify for the government-sponsored insurance programs, each community must have in place a floodplain land-use plan with zoning regulations approved by the state government.

The conceptual model shown in Figure 4–6(d) includes only a limited range of human response. Findings from empirical research have resulted in further refinements as described in Section 4.6 in which a "multiple-means" set of alternative adjustments is specified. This is depicted in Figure 4–7(a).

Consideration of the factors that enter into the selection process has led to the adoption of a cognitive view of hazards (White, 1964; Burton and Kates, 1964) in which the perception of hazard and adjustments by the decision maker becomes a significant variable. Recent elaborations of the model (Figure 4–7(b)) include this perceptual component and also describe the hazard not as a separate event in nature, but as a product of the joint probability of occurrence of the extreme event in the natural environment and a given level of adjustment in the human use system.

In this form, the model still remains unsatisfactory, however, in several respects. While subcomponents (flood frequency, flood damages, and the effect of some adjustments) have been put into operation (White, 1964), most of the model remains at the conceptual level. For example, the model fails to relate hazard adjustment to everyday places and work activities, or to take into account choices about livelihood and location, or to account for differences in individual as opposed to collective decision making. On the last score, human perceptions of natural hazards are recognized as an important element in the process of hazard control and management, but knowledge is lacking. This is not so much a failure of hazard research as a commentary on the complexity of human behavior and the general failure of economic and psychological models of humans to account for that behavior.

Both empirical and theoretical evidence suggest that despite a powerful and growing

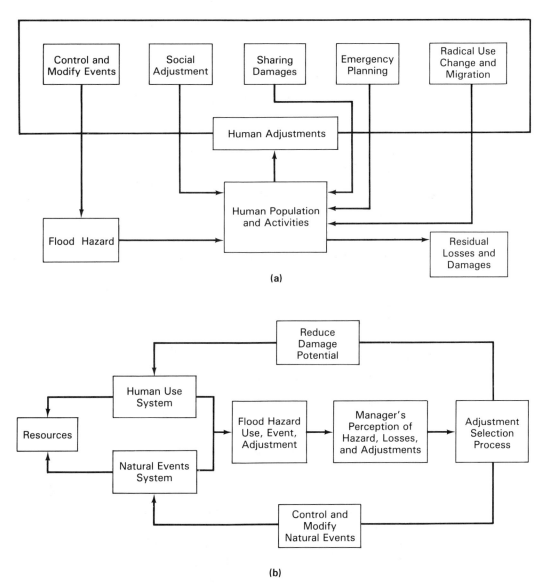

Figure 4–7 Hazard models—II

ability to exercise control over nature by technological means, deaths and damages as well as other impacts of natural hazards will continue to occur in the future. In fact, damages may well increase and take more catastrophic forms unless management improves. As modern societies seek to exercise more control over their environment, the fact that success still eludes us in relation to some of the most basic natural hazards should be a protection against overconfidence and naive optimism in attempts to control the environment in all its varied aspects.

PROBLEMS

4.1. Select a natural hazard that occurs in the region where you live and work. Obtain the following data on the hazard:
  (a) The frequency and magnitude of occurrence of the natural event.
  (b) The impacts of the hazard in terms of the categories of Figure 4–3.
  (c) A list of the most commonly used adjustments to the hazard, classified according to the categories given in Section 4.6.
  (d) A list of the theoretically possible adjustments, whether they are used or not.
  (e) The way in which the hazard is perceived by those exposed to it, by the public, and by the scientists and engineers who manage or control the hazard.

4.2. **(a)** Prepare an assessment of the chosen hazard in your community with particular reference to the following questions:
  (1) Is the hazard being controlled adequately?
  (2) Are the actual and potential impacts of the hazard increasing or decreasing?
  (3) How could the adjustments to the hazard be increased in number and effectiveness?
  (4) How do the perceptions of those at risk compare with those of the general public and scientific and technical hazard managers?
  (b) Prepare a conceptual model of the hazard in the manner of Figures 4–6 and 4–7 to show how the hazard is generated and to model the process of adjustment.

REFERENCES

ALGERMISSEN, S. T., et al. *A Study of Earthquake Losses in the San Francisco Bay Area: Data and Analysis.* Washington, D.C.: National Oceanic and Atmospheric Administration, U.S. Department of Commerce, 1972.

AULICIEMS, A., and BURTON, I. "Trends in Smoke Concentrations Before and After the Clean Air Act of 1956." *Atmospheric Environment* 7 (1973), pp. 1063–1070.

BLAIR, J. P. "Home to Tristan da Cunha." *National Geographic* 125 (1964), pp. 60–81.

BURTON, I. "A Preliminary Report on Flood Damage Reduction." *Geographic Bulletin* 7 (No. 3). Ottawa: Department of Mines and Technical Surveys, 1965.

BURTON, I., and KATES, R. W. "The Perception of Natural Hazards in Resource Management." *Natural Resources Journal* 3 (1964), pp. 412–441.

BURTON, I., KATES, R. W., and WHITE, G. F. *The Environment as Hazard.* New York: Oxford University Press, 1978.

BURTON, I., VICTOR, P., and WHYTE, A. V. *Final Report on the Mississauga Evacuation: A Report to the Solicitor-General of Ontario.* Toronto: Ontario Ministry of the Solicitor-General, 1981.

COCHRANE, H. C., et al. *Social Science Perspectives on the Coming San Francisco Earthquake: Economic Impact, Prediction, and Reconstruction*. Natural Research Working Paper No. 25. Boulder, Colo., Institute of Behavioral Science, 1974.

HAAS, J. E., KATES, R. W., and BOWDEN, M. J. (eds.). *Reconstruction Following Disaster*, Cambridge, Mass.: MIT Press, 1977.

KUNREUTHER, H. *Limited Knowledge and Insurance Protection: Implications for Natural Hazard Policy*. Philadelphia: University of Pennsylvania Press, 1977.

LARSEN, R. I. "Relating Air Pollutant Effects to Concentration and Control." *Journal of the Air Pollution Control Association* 20 (1970), pp. 214–225.

PALMER, WAYNE C. *Meteorological Drought*, U.S. Weather Bureau, Office of Climatology Research Paper 45. Washington, D.C.: U.S. Weather Bureau, 1965.

POST, J. D. *The Last Great Subsistence Crisis in the Western World*, Baltimore: Johns Hopkins Press, 1977.

THOMPSON, STEPHEN A. *Trends and Developments in Global Natural Disasters, 1947–1981*. Natural Hazard Research Working Paper No. 45. Boulder, Colo., Institute for Behavioral Science, 1982.

THORNTHWAITE, C. W., and MATHER, J. R. *The Water Balance*. Publications in Climatology, vol. 8. Centerton, N.J.: Laboratory of Climatology, 1955.

WHITE, G. F., et al. *Changes in Urban Occupance of Flood Plains in the United States*. Research Paper No. 57. Chicago: University of Chicago Press, 1958.

WHITE, G. F. *Choice of Adjustment to Floods*. Research Paper No. 93. Chicago: University of Chicago Press, 1964.

WHITE, G. F., and HASS, G. *Assessment of Research on Natural Hazards*. Cambridge, Mass.: MIT Press, 1975.

WHITE, G. F., and TOLBA, M. *Global Life Support Systems*. United Nations Environment Programme Information No. 47. Nairobi, Kenya: United Nations Environment Programme, 1979.

5

Human Environmental Disturbances

F. Kenneth Hare

Thomas C. Hutchinson

5.1 OVERVIEW

In Chapter 1, we reminded ourselves of the many technological improvements made for human existence and the enjoyment of life. But there is no denying that destructive impacts on the environment have occurred even in the most remote places. High-flying research aircraft have identified polluting gases 18 km above the Antarctic continent. Synthetic chemicals unknown in nature are often detected in remote places. Lead compounds, released into the atmosphere from car exhausts, can be found in the glacial ice of Greenland and Antarctica. Untouched forests or grassland can hardly be located. Human beings are the most powerful disturbers of their own environment, even though their health and perhaps their survival on earth depend on its condition.

The oceans show some disturbances very clearly. Oil seeping from ships has spread hydrocarbons over much of the surface. Some of these occur as small nodules that are washed up on beaches worldwide, especially near shipping lanes. Here and there much larger spills have devastated communities living along shorelines. Given the huge size of modern tankers, it is surprising that these spills have not been even more extensive. Tritium, radioactive hydrogen-3, from airborne nuclear bomb testing in the 1950s and early 1960s has penetrated several hundred meters into the ocean waters. As yet, the deep waters are largely unaffected, but they, too, will slowly absorb persistent pollutants unless we change our ways.

Forests and prairies show a different kind of disturbance. At least half of the world's original forest cover has been cleared to make way for agriculture or pastureland. In North America, for example, it is very hard to find surviving areas of prairie grassland on the high plains, or of deciduous forest that truly resembles what the eastern pioneer settlers saw when they colonized the land. The soil, too, has been drastically changed. Over a quarter of the carbon stored in the world's soil has been oxidized, and returned to the atmosphere as carbon dioxide—because of plowing by farmers and overgrazing by farm animals. Obviously, people must feed themselves and find firewood, timber, and minerals. But doing so has badly damaged the natural environment, and the damage is accelerating.

Clearly, there are two reasons for this damage. One is that we have no choice but to exploit the natural environment: food, minerals, and shelter are essential to our lives. With the continuing increase in world population for at least the next 50 to 100 years, this legitimate pressure will only get more intense. All we can do is to make sure that our technology of exploitation and use of resources are efficient and create as little damage as possible. The second reason, however, is carelessness, or even wanton destruction. Too many of our cities and industries continue to pour their waste products into the air or water, our agriculture is often unnecessarily destructive of the soil, and many of us are guilty of excessive resource consumption and wastage as individuals.

The duty of the engineer and scientist is thus obvious: to raise the level of technology to the point where the real needs of humanity can be met, while the environment can still be protected. But how can this be done?

The problems to be dealt with occur on all scales. Some problems, such as smoky or malodorous industries, are local and can be readily controlled; the trouble is easily located and can usually be corrected by better methods of combustion or waste disposal, albeit at considerable expense. Other cases affect large regions and involve thousands of polluters and millions of victims. Acid rain (Section 5.3) is like this. It results from emissions of sulfur dioxide and nitrogen oxides from chimneys and exhaust pipes, and now affects all of northeastern North America and northwestern Europe. Still other problems are literally worldwide. The carbon dioxide effect (Section 5.2) is of this sort. All of us contribute to it whenever we light a fire or drive a car. And everyone is affected by the result. Many other human environmental disturbances attributable to "progress" are covered in detail in Part 3.

The actions needed to remedy these human environmental disturbances will be dictated by questions of scale, and also by the kind of technology involved. The scientist's and engineer's first job is to understand the problem. This cannot be overemphasized: the physician cannot cure the patient before he or she has accurately diagnosed the illness. In particular, one must see the problem as a whole and understand how it affects other things. In environmental management it is usually best to go for as comprehensive a solution as possible, because so many things are connected. Unfortunately, this is often made difficult because of constraints of a political, legal, or jurisdictional nature. Air pollution, for example, is often found to be relevant to water quality, the health of crops and humans, the corrosion of buildings, and even aesthetic appeal. Controlling air pollution itself may be quite simple, even if expensive. But repairing the damage already done in these other sectors will be much more difficult and certainly expensive.

To bring home the interaction between the air, water, and land systems, the next two sections will deal with two major environmental issues of today: the buildup of carbon dioxide in the atmosphere, and the problem of acid rain. Acid rain is fairly well understood, and the means of control exist. But controlling it will be costly, and so far there have been few effective steps to bring control about. The carbon dioxide buildup is less well understood, although it is almost certainly attributable to human activities. Control may be technically possible, but will be very costly to undertake. Much will be learned from a study of the two issues, as representative of the many other kinds of environmental disturbances which could have been chosen here.

5.2 CARBON DIOXIDE BUILDUP AND THE GREENHOUSE EFFECT: A GLOBAL ENVIRONMENTAL ISSUE

5.2.1 The Present Situation

One of the most important environmental changes now in progress is a buildup of atmospheric carbon dioxide (CO_2). Undoubtedly, the added CO_2 in the atmosphere is coming from the burning of fossil fuels, the cutting of forests, and the wastage of soil humus (the colloidal organic complex in the soil). It is thus an artificial disturbance of the environment. In this section we shall assess the character and probable consequences of the CO_2 buildup and consider whether there is any need to take control measures, now or in the future. The main outcome of the CO_2 buildup is likely to be a change of climate, notably towards greater warmth. This may well affect the world economy.

The atmospheric CO_2 content is usually measured in terms of its concentration relative to all other gases in parts per million by volume (ppmv). Though there are large diurnal variations in the CO_2 concentration near the ground (because of the action of green plants or fuel consumption), the gas is well mixed in the lower atmosphere. Moreover, concentrations are much the same at all levels in both hemispheres. Annual average values in 1987 were near 347 ppmv. Figure 5–1 shows how the concentration of CO_2 has changed since serious monitoring began in 1958. An unsteady, but persistent, increase from year to year is recorded for all monitoring stations throughout the world.

Unfortunately we have no systematic records prior to 1958, so we do not know when the increase started. Preindustrial atmospheric CO_2 was probably in the range of 270–300 ppmv. If we assume that 100 years ago the concentration was near 290 ppmv, the subsequent increase has been about 50 ppmv, an average of 0.5 ppmv per annum. At the Mauna Loa Observatory in Hawaii, the annual increase since 1958 has varied from 0.5 ppmv in 1962–63 and 1974–75 to 2.2 ppmv in 1972–73. The recent rate of increase is clearly higher than earlier in the century, but has itself fluctuated considerably. Over the 1970s as a whole, the increase has been at the rate of 3.8 percent per decade. Few other global environmental changes of such a magnitude have actually been measured.

Since the mass of carbon in the planet is virtually constant, the increase must be coming from another storage reservoir. Figure 5–2 shows an estimate of the identified reservoirs and transfer pathways, with storages and transfer rates in gigatonnes (1 Gt =

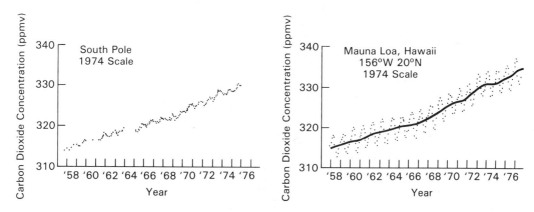

Figure 5–1   Trends in carbon dioxide concentration in the past two decades

10^9 t $= 10^{12}$ kg) of carbon per annum (Gt a^{-1}). The atmospheric store of carbon in 1987 was believed to be about 737 Gt, as against 610 Gt in 1860. Three net transfers are indicated:

1. An addition *to* the atmosphere of about 5 Gt a^{-1} due to the burning of fossil fuels (coal, oil, gas, and peat), whose storage in the earth's crust exceeds 5,000 Gt.

2. An addition *to* the atmosphere of 0 to 2 Gt a^{-1} of carbon from the oxidation of plant

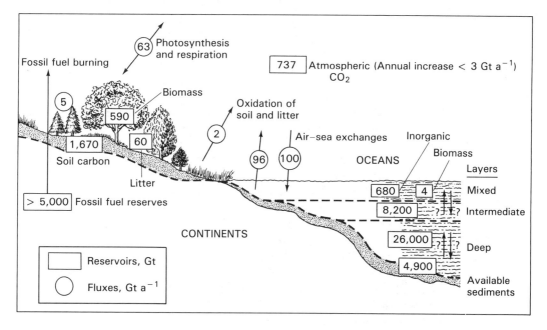

Figure 5–2   A sketch of the main storage reservoirs and transfer processes in the carbon cycle. Data updated from U.S. Department of Energy Report 008, 1980. All values in gigatonnes (Gt) or in gigatonnes per annum (Gt a^{-1}) of carbon

tissues, litter, and soil carbon, due mainly to the cutting of forests. Total storage is assumed to be 590 Gt of living biomass, 60 Gt of litter, and 1,670 Gt of soil humus. Photosynthesis and respiration transfer are assumed equal at 63 Gt a^{-1}.

3. A net transfer *from* the atmosphere *to* the oceans of 4 Gt a^{-1} of carbon (the difference between very large two-way exchanges).

These estimates suggest an annual increase of 2–4 Gt of carbon in the atmosphere. The average observed increase is equivalent to a little under 3 Gt a^{-1}. Hence, the additions under (2) or the transfers to the ocean under (3) may be wrong, since the fossil fuel consumption in (1) is thought to be reasonably accurate. In fact, there is wide disagreement about the transfers from living biota, and from soils, into the ocean. Until this issue is resolved by research, however, we have to be content with the notion that the atmospheric carbon content of 737 Gt is being added to at a rate a little below 3 Gt a^{-1}, and that this is about half the release of carbon by fossil fuel burning (the *retention rate*). The oceans are the only identified major sink for atmospheric carbon.

Answers to the question ''Will the increase continue?'' obviously depend upon the future use of fossil fuels, and, to a lesser extent, the future use of soil and forest resources. The world is currently shifting away from high-quality crude oil into less efficient sources, most of all coal, as discussed in Chapter 3. If this shift continues, CO_2 production per unit energy produced will increase sharply. The same will be true if there is a shift into heavy oils, tar sands, and oil shales, or into coal liquefaction (synfuels). Of the available energy options, only nuclear fission (and, in the distant future, fusion), hydroelectricity, and solar power offer any relief to the CO_2 buildup. Even natural gas, the most efficient of the fossil fuels, adds CO_2 to the atmosphere.

Energy use, moreover, will continue to increase, in spite of price rises and political uncertainties. The annual growth in the commercial world energy supply from 1890 onwards was about 5.5 percent, except for hesitations during the two World Wars and the slowdown of the 1930s. Because of steadily increasing efficiency, due to the shift from coal to oil and natural gas, the annual rate of CO_2 emission during this period grew only at 4.3 percent. Neither of these figures, however, is likely to apply in the future. The rate of expansion in energy consumption will probably diminish due to high prices, but the shift back to coal and away from oil will increase the carbon dioxide released for each unit of energy produced.

Furthermore, we are now faced with a new factor in the problem—the realization that other gases have a similar effect to CO_2. That is, the greenhouse effect is due, not just to CO_2, but to the buildup of these other gases as well, whose accumulation in the atmosphere is becoming increasingly obvious. These other greenhouse gases include methane (CH_4), nitrous oxide (N_2O), and various synthetics, notably the chlorofluorocarbons (compounds of carbon, chlorine, and fluorine). Chlorofluorocarbons are in widespread use as refrigerants, as propellants in spray cans, and for expanding plastic foam. All are similar to CO_2 in radiative behavior. Although they are present only in minute quantities, they are believed to rival CO_2 in warming effect. Hence, the global greenhouse warming may well be *double* the effect expected from CO_2 alone.

5.2.2 Speculations on the Future

Various visions of the future have been presented for future fossil fuel use at selected levels of total demand. The International Institute for Applied Systems Analysis (IIASA) at Laxenburg, Austria, has been most active. The world power consumption rate in 1987 was over 8 terawatts (TW). The most likely scenario for 2030 A.D. visualizes a total demand of 35 TW, with fossil fuel use of 17 TW. The U.S. Department of Energy has estimated a world demand of 27 TW in 2025 A.D., with fossil fuel contributing 21 TW. This U.S. estimate would lead to carbon dioxide emissions equivalent to 13.6 Gt a^{-1} of carbon compared to the 1987 emission of about 5 Gt a^{-1}. Great uncertainties attach to such estimates, especially since the future of nuclear fission may now be doubtful. In 1979 there was speculation that a doubling of the atmospheric concentration of 340 ppmv of CO_2 then existing might occur as early as 2020. This now looks unlikely, but not impossible. If more conservative estimates are realized, CO_2 levels in the atmosphere are likely to lie in the vicinity of 435 ppmv in 2025, although with a wide margin of error. Figure 5–3 shows the U.S. Department of Energy (1980) estimates, given various assumptions, concerning the retention rate of CO_2 in the atmosphere. It is clear that a doubling of CO_2 before midcentury is unlikely, and may not even happen before 2100 A.D. Nevertheless, remaining reserves of fossil fuel are sufficient to permit a four- to sixfold increase in atmospheric CO_2 at some later time. It is therefore important to attempt to estimate the probable climatic consequences of a doubling and a quadrupling of CO_2.

On the other hand, the doubling effect expected from the other greenhouse gases may well accelerate the warming. An international assessment suggests that the *total* greenhouse effect may be the equivalent of a doubling of CO_2 by the year 2030. (World Meteorological Organization, 1981).

These changes will be effectively worldwide, since both released CO_2 and the other greenhouse gases are rapidly spread by the wind systems, and are only slightly soluble in rain. Although the release of CO_2 and the other gases to the atmosphere is strongly concentrated in the industrial regions, the effect promptly becomes global. The acid rain problem, discussed in Section 5.3, is different, because sulfur and nitrogen oxides and other derived chemical species are removed fairly near their sources. The greenhouse effect is truly global, and calls for international monitoring and action.

5.2.3 Effects of Greenhouse Gas Buildup

The buildup of the greenhouse gases inevitably influences the temperature of the atmosphere and the earth's surface. Carbon dioxide emits and absorbs radiation at wavelengths typical of the earth and atmosphere. If its concentration increases, the atmosphere offers increased resistance to the necessary escape of radiation to space. Since incoming solar radiation is not much affected by the change in the concentration of CO_2, surface temperatures must rise as a result of the increased resistance to the return flow. Though not identical, the influence of the other greenhouse gases is similar. The height in the atmosphere from which the radiation eventually escapes is also raised slightly.

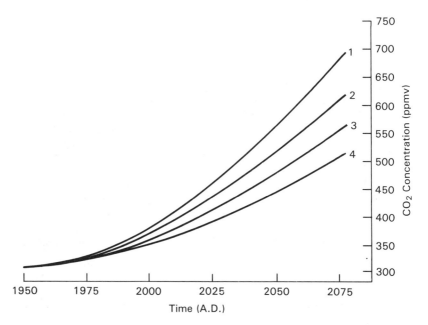

Figure 5–3 Estimated future fossil fuel consumption and CO_2 concentrations to 2075 A.D. Source: U.S. Department of Energy (1980)

Note: The four curves show possible trends of CO_2 concentrations under the following assumptions:

Curves 2 and 3: Based on "best estimate" of fuel consumption made by U.S. Dept. of Energy in 1980.

Curve 2:   Applies if airborne fraction of CO_2 emitted will be 53.5 percent.

Curve 3:   Applies if airborne fraction of CO_2 emitted will be 43 percent.

There is uncertainty about the relative retention of emitted CO_2 by the atmosphere and that which is transferred to the ocean. Current best information shows an emission from burning fuel of 6 Gt a^{-1}, a retention in the air of somewhat less than 3 GT a^{-1}, and therefore a transfer of somewhat more than 3 GT a^{-1} to the oceans. Since there is uncertainty in the air retention rate, the two curves perhaps indicate upper and lower limits on CO_2 concentrations under the two assumptions.

Curves 1 and 4: Based on a 25% higher fuel consumption (Curve 1 over 2) and 25% lower fuel consumption (Curve 4 compared to 3).

To predict the consequences of this radiative change, we must take into account the redistribution of available energy by winds and, if possible, by ocean currents. Early attempts to do this by means of simple one-and-two dimensional models led to the qualitative conclusion that the earth's surface would definitely become warmer as CO_2 increased, but the calculated warming varied from estimate to estimate.

The most convincing present-day answers are those obtained from experiments using three-dimensional general circulation models (GCMs) of the atmosphere. Manabe and Stouffer (1980) reported an experiment in which they introduced into a nine-level GCM a realistic earth's surface—continental coastlines and topography, and an ocean with variable heat capacity (but not currents to redistribute heat). Cloudiness was held constant, since another study with simpler geography indicated that the cloud feedback process was not important. The experiment indicated the following main results (Manabe and Stouffer, 1980; Manabe and Wetherald, 1980):

1. A *quadrupling* of CO_2 concentration (or of the other greenhouse gases) should raise the world annual surface air temperature by 4.5K in the northern hemisphere and by 3.6K in the southern hemisphere. For a *doubling*, the likely planetary increase is about 2K. The warming of the northern hemisphere will be greatest in early winter, and least in summer.

2. High-latitude areas will be more affected than others, because the warming pushes the sea-ice limit poleward. In winter the rise in temperature in Alaskan and Canadian latitudes will be on the order of 5–12K, and in summer it will be 4–5K, in each case for a quadrupling, with half these amounts for a doubling. North Americans may not find such warming a problem, but there may be other effects that are not so favorable (see item 4).

3. The permanent pack ice that covers the Arctic Ocean and the Canadian Arctic channels will disperse if a quadrupling occurs, and will be replaced by prolonged annual winter ice of the sort now typical of large areas around Antarctica, Hudson's Bay, and the Baltic Sea. A doubling will reduce the thickness and extent of the permanent pack ice, but will not disperse it (although another calculation reported by Kellogg and Schware (1981) finds that a doubling *will* disperse the pack ice).

4. With CO_2 quadrupled, a significant decrease in available soil moisture is indicated for a narrow band of latitude near 45°N. This is attributed to altered precipitation and increased evaporation arising from a poleward shift of cyclonic activity. This latitude is close to North America's main belt of agriculture. Unfortunately, the longitudinal "resolution" of the model does not indicate whether the increased dryness really will affect our farming system. Nor is it clear whether a doubling of CO_2 will produce similar, though less marked, effects. Another study, carried out at the Geophysical Fluid Dynamics Laboratory (GFDL) of the National Oceanographic and Atmospheric Administration in Princeton, New Jersey, with simpler geography, does indicate such an effect, and puts the most affected latitudes near 40°N (Manabe and Wetherald, 1980).

These predictions, from the model, of rising temperatures have major implications for North America, some good and some bad. The good effects include the following:

1. A sharp decrease (perhaps 15–25 percent) in space heating costs due to warmer winters, partly offset by increases in air conditioning costs (possibly 10 percent).

2. A longer growing seasons for crops, and hence the possibility of better harvests in northern regions, again partially offset by decreased yields further south (where summers are often too warm for good crop performance).

3. Much easier navigation, and for a longer period during the summer, in the Arctic

seas (e.g., Beaufort, Bering, Baffin Bay), and in the Canadian Arctic (Hudson's Bay and Strait, Lancaster Sound, Barrow Strait, and other straits), together with easier conditions for offshore oil and gas development.

Less welcome effects are as follows:

4. Drier crop conditions in many parts of the midwest and Great Plains, including the Canadian Prairies, requiring still greater use of irrigation water, already expensive and in short supply.

5. Widespread melting of the permanently frozen ground (*permafrost*) now underlying many parts of Alaska and northern Canada. This will greatly alter building technology and conditions for road and pipeline construction in these areas.

Such dramatic changes are at present hypothetical and may never take place. However, most other model calculations of the greenhouse effect tend to confirm the Princeton results. Nonetheless, still other studies suggest a smaller effect on atmospheric temperatures. Such conflicts are common in the early days of studying natural systems. They arise from the difficulty of incorporating adequate detail into the boundary conditions of the models, and their different systems of equations.

Again, the possible changes are of a global nature; the reason for considering the possible effects only on North America is that this is where the known studies were made.

5.2.4 Control Measures

Although it has not yet been demonstrated that the greenhouse effect will be, on balance, harmful to humans, the possibility remains that the world will decide that the risk is too great to take. In such a case it may be that international action to contain the problem will be sought. The fact that CO_2 emission comes mainly from advanced industrial powers makes such action conceivable. Agreement between the U.S.A., the U.S.S.R., and the European Economic Community would be the necessary first step. It seems unlikely, however, that the world will be able to move quickly away from its present dependence on fossil fuels. Restrictions on oil supplies may even accelerate the use of less efficient combustion processes which would release even greater quantities of CO_2 to the atmosphere. Hence, agreement, if it comes, is more likely to focus on the possibility of control than on the abandonment of carbonaceous fuels.

The technological removal of CO_2 from flue gas and exhaust pipes is feasible, but prohibitively energy intensive and costly. An analysis by the U.S. Department of Energy (1980) argues that it is feasible only in an all-electric economy, since otherwise capital investment requirements are out of the question. For centralized flue emissions, removal by monoethanolamine (MEA) absorption/stripping is the least energy-intensive technology available, but net power plant efficiency is reduced from an assumed 38 percent for zero removal to about 20 percent for complete removal—a huge burden. Disposal of the immense volume of carbon removed would have to be in the deep ocean, which would both add further costs and pose a threat to the ocean carbon cycle. Any attempt to increase CO_2 absorption by the oceans seems in the realm of science fiction, and is in any case dangerous given the present unsatisfactory state of knowledge of the oceanic carbon cycle.

It may well prove easier to control some of the greenhouse gases other than CO_2, notably the chlorofluorocarbons (CFCs). A tentative 31-nation agreement was reached in 1987 to limit the use of chlorofluorocarbons, but even if all nations adhere to the agreement, it will be decades before the slowdown becomes effective.

Of other possible measures, only the control of biotic and soil carbon exchanges seems useful, on other grounds as well as for the removal of CO_2. The present storage of carbon in biota, chiefly in the woody stems, branches, trunks, and roots of shrubs and trees, has been variously estimated from about 500 Gt to over 900 Gt, with annual exchange rates due to photosynthesis and respiration in the range 45–70 Gt a^{-1}. This storage is thus comparable to the 737 Gt in the atmosphere. Forest clearance, with subsequent use of the land for less efficient carbon storage, obviously transfers carbon to the atmosphere, chiefly due to burning. A recent careful estimate puts the net transfer to the atmosphere as 0–2 Gt a^{-1}, though much higher estimates are also current. Storage in soil is estimated to be in the range 1,450 to 1,730 Gt, about one-quarter less than before the agricultural revolution. Present-day transfers of soil carbon to the atmosphere are variously estimated from negligible values to one remarkable figure of 4.6 Gt a^{-1}. We thus have only a hazy notion of the size of the biotic-soil reservoir of carbon and of the net exchanges with the atmosphere.

Forest clearance for agriculture and the reduction of forest biomass because of poor forestry practices are clearly within the realm of possible management. Most of the storage is in the tropical rain forest, which is being rapidly converted to other uses. Large amounts are also stored in northern forest formations (about one-eighth of the world total). It has been estimated that total forest storage in Canada alone is 44 Gt, and that a further 38 Gt are stored in the huge northern muskegs. One way in which nations can at least slow down the CO_2 buildup is to prevent further wastage of forest biomass. Good forest management should aim at a high level of stored biomass, or standing crop. Much forest exploitation works today in the other direction. In Canada the biomass storage in forests must have declined substantially since cutting began, in spite of regeneration. In the United States this decline has been halted and even reversed: the standing crop of merchantable timber has recently been slowly increasing. In fact, the continent now faces an era of fully managed forestry in which it should be possible for us to rebuild some of the biomass losses of the past two centuries. It may be more difficult, however, to reverse the loss of soil humus that affects our agricultural areas. Nevertheless, every effort should be made to do so if there is a need to slow down carbon accumulation in the atmosphere. Though technological control of the CO_2 buildup seems impossible, there are avenues of land and natural resource management along which North America can proceed and which we can advocate internationally.

5.2.5 Conclusions

The economic impact of the greenhouse effect is not yet well understood, but this much can be said: there will be benefits as well as costs for most societies of our planet. It is not clear whether the United States and Canada will lose or gain from the changes, if they occur, and it is by no means certain that they will happen. If they do, we do not know in detail how drastic they will be or how they will distribute themselves across the national maps. All that scientists can advise at present is to be wary: the nations should be on yellow alert.

This means that we should monitor the atmosphere, oceans, and biota and work hard to understand and predict the impending changes, so as to detect their progressive arrival. We should be ready with strategies to offset any hardships the changes may bring, and to exploit any opportunities created. If these things are done, on balance we may profit from the greenhouse effect.

Since the problem is worldwide, and since prosperity depends on world trade, we must also involve ourselves in international efforts to resolve the uncertainties. Participation in the World Climate Programme and similar efforts is not merely a duty; it is in the interest of each nation.

The time scale of the impending changes is so long that they lie outside the ordinary framework of national policies and international statesmanship. If they happen, however, they may well achieve a transformation of the environment greater than anything experienced thus far in the history of civilization.

5.3 ACID RAIN: A MAJOR REGIONAL ENVIRONMENTAL ISSUE

5.3.1 The Nature of the Problem

Acid Rain, the label regularly given to both wet and dry acidic deposition, is a fairly recent addition to our language. Although the term was actually coined 110 years ago by the British chemist Angus Smith from his studies of air in Manchester, England, it was not until a rain quality monitoring network was developed in northern Europe in the 1950s that the widespread occurrence of acid rain was recognized. For the last decade, acid rain has been a major concern as it continues to defile major areas of our planet.

Acid rain occurs within, and downwind of, areas of major industrial emissions of sulfur dioxide (SO_2) and the oxides of nitrogen (NO_x). After SO_2 and NO_x are emitted into the atmosphere, they are transformed into sulfate or nitrate particles and, by combining with water vapor, into mild sulfuric or nitric acids. These acids then return to earth as dew, drizzle, fog, sleet, snow, and rain.

Normal, clean rain is slightly acidic, with a pH level of about 5.6. This is due to the equilibrium between rainwater and the CO_2 in the air, which dissolves to a sufficient extent in the droplets to give a weak carbonic acid solution. Today, over wide areas of eastern North America and northern Europe where heavy rainfalls predominate, rain falls with a pH value close to 4.0 and, on rare occasions, 3.0. The concern relates largely to the effects of the acidity on fish populations and other aquatic animals, to potential damages to crops and forests, and to accelerating deterioration of building materials. It even seems likely that acidified rains may enter the groundwater storage and increase the solubility of toxic metals. Acid waters can also dissolve metals, such as lead and copper, from hot and cold water pipes. These problems are now very widespread. The potential effects on tourism and on the recreational uses of lakes and rivers are enormous. Estimates for Ontario alone are for a multimillion dollar loss of tourist dollars per year unless the problem is ameliorated.

5.3.2 Sources and Distribution of Acid Rain

The pollutant material that comes down with rain is called **wet deposition,** and of course includes particulates and gases scavenged from the air by raindrops. The material reaching the ground by gravity during dry intervals is called **dry deposition,** and includes particulates, gases, and aerosols. Pollutants may be carried hundreds or even thousands of kilometers by the prevailing winds. This phenomenon is known as the long range transport of airborne pollutants (LRTAP). In 1968, Svante Oden from Sweden demonstrated that the precipitation over the Scandinavian countries was gradually becoming more and more acidic, that sulfur compounds in the polluted air masses were primarily responsible, and that large quantities of the acidifying substances came from emissions in the industrial areas of Central Europe and Britain. Soon afterwards, data on changes in lake acidity with time were developed. Trajectory studies in North America have demonstrated that more than 50 percent of the deposition of acid rain in Central Ontario is due to air masses passing over the major sulfur-emitting sources in the midwestern states of the United States, especially Ohio and Indiana. Acid rains in the Adirondacks and southern Quebec, on the other hand, often appear to have their origins in the industrial eastern seaboard states of New York, Massachusetts, and Maryland, and from Pennsylvania and other states over which the air has previously passed (Figure 5–4). The Canadian maritime provinces are affected by emissions from the U.S. eastern seaboard and also on occasions from smelter sources in Ontario and Quebec. More than ten percent of the acid rain that falls in the northeastern United States comes from Canadian sources.

Figure 5–5 depicts the distribution of emissions of SO_2 and NO_x in North America. Figure 5–6 shows the areas vulnerable to acidification. The lines indicate the incidence of wet sulfate deposition, and the numbers show the levels of deposition in kg/ha/yr. Levels of deposition exceeding 20 kg/ha/yr (18 lb/acre/year) are generally regarded as threatening in vulnerable areas. Nitrate depositions, not shown on the map, threaten the same areas.

A comparison between sulfur and NO_x sources in several states and in Ontario is given in Table 5–1. The major contribution from coal-fired generating plants in the major sulfur-emitting states is apparent in Ohio, Pennsylvania, Indiana, Illinois, and Kentucky, while the predominance of the nickel-copper smelter at Sudbury as a sulfur source for Ontario (and beyond) is also noted. In all cases, these very-large-source emissions have been dispersed in the atmosphere as a result of the development of the tall-stack pollution dilution technology of the late 1960s and 1970s and now contribute to regional acid rain problems.

Acidic materials accumulate in the frozen snowpack in regions subjected to acid deposition. The first major melt of the spring releases the majority of the acidic accumulation, which runs off as meltwater over still-frozen soil and quickly enters the rivers and streams. One consequence is the sudden intrusion into lakes of a ''plug'' of acidic waters, especially

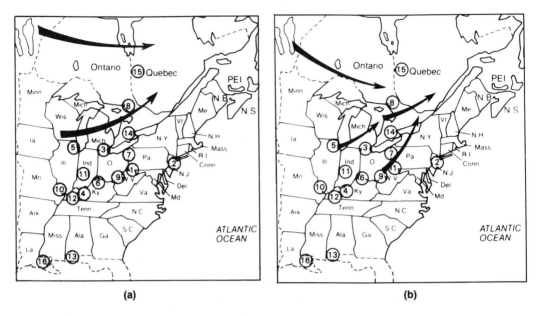

(a)   (b)

Figure 5–4   Storm trajectories over major SO$_2$- and NO$_x$-emitting areas in (a) summer and (b) winter. Source: Ontario Ministry of the Environment, 1980

Note: U.S. emission rates from the SURE II data base are 1977–78 emission rates for area sources. Canadian data from Environment Canada are estimated 1978 emission rates for major SO$_2$ point sources and 1974 emission rates for other area sources. Storm trajectories by J. Kurtz, meteorogical scientist. Environment Ontario based on 40 years of data. U.S. Weather Bureau.

Eastern North America—Major SO$_2$- and NO$_2$-Emitting Areas

Geographical Area	Grams/Sec.	Geographical Area	Grams/Sec.
1. East and West Pittsburgh: Upper and Central Ohio River Valley	98,718.7	8. Sudbury, Ontario	43,915.3
2. New York, New Jersey	81,892.2	9. Lower & central Ohio River Valley; Clarksburg, West Virginia	42,401.3
3. Toledo, Ohio; Detroit, Michigan	65,421.6	10. Eastern Missouri, Illinois	41,298.8
4. Western Kentucky; southern Indiana	53,623.7	11. Indianapolis, Indiana	30.202.9
5. Chicago Illinois	53,040.7	12. Western Kentucky	25,849.3
6. Cincinnati, Ohio; northern Kentucky	50,051.0	13. Mobile: southern Alabama	24,138.5
7. Cleveland, Ohio; western Pennsylvania	47,997.7	14. Toronto, Ontario	18,584.7
		15. Rouyn-Noranda, Quebec	16,404.2
		16. Southern Louisiana	14,596.8

in their shallow inshore areas. Figure 5–7 shows the "spring pH depression" in one of the six inflowing streams to Harp Lake, a study lake in Muskoka, Ontario. As the spring runoff increases the amount of water in the stream, the acidic melted snow causes the pH to drop, producing severe chemical "shock effects" on aquatic life. The disastrous effects of this type of phenomenon on shallow-water spawning fish are considered later.

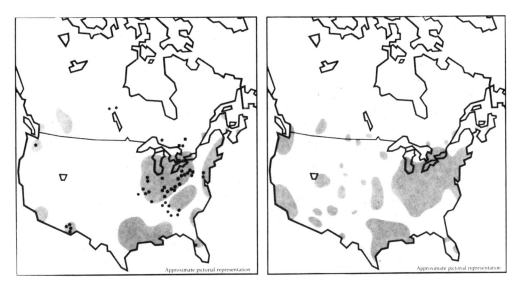

Figure 5–5   (a) Distribution of emissions of sulfur dioxide (SO_2); (b) Distribution of emissions of nitrogen oxides (NO_x). Source: Environment Canada (1984)

TABLE 5–1   QUANTITIES OF SO_2 AND NO_x EMITTED ANNUALLY FROM VARIOUS STATES IN THE U.S.A. AND ONTARIO

	Sulfur dioxide (10^3 tons/year)	Nitrogen oxides (10^3 tons/year)
Ohio	3,259 (Reduced to 2,700 in 1980)	1,187
Pennsylvania	2,495	1,023
Ontario	2,100*	Unknown
Indiana	1,891	960
Illinois	1,707	1,274
Kentucky	1,631	569
Texas	1,541	2,117
Missouri	1,507	618
Tennessee	1,277	560
Arizona	1,239	276
West Virginia	1,226	471
Michigan	1,225	742
Alabama	1,038	511
New York	1,022	906
California	675	1,284

* Includes emissions from Ontario Hydro and smelters from the Sudbury complex.

Source: Government of Ontario report, 1981

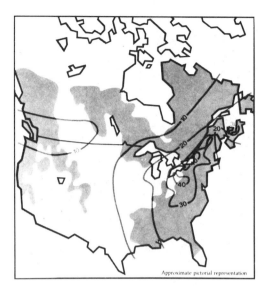

Figure 5–6 Areas vulnerable to acidification, based on bedrock geology and surficial deposits. Source: Environment Canada, 1984

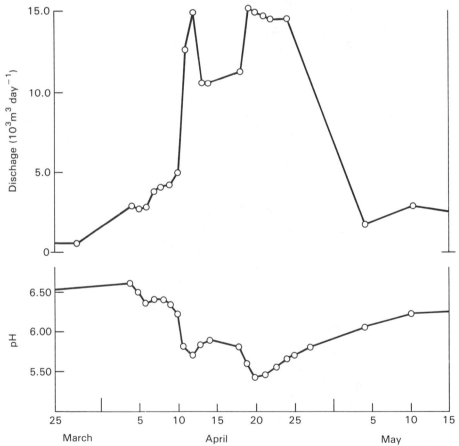

Figure 5–7 "Spring pH depression" of a stream. Source: Ontario Ministry of the Environment, 1980

5.3.3 Effects of Acid Rain on Aquatic Systems

The most important effect of acid rain on aquatic systems is the decline in fish populations, which is especially damaging to sports fishing. The indirect result on tourism is economic. Other aquatic effects of acid rain include those on humans who eat fish having an increased concentration of metal in their flesh and the reduction of certain groups of zooplankton, algae, and aquatic plants, which disrupts the overall food chain in lakes, thereby causing potential ecological imbalances.

Studies have clearly demonstrated that trout and Atlantic salmon are particularly sensitive to low pH levels, which interfere with their reproductive processes and frequently lead to skeletal deformities.

High aluminum concentrations in acidifying waters are often the actual trigger which kills fish—and probably other sensitive biota such as planktonic crustaceans. In alkaline or near neutral lakes, aluminum concentrations are very low. As the pH decreases, however, the previously insoluble aluminum which is present in very high concentrations in rocks, soils, and river and lake sediments begins to go into solution (Figure 5–8). Once in solution, aluminum is remarkably toxic to many forms of aquatic life at low concentrations, i.e., from 0.1 to 1 mg/L. Although aluminum concentration increases exponentially below a pH of about 4.5–4.7, toxicity to fish occurs at a higher pH than this. Studies at Cornell University by Baker and Schofield (1980) show that the maximum toxicity of aluminum to fish occurs around pH 5.0. This is due to the rather complex chemistry of aluminum, for which the chemical forms and their ratios in solution change with pH. Free ionic aluminum occurs mainly below pH 4.2 and is extremely toxic. At a pH of around 5.0 the hydroxyl forms predominate, and the toxicity declines above and below this pH level. At

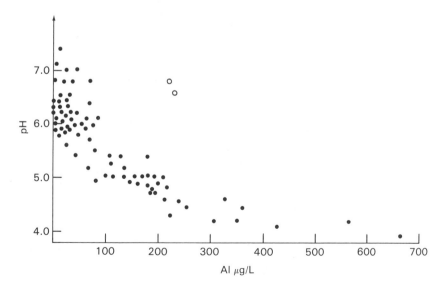

Figure 5–8   The relationship between lake pH and total aluminum in some Swedish clearwater lakes. Source: Dickson, 1980

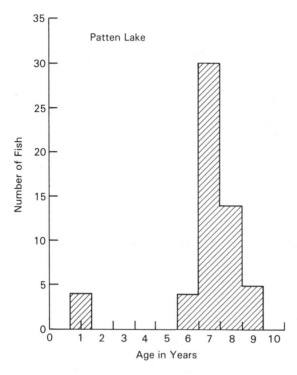

Figure 5–9 Age composition of yellow perch from Patten lake, Ontario. Source: Ryan and Harvey, 1980. Copyright © 1980 by Dr. W. Junk Publishers. Reprinted by permission of Kluwer Academic Publishers.

pH 5.0, aluminum concentrations of 0.2 mg/L or greater cause fish gill damage and secretion of mucus onto the gills in brown trout and in whitesuckers. The slimy mucus appears to plug the gills, causing respiratory problems. In addition, the essential integrity of the semipermeable gill membranes, through which exchanges of gases and salts take place, is altered.

It thus seems that not only can an increase in H^+ ions cause fish kills and declines in populations, but also that aluminum can be an additional and perhaps crucial toxic factor in waters around pH 5.0, and certainly at pH 4.0.

Although fish can die of acidification, more commonly they simply fail to reproduce. Yearlings fail to enter the stock or else enter in low numbers, and after a number of years of this reproductive failure, which produces an increasingly old population, the species eventually disappears from the lake or stream. This aging of a population and failure of year classes is illustrated by the data on yellow perch for Patten Lake, Ontario (Figure-5–9).

Some of the areas affected by acid rain are as follows:

- About a dozen rivers in Nova Scotia, far removed from local upwind pollution sources, no longer support healthy populations of Atlantic salmon.
- About 200 lakes in the Adirondacks of upper New York State no longer support brook trout and smallmouth bass. Thousands more lakes in the area are losing their capacity to buffer acid rain.

- Of the 4,016 lakes tested in the Province of Ontario, 155, or 4 percent, have been found to be acidified, with their ability to support aquatic life extremely limited. A total of 2,896 lakes had some susceptibility to acidification. Schindler (1987) suggests that the above estimates substantially understate the magnitude of the problem.

Similar phenomena have occurred in southern Norwegian rivers, in a good many lakes in Galloway, Scotland, and in the Erzgebirge region in East Germany, where fish populations have either vanished or suffered marked reductions over the past 30 years.

Many species of amphibians, i.e., frogs, toads, and salamanders, breed in temporary pools formed by spring rains and melted snow. The eggs and developing embryos are exposed to the springtime acid shock, and deformity or death occurs. Fieldwork has established that 80 percent of salamander eggs failed to hatch in waters with a pH level below 6.0. For the cricket frog and northern spring peeper, an exposure to waters with a pH level of about 4.0 resulted in more than 85 percent mortality. Amphibians are significant members of both the water and land ecosystems. As both major predators of aquatic insects and high-protein food for many birds and mammals, they are important links in the food chain.

Some groups of biota, such as the molluscs, which include animals with shells, e.g., snails, limpets, mussels, and oysters, are strongly dependent upon calcium for their outside skeletal protection. Since acid water readily dissolves calcium carbonate and interferes with calcium uptake by these organisms, they cannot survive in such waters.

Many of the crustaceans (lobster family) in the small free-swimming group known as zooplankton (microscopic animals in the water column) are also very sensitive to increased acidity of fresh waters. Since many of these zooplankton are very important sources of food for fish, their loss could eliminate certain fish species without any direct effect of the acidity on the fish themselves.

Finally, in considering food-chain effects, the key role of the green plants has to be recognized. Green plants are the support system for the entire aquatic biota, since they are the only organisms able to fix carbon (in the presence of light) and so produce the essential carbohydrates, fats, and proteins. Their demise would cause a direct collapse of the food chain.

5.3.4 Effects of Acid Rain on Terrestrial Ecosystems

Effects on forests.   The forests of Canada, the United States, and Scandinavia are of enormous economic importance. Hundreds of thousands of people are employed in the various wood- and forest-associated industries. One in 10 Canadians is employed directly or indirectly in such industries, and Sweden and Norway have a similar employment profile. In addition, the forests and lakes in these countries are major tourist and recreational areas. Acid rain poses an insidious and potentially devastating threat to our forests. It has been shown that seedlings can be damaged by moderately acidic rain (pH 4.6). Researchers are beginning to evaluate the role of acid rain in increasing the vulnerability of trees to disease and insects. Direct, visible acid rain damage to foliage is not being seen, but the dramatic and striking death and dieback of trees in Central Europe is a catalyst for such concerns. Thousands of hectares of spruce and fir forests in Czechoslovakia and East Germany have died in the past 15 years. The forests of the Hartz Mountains and the Black Forest in West Germany are also in trouble, with beech and spruce dying or in a growth decline on the less well-buffered soils.

According to Professor Bernhardt Ulrich of the University of Göttingen in West Germany, the increased acidity of rain in Germany over the past 25 years, combined with the high and acidic snowfall in mountainous areas, has caused leaching of calcium and magnesium from the soils, and at the same time has increased aluminum concentrations in the soil solution. Thus, the Ca:Al ratio has been reduced. When this molar ratio falls below 1.0, aluminum uptake into the fine absorbing roots is favored, and this results in aluminum toxicity in these fine roots, whereupon the roots may die or have reduced vigor. This in turn allows the entry of pathogenic (disease-causing) bacteria and fungi which infect the trees and gradually play a role in their decline. The occurrence of increased aluminum concentrations in the soil solution has been especially prominent in years in which severe summer droughts have occurred, such as in 1975 and 1976 in Europe. Under those circumstances, the aluminum concentrations are increased as a result of the drought-induced concentration of the soil solution. Certainly, the rates of forest decline in West Germany have accelerated markedly since 1975. The high level of industrial activity in and around West Germany is believed to be a key factor in this, as is the high rainfall of the mountain areas, where effects are most severe. In higher altitude forests, acidic cloud waters bathe the trees in fog for long periods each year. Photos 5–1(a) and (b) are examples of the effects of acidic air pollution.

One of the greatest difficulties we face in studying forest growth and the possible effects of acid rain on it is the very considerable variation in growth from year to year caused by normal climatic fluctuations and by insect attack. Growth can differ severalfold from year to year. It is extremely difficult, therefore, to pick up small trends in forest growth decline over a short period of time. Assessments of this type have usually made use of the annual width of wood laid down in tree trunks as annual rings. Such studies have been done in the United States and Norway. All have made use of a limited amount of data, have had difficulties in taking account of the differential normal growth at different ages within a species, and have been inconclusive. One American study suggests that "acid rain merits strong consideration as a factor suppressing tree growth in the Pine Barrens of New Jersey," but others suggest that no clear conclusions can be inferred from the data. We thus have a most frustrating situation in which we *might* be facing a serious decline in forest vigor, but are unable at this time to sort out the various alternative explanations.

In experiments, acid has been sprayed in the field or in controlled laboratory (greenhouse) conditions. Several of these studies have shown an enhanced growth with increasing acidity of the spray down to pH 3.0. In a study by the U.S. Environmental Protection Agency (Lee and Neely, 1980), increased seedling growth occurred in four species, while seven others were unaffected down to pH 3.0. It was suggested that, owing to soil properties, the growth effect was a fertilization effect of sulfur uptake through the foliage.

The most detailed study was done on Scots pine stands in Norway (Drablos and Tollan, 1980). Applications were made above the canopy and at pH values from 5.6 down to 2.0. In Scots pine saplings, increased height and diameter growth were observed in the first four years of the experiment, even at pH 2.0. The effect was attributed to nitrogen

Photos 5–1 (a) and (b)   The effects of acidic air pollution.

(a) A severely affected region located about 8 km from two of the Sudbury-area nickel-copper smelters in the Province of Ontario. The forest destruction has largely been brought about by sulfur dioxide fumigations over many years. Conifers are absent and only stunted birch and red maple remain here. Soils have been acidified and contaminated with heavy metals. Aluminum is solubilized in the strongly acidic soils. (Photo courtesy T.C. Hutchinson)

(b) Die-back and decline of spruce in the Adirondack region of New York State. The gradual loss of needles from the top of the tree and from the tips of the branches can be seen. This is similar to much of the forest damage in Germany (Photo courtesy T.C. Hutchinson)

fertilization from the nitric acid additions. These data have been seized upon by some as ''proof'' that acid rain will only benefit the forests. Unfortunately for this hypothesis, the last two years of the Norwegian data reversed this trend, showing a decline in growth in the plots treated with acid as compared with the control plots. The beneficial fertilization effect was apparently overcome by the detrimental acidity-aluminum effects.

It is important to realize that acidic soils, by themselves, are not harmful to plant growth. Acidification of soils and leaching of nutrients from them, especially of calcium, magnesium, and other bases, are normal soil processes. The vast boreal forests extending worldwide in high latitudes of the northern hemisphere are growing on acid soils developed since the last major glaciation of 10–12,000 years ago. The plants are, therefore, adapted to the acid soil. The question we face in evaluating the acid rain hazard is whether the increases in acidity will push these forests over a threshold to which they are not physiologically adapted. The answer is crucial, but not presently known. Certainly the decline of red spruce in eastern North America is well documented (Klein and Perkins, 1987), and the spread of sugar maple dieback in Quebec and adjacent areas since 1982 is cause for great concern.

Another effect of acid rain on forests includes the leaching of easily acid-soluble components from the foliage, from the trunks of trees, and from the upper layers of the soils. Some of these are redeposited in the soil, or else leached into the drainage basin or the groundwater. The increased levels of K, Ca, Mg, Al, and SO_4 appearing in streams in areas affected by acid rain are believed to be derived from the soils. It is possible that with time, the base components of such soils will be so depleted that nutrient deficiencies will occur. Aluminum toxicity may also be induced. For two reasons, high aluminum concentrations appear to be harmful to many higher plant species through effects on their root systems. First, cell division in the roots is inhibited, and the roots lose their flexibility and plasticity, becoming short and brittle. Second, aluminum has a number of effects on other ions, amongst which are interference with phosphorus uptake and precipitation as aluminum phosphate.

Effects on crops.   While the sensitivity of many crops appears to be much greater than that of many tree species to direct foliar damage by acid rain, no solid evidence exists that the leaves of crops have yet been damaged by acid droplets in the field. A number of detailed studies, however, have begun to suggest that even in a well-buffered agricultural system acid rain may be detrimental. In a study by Lee and Neely (1980), of 27 crop plants grown in pots and exposed to simulated acid rain over a pH range from 2.5 to 5.7, visible, unsightly foliage lesions appeared in 21 crops at a pH of 3.0 (which occurs with a rainfall frequency of 0.5–1.0 percent in affected regions of North America). Studies of major Ontario crops by Hutchinson, 1981 (unpublished) showed that lettuce, beets, onions, soybeans, pinto beans and tobacco were all severely affected in rains of pH 2.5 and 3.0. Such crops as tobacco, lettuce, and spinach depend upon healthy foliage for their sale. Studies at the Brookhaven National Laboratory in the United States (Evans et al., 1983) demonstrated that plants exposed to simulated acidic rainfalls of pH 4.2, 3.8, and 3.5 had decreased seed yields 2.6, 6.5, and 11.4 percent, respectively, compared to plants exposed to ambient rainfall only. Such seed losses in a major crop, such as soybeans, would amount to losses of many millions of dollars per year in the United States.

Experiments have shown that the critical stage in the life cycle of plants at which pollen is transferred to the female flower and germinates to produce a long fertilization (pollen) tube is very sensitive to low pH (Sidhu, 1983). Generally, apple and grape pollen germination and tube growth are reduced at a pH of 3.5 and below. In studies of boreal forest species (Cox, 1983), birch pollen was found to be very sensitive while pine pollen was not. For fruit crops, which are obviously dependent upon a good fruit set at pollination time, acid rain poses a hazard which has not been evaluated.

In summary, it seems clear that terrestrial systems are less sensitive to acid deposition than are aquatic systems. Some of the short-term effects of acid rain may even be beneficial, probably because of the fertilizing nitrogen inputs. Over the longer term, however, it is quite possible that damaging effects will occur. Nutrient cycling and balances in the forest will undoubtedly be affected, and tree growth may decline.

5.3.5 Effects of Acid Rain on Groundwater, Materials, and Buildings

Groundwater and drinking water quality.   Groundwater accumulates very slowly by the percolation of surface waters through the soil and bedrock to the water table. If groundwaters became acidified, then those municipalities depending on groundwater as a drinking water supply might have to chemically adjust the water to acceptable standards. However, in many rural areas and especially in cottage areas, wells are driven below the water table and the water is pumped to the surface for direct consumption, with no provision for treatment. Scientific evidence indicates that groundwater acidification, and its consequent contamination by acid-soluble metals, is occurring in some areas. The major metals of concern are lead, copper, and zinc, which may be leached from water pipes and containers, and aluminum coming from the bedrock itself.

Studies by the Geological Survey of Sweden (Swedish Ministry of Agriculture, 1982) have shown that pH values between 5 and 6 occur in groundwater all over Sweden except in the southwestern parts, where pH values between 4 and 5 have been recorded in shallow wells, with a few such wells even being below 4.0. The water in the ground is quite susceptible to infiltration and replacement by acid waters percolating from above. The risk that this groundwater will become contaminated by heavy metals and aluminum increases substantially as the pH of rain and percolated soil water falls to 4.0, leading to adverse health effects. The fact that over one million Swedes use water drawn from a well of their own, and about half of these live in areas where acidified lakes are common, emphasizes the importance of the problem in such areas.

Corrosion of water pipes.   Acidic water corrodes water pipes. This leads of course to more frequent pipe replacement, but, more importantly, to the risk that metals leached from the pipe walls can reach humans directly through water consumption. In Sweden copper pipes predominate, as they do in Canada and eastern North America as a

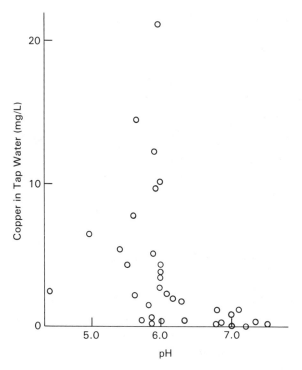

Figure 5–10 Copper content in tap water in Western Europe. Source: Swedish Ministry of Agriculture, 1982

whole. Old lead pipes still occur in the older residences, and are also common in old houses in Britain and Ireland. Galvanized pipes, with a high zinc content, are also widely used. Copper solubility increases sharply below pH 5.0, and also with increasing temperature. Hot water pipes with acid water in them are the most susceptible to dissolution. Concentrations as high as 20 mg/L Cu have been reported in cold water, and 45 mg/L in hot water that has stood in the pipes overnight (Figure 5–10). The World Health Organization (WHO) recommends that copper content be a maximum of 1.5 mg/L. Cases have been reported from Sweden of people's hair turning green after rinsing in warm water with high copper levels, and of children contracting diarrhea due to levels of copper as low as 0.5 mg/L. Cadmium and lead can also be dissolved from soldered joints, and zinc from galvanized pipes. In the Adirondacks of New York State, similar problems have occurred. In some acidified groundwater areas, high levels of copper, lead, and aluminum have been found in water in the house.

Effects on buildings, materials, and paint. Stone buildings, statues, and monuments are eroded by a number of airborne pollutants, including acid rain. Building materials such as steel, paint, plastics, cement, masonry, galvanized steel, limestone, sandstone, and marble also risk damage. The frequency with which structures need to have new protective coatings replaced is increasing with resulting additional costs, estimated at billions of dollars annually.

The effects of the various pollutants cannot yet be reliably separated from each other. However, it is generally accepted that the major single corrosive agent of building materials is sulfur dioxide and its by-products.

Sandstones and limestones have often been employed as materials for monuments and sculptures. Both corrode more rapidly in sulfur-laden city air than in the countryside. When sulfur pollutants are deposited on a sandstone or limestone surface, they react with the calcium carbonate in the material, converting it into the readily soluble calcium sulfate (gypsum), which washes off in the rain. In the Acid Rain Report commissioned by the governor of Ohio in 1980 (Scientific Advisory Task Force, 1980), the committee states that "acid rain is of special concern because of its effects on structures having archaeological or historical significance." The disfiguration and dissolution of famous statues and monuments such as the Acropolis in Athens, famous art treasures in Italy, etc., has accelerated greatly over the past 30 years, often after they have stood for centuries. This is a tragedy which defies economic analysis.

5.3.6 Remedial and Control Measures

Since it is apparent that very substantial harm is being done to our environment, it is clear that remedial action is needed. We have to be aware of the complexity of the problem with its ramifications and interactions in air, soil, water, and sediments, and its effects on plants, animals, and microbes. If large costs are likely to be associated with certain types of remedial action, we need to be sure that these costs are justified and that the action will be effective.

There can be no quick solutions. The cleanup may take decades, even if we start today. In the past few years we have established the fundamental requirements for action, viz.,

- The recognition that acid rain is a serious problem
- The knowledge that reduction of emissions is the best solution

Sulfur oxides are produced from the burning of fuels, smelting of ores, and other industrial processes. Sulfur oxide emissions can be reduced by taking the following measures before, during, and after combustion.

BEFORE COMBUSTION

Fuel switching	Changing from fuels with higher sulfur content to those with lower sulfur content
Fuel blending	Blending fuels with higher and lower sulfur content to produce a fuel with a medium-level sulfur content
Oil desulfurization	Removing sulfur during the oil refining process by hydrogenation (adding hydrogen)

| Coal washing (physical cleaning of coal) | Crushing and removing sulfur and other impurities from coal by placing the coal in liquid. The clean coal floats, the impurities sink |
| Chemical cleaning of coal | Dissolving the sulfur in coal with chemicals |

DURING COMBUSTION

| Fluidized bed combustion (FBC) | Mixing finely ground limestone with coal and burning it in suspension |
| Limestone injection in multistage burners (LIMB) | Injecting finely ground limestone into a special burner |

AFTER COMBUSTION

| Flue gas desulfurization (FGD), or scrubbing | Mixing a chemical absorbent such as lime or limestone with the flue gas to remove sulfur dioxide |

Sulfur oxide emissions from nonferrous smelters can be reduced by a variety of means, including

Mineral separation	Removing some of the sulfur-bearing minerals from the metal-bearing minerals before smelting
Process change	Using smelting processes producing less SO_2 or producing waste streams that are more easily controlled
By-product production	Capturing SO_2 after the smelting process, to produce sulfuric acid (used in many industrial processes and in making fertilizer), liquid SO_2 (used in pulp and paper processing), or elemental sulfur (used in industrial processes)

The Organization for Economic Co-operation and Development (OECD), which includes most of the western industrialized nations, has carried out extensive research into the consequences of long-distance transport of acid rain (OECD Report, 1981). It has also looked in detail at the control technologies and the costs associated with their implementation. The research leaves no doubt that the political and socioeconomic consequences will be severe if nothing is done.

The OECD's calculations apply to the northwestern and southern parts of Europe and are based on the 1980 U.S. dollar. The average cleaning cost is put at $780 per ton of sulfur. Applied to the whole of Europe exclusive of the Soviet Union, the cost of a 50 percent reduction in SO_2 would be about $8 billion per year.

The General Accounting Office of the U.S. government has suggested in its 1982 report on acid rain that an SO_2 emissions reduction of 10 million tons per year in the eastern U.S. would cost \$3–4.5 billion 1980 dollars. (GAO Report 1982). (Total 1980 SO_2 emissions in the U.S. were 24.1 million tons). Clearly, the U.S. will not risk implementing such a costly program until it is sure it has appropriate technology and a certainty of success in mitigating the effects of acid rain.

Over the next few years decisions will have to be made on these technologies. Meanwhile, tensions between the United States and Canada and between the Scandinavians and some of their European neighbors are likely to increase as their environments are further degraded and tourism, recreation, and perhaps soon agriculture and forestry are affected.

5.4 LESSONS LEARNED

Although the two issues of human environmental disturbances examined in Sections 5.2 and 5.3 are very different, they can teach us important lessons about controlling these pervasive environmental problems. Among these lessons are the following.

Human technology can be the cause of serious economic impacts over very large areas of the world, including areas hundreds or thousands of kilometers from the emitters of the pollution. This is so because the atmosphere is a most effective carrier of gases and particles. Poorly soluble gases such as CO_2 and various synthetics like the halocarbons are spread worldwide, and become long-lasting or permanent parts of the atmosphere. More soluble gases such as SO_2 and NO_x can affect large parts of continents and cause serious damage to ecological systems, tourism, agriculture, and forestry, and to buildings and materials.

It follows from this that corrective action can be taken only if the governments concerned agree to cooperate. In North America, any attempts to solve the acid rain problem involve the U.S. and Canadian governments as well as the states and provinces in much of the eastern half of the continent. Since high costs are involved, and since the victims are far distant from the emitters, it will be very difficult for the governments to act jointly.

Nevertheless, public pressure is growing for remedial action. This implies that future technology will have to be much cleaner than before, and that there will be a continual search for sources of energy and raw material that create less waste which can escape to the atmosphere.

Control of such problems has two main components: regulation and technical control. *Regulation by government* is a long-established measure in the case of air and water pollution on the local scale. On this scale it is easy to identify the polluters, to trace the pathway followed by the pollutant, to identify real effects, and to relate these on the dose-response basis to the concentrations of the pollutants. It is correspondingly easy to design suitable regulations for emission or ambient standards, and to monitor their usefulness.

It is much more difficult to proceed in this manner in the case of regional or worldwide problems like those of acid rain and CO_2 buildup. Opponents of remedial action can plead, often successfully, that the costs outweigh the benefits, that the true culprits cannot be identified, or that there is no hard, scientific proof that the adverse effects are truly the result of the pollution; and there may even be benefits. As we saw in Sections 5.2 and 5.3, there is much uncertainty in both cases, and this uncertainty is not likely to be quickly removed.

Technical control measures are the special province of the scientist and engineer. The best of these measures is the choice of pollution-free methods in the originating industry, e.g., in power generation and metal smelting. But even this may still cause problems. For example, nuclear generating stations are a much "cleaner" way of producing energy than are coal-fired generating stations. However, the public is alarmed at the possible danger of accidents in the reactors, and at the potential difficulties of future waste disposal. Hence, although it reduces SO_2, NO_x, and CO_2 emissions, the substitution of nuclear reactors for coal or oil furnaces may not be acceptable as a permanent solution to some people. Nevertheless, the scientist and engineer must constantly seek cleaner and more efficient methods to remove the problems at their sources.

Removal of such pollutants as SO_2, NO_x, and CO_2 from flue gases is technically feasible, but very expensive. The technical means to achieve this are presented in Chapter 13. In the case of CO_2, the removal of carbon would reduce the efficiency of the net heat conversion so greatly that the cost would be prohibitive. The acid rain precursors are more easily dealt with, but even here the cost of removal will be high.

A political value judgment has to be made on such questions: the representatives of the public must decide whether environmental protection justifies the cost. The probability is that they will gradually move toward higher levels of protection; hence, the specialists in pollution control must be ready with the technical solutions.

PROBLEMS

5.1. In Section 5.2.2, world power consumption in 1987 was estimated at over 8 terawatts (TW). In Section 3.2 of Chapter 3, it is stated that "the world's consumption of commercially provided energy in 1985 was about 330 exajoules (EJ)." Are these statements compatible?

5.2. In this chapter examples of global (CO_2), regional/continental (acid rain), and local disturbances have been presented. From your current knowledge of environmental disturbances, natural or manufactured, list other examples which in your opinions are of (a) a global, (b) a regional/continental, and (c) a local scale.

5.3. Select one of the examples listed in Problem 2, and prepare a short (two-three page) statement of your understanding of that particular environmental problem.

5.4. Several types of evidence have suggested that increased acidity of lake and river waters is responsible for a decline in their fish populations. Explain.

5.5. If SO_2 emissions are controlled by an appropriate technology, what will be the effect of this on rain in such regions as California, Michigan, and Ontario (see Table 5–1)?

5.6. Trees require adequate sulfur (S) and nitrogen (N) for healthy growth. Why, then, should acid rain be a problem for them?

5.7. Why are the gases SO_2 and NO_x believed responsible for regional acidification? Why is it that long-distance transport of pollutant air masses is believed responsible for the regional acid rain problem?

5.8. Old and aging fish populations in a lake may indicate the presence of acid rain. Why?

5.9. Why is it difficult to use annual ring increments as an indication of acid rain damage?

REFERENCES

On Carbon Dioxide Buildup

KELLOGG, W. W., and SCHWARE, R. *Climate Change and Society: Consequences of Increasing Atmospheric Carbon Dioxide*, Boulder, Colo: Westview Press, 1981.

MANABE, S., and STOUFFER, R. J. "Sensitivity of a Global Climate Model to an Increase of CO_2 Concentration in the Atmosphere." *Journal of Geophysical Research* 85 (1980): 5529–5554.

MANABE, S., and WETHERALD, R. T. "On the Distribution of Climatic Change Resulting from an Increase in CO_2 Content of the Atmosphere." *Journal of the Atmospheric Sciences* 37 (1980): 99–118.

National Academy of Sciences. *Understanding Climatic Change.* Washington, D.C.: N.A.S., 1975.

U.S. Department of Energy. *A Comprehensive Plan for Carbon Dioxide Effects Research and Assessment, Part I, The Global Carbon Cycle and Climatic Effects of Increasing Carbon Dioxide, Carbon Dioxide Effects Research and Assessment Program.* Report 008. Washington, D.C.: U.S. Department of Energy, 1980.

U.S. Department of Energy. *Environmental Control Technology for Atmospheric Carbon Dioxide, Carbon Dioxide Effects Research and Assessment Program.* Report 006. Washington, D.C.: U.S. Department of Energy, 1980.

World Meteorological Organization. *World Climate Programme, On the Assessment of the Role of CO_2 on Climate Variations and their Impact.* (Villach, Austria and Geneva, Switzerland: World Meteorological Organization, 1980 and 1981.

On Acid Rain

BAKER, J. P., and SCHOFIELD, C. L. "Aluminum Toxicity to Fish as Related to Acid Precipitation and Adirondack Surface Water Quality." In *Ecological Impact of Acid Precipitation*, edited by D. DRABLOS and A. TOLLAN, pp. 292–296. Sandefjord, Norway: SNSF, Oslo, 1980.

BEAMISH, R. J., and HARVEY, H. H. "Acidification of the La Cloche Mountain Lakes, Ontario, and Resulting Fish Mortalities." *Journal of Fisheries Research Board of Canada* 29 (1972): 1131–1143.

BEAMISH, R. J., LOCKHART, W. L., VANHOON, J. C., and HARVEY, H. H. "Long-term Acidification of a Lake and Resulting Effects on Fishes." *Ambio.* 4 (1975): 98–102.

COX, R. M. "Sensitivity of Forest Plant Reproduction to Long Range Transported Air Pollutants." *New Phytologist* 95 (1983): 269–276.

DICKSON, W. "Properties of Acidified Waters." In *Ecological Impact of Acid Precipitation*, edited by D. DRABLOS and A. TOLLAN, pp. 75–83. Sandefjord, Norway: SNSF, Oslo, 1980.

DRABLOS, D., and TOLLAN, A. *Ecological Impact of Acid Precipitation.* Sandefjord, Norway: SNSF, Oslo, 1980.

Environment Canada. *The Acid Rain Story.* Ottawa: Environment Canada, 1984.

Environment Canada. *Downwind: The Acid Rain Story.* Ottawa: Environment Canada, 1981.

Evans, L. S., Lewin, K. F., Petti, M. J., and Cunningham, E. A. "Productivity of Field-Grown Soybeans Exposed to Simulated Acidic Rain." *New Phytologist* 93 (1983): 377–388.

GAO Report. "General Accounting Office Report on Acid Rain." Washington, D.C., U.S. Government, 1981.

Harvey, H. H. "Widespread and Diverse Changes in the Biota of North American Lakes and Rivers Coincident with Acidification." In *Ecological Impact of Acid Precipitation*, edited by D. Drablos and A. Tollan, pp. 93–98. Sandefjord, Norway: SNSF, Oslo, 1980.

Harvey, H. H., and Pierce, R. C., eds. *Acidification in the Canadian Aquatic Environment*. Ottawa: National Research Council of Canada, 1981.

Hutchinson, T. C. "Unpublished Report to the Ontario Ministry of Environment on the Relative Sensitivity of Ontario Crops to Acid Rain Spray." 1981.

Klein, R. M., and Perkins, T. D. "Cascades of Causes and Effects of Forest Decline." *Ambio*. 16 (1987): 86–93.

Lee, J. J., and Neely, G. E. *Sulfuric Acid Rain Effects on Crop Yield and Foliar Injury*. Corvallis, Or.: EPA, 1980.

NATO. *Effects of Acid Precipitation on Terrestrial Ecosystems*. New York: Plenum Press, 1980.

OECD, Report, "The Costs and Benefits of Sulfur Oxide Control. A Methodological Study." Paris, Organization for Economic Co-operation and Development, 1981.

Ontario Ministry of Environment. *The Case Against Acid Rain*, Toronto: Ontario Ministry of the Environment, 1980.

Ryan, P. M., and Harvey, H. H. "Growth Responses of Yellow Perch to Lake Acidification in the La Cloche Mountain Lakes of Ontario." *Environmental Biology of Fish* 5 (1980): 97–108.

Scientific Advisory Task Force. *Acid Rain, Report to the Governor of Ohio*. Columbus: State of Ohio, 1980.

Schindler, D. W. Personal Communication, 1987.

Sidhu, S. S. "Effects of Simulated Acid Rain on Pollen Germinations and Pollen Tube Growth of White Spruce (*Picea Glauca*)." *Canadian Journal of Botany* 61 (1983): 3095–3099.

Swedish Ministry of Agriculture. *Acidification Today and Tomorrow*. Stockholm: Swedish Ministry of Agriculture, 1982.

6

Physics
and Chemistry

Gary W. Heinke

J. Glynn Henry

6.1 INTRODUCTION

Much of this book deals with water, with air, and with mixtures of solids such as refuse or sludges. When we say water, normally we mean not H_2O, but water in the form of rain, rivers, lakes, groundwater, or seawater. In each of these cases, we are dealing with very dilute systems which have particles dispersed and solutes dissolved in water, the universal solvent. When we say air, we generally mean not just the pure mixture of nitrogen, oxygen, and trace gases, but also the gaseous pollutants, as well as the liquid and solid particles suspended in the air.

This chapter therefore summarizes some fundamentals concerning particles and particle dispersions, and then presents some basic information from physics, chemistry, physical chemistry, and reaction kinetics that are relevant to water and air systems. Some of this material is covered in elementary courses in the various disciplines, and some will be new here.

6.2 PARTICLE DISPERSION

In order to be able to describe and treat natural waters, wastewaters, air, solid wastes, and sludges, knowledge about the medium and the particles and solutes in it is essential. Some properties of the medium may be greatly unaffected or only slightly affected by the presence of particles or solutes. For example, the density of various waters, including heavily polluted wastewaters, is so close to that of water that the small differences can usually be ignored. Even seawater, with a total dissolved solids concentration of about 34,500 mg/L, has a density only 2.5 percent greater than that of pure water. Other properties of the medium may be greatly affected by the presence of particles or solutes. A familiar example would be the loss of visibility in air when fine liquid particles are present (fog) or fine solid particles are present (smoke).

6.2.1 Particle Size, Shape, and Distribution

A particle can be defined as any distinct, i.e., particulate, portion of solid, liquid, or gaseous matter larger than a single small molecule (larger than 1 nanometer in diameter). Water, air, and solid wastes contain many particles which vary considerably in size. For many situations, it will be important to find a convenient way to express the size of particles, their shape, and their size distribution.

Figure 6–1 presents a schematic diagram on a logarithmic scale to cover the range of sizes of particles of importance in environmental engineering. The boundaries shown are flexible. In water, particles are said to be in suspension when they can be removed by settling or by filtration through filter paper. The lower limit for this is about 0.4 μm. Particles smaller than that are called colloids. They normally range in size from between 1 to 400 nanometers (nm), so they are not visible with an ordinary high-powered microscope. Because of their importance in environmental engineering, colloids are discussed separately in Section 6.2.2. Below about 1 nm, particles are considered to be dissolved, with diameters ranging from those of a single atom (about 0.2 nm) to the size of a molecule (about 1 nm). They are dispersed in the solvent to form a solution.

Seldom will the particles in a mixture or suspension be of uniform size or spherical in shape. To describe the mixtures of particles analytically is difficult, especially for solid wastes, where particles differ greatly in size and shape. Small particles in air or liquid suspensions are more easily characterized. Irregular shape as defined by the length, width, and height of the particle is normally related mathematically to an equivalent diameter. Other methods are based on shape factors, which may compare the particle surface area or its settling velocity to that of an equivalent sphere. One such example, the sphericity Ψ, is presented in Section 6.2.4 in Equation 6.8.

The size distribution of spherical or equivalent spherical particles generally cannot be expressed by a single parameter or function, as, for example, average particle diameter. Instead, classification may depend on the way particle diameters are actually measured. Methods range from the viewing of material for coarse sizes to particle counting through a microscope for small sizes. In any case, particles are classified into an arbitrary number of size ranges which may, for example, be sieve sizes. This information can be plotted as shown in Figure 6–2, which is a plot of cumulative weight less than a given diameter d_p versus particle diameter. Mathematical expressions are also used to describe particle size distribution, but these are beyond the scope of this book.

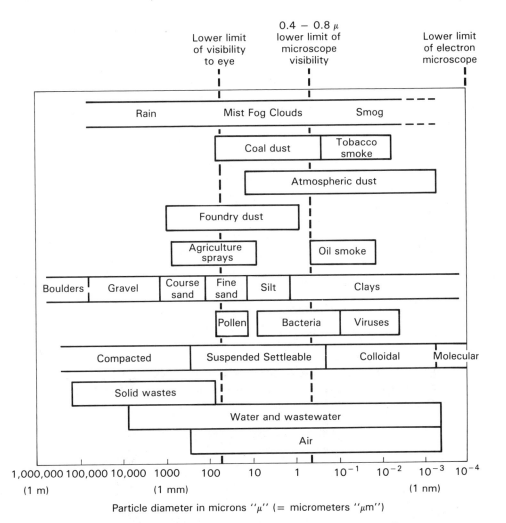

Figure 6–1 **Typical range of particle size.** Source: Adapted from several sources including Williamson (1973), Hidy and Brock (1970), and Perkins (1974).

Note: *Common Units*:   μ, microns (1 × 10⁻⁶ m) and
              mμ millimicrons (1 × 10⁻⁹ m)
              were common terms before SI was adopted.

     SI Units: μ is replaced by μm, micrometers (1 × 10⁻⁶)
                mμ is replaced by nm, nanometers (1 × 10⁻⁹)

(Both systems are still in use, and both may be found in this book).

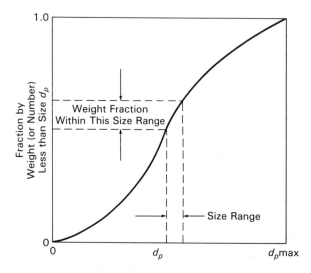

Figure 6–2   Particle size distribution

6.2.2 Colloidal Dispersions

Colloidal dispersions consist of very small particles ranging in size from about 1 to 400 nm, separated by the dispersion medium. The dispersed colloidal particles can be solid, liquid, or gaseous. Although dispersions in a solid medium occur, the dispersion medium of interest in the environmental field is either a liquid or a gas. Common names for dispersions are given in the following table:

DISPERSED PHASE	DISPERSION MEDIUM	COMMON NAME
Solid	Liquid	Suspension
Liquid	Liquid	Emulsion
Gas	Liquid	Foam
Solid	Gas	Smoke, Aerosol
Liquid	Gas	Fog, Aerosol

The characteristic properties of colloidal particles can be attributed to their small size, which provides a very large surface area per unit volume. To demonstrate just how large a surface area they possess, consider a cube 1 cm to a side. Such a cube has a surface area of 6 cm². If we divide it into eight cubes $\frac{1}{2}$ cm to a side, we double the total surface area to 12 cm², but the volume remains the same. By continuing to make smaller and smaller cubes until they are well into the colloidal size range, or about 10^{-5} mm to a side, the total volume will still be the same but the total surface area will have increased from 6 cm² to 600 m², a millionfold increase! Because of this high surface-to-volume ratio, colloids have tremendous adsorptive capacity relative to their small mass. Also, with the large surface area the weak atomic surface changes on colloids become a significant factor in their behavior.

The two surface-related properties of colloids are therefore their adsorptive capacity and electrokinetic charge. **Adsorption** refers to the ability of certain solids to concentrate on their surface, substances from the surrounding medium. The **electrical charge** which all colloidal particles carry may be positive or negative and varies in magnitude depending on the type of material the colloid is made of. Like-charged particles repel each other and thereby prevent the formation of larger particles through agglomeration. This repulsion between particles makes it difficult to separate the particles from their dispersion medium. Coagulation, a treatment process which overcomes the problem by adding particles (ions) of opposite charge, is discussed in Chapter 11. Coagulation is also used to remove particles from polluted air (Chapter 13).

6.2.3 Methods of Expressing Particle Concentrations

The mass of particles in a unit volume (or mass) of dispersion medium is called **particle concentration.** There are several different ways of expressing particle concentrations in air, water, and wastewater.

Air.   The usual units for expressing the concentration of small particles suspended in air are grams of particulates per m^3 of air. The concentration is determined by drawing a known volume of air through a preweighed filter and weighing the amount of particulates which have been trapped. For dust, which consists of larger particles which settle quickly, measurements are made by collecting the settleable material in dustfall jars for a specific time and determining the accumulated weight. The concentration is then expressed in weight collected per unit area over a given period. Examples of such units are ton/mile^2 $\times$ month, or kg/m^2 $\times$ month. Several other methods of expressing the concentration of particles in air are also in use (see Chapter 13).

Water and wastewater.   Concentrations of particles in water are normally expressed differently than in wastewater. Because there are only small amounts of particles in most natural waters, and particularly in drinking water, an optical method is used rather than a gravimetric method. **Turbidity** is an expression of the optical property that causes light to be scattered and absorbed rather than transmitted in straight lines through the sample. Turbidity in water is caused by suspended matter such as clay, mud, algae, silica, rust, bacteria, and other particulates. The standard unit of turbidity measurement adopted by the water industry is the nephelometric turbidity unit (NTU) named after the instrument used to measure turbidity, the nephelometer (Gk: nephos = cloud). It is based on a specified concentration of a formazin polymer suspension.

For wastewaters and sludges, the suspended solids (SS) concentrations are generally sufficiently high that gravimetric methods are best, and the units normally used are mg/L. A small sample (100 mL) of wastewater or sludge is filtered through a glass-fiber filter which is dried and weighed before and after filtration in order to determine the SS concentration. An alternative, gravimetric procedure, useful when filtration is difficult, is to weigh a known volume of sample before and after evaporation of the liquid (at 103°C),

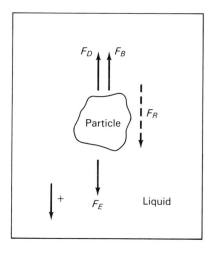

Figure 6–3 Forces acting on a particle settling in a quiescent fluid

with the residue representing the SS. Solids concentrations that exceed 10,000 mg/L may be expressed in percent (1% = 10,000 mg/L).

6.2.4 Settling of a Particle in a Fluid

The principles involved in the settling of a particle in a fluid can be applied to the removal of suspended solids in a river or lake (Chapter 9), the design of clarifiers for water treatment (Chapter 11) or wastewater treatment (Chapter 12), and the settling of particulates from air (Chapter 13).

Consider the situation of a single sand particle settling at velocity u in a quiescent fluid. A particle falling under the action of gravity will accelerate until the frictional drag of the fluid just balances the gravitational acceleration, after which it will continue to fall at a constant velocity known as the terminal settling velocity u_t. This velocity can be calculated by making a force balance on the particle (see Figure 6–3), viz.,

$$F_R = F_E - F_B - F_D \tag{6.1}$$

where F_E = external force on particle, in this case gravity (but could be another external force such as centrifugal force in centrifugation)
F_B = buoyancy force
F_D = friction or drag force, opposing settling of particle
F_R = resultant force (= 0 when terminal velocity is reached)

These forces can be expressed as

$$F_R = M\frac{du}{dt}$$

$$F_E = Ma_E = Mg$$

$$F_B = \frac{\rho}{\rho_p} MA_E = \frac{\rho}{\rho_p} Mg$$

where M = mass of particle

u = settling velocity of particle

a_E = acceleration, equal to g for gravitational settling (for centrifugation, a_E = $r\omega^2$, where ω = angular velocity)

ρ = density of fluid

ρ_p = density of particle

Substituting the values of F_R, F_E, F_B, and F_D into Equation 6.1 and rearranging, we get

$$\frac{du}{dt} = \frac{\rho_p - \rho}{\rho_p} g - \frac{F_D}{M} \qquad (6.2)$$

Experimentally, it has been found that

$$F_D = \frac{C_D A_p u^2}{2}$$

where C_D = coefficient of drag or friction (dimensionless)

A_p = projected area of particle at right angles to direction of settling

Substituting these values into Equation 6.2, we get

$$\frac{du}{dt} = \frac{\rho_p - \rho}{\rho_p} g - \frac{C_D A_p \rho u^2}{2M} \qquad (6.3)$$

Assuming that the particle is spherical,

$$\frac{A_p}{M} = \frac{\pi d_p^2 / 4}{(\pi d_p^3 / 6)\rho} = \frac{3}{2\rho_p d_p} \qquad (6.4)$$

where d_p is the particle diameter. Therefore, substituting this value of A_p/M into Equation 6.3 yields

$$\frac{du}{dt} = \frac{\rho_p - \rho}{\rho_p} g - \frac{3C_D \rho u^2}{4\rho_p d_p} \qquad (6.5)$$

The terminal settling velocity u_t is reached very quickly. At that point,

$$\frac{du}{dt} = 0$$

so solving for u_t, we get

$$u_t = \left[\frac{4g(\rho_p - \rho)d_p}{3C_D \rho} \right]^{1/2} \qquad (6.6)$$

Equation 6.6 applies to settling particles ($\rho_p > \rho$) as well as rising particles ($\rho_p < \rho$) such as oil or air in water.

It has been found experimentally that C_D is a function of the dimensionless Reynold's number* Re, i.e.,

$$C_D = \frac{b}{Re^n}$$

where the values of b and n are as given in the following table:

Flow	Re	b	n	Remarks
Laminar	<2	24	1	Friction drag predominates
Intermediate	2–500	18.5	0.6	Friction and form drag both important
Turbulent	500–200,000	0.44	0	Form drag predominates

* Reynold's number Re is defined as

$$Re = \frac{u \cdot d_p \cdot \rho}{\mu}$$

where μ is the dynamic (absolute) viscosity of the medium.

For laminar flow conditions,

$$C_D = \frac{24}{Re} = \frac{24\mu}{\rho u d_p}$$

By substituting into Equation 6.6, we obtain

$$u_t = \frac{(\rho_p - \rho)g d_p^2}{18\mu} \tag{6.7}$$

Equation 6.7 is known as Stokes's terminal velocity equation. It applies only if laminar flow occurs and if the particles are spherical.

Both the suspended matter in natural waters and wastewaters and particulate emissions settling in air are seldom spherical. They have a greater surface area per unit volume than a sphere, and will therefore settle more slowly than spheres of equivalent volume. For particles of irregular shape, an index called **sphericity** has been defined, in such manner that

$$\frac{A_p}{V_p} = \frac{6}{\Psi d_p} \tag{6.8}$$

where the sphericity Ψ is a relative index of the roundness of the particle and is dimensionless. Note that for a sphere $\Psi = 1$, whereas, for nonspherical particles Ψ will always be less than 1. For example, for mica flakes $\Psi = 0.28$, for jagged flint sand $\Psi = 0.66$, for pulverized coal $\Psi = 0.73$, and for nearly spherical Ottawa sand $\Psi = 0.95$. The practical application of Stokes's terminal velocity equation is more complicated, since particles in water or air are not normally discrete particles and the water or air medium is not quiescent.

Therefore, experimental or empirical methods must normally be used for the design of settling tanks for water or wastewater (Chapters 11 and 12) or treatment units for cleaning polluted air (Chapter 13).

Example 6.1

Calculate the settling velocity of two spherical particles of diameter (a) 0.1 mm, and (b) 0.001 mm, in still water at a temperature of 20°C. The specific gravity of the particles is 2.65. Assume laminar flow conditions. A common detention time for sedimentation tanks is 2 hr. Will these particles settle to the bottom of a 3.5-m-deep tank in that time?

Solution   From Appendix A-4,

$$\rho_{20°C, \text{ water}} = 998 \text{ kg/m}^3$$
$$\rho_p = 2.65 \times 998 \text{ kg/m}^3$$
$$\mu_{20°C, \text{ water}} = 1.00 \times 10^{-3} \text{ N-s/m}^2 = 1.00 \times 10^{-3} \text{ kg/m-s}$$

Therefore, using Equation 6.7, we obtain

(a) $u_t = (2.65 - 1.00) \, 998 \text{ kg/m}^3 \times 9.81 \text{ m/s}^2 \times \dfrac{(0.1 \times 10^{-3})^2 \text{ m}^2}{18 \times 1.00 \times 10^{-3} \text{ kg/m-s}}$

$= 9.0 \times 10^{-3} \text{ m/s} = 9.0 \text{ mm/s}$

So the time to settle 3.5 m $= 3.5/9.0 \times 10^{-3} = 390 \text{ s} = 6.5 \text{ min} \ll 2 \text{ hr.}$
  This settling rate is certainly practical for particle removal by sedimentation.

(b) $u_t = 9.0 \times 10^{-7} \text{ m/s} = 9.0 \times 10^{-4} \text{ mm/s}$

So the time to settle 3.5 m $= 390 \times 10^4 \text{ s} = 1{,}083 \text{ hr.}$
  This settling rate is much too slow for particle removal by sedimentation alone, so coagulation/flocculation must be employed to increase the particle size and thus achieve greater settling velocity (see Chapter 11).

6.3 SOLUTIONS

6.3.1 Solutions and Solubility

Gases, liquids, and solids may dissolve in water to form true solutions. The substance that dissolves is called the **solute,** and the substance or medium in which it is dissolved is called the **solvent.** A solution may have any concentration of the solute below a certain limit, called the **solubility** of that substance in that medium. A solution which contains, at a given temperature, as much solute as it can hold in the presence of the dissolving substance is called a **saturated solution.** Solutions that contain less solute are called **unsaturated,** and those that (under special conditions) contain more are called **super-saturated.**

Several factors affect solubility. Temperature and the chemical character of the substances involved are the most important. Pressure is relatively unimportant for liquids, except for deep underground supplies or deep ocean water. The solubilities of most substances increase as temperature increases, but there are important exceptions. In general, if a substance dissolves at saturation with absorption of heat, then the solubility will increase as the temperature goes up. On the other hand, if heat evolves in the solution process, solubility will decrease with an increase in temperature of the solvent. A number of the calcium compounds, including $CaCO_3$, $CaSO_4$, and $Ca(OH)_2$, decrease in solubility as the temperature goes up. The solubility of oxygen in water also decreases with rising temperature. This has important consequences, because of the decreased capacity of natural waters to supply oxygen for aquatic life and for oxidation of organic pollution during the summer months.

The chemical character of the solute and solvent may be such that a solution forms readily. For example, alcohol (C_2H_5OH) and water (H_2O) mix easily, with no saturation limit. That is, they are completely miscible. On the other hand, water and mercury are almost completely immiscible. There is a wide range of miscibility between these extremes. Handbooks provide information on the solubilities of substances (solubility constants) in various solvents. The solubilities of many substances are affected by chemical reactions with water or other solvents. For example, calcium carbonate is only slightly soluble in pure water, but has a much higher apparent solubility if the water contains carbon dioxide (CO_2), because of the chemical reaction between $CaCO_3$ and CO_2. In general, substances such as salts, acids, and bases dissolve in water to form solutions containing **ions.** In such solutions the presence of excess amounts of any of the ions can greatly affect solubility.

In precipitation reactions, a form of the equilibrium constant is called the **solubility product.** For example, in the reaction of calcium carbonate ($CaCO_3$) in water, where $CaCO_3$ (solid) $\rightleftarrows Ca^{+2} + CO_3^{-2}$, the solubility product is defined as $[Ca^{+2}]\,[CO_3^{-2}] = K_{sp}$. K_{sp} is the solubility product constant, and [ ] indicates the concentration of the substance in mol/L. Numerical values of K_{sp} for precipitation reactions can be obtained from handbooks.

The solubility of gases in liquids depends on the nature of the gas, on the nature of the solvent, and on pressure and temperature. For example, nitrogen (N_2), hydrogen (H_2), and oxygen (O_2) are relatively insoluble in water, whereas ammonia (NH_3) and hydrogen sulfide (H_2S) are quite soluble. A discussion of the gas laws, including those of importance to gas-liquid transfer, is presented in Section 6.4.

Natural waters always contain dissolved ions, which come from the contact of water with minerals such as limestone, magnesite, gypsum, and salt beds. The most common cations found in natural waters are calcium (Ca^{+2}), magnesium (Mg^{+2}), sodium (Na^+) and potassium (K^+). The most common anions are the bicarbonates (HCO_3^-), chlorides (Cl^-), sulfates (SO_4^{-2}), and, to a lesser extent, the nitrates (NO_3^-). In any water, electroneutrality is maintained, so that the sum of the cations must always equal the sum of the anions. Minor ionic species found in water include, for example, the following:

CATIONS	ANIONS
Iron (Fe^{+2} or Fe^{+3})	Carbonate (CO_3^{-2})
Manganese (Mn^{+2})	Hydroxide (OH^-)
Aluminum (Al^{+3})	Sulfur (S^{-2}, SO_3^{-2})
Ammonium (NH_4^+)	Phosphates (PO_4^{-3}, HPO_4^{-2}, $H_2PO_4^-$)
Copper (Cu^{+2})	

Human activity, principally through industrial waste discharges, may add ions to natural waters, sometimes resulting in significant pollution problems. The heavy metal ions, which may be toxic to microorganisms, plants, and animals, are prime examples.

6.3.2 Methods of Expressing the Composition of Solutions

The following two systems for expressing the composition of solutions are commonly used:

1. *Mass/mass* (commonly stated as weight/weight), or more explicitly, the mass of solute per mass of solution. A typical unit used is mg/kg, also expressed as ppm* (parts per million). This method is not temperature dependent.
2. *Mass/volume* (commonly stated as weight/volume), or more explicitly, the mass of solute per volume of solution. A typical unit used is mg/L.* This method is temperature dependent, since volume varies with temperature. Therefore, temperature should be reported when stating concentration by this method.

Within each of these two systems, there are several methods for expressing concentrations:

1. *Mass/mass (or weight/weight)*

    (a) *Percent by weight*   Example: Solute 1% NaCl
                                            Solvent 99% H_2O
    (b) *Parts per million*   Example: 10,000 ppm of NaCl
                                            10,000 mg/kg NaCl
    (c) *Molality, m* = number of gram moles** of solute per 1,000 g of solvent

---

* The units ppm and mg/L are often used interchangeably. This is justified if the specific gravity of the solution is unity, which is approximately true for most waters and wastewaters. If the specific gravity is not unity, conversion can be calculated from

$$\text{Concentration in ppm (mg/kg)} = \text{Concentration in (mg/L)} \times \frac{1}{\text{specific gravity of solution}}$$

For very dilute solutions, it may be more convenient to express the concentration in micrograms per liter (μg/L). This is equivalent to parts per billion (ppb), where a billion is understood to be 10^9.

** One mole is the amount of a substance which contains Avogadro's number of molecules. Therefore, the mole refers to a fixed *number* of any type of particles rather than a weight. In practice, we do not count molecules or particles; we weigh them. Therefore, an engineering definition of a mole is that mass of a substance (in appropriate units—g, kg, lb, ton, etc.) which is numerically equal to the molecular weight of the substance. For example, 1 gram-mole of oxygen = 32 g of oxygen.

Example: NaCl solution in water with a molality of 1

1 gram mole of NaCl (solute) =		58.5 g
H_2O (solvent)	=	1,000.0 g
		1,058.5 g of solution

(d) *Mole fraction, X_i* $= \dfrac{\text{number of moles of solute, } n}{\text{total number of moles, } n_t}$

The following example will illustrate the conversion from one method to the others.

Example 6.2

Express the composition of a 2%-by-weight NaCl solution in water in terms of the other mass/mass units.

Solution Choose 100 g of solution as a basis. (Any other weight would also do.) We have 2 g NaCl and 98 g H_2O. Thus,

$$for\ (b),\ \frac{2}{100} = \frac{2 \times 10^4}{10^6} = 20{,}000 \text{ ppm}$$

$$for\ (c),\ \text{molality } m = \frac{2/58.5}{98/1{,}000} = 0.349$$

$$for\ (d),\ \text{mole fraction } X_{\text{NaCl}} = \frac{2/58.5}{2/58.5 + 98/18} = 0.0062$$

$$\text{mole fraction } X_{H_2O} = \frac{98/18}{2/58.5 + 98/18} = \frac{0.9938}{1.0000}$$

(The sum of mole fractions must equal unity.)

2. *Weight/volume*

(a) mg/L Example: 1,000 mg of NaCl in 1 L of solution

(b) Molarity M = number of gram moles of solute per 1 liter of solution

Example: NaCl solution of molarity = 1 M, contains 1 g mole or 58.5 g NaCl per liter of solution.

Solutions of equal molarity have equal numbers of molecules of dissolved substance per liter or per any other unit volume.

(c) Normality N = number of gram-equivalent weights of solute per 1 liter of solution (eq/L)

Because a given substance can have more than one gram-equivalent weight depending on the reaction it undergoes, it is necessary, in expressing a concentration as a normality, to specify in what reaction or type of reaction the solution is going to be used. In general,

$$\text{Equivalent weight in g/eq} = \frac{\text{atomic or molecular weight (g)}}{n \text{ (equivalents)}} \tag{6.9}$$

where n is a positive integer and

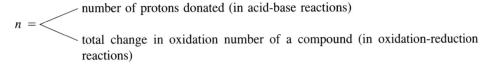

$$n = \begin{cases} \text{number of protons donated (in acid-base reactions)} \\ \text{total change in oxidation number of a compound (in oxidation-reduction reactions)} \end{cases}$$

For very dilute systems, it is often more convenient to use milliequivalents per liter (meq/L) instead of (gram) equivalents per liter (eq/L). Note that 1 eq/L = 1,000 meq/L, and 1 eq/m³ = 1 meq/L.

For example, in the acid-base reactions

$$H_3PO_4 + NaOH \rightarrow NaH_2PO_4 + H_2O$$

$$H_3PO_4 + 2NaOH \rightarrow Na_2HPO_4 + 2H_2O$$

and

$$H_3PO_4 + 3NaOH \rightarrow Na_3PO_4 + 3H_2O$$

n = 1, 2, and 3, respectively.

The advantage of *normal* solutions is that if the normality of two solutions A and B are the same, then 1 mL of A will react with exactly 1 mL of B. This is because

$$V_A N_A = V_B N_B \tag{6.10}$$

where N_A, N_B = normality of solutions A, B, which is the number of gram-equivalent weights in one liter of solutions of A and B

and V_A, V_B = volume of solution A (B) of normality N_A (N_B) that reacts with volume V_B (V_A) of normality N_B (N_A)

Therefore, if $N_A = N_B$, then $V_A = V_B$. This relationship can be used to find the normality of an unknown solution.

Example 6.3

If a solution contains 5 g/L of NaOH, calculate the concentration of NaOH in terms of weight/volume units expressed as (a) mg/L, (b) molarity, and (c) normality. The temperature of the solution is 20°C. The reaction is

$$H_3PO_4 + 3NaOH \rightarrow Na_3PO_4 + 3H_2O$$

Solution

(a)   Concentration in mg/L   = 5,000 mg/L

(b)   Concentration in molarity $= \dfrac{5/40}{1} = 0.125$ M

Note: To convert molarity to mg/L, use

$$mg/L = \text{molarity} \times \text{gram molecular weight} \times 10^3$$
$$= 0.125 \times 40 \times 10^3 = 5,000 \text{ mg/L}$$

(c)   Equivalent weight for NaOH in this reaction $= \dfrac{40 \text{ g}}{3 \text{ eq}} = 13 \cdot 33 \text{ g/eq}$

Number of equivalent weights of NaOH in 1 L solution $= \dfrac{5 \text{ g/L}}{13.33 \text{ g/eq}}$

$$= 0.375 \text{ eq/L}$$

Therefore,

$$\text{Normality N} = 0.375 \text{ eq/L}$$

Example 6.4

Determine the normality of a solution of NaOH, of which 17.5 mL is required in the titration of 35.0 mL of a 0.2-N HCl solution. A drop of methyl orange indicator in the acid solution serves to indicate the end point of titration, by a change in color.

Solution From Equation 6.10, $V_A N_A = V_B N_B$. Let A be the HCl solution and B the NaOH solution. Then

$$N_{\text{NaOH}} = \frac{V_{\text{HCl}} \times N_{\text{HCl}}}{V_{\text{NaOH}}} = \frac{35.0 \times 0.2}{17.5} = 0.4N$$

Therefore, the normality of the NaOH solution is 0.4N.

Concentration in terms of a common constituent.   It has been found useful in water chemistry to express concentrations in terms of a common constituent when different chemical forms, all containing the common constituent, are present in water. This method is not used in general chemistry and therefore requires explanation. For example, nitrogen compounds can be present in the following forms in wastewater:

Ammonia nitrogen	NH_4^+, NH_3
Organic nitrogen	various forms
Nitrite nitrogen	NO_2^-
Nitrate nitrogen	NO_3^-

It is customary to report all results in terms of nitrogen N so that values can be directly compared. For example, Figure 6–4 shows the changes occurring in the forms of nitrogen in a wastewater under aerobic conditions. All concentrations are expressed as mg/L N. The expression 10 mg/L NO_3^- N means that the nitrate (NO_3^-) concentration is 10 mg/L expressed as N.

Example 6.5

A nitrogen analysis of a wastewater sample gave the following results:

Ammonia	30.0 mg/L NH_3
Nitrite	0.10 mg/L NO_2^-

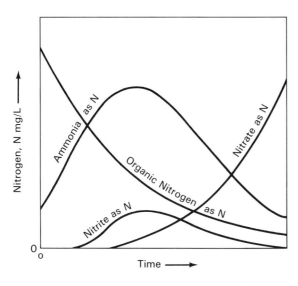

Figure 6–4 Forms of nitrogen compounds in wastewater under aerobic conditions
Source: Sawyer and McCarty, 1978

Nitrate	1.50 mg/L NO_3^-
Organic nitrogen (various forms)	15.0 mg/L N

Find the total concentration of nitrogen.

Solution

$$\text{Ammonia} \qquad NH_3 = \frac{14}{17} \times 30.0 = 24.70 \text{ mg/L} \quad NH_3^- \text{ N}$$

$$\text{Nitrite} \qquad NO_2^- = \frac{14}{46} \times 0.10 = 0.03 \text{ mg/L} \quad NO_2^- \text{ N}$$

$$\text{Nitrate} \qquad NO_3^- = \frac{14}{62} \times 1.50 = 0.34 \text{ mg/L} \quad NO_3^- \text{ N}$$

$$\text{Organic nitrogen} \qquad\qquad\qquad = 15.0 \text{ mg/L} \quad \text{N}$$

$$\text{Total concentration of nitrogen} \qquad = 40.1 \text{ mg/L} \quad \text{N}$$

The common-constituent method is also frequently used for phosphorus concentration and for expressing the hardness and alkalinity of water, both of which are part of the carbonate system. Hardness is caused by divalent metallic cations, principally Ca^{+2} and Mg^{+2}, while alkalinity is contributed to by the anions HCO_3^-, CO_3^{-2}, and OH^-. It has been common practice to express hardness and alkalinity in terms of mg/L of $CaCO_3$, but the use of "equivalents" is increasing. We have

$$\text{Hardness (in mg/L as } CaCO_3) = M^{+2} \text{ (in mg/L)} \times \frac{50^* \text{ mg/meq}}{\text{equiv wt } M^{+2}} \qquad (6.11)$$

* For conversion between milliequivalents per liter of Ca^{+2} or Mg^{+2} and milligrams per liter as $CaCO_3$, use

$$\text{Milliequivalent weight of } CaCO_3 = \frac{100 \text{ mg/mol}}{2 \text{ meq/mol}} = 50 \text{ mg/meq}$$

where M^{+2} represents a divalent metallic ion. Note that n in Equation 6.9 is now equal to 2 based on the reaction

$$CaCO_3 \rightarrow Ca^{+2} + CO_3^{-2}$$

Example 6.6

Calculate the hardness, in mg/L $CaCO_3$, of a water sample with the following analysis:

cation	conc.	Equiv. wt
Na^+	40 mg/L	23
Mg^{+2}	10 "	12.2
Ca^{+2}	55 "	20.0
K^+	2 "	39.0

Solution Only the divalent ions Ca^{+2} and Mg^{+2} contribute to hardness. Thus, from Equation 6.11,

$$\text{Hardness, mg/L } CaCO_3 = 55 \times \frac{50}{40/2} + 10 \times \frac{50}{24.3/2}$$

$$= 138 + 41 = 179 \text{ mg/L as } CaCO_3$$

6.3.3 Acid-Base Reactions

Acid-base reactions, perhaps the most important class of chemical equilibria, are particularly important in water chemistry. Examples include the carbonate system and its relationship to pH, acidity, and alkalinity; the concentration of metal ions in water; water softening; and certain precipitation reactions and oxidation-reduction reactions.

Lowry-Bronsted definition of acid-base. There are several definitions of an acid and a base. The most common is that of Lowry-Bronsted. It states that:

> An **acid** is a substance having the tendency to lose or donate a proton (H^+), and a **base** is a substance having the tendency to add or accept a proton.

Therefore, in an acid-base reaction, we must always have, in addition to the acid (or base), some substance which will accept (or donate) the proton.

Two reactions, both involving water, further illustrate the definition. The first is

$$\underset{\text{acid}}{HCl} + \underset{\text{base}}{H_2O} \rightleftarrows \underset{\substack{\text{conjugate} \\ \text{acid}}}{H_3O^+} + \underset{\substack{\text{conjugate} \\ \text{base}}}{Cl^-} \tag{6.12}$$

HCl is an acid because it can donate a proton (H^+) to the base H_2O, which can accept the proton to become H_3O^+. Once an acid has donated a proton, it is able to accept another proton and is therefore termed a **conjugate base,** here Cl^-. Similarly, a base, having accepted a proton, is now in a position to donate a proton and is therefore called a **conjugate acid,** here H_3O^+. HCl and Cl^- differ only by a proton. They are called a **conjugate acid-base** pair, as are H_3O^+ and H_2O.

When water reacts with ammonia, NH_3, a base, the water behaves as an acid rather than as a base as in Equation 6.12. We have

$$NH_3 \; + \; H_2O \; \rightleftarrows \; NH_4^+ \; + \; OH^-$$

| base | acid | conjugate acid | conjugate base |

(6.13)

Here we are dealing with the conjugate acid-base pairs, NH_4^+ and NH_3, and H_2O and OH^-

The strength of an acid, that is, how large or small its tendency is to lose a proton, can be measured by comparing this tendency against a common base. Water is such a base, as shown in the general reaction

$$HA + H_2O \rightleftarrows H_3O^+ + A^- \qquad (6.14)$$

where HA stands for an acid. The equilibrium constant for this reaction is

$$K_A = \frac{[H_3O^+][A^-]}{[HA]}$$

where [ ] means "the concentration of," in mol/L. K_A is then also called the acid dissociation constant. When an acid is strong, its conjugate base will be weak, and vice versa. Table 6–1 lists a number of acids, their conjugate bases, and their dissociation constants at a common temperature of 25°C. The use or occurrence of these acids is also indicated. Information on other dissociation constants at different temperatures can be obtained from handbooks of chemistry.

All strong acids are completely dissociated. Examples of strong acids are provided in the following table:

	Strong acid	Weak conjugate base
Hydrochloric acid	HCl	Cl^-
Sulfuric acid	H_2SO_4	HSO_4^-
Nitric acid	HNO_3	NO_3^-
Perchloric acid	$HClO_4$	ClO_4^-

Ionization of water.   As we have learned, water can act either as an acid or a base, depending on the other reacting substance. It is itself weakly and reversibly ionized, as shown by

$$H_2O \; + \; H_2O \; \rightleftarrows \; H_3O^+ \; + \; OH^-$$

| acid | base | conjugate acid | conjugate base |

(6.15)

TABLE 6-1 ACIDS/CONJUGATE BASES AND DISSOCIATION CONSTANTS

Occurrence	Acid	Conjugate base	Reaction	[2]Dissociation constant K_a @ 25°C mole/L
• pH control, Coagulation	Hydrogen sulfate ion, HSO_4^- (relatively strong acid)	SO_4^{-2} (very weak conj. base)	$HSO_4^- + H_2O \rightleftarrows H_3O^+ + SO_4^{-2}$	1.2×10^{-2} (highest)
• Softening, Buffers	[3]Phosphoric acid, H_3PO_4 (K_1)	$H_2PO_4^-$	$H_3PO_4 + H_2O \rightleftarrows H_3O^+ + H_2PO_4^-$	7.5×10^{-3}
• Fluoridation	Hydrofluoric acid, HF	F^-	$HF + H_2O \rightleftarrows H_3O^+ + F^-$	6.8×10^{-4}
• Nitrification, Chlorine demand	Nitrous acid, HNO_2	NO_2^-	$HNO_2 + H_2O \rightleftarrows H_3O^+ + NO_2^-$	4.5×10^{-4}
• Sludge digestion	Acetic acid, CH_3COOH	CH_3COO^-	$CH_3COOH + H_2O \rightleftarrows H_3O^+ + CH_3COO^-$	1.8×10^{-5}
• Coagulation, pH control	[1]Carbonic acid, H_2CO_3 (CO_2) (K_1)	HCO_3^-	$CO_2 + 2H_2O \rightleftarrows H_3O^+ + HCO_3^-$	4.2×10^{-7}
• Odor control, Aeration	Hydrogen sulfide, H_2S	HS^-	$H_2S + H_2O \rightleftarrows H_3O^+ + HS^-$	1.1×10^{-7}
• Softening	Dihydrogen phosphate ion, $H_2PO_4^-$ (K_2)	HPO_4^{-2}	$H_2PO_4^- + H_2O \rightleftarrows H_3O^+ + HPO_4^{-2}$	6.2×10^{-8}
• Disinfection	Hypochlorous acid, HOCl	OCl^-	$HOCl + H_2O \rightleftarrows H_3O^+ + OCl^-$	2.9×10^{-8}
• Disinfection, Sludge digestion	Ammonium ion, NH_4^+	NH_3	$NH_4^+ + H_2O \rightleftarrows H_3O^+ + NH_3$	5.6×10^{-10}
• Waste toxicity	Hydrogen cyanide, HCN	CN^-	$HCN + H_2O \rightleftarrows H_3O^+ + CN^-$	4.8×10^{-10}
• Coagulation, pH control	Bicarbonate ion, HCO_3^- (K_2)	CO_3^{-2}	$HCO_3^- + H_2O \rightleftarrows H_3O^+ + CO_3^{-2}$	4.8×10^{-11}
• Softening	Hydrogen phosphate ion, HPO_4^{-2} (K_3) (very weak acid)	PO_4^{-3}	$HPO_4^{-2} + H_2O \rightleftarrows H_3O^+ + PO_4^{-3}$	4.7×10^{-13}
	Water, H_2O	OH^- (relatively strong conjugate base)	$H_2O + H_2O \rightleftarrows H_3O^+ + OH^-$	1.8×10^{-16} (lowest)

Note 1: The concentration of CO_2 dissolved in water is about 600 times the concentration of H_2CO_3, but the conversion from CO_2 to H_2CO_3 is very rapid, and therefore the assumption that all of the dissolved CO_2 exists as H_2CO_3 is all right for us in calculating pH in CO_2 water solutions.

Note 2: The dissociation constant for the conjugate base is $K_b = K_w/K_a$, where $K_w = $ ion product for water.

Note 3: Some acids, such as phosphoric acid (H_3PO_4) or carbonic acid (H_2CO_3), have more than one hydrogen ion (proton) to donate. The stepwise ionization leads to separate dissociation constants, which are indicated by a subscript—K_1, K_2, etc.

The dissociation constant K is given by

$$K = \frac{[H_3O^+] \cdot [OH^-]}{[H_2O]} = 1.8 \times 10^{-16} \text{ mol/L (@ 25°C)}$$

which is usually simplified to

$$[H_3O^+] \cdot [OH^-] = K[H_2O]$$

where

$$[H_2O] = \frac{1{,}000 \text{ g/L}}{18 \text{ g/mol}} = 55.5 \text{ mol/L}$$

so that

$$K_w = 1.8 \times 10^{-16} \times 55.5 = 1.0 \times 10^{-14}$$

K_w is called the ion product constant for water.

As in the case of the acid dissociation constants K_a, temperature affects K_w, the ion product constant for water, as shown in the following table:

T, °C	K_w
0	1.1×10^{-15}
10	2.9×10^{-15}
20	6.8×10^{-15}
25	1.0×10^{-14}
30	1.5×10^{-14}
100	7.0×10^{-13}

The dissociation at 100°C is more than two orders of magnitude greater than at 0°C.

The pH scale.   The strength of an acid or base can be indicated by its molar concentration of hydrogen ions. However, because this is awkward, the convention has been established to express the hydrogen ion concentration in terms of its negative logarithm, called the **pH** of the solution. Thus

$$pH = -\log[H_3O^+] \tag{6.16}$$

and similarly pOH is used to signify the negative logarithm of the hydroxide-ion concentration, so

$$pOH = -\log[OH^-] \tag{6.17}$$

Because $[H_3O^+][OH^-] = K_w = 10^{-14}$ at 25°C, it follows that

$$pH + pOH = 14 \text{ at } 25°C$$

An aqueous solution which is **neutral**, i.e., neither acidic nor basic, has by definition equal concentrations of H_3O^+ and OH^- ions, and at 25°C its pH = pOH = 7. Aqueous solutions with a pH less than 7 are referred to as acidic, and those with a pH greater than 7 are called basic or alkaline.

Example 6.7

(a) Find the pH of a solution with $[H_3O^+] = 3.4 \times 10^{-4}$ mol/L
(b) Find the $[H_3O^+]$ if the pH of a solution is 6.7.

Solution   From Equation 6.16,

(a)
$$pH = -\log [H_3O^+] = -\log (3.4 \times 10^{-4})$$
$$= -\log 3.4 - \log 10^{-4}$$
$$= -0.53 + 4$$
$$pH = 3.47$$

(b)
$$6.7 = -\log [H_3O^+]$$

or

$$10^{-6.7} = [H_3O^+]$$
$$10^{0.3} \times 10^{-7} = H_3O^+$$

or

$$[H_3O^+] = 2 \times 10^{-7} \text{ mol/L}$$

The carbonate system.   The most important acid-conjugate base system in air-water interactions is the carbonate system. It controls the pH of most natural waters and consists of the following species:

Carbon dioxide, CO_2, in gaseous form $CO_2(g)$ or dissolved in water $CO_2(aq)$
Carbonic acid, H_2CO_3
Bicarbonate ion, HCO_3^-
Carbonate ion, CO_3^{-2}
Carbonate-based solids, principally calcium and magnesium

Examples of the importance of the carbonate system in the environmental field include

- The production of CO_2 in biological respiration
- The consumption of CO_2 in photosynthesis
- The air-water interchange of CO_2
- The dissolution of carbonate minerals, principally $CaCO_3$ and $MgCO_3$, by groundwater
- The buffering capacity of natural waters, principally due to the carbonate system (acidity and alkalinity)
- Water softening
- Several water and wastewater treatment processes
- The interchange between solid and dissolved forms of $CaCO_3$ ($MgCO_3$) at the bottom of lakes.

The nature of the carbonate system is difficult to establish. It may be involved in homogeneous (one phase) solution equilibria as well as heterogeneous air-water and water-solid equilibria.

Snoeynik and Jenkins (1980) have identified four systems:

1. An open system with no solid present
2. An open system with a solid present
3. A closed system with no solid present
4. A closed system with a solid present

Some examples of real systems are shown and described in Figure 6–5. The calculations of carbonate species concentration in open and closed systems, and in the presence of metal ions and of carbonate-containing solids, is beyond the scope of this book. A detailed treatment can be found in Snoeynik and Jenkins (1980) and Butler (1982).

The equilibria for the carbonate system are as follows:

Equilibrium	Equation	Equilibrium constant	Equation number
Air ⇌ Water:	$CO_{2(g)}$ ⇌ $CO_{2(aq)}$	$K_H = 3.2 \times 10^{-2}$ (see Henry's Law, Section 6.4)	(6.18)
In water:	$CO_{2(aq)} + H_2O$ ⇌ $H_2CO_3^*$	$K_m = 1.6 \times 10^{-3}$ (dimensionless)	(6.19)
	$H_2CO_3 + H_2O$ ⇌ $H_3O^+ + HCO_3^-$	$K_{a1} = 4.2 \times 10^{-7}$ mol/L	(6.20)
	$HCO_3^- + H_2O$ ⇌ $H_3O^+ + CO_3^{-2}$	$K_{a2} = 4.8 \times 10^{-11}$ mol/L	(6.21)
Solid ⇌ Water:	$CaCO_{3(s)}$ ⇌ $Ca^{+2} + CO_3^{-2}$	$K_{so} = 5.0 \times 10^{-9}$ mol²/L²	(6.22)
	$CaCO_{3(s)} + H_3O^+$ ⇌ $Ca^{+2} + HCO_3^- + H_2O$	$K = K_{so}/K_{a1}$	(6.23)

* $K_m = 1.6 \times 10^{-3} = \left[\dfrac{H_2CO_3}{CO_{2(aq)}} \right]$, indicating that $[H_2CO_3] \ll [CO_{2(aq)}]$, a ratio of about 1:600. However, it is customary to let $[H_2CO_3]$ represent the sum of $[CO_{2(aq)}] + [H_2CO_3]$.

Note: Constants shown above are for T = 25°C. For values at other temperatures, refer to handbooks.

Buffering capacity of natural waters.   A knowledge of the carbonate system helps us to understand how most natural waters are able to resist changes in pH upon the addition or formation of acidic or alkaline material. In natural waters this buffering capacity is attributable mostly to the presence of species of the carbonate system. Bases such as HCO_3^-, CO_3^{-2}, and OH^- give the water the ability to resist changes in pH when a strong acid is added. Acids such as $H_2CO_3(CO_2)$, HCO_3^-, and H_3O^+ provide buffering against the addition of strong bases.

A **buffer** is a substance in solution which offers resistance to changes in pH when acidic or alkaline material is added to or formed in the solution. Within the pH range of 6 to 9 which is characteristic of most natural waters, only weak acids and bases have this capacity. Figure 6–6 shows what happens to pH when a strong base (NaOH) is added to a weak acid (H_2CO_3) and to a strong acid (H_2SO_4). Note, from the sloping part of the titration curve of carbonic acid, that buffering in the pH range of 6 to about 8.5 is provided,

Titration: Open System, No Solids

A water sample containing carbonate species is titrated with a strong acid. During titration the sample is warmed and vigorously bubbled with air to equilibrate the sample with the atmosphere. The system can be sketched as in the following figure. The system is *open* to the atmosphere and not in equilibrium with a carbonate-containing solid.

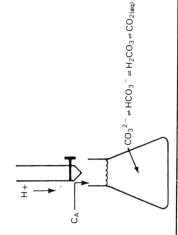

$$CO_3^{2-} \rightleftharpoons HCO_3^- \rightleftharpoons H_2CO_3 \rightleftharpoons CO_{2(aq)} \rightleftharpoons CO_{2(g)}$$

Titration: Closed System, No Solids

A water sample containing carbonate species is titrated with strong acid. The titration is conducted rapidly in a nearly full container with little shaking. The system can be sketched as in the following figure. The system is considered to be *closed* to the atmosphere (although some CO_2 may escape) and is not in equilibrium with any carbonate-containing solid.

$$CO_3^{2-} \rightleftharpoons HCO_3^- \rightleftharpoons H_2CO_3 \rightleftharpoons CO_{2(aq)}$$

Lake: Closed System with Solids

The hypolimnion of a stratified lake is in contact with sediment containing calcite, $CaCO_{3(s)}$. The system can be sketched as in the following figure. The system is *closed* to the atmosphere and tends to be in equilibrium with a carbonate-containing solid.

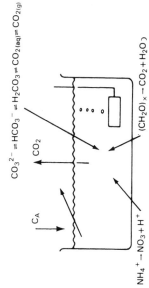

$$CO_3^{2-} \rightleftharpoons HCO_3^- \rightleftharpoons H_2CO_3 \rightleftharpoons CO_{2(aq)}$$

Water Treatment: Closed System with Solids

Chemicals (e.g., lime, soda ash, strong acid, strong base, and CO_2) are added to the water in a water treatment plant to soften the water and adjust its pH. The system can be sketched as in the following figure. Because the surface area to volume ratio of the treatment basin is small, the residence time is short, and solid $CaCO_3$ is in abundance, the system is considered to be *closed* to the atmosphere with carbonate-containing solids. (A closed system assumption is not entirely accurate because some atmospheric CO_2 exchange may occur.)

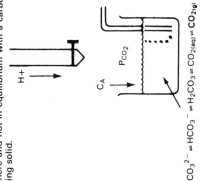

$$CO_3^{2-} \rightleftharpoons HCO_3^- \rightleftharpoons H_2CO_3 \rightleftharpoons CO_{2(aq)}$$

Wastewater Treatment: Open System without Solid

Nitrification (oxidation of ammonia to nitrate) and oxidation of organic matter to CO_2 and water take place in the vigorously aerated aeration basin of an activated sludge plant. The system can be sketched as in the following figure. The system is *open* and in equilibrium with the atmosphere with a P_{CO_2} of the aerating gas. It is not in equilibrium with a carbonate-containing solid.

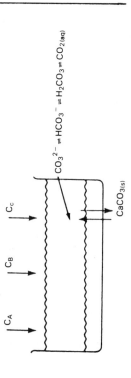

$$CO_3^{2-} \rightleftharpoons HCO_3^- \rightleftharpoons H_2CO_3 \rightleftharpoons CO_{2(aq)} \rightleftharpoons CO_{2(g)}$$

$$NH_4^+ \rightarrow NO_3 + H^+$$

$$(CH_2O)_x \rightarrow CO_2 + H_2O$$

Figure 6–5 Examples of carbonate systems. Source: Snoeynik and Jenkins, 1980. Reprinted by permission of John Wiley & Sons Inc. Copyright © 1980.

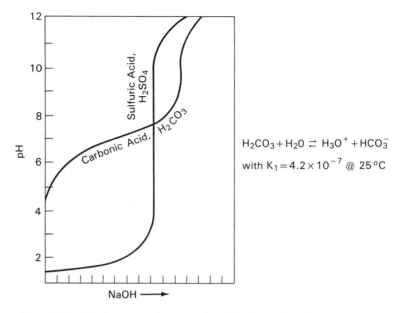

Figure 6–6   Titration curves for carbonic acid and sulfuric acid

whereas the steepness of the sulfuric acid titration curve indicates no buffering capacity in this pH range. This resistance to change in pH upon the addition of alkaline or acidic material is explained shortly.

In a natural water (pH about 7) containing free CO_2 and bicarbonate alkalinity, (HCO_3) the reactions of the CO_2 and HCO_3 (Equations 6.19 and 6.20) and the dissociation of the water itself (Equation 6.15) illustrate how buffering occurs. Let us restate these equations as

$$CO_2 + H_2O \rightleftarrows H_2CO_3 \tag{6.24}$$

$$H_2CO_3 + H_2O \rightleftarrows H_3O^+ + HCO_3^- \tag{6.25}$$

$$H_2O + H_2O \rightleftarrows H_3O^+ + OH^- \tag{6.26}$$

If we add a small amount of NaOH, a highly alkaline material (that is, we add OH^-), then because $[H_3O^+][OH^-] = K_w = 10^{-14}$, an increase in OH^- causes a decrease in $[H_3O^+]$. But a decrease in $[H_3O^+]$ will shift Equation 6.25, and therefore also Equation 6.24, to the right, producing more $[HCO_3^-]$ and $[H_3O^+]$ and reducing CO_2. The net result is a slight increase in pH. This addition of hydroxyl ions can be continued without any marked increase in pH, until all free CO_2 has been converted to HCO_3^- at a pH of about 8.3. Similarly, if we add a small amount of a strong acid, say, HCl, the H_3O^+ will increase. This will shift the equations to the left, the end result being an increase in free CO_2 and a slightly lower pH. This buffering capacity of natural waters is an extremely important property, because it prevents large shifts in the pH of the water upon the addition of acidic or alkaline contaminants. Many bacteria and other aquatic life have a relatively narrow range of pH tolerance and would be destroyed if they were not protected by the carbonate system.

Acidity and alkalinity.   The **acidity** of a water is a measure of its capacity to neutralize bases; **alkalinity** is a measure of the water's capacity to neutralize acids.

From an examination of the titration curves of a strong acid (H_2SO_4) and a relatively weak acid ($H_2CO_3(CO_2)$) in Figure 6–6, it is apparent that below pH 4.5, acidity is due to the presence of a strong mineral acid (H_2SO_4), whereas between pH 4.5 and 8.5, $H_2CO_3(CO_2)$ is the source of the acidity tending to neutralize the strong base (NaOH). In natural waters, carbon dioxide from the atmosphere and from the bacterial oxidation of organic matter and mineral acidity from industrial wastes, from mine drainage, and from acid rain are the principal sources of acidity. Acid waters are not a threat to human health, but they are of great concern because of their corrosiveness and because they upset the ecology of lakes. Acidity is determined in the laboratory by titrating a known volume of the sample with a standard solution of an alkaline reagent until the end point, at pH 4.5 or 8.5 (depending on the type of acidity present), is reached. End points can be indicated by a pH meter or by chemicals which change color at pH 4.5 or 8.5 (methyl orange or phenolphthalein).

Bicarbonates, formed by the action of CO_2 on basic materials (Equation 6.27), enter surface waters where they represent the major form of alkalinity.

$$CO_2 + H_2O + CaCO_3 \rightarrow Ca(HCO_3)_2 \tag{6.27}$$

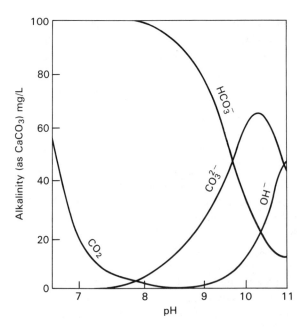

Figure 6–7 Relative amounts of CO_2, HCO_3^-, CO_3^{-2} and OH^- at various pH levels (values calculated for water with a total alkalinity of 100 mg/L at 25°C. Source: Sawyer and McCarty (1978)

At higher pH levels, natural waters may also contain considerable amounts of carbonate and hydroxide alkalinity, as shown in Figure 6–7, which indicates the relative amounts of carbonate in water. Higher pH may be caused by algae, which remove CO_2 from water through photosynthesis and thereby increase pH. The alkalinity of water, either high or low, has no ill effects on humans, but highly alkaline waters are unpalatable. Total alkalinity is measured by titration with dilute sulfuric acid to the end point at about pH 4.5. The use of $N/50$ H_2SO_4 is convenient for expressing alkalinity in terms of mg/L as $CaCO_3$, since the equivalent weight of $CaCO_3$ is 50. We have

Total alkalinity (mg/L as $CaCO_3$)

$$= \text{total mL } N/50 \text{ H}_2\text{SO}_4 \text{ to pH } 4.5 \times \frac{1,000}{\text{mL sample}} \quad (6.28)$$

Example 6.8

A waste sample of 100 mL requires 7.5 mL of N/50 H_2SO_4 to titrate to pH = 4.5. What is its total alkalinity in mg/L $CaCO_3$?
From Equation 6.28,

$$\text{Total alkalinity} = 7.5 \times \frac{1,000}{100} = 75 \text{ mg/L as CaCO}_3$$

The forms and concentrations in which alkalinity is present in a water can be found from the preceding alkalinity measurements, i.e., by titration with N/50 H_2SO_4, or, more accurately, by calculations based on the equilibrium equations and the fact that the sum of the cation concentration must equal that of the anion concentration.

For the calculational method, the applicable equations are Equation 6.15 relating H_3O and OH, namely,

$$[\text{H}_3\text{O}^+][\text{OH}^-] = K_w = 1 \times 10^{-14} @ 25°C$$

from which, by measuring pH, the hydroxide alkalinity [OH] can be calculated, and Equation 6.21 relating CO_3 and HCO_3, namely,

$$\frac{[\text{H}_3\text{O}^+][\text{CO}_3^{-2}]}{[\text{HCO}_3^-]} = K_{a2} = 4.8 \times 10^{-11} @ 25°C \quad (6.29)$$

Now, because the cations and anions must balance, and since alkalinity is equivalent to all the cations being measured except for the H_3O^+ ion, we can write

$$[\text{Alkalinity}] + [\text{H}_3\text{O}^+] = [\text{HCO}_3^-] + 2[\text{CO}_3^{-2}] + [\text{OH}^-] \quad (6.30)$$

Note that the carbonate concentration $[CO_3^{-2}]$ is multiplied by 2 since it combines with 2 hydrogen ions in forming carbonic acid, and that the ion concentrations are in mol/L. Concentrations in mg/L as $CaCO_3$ are 50,000 times these values, since the gram equivalent weight of $CaCO_3$ is 50.

The unknowns in Equations 6.29 and 6.30 are $[HCO_3^-]$ and $[CO_3^{-2}]$. These can be found by solving these equations simultaneously. The result (Sawyer and McCarty, 1978) is

Carbonate (CO_3^{-2}) alkalinity (mg/L as $CaCO_3$)

$$= \frac{\text{Total alkalinity (mg/L as } CaCO_3) - 50{,}000\,[H_3O^+] - 50{,}000\,\dfrac{K_w}{[H_3O^+]}}{1 + \dfrac{[H_3O^+]}{2K_{a2}}} \quad (6.31)$$

Bicarbonate (HCO_3^-) alkalinity (mg/L as $CaCO_3$)

$$= \frac{\text{Total alkalinity (mg/L as } CaCO_3) - 50{,}000\,[H_3O^+] - 50{,}000\,\dfrac{K_w}{[H_3O^+]}}{1 + \dfrac{2K_{a2}}{[H_3O^+]}} \quad (6.32)$$

Example 6.9

The following information is available on a wastewater sample.

$$\begin{aligned}
\text{Total alkalinity} &= 75 \text{ mg/L as } CaCO_3 \text{ (by titration)} \\
\text{Temperature} &= 25°C \\
\text{pH} &= 10.1 \text{ (by pH meter)}
\end{aligned}$$

Calculate the bicarbonate, carbonate, and hydroxide alkalinities.

Solution From Equation 6.16,

$$\begin{aligned}
10.1 &= -\log\,[H_3O^+] \\
10^{-10.1} &= [H_3O^+] \\
10^{0.9} \cdot 10^{-11} &= [H_3O^+] \\
[H_3O^+] &= 7.9 \times 10^{-11} \text{ mol/L}
\end{aligned}$$

From Equation 6.31,

$$\text{mg/L } CO_3^{-2} \text{ (as } CaCO_3) = \frac{75 + 50{,}000 \times 7.9 \times 10^{-11} - 50{,}000\left(\dfrac{10^{-14}}{7.9 \times 10^{-11}}\right)}{1 + \dfrac{7.9 \times 10^{-11}}{2 \times 4.8 \times 10^{-11}}}$$

$$= \frac{75 + 3.9 \times 10^{-6} - 6.3}{1 + 0.82} = \frac{68.7}{1.82} = 37.8$$

From Equation 6.32,

$$\text{mg/L } HCO_3 \text{ (as } CaCO_3) = \frac{75 + 3.9 \times 10^{-6} - 6.3}{1 + \dfrac{1}{0.82}} = \frac{68.7}{2.22} = 30.9$$

From Equation 6.15,

$$[OH^-] = \frac{1 \times 10^{-14}}{[H_3O^+]}$$

or

$$\text{mg/L } [OH^-] \text{ (as } CaCO_3) = 50{,}000 \times \frac{10^{-14}}{7.9 \times 10^{-11}} = 6.3$$

Check: Total alkalinity $= 37.8 + 30.9 + 6.3 = 75$ (by calculation).

6.4 GASES, GASEOUS MIXTURES, AND GAS-LIQUID TRANSFER

A knowledge of the behavior of gases and gaseous mixtures under varying environmental conditions is necessary for air pollution control as well as for the control of water and land pollution. For example, in the anaerobic digestion of wastewater, three main gases, CO_2, CH_4, and H_2S are produced, which are respectively corrosive, energy rich, and poisonous. Gas from organic decomposition in landfill sites can lead to fires and explosions, which can be dangerous to any development on or near the site. In addition, the dissolution of gases in liquids of all kinds and the removal of dissolved gases from liquids are of particular significance to the environmental engineer and scientist.

In this section we shall review ideal gases and the laws describing their behavior. In most environmental situations, gases are at low enough pressures that they behave almost like ideal gases. Through the experimental study of gases, certain "laws" or generalizations have been evolved: Boyle's law, Charles's law (also known as Gay-Lussac's law), the ideal gas law, and Dalton's law. For the reasons noted in Section 6.3, Raoult's law and Henry's law, which deal with gas-liquid systems, are presented after the gas laws.

6.4.1 Gas Laws

Boyle's law.   Boyle's law states:

The volume V of a gas varies inversely with its pressure at constant temperature.

That is to say,

$$V_{\text{at } T \text{ constant}} = \frac{1}{P} \tag{6.33}$$

or

$$P_{\text{at } T \text{ constant}} = \frac{1}{V} \tag{6.34}$$

or

$$P \cdot V_{\text{at } T \text{ constant}} = \text{Constant} = K \tag{6.35}$$

Figure 6–8(a) is a plot of pressure versus volume for a gas at three temperatures T_1, T_2, and T_3 with the experimentally measured P-V data points shown. Equations for these curves would be the same as Equation 6.35, of the form $xy = $ constant, which is the mathematical expression for a hyperbola. A plot of P vs. $1/V$ (Figure 6–8(b)) at a constant

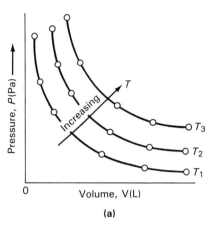
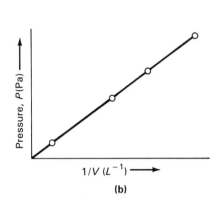

(a)   (b)

Figure 6–8   Relationship of pressure to volume

temperature should yield a straight line; therefore, a plot of the experimental values enables us to judge how closely the gas follows Boyle's law.

Boyle's law has direct application in converting measured gas volumes at various pressures (i.e., altitudes) to standard conditions.

Charles's law (or Gay-Lussac's law).   Charles's law states:

The volume of a gas at constant pressure varies in direct proportion to the absolute temperature of the gas.

Experimentally, it can be shown that for all gases maintained at a constant low pressure, the increase in volume for each degree Celsius rise in temperature is $\frac{1}{273}$ the volume of the gas at 0°C, or

AT CONSTANT P:

$$V = V_{0°C} + \frac{T_C}{273} V_{0°C} = V_{0°C} \frac{273 + T_C}{273} \tag{6.36}$$

If we define a new temperature scale T_K such that

$$T_K = T = 273 + T_C$$

where T_C = temperature in degrees Celsius; and at $T_C = 0°C$, $T_K = T_0 = 273$ (actually, 273.15); and $T_K = T$ = absolute temperature in Kelvin, then Equation 6.36 becomes

AT CONSTANT P:

$$\frac{V}{V_0} = \frac{T}{T_0}$$

or

$$V \text{ at constant } P = \frac{V_0}{T_0} T = \text{constant} \times T \tag{6.37}$$

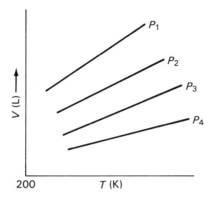

Figure 6–9 Relationship of volume to temperature

Figure 6–9 is a plot of Equation 6.37 relating the volume of any gas (since all gases have the same volume at the same temperature and pressure) to the absolute temperature at different constant pressures. Charles's law applies over a limited range of temperature, and straight lines result. Theoretically, at $T_0 = 0K$, $V = 0$. In fact, however, gases will liquefy or solidify long before the absolute zero temperature is reached, and the straight lines would be shorter than shown. Nonetheless, extrapolation of the lines for different gases to the point where they intersect the temperature axis would show that the point of intersection is the same for all gases and near $0K$ ($-273°C$).

Charles's law can be used to calculate pressure in rigid containers as the temperature varies. Determination of the required size of gas tank and the pressures to be expected over a range of temperatures would be a practical application of the combination of Boyle's and Charles's laws.

Ideal gas law.   Let us now examine the situation when the pressure P and temperature T of a gas vary between state 1 (P_1, T_1, and V_1) and state 2 (P_2, T_2, and V_2). Experimentally, we have measured P_1, T_1, V_1, P_2, and T_2, and we wish to calculate V_2. Then, from Boyle's law,

$$\text{at } T_{\text{constant}} = T_1: \frac{V_x}{V_1} = \frac{P_1}{P_2} \qquad (6.38)$$

where V_x is the volume of the gas at T_1 and P_2.
From Charles's law,

$$\text{at } P_{\text{constant}} = P_2: \frac{V_2}{V_x} = \frac{T_2}{T_1} \qquad (6.39)$$

or

$$V_2 = V_x \frac{T_2}{T_1} = V_1 \frac{P_2}{P_1} \times \frac{T_2}{T_1}$$

which, upon rearranging, becomes

$$\frac{V_1 P_1}{T_1} = \frac{V_2 P_2}{T_2} = \text{constant} = K$$

In general, then,

$$PV = KT \qquad (6.40)$$

The numerical value of K is determined by the number of moles of gas involved and the units of P, V, and T. It is independent of the type of gas. Let

$$K = nR$$

where n = number of moles of gas
$\quad\quad R$ = universal gas constant (per mole of gas)

Then

$$\boxed{PV = nRT} \qquad (6.41)$$

Equation 6.41 is called the **ideal gas law.** It is the simplest form of the general equation of state which applies to real gases. For low pressures and normal temperatures, most gases behave like ideal gases. The numerical evaluation of R can be obtained from the experimental fact that 1 gram mole of any ideal gas at standard conditions [standard temperature and pressure, (STP)] of 0°C (273.15 K) and 101,325 Pa occupies a volume of 22.414 liters. Therefore,

$$R = \frac{Pv}{nT} = \frac{101{,}325 \, \text{Nm}^{-2} \times 22.414 \times 10^{-3} \, \text{m}^3}{1 \, \text{g mole} \times 273.15 \, \text{K}}$$

$$= 8.31 \, \text{NmK}^{-1} \, \text{mol}^{-1} = 8.31 \, \text{JK}^{-1} \, \text{mol}^{-1}$$

(Exact value 8.31441)

The dimensions of R are

$$\left[\frac{ML^2}{t^2 \, T \, \text{mole}} \right]$$

If P is in atmospheres and V is in liters, then $R = 0.082056 \text{L atm K}^{-1} \, \text{mol}^{-1}$. Other values of R in different systems of units can be obtained from handbooks.

Dalton's law of partial pressures.   Dalton's law states:

In a mixture of gases, each gas exerts pressure independently of the other gases. The partial pressure of each gas is proportional to the amount, as measured by percent volume or mole number, of that gas in the mixture.
That is to say,

$$P_{\text{total}} = P_1 + P_2 + P_3 + \cdots = \Sigma P_i$$

where P_i is the partial pressure which gas i would exert if it occupied the total volume alone.

If the pressure and temperature of a gaseous mixture are not extreme, we may use the ideal gas law for the mixture, which contains n_A moles of gas A, n_B moles of gas B, and n_C moles of gas C, with a total volume V and at temperature T. In that case, the partial pressures of the three gases are given by

$$P_A = \frac{n_A RT}{V} = \frac{n_A}{n_{\text{total}}} P_{\text{total}}$$

$$P_B = \frac{n_B RT}{V} = \frac{n_B}{n_{\text{total}}} P_{\text{total}} \qquad (6.42)$$

$$P_C = \frac{n_C RT}{V} = \frac{n_C}{n_{\text{total}}} P_{\text{total}}$$

and the total pressure is

$$P_{\text{total}} = P_A + P_B + P_C = (n_A + n_B + n_C) \times \frac{RT}{V} = n_{\text{total}} \cdot \frac{RT}{V} \qquad (6.43)$$

Example 6.10

The pressure gauge on a watermain indicates 80 psig. The atmospheric pressure is 100 kPa. Calculate the absolute pressure in pascals in the watermain.

Solution

$$\text{Absolute pressure} = \text{gauge pressure in main} + \text{atmospheric pressure}$$
$$= 80 \text{ psi} \times 6{,}895 \text{ Pa/psi} + 100{,}000 \text{ Pa}$$
$$= 551{,}600 + 100{,}000$$
$$= 651{,}600 \text{ Pa}$$
$$= 651.6 \text{ kPa}$$

Note: The Pascal as a unit of pressure will take some time to get used to. For those familiar with pound per square inch (psi) or pound per square foot the conversion to Pascals seems awkward. However, one simple relationship is that, for elevations near sea level, 1 atm (14.7 psi) is approximately 100 kPa, which is called a bar. The bar is used in Germany and other countries, and the millibar (100 Pa) is a basic unit in meteorology (Chapter 7).

Example 6.11

Calculate the required volume of a gas tank which must hold at least seven days of CH_4 gas produced in a digestion process. Daily gas production is 500 kg. The temperature of the gas is 25°C, and the pressure in the tank is 200 kPa.

Solution

$$\text{Molecular mass of } CH_4 = 12 + (4 \times 1) = 16 \text{ g}$$
$$\text{Number of g moles/week} = \frac{7 \times 500 \times 1{,}000}{16} = 218{,}750$$

From Equation 6.11,

$$V = \frac{nRT}{P}$$
$$= \frac{218{,}750 \text{ (g moles)} \times 8.31 \text{ (J-K}^{-1}\text{ g mole}^{-1}\text{)} \times 298 \text{ (K)}}{200{,}000 \text{ (Pa)}} = 2708.5 \text{ J/Pa}$$

Since

$$Pa = N \cdot m^{-2} = (kg \cdot m \cdot s^{-2})m^{-2}$$
$$= kg \cdot m^{-1} s^{-2}$$

and

$$J = N \cdot m = kg \cdot m^2 \cdot s^{-2}$$

it follows that $J/Pa = m^3$ and the required tank volume $= 2708.5$ m^3.

Example 6.12

The composition of digester gas from the anaerobic digestion of wastewater sludge is 68% CH_4, 30% CO_2, and 2% H_2S. If 1,000 kg of the gas mixture is stored in a tank at a pressure of 300 kPa, calculate the partial pressure of each component present.

Solution   The molar masses of the components are

$$\text{for } CH_4 = 16 \text{g/mol}$$
$$\text{for } CO_2 = 44 \text{g/mol}$$
$$\text{for } H_2S = 34 \text{g/mol}$$

The amount of each component present is given as follows:

$$n_{CH4} = \frac{1,000 \text{ kg} \times 1,000 \text{ g/kg} \times 0.68}{16 \text{ g/mol}} = 42,500 \text{ mol}$$

$$n_{CO2} \qquad\qquad = \quad 6,820 \text{ mol}$$

$$n_{H2S} \qquad\qquad = \qquad 590 \text{ mol}$$

and therefore

$$n_{total} = 49,910 \text{ mol}$$

The partial pressures are then

$$P_{CH4} = \frac{n_{CH4}}{n_t} p_t = \frac{42,500}{49,910} \times 300 \text{ kPa} = 255 \text{ kPa}$$

$$P_{CO2} \qquad\qquad = \quad 41 \text{ kPa}$$

$$P_{H2S} \qquad\qquad = \quad 4 \text{ kPa}$$

$$P_{total} = 300 \text{ kPa}$$

6.4.2 Gas-Liquid Transfer

Raoult's law and vapor pressure.   Raoult's law deals with the vapor pressure of an ideal solution, which is defined as one whose properties are a molar average of the corresponding properties of the components of that solution.

Raoult's law can be stated as follows:

If a solution obeys Raoult's law, then the partial pressure of any component depends first on how volatile it is, and second on how much of it is present in the solution. The vapor pressure of the component measures the first property, while the mole fraction of the component measures the second.

Mathematically, this can be expressed as

$$P_a = X_A P_A \tag{6.44}$$

where P_a = partial pressure of component A in equilibrium with the solution
P_A = vapor pressure of substance A when pure at the temperature of the solution
X_A = mole fraction of component A in the solution.

Note that Raoult's law is different from Dalton's law. Raoult's law allows the calculation of partial pressures in the gas phase on the basis of the composition of the liquid. Dalton's law, on the other hand, defines the partial pressure on the basis of the composition of the gas phase. The latter may be quite different from the composition of the liquid phase.

The separation of components with different vapor pressures through repeated evaporation/distillation and condensation is achieved in a number of industrial waste and processing operations. This technique is a practical example of Raoult's law.

Gases dissolved in liquids: Henry's law.   Many situations encountered in environmental science and engineering involve the transfer of gases into and out of liquids. For example, the aeration of rivers and lakes involves the transfer of oxygen from the air to the water, thus supplying the dissolved oxygen essential for fish and many forms of aquatic life. The aeration of waters and wastewaters to remove odorous gases and the aeration of wastewater for biological oxidation are other examples.

The degree of solubility of a gas in a liquid depends on the kind of gas it is, the nature of the solvent liquid, the pressure, and the temperature. The solvent liquid in many environmental applications will be water. Slightly soluble gases include N_2, H_2, O_2, and He. NH_3, on the other hand, is a very soluble gas. The nature of the solvent is important. For example, N_2, O_2, and CO_2 are much more soluble in alcohol (ethyl alcohol) than in water, whereas H_2S and NH_3 are much more soluble in water than in alcohol.

Many of the solutions occurring in the environmental field are very dilute mixtures. Henry's law is a special case of Raoult's law applied to dilute solutions. In such solutions the partial pressure of the solute, which is present in small quantities, may be different from that predicted by Raoult's law, but it will nevertheless be proportional (i.e., linearly related) to its mole fraction.

Suppose that we are dealing with a solution of a small quantity of ideal gas B in ideal solvent A, as for example oxygen dissolved in water. Then, mathematically, Henry's law can be expressed as

$$P_B = X_B K_H(A,B) \tag{6.45}$$

where P_B = partial pressure of the solute B in the gas
X_B = mole fraction of B in the solution
$K_H(A,B) = K_H$ = Henry's constant, which depends on the properties of both the solute B and the solvent A.

Quite often, Henry's law is stated as

$$X_B = K_H^* p_B \tag{6.46}$$

which is the same as Equation 6.45, except that $K_H(A,B) = 1/K_H^*$. And sometimes it is stated as

$$C_B = K_H^{**} p_B \tag{6.47}$$

where C_B = concentration of the gas dissolved in the liquid at equilibrium, in mL/L (or mg/L)

and K_H^{**} will be numerically and dimensionally different from K_H^* and $K_H(A,B)$.

When looking up values for Henry's constant in handbooks, it is important to know to which of the three equations the values apply.

Henry's constant is also temperature dependent. Table 6–2 provides values of K_H for a number of gases of importance in the environmental field and at selected temperatures over the range normally encountered. Values for other gases can be found in engineering handbooks, such as Perry (1984). Through the combination of Dalton's, Raoult's, and Henry's laws, gas-liquid transfer problems at equilibrium can be solved.

Note carefully that Henry's law is an *equilibrium* law. For example, it allows the calculation of the dissolved oxygen equilibrium (saturation) concentration in a river at a certain temperature. If this river receives organic waste, it uses up a portion or all of the dissolved oxygen in the water, thereby creating an undersaturated dissolved oxygen concentration in the water. The magnitude of the differential between the equilibrium concentration and the actual concentration will govern the rate at which oxygen is transferred from the air to the river water. This can be expressed mathematically as

$$\frac{dC}{dt} \propto C \propto (C_{\text{equil}} - C_{\text{actual}}) \tag{6.48}$$

The kinetics of this gas-liquid transfer are not considered in this book.

Example 6.13

Calculate the amount of dissolved oxygen (abbreviated DO) in mg/L present in river water at 20°C and at 0°C under saturated conditions. Assume atmospheric pressure is 100 kPa.

Solution From Table 6–2,

$$K_{H 20°C} = 4,060 \text{ MPa}$$
$$K_{H 0°C} = 2,580 \text{ MPa}$$

From Equation 6.45,

$$P_{O_2} = X_{O_2} \times K_H(H_2O, O_2)$$

From Equation 6.42,

$$P_{O_2} = 0.209 \times 100 \text{ kPa, since air contains 20.9\% oxygen}$$
$$= 20,900 \text{ Pa}$$

Therefore,

$$X_{O_2 20°C} = \frac{20,900}{4,060 \times 10^6} = 5.15 \times 10^{-16}$$

TABLE 6–2 VALUES OF HENRY'S CONSTANT K_H FOR SELECTED GASES DISSOLVED IN WATER (use with equation 6.45)

T, °C	O₂		H₂		N₂		H₂S		CH₄		CO₂	
	MPa	10⁴ atm	MPa	10⁴ atm	MPa	10⁴ atm	MPa	10⁴ atm	MPa	10⁴ atm	MPa	10⁴ atm
0	2,580	2.55	5,870	5.79	5,360	5.29	27.2	0.0268	2,270	2.24	73.8	0.0728
10	3,310	3.27	6,450	6.36	6,770	6.68	37.2	0.0367	3,010	2.97	105	0.104
20	4,060	4.01	6,920	6.83	8,150	8.04	48.9	0.0483	3,810	3.76	144	0.142
30	4,810	4.75	7,390	7.29	9,365	9.24	61.7	0.0609	4,550	4.49	189	0.186
40	5,420	5.35	7,610	7.51	10,540	10.4	75.5	0.0745	5,270	5.20	236	0.233

Source: *National Research Council International Critical Tables*, Vol. III, New York: McGraw-Hill, 1929. Conversion to SI by authors

Note: Henry's Law: $p_B = X_B K_H(A,B)$ (in atm or Pa), where X_B = mole fraction of B in the solution and $K_H(A,B) = K_H$ = Henry's constant (in atm or Pa).

and

$$X_{O_{2^0°C}} = \frac{20,900}{2,580 \times 10^6} = 8.10 \times 10^{-16}$$

The mass of oxygen at each temperature is

$$5.15 \times 10^{-6} \text{ mol } O_2 \times 32 \text{ g/mol} = 1.65 \times 10^{-4} \text{ g @ } 20°C$$
$$8.10 \times 10^{-6} \text{ mol } O_2 \times 32 \text{ g/mol} = 2.59 \times 10^{-4} \text{ g @ } 0°C$$

The mass fraction of O_2 is

$$\frac{1.65 \times 10^{-4}}{1.65 \times 10^{-4} + 18.0} = 9.2 \times 10^{-6} \text{ @ } 20°C$$

and

$$\frac{2.59 \times 10^{-4}}{2.59 \times 10^{-4} + 18.0} = 14.4 \times 10^{-6} \text{ @ } O°C$$

Therefore, the dissolved oxygen concentration at the two temperatures is

$$DO_{20°C} = 9.2 \text{ mg/L (ppm)}$$

and

$$DO_{0°C} = 14.4 \text{ mg/L (ppm)}$$

Note that oxygen is a rather insoluble gas in water and that an increase in temperature decreases its solubility in water.

6.5 MATERIAL BALANCES

6.5.1 Concept of Material Balance

When rain falls, some of it will evaporate directly into the atmosphere, more will evaporate from the land and water surfaces back to the atomsphere, a portion will be absorbed by vegetation and transpirate back to the atmosphere, and the remainder will run off directly to rivers and lakes or infiltrate into the groundwater, to eventually rejoin the surface water. This water system is called the hydrologic cycle. Figure 6–10(a) is a simplified schematic diagram of the hydrologic cycle for a small land/lake region. A number of separate "systems" can be identified: (1) the atmospheric clouds over (a) the lake, (b) the land, and (c) the lake and land; (2) the land; (3) the lake; and (4) the entire system. An examination of Figure 6–10(b) shows that the quantities of water crossing the boundaries of each subsystem balance. (For a similar water balance on a *global* basis, see Figure 7–10.)

This example illustrates the law of conservation of matter. In chemistry, scientists found that the sum of the weights of the substances entering into a reaction always equaled the sum of the weights of the products of the reaction. The general concept of the law of conservation of matter can be illustrated by three equations, applied to an enclosed, isolated system. First, there is

$$\text{INPUT} = \text{OUTPUT} \tag{6.49}$$

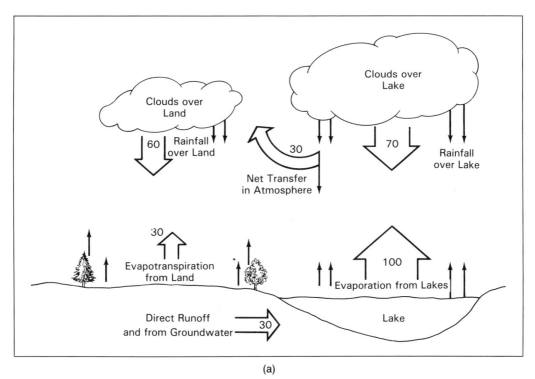

(a)

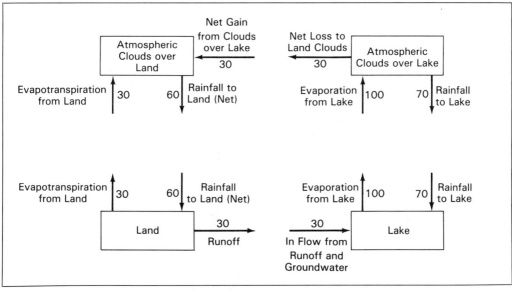

(b)

Figure 6–10 Hydrologic cycle in a small land/lake region. (a) Schematic of the Hydrologic Cycle (b) Material Balance on the Hydrologic Cycle

or, simply stated, "What comes in must go out."

If material accumulates within the system, then

$$\text{ACCUMULATION} = \text{INPUT} - \text{OUTPUT} \qquad (6.50)$$

Furthermore, if material is produced or consumed within the system, then the most general case can be described as

$$\begin{aligned}\text{(Rate of) ACCUMULATION} = \;&\text{(Rate of) INPUT} - \text{(Rate of) OUTPUT} \\ &+ \text{(Rate of) PRODUCTION} - \text{(Rate of) CONSUMPTION} \quad (6.51)\end{aligned}$$

where the parenthetical (Rate of) allows for changes with time in flowrates, production rates, and consumption rates, and therefore in accumulation rates. Material balances, also often referred to as mass balances, are a very useful tool to examine a process or parts of a process. They are used extensively in chemical engineering, and can be also very useful in the environmental field. Material balances are useful as a check on measurements of all streams which may be difficult or impossible to measure directly. They are also helpful in the design of a process and in the accounting of all materials of production and consumption (including waste products) in a process. When there is no accumulation in a system, it will be called **steady state.** For the **unsteady state,** the rate of accumulation is changing with time. The filling or emptying of a storage tank would be an example of an unsteady material balance. The following illustrative examples will demonstrate applications of single material balances.

Example 6.14: Settling of Suspended Solids From Wastewater

A settling tank is used to remove suspended solids from wastewater. The rate of flow of wastewater into the tank is 10 L/s, and the influent concentration of suspended solids (SS) is 200 mg/L. The removal efficiency of the settling tank for suspended solids is 60%. Calculate the amount of suspended solids (sludge) accumulating in the sludge zone each day.

Solution It will be helpful to draw a diagram of the process and to mark all the known data on it, as well as the unknown, identified by a question mark.

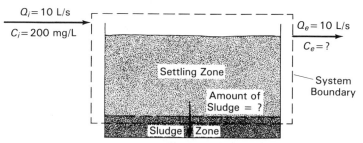

Longitudinal Section

Now, draw the appropriate system boundaries if there is more than one system. In this case the boundary will be around the settling zone. Make assumptions, if any are necessary, in order to solve the problem. In this case a reasonable assumption is that the amount of water which will be withdrawn when pumping out the sludge from the sludge zone is very small as compared to the inflow of wastewater, and can therefore be neglected. Therefore, $Q_e = Q_i$.

The concentration C_e can easily be calculated:

$$C_e = C_i \times \frac{100 - (\text{removal efficiency})}{100} = 200 \left(\frac{100 - 60}{100} \right)$$

$$= 200 \times 0.40 = 80 \text{ mg/L}$$

The material to be balanced in this case is the mass of suspended solids. There is no accumulation of suspended solids in the settling zone, the system around which we have drawn the boundary. There is also no production or consumption of suspended solids within the settling zone. Therefore, Equation 6.49 applies, and we have

Solids Balance: $\text{INPUT} = \text{OUTPUT}_{\text{in effluent}} + \text{OUTPUT}_{\text{to sludge zone}}$

or

$$\text{OUTPUT}_{\text{to sludge zone}} = \text{INPUT} - \text{OUTPUT}_{\text{in effluent}}$$

In solving problems of material balances, it is helpful to use a fixed time interval or an assumed quantity of materials as a basis for calculations. For this flow problem, one day is a convenient time period to use. We obtain

$$\text{INPUT SS} = 10 \text{ L/s} \times 60 \text{ s/min} \times 60 \text{ min/hr} \times 24 \text{ hr/d} \times 200 \text{ mg/L} \times 10^{-6} \text{ kg/mg}$$
$$= 172.8 \text{ kg/d}$$

Similarly,

$$\text{OUTPUT SS in effluent} = 69.1 \text{ kg/d}$$

Therefore,

$$\text{OUTPUT}_{\text{to sludge zone}} = 172.8 - 69.1 = 103.7 \text{ kg/d}$$

Example 6.15: Dilution

An industry discharges its liquid waste into a river which has a minimum flowrate of 10 m^3/s. The major pollutant in the waste is a nonreactive organic material called P. The waste stream has a flowrate of 0.1 m^3/s, and the concentration of P in the waste stream is 3,000 mg/L. Upstream pollution has caused a concentration of 20 mg/L P in the river upstream of the industrial discharge under the minimum flowrate conditions. The state regulatory agency has set a maximum limit of 100 mg/L P in the river. Assume that complete mixing occurs in the river. Will the industry be able to discharge the waste without treatment?

Solution   A diagram of the process for minimum flow conditions in the river is as follows:

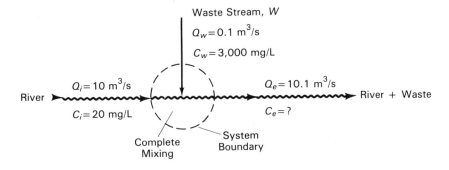

A material balance on P, for a time interval of one second, is

$$\text{INPUT} = \text{OUTPUT}$$

or

$$\text{Input}_{\text{upstream river}} + \text{Input}_{\text{waste}} = \text{Output}_{\text{downstream river}}$$

$$10 \text{ m}^3 \times 10^3 \text{ L/m}^3 \times 20 \text{ mg/L} + 0.1 \text{ m}^3 \times 10^3 \text{ L/m}^3 \times 3{,}000 \text{ mg/L}$$

$$= 10.1 \text{ m}^3 \times 10^3 \text{ L/m}^3 \times C_e \text{ mg/L}$$

So

$$200 + 300 = 10.1 C_e$$

or

$$C_e = \frac{500}{10.1} = 49.5 \text{ mg/L}$$

Therefore, no treatment is required.

6.5.2 Guidelines for Making Material Balances

A few general guidelines for solving problems can be stated as follows (adapted from Himmelblau, 1982):

1. Draw a diagram or flowchart of the process.
2. Calculate all weights, flowrates, concentrations, etc., which can be determined from the information provided without making material balances.
3. Show all known data (flowrates, concentrations, etc.) on the diagram.
4. Give appropriate symbols to any unknown quantities, and indicate each unknown by a question mark.
5. Select a convenient basis on which to carry out all calculations, for example, a suitable time interval, such as a day or a second, or a fixed quantity of material such as 100 kg or 1 lb.
6. Select the appropriate system boundaries for the material balance(s) to be made. Choose boundaries in such a way that calculations are kept as simple as possible.
7. Write the material balances. These may include a balance on the total material and a balance for each of the component materials involved in the problem. From algebra, we know that we must have as many independent equations as we have unknowns.
8. Make assumptions, if any are necessary, that may make the problem simpler. Experience will be required to do this wisely.

The following examples will provide practice in solving problems. Additional problems on material balances are given at the end of the chapter, and several occur in later chapters where such problems are appropriate.

6.5.3 Examples of Material Balances

Example 6.16

The sludge removed from the sludge zone of the settling tank in Example 6.14 has a solids concentration of 3%. In order to be able to burn the sludge in an incinerator, it must be dewatered. This is to be carried out by a gravity thickener which can achieve an underflow concentration of 8%, and then the sludge will be concentrated further in a vacuum filter which will remove 75% of the water from the feed stream. The density of wet sludge is approximately equal to that of water. Calculate (a) the flowrate of thickened sludge which must be handled by the vacuum filter, and (b) the composition of the filter cake produced by the vacuum filter.

Solution   The process can be broken into its two components, and a diagram can be drawn showing all data and system boundaries:

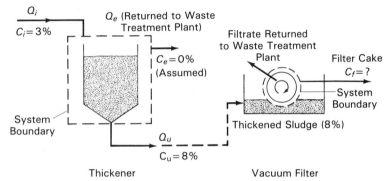

(a) *Thickener*

The rate of flow Q_i associated with the withdrawal of 103.7 kg/d of solids at a concentration of 3% is

$$Q_i = \frac{103.7 \text{ kg/d} \times 10^6 \text{ mg/kg}}{86,400 \text{ s/d} \times 30,000 \text{ mg/L}}$$

$$= 0.040 \text{ L/s}$$

Q_i is thus 0.4% of the inflow to the settling tank (10 L/s), which shows that the assumption that $Q_e = Q_i$ made in Example 6.14 was reasonable. Similarly,

$$Q_u = \frac{103.7 \text{ kg/d} \times 10^6 \text{ mg/k}}{86,400 \text{ s/d} \times 80,000 \text{ mg/L}}$$

$$= 0.015 \text{ L/s}$$

Note that we have assumed that all of the solids are settling to the bottom of the thickener, and therefore the effluent concentration $C_e \simeq 0$. This may not always be a good assumption, but certainly $C_e \ll C_u$. Therefore, this part of the question can be answered without making a material balance.

(b) *Vacuum Filter*

Choose 1 kg of thickened sludge as a basis for calculation. Assume that the amount of solids in the filtrate is negligible. The composition of the three streams is shown in the boxes in the following diagram:

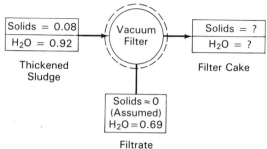

Thickened
Sludge

Filter Cake

Filtrate
(75% of Water Removed)

We have

$$H_2O \text{ removed} = 0.75 \times 0.92 = 0.69 \text{ kg per kg of thickened sludge}$$

The final amount of solids in the filter cake is 0.08 kg, and the associated water is

$$\text{IN} - \text{REMOVAL} = \text{REMAINDER}$$
$$0.92 - 0.69 \qquad = 0.23 \text{ kg } H_2O$$

The composition of the filter cake is as follows:

	kg	%
Solids	0.08	25.8
H_2O	0.23	74.2
Total	0.31	100%

Note that although the filter cake still contains about 75% water, it has the consistency of a piece of wet felt and cannot be pumped. It is therefore transported by a conveyor belt to a storage/loading area.

Example 6.17: Sludge Drying

The filter cake from Example 6.16 is fed to a rotary kiln dryer. After 500 kg of water are removed in the drying operation, the dried sludge is found to contain 30% water. What is the weight of the filter cake fed to the dryer?

Solution   A diagram of the process is as follows:

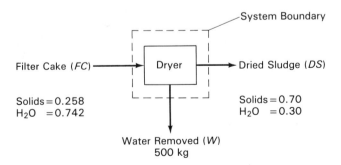

Choose 500 kg of water removed as a basis for calculation. Then the total material balance (solids and water) is given by

$$INPUT = OUTPUT$$
$$FC = DS + W = DS + 500 \text{ kg}$$

The solids balance is $0.258FC = 0.70DS$

or
$$DS = \frac{0.258}{0.70} FC = 0.369FC$$

Substituting into the total materials balance gives

$$FC = 0.369FC + 500 \text{ kg}$$
$$0.631FC = 500$$
$$FC = \frac{500}{0.631} = \underline{792 \text{ kg}}$$
$$DS = \qquad \underline{292 \text{ kg}}$$

Example 6.18: Mixing with Accumulation*

A mixing tank contains 30 ft³ of water. A waste stream containing 2 lb per ft³ of pollutant A flows into the tank at a flowrate of 3 ft³ per minute. Liquid flows from the tank at a rate of 1 ft³ per minute. Assume that the tank is completely mixed, i.e., the effluent concentration is equal to the tank concentration of pollutant A. Calculate the concentration of A in the effluent when the tank contains 50 ft³ of solution. Assume the pollutant is nonreactive.

Solution   This problem involves the accumulation of water and pollutant A in the tank. A diagram of the process is made, showing all data. It is an unsteady case.

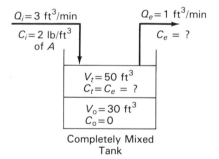

Completely Mixed
Tank

Accumulation of Water
The rate of accumulation of liquid in the tank is

$$Q_i - Q_e = 3 - 1 = 2 \text{ ft}^3/\text{min}$$

So the time to reach $V_t = 50$ ft³ is

$$t = \frac{V_1 - V_0}{Q_i - Q_e} = \frac{50 - 30}{2} = 10 \text{ min}$$

---

* (Adapted from a similar problem by Jenson and Jeffries, *Mathematical Methods in Chemical Engineering*, New York: Academic Press, 1963, pp. 11.)

Accumulation of Pollutant A

Initially there is no pollutant A in the tank, so $C_0 = 0$. After 10 minutes, the unknown concentration of A is given by $C_{10} = C_e$.

Assume that a linear variation occurs with time in the concentration of A. Then a material balance on A can be made over the time period of 10 minutes as follows:

$$\text{IN} - \text{OUT} = \text{ACCUMULATION}$$

$$(Q_i \times t)C_i - (Q_e t)\frac{C_{10} - C_0}{2} = V_{10} \times C_{10}$$

$$(3 \times 10) \times 2 - (1 \times 10)\frac{C_{10} - 0}{2} = 50 \times C_{10}$$

$$60 - 5C_{10} = 50 \times C_{10}$$

$$C_{10} = \frac{60}{55} = \underline{1.09 \text{ lb/ft}^3}$$

A more detailed treatment, not assuming a linear increase in the concentration of A, can be made as follows.

Let both the volume V (ft³) and the pollutant concentration in the effluent/tank, C_e (lb/ft³), be a function of time t (min). A systematic listing of all properties of the system at times t and $t + dt$ can be made thus:

Properties of the system		at t
Input rate of A (ft³/min)	$= Q_0$	$= 3$
Input concentration of A (lb/ft³)	$= C_0$	$= 2$
Output rate of solution (ft³/min)	$= Q_e$	$= 1$
Output concentration of A (lb/ft³)	$= C_e$	$= f(t)$
Volume of liquid in tank (ft³)	$= V$	$= f(t)$
Content of A in tank (lb)	$= VC_e$	$= f(t)$

Volume balance over time interval dt

$$\text{INPUT} - \text{OUTPUT} = \text{ACCUMULATION}$$

$$3dt - 1dt = \frac{dV}{dt}\,dt$$

or

$$\frac{dV}{dt} = 2$$

Pollutant A balance over time interval dt

$$\text{INPUT} - \text{OUTPUT} = \text{ACCUMULATION}$$

$$3 \times 2dt - 1C_e dt = \frac{d(VC_e)}{dt}\,dt$$

By simplification,

$$6 - C_e = C_e\frac{dV}{dt} + V\frac{dC_e}{dt}$$

Substituting for $dV/dt = 2$ and $V = 30 + 2t$, and rearranging gives

$$\frac{dC_e}{6 - 3C_e} = \frac{dt}{30 + 2t}$$

Integrating yields

$$-\frac{1}{3} \ln (6 - 3C_e) = \frac{1}{2} \ln (30 + 2t) + I$$

where I is the integration constant.

Now, we know that at $t = 0$, $C_e = I$; therefore,

$$\frac{1}{3} \ln 6 = \frac{1}{2} \ln 30 + C$$

whence

$$I = -\frac{1}{3} \ln 6 - \frac{1}{2} \ln 30$$

By substituting and combining, we find that

$$-\frac{2}{3} \ln (1 - 0.5C_e) = \ln (1 + 0.067t)$$
$$1 - 0.5C_e = (1 + 0.067t)^{-3/2}$$
$$C_e = 2 - 2 (1 + 0.067t)^{-3/2}$$

Note that the assumption of a linear increase in the concentration of pollutant A is not quite correct, since, for $t = 10$ min, $C_e = 2 - 2 \times 1.67^{-3/2} = 1.07$ lb/ft³. Whether or not a linear assumption is justified depends on the accuracy of the data and the use of the results. Many times, complex problems can be greatly simplified by making certain assumptions. Whether to do so and what assumptions to make depends on the experience of the engineer or the scientist.

In the material balance examples so far, there has been no production or consumption of a pollutant involved. In practice, many pollutants will undergo change in the medium, the rate of change being described by a rate equation. This topic will be introduced in Section 6.6, and a material balance problem with chemical reaction will be illustrated there.

6.6 REACTION KINETICS AND REACTORS

6.6.1 Reaction Kinetics

Not all chemical reactions reach equilibrium quickly. Reactions which are time dependent are called **kinetic reactions.** There are many cases in the environmental field in which a reaction of a pollutant or other substance in a medium is time dependent. Some examples are

- The removal of organic matter in water
- The growth of biological masses

- Radioactive decay
- Chemical disinfection
- Gas-water transfer
- Industrial waste reactions.

Reaction kinetics can be defined as the study of the effects of temperature, pressure, and concentration on the rate of a chemical reaction. The brief introduction to reaction kinetics presented here can be supplemented by reference to Levenspiel (1972) or other similar texts.

The **rate of reaction, r_i,** is a term used to describe the rate of formation or disappearance of a substance (or chemical species). Reactions such as biological oxidation and disinfection, which occur within a single phase (i.e., liquid, solid, or gaseous), are called homogeneous reactions. Those like ion exchange and adsorption, which occur at surfaces between phases (the solid-water, or air-water interface) are referred to as heterogeneous reactions. There are other classifications between homogeneous and heterogeneous systems and those where a catalyst affects the rate, but homogeneous reactions are the most common and will be emphasized here.

For homogeneous reactions,

$$r_i = \frac{\text{moles (or mass)}}{\text{unit volume} \times \text{unit time}} \tag{6.52}$$

For heterogeneous reactions,

$$r_i = \frac{\text{moles (or mass)}}{\text{unit surface} \times \text{unit time}} \tag{6.53}$$

The sign convention is positive ($+$) for the formation of a substance and negative ($-$) for the disappearance of a substance. The rate at which these reactions occur is a function (f) of temperature (T) and pressure (P) and also of the concentration of the reactant(s). The rate relation is therefore

$$r_i \propto f(T_1 P_1 [A], [B], \ldots) \tag{6.54}$$

Usually, the temperature (T) and pressure (P) effects are separated from the effects of the concentration; therefore,

$$r_i = k f_1(T,P); f_2([A], [B], \ldots) \tag{6.55}$$

where k is the rate constant, normally a function of temperature only, and f_1 and f_2 mean "function of ( )."

Assuming that the pressure and temperature are kept constant, we can examine how the concentration of one or more of the reactants affects the reaction rate. For the stoichiometric equation

$$aA + bB \rightarrow cC \tag{6.56}$$

where a, b, and c are the stoichiometric equation coefficients for the reactants A and B and the product C, the rate equation is

$$r_A = -k[A]^\alpha[B]^\beta = k[C]^\gamma \tag{6.57}$$

where $[A]$, $[B]$, and $[C]$ are the respective concentrations and α, β, and γ are empirically found exponents. The negative sign indicates that A and B are disappearing while C is increasing.

The **order of reaction** is defined as the sum of the empirically found exponents α + β, and the order with respect to reactant A is α, to B is β, and to product C is γ. The exponents are often whole numbers, i.e., 0, 1, 2, etc., but fractional exponents also occur. In many cases, reactions will be zero, first, or second order. Expressed mathematically,

$$r_A = -k \qquad \text{zero-order reaction} \tag{6.58}$$
$$r_A = -kA \qquad \text{first-order reaction} \tag{6.59}$$
$$r_A = -kA^2 \qquad \text{second-order reaction} \tag{6.60}$$
$$r_A = -kAB \qquad \text{second-order reaction} \tag{6.61}$$

A more complex example is

$$r_A = \frac{k[A]}{1 + k^1[A]} \tag{6.62}$$

- At a low concentration of A (at the end of the reaction), $k^1[A] \ll 1$; therefore, the reaction reduces to first order.
- At a high concentration of A (at the beginning of the reaction), $k^1[A] \gg 1$; therefore, the reaction reduces to zero order.

Equation 6.62 is an example of a **saturation-type** reaction, which is quite common in environmental problems. It has a maximum rate near the beginning of the reaction, where the rate is independent of the concentration of the reactant(s), and then decreases as a reactant becomes limiting.

Figure 6–11 is a graphical representation showing how the reaction rate r $(d[A]/dt)$ varies with time for different orders of reaction.

Types of reactions.   **Elementary Reactions** are defined as those reactions occurring in a single step where the stoichiometric equation represents not just a mass balance, but also what actually happens on a molecular scale. In these cases, $a = \alpha$, $b = \beta$, and $c = \gamma$, and the rate equation can be written from stoichiometry. For the elementary reaction of Equation 6.56, the rate equation becomes

$$r = -k[A]^a [B]^b = k[C]^c \tag{6.63}$$

and the overall rate of reaction (and the rates of r_A, r_B, r_C for reactants A, B and product C) are

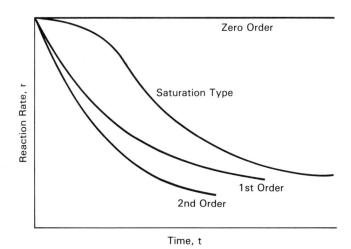

Figure 6–11   Graphical representation of rate equations

$$r = \frac{r_A}{a} = \frac{r_B}{b} = \frac{r_C}{c} \tag{6.64}$$

With nonelementary reactions, there is no direct relation between the stoichiometric equation and the reaction rate. It is assumed that a series of elementary reactions is taking place, and, consequently, rate constants must be determined experimentally.

Elementary reactions in the environmental field may be single, as in

$$A \rightarrow C \tag{6.65}$$

or multiple, as in

$$A \rightarrow B \rightarrow C \tag{6.66}$$

and either type may be reversible. For example, for the elementary, multiple, irreversible reaction

$$aA \ \xrightarrow{r_1} \ bB \ \xrightarrow{r_2} \ cC \tag{6.67}$$

the rates of reaction are

$$r_1 = \frac{r_A}{a} = \frac{r_{B1}}{b} \tag{6.68a}$$

$$r_2 = \frac{r_{B2}}{b} = \frac{r_C}{c} \tag{6.68b}$$

and

$$r_A = ar_1 \tag{6.69a}$$
$$r_B = br_1 + br_2 \tag{6.69b}$$
$$r_c = cr_2 \tag{6.69c}$$

Example 6.19

 (a) A reaction has the stoichiometric equation $A \rightarrow C + D$. What is its order of reaction?
 (b) If it is known that the reaction is elementary and irreversible, what is the order of reaction with respect to A?

Solution

 (a) The question cannot be answered, since it is not stated whether the reaction is elementary and irreversible.
 (b) $r_A = -kA$. Therefore, the reaction is first order.

Example 6.20

An elementary, irreversible reaction has the stoichiometric equation $2A + \frac{1}{2}B \rightarrow C$. Calculate the rates of formation and disappearance of the three components of the reaction and their relationship to one another.

Solution

$$r_A = -k_1[A]^2 [B]^{1/2}$$
$$r_B = -k_2[A]^2 [B]^{1/2}$$
$$r_C = +k_3[A]^2 [B]^{1/2}$$

From stoichiometry, we know that

$$\frac{r_A}{-2} = \frac{r_B}{-\frac{1}{2}} = \frac{r_C}{+1}$$

Therefore,

$$r_C = -\tfrac{1}{2}r_A = -2r_B$$

or

$$k_3 = \frac{k_1}{2} = 2k_2$$

Effect of temperature on reaction rate constant.   It has been found experimentally that most reaction rates increase with increasing temperature, as shown in Figure 6–12(a) doubling (approx.) for a 10°C increase at lower temperature. A plot of ln k against $1/T$ (Figure 6–12(b)) provides a straight line and a means to predict reaction rates at different temperatures. Therefore,

$$\frac{d(\ln k)}{d(1/T)} = \text{constant (slope of line)} = -\frac{E_a}{R} \qquad (6.70)$$

where E_a = Arrhenius activity energy
      R = universal gas constant
      T = temperature, K
      k = reaction rate constant, various units.

E_a and R have to have consistent units.

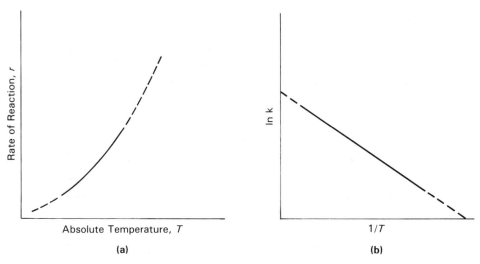

Figure 6–12   Effect of temperature on reaction rate constant. (a) Rate of reaction vs. absolute temperature. (b) Natural logarithm of reaction rate constant vs. reciprocal of absolute temperature.

Equation 6.70 can be integrated to give

$$k = Ae^{(-E_a/RT)} \tag{6.71}$$

where A is the van't Hoff-Arrhenius coefficient, in appropriate units.

Equation 6.71 is known as the Arrhenius temperature-dependence equation. It is often convenient to rearrange it as

$$\frac{k_2}{k_1} = (e^{E_a/RT_1T_2})^{T_2 - T_1} \tag{6.72}$$

to facilitate comparison of the reaction rate constants at two temperatures. In environmental engineering the range of temperatures is usually small, so the product $T_2 T_1$ is approximately constant.

Let $e^{E_a/RT_1T_2} = \theta$, where θ is the temperature coefficient. Then Equation 6.72 can be rearranged to yield

$$k_2 = k_1\theta^{(T_2 - T_1)} \tag{6.73}$$

Equation 6.73 is frequently used in biochemical reactions as well as physicochemical reactions for easy calculation of temperature effects, provided that information on θ is available.

Example 6.21

The rate of growth of a biochemical system at a temperature of 20°C has the reaction rate constant k_{20}. Calculate the relative rate at 30°C if the temperature coefficient $\theta = 1.072$.

Solution   From Equation 6.73, $k_{30} = k_{20} \cdot 1.072^{(30 - 20)} = 2k_{20}$, that is, a doubling of the rate for a 10°C temperature rise.

6.6.2 Types of Reactors

A number of physical (sedimentation, filtration, equalization, etc.), chemical (precipitation, coagulation, softening, etc.), and biochemical (activated sludge, anaerobic digestion, etc.) treatment methods are used in the environmental engineering field. They are generally carried out within tanks. When a reaction of a chemical or biochemical nature occurs within a tank, the tank is usually referred to as a reactor.

Reactors can generally be divided into two types: **batch reactors** and **flow reactors.** In a batch reactor, the materials are added to the tank, thoroughly mixed, and then left for a sufficient time for the reaction to occur. At the end of the given time, the mixture is removed from the tank. Because the material in the tank is normally well mixed, the composition within the reactor is uniform at any instant of time. However, as the reaction proceeds, the composition changes with time. A **batch reaction** is therefore referred to as an unsteady-state operation.

In a **flow reactor,** material flows into, through, and out of the reactor. Depending on the mixing conditions and flow patterns within the tank, we speak of ideal and real reactors. Figure 6–13 shows the spectrum of flow reactors, with the two ideal ones at each end. The ideal reactor of part (a) is called a **plug flow tubular reactor (PFTR),** or sometimes just a plug flow, piston flow, or tubular flow reactor. The flow pattern within the tank is characterized as uniform. That is, the fluid particles pass through the tank and are discharged in the same sequence as they entered the tank. The particles remain in the tank for a period equal to the theoretical detention time. The situation is equivalent to forcing a fluid through a long tube, as when water flows through a garden hose. There is no mixing of the fluid in a longitudinal direction, although there may or may not be some lateral mixing. The operation can be steady, if the rate of flow is constant with time, or unsteady, if it changes with time. At the other end of the spectrum of flow reactors (Fig. 6–13c) is the other ideal flow reactor, called the **completely stirred tank reactor (CSTR),** or sometimes just the stirred tank or backmix reactor. It has the characteristic that the contents of the tank are

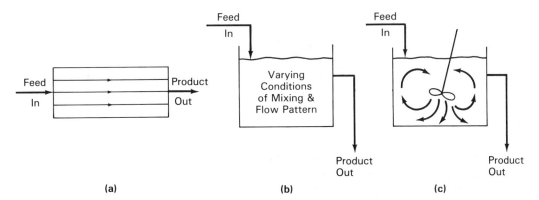

Figure 6–13 Flow reactors. (a) Plug flow tubular reactor (ideal reactor). (b) Real reactors. (c) Completely stirred tank reactor (ideal reactor).

so completely mixed that the composition is uniform throughout. Therefore, the composition of the effluent is the same as that of liquid in the tank. Real flow reactors have mixing conditions (and thereby flow patterns) which fall between the PFTR (no mixing) and CSTR (complete mixing). Some real flow reactors can be approximated by one or the other of the two ideal flow reactors. In other cases correction factors have to be applied to the solutions developed for reactions occurring in ideal reactors, which are simpler to develop than for real reactors.

A comparison of the batch, PFTR, and CSTR reactors is made in Table 6–3. In an industrial operation which produces various waste products of relatively small quantity but perhaps high strength, a batch operation for the treatment of waste may be useful. It allows intermittent operation whenever there is a sufficient volume of waste produced, and may allow easy change from one waste to another. Waste treatment in the metal plating industry, or in certain operations of the textile industry, are examples where batch treatment is used. Certainly, it is untrue that batch operations are old fashioned. On the other hand, where waste streams are large and being produced continuously, a flow reactor is much more sensible. Examples include municipal waste treatment of domestic and liquid industrial waste. Another variation called a semibatch reactor is used, when intermittent feed to a reactor with long detention time occurs. An example of this is the anaerobic digestion of sewage sludges in a municipal waste treatment plant (see Chapter 12).

Detention time. In a PFTR, by definition, each fluid particle spends exactly the same amount of time flowing through the reactor. This **flow-through time** is generally called the **detention time.** For a PFTR, the detention time for each particle can be obtained from the equation

$$t = \frac{V}{q} \tag{6.74}$$

where t = detention time, $[t]$
       V = volume of liquid in the ideal reactor $[L^3]$
       q = volumetric flowrate of feed (inflow) q_0 or product (effluent) q_f $[L^3/t]$ and
           $[L]$ indicates dimensions of length

For liquids, $q_0 = q_f$; but with gases there may be a volume change, and q_0 may not be equal to q_f. In this case, q_f should be used to calculate detention time.

For a CSTR some fluid particles may move through the reactor in a much shorter or longer time than the average detention time, but the latter can still be calculated from Equation 6.74.

Nonideal flow and tracer studies. In practice, most real tanks do not behave like the two ideal flow reactors. Deviations from the ideal flow pattern occur as a result of (1) "channeling" of parts of the liquid through the reactor because of density differences caused by temperature variations; (2) short circuiting, perhaps because of uneven weir outlet elevations; (3) the existence of stagnant regions, and (4) dispersion caused by turbulence and local mixing. Because of these deviations, the effective average detention time

TABLE 6-3 COMPARISON OF TYPES OF REACTORS

	Batch reactor	Flow reactors	
		PFTR	CSTR
Type of vessel Dimensions	Tank $\dfrac{\text{Length}}{\text{Diameter}} \cong 1$	Tube or long tank $\dfrac{\text{Length}}{\text{Diameter}} \gg 1$	Tank $\dfrac{\text{Length}}{\text{Diameter}} \cong 1$
Types of Fluids	Liquid Solid	Liquid Gas Gas/solid	Liquid Gas Fluidized bed
Mixing between reactants and products	None (separated in time)	None (separated in distance)	Complete
Conversion[1] of single irreversible reaction	Highest	Highest	Lowest
Advantages	Very flexible, can change from one process to another from day to day Capital cost is low Highest conversion, therefore smaller reactor	Operating in steady state (except for startup and shutdown), easy to check on operation through T, P, and concentration checks. Lower capital cost than CSTR Highest conversion, therefore smaller reactor Lower operating cost than batch reactor	Constant with time (no differential equations involved in solutions) Easier to control process since T, P are constant Better quality control Lower operating costs than PFTR
Disadvantages	Always operating in unsteady state Operating costs are high since more manpower is required Harder to control quality of products	Less flexible than batch reactor if one wishes to change from one process to another Need long reactor (tube) if the reaction is slow Difficult to control local temperature (hot spots)	Lowest conversion requires bigger reactor Higher capital costs since mixers are required

[1] The extent of conversion can be explained as follows. In general, the rate of a reaction (see Section 6.6.1)

$$r = \frac{dC}{dt} = k f_1(T, P) f_2(C_A, C_B)$$

where C_A and C_B are the concentrations of the reactants (or products). If, for a given reactor, the temperature, T and pressure, P are held constant, then dC/dt, the change of concentration with time, is only a function of the concentration in the tank, i.e.,

$$\frac{dC}{dt} = k_1 f(C_A, C_B)$$

In a CSTR, the concentration of materials A and B in the feed stream is immediately lowered to the concentration in the tank (and effluent); therefore, the overall rate of reaction, and consequently the conversion of reactant to product, will be less than for a PFTR, where no such lowering of concentration upon entering the reactor occurs. Put more simply, in a CSTR the reactants in the feed stream are diluted immediately with products, which lessens the rate of reaction and therefore the conversion.

is less than that calculated for the ideal reactor. Some fluid particles may flow through the tank very quickly, others may take several average detention times, and still others may reach stagnant or "dead volume" regimes, thereby reducing the useful volume of the tank. In order to obtain a complete picture of the flow of fluid in the tank, velocities would have to be measured throughout the tank. This is a very time-consuming task. A less arduous task is to find out how much of the fluid that was in the tank at time t_0 remains at a later time. A technique using tracers can provide a picture of the "residence time distribution" of the flow through the tank. A dye, salt solution, or other nonreactive material is introduced into the inlet to the tank and continuously monitored at the outlet of the tank. Two common methods of introducing the tracers are the following:

Continuous Input   A tracer is put into the inlet continuously to provide a tracer concentration of C_0 until the end of the experiment. Also called step input.

Pulse Input   All the tracer is dumped into the inlet in as short a time as possible so that the initial concentration C_0 in the reactor = Q/V, the quantity of tracer added/reactor volume. Also called slug input.

The response of the outlet in each of the two cases for a real reactor is shown in Figure 6–14. In both cases C is the concentration of tracer in the outlet stream.

Suppose that red dye is used as a tracer, and that the fluid flowing through the reactor is white. In the case of a continuous input, at time $t = 0$ a steady flow of red liquid is added at the inlet. Some red particles will flow in a very short time to the outlet stream, making it slightly pink. As time goes on, the outlet stream will become steadily pinker until a steady color of pink appears in the outlet, the degree of which is determined by the relative flow volumes of white and red liquid.

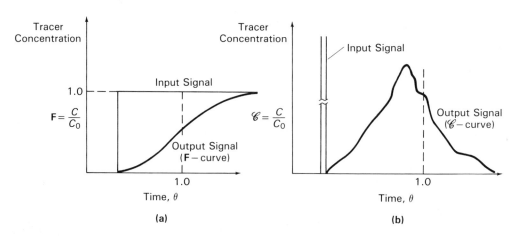

Figure 6–14 Response to tracer input at the outlet of a reactor. (a) Continuous input. (b) Pulse input where C/C_0 = proportion of tracer in the outlet stream = F for a continuous input and = $\mathscr{C}$ for a pulse input. Time θ (dimensionless) = $t/\bar{t}$, the actual detention time/theoretical detention time.

For the pulse input, the same early appearance of pink as in the continuous case would occur at the outlet, but since all of the red liquid would have been dumped in at once, with no further additions, the color at the outlet will reach a peak intensity of pink and then begin to fade until eventually the liquid will be all white again.

Example 6.22

Describe mathematically the response at the reactor outlet to a continuous nonreactive tracer input to the inlet of a CSTR. The volumetric flow rate is q.

Solution

C_0 = concentration of tracer in continuous input
C = concentration of tracer in tank (and in outlet stream of a CSTR)

By a material balance on the tracer,

$$\text{Input} = \text{output} + \text{accumulation}$$

$$qC_0 = qC + \frac{d}{dt}(VC)$$

Dividing by qC_0, we obtain

$$1 = \frac{C}{C_0} + \frac{V}{q} \times \frac{d}{dt}\frac{(C)}{C_0}$$

From before,

$$\frac{C}{C_0} = F, \qquad \frac{V}{q} = \bar{t}$$

and since $\theta = t/\bar{t}$ (see Figure 6–14)

$$\frac{1}{d\theta} = \bar{t} \cdot \frac{1}{dt}$$

(By convention, we call the response at the outlet to a continuous input an F-curve, and to a pulse input a $\mathscr{C}$-curve.) Therefore,

$$1 = F + \frac{dF}{d\theta}$$

$$\int_0^\theta d\theta = \int_0^1 \frac{dF}{1 - F}$$

$$-\theta = -\ln(1 - F)$$

$$e^{-\theta} = 1 - F$$

or

$$F = 1 - e^\theta$$

In a manner similar to that of Example 6.22, it can be shown that for a pulse input to a CSTR, the response at the outlet is

$$\mathscr{C} = e^{-\theta}$$

If the proportion F or $\mathscr{C}$ of tracer present in the effluent is known, then the proportion I of tracer still in the tank can be determined. For a PFTR, the response at the outlet will

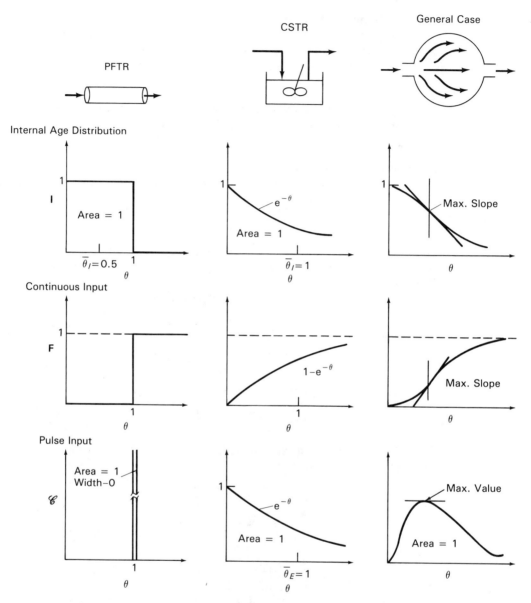

Figure 6–15 *I, F,* and $\mathscr{C}$ curves for various reactor types (adapted from Levenspiel, 1972)

be identical to the input, delayed by the detention time $\bar{t}$. For a CSTR, inlet and outlet concentrations are identical by definition.

Figure 6–15 provides information for a PFTR, CSTR, and real reactor on the *I, F,* and $\mathscr{C}$ curves for tracer inputs. Various names other than "fraction of tracer present" are sometimes given to these curves. For example *I* may be called the age distribution of molecules within the tanks, or residence time distribution, while F or $\mathscr{C}$ may be referred to as effluent age distribution, output tracer distribution, or output residence time distribution. Although such expressions for C/C_0 occur frequently in the literature, they tend to be confusing and have therefore not been used here.

For reasons stated earlier, reactors deviate from the flow regime of either a PFTR or CSTR, and various mathematical and physical models have been developed to approximate nonideal behavior. Further discussion of these models can be found in Levenspiel (1972).

Typical tracer studies.   The use of tracer studies to evaluate the performance of three primary settling tanks is illustrated in Figure 6–16 and Table 6–4. The figure shows the $\mathscr{C}$ curves for the tanks at the Windsor, Sarnia, and CCIW (Burlington), Ontario, sewage treatment plants. The table lists the hydraulic efficiency parameters obtained from the $\mathscr{C}$ curves under various flow conditions expressed as an "overflow rate." (The concept of overflow rate, that is, ratio of the inflow Q to the surface area of the tank, is considered in section 12.5.2.)

The hydraulic efficiency of a tank can be defined as the ratio of the actual mean detention time t_g to the theoretical detention time T expressed as a percent. For the ideal settling tank, this ratio will be unity (or 100 percent). For an actual settling tank, it will always be considerably less than unity because of the presence of stagnant zones within the tank. The efficiency of the tank at Sarnia (about 73 percent) is much higher than it is at Windsor (30–42 percent). Also, the time to the initial appearance of the tracer and to

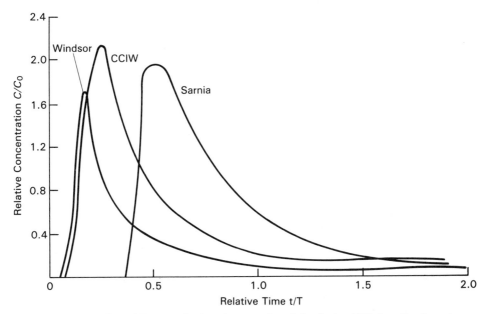

Figure 6–16 Typical $\mathscr{C}$ curves for the primary sedimentation tanks at Windsor, Sarnia, and CCIW (Burlington), Ontario, sewage treatment plants. (G. W. Heinke, A. J. Tay and M. A. Qazi, *Journal Water Pollution Control Federation* 52, (1980):2946.

TABLE 6-4 HYDRAULIC EFFICIENCY PARAMETERS FOR WINDSOR, SARNIA, AND CCIW (BURLINGTON), ONTARIO, PRIMARY SEDIMENTATION TANKS

Overflow rate, m³/m² · d		Hydraulic efficiency parameters, min.				
		t_i	t_p	t_g	T	t_g/T %
Sarnia						
	36	29	48	79	110	72
	49	24	39	59	80	74
	73	19	31	39	53	73
	98	16	25	29	40	73
Windsor						
	24	20	46	60	200	30
	49	10	25	33	100	33
	73	7	18	27	67	40
	98	5	15	21	50	42
CCIW						
	29	8	28	42	110	38
	49	5	25	36	66	55
	73	5	20	30	50	64
	98	5	16	24	33	73

where C_0 = dose of tracer per unit volume of settling tank

C = tracer concentration in the effluent at time t

t_i = time interval for initial detection of the tracer in the effluent, in minutes

t_p = time interval to reach peak concentration of tracer in effluent, in minutes

t_g = actual mean detention time, in minutes (centroid of the $\mathscr{C}$ curve)

T = theoretical detention time, in minutes.

Source: G. W. Heinke, A. J. Tay and M. A. Qazi, Journal Water Pollution Control Federation 52, (1980):2946.

the peak concentration occurs much earlier at Windsor than at Sarnia, indicating that severe short circuiting occurs at Windsor. The hydraulic parameters at the CCIW plant fall in between those of Windsor and Sarnia. It can be concluded, supported by the suspended solids removal data, that the Sarnia plant has a very efficient settling tank, whereas the Windsor plant provides poor removal of suspended solids.

6.6.3 Determination of Reaction Rates

In order to develop expressions for the rates of chemical or biochemical reactions—that is, to determine the order of a reaction and the reaction rate constants—laboratory or pilot plant experiments are conducted. The objective is to develop data on the concentration of the reactants and/or products versus time for a batch reactor or versus flowrate (which amounts to a time scale) for a continuous-flow reactor.

Either a batch or a continuous system could be used, but batch reactors are more common because they are simpler. Of the several methods available for determining rate constants, the method of integration is the most popular and is presented here. For simplicity, only irreversible reactions involving one reactant will be considered.

By integrating the rate equations (Equations 6.58 to 6.61) for zero-, first- and second-

TABLE 6–5 PLOTTING PROCEDURE TO DETERMINE ORDER OF REACTION BY METHOD
OF INTEGRATION FOR A PFTR

Order	Rate equation	Integrated equation	Linear plot	Slope	Intercept
0	$\dfrac{d[A]}{dt} = -k$	$[A] - [A_0] = -kt$	$[A]$ vs. t	$-k$	$[A_0]$
1	$\dfrac{d[A]}{dt} = -k[A]$	$\ln\dfrac{[A]}{[A_0]} = -kt$	$\ln [A]$ vs. t	$-k$	$\ln [A_0]$
2	$\dfrac{d[A]}{dt} = -k[A]^2$	$\dfrac{1}{[A]} - \dfrac{1}{[A_0]} = kt$	$\dfrac{1}{[A]}$ vs. t	k	$\dfrac{1}{[A_0]}$

order reactions from $[A] = [A_0]$ at time $t = 0$ to $[A] = [A]$ at time $t = t$, we obtain expressions which, can be plotted as a straight line if we choose suitable ordinates.

The results for an ideal plug flow reactor are shown in Table 6–5.

The data for the integration method are obtained by measuring the decreasing reactant concentration at various times and calculating the rate constants for zero-, first- and second-order reactions. (Higher order reactions are not common in environmental engineering.) We could also plot the data to see which order of reaction gives the best fit. Example 6.23 illustrates the procedure.

Example 6.23

The following data were obtained for the reaction $A \rightarrow B + C$. Determine the order of the reaction and the value of the reaction rate constant k.

t, min	0	10	20	40	60
A, mg/L	90	72	57	36	23

Solution Assume that the reaction is zero or first order. The following table gives the appropriate calculations:

t (min)	A (mg/L)	$[A]/[A_0]$	$\ln\dfrac{[A]}{[A_0]}$	zero order k_0 (mg/L · min)	first order k_1 (min^{-1})
0	90	1.00	0.0		
10	72	0.80	-0.223	$+1.80$	-0.0223
20	57	0.63	-0.457	$+1.65$	-0.0228
40	36	0.40	-0.916	$+1.35$	-0.0229
60	23	0.26	-1.347	$+1.12$	-0.0225

A sample calculation for the period from zero to 10 minutes is as follows:

$$k_0 = \frac{[A_{10}] - [A_0]}{-t} = \frac{72 - 90}{-10} = +1.80$$

$$k_1 = \frac{\ln([A_{10}]/[A_0])}{t} = \frac{-0.223}{10} = -0.0223$$

Since the k_1 rate is much more consistent than the k_0 rate calculated for the various intervals of time, the reaction is judged to be first order and the average rate constant $k_1 = -0.0226$.

We could also have found k graphically by plotting $[A]$ versus t and $\ln[A]$ versus t. It would then have been clear that the data fit a straight line better on the graph of $\ln[A]$ versus t than on the graph of $[A]$ versus t. Therefore, first order is selected, and the slope of the straight line is $-k_1$.

Note that with the method of integration only whole numbers of the exponents α, β, γ, etc., can be determined. If it is desirable or necessary to determine fractional exponents, because the data appear to justify it, then the method of differentiation (not described here) must be used. For further information, see Levenspiel (1972).

6.6.4 Principles of Reactor Design

Reaction rate equations determined from an analysis of a batch reactor can provide a basis for the design of a continuous-flow reactor. If the temperature change and pressure drop through the reactor can be neglected, a mass balance on the change in the quantity of reactants relates the residence time, degree of reactant conversion, and the reaction rate. The results will vary with the type of reactor and the order of the reaction taking place.

The general material balance for any component A in an element of volume in the reactor will be

$$\text{INPUT} = \text{OUTPUT} - \text{LOSS BY REACTION} + \text{ACCUMULATION} \quad (6.75)$$

$$
\begin{array}{llll}
\text{[rate of flow} & \text{[rate of flow} & \text{[rate of loss} & \\
 & & \text{of } A \text{ due to} & \text{[rate of accumulation} \\
\text{of } A \text{ into} & = \text{ of } A \text{ out of} & - \text{ chemical} & + \text{ of } A \text{ in the} \\
\text{the element]} & \text{the element]} & \text{reaction in} & \text{element]} \\
 & & \text{the element]} &
\end{array}
$$

The procedure for a first-order reaction in an ideal CSTR is as follows. In a CSTR at a steady state, reactant A is at a uniform concentration throughout the reactor and does not accumulate. A mass balance can therefore be made on reactant A over the whole reactor, and Equation 6.75 becomes

$$\text{INPUT} = \text{OUTPUT} - \text{LOSS BY REACTION}$$

or

$$Q[A_0] = Q[A] - r_A V \quad (6.76)$$

where Q = rate of inflow
$\quad\quad$ = rate of outflow (m³/hr)
$\quad[A_0]$ = concentration of reactant A in feed (mol/m³)
$\quad\;[A]$ = concentration of reactant A in tank and in effluent (mol/m³)
$\quad\;r_A$ = rate of reaction of reactant A at concentration $[A]$
$\quad\quad$ = $-k[A]$ for a first-order reaction (mol/m³ · hr)
$\quad\;\;V$ = liquid volume in tank = $Q \cdot t$ (m³)
$\quad\;\;t$ = mean hydraulic detention time (hr), or residence time $\bar{t}$

Dividing Equation 6.76 by $Q[A_0]$ gives

$$1 = \frac{[A]}{[A_0]} - \frac{r_A V}{[A_0]Q}$$

and since $V/Q = t$ and $r_A = -k[A]$, it follows that

$$1 = \frac{[A]}{[A_0]} + \frac{k[A]t}{[A_0]}$$

$$= \frac{[A]}{[A_0]} (1 + kt)$$

so that

$$kt = \frac{[A_0]}{[A]} - 1$$

A similar procedure for other reactors and/or orders of reaction results in the kinetic equations summarized in Table 6–6.

TABLE 6–6 KINETIC EQUATIONS RELATING MEAN RESIDENCE TIME t AND REACTANT CONCENTRATION [A] IN PFT AND CST REACTORS

Type of reaction			Type of reactor	
Equation	Order	Rate	Plug flow	Mixed flow
$A \rightarrow C$	0	$-k$	$kt = [A_0] - [A]$	$kt = [A_0] - [A]$
$A \rightarrow C$	1	$-k[A]$	$kt = \ln \dfrac{[A_0]}{[A]}$	$kt = \dfrac{[A_0]}{[A]} - 1$
$2A \rightarrow C$	2	$-k[A]^2$	$kt = \dfrac{1}{[A_0]}\left(\dfrac{[A_0]}{[A]} - 1\right)$	$kt = \dfrac{1}{[A]}\left(\dfrac{[A_0]}{[A]} - 1\right)$

Example 6.24 shows the application of these kinetic equations.

Example 6.24

A reactor is to be used to carry out conversion of component A to product C. Specifications call for a 99% conversion of component A. The first-order rate constant $k = 1.0\ \mathrm{hr}^{-1}$. Because the reactor is relatively long and narrow, the engineer assumes plug flow conditions. However, because there are powerful mixers situated at intervals along the reactor, the actual mixing conditions are those of a completely stirred tank reactor. By assuming plug flow conditions, what error in required volume of reactor will the engineer make if the feed rate is 1,000 ft³/ hr? What will the actual conversion be? Assume (1) a constant-volume reaction for the fluid of density $\rho = 1.00$, (2) that steady state conditions apply, and (3) that all conversion takes place in the reactor.

Solution

PFTR: For a first-order reaction in a PFTR, the applicable kinetic equation from Table 6–6 is

$$kt = \ln \frac{[A_o]}{[A]}$$

For 99% conversion,

$$\frac{[A_o]}{[A]} = \frac{100}{1}$$

and for $k = 1.0$ hr^{-1},

$$1 \times t = \ln \frac{(100)}{1} = 4.6 \text{ hr}$$

Since $Q = 1,000$ ft^3/hr, the required $V = Q \times t = 1,000 \times 4.6 = 4,600$ ft^3.

CSTR: For a first-order reaction in a CSTR, the applicable equation from Table 6–6 is

$$kt = \frac{[A_o]}{[A]} - 1$$

For 99% conversion,

$$\frac{[A_o]}{[A]} = \frac{100}{1}$$

and for $k = 1.0$ hour^{-1},

$$t = \frac{100}{1} - 1 = 99 \text{ hr}$$

The required volume $V = Qt = 1,000 \times 99 = 99,000$ ft^3. Therefore, the error in the required volume is

$$99,000 - 4,600 = 94,600 \text{ ft}^3$$

Actual Conversion

If the actual volume of the reactor is 4,600 ft^3, but the working conditions are those of a CSTR, that is,

$$kt = \frac{[A_o]}{[A]} - 1$$

and

$$t = \frac{V}{Q} = \frac{4,600 \text{ ft}^3}{1,000 \text{ ft}^3 \text{ hr}^{-1}} = 4.6 \text{ hr}$$

then

$$1 \times 4.6 = \frac{100}{[A]} - 1$$

so that

$$[A] = \frac{100}{5.6} = 17.9\% \text{ remaining}$$

and the actual conversion is

$$\frac{100 - 17.9}{100} = 82.1\%$$

Example 6.24 shows that the removal efficiency of a CSTR for a first-order reaction is about $\frac{1}{6}$ less than for a PFTR. We also saw that for the same 99 percent conversion, the CSTR had to be $99{,}000/4{,}600 = 21.5$ times larger than the PFTR. From this evidence, we might conclude that because they are so much more efficient, we should design all reactors as PFTRs. However, the differences between the two types of reactors decrease with lower orders of reaction and lower removal requirements. Also, in practice, full-scale PFTR units do not perform close to ideal conditions, but rather perform somewhere between a PFTR and a CSTR. Moreover, reactions may not be exactly first order, but in between zero and first order. In any case, the volumes provided are normally well above the theoretical requirements for conversion and the slightly lower efficiency of the CSTR may be offset by its greater stability and more uniform effluent characteristics under varying loading conditions.

Information on more complex systems with reactors in series or in parallel, variations in inflow, recirculation, and heterogeneous reactions is available elsewhere (Levenspiel, 1972).

PROBLEMS

6.1. Balance the following equations:
    (a) $FeS + HCl \rightarrow FeCl_2 + H_2S$
    (b) $Cl_2 + KOH \rightarrow KCl + KClO_3 + H_2O$
    (c) $MnO_2 + NaCl + H_2SO_4 \rightarrow MnSO_4 + H_2O + Cl_2 + Na_2SO_4$
    (d) $H_2C_2O_4 + KMnO_4 + H_2SO_4 \rightarrow CO_2 + MnSO_4 + K_2SO_4 + H_2O$
    (e) $Fe(OH)_2 + H_2O + O_2 \rightarrow Fe(OH)_3$

6.2. Calculate the rise velocity of an air bubble 100 microns in diameter in a tank of water at 20°C.

6.3. Calculate the length of time it will take for a jagged flint sand particle with an equivalent diameter of 0.8 mm to settle to the bottom of a 4-m-deep tank.

6.4. Find the terminal settling velocity, in 20°C water, of a spherical sand particle of diameter 0.07 mm.

6.5. The following information is from a laboratory test to determine the suspended solids concentration of a sample of untreated wastewater. A 100-mL sample is filtered through a filter pad. The weight of the clean filter pad and crucible, both of which were dried, cooled, and then weighed, is 48.610 g. After filtration, drying at 104°C, and cooling, the weight of the crucible, filter pad, and dried solids is 48.903 g. What is the concentration of suspended solids in the wastewater sample, in mg/L?

6.6. How many moles of H_2SO_4 are required to form 65.0 g of $CaSO_4$ from $CaCO_3$?

6.7. A sample of 25.26 g of hydrated magnesium sulfate ($MgSO_4$ X H_2O) is heated to 400°C to remove the water of crystallization. It is found that 12.34 g of anhydrous magnesium sulfate is left. What is the value of X?

6.8. Ethanol is accidentally spilled into a river, where it is degraded by microbial action according to the reaction equation

$$C_2H_5OH + 3O_2 \rightarrow 2CO_2 + 3H_2O$$

(a) How many kilograms of oxygen are consumed in the process if 500 lb of ethanol were spilled?

(b) How many kilograms of CO_2 are produced?

6.9. How many grams of magnesium will be necessary to form 1 kg of magnesium carbonate?

6.10. Air is a solution whose major components are nitrogen, oxygen, and argon, with mole fractions of 0.781, 0.210, and 0.009, respectively. Calculate the mass fractions of each.

6.11. A sample of 7.14 g of potassium iodide is dissolved in 145 g of water. What are (a) the molality, and (b) the mole fraction, of KI in the solution?

6.12. Calculate the pH of a solution containing

(a) 25 mg/L hydrochloric acid

(b) 25 mg/L sodium hydroxide

(c) 100 mg/L acetic acid

(d) 100 mg/L hypochlorous acid

6.13. A 1-liter aqueous solution contains 100 mg/L of HCl. Its pH is to be altered by the addition of NaOH at a concentration of 1 mol/L, also in aqueous solution. Calculate

(a) The initial pH of the solution (HCl only)

(b) the pH after the addition of 1 mL of NaOH

(c) the pH after the addition of 2 mL of NaOH

(d) the pH after the addition of 3 mL of NaOH

6.14. A water sample has been analyzed with the following results:

Cation	Amount
Ca^{+2}	104 mg/L
Mg^{+2}	38 mg/L
Na^{+}	19 mg/L

Calculate the number of milliequivalents per liter of each cation, the total hardness, and the alkalinity as mg/L $CaCO_3$.

6.15. A useful test of the accuracy of a water analysis is to compare the reported cation and anion concentrations in meq/L. In a perfect analysis, all cations and anions contained in the water would balance. The reported results of a water sample are as follows:

Cations		Anions	
Na^{+}	90	Cl^{-}	102
Ca^{+2}	60	HCO_3^{-}	220
Mg^{+2}	20	SO_4^{-}	64
Fe^{+2}	2	NO_3^{-}	1

Does this analysis fall within a maximum acceptable error of 10%?

6.16. An analysis of a wastewater sample yields the following results:

Total alkalinity = 72 mg/L as $CaCO_3$
Temperature   = 25°C
pH   = 9.8

Calculate the carbonate, bicarbonate, and hydroxide alkalinities.

6.17. If Lake Ontario water has concentrations of Ca^{+2} and Mg^{+2} of 0.00096 and 0.00022 mole/L, respectively, what is the hardness of the water, expressed in mg/L of $CaCO_3$?

6.18. The following concentrations of cations were obtained from a water analysis:

Cations	mg/L
Ca^{+2}	60
Mg^{+2}	20
Na^+	15.5
K^+	8

Calculate the hardness and alkalinity as mg/L $CaCO_3$.

6.19. Calculate the alkalinity of water containing 20 mg/L of Ca^{+2} and 15 mg/L of Mg^{+2}, expressed as $CaCO_3$. Use (a) the formulae provided in Section 6.3.3, and (b) the method of equivalent weights.

6.20. Upon analysis, a sample of water is found to contain the following elements at the concentrations indicated:

Carbon dioxide (CO_2)	8.8 mg/L
Calcium bicarbonate ($Ca(HCO_3)_2$)	186.3 mg/L
Calcium sulfate ($CaSO_4$)	81.6 mg/L

Lime ($Ca(OH)_2$) is used to precipitate the CO_2 and $Ca(HCO_3)_2$, and soda ash (Na_2CO_3) is able to precipitate the calcium sulfate according to the following equations:

$$CO_2 + Ca(OH)_2 \rightarrow CaCO_3 \text{ (ppt)} + H_2O$$

$$Ca(HCO_3)_2 + Ca(OH)_2 \rightarrow 2CaCO_3 \text{ (ppt)} + H_2O$$

$$CaSO_4 + Na_2CO_3 \rightarrow CaCO_3 \text{ (ppt)} + Na_2SO_4$$

Calculate the mass of lime and soda ash required to theoretically soften one liter of the water completely.

6.21. A cylinder storing oxygen at 20 MPa at 0°C might explode if the pressure exceeds 50 MPa. At what maximum temperature (°C) may this cylinder be safely stored, allowing for a safety factor of 2.0?

6.22. An engineer wishes to store methane gas (CH_4) produced in an anaerobic sludge digester at a sewage treatment plant. If the gas, produced at a rate of 200 kg/day, is to be stored at 20°C and 4 MPa, what volume of tank is required for a 10-day storage period.

6.23. A 50-mL sample of oxygen at a pressure of 0.1 MPa is mixed with a 250-mL sample of nitrogen at the same temperature and at a pressure of 0.0667 MPa. The mixture is placed in a 150-mL vessel, with no change of temperature. Calculate the partial pressure of each gas and the total pressure in the vessel.

6.24. A sample of 1.002 g of graphite (C) is completely burned in a steel vessel containing 250 mL of oxygen at a pressure of 1.00 MPa at 27°C. Calculate the mole fraction of each gas and the total pressure after combustion, assuming that all gases are ideal and that the temperature increases 2.5°C.

6.25. What volume of oxygen at 27°C and 0.21 atm is required for the combustion of 25 g of methane gas?

6.26. Anaerobic digestion of an industrial waste, largely acetic acid, produces carbon dioxide and methane gas. Calculate the volume of CO_2 and CH_4 gas produced daily at 20°C for an average daily waste production of 500 kg of CH_3COOH.

6.27. A new power plant without SO_2 removal facilities is going to be built on the outskirts of a city. Coal with a 1.8% by weight sulfur content will be used. It is estimated that an area of about 5 km in each direction from the plant will be affected by the smokestacks and will, under the worst circumstances, contain about one day's production of SO_2 from the plant. The precipitation record indicates a typical rainfall of 5 cm in 24 hours with an average pH of 6.2. Calculate the maximum permissible daily use of coal without lowering the pH in the rainfall to less than 5.0. The following equations apply (assume that the rainfall in the affected area dissolves all the SO_2 present there):

$$S + O_2 \rightarrow SO_2$$

$$2SO_2 + O_2 + 2H_2O \rightarrow 2H_2SO_4$$

6.28. The acid rain problem was discussed in Chapter 5. From this general information and the knowledge gained in Chapter 6, consider the following situation:

An industrial plant emits SO_2 into the atmosphere on a steady basis. Rainfall records indicate that the annual precipitation is 80 cm/yr, at a pH of 4.5. A nearby lake has the following characteristics: Surface area = 8 km², average depth = 10 m, pH = 5.5, alkalinity = 25 mg/L as $CaCO_3$. The area from which runoff drains directly to the lake is 25 km². Assume that only 20% of the rain falling on the land reaches the lake. Assume also that on a yearly basis the lake is completely mixed, and that the small river flowing into and out of the lake can be neglected as far as acidification is concerned. How many years will it be before the lake will reach a pH of 5.0?

6.29. A city situated on a large river disposes of its treated wastes to the river on a continuous basis. The minimum flow in the river is 210 m³/s, and the discharge rate from the treatment plant is 12.5 m³/s. If the maximum acceptable limit for a certain pollutant is 1.0 ppm (1 mg/L) in the downstream river, and the "background" concentration of this pollutant upstream is 0.4 ppm, what is the maximum concentration of the pollutant, in mg/L, that can be safely released from the water pollution control plant?

6.30. In the situation described in Problem 6.29, assume that one factory is responsible for the release of the pollutant, and that overall treatment plant efficiency for removal of this material is 60%. If the average waste flow out of the factory is 0.05 m³/s, then, neglecting volumes removed from flows (that is, in sludges, etc.), what is the maximum concentration of the pollutant, in mg/L, that may be released to the sewer system?

6.31. A domestic wastewater contains 350 mg/L of suspended solids. Primary sedimentation facilities remove 65% of these solids. Approximately how many gallons of sludge containing 5.0% solids will be produced per million gallons of wastewater handled?

6.32. A gravity thickener receives 33,000 gpd of wastewater sludge and increases the solids content from 3.0% to 7.0% with 90% solids recovery. Calculate the volume of thickened sludge.

6.33. Dust removal from an airstream of a municipal incinerator is accomplished by four dust collectors operating in parallel, each handling $\frac{1}{4}$ of the total airflow of 200 m³/min. The airstream contains 10 g/m³ of suspended solids. The efficiency of the dust collectors is 98%. One of the dust collectors has to be taken out of service for one month. The maximum permissible solids concentration from the combined stack discharge is 1.0 g/m³. Will the plant be able to meet this standard during the shutdown period?

6.34. The rates of enzyme-catalyzed reactions sometimes follow a rate equation such as

$$r_A = \frac{-k[A]}{1 + k_1\,[A]}$$

(a) What is the order of this reaction?

(b) Indicate an approximate method of plotting the experimentally obtained data of $[A]$ and t by the method of integration, so that two straight lines are obtained if the rate equation is followed.

(c) Comment on the defects of fitting a straight line to the two plots. (Courtesy of C. Crowe, McMaster University.)

6.35. Benzene diazonium chloride decomposes according to the equation

$$C_6H_5N_2Cl \rightarrow C_6H_5Cl + N_2$$

At 50°C, with an initial concentration of 10 g/L of $C_6H_5N_2Cl$, the following results were obtained:

Time (min)	6	9	12	14	18	22	24	26	30	∞
N₂ evolved (cm³)	19.3	26.0	32.6	36.0	41.3	45.0	46.5	47.4	50.4	58.3

(a) By inspection, suggest the most likely order of reaction, with reasons for your choice.

(b) Use the method of integration to determine the order of the reaction and the rate constant. (Courtesy of C. Crowe, McMaster University.)

6.36. A chemical reaction is carried out in a CSTR. Component A is converted to product C, the rate equation being reported as

$$r_A = -0.15(s^{-1})\,[A]$$

(a) Calculate the volume required for a 90% conversion of A for a volumetric flow rate of 100 L/s, assuming that $[A_0] = 0.10$ mol/L.

(b) After the design is completed, the engineer finds out that an error has been made in the order of reaction. It turns out not to be first order, but zero order, the correct equation being

$$r_A = -0.15\,(s^{-1})$$

What effect will this have on the design?

6.37. A wastewater is to be treated in a CSTR. Assume an irreversible, first-order reaction $r_A = -k[A,]$ where $k = 0.15d^{-1}$. Determine the flowrate which can be handled if the reactor volume is 20 m^3 and 98% treatment efficiency is required. What volume would be required for the same flowrate if the treatment efficiency need only be 92%?

6.38. A liquid-phase reaction takes place in two CSTR reactors operating at steady state in parallel at the same temperature. One reactor is twice the size of the other. The total feed stream is split appropriately between the two reactors to achieve the highest fractional conversion of reactant, which is 0.70. The smaller reactor needs to be taken out of service for repair. If the total feed rate stays the same, what is the resulting fractional conversion in the larger reactor? Assume the reaction is first order. (Courtesy of R. Missen, University of Toronto.)

6.39. A second-order, liquid-phase reaction ($A \rightarrow$ products) is to take place in a batch reactor at constant temperature. The rate constant is 0.05 L mol^{-1} min^{-1}. The initial concentration $[A_0]$ is 2 mol L^{-1}. If the downtime t_D between batches is 20 minutes, what should be the reaction time t_R for each batch so that the rate of production is maximized on a continuing basis? Note that the total batch time is $t_R + t_D$. (Courtesy of R. Missen, University of Toronto.)

6.40. A liquid reaction is carried out in a batch reactor at constant temperature. A 50% conversion is achieved in 20 min. How long will it take to achieve the same conversion (a) in a PFTR and (b) in a CSTR?

6.41. The required detention time of a plug flow tubular reactor, used for wastewater treatment, is a minimum of 3 hr. The dimensions of the tank are: Length = 100 m, width = 10 m, depth = 4 m. Calculate the maximum flowrate, in m^3/s, which can be accommodated, and the velocity of flow, in m/s.

REFERENCES

APHA, AWWA, and WPCF. *Standard Methods for the Examination of Water and Wastewater.* 16th ed. Washington D.C.: American Public Health Association, American Water Works Association, Water Pollution Control Federation, 1985.

BIRD, R. B., STEWART, W. E., and LIGHTFOOT, E. N. *Transport Phenomena.*, New York: Wiley, 1960.

BRECK, W. G., BROWN, R. J. C., and McCOWAN, J. D. *Chemistry for Science and Engineering.* Toronto: McGraw-Hill Ryerson, 1981.

BUTLER, J. N. *Carbon Dioxide Equilibria and Their Applications.* Reading, Mass.: Addison-Wesley, 1982.

FAIR, G. M., GEYER, J. C., and OKUN, D. A. *Water and Wastewater Engineering.* Vol 2 New York: Wiley, 1982.

HIDY, G. M., and BROCK, J. R. *The Dynamics of Aero-colloidal Systems.* Oxford: Pergamon Press, 1970.

HIMMELBLAU, D. M. *Basic Principles and Calculations in Chemical Engineering.* Englewood Cliffs, N.J.: Prentice Hall, 1982.

LEVENSPIEL, O. *Chemical Reaction Engineering.* New York: Wiley, 1972.

MAHAN, B. H. *University Chemistry.* Reading, Mass.: Addison-Wesley, 1975.

PERKINS, H. C. *Air Pollution*. New York: McGraw-Hill, 1974 (originally from C. E. LAPPLE, Stanford Research Institute Journal 5(94), 1961.

PERRY, R. H. *Chemical Engineer's Handbook*. New York: McGraw-Hill, 1984.

RICH, L. G. *Unit Operations of Sanitary Engineering*. New York: Wiley, 1980.

SAWYER, C. N., and MCCARTY, P. L. *Chemistry for Environmental Engineering*. New York: McGraw-Hill, 1978.

SNOEYNIK, V. L., and JENKINS, D. *Water Chemistry*. New York: Wiley, 1980.

WILLIAMSON, J. *Fundamentals of Air Pollution*, Reading, Massachusetts: Addison-Wesley, 1973.

7

Climatology
and Meteorology

F. Kenneth Hare

7.1 INTRODUCTION

The atmosphere is a vital component of the human environment. It transmits and alters the solar energy that controls our climate. It acts as a shield, protecting us from damaging meteoritic impacts and from penetrating radiation, such as ultraviolet rays from the sun. It supports the flight of birds and insects and transports seeds and spores. In fact, its gases provide the raw materials for life itself: without them, we could not exist. Therefore, its chemistry is very important to us.

Weather and climate are the two aspects of the atmosphere of which we are most aware. **Weather** is the name we give to the states of the sky, air, wind, and water. Weather elements are rain, snow, heat, wind, thunder, and fog. These elements are often present, but usually pass away in a few hours or days. Our integrated experience of weather is the **climate,** the characteristic annual cycle of weather. Depending on our expectation of the weather, we make decisions from a choice of clothing to the date of the harvest.

Despite its importance, most of us are aware of the atmosphere only when cloud cover, fog, or haze is present, or when we look at the blue of the daytime sky, which is shortwave sunlight scattered by atmospheric gases. However, although it is not usually visible to the naked eye, no other part of the environment is so thoroughly monitored. Weather satellites look downwards at the atmosphere in several different wavelength bands. Radiosonde balloons, measuring temperature, pressure, and humidity, are sent up to about

30 km once or twice daily from over 1,500 stations around the globe. A close network of ground observing stations also measures atmospheric properties, in some cases hourly and in some cases continuously, by recording instruments. All this effort, the product of 150 years of evolution, is coordinated internationally by the World Meteorological Organization, with headquarters in Geneva.

The scientific study of weather gave birth to the science of **meteorology,** which is an investigation of the physics and chemistry of the lower atmosphere. The upper atmosphere—which begins 100 km above sea level—behaves differently. The scientific study of it is sometimes called **aeronomy.** Electromagnetic forces and chemical activity are more important in this field than in meteorology. **Climatology,** the study of climate, is concerned with how the earth's atmosphere behaves over long periods of time. All these areas of study, together with atmospheric chemistry, form the **atmospheric sciences.**

Our activities have already affected the behavior of the atmosphere. Pollution has changed its composition. The cutting of forests has reduced the surface roughness of the earth and altered the turbulent flow of the winds. The energy balance of the earth's surface and lower atmosphere has been significantly changed, especially in cities. So far the human impact has been small, but it is growing rapidly. Technology can help to reduce that impact.

The engineer and scientist, therefore, need to know the basic facts about weather and climate for several reasons. For example, not only does the control of air pollution require knowledge of how the lowest layers of the atmosphere behave, but recently we have discovered that some pollutants spread through the entire atmosphere, so that the higher layers also need to be studied. An understanding of world climate is useful as well to those involved in projects far from their home base. And in the poorer countries, the vagaries of climate can make supplies of food and water unreliable and thus contribute to economic difficulties. Similarly, a knowledge of rainfall occurrence and intensity is essential for the design of public works such as drainage improvements, reservoirs, dams, and water supply facilities. Also, snow and wind conditions must be considered in the design of structures; and the range in temperature conditions relates directly to the design of heating and air conditioning systems and insulation requirements.

There are many aspects of the atmospheric sciences which will not be dealt with in this chapter. An example of an omitted field is bioclimate, the relation of climate to living organisms (including humans). Also, very little attention has been given to the weather map and to weather forecasting. Synoptic meteorology, as the study of these matters is called, is very important to aerial navigators and aerospace engineers.

7.2 BASIC ATMOSPHERIC PROPERTIES

7.2.1 Composition and Physical State

The atmosphere is made up of a mixture of gases, with numerous suspended particles, some solid and some liquid. The lower atmosphere is electrically neutral, containing few free ions; for the most part, it is composed of molecules. The upper atmosphere, by contrast, is extensively ionized: many gases of the lower atmosphere are broken up at the high levels

TABLE 7–1 COMPOSITION OF PURE, DRY AIR (i.e., WITHOUT WATER VAPOR) IN LOWER ATMOSPHERE, WITH MOLECULAR WEIGHTS AND ENVIRONMENTAL ROLES

Gases	Formulae	Concentration (% by volume)	Molecular weights (kg mol^{-1} × 10^3)	Environmental roles
Active gases				
Nitrogen	N_2	78.09	28.0	Inert as N_2; essential to life as N
Oxygen	O_2	20.95	32.0	Essential to life; chemically active
Hydrogen	H_2	5.0×10^{-5}	2.0	Important in atmospheric chemistry
Inert gases				
Argon	Ar	0.93	39.9	Inert
Neon	Ne	1.8×10^{-3}	20.2	Inert
Helium	He	5.2×10^{-4}	4.0	Inert; escapes from earth's crust
Krypton	Kr	1.0×10^{-4}	83.7	Inert
Xenon	Xe	8.0×10^{-6}	131.3	Inert
Radon	Rn	6.0×10^{-18}	222.0	Radioactive; variable in height and time, because of decay
Variable gases				
Carbon dioxide	CO_2	3.4×10^{-2}	44.0	Essential to life; optically active
Ozone	O_3	1.0×10^{-6}	48.0	Toxic, radiatively and chemically active

Other trace constituents include sulfur dioxide (SO_2), carbon monoxide (CO), oxides of single nitrogen (NO_x), and various pollutants

Source: R. J. List; Smithsonian Meteorological Tables, Washington, D.C.: Smithsonian Institute, Table 110, p. 389, 1951.

of the upper atmosphere into single atoms, or into free radicals such as hydroxyl (OH). Because of its special role, water vapor (H_2O) is often dealt with separately. The atmosphere is then said to be made up of **dry air** and **water vapor,** together with suspended particles. Table 7–1 shows the main constituents of dry air.

As Illustrated in Table 7–1, the atmosphere consists mainly of a mixture of gases of constant concentration. Oxygen and nitrogen molecules, both with two atoms, form 99.04 percent by volume, and inert argon atoms a further 0.93 percent. All the remaining constituents of dry air make up only 0.03 percent. Yet they are quite important. Carbon dioxide, for example, is essential to life, and is critical to climatic control. Ozone, present chiefly above 15 km, is very toxic, and also has a marked effect on climate. In addition, it shields us from damaging ultraviolet radiation. Carbon dioxide, ozone, and radon (a radioactive decay product escaping from the solid earth) vary in concentration.

Dry air is so nearly fixed in composition that we can treat it physically as a single gas with a molecular weight of 0.028964 kg mol^{-1}. At the range of temperatures and pressures observed in nature, dry air obeys the law for a perfect gas, that is, $PV = nRT$ (Equation 6.41), or, slightly differently stated,

$$p = R\rho T \qquad\qquad (7.1)$$

where p = pressure (N m^{-2} = Pascal)
   ρ = density (kg m^{-3})
   T = Kelvin temperature (K)
   R = gas constant for dry air (287.0 J kg^{-1}K^{-1})

This equation of state is one of the laws governing atmospheric behavior. Only two parameters of state are really needed, so in practice we use temperature and pressure, which are easy to measure. The mean surface temperature of the earth is 288K, and the mean sea-level pressure is 1,013.2 mbar. (For convenience, the meteorologist uses the *millibar* (mbar) as a unit. One mbar is equal to 100 Pascals (Pa).) Note that pressure does not determine temperature, or vice versa: cold air can have low pressure, and warm air high pressure.

Water may be present in the atmosphere as gas (vapor), solid, or liquid. Water vapor is always present. Its partial pressure may be as high as 30 mbar, equivalent to about 3 percent by volume. The precipitable water is the liquid water equivalent of the water vapor present in any column of the atmosphere. In practice, most of the water vapor is in the lowest 5 km. The precipitable water for the entire vertical column of area of 1 m² varies from almost zero in very cold air to about 60 kg m^{-2} in the most humid parts of the tropical countries. It occasionally reaches 50 kg m^{-2} in temperate North America. The molecular weight of natural water is 0.018016 kg mol^{-1}. Natural water contains two stable isotopes of hydrogen, ordinary ^{1}H, and deuterium (^{2}H or D), which contains an extra neutron.* Very small quantities of radioactive tritium (^{3}H or T) are present, chiefly because of earlier bomb testing.

The smaller particles within the atmosphere form an aerosol (i.e., colloidal-size particles in a gas) which are too small to fall out rapidly. The smallest detectable of these particles is 10^{-1} to 10^{-3} micrometer (μm) in radius. Such particles are very numerous inland, especially in cities. At sea there are fewer of them, but maritime air contains much larger chloride particles that play a key role in condensation. Certain sizes of particles reflect sunlight diffusely and form the reduced visibility conditions called haze, mist, or fog. Many atmospheric particles are, in fact, liquid, since they attract water vapor condensation and go into solution.

Without the aerosol, clouds, rain, and snow could not form. But too many particles from chimney smoke, car exhausts, or loose soil reduce visibility and may cause health problems. Most air pollution is of this variety.

---

* The notation ^{1}H and ^{2}H is used here to differentiate between an ordinary hydrogen atom of mass 1 (1 proton) and a hydrogen atom of mass 2 (1 proton and 1 neutron).

7.2.2 Thermal and Electrical State

Figure 7–1 shows the permanent layers of the atmosphere. At the base is the **troposphere,** capped by a surface of minimum temperature called the **tropopause** at levels between 10 and 17 km above the sea. Temperature decreases with height in the troposphere, since the main heat source is solar radiation absorbed at ground level. The rate of decrease, called the **lapse rate,** is about 5.0K km^{-1}. The troposphere contains most of the water vapor, clouds, and storms of the atmosphere. Winds tend to be strongest at the tropopause, the level of the jet streams. This is also the level at which most aircraft cruise.

Above the tropopause, temperature increases with height in the **stratosphere,** reaching a maximum at 50–55 km, the so-called **stratopause,** where temperature is about as high as at ground level. The warmth is due to the absorption of ultraviolet radiation from the sun by oxygen (O_2) and ozone (O_3). Most of the world's ozone is found in the stratosphere, where it may exceed 5 parts per million by volume (ppmv). Hence, stratospheric air is lethal to human beings. There is very little water vapor at these heights.

The **mesosphere** extends from the stratopause at 50–55 km to another temperature minimum at 80 km, the **mesopause.** The mesosphere is a windy and turbulent region, but

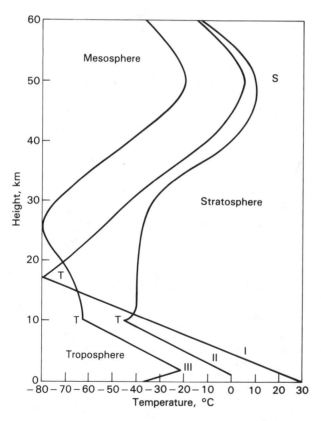

Figure 7–1 Typical temperature distributions with height, key: I—hot, tropical areas, all year; II—polar regions, summer; III—polar regions, winter (T marks the tropopause on each curve; S is the stratopause)

there is too little water vapor for clouds to form. Above the mesopause, temperature increases indefinitely upwards into the **thermosphere,** the hot upper atmosphere.

The air nearest the earth's surface is called the **boundary layer.** The **planetary boundary layer** (below 1,000 m) is the layer in which the wind is affected by friction with the earth's surface. The bottom 50 m is often called the **surface boundary layer.** These layers are very important to the engineer, most of whose work is done at such levels.

The temperatures of the air, sea, and land surface are controlled by unequal heating and cooling by the sun or outgoing radiation. This accounts for the familiar changes of heat and cold during a typical day, and between seasons. It also explains why the tropics are warm and polar regions cold. The transport of heat by winds and ocean currents also affects temperature.

The lower atmosphere is usually electrically neutral, unlike the ionized upper atmosphere. Nevertheless, strong potential gradients do exist, especially in and around thunderstorms. In a thunderstorm, gradients of 50,000 volts m^{-1} are sometimes observed near the ground. Lightning (a discharge) occurs when gradients on the order of 100,000 to 300,000 volts m^{-1} are generated in the towering clouds of thunderstorms. Engineered structures like steel towers or tall buildings are often struck by such discharges.

7.3 ENERGY OUTPUTS AND INPUTS

7.3.1 Solar Radiation

The sun provides 99.97 percent of the energy used at the earth's surface for all natural processes. The only other sources are (*i*) geothermal energy, the source of which is nuclear disintegrations in the earth's interior, and (*ii*) starlight from space. Both are tiny by comparison with the energy of the sun. The energy we use in our economy is also mainly solar. Coal, oil, and natural gas contain solar energy stored in plant tissues as a result of photosynthesis in the remote past. Burning them releases ancient solar energy and carbon dioxide into the atmosphere. Currently, we burn these fuels at a world rate under 10^{13} W, which is small by comparison with the rate at which the earth receives solar energy (1.74 $\times$ 10^{17} W).* The latter means that the annual mean energy received is 5.5 $\times$ 10^{24} J, or 1.5 $\times$ 10^{18} kilowatt-hours.

The sun is a fairly constant star. We can detect only small variations in the nature and intensity of the radiation it emits, and even these may be due to faulty measurement. Hence, we speak of the **solar constant,** which is the intensity of solar radiation reaching the top of the earth's atmosphere.

Measured at right angles to the solar beam, the solar constant is now estimated to be 1,368 W on each square meter of the circular outline (disk) of the earth as it faces the sun. The spin of the earth distributes this power over the whole surface of the sphere,

---

* The watt is the unit of power, the rate at which energy is produced, consumed, or transmitted. It is 1 joule per second (J s^{-1}). A kilowatt (kW) is 10^3 watts. The kilowatt-hour is commonly used as a unit of energy. It equals 36×10^5 joules.

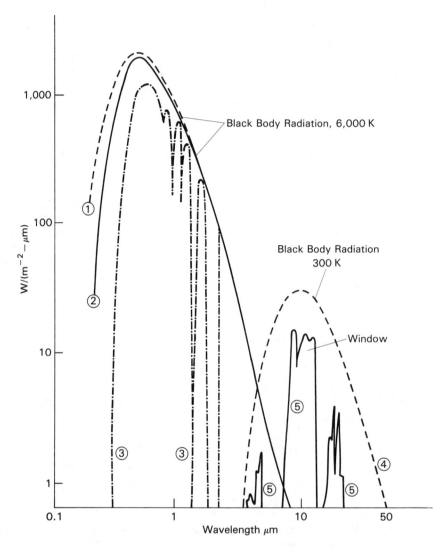

Figure 7–2 Spectra of solar (shortwave) and terrestrial (longwave) radiation.
Source: Adapted from Sellers, 1965

Power per unit area (W m^{-2}) per micrometer wavelength emitted by a black body at 6,000K
(curve 1) and another at 300K (curve 4), roughly the surface temperatures of the sun and
earth, respectively. Other curves have the following meaning: 2, actual power of solar
radiation at the top of the atmosphere; 3, the same at the base of the atmosphere; 5,
power of radiation passing *directly* from earth's surface to space, showing atmospheric
window.

whose area is four times as great as that of the disk. Hence, the mean solar constant per
unit area of the earth's surface is 342 W m^{-2}. (Surface area of a sphere $= 4\pi r^2$)

Solar radiation resembles that of a black body (perfect radiator) near 6,000K, which
is henceforth assumed to be the temperature of the sun's surface. The highest radiation
intensity occurs near a wavelength of 500 nm (see Figure 7–2), with most of the power

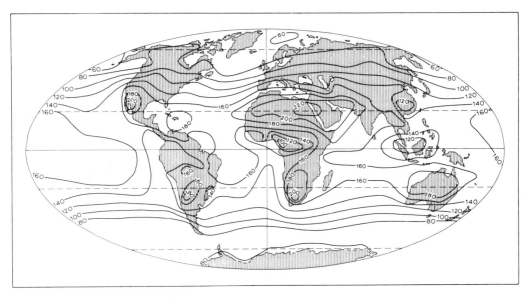

Figure 7–3 Average annual solar radiation on a horizontal surface at ground level (K cal/ cm^{-2} yr^{-1}). Source: M. I. Budyko. *The Heat Balance of the Earth's Surface*. Translated by N. S. Stepanova. Washington, D.C.: U.S. Dept. of Commerce, 1958

contained in the range 200–5,000 nm. Meteorologists call this **shortwave radiation,** because it is of shorter wavelength than radiation emitted by the earth itself. The human eye detects light between about 400 and 700 nm, which is called the **visible-light** wave band. Shorter radiation (200–400 nm) is called **ultraviolet,** and longer radiation **infrared.**

Figure 7–3 shows the mean annual solar radiation received at the earth's surface. This amount is well below the areal average solar constant, because the sun's rays are weakened on their way down through the atmosphere. On average, the radiation is distributed as follows:

1. About 17 percent is absorbed by clouds, water vapor, and carbon dioxide, heating the atmosphere directly.
2. About 30 percent is reflected back to space from clouds (which accordingly appear white to an observer on a spacecraft) and from atmospheric gases or particles.
3. About 53 percent reaches the ground. About two-thirds of this is in the form of direct sunlight, capable of casting shadows. The remainder is diffuse—the blue light of the sky and the gray of a cloudy day.

The actual mean intensity (averaged over 24 hours) of solar radiation at ground level varies from about 250 W m^{-2} in subtropical deserts to as little as 80 W m^{-2} in cloudy subpolar areas. Obviously, it is near zero at night, and day values are considerably higher than average. At times in clear weather, when the sun is nearly vertically overhead, values approaching the solar constant (1,368 W m^{-2}) are observed for short periods.

In midlatitude areas, mean values are close to 130–160 W m^{-2} on a 24-hour basis. To accumulate 1 kW of solar power, it is therefore necessary to collect it over an area of 6–8 m^2, even if perfect absorption is achieved. Solar radiation has low power per unit area, and is thus expensive to convert for high-temperature uses. Hence, large collectors as well as extensive storage capacity are needed if the energy is to be used as heat.

7.3.2 Terrestrial Radiation

Year in and year out the sun goes on heating the earth, whose temperature nevertheless remains almost the same. Therefore, the earth must be sending the same amount of energy back to space. It can only do this by means of radiation. The earth's surface acts much like a black body with a temperature of 288K. At such a temperature, a black body emits energy at wavelengths between about 4,000 and 50,000 nm. Peak intensity is at almost exactly 10,000 nm. Thus, terrestrial radiation is often called **long-wave radiation.** Clouds also radiate like black bodies, and are only slightly cooler than the earth. The necessary return flow to space takes the form of long-wave radiation from (a) the earth's surface or (b) the atmosphere, especially the tops of clouds.

The atmosphere is chiefly transparent: most of its gases neither absorb nor emit much radiation, with three important exceptions:

1. Water vapor (H_2O) absorbs and emits radiation very strongly between 5,000 and 7,000 nm, and above 17,000 nm.
2. Carbon dioxide (CO_2) absorbs and emits strongly near 4,500 nm and above 13,500 nm.
3. Ozone (O_3) absorbs and emits near 9,600 nm.

Thus, in cloud-free weather there is a gap or "window" between 7,000 and 13,500 nm in which long-wave radiation emitted by the earth's surface or clouds escapes freely to space. In addition, the gases H_2O, CO_2, and O_3 send radiation upward to space in the wave bands listed. An observer or satellite looking down at the earth sees upcoming long-wave radiation that has originated from various levels of the atmosphere or from the earth's surface. By the right choice of wave band, the observer can identify the level from which the radiation comes, and by means of its intensity, can estimate the temperature of the emitting layer or surface. In this way, satellites can scan the earth even at night, and can identify clouds and measure their heights. Vertical profiles of temperature can also be measured. Movies of the motion of clouds can be made from satellite data of both reflected visible and long-wave emitted radiation. It is important to note that the long-wave absorption and emission by these gases and by clouds have the effect of making the escape of energy to space more difficult than it would be in a clear, dry atmosphere. This action by the gases and clouds is often called the **greenhouse effect.** It makes the earth's surface about 35°C warmer than it otherwise would be.

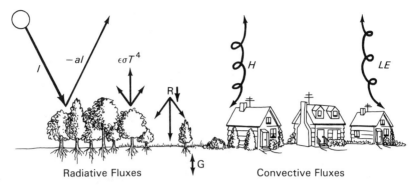

Figure 7–4 The surface heat balance

Radiative heating and cooling include:

I, the incoming solar (shortwave) radiation
$-aI$, the fraction of I reflected back unused
$\epsilon\sigma T^4$, the long-wave radiation leaving the surface
$R\downarrow$, the long-wave radiation back from the air (from CO_2, H_2O vapor, O_3, and clouds)

Convective heating and cooling include:

H, the flux of sensible heat (usually upward) due to eddies
LE, the flux of latent heat, also due to eddies, associated with evaporation or condensation
at the surface
G, the flux of heat into and out of the soil, due primarily to conduction

7.3.3 The Surface Radiation Balance

The rate of net radiative heating or cooling at the earth's surface is called the **net radiation,** or **radiation balance.** It is the sum of all the gains and losses of radiant power at the earth's surface (see Figure 7–4), given by

$$R_n = I(1 - a) + R\downarrow - \epsilon\sigma T^4$$
$$\text{Terms:} \quad (1) \qquad (2) \qquad (3) \tag{7.2}$$

where R_n = net radiation ($W\ m^{-2}$)
$\quad I$ = solar radiation at surface ($W\ m^{-2}$)
$\quad a$ = albedo for shortwave radiation (dimensionless)
$R\downarrow$ = downward long-wave radiation from atmosphere ($W\ m^{-2}$)
$\quad \sigma$ = Stefan-Boltzmann constant ($5.67 \times 10^{-8}\ W\ m^{-2}\ K^{-4}$)
$\quad T$ = temperature of the surface (K)
$\quad \epsilon$ = emissivity of surface (ratio of actual to black-body radiation)
$\quad$ (dimensionless).

A plus sign means an energy gain at the surface. Since all fluxes are measured or computed with respect to a horizontal surface, they can be thought of as *vertical* energy transfers.

Term (1) on the right-hand side of the equation is the **absorbed solar radiation,** with a the **albedo,** i.e., the fraction of the solar radiation that is reflected back. The surface

TABLE 7–2   RADIATIVE PROPERTIES OF NATURAL SURFACES

Surface	Remark	Albedo a	Emissivity (all-wave) ϵ
Soils	Dark, wet	0.05–0.40	0.90–0.98
	Light, dry		
Desert		0.20–0.45	0.84–0.91
Grass	Long (1.0 m)	0.16	0.90
	Short (0.02 m)	0.26	0.95
Agricultural crops and tundra		0.18–0.25	0.90–0.99
Orchards		0.15–0.20	—
Forests			
Deciduous	Leaves fallen	0.15	0.97
	Leaves on	0.20	0.98
Coniferous		0.05–0.15	0.97–0.99
Water	Small zenith angle	0.03–0.10	0.92–0.97
	Large zenith angle	0.10–1.00	0.92–0.97
Snow	Old	0.40	0.82
	Fresh	0.95	0.99
Ice	Sea	0.33–0.45	0.92–0.97
	Glacier	0.20–0.40	

Source: Oke, 1978

albedo depends on the nature of the material (for example, soil, plants, or water). Typical values are given in Table 7–2. Most land surfaces have albedos in the range 0.1 to 0.3. Snow usually exceeds 0.8, and water is well below 0.1.

Term (2), $R \downarrow$, is the warming of the surface due to the long-wave radiation from clouds, water vapor, carbon dioxide, ozone, and aerosols. It is usually smaller than $\epsilon \sigma T^4$ (Term 3), the escape of long-wave radiation from the surface, so that the sum of the long-wave gains and losses is usually a net cooling. Most natural materials radiate almost as black bodies at these temperatures, that is, emit energy at a flux of σT^4 according to the Stefan-Boltzmann law. However, actual losses are normally a bit less than black-body values, with the emissivity ϵ usually in the range of 0.90 to 0.98. Representative values of ϵ are given in Table 7–2 also.

Example 7.1

Calculate the net radiation R_n under the following conditions:
- Incoming solar radiation at midday, $I = 1,000$ W m^{-2}
- Albedo of surface, $a = 0.20$
- Emissivity, $\epsilon = 0.95$
- Temperature of surface, $T = 300$K
- Downward long-wave radiation from atmosphere, $R \downarrow = 250$ W m^2 (measured)

Solution   From Equation 7.2,

$$R_n = 100 (1 - 0.2) + 250 - 0.95 \times 5.67 \times 10^{-8} \times 300^4$$
$$= 800 + 250 - 440$$
$$= 610 \text{ W m}^{-2}$$

Instruments called **net radiometers** are available to measure R_n.

Each of the streams of radiation varies daily and annually. Therefore, so does R_n. Figure 7–5 illustrates a typical summer day's radiation, with its wide variation. Included are the following:

1. Solar radiation is near zero at night. It rises to a peak near local noon, and then returns to zero just after sunset. The total daily influx depends on how close the sun is to vertical at noon, a function of season and latitude. It is readily measurable. Albedo can

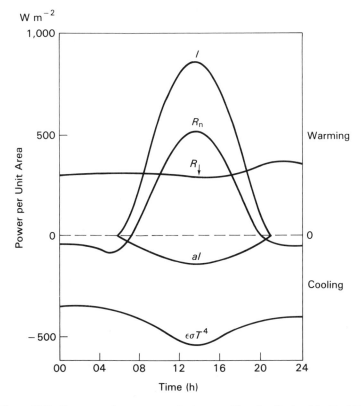

Figure 7–5   Energy exchanges over short grass, Matador, Sask., July 30, 1971.
Source: Oke, 1978

The curves are typical of a clear, sunny day. The surface is heated by I, the solar radiation, and $R\downarrow$, the long-wave radiation from the atmosphere. It is cooled by aI, the reflected solar radiation (a is the albedo), and by $\epsilon\sigma T^4$, the long-wave radiation emitted. Note that the long-wave exchanges vary little during the 24 hours.

also be determined quite easily, simply by measuring the incoming radiation and then reversing the instrument to measure the sunlight reflected back. The ratio of outgoing to incoming radiation gives the albedo.

2. Long-wave heating from the atmosphere and long-wave cooling of the surface are much more difficult to observe. They are usually calculated from measurements of air temperature and humidity. Unlike solar radiation, they vary rather slowly. Usually they add up to a cooling both by day and by night. During the day the heating by the sun offsets the net long-wave cooling, but at night the latter is in control. Temperature falls until dawn.

3. A considerable variation in the amounts of all the types of radiation is caused by weather changes. An overcast sky, for example, largely prevents long-wave cooling, because downward long-wave radiation is strong during such conditions. The clouds also reduce solar radiation. So clear and cloudy days have quite different temperatures.

7.3.4 Energy Use at the Surface

How is the energy of Equation 7.2 used? A simple heat balance equation (see Figure 7–4) provides the answer. We have

$$\underset{\text{(heat gained)}}{R_n - G} = \underset{\text{(heat used)}}{H + LE + MS + Q} \tag{7.3}$$

where R_n = net radiation, as in Equation 7.2 (W m^{-2})

    G = loss of heat by conduction into soil (soil heat flux) (W m^{-2})

    H = loss of heat due to upward flux of heated eddies (turbulent heat flux) (W m^{-2})

    E = evapotranspiration of water (evaporation plus transpiration through plant tissues) (kg m^{-2}s^{-1})

    S = snowfall to be melted (kg m^{-2}s^{-1})

    Q = energy conversion by photosynthesis in green plants (W m^{-2})

L and M = latent heats of vaporization and melting (fusion) of water and ice (nearly constant at 2.44×10^6 and 3.33×10^5 J kg^{-1}, respectively)

Heat flows into and out of the soil mainly by conduction at a rate G that is readily measured by soil heat flux transducers. The flux, i.e., the rate of flow per unit area, is downward when solar heating is strong, and upward at other times. It tends to vanish over a day or a year, and is in any case small (of the order a tenth to a hundredth of R_n). It is usually ignored, or else regarded as a small reduction of the net radiation (which is why it appears on the left of Equation 7.3). The even smaller geothermal heat flux from the earth's interior can also be ignored.

The net heating ($R_n - G$) is the heat source for the processes on the right-hand side of Equation 7.3. These are as follows:

1. H and LE are the sensible and latent heat fluxes, respectively, that are carried to and from the surface by turbulent eddies in the wind—the gusts and up-and-down movements typical of windy weather.

2. *MS* is the heat needed to melt snow. This is usually small, but in regions of heavy snowfall as much as 10 percent of the annual net radiation influx may be used in this way.

3. *Q* is the very small amount of heat used by green plants during the manufacture of tissues by photosynthesis. It is rarely more than 1 percent of the net radiation.

Example 7.2

A typical summer value of net radiation by day is $+500$ W m^{-2}, and of the soil heat flux is -10 W m^{-2}. Hence, net heating is 490 W m^{-2}. Over a moist, plant-covered soil, typical values of the other terms in Equation 7.3 would be as follows:

Photosynthesis (*Q*)	5 W m^{-2}
Sensible heat flow (*H*)	85 W m^{-2}
Evapotranspiration (*LE*)	400 W m^{-2}

If the soil is dry, however, most of the available heat will produce sensible heat flux, and not evaporation. The ratio *H/LE* is called the **Bowen ratio.** In this example it is 85/400, or about 0.21. In cold or dry conditions it is much higher. The Bowen ratio plays an important role in hydrology and climatology.

7.4 WIND, STABILITY, AND TURBULENCE

7.4.1 Motion of the Lower Atmosphere

If air moves relative to the ground, we feel it or see it as wind—just air in motion. It is set in motion (i.e., accelerated) by a series of forces:

1. **The pressure gradient force** tending to impel air motion from areas of high to areas of low pressure.

2. **Gravitation,** which tends to accelerate the air downwards at a rate close to 9.8 m s^{-2}.

3. **Friction,** acting opposite to the wind direction, and proportional roughly to the square of the wind speed.

4. **The Coriolis force,** caused by the rotation of the earth, often called the deflecting force of the earth's rotation. It acts at right angles to the wind direction and is proportional to the wind speed. It thus acts toward the right in the northern hemisphere and toward the left in the southern hemisphere (when viewed from above).

In practice, the wind tends to blow with constant velocity (i.e., no change in either speed or direction) relative to the earth. Newton's Second Law says that if a force is applied, a proportional acceleration will occur. It follows that if there is no acceleration, there can be no forces acting. Since the preceding forces do exist, evidently they must usually be in balance, that is, they cancel one another. Only in this way can the wind blow with constant velocity. In other words, the air usually moves under balanced forces.

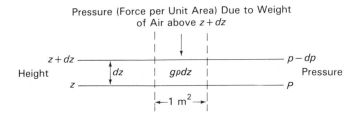

Figure 7–6 The hydrostatic equation: Variation of pressure with height

The decrease in pressure-*dp* due to the layer *dz* reflects the weight of the 1 m² column between *z* and *z* + *dz*, which is *gρdz*. The pressure *p* at level *z* is due to the weight of all such layers above.

We can see this most readily if we consider a unit kilogram of air somewhere in the atmosphere. It is accelerated downwards at 9.8 m s⁻² by gravitation. Yet in practice, it usually remains at the same level. Thus, the downward gravitational force must be balanced by some equal and opposite force or forces. Actually, the upward pressure gradient force does the balancing. Atmospheric pressure in air at rest is due simply to the weight of the overlying air (Figure 7–6). If we consider the air column to be made up of thin slabs 1 m² in area, and *dz* m thick, the weight of each slab is, by definition, *gρdz*, where *g* is 9.8 m s⁻², ρ is the density in kg m⁻³, and *z* is the height in m. At any level *z*, the pressure is given by

$$p = \int_z^\infty g\rho \, dz$$

the sum of weights of all the layers above the level *z*. The atmosphere distributes its mass so that everywhere the upward thrust due to the decrease of pressure with height balances the downward acceleration of gravity. This is expressed by the **hydrostatic equation**

$$dp = -g\rho \, dz \qquad (7.4)$$

Solving Equation 7.1 for ρ and substituting into Equation 7.4, we get

$$\frac{dp}{p} = -\left(\frac{g}{RT}\right) dz \qquad (7.5)$$

which can be used to set altimeters, or to calculate pressure differences over a small range of heights. The hydrostatic pressure *p* is the same in all directions.

Usually we restrict the word "wind" to mean the horizontal movement of air, which we can feel on our faces or our backs. The balance of forces for steady, straight motion is expressed by

$$2\omega \sin \phi \rho V_g = \frac{dp}{dn} \qquad (7.6)$$

where ω = rate of rotation of the earth (7.3 × 10⁻⁵ rad s⁻¹)
     φ = latitude

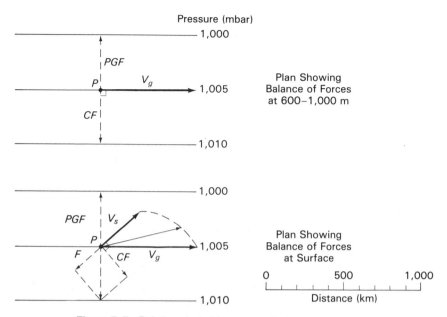

Figure 7–7   Relation of wind to pressure in the horizontal plane

V_g = wind velocity, parallel to isobars (lines of constant pressure) (m s^{-1})
dp/dn = pressure gradient (i.e., rate of change of pressure p with respect to distance n at right angles to isobars) (N m^{-3})

In this case, the pressure gradient force per unit mass is balanced by the Coriolis force, the term on the left of Equation 7.6. The air flows along the isobars at a velocity V_g which is just fast enough to make the left-hand term equal in magnitude to the pressure gradient term on the right-hand side. Equation 7.6 defines the **geostrophic wind**, an idealized wind that is quite close to the real wind at all heights between 600 to 1,000 m above the ground.

Figure 7–7 shows how the balance works. At the point P, using the distance scale we can estimate the pressure gradient at 15 mbar/1,000 km, which is equal to 15×10^2 N m$^{-2}/10^6$ m. The wind blows along the isobars at a speed V_g, keeping the low pressure on its left, since we are hypothetically dealing with the northern hemisphere. If we insert the preceding value in Equation 7.6, we get $V_g \approx 20$ m s^{-1} at 30° latitude, for example. In other words, to make the deflecting force equal and opposite to the pressure gradient force, the wind must blow along the isobars at 20 m s^{-1} in the case shown in Figure 7–7.

We can, of course, easily measure the wind by tracking floating balloons, or by mounting a wind vane with a speed sensor attached. But it is even easier to measure pressure by means of a barometer. Hence, weather maps use pressure distribution as their main indicator of wind. Pressure is measured simultaneously at thousands of stations worldwide, the values are standardized to sea level using Equation 7.5, and then the isobars are drawn in. The resulting map is a good approximation to the distribution of wind, as well as of pressure. Equation 7.6 allows us to calculate wind speed, the isobars being roughly streamlines.

The two sketches in Figure 7–7 show how wind is related, in an idealized fashion, to isobars of constant pressure. At 600–1,000 m, the wind flows along the isobars at a speed V_g (the geostrophic wind) that makes the Coriolis force (*CF*) equal and opposite to the pressure gradient force (*PGF*). At the surface, friction (*F*) with the ground slows down the wind, so that the Coriolis force is decreased and no longer balances the pressure gradient force (PGF). To balance the PGF, the wind turns across the isobars (lower diagram) so that *CF* and *F* together balance *PGF* (see the parallelogram of forces). The angle of the surface wind V_s to the isobar is on the order of 40°, and the speed is about half that of the geostrophic wind V_g. Between the two levels, as one ascends, the wind increases and turns towards the isobar. The wind arrow point traces out a spiral curve on the way up (see the lower diagram, where a sample midlevel wind is shown). The system is reversed in the southern hemisphere.

On real weather maps the isobars are rarely straight. Instead, the maps show oval or circular areas of high and low pressure hundreds or even thousands of kilometers across. The free air flows clockwise along the isobars of a high pressure system (an **anticyclone**), and anticlockwise in a low pressure system (a **cyclone**). This rule for the northern hemisphere is reversed in the southern. Even though isobars are in fact curved, and the wind thus changes its direction, the speed is still close to the geostrophic value in most cases. Anticyclones and cyclones represent gigantic horizontal eddies in the atmosphere. They migrate slowly, bringing with them fair (anticyclones) or foul (cyclones) weather. Near the equator the relation between wind and pressure breaks down, and pressure differences are small.

7.4.2 Turbulence and Stability

The rules just described allow many kinds of motion. In practice we think of wind as both organized motion, which is described by Equation 7.6, and unorganized motion, or turbulence, which is not described by that equation. Both organized winds and turbulence are important environmentally because they transport water vapor, carbon dioxide, heat, and pollutants. In this section, we look at turbulence, which is most important in bringing about vertical transport. It is very important in the study of air pollution, discussed in Chapter 13. It is also crucial in the design of buildings, aircraft, bridges, and other structures exposed to the wind. The gusts and eddies in turbulent air cause the worst stress that these structures have to withstand.

Turbulence includes the gusts that assault us as we walk into a strong wind. In the atmosphere, it is of two kinds, forced and free. Forced turbulence occurs when wind encounters obstacles, as it always does at the earth's surface. Free turbulence develops when the atmosphere encourages the growth of small initial disturbances. Usually this is because the motion releases buoyancy. Both kinds of turbulence occur near ground level as well as in the higher layers.

Free turbulence, or free convection as it is better called, is encouraged by certain distributions of temperature in the atmosphere. In Figure 7–1, we see that temperature in the troposphere usually decreases with height; i.e., dT/dz is negative. The **environmental lapse rate** is defined as the rate of decrease of temperature as one ascends through the atmosphere. It is usually close to 5K per kilometer of ascent. If a bubble or slab of air rises, either freely or forced, its pressure falls, and the volume expands. Work is done against the surrounding air. This leads to a cooling of the rising air, given by the equation

$$\frac{dT}{dz} = -\frac{g}{c_p} \tag{7.7}$$

where g is the acceleration due to gravity (9.8 m s^{-2}), and c_p is the specific heat of air at constant pressure (1.0 kJ kg^{-1} K^{-1}). This equation defines the **dry adiabatic cooling rate,** valid only as long as condensation does not occur. The word "adiabatic" means "without communication of heat," signifying that the change of temperature is due to energy conversion, not loss. Since g and c_p are almost constant, dT/dz is also constant at 9.8K km^{-1}. Air becomes buoyant if the temperature in the surrounding air decreases with height at a rate exceeding this figure. This usually happens during the day when the sun heats the ground, rather than at night. Hence, buoyant convection near the ground is largely confined to the daytime.

If the air is moist and becomes saturated as it ascends, a cloud forms. The latent heat of vaporization is released into the ascending air, which then cools less rapidly. A typical cooling rate for saturated air is 4 to 5K km^{-1}, which is quite similar to the usual values of the environmental lapse rate. Hence, rising saturated air—the cloud systems of the atmosphere—can usually ascend further than can dry air, because it remains buoyant longer.

Stability is the condition in the air that tends to damp down any convection that may be present. **Instability** is the opposite—the condition that encourages convection to grow. For unsaturated air, an environmental lapse rate of less than 9.8K km^{-1} is stable. Saturated air is stable if the environmental lapse rate is less than the reduced cooling rate in the ascending currents.

The preceding relationships are visible on most sunny days in the warmer seasons. In the early morning, the air is stable and the sky is cloudless. As the sun heats the ground, however, the environmental lapse rate increases rapidly, until a shallow layer with lapse rates greater than the dry adiabatic cooling rate (9.8K km^{-1}) is created (see Figure 7–8). This layer is turbulent, and as the day progresses, it may become 1 or 2 km deep. Flight in light aircraft in this layer will be very bumpy. If the air is moist, clouds may form on many of the ascending "thermals" and then may rise to considerable heights, especially if the upper troposphere is cool. The cauliflower-head-like **cumulus clouds** mark such rising columns. Under very unstable conditions—warm near the ground, very cool above—they may rise far above the freezing level. The cloud tops then spread out to form the familiar anvil-shaped thunderheads called **cumulonimbus** clouds (see Photograph 7–3), from which heavy showers originate. As evening progresses, however, the solar heating at ground level ceases, and conditions become more stable again. The clouds stop growing and then disperse. Nocturnal thunderstorms, common in very humid air masses, especially

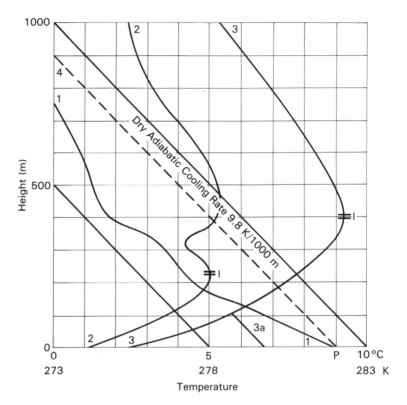

Figure 7–8 Lapse rates of temperatures in the boundary layer

The dry adiabatic cooling rate of 9.8K/1,000 m is shown for air at 10°C (283K) and 5°C (278K). Air rising from the ground will cool at this rate unless it is saturated. Curve 1 shows conditions typical of midday hours with strong heating. Heated air at point *P* will rise and cool along dashed curve 4, whereas the actual temperature of the surroundings, as measured by actual sounding (curve 1), is colder, because the environmental lapse rate exceeds the dry adiabatic rate. Hence, the rising air will be buoyant, because warm air is less dense. Curve 2 shows an example of a surface inversion (marked I), with air at 220 m 4K warmer than at the ground. A second inversion occurs at 450 m. Curve 3 shows a much stronger inversion at 400 m, typical of dawn, with the temperature at the inversion 7K warmer than at the surface. Later in the morning, with the sun heating the ground and a wind stirring the lower layers, a shallow layer of normal, near-adiabatic lapse rate develops (curve 3a). Air rising from the ground in any of these cases—curves 2, 3 and 3a—will soon become colder than its surroundings and will thereby cease to be buoyant. Smoke, haze, and pollutants are trapped by such inversions.

in the Great Plains of the U.S., also require unstable conditions, but in this case a different mechanism triggers them.

Negative environmental lapse rates, i.e., an increase in temperature with height, are common at night, and in winter may persist all day. The top of the layer of negative lapse rates is called an **inversion** (Figure 7–8). Such conditions are very stable. There is usually little wind, and pollutants emitted into the inverted layer are slow to disperse. Dense fog may be present if the air is moist. Inversions often occur over cool sea or lake waters, and may be drawn short distances inland. This occurs in the coastal valleys of California, and at many localities along the shorelines of the Great Lakes.

7.5 WATER IN THE ATMOSPHERE

7.5.1 Humidity and Precipitation

Water has the special property that it exists in the atmosphere in all three phases—liquid, gaseous, and solid. It is called a **vapor** when in the gaseous phase.

There is an upper limit to the concentration that water vapor can attain, called **saturation.** At this limit it tends to condense to liquid or solid forms, provided that suitable surfaces exist on which this can happen. The saturated vapor condenses on small hygroscopic nuclei within the aerosol called **condensation nuclei.** In the atmosphere, condensation usually forms liquid water at temperatures well below the melting point of pure ice (273.2K); the droplets are then said to be supercooled. Only below 233K does condensation always form ice crystals. The water and ice particles are very small (usually below 2,000 nm in diameter), and remain suspended as cloud or fog. At ground level the vapor may also condense as dew or hoarfrost. Rime is the name given to the clear ice formed when supercooled droplets freeze on contact with solid surfaces.

Saturation vapor concentration depends only on the temperature of the vapor (or of a plane water or ice surface with which it is in equilibrium). The partial pressure due to water in the atmosphere is called the **vapor pressure** e, which is always less than about 40 mbar (4 kPa), and may approach zero in very cold air. The saturation vapor pressure over water, e_s, is given to a good approximation by

$$\log_{10} e_s \text{ (in mbar)} = 9.4051 - \frac{2,345}{T} \tag{7.8}$$

where T is taken to be the temperature of the air in which the vapor is mixed. Figure 7–9 is the curve of this function, compared with that for ice below 273.2K. Note that there is much lower vapor pressure over ice than over supercooled water. Saturation vapor pressure increases rapidly as temperature rises. If unsaturated air is cooled, the water vapor becomes saturated at some temperature (K) called the **dew point.** Condensation then begins on any suitable surface.

Several other measures are used to express the actual humidity of air. The vapor pressure e has already been mentioned. The **relative humidity** is the ratio e/e_s, or the actual vapor pressure divided by the saturation vapor pressure at the air temperature. The **humidity mixing ratio** x is the mass of water vapor mixed with a unit mass of dry air, and the **specific humidity** q is the mass of water vapor in a unit mass of moist air. The

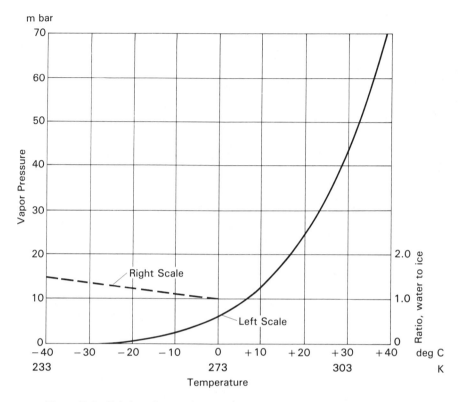

Figure 7–9 Relation of saturation vapor pressure (over a plane water surface) to temperature (deg C and K).

Dashed curve (use right scale) is the ratio of saturation vapor pressure over water to that over ice at the same temperature. At 0°C (273K), the two are the same. At −40°C (233K), the saturation vapor pressure over water exceeds that over ice by a factor of 1.47.

absolute humidity σ_w, or the **vapor density,** is the mass of water vapor per unit volume of moist air. Some useful formulae are

$$q = \frac{x}{1 + x} \tag{7.9}$$

$$x = \frac{q}{1 - q} = \frac{0.622e}{p - e} \approx \frac{5e}{8(p - e)} \tag{7.10}$$

$$r = \frac{e}{e_s} \approx \frac{x}{x_s} \approx \frac{q}{q_s} \tag{7.11}$$

where p is the atmospheric pressure in mbars and q, x, and r are all dimensionless quantities. Nevertheless, x and q are usefully measured in g kg^{-1}, and r is usually given as a percentage of saturation (that is, a relative humidity of 0.7 is referred to as 70 percent).

Because cloud droplets are so numerous, they can easily accommodate increased condensation with only a small increase in their radius. At these still small sizes, the droplets are kept suspended by the turbulence in the air. Yet rain or snow often falls from thick clouds. How does this come about? There are at least two known explanations:

1. If the cloud's temperature is below 273K but above 233K, supercooled water droplets are usually dominant. If small ice crystals form in or fall into such clouds, they will be in an environment that is supersaturated with respect to ice (see Figure 7–9). Hence, they grow rapidly by direct solid condensation. This is known as the **Bergeron-Findeisen process.** Eventually the ice crystals may fall as snow, often after coalescing with other crystals or with supercooled droplets that freeze on to them. A significant hazard for aircraft occurs when such supercooled droplets freeze on to leading edges, such as wings, propellers, and struts. This is called rime icing.

2. As there are always differences in size among droplets and crystals, differences in fall velocity lead to growth of the larger droplets by accretion. Under favorable circumstances, the droplets may grow to the point where they can fall to the ground. Typical rain and drizzle have drop sizes in the range 0.5 to 2.5 mm.

Experience has shown that both of the preceding processes are usually at work in the cold clouds of middle- and high-latitude rainstorms and snowstorms, in that significant falls originate in clouds that reach well above the freezing level. Substantial rain may, however, fall from warm cumulus clouds in the marine tropics.

Photograph 7–1   Carl Milles's statue, *The Hand of God*, silhouetted against fair-weather Cumulus, Stockholm, Sweden

These clouds form when shallow convection currents carry moisture upwards to the condensation level, but are then stopped by an inversion. (Photo courtesv F. K. Hare)

Photograph 7-2   Stratus (below) and altostratus cloud on a winter's day, Rakaia
Gorge, New Zealand

Note also the river terraces cut by the Rakaia River in the past 10,000 years of a warmer
climate. (Photo courtesy F. K. Hare).

All significant **precipitation** (i.e., rain, snow, and hail) falls from clouds formed by
uplift. Rising moist air cools adiabatically below its dewpoint. Two broad families of clouds
are recognized: (*i*) those due to slow, slanting uplift, called **layer clouds,** that are typical
of cooler season cyclonic storms in both hemispheres, and (*ii*) those due to rapid convective
uplift of air columns in unstable air, called **cumulus** and **cumulonimbus clouds.** Both of
the situations described exist when moist air is forced to climb over mountains. Different
types of clouds are shown in Photographs 7–1 through 7–3.

The layer cloud family includes stratus (usually formed by surface chilling of moist
air), altostratus or nimbostratus (rain and snow clouds), and cirrus, cirrostratus, and
cirrocumulus (the thin, wispy clouds common against the blue of the sky). All are formed
when warm, moist air rises gradually (at rates typically about 0.1 m s^{-1}) as it moves
eastward or northeastward over cooler air, usually in developing cyclones. Such cloud
masses often become thick enough and cool enough to start snowfall via the Bergeron-
Findeisen process. As the snow falls into the lower layers, it melts to rain, except in cold,
wintry conditions. Much of the prolonged rain and snow of the cooler seasons in middle
and high latitudes is formed in this way. The cloud systems responsible may cover tens of
thousands of square kilometers and be 5–8 km deep.

The cumulus type of cloud, arising from rapid uplift in small columns (on the order
of 0.5 to 100 square kilometers in cross section), ranges from the small puffs of cumulus
typical of a fair day to huge cumulonimbus masses rising to over 15 km in height. These
clouds produce short, violent showers. They are the dominant rainclouds of the tropics and
of the midlatitude summer. Thunder and lightning often accompany the heavier falls.
Cumulonimbus cloud is also common over the warm oceans when cold air flows toward

Photograph 7–3 Thunderheads (cumulonimbus clouds) of tropical disturbance over the Pacific Ocean between Hawaii and Fiji.

These thunderheads are caused by the convergence of the trade winds of the two hemispheres over warm ocean surfaces. (Photo courtesy J. S. Simpson and H. Riehl)

the equator. In the most violent of these storms, the rising column of air may acquire a rapid rotation about a vortex core with very low pressure. These **tornadoes,** as they are called, are the most destructive of all storms. They are small (less than 1 km across), but may travel long distances, and are unrelated to cyclonic storms.

7.5.2 The Hydrologic Cycle

The movement of water between air, sea, lakes and rivers, land, soils, glaciers, and living organisms forms the **hydrologic cycle,** sketched in Figure 7–10. This cycle was used earlier (Figure 6–10) to illustrate a water balance in a small region. Here, a similar balance on a *global* basis is shown. This movement almost balances out. Water is chemically active and enters into compounds very easily. As far as we know, these chemical exchanges release as much water as they remove. Overwhelmingly, the main mass of water is in the oceans, which cover two-thirds of the earth's surface. A small part cycles annually through the storage reservoirs shown in Figure 7–10.

   The atmospheric part of the cycle begins when evaporation occurs. This takes place off open water surfaces and also off ice, plants, soil, and all other surfaces recently wetted by precipitation. Evaporation requires:

1. *An energy source*, which is mainly the net radiation at the evaporating surface, but may also be heat from the soil, the turbulent heat flux in the atmosphere (common when warm air moves over cold surfaces), or heat from warm ocean waters. The heat required to evaporate 1 kg of water varies from 2.50×10^6 J kg^{-1} at 273K to 2.43×10^6 J kg^{-1} at 303K. To melt ice at 273K requires 3.33×10^5 J kg^{-1}. These are very large amounts of heat. On a typical day much of the available net radiation is used to evaporate water from moist surfaces such as wet soil, lakes, or growing crops (which deliver water to their leaf surfaces through the physiological process of transpiration). Off a plant-covered land surface, the latter may be dominant, and can rarely be separated from evaporation

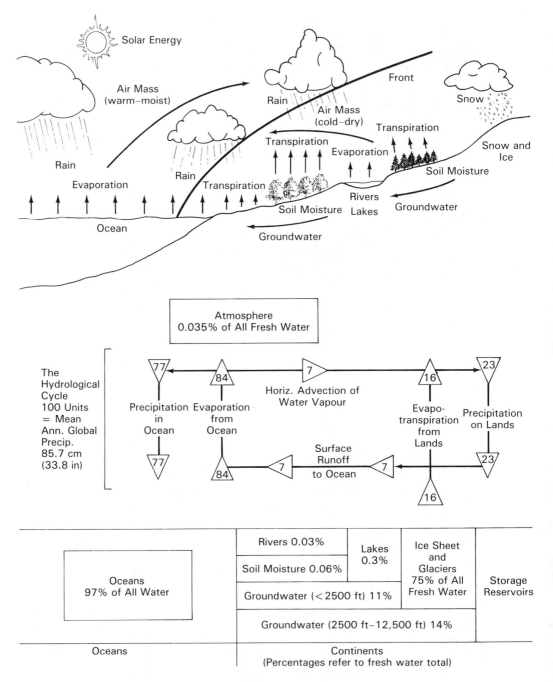

Figure 7–10 The hydrologic cycle. Adapted from: R. G. Barry and R. J. Chorley, *Atmosphere, Weather and Climate*, London: Methuen, 1982

Shown is the hydrologic cycle and water storage of the globe. The exchanges in the cycle are given as percentages of the mean annual global precipitation of 85.7 cm (33.8 in). The percentage storage figures for atmospheric and continental water are percentages of all fresh water. The saline ocean waters make up 97 percent of all water.

off the soil. Hence, the word **evapotranspiration** is often used for the collective process, which requires the same energy as evaporation.

   2. *A sink for the vapor produced*, which is simply the capacity of the atmosphere to transport it away, mainly by turbulent diffusion. Strong winds and unstable lapse rates favor rapid evaporation.

   The measurement of evapotranspiration is very difficult. The standard instrument is the evaporation pan, in which the loss of water from a confined surface is measured mechanically. For turfed, cropped, or bare soil, the lysimeter can be used. This is an isolated block of soil, in which grass or crops can be planted, from which one can measure the percolation from the base. Knowing the precipitation into the block, and allowing for storage changes, one can compute the evaporation or evapotranspiration. All such gauges suffer from the drawback that, unless great care is used, they will have different net radiation and turbulent characteristics from natural exposures. This is especially true of evaporation pans. Except at research stations with sophisticated instruments, the usual procedure is to rely on semiempirical formulae. A selection of these may be found in Bruce and Clark (1966), Hare and Thomas (1979), Mather (1974), and Oke (1978). Over a wide range of surface conditions, and over periods of several hours, a simple relation between net radiation and evaporation can be used, viz.,

$$E = \frac{\alpha s R_n}{L(s + \gamma)} \tag{7.12}$$

where $s = de_s/dT$ (mbar K^{-1}) from Figure 7–9
$\quad \gamma = 0.64$ mbar K^{-1}
$\quad \alpha = $ a proportionality constant whose value is near 1.0 over a wide range of surfaces, except under very dry conditions; it rises to a maximum of 1.26 over fully moist surfaces
$\quad R_n = $ net radiation, as before.

   The main source of evapotranspiration is rainfall that has been stored in the soil. Much of this water is held too tightly to be easily removed, but all soils contain some available water that can be extracted by plants and released to the atmosphere by transpiration, the flow of water through plant tissues and its loss as moisture through leaves and stems. The root systems of plants, which tap the soil water, are usually shallow, the top meter of soil normally providing most of the evaporated and transpired water. Once this soil water is removed, evapotranspiration ceases and plants tend to wilt. A recently wetted soil that contains all its available water but has drained all its surplus water is said to be at field capacity.

   Precipitation and evaporation are normally measured in millimeters of depth, i.e., the depth of rain or melted snow which would accumulate on or be lost from an impervious flat surface per unit time. The usual SI unit is kg m^{-2} per unit time, which is equivalent to 1 mm of rain for the same period. Rain and snow gauges simply catch the precipitation as it falls, and the depth is measured and recorded at fixed intervals.

A water balance must exist at any point. Much like the heat balance equation (Equation 7.3) is the **water balance** equation,

$$P = N + E + \text{storage change} \qquad (7.13)$$

where P is precipitation, N is percolation and runoff, E is evaporation, and the storage-change term indicates that water may be either stored in plant soil or subsoil or released from storage. All units are kg m^{-2} (millimeters of depth) per unit time.

Note that E occurs in both the heat balance and the water balance equations. For annual totals it is common to use the **runoff ratio** $C = N/P$, which varies from nil in very dry areas to as high as 0.9 in extremely wet, cool climates. Runoff, the surplus precipitation that finds its way into the streams, is the main subject of study of the science of **hydrology.**

7.6 CLIMATE

7.6.1 World Distribution

Figures 7–11 and 7–12 show the average or expected distribution of precipitation and temperature around the globe. Hundreds of other world distributions could be added. We have a good idea of the distribution of pressure, solar radiation (see Figure 9–1), temperature, precipitation, wind, humidity, and clouds at many levels in the atmosphere, especially below 30 km (a pressure of 10 mbar). Many other climatic elements can be mapped as well. Particularly important are the temperature and salinity of the sea surface waters.

These distributions have a long history. For the past 10,000 years climatic changes, though significant in human history, have not been large. Before that, however, there were many times when the world's climate was colder and drier than at present. In the past one million years, at least nine phases—the **glacial epochs**—are known when ice and snow were more widespread than now. Other glacial ages occurred in the remote past, one of them 650 million years ago. Thanks to much geophysical and geochemical work on the sediments of the deep oceans, we now have a detailed knowledge of the most recent age of glacial epochs, the Quaternary period. Since rapid fluctuations of climate characterized this period, it is natural to ask the question: Can the present climate change?

The fundamental controls of climate are the solar constant, which influences temperature on earth; the composition of the atmosphere; the distribution of land, sea, and mountains; and the rate of the earth's rotation. All these have changed over the long history of the earth but, with the possible exception of the atmosphere, now seem relatively stable.

7.6.2 Climatic Variability

Despite the stability of the climate-controlling factors, we do have direct evidence of fluctuations of climate during the past century, for which good records are available. Surface air temperatures averaged over the world vary from year to year by as much as ±0.5K.

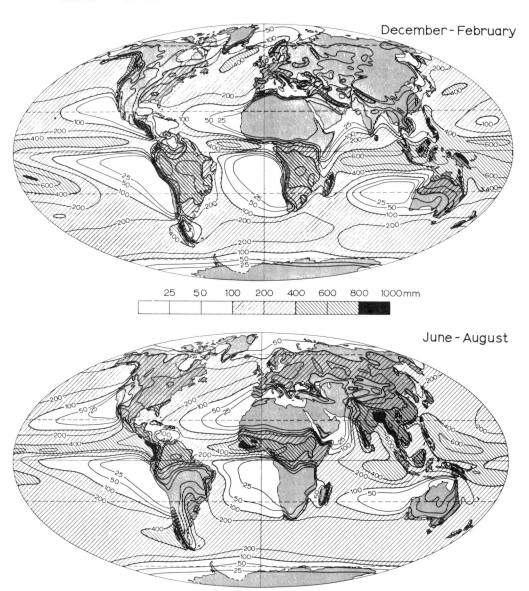

Figure 7–11 World distribution of precipitation (annual mean). Source: F. Möller, Petermanns Geographische Mitteilungen, 95 Jahrgang, 1–7, 1951.

Mean global precipitation (in mm) for the periods December–February and June–August.

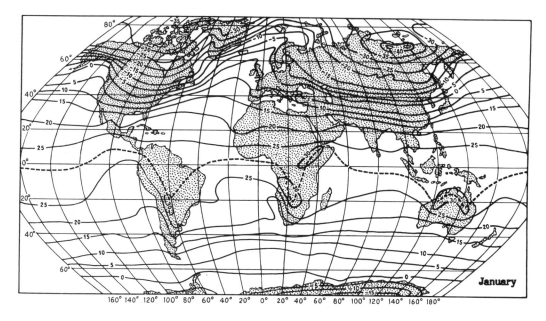

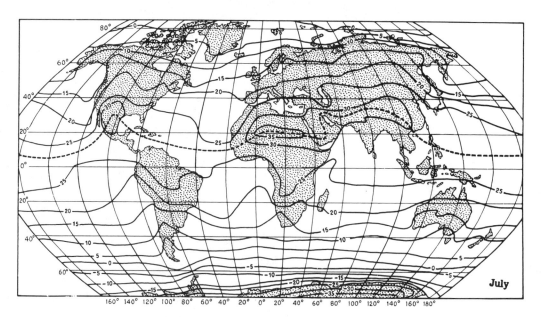

Figure 7–12 World distribution of temperature (January and July means). Source: R. G. Barry and R. S. Chorley, *Atmosphere, Weather and Climate*, London; Metheun, 1982

Mean sea-level temperatures in January and July (°C). The positions of the thermal equator are approximately shown by dashed line.

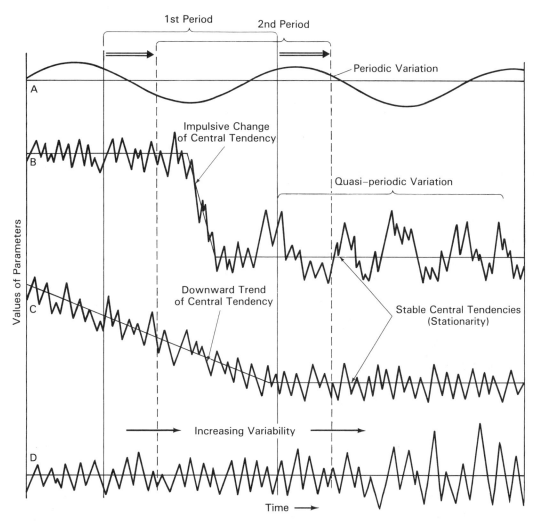

Figure 7–13 How climatic parameters vary with time

Idealized time series (curves *A* to *D*) of a representative parameter of a climatic element that is continuous in time (such as temperature or pressure). Vertical bars indicate an arbitrary averaging or integrating period (usually 30 years) which is recalculated each decade (see dashed bars).

Between 1860 and 1985, however, there has been a slow, fairly uniform background warming of 0.6K (a trend). Small though these changes are, they have considerable economic effects. There have also been significant changes in precipitation in some regions.

In assessing the meaning of such changes or fluctuations, it is important to remember that climate is inherently variable. Figure 7–13 shows an idealized set of **climatic time series**, i.e., values of climatic elements continuously plotted against time. To define climate, we average the values for a particular element (temperature, rainfall, etc.) over some arbitrary period, usually 30 years. Present practice is to recalculate these averages every ten years. The set for 1951–1980 is currently in standard use. The vertical solid and dashed bars in Figure 7–13 illustrate two such periods. Curves *A* through *D* show the different variations that occur in climatic time series, viz.,

1. *Periodic variations* (curve *A*) with various periods. The only common ones in climate are *daily* variations (for example, warm days and cold nights) and *seasonal or annual* variations (for example, warm summers and cold winters). Most other periodic variations ("cycles") allegedly found in climate turn out to be spurious.

2. *Quasi-periodic variations* (curve B), where a few years of high values are followed by a few years of lower values, but without any regular period.

3. *Upward or downward trends*. Curve *C* shows a downward trend followed by a period of *stationarity* (no trend). Curve *D* shows a trend toward increasing year-to-year differences of value (variability).

4. *Impulsive changes* of central tendency (curve *B*), in which the mean value changes suddenly to a new, stable level.

5. *Short-period variations* on curves *B* to *D*, without apparent pattern, resembling "noise" cluttering a radio signal. These include the familiar weather changes, but also such phenomena as warm and cold or wet and dry years.

On this basis, we say that climatic change occurs only if there is a statistically real difference between successive averaging periods. The changes that occur *within* an averaging period are called the *variability* of climate. Good measures of this variability are as important as the average values. Scientists and engineers, dealing with climate-related environmental problems, need at least the following information:

1. Good averages, such as mean monthly and annual temperature, rainfall, and solar radiation.

2. Some measures of variability. These can include standard deviations, frequency data, and extreme values likely to recur every decade or century. Data on return periods (probable intervals between specified values) are also needed, especially by the civil engineer and hydrologist.

3. Some estimate of *future* trends, or impulsive changes. This is especially important to civil engineers and agricultural scientists, but also affects heating and ventilating specialists.

Most national weather services can provide up-to-date data of these kinds. Estimates of future trends and changes are, however, harder to come by. They form the main subject of research in modern climatology.

Will the present climate change? There is no sure evidence that it is now doing so. Detecting climatic change is like trying to tune in a very faint AM radio signal on a thundery day: the noise overwhelms the signal. Climate is so variable that it is very hard to detect the "signal" of lasting change through the "noise" of short-term variations. For example, temperature usually rises more between 10 A.M. and 12 N. on a summer day than annual mean temperature is believed to have changed in the past 10,000 years.

Of the external controls of climate which have been discussed, only the composition of the atmosphere (Section 7.2) and the solar constant (Section 7.3) seem capable of rapid change. The solar constant undergoes small short-term changes, on the order of 0.1 or 0.2 percent, and a striking 11-year sunspot cycle occurs that alters the ultraviolet and particle radiations reaching the earth. Most authorities feel, however, that these solar variations are too small to affect climate significantly. Three compositional changes are suspected of being more important: the buildup of carbon dioxide and certain other trace gases, in progress at least since 1880; a possible increase in the particle load; and the effects of a decrease in ozone in the stratosphere because of pollution.

The carbon dioxide (CO_2) buildup and the greenhouse effect have been dealt with in Chapter 5. It was shown that the CO_2 concentration in the atmosphere is expected to double in the mid or late twenty-first century, and that this may induce a general rise in world temperature of 2 to 3K. Striking effects on precipitation and ice distribution are also expected, but substantial changes in sea level are not anticipated. Most atmospheric scientists expect these changes to happen, but they have not yet been detected in the present climate. The CO_2 increase will strengthen the greenhouse effect described in Section 7.3.

A second possible influence on temperature is a change in the aerosol and particle load in the atmosphere. Unfortunately, we cannot be sure that the atmospheric aerosol or particle load is increasing, nor do we know much about the probable climatic effect of such an increase. One school says that an increase is in progress, due partly to human activity and partly to increased explosive volcanic eruptions such as those of El Chichón, the Mexican volcano that killed 2,000 people in 1982. Some climatologists believe that the net effect of such an increase will be a lowering of mean temperature, because of increased reflection of solar radiation back to space. Others challenge this view.

The ozone (O_3) effect on climate is more complex, and is only one of the possible environmental consequences of the decrease in ozone now in progress. The changes occurring in the ozone layer and their consequences are as follows:

1. Ozone is created in the stratosphere when "hard" solar ultraviolet radiation of wavelength less than 242 nm breaks up molecular oxygen into atoms, which can then combine with O_2 to form O_3. This process is continuous in sunlight, and would lead to a progressive increase in O_3 in the stratosphere unless balanced by processes of decay.

2. These decay processes include the breakup of ozone by longer wave (> 320 nm) ultraviolet radiation and visible light; chemical combination of ozone with oxygen; and interaction with oxides of odd nitrogen, which are derived from nitrous oxide (N_2O) diffusing upward from the soil. The last process accounts for 60–70 percent of the required destruction.

3. The balance between (1) and (2) creates a continuous **ozone layer** in the stratosphere. The ozone strongly absorbs solar radiation, leading to the high temperature of the stratopause and also robbing the downward solar beam of its power that might otherwise heat the earth's surface. Moreover, it removes the more damaging wavelengths of ultraviolet, thereby protecting exposed flesh and plant tissues. Hence the term "ozone shield."

4. Pollutants now being released include certain gases that can upset the aforementioned chemistry. In particular, the chlorofluoromethanes (CFMs) now referred to as chlorofluorocarbons, or CFCs, used in spray-can propellants and as refrigerants, all eventually accumulate in the atmosphere. They are quite stable at low heights, but are broken up in the stratosphere, releasing chlorine, which attacks ozone. Since 1978 remarkable decreases in ozone concentrations have been observed over Antarctica which may be due to the CFC effect. The continued release of CFCs at current levels would significantly reduce the ozone protective layer, thus increasing the ultraviolet flux at ground level and the incidence of skin cancer.

5. A decrease in O_3 concentration would cool the stratosphere and warm the earth's surface, but the effect would be masked by other changes. The main impact on climate would be due to the greenhouse effect of the CFCs, which absorb and emit long-wave radiation much as CO_2 does. One estimate is that the effect of the CFCs and other pollutants on the atmosphere will be to reinforce the expected CO_2 warming—perhaps doubling its effect.

The combined impact of the CO_2 and pollutant buildups may therefore be to raise world temperatures appreciably in the next century, perhaps at a rate of 0.04 to 0.06K per annum. This is much faster than the climatic fluctuations recently observed. It will have considerable effect on world agriculture (mainly because of associated rainfall changes) and on sea-ice distribution. Navigation will become easier in the arctic seas. Efforts are now being made internationally, under the UN-sponsored World Climate Program, to see how important these effects will be.

7.6.3 The Climatic System

Climate interacts with soil, rock, plants, animals, surface water, and ice. The **climatic system** is the climatologist's name for this interaction. Biologists talk of **ecosystems,** by which they mean the relation of living organisms and communities to their total physical environment. Whichever term is used, one must try to see the linkages. It is exchange, disturbance, recovery, and lasting change within these systems that are at the root of environmental science. Environmental engineers and scientists who forget that things are interconnected court environmental disaster.

Our climate and the oceans are closely linked. Water, energy, carbon dioxide, and particles are exchanged between sea and air at prodigious rates. Actually, most of the exchanges take place within the water layer near the ocean surface—a layer less than 100 m deep. The deeper ocean waters are largely isolated from the atmosphere, taking as much as 1,000 years to exchange water and materials with the ocean surface. Ocean currents are in part driven by the winds, which exert a powerful drag force. They transport large amounts of heat. Without such transport, world temperature contrasts would be much greater.

Similar links exist between climate and the great continental glaciers that cover Greenland and Antarctica. For many millennia, much of the snow that has fallen on these land areas has accumulated as glacial ice, which now lies several kilometers deep over their central areas. The sun and atmosphere are unable to deliver enough energy to melt the ice. Antarctica discharges most of its surplus ice into the ocean as gigantic bergs or floating shelf ice. Greenland loses about half its annual surplus as meltwater, and the other half as icebergs that drift down toward Newfoundland. In each of the glacial epochs of the Quaternary period, similar ice sheets covered North America and northern Europe. The most recent vanished 6,000 to 9,000 years ago. But there is little prospect that the predicted warming of climate in the next century will do likewise for Greenland and Antarctica. Sea level is unlikely to rise more than one meter in the next century, because any marginal melting of the ice will be compensated, fully or in part, by increased snowfall on the upper slopes of the glaciers. Melting of sea ice leaves the sea level unaffected.

7.6.4 Urban Climates

Engineers and meteorologists do much of their work in cities or heavily industrialized areas. The climates of such areas differ from those of open country in many ways. Also, serious air pollution (dealt with in Chapter 13) is mainly found in cities.

The climates of cities are modified by several factors:

1. Cities are rougher than open country, so that wind flowing across them is made more turbulent by contact with obstacles like buildings or power lines.
2. The surface materials of cities are quite unlike natural soil or vegetation. There are many tall concrete, brick, or steel structures. Vegetation, which in nature pumps much water into the atmosphere, keeping the plants cool, is absent over much of the area.
3. In the city, the heat and water balances are changed from what they are in the countryside. The buildings, streets, parking lots, and industrial plants of a city have quite different properties from open country as regards (a) storage of heat, (b) storage of water, (c) absorption of solar radiation, and (d) all components of the hydrologic cycle, viz., evaporation, percolation, runoff, and water storage.
4. In addition, cities release a great deal of heat to the atmosphere from furnaces, automobiles, and other fuel-consuming activities.

A direct consequence of these factors is that the boundary layer over a city is quite unlike that over the surrounding country. Figure 7–14 depicts the main effects. The modified boundary layer over the city forms a sort of dome, which is higher by day than by night. At night a strong inversion may occur, trapping the pollutants. Most of the pollution released by the city stays in this dome, although some is carried off by the wind. The various transfers of water and energy in Figure 7–14 carry the same symbols as in the heat and water balance equations (7.3 and 7.13). The main differences are that:

1. During both day and night there is an extra heat source in the city, marked F, which is due to the use of fuel and mechanical energy in buildings and vehicles.
2. The structures of the city have a high heat storage capacity. They warm up by day as the sun strikes them, and release the heat at night.

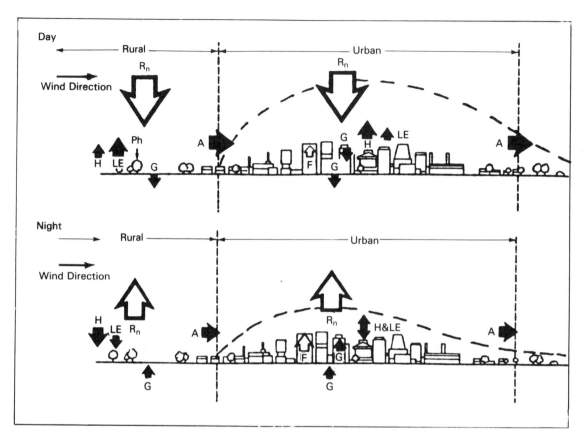

Figure 7–14 The urban boundary layer by day and night. Source: Oke, 1978.
Schematic representation of the urban and rural heat balances. Directions of *H* and *LE* at night are uncertain. Dashes indicate dome of urban boundary layer.

The result is that the city is usually warmer than the country. Figure 7–15 shows, as an example, the mean temperature on twelve nights at Winnipeg, Manitoba, a city of 650,000 population. The temperatures are shown as deviations from those measured at the airport on the edge of the built-up area. Over the central business district, temperatures are on the average 5–6K above those in the open country. The lines of equal temperature differences clearly follow the outline of the city. The warm city area is called an *urban heat island*. All large cities have them. In extreme cases, temperatures in a city center at night may be over 10K warmer than the surrounding country.

Heat islands form most readily in calm weather. Strong winds tend to disperse them, and with the heat the pollutants, too. A windy day gives the city much the same weather as the surrounding country. Interference by high buildings makes the wind even gustier in

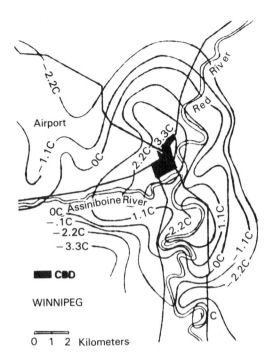

Figure 7–15 The urban heat island at Winnipeg, Manitoba. Source: Hare and Thomas, 1979

The urban heat at Winnipeg, based on mean deviation from airport temperature during twelve experimental runs (°C). "CBD" means Central Business District.

the city. All of us have observed the violent wind gusts in the city center near skyscrapers. With light winds, however, the buildings have the opposite effect. The air over the inner city is trapped by the obstacles and remains in the urban canyons. Pollutants accumulate, and, if fog forms, the pollutants combine with the water or ice droplets to form smog.

As cities have grown bigger, their heat islands have become more intense. The bigger the city, the bigger is the heat island. For city dwellers, this has been a real climatic change. In Paris, France, for example, mean annual temperature rose almost 1K between 1880 and 1965. Mean annual temperature in Tokyo, Japan, rose by 1.4K between 1915 and 1970. These effects are not surprising if one remembers that energy use in a city may compare with natural radiative heating. Table 7–3 shows the power used per unit area in selected cities. Note that the figures shown are not in all cases for whole cities. For example, Manhattan is a borough of New York City, and the data for Sydney are for its inner area only. Observe that in Manhattan and Moscow the artificial heating exceeds the natural annual net radiation, i.e., the sum of all the natural inputs and outputs of power by radiation of various kinds. In Manhattan, the man-made power is actually about eight times the natural source.

From experience, we find that each city is different from its neighbors. Differences of topography, for example, have marked effects. Many cities, such as Los Angeles, Vancouver, Fairbanks, and Milan, are built in valleys or basins that tend to trap the urban boundary layer within walls of high ground. These areas have much more severe air pollution

TABLE 7–3   POWER USE PER UNIT AREA, SELECTED CITIES

City	Population (millions)	Area (km²)	Power use per unit area (W m⁻²)
Manhattan (N.Y.)	1.7	234	630
Moscow (U.S.S.R.)	6.4	878	127
Sydney (Australia)	0.1	24	57
West Berlin (F.R.G.)	2.3	233	21
Los Angeles (Calif.)	7.0	3,500	21

Source: Landsberg, 1981

problems than do cities in open sites, like Chicago, Illinois. Even quite shallow basins can produce marked effects on calm nights. Cities like London and Paris are examples of this.

PROBLEMS

7.1 Warm air may be as dense as cold air. What pressure is needed to make air at 30°C (303K) as dense as air at 0°C (273K) whose pressure is 1,000 mbar?

7.2 Solar radiation at noon on a particular day is measured to be 900 W m⁻². The surface is short grass and has a measured temperature of 15°C (288K). The downward long-wave radiation from the atmosphere is measured at 87% of the upward long-wave radiation from the surface. Calculate the probable value of the net radiation.

7.3 Using the hydrostatic equation (Equation 7.5), calculate the rate of decrease of pressure with height at a pressure of 700 mbar if the temperature is −3°C (270K). Hint: Assume that dz is a layer 10 m thick.

7.4 The pressure gradient on a particular occasion is 10 mbar/1,000 km. Calculate a probable value for the wind speed at 600 meters above the ground. (Assume that the air temperature is +7°C (280K) and the air pressure is 1,000 mbar. Latitude is 30°.)

7.5 Air with a relative humidity of 50% at 12°C (285K) is heated without change of pressure to 20°C (298K). What will be its specific humidity at the new temperature?

7.6 What is meant by the periodic variation of a climatic element? What examples can you give of such variation? Use your own experience as well as the text.

7.7 Using library sources, locate the world's main desert regions. What can you say about their distribution? Can you suggest a reason for their dryness?

7.8 Near the North Pole, winter temperatures fall quickly to −35°C (238K) and then remain near that level, even though several months of complete darkness remain. Can you suggest why the temperature does not continue to fall?

7.9 Prepare a rough checklist of design problems in the field of civil engineering in which weather data are likely to be important. Where would you get these?

7.10 Suppose that there is a long series of major volcanic eruptions over a period of several years. What effect might such a series have on world climate?

LIST OF SYMBOLS AND UNITS

Symbol	Meaning	Units	Numerical values
a	albedo (fractional reflectivity)	dimensionless	
c_p	specific heat of air at constant pressure	kJ kg^{-1} K^{-1}	1.0
e	vapor pressure of water	N m^{-2} (Pascal)	
e_s	saturation vapor pressure of water	N m^{-2} (Pascal)	
E	evaporation or evapotranspiration per unit time	kg m^{-2} (mm depth of water)	
F	release of heat by city fuel burning	W m^{-2}	
G	soil heat flux	W m^{-2}	
H	turbulent (convective) heat flux	W m^{-2}	
I	solar radiation flux	W m^{-2}	
L	latent heat of vaporization of water	J kg^{-1}	2.44×10^6
M	latent heat of melting ice	J kg^{-1}	3.33×10^5
n	distance (horizontal)	m	
N	runoff or water surplus per unit time	kg m^{-2} (mm depth of water)	
p	atmospheric pressure	mbar (mean sea level averages)	1013.25
P	precipitation per unit time	kg m^{-2} (mm of precipitation)	
q	specific humidity	g kg^{-1} (dimensionless)	
Q	photosynthetic energy conversion	W m^{-2}	
r	relative humidity	percent of saturation (dimensionless)	
R_n	net radiation flux	W m^{-2}	
R	universal gas constant (dry air)	J kg^{-1} K^{-1}	287.0
$R \downarrow$	long-wave radiation flux from atmosphere	W m^{-2}	
s	de_s/dT	mbar K^{-1}	
S	snowfall per unit time	kg m^{-2} (mm depth of snow)	
T	temperature	K (Kelvin)	
V	wind velocity or speed	m s^{-1}	
V_g	geostrophic wind	m s^{-1}	
V_s	surface wind	m s^{-1}	
x	humidity mixing ratio	g kg^{-1} (dimensionless)	
z	height	m	

GREEK SYMBOLS

Symbol	Meaning	Units	Numerical values
α	parameter in Equation 7.12	dimensionless	
γ	psychrometric constant	mbar K^{-1}	0.64
ϵ	emissivity of surface	dimensionless	
μ	micro	—	1/1,000
ϕ	latitude	deg or rad	
ρ	air density	kg m^{-3}	
σ	Stefan-Boltzmann constant	W $m^{-2} K^{-4}$	5.6×10^{-8}
ω	rate of rotation of earth	rad s^{-1}	7.3×10^{-5}

Note: The S.I. system is used where possible. For lengths (including wavelength), measures are in meters (m), micrometers (μm) or nanometers (nm). Precipitation, runoff, and evaporation are usually measured in terms of depth accumulated (mm). Snowfall is usually measured in mm of snow melt; if measured fresh, the centimeter (cm) is preferred. Snow melt has a mean density of 0.1; hence, 1 cm of snow is roughly equal to 1 mm of snow-melt water.

REFERENCES

BRUCE, J. P., and CLARK, R. N. *Introduction to Hydrometeorology*. Oxford Eng: Pergamon Press, 1966.

FLEAGLE, R. G. and BUSINGER, J. A. *An Introduction to Atmospheric Physics*. New York: Academic Press, 1963.

HALTINER, G. J., and MARTIN, F. L. *Dynamical and Physical Meteorology*. New York: McGraw-Hill, 1957.

HARE, F. K., and THOMAS, M. K. *Climate Canada*. 2d ed. Toronto: Wileys Publishing of Canada, 1979.

IRIBARNE, J. V., and GODSON, W. L. *Atmospheric Thermodynamics*. Dortmundt FRG: Reidel, 1973.

JUNGE, C. E. *Air Chemistry and Radioactivity*. New York: Academic Press, 1963.

LANDSBERG, H. *The Urban Climate*. New York: Academic Press, 1981.

MATHER, J. R. *Climatology, Fundamentals and Applications*. New York: McGraw-Hill, 1974.

MONTEITH, J. L. *Principles of Environmental Physics*. London: Edward Arnold, 1973.

OKE, T. R. *Boundary Layer Climates*. London: Methuen, 1978.

SELLERS, W. D. *Physical Climatology*. Chicago: University of Chicago Press, 1965.

World Meteorological Organization. *Proceedings of the World Climate Conference*. Geneva: World Meteorological Organization, 1979.

8

Microbiology
and Epidemiology

Gary W. Heinke

8.1 INTRODUCTION

Although the word "health" does not appear in the title, this chapter deals with health—
the protection of human health from environmental influences. Epidemiology—the science
concerned with the study of epidemics—was the basis for environmental sanitation and
preventive medicine for the past century and a half and is worthy of a brief discussion here.
Because of the great importance of microorganisms in environmentally transmitted diseases
of humans, and because of their importance in ecology and in the technology of environmental control, microbiology—the study of microorganisms and their activities—is also
introduced in this chapter.

   Microbiology (Greek **micros**, "small," **bios**, "life," and **logos**, "study of") is the
study of microorganisms and their activities. Environmental or sanitary microbiology concerns itself with microorganisms commonly found in water, wastewater, air, and in some
cases soil that may affect public health, decompose organic matter, or perform a useful
function.

   Epidemiology (Greek **epi**, "upon," **demos**, "people," and **logos**, "study of")
means "the study of what has come upon the people"; taken in the context of disease, it
means the study of the causes of disease among a population. **Epidemic** describes the
widespread outbreak of an infectious disease in a community. **Endemic** refers to diseases
which are continuously present in a particular population. Since the objective of epide-

miological studies is to control the spread of disease, the determination of the etiologic agent (that which causes the disease) and the mode of transmission of the disease are of prime importance for successful control.

Only recently have we realized that many noninfectious diseases are caused by the toxic substances in industrial wastes. Both inorganic and organic contaminants are implicated, and long-term epidemiological studies are needed to determine the ''safe'' concentrations and exposure times that can be tolerated without adverse environmental effects.

Endemic refers to a disease prevalent in, and confined to, a particular population.

An **epidemic** is an outbreak of an infectious disease spreading widely in an area.

Epidemiology is the study of the causes of a disease in a community.

Microbiology is the study of microorganisms and their activities.

8.2 FUNDAMENTALS OF MICROBIOLOGY

8.2.1 Classification of Microorganisms

Most living things were originally classified as belonging to either the plant kingdom or the animal kingdom. However, many microorganisms did not fit unequivocally into either of these two categories, and Haeckel proposed in 1866 that a third kingdom be recognized—the **protista.**

The protista included protozoa, algae, fungi, and bacteria (viruses were unknown in 1866). With advances in knowledge of cell ultrastructure, the Protista were further subdivided into two categories: the Higher Protista (the **eukaryotes**), consisting of either unicellular or multicellular organisms that have a true nucleus, and the Lower Protista (the **prokaryotes**), consisting of organisms which have no true nucleus. In the prokaryotes, which include only bacteria and blue-green algae, the genetic material of the cell—the DNA—is not organized into structures recognizable as chromosomes and is not separated from the cytoplasm by a nuclear membrane. The blue-green algae are now generally referred to as blue-green bacteria or **cyanobacteria;** thus, ''prokaryotes'' and ''bacteria'' are synonymous terms. Protozoa, algae, and fungi are grouped as eukaryotic protists. Viruses, which are noncellular, are included in neither of the above groupings. Based on this classification, microorganisms can be grouped as the eukaryotic protists, the prokaryotes, and the viruses (Gaudy and Gaudy, 1980).

Bacteria are the most important group of microorganisms. They are essential to the nutrient cycle of the ecosystem. Pathogenic (disease-causing) bacteria have received the most attention, and are discussed further in the section on epidemiology. Many other bacteria are important in water and wastewater treatment processes, in the natural self-purification of streams and lakes, and in the decomposition of materials in landfills, soils, and compost heaps. **Viruses,** which are smaller than bacteria, may also cause diseases in plants and animals as well as in humans.

Algae are a group of photosynthetic plant-like microorganisms. They can cause problems in water supplies by imparting tastes and odors and by clogging filters. They are beneficial in oxidation ponds, providing oxygen for low-cost wastewater treatment. On the other hand, excessive amounts of nutrients in water can lead to algal blooms which, when they decompose, remove dissolved oxygen from lakes. The process of nutrient enrichment called eutrophication is discussed in Chapter 9.

Fungi are unicellular or multicellular nonphotosynthetic protists which are able to survive under low pH conditions. They are useful in the biological treatment of some industrial wastes and in the composting of solid organic wastes.

Protozoa are generally an order of magnitude larger than bacteria and are useful in the biological treatment processes discussed in Chapter 12.

Rotifers are multicellular microorganisms which are sometimes present in the effluent of biological waste treatment plants. They perform a ''polishing'' function by consuming organic colloids, bacteria, and algae.

Crustaceans are multicellular organisms with a hard body or shell. Some of them are microscopic in size and serve as food for fish. They are considered indicators of normal, unpolluted conditions in receiving waters. Figures 8–1 and 8–2 show some of the afore-mentioned microorganisms.

According to **microbial nomenclature,** microorganisms are given two names to indicate their *genus* (plural: genera) and *species*. For example, *Escherichia coli* is the combination of two names: *Escherichia* indicates the genus and *coli* the species. The generic name is written with a capital letter and the species name with a small letter.

8.2.2 Bacteria

Bacteria (singular:.bacterium) are unicellular microscopic organisms. They are found in water, wastewater, soil, air, and milk, on plants (fruits, vegetation), animals, and humans (skin, intestinal tract). Bacteria reproduce by binary fission and are characterized by their shape, size, structure, and arrangement of cells. Individual bacteria have one of three general shapes: spherical (called cocci—singular, coccus), cylindrical or rod-like (called bacilli—singular, bacillus), and spiral-shaped (spirilla—singular, spirillum). Bacterial cells may be arranged in groups such as pairs, clusters, or chains (Figure 8–1(b) and (c). Some examples of important bacteria in the environmental field are listed in Table 8–1. Most bacteria range in size from 0.5 to 5.0 μm long and 0.3 to 1.5 μm wide. Cocci are about 0.1 μm in diameter.

Figure 8–3 is a schematic diagram of a typical bacterial cell. All bacteria have a rigid cell wall which maintains the shape of the cell and protects the contents from osmotic pressure. If the cell wall were removed, the cell would quickly collapse or burst from the pressure of its contents. The wall is usually 0.02 μm to 0.03 μm thick and accounts for 10 to 40 percent of the dry weight of the organism.

Some bacteria are covered by a layer of viscous substance called the **capsule** or slime

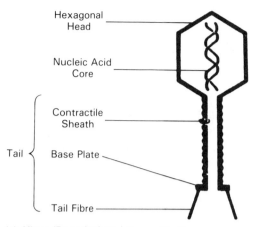

- Hexagonal Head
- Nucleic Acid Core
- Tail
  - Contractile Sheath
  - Base Plate
  - Tail Fibre

(a) **Virus** (Bacteriophage) Source: Ward's Natural Science Establishment, Inc. Rochester, N.Y. 1964.

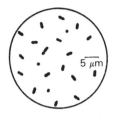

(b) **Bacteria** *Streptococcus pneumoniae*. One of the causative agents of pneumonia showing the typical arrangement of pairs of spherical bacteria cells. Size ranges from 0.5 to 1.25 μm in diameter. Source: Pelzar, M. J. and E. C. S. Chan, *Elements of Microbiology*, New York: McGraw Hill, 1981.

(c) **Bacteria** *Salmonella typhi*. The causative agent of typhoid are typical rod-shaped bacteria (bacilli). Source: Pelzar, M. J. and E. C. S. Chan, *Elements of Microbiology*, New York: McGraw Hill, 1981.

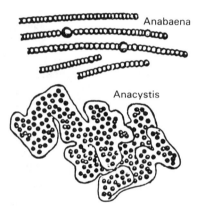

Anabaena

Anacystis

(d) **Algae.** Two forms of algae suspected to be responsible for tastes and odors in drinking water. Source: Palmer, 1959.

(e) **Fungi** (Mycelium). Source: Buckman and Brady, 1960.

Figure 8–1   Virus, bacteria, algae, and fungi

(a) **Protozoan** *Vorticella*. A protozoan
covered with hairlike cilia. Source: Ontario
Ministry of the Environment, "Activated Sludge
Process Workshop Manual," Fifth Edition (2nd
Revision) Ministry of Government Services,
Publication Centre, Toronto, August, 1978.

(b) **Rotifer.** A multi-cellular animal that
feeds on bacteria and organic matter.
Two rows of cilia surround the head of
the organism and appear to be rotating
as they sweep food into the oral cavity.
Source: Clark, Viessman, and Hammer, 1977.

Daphnia   *Cyclops*

(c) **Crustaceans.** Very small microscopic multicellular organisms with hard shells.
They feed on other microorganisms and organic matter and are in turn food for
small fish. Source: Clark, Viessman, and Hammer, 1977.

Figure 8–2   Protozoan, rotifer, and crustaceans

TABLE 8–1 SOME BACTERIA OF SIGNIFICANCE IN THE ENVIRONMENT

Group of Bacteria	Genus	Environmental Significance
Pathogenic bacteria	*Salmonella*	Cause typhoid fever
	Shigella	Cause dysentery
	Mycobacterium	Cause tuberculosis
Indicator bacteria	*Escherichia*	Fecal pollution
	Enterobacter	"
	Streptococcus	"
	Clostridium	"
Decay bacteria	*Pseudomonas*	Degrade organics
	Flavobacterium	Degrade proteins
	Zooglea	Floc-forming organism in activated sludge plants
	Clostridium	Produce fatty acids from organics in anaerobic digester
	Micrococcus	" " " " " " " "
	Methanobacterium	Produce methane gas in anaerobic digester from fatty acids
	Methanococcus	" " " " " " " " "
	Methanosarcina	" " " " " " " " "
Nitrifying bacteria	*Nitrobacter*	Oxidize inorganic nitrogenous compounds
	Nitrosomonas	" " " "
Denitrifying bacteria	*Bacillus*	Reduce nitrate and nitrite to nitrogen gas or nitrous oxide
	Pseudomonas	" " " " " " " " " "
Nitrogen-fixing bacteria	*Azotobacter*	Capable of fixing atmospheric nitrogen to NH_3
	Beijerinckia	" " " " " " "
Sulfur bacteria	*Thiobacillus*	Oxidize sulfur and iron
	Desulfovibrio	Involved in corrosion of iron pipes
Photosynthetic bacteria	*Chlorobium*	Reduce sulfides to elemental sulfur
	Chromatium	" " " " "
Iron bacteria		
filamentous	*Sphaerotilus*	Responsible for sludge bulking in activated sludge plants
iron oxidizing	*Leptothrix*	Oxidize ferrous iron

layer. It is believed that capsular material is excreted from the cell, but because of the viscosity of the slime, it is not readily diffused away. The presence of capsules on some pathogenic bacteria increases their infective capacity. Loss of the capsule in some other cases results in loss of the ability to cause disease. Capsules have also been blamed for the production of slimes in some industrial processes.

Many bacteria are **motile,** or capable of rapid movement in liquids by rapid lashing of one or more of their whip-like **flagella.** These are long thread-like appendages found mainly on bacilli. There may be one or more flagella attached at one end of the cell, or there may be many distributed along the length of the bacillus. Bacteria without flagella are usually nonmotile. The existence and form of flagella help to differentiate between various bacterial groups.

Immediately beneath the cell wall is the semipermeable **cytoplasmic membrane** (about 7.5×10^{-3} μm thick). It serves the very important function of providing a semi-

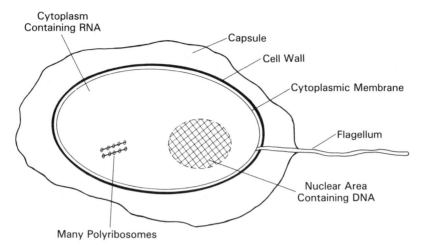

Figure 8–3   A schematic diagram of a typical bacterial cell

permeable boundary separating the protoplasm from the external environment while allowing the passage of nutrients into the cell and waste products out. Damage to this membrane by chemical or physical agents can cause the death of the cell.

The protoplasm, or the internal contents of the cell, can be divided into three different areas: the cytoplasm, the nuclear area, and the polyribosomes. The **cytoplasm** is granular in appearance, due in part to the abundance of RNA. The fluid portion of the cytoplasm contains dissolved nutrients. The **nuclear area** contains the DNA or chromatin which is diffused throughout the cell in prokaryotes.

RNA (ribonucleic acid) is a long-chained, single-helix molecule containing phosphoric acid, D-ribose (a sugar), adenine, guanine, cytosine, and uracil. It is indispensable for the biosynthesis of protein, helping to arrange the order of the amino acids that make up the specific proteins required by the cell. In conjunction with protein, RNA forms densely packed particles called **polyribosomes.** These produce the enzymes which are complex organic catalysts, generally specific to each biochemical reaction.

DNA (deoxyribonucleic acid) is a very long-chained, double-helix molecule occurring in the nuclear area of all cells. It contains phosphoric acid, 2-deoxy-D-ribose (a sugar), the purine bases adenine and guanine, and the pyrimidine bases cytosine and thymine. Although the DNA of a bacterial cell is diffused and not contained in a nuclear membrane, it is confined to certain areas within the cell, and these can be considered a primitive form of nucleus. DNA is responsible for the genetic stability of the species.

Some bacteria, for example *Bacillus* and *Clostridium*, form **spores. Spores** represent a dormant or resting phase of the cell and in this respect are analogous to the seeds of plants. A normal, active cell is called a vegetative cell. Bacteria capable of forming spores may exist as vegetative cells for many generations. When the cell is exposed to adverse growth conditions, spores may form within the cytoplasm. The spore can be smaller or larger than the vegetative cell and may occur at the end of the cell or near the center. These features are useful in characterizing spore-forming bacteria.

Spores are extremely resistant to adverse chemical or physical environments. Spore-forming bacteria are common in the air, soil, and water. Their resistance is due to an impermeable spore wall made up of a dipicolinic acid-calcium complex and also to the dehydration of the cell contents. Under conditions conducive for growth, the spore germinates and a new vegetative cell emerges. This survival ability makes spore-forming bacteria difficult to destroy, but is of obvious benefit to the bacterium.

8.2.3 Growth and Death of Bacteria

All living organisms have nutritional and physical requirements that must be met in order to sustain their life. Among the many species of bacteria, there are wide variations in nutritional requirements and the physical conditions they can withstand. Certain bacteria grow at temperatures below 0°C, others at temperatures as high as 99°C. Some bacteria require atmospheric oxygen, whereas others are hindered by its presence.

Bacteria are divided into two broad groups with respect to their energy and carbon sources: **heterotrophic** and **autotrophic.** Heterotrophic bacteria obtain both their energy and carbon from an organic compound or organic matter. Autotrophic bacteria require carbon dioxide as their carbon source and obtain their energy from sunlight or by the oxidation of inorganic materials. If autotrophs require sunlight as their energy source, they are called **photoautotrophs.** If they obtain their energy by oxidizing inorganic chemical compounds, they are called **chemoautotrophs.**

In addition to carbon, other nutrient requirements include nitrogen, sulfur, phosphorus, and traces of metallic elements such as magnesium, calcium, and iron. Again, bacteria vary widely in the way they obtain these nutrients. Some bacteria can ''fix'' or obtain nitrogen from the atmosphere, others obtain nitrogen from inorganic sources such as ammonia or nitrates.

A number of bacteria are very specific in their nutrient requirements, whereas others are able to utilize a variety of sources for their needs. For example, *Escherichia coli* can manufacture their own vitamin requirements from other compounds, but the **lactobacilli** will not grow unless specific nutrients are immediately available. The latter organisms are called **fastidious heterotrophs.**

Many factors affect bacterial growth. The major physical factors are temperature, the gaseous environment, and pH. Bacteria can be grouped according to the temperature range within which their growth occurs. **Psychrotrophs** are those bacteria which can grow at temperatures as low as 0°C and up to 25–30°C. Those psychrotrophs which have an optimal temperature of 15°C or lower and a maximal temperature for growth at about 20°C are called **psychrophiles. Mesophiles** grow best within a temperature range of 30–40°C, while **thermophiles** can grow at temperatures up to 99°C and have temperature optima of 40°C or higher.

The most important gases involved directly in bacterial growth are oxygen for aerobic biological oxidation and carbon dioxide as a source of carbon for autotrophs. Because of the importance of oxygen, it is often useful to divide bacteria into the following groups on the basis of their need for oxygen:

- Aerobic bacteria require free oxygen for growth.
- Anaerobic bacteria can grow without free oxygen.
- Facultative bacteria can grow with or without oxygen.
- Microaerophilic bacteria grow in the presence of minute quantities of free oxygen.

The adjectives **facultative** and **obligate** describe the degree of dependence on a particular condition. For example, an obligate anaerobe is a bacterium that will not grow at all in the presence of free oxygen. A facultative autotroph is an organism that normally uses CO_2 as a source of carbon, but can also grow heterotrophically with organic compounds as energy sources.

The third major factor influencing bacterial growth is pH. Most bacteria exhibit optimum growth at a pH range from 6.5 to 7.5, with maximum limits for growth between pH 4.0 and pH 10.0. The metabolic activities of bacteria can cause shifts in the pH of their environment. Therefore, the environment must have a buffering capacity in order to neutralize these shifts if growth is to continue for an extended period of time. Other physical conditions are important for some species of bacteria. For example, phototrophic bacteria require light as their source of energy and a few bacteria require unusually high salt concentrations, like those found in locations such as the Dead Sea or Utah's Great Salt Lake. All bacteria require moisture for growth, since all nutrients must be dissolved in order to penetrate the cell membrane.

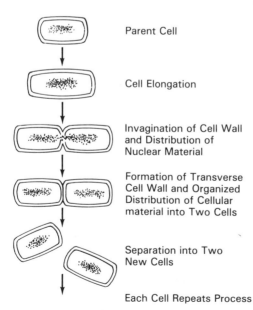

Parent Cell

Cell Elongation

Invagination of Cell Wall
and Distribution of
Nuclear Material

Formation of Transverse
Cell Wall and Organized
Distribution of Cellular
material into Two Cells

Separation into Two
New Cells

Each Cell Repeats Process

Figure 8–4 Binary fission of bacterial cells. Source: Pelczar, Reid, and Chan, 1977

Growth and reproduction of the cell occur as nutrients are taken into the cell and processed into new cell material. Nuclear material is reproduced and distributed in the cell; a cell wall or septum develops which divides the bacteria and separates it into two viable cells. The reproductive process of bacteria called **binary fission** is a characteristic of bacterial growth (Figure 8–4).

Bacterial populations can reach high densities very quickly. The individual cells double at a rate characteristic for each organism under a given set of conditions. This time interval is known as the generation time. Generation times at 20°C range from 15 to 20 minutes for *Escherichia coli* to several hours for other species (11 hours for *Nitrosomonas europae*). Although bacteria can grow in a wide variety of conditions, optimum growth requires a specific environment for each species.

The rate of growth of a bacterial population is directly proportional to the number of bacteria present. This can be expressed mathematically as

$$\frac{dB}{dt} = kB \tag{8.1}$$

where dB/dt = rate of growth
  B = concentration of bacteria at time t
  k = first-order growth-rate constant.

Integrating Equation 8.1 yields

$$\ln \frac{B}{B_0} = kt \tag{8.2}$$

where B_0 is the initial population concentration.
If G is the generation (doubling) time then $B = 2B_0$, and Equation 8.2 can be rewritten as

$$k = \frac{\ln 2}{G} \tag{8.3}$$

Substituting this value of k into Equation 8.2, we can express the bacterial population B as a function of time t as

$$B = B_0 \cdot 2^{t/G} \tag{8.4}$$

Taking the logarithm of this equation, we obtain

$$\log B = \log B_0 + \left(\frac{t}{G}\right) \log 2 \tag{8.5}$$

A plot of B against t on semilog paper would produce a straight line with a slope of $0.3/G$ ($\log 2 = 0.3$) and a y-intercept of B_0.

This type of logarithmic growth is in reality typical of only a small portion of the normal growth pattern of a bacterial population in a batch culture, as shown in Figure 8–5. There is an initial period of what appears to be no growth, followed by rapid growth and then stabilization at a maximum amount ending in a declining or death phase.

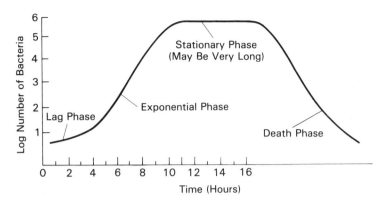

Figure 8–5 Typical bacterial growth curve. Source: Mitchell, 1974

During the initial period, called the **lag phase,** the cells adjust to their new environment. They may be deficient in certain enzymes or coenzymes required to metabolize the surrounding nutrients. These enzymes, therefore, have to be synthesized by the cells. The individual cells also increase in size beyond their normal limits as new protoplasm is developed. When this adjustment period is complete, the cell can divide and start reproducing normally. At the end of the lag phase, there is a gradual transition into the exponential phase as described previously. This is the **log** (logarithmic) or **exponential growth** phase in which the population doubles at regular intervals. It is the period of most rapid growth under optimal conditions. The bacterial population is the most uniform during this period in terms of chemical composition, metabolic rates, and other physiological characteristics. This phase of rapid growth obviously cannot continue indefinitely because of food limitation and the cells start to die off. This results in a decrease in the growth rate until zero growth is achieved.

When the number of new cells being produced equals the number of cells dying, a dynamic equilibrium is reached at which there is no further increase. This is called the **stationary phase.** The reason for cessation of the growth phase is usually due to the exhaustion of one or more nutrients.

The **death** or **declining** phase is reached when the death rate starts to exceed the growth rate. In addition to the depletion of nutrients, toxic by-products of cell metabolism can build up in the environment, inhibiting further growth.

In continuous biological waste treatment processes (Chapter 12), the bacterial population degrading the waste organic matter is predominantly in the stationary to declining phase.

Just as conditions can be created for the optimum growth of bacteria, so unfavorable conditions can be used to eliminate bacteria. Complete destruction of microbial life is called **sterilization. Disinfection,** on the other hand, implies the destruction of pathogens (disease-causing organisms) only. The simplest means of sterilizing a small amount of material is to heat it to a temperature at which the cell protein is destroyed. Living organisms are destroyed at 100°C. However, bacterial spores require a higher temperature for destruction.

At a temperature of 120°C, complete sterilization with steam at about 210 kPa (15 psig) is generally achieved in less than 20 minutes.

Microorganisms can also be destroyed by shortwave radiation (200–400 nm) or by high-frequency sound. Shortwave ultraviolet irradiation can be used to sterilize surfaces or large enclosed areas. The radiation destroys the cells' nucleic acids. Ultrasonic waves at frequencies in the range of 200,000 cycles per second can effectively rupture cell walls of bacteria. Usually, however, ultrasonic energy (**sonification**) is not employed to control microbial populations, but it is useful as a method for disrupting cells to extract their intracellular constituents.

Disinfection is most often achieved by the use of chemical bactericidal agents. Oxidizing chemicals such as the chlorine compounds, iodine, and ozone are very effective for killing microorganisms in water and wastewater. Chlorine and, to a lesser extent, ozone are the most widely used disinfection agents. These agents permanently destroy the cell or parts of it so that it can not reproduce even after the removal of the agent. They act by oxidizing the enzymes and other material in the cytoplasm. The nonspecific action of chlorine and ozone makes it unlikely that resistant bacterial strains will develop. The rate of disinfection is dependent on the nature of the disinfectant, the cell physiology, and the environment. Some important variables are the concentration of the disinfectant, the contact time between the microorganisms and the disinfectant, the temperature, and the pH. The application of disinfection in water treatment is discussed in Chapter 11.

8.2.4 Viruses, Algae, Fungi, and Protozoa

Other microorganisms of importance in environmental science and engineering include viruses, algae, fungi, and protozoa.

Viruses. The smallest of these range in size from 10 nm to 250 nm (1 nm = 10^{-3} μm). By comparison, the size of a small bacterium, *Salmonella typhi*, is 1,000 nm, or 1 μm. Viruses are unique in that they contain no internal enzymes and therefore cannot grow or metabolize on their own. They are obligate parasites, infecting the tissues of bacteria, plants, and animals, including humans. Some examples of human pathogenic viruses are those that cause smallpox, infectious hepatitis, influenza, and poliomyelitis.

Figure 8–6 is a sketch of mumps and influenza viruses. In general, viruses are composed of a nucleic acid core, either DNA or RNA, surrounded by a protein covering called a **capsid.** This capsid is made up of smaller units called **capsomeres.** A complete

Mumps

Influenza

Figure 8–6 Sketch of mumps and influenza viruses (80–120 nm in size). Source: Pelczar, Reid, and Chan, 1977 (Suggested by a drawing by R. M. Chapman Jr., Time, Nov. 17, 1961)

virus unit is called a **virion.** Viruses are formed according to geometric rules of symmetry. The types of protein making up the capsid help to distinguish viruses from each other. Each type of virus can infect only a specific type of host cell so that, for example, an animal viral disease cannot be transmitted to humans.

According to cell theory, viruses are not living organisms. They do, however, have the ability to reproduce or replicate themselves within their specific host cells. Because they are not really alive outside of a host cell, they survive for a long time between infections and can only be ''killed'' by alteration of their molecular structures.

The algae, fungi, and protozoa are much more complex and have more specialized structures than the viruses or bacteria.

Algae.   Except for the blue-green algae they have a discrete nucleus surrounded by a nuclear membrane and are therefore classified as **eukaryotic,** meaning ''having a true nucleus.'' They have thick cell walls.

Algae, therefore include members of both the higher and lower protista. Their size

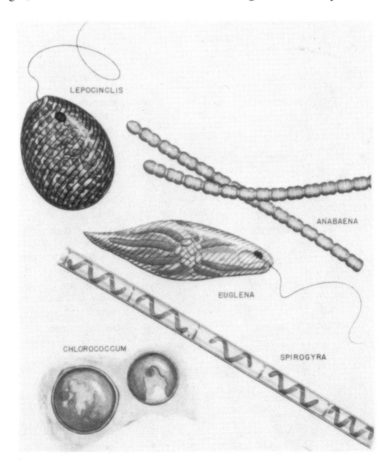

Figure 8–7 Some types of algae found in polluted water. Source: Palmer, 1959.

ranges from the microscopic unicellular phytoplanktons to the large multicellular seaweeds. The shapes of unicellular algae can be spherical, cylindrical, clublike, or spiral. Multicellular colonies can grow in filaments or long tubes or simple masses of single cells that cling together (Figure 8–7). The filamentous or tube-like growths can be branched or bundled together and may even contain cells that perform special functions. These appear superficially like the higher plants. Regardless of the size or complexity, all algal cells contain photosynthetic pigments and are thus capable of photosynthesis. The pigments are found in distinct bodies called **plastids, chloroplasts,** or **chromatophores.** Algae are important primary producers in the aquatic food chain, although they can be a problem in water supplies, since they contribute to tastes and odors, clog water intakes, shorten filter runs, and cause high chlorine demand. Excessive growth of algae, known as **algal blooms,** forms a blanket of organic material which interferes with the recreational use of waters. Algae are classified on the basis of their pigments. Seven general groups are shown in Table 8–2. Groups I, II, IV, and VII are of interest in the environmental field because of their appearance in both clean and polluted water. The others are mostly marine algae.

TABLE 8–2   CLASSIFICATION OF ALGAE

Division	Color	Environment/Cell arrangement/Comments
I. Chlorophyta	grass green	Fresh water; mainly clean-water algae except *Chlorella, Scenedesmus,* mostly colonial, filamentous.
II. Chrysophyta	yellow green	Clean, cold water; mainly cellular, some colonial. Diatoms have silica in cell walls.
III. Pyrrophyta	yellow brown	Mostly marine; 90% unicellular, two flagella.
IV. Euglenophyta	green	Fresh water; requires organic nitrogen; will grow as a protozoan in absence of light; unicellular, motility by flagellum.
V. Rhodophyta	red	Mostly marine; very clean, warm-water, colonial; sheets are common.
VI. Phyophyta	brown	Marine; cool-water; colonial, large. Example: *Macrocystis,* giant kelp.
VII. Cyanophyta	blue-green*	Fresh water, warm, often polluted; unicellular, gelatinous clumps; no chloroplasts or true nucleus; nitrogen fixers, often responsible for algal blooms.

*.Blue-green algae are now generally referred to as blue-green bacteria or cyanobacteria.

The hyphae form a mycelium on the surface and extract nutrients from it. New surfaces are attached by stolons. The spores are in a sporangium at the tip of a specialized hypha, called a sporagiophore.

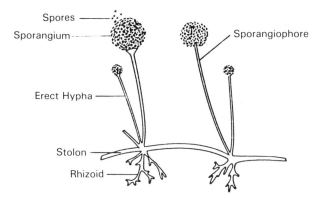

Spores

Sporangium

Sporangiophore

Erect Hypha

Stolon

Rhizoid

Figure 8–8   Sketch of a fungus.
Source: Mitchell, 1974

Fungi.   Fungi are nonphotosynthetic higher protists (eukaryotes), and may be divided into three groups: **molds,** which are filamentous fungi; **yeasts,** which are nonfilamentous fungi; and **mushrooms,** which are macroscopic fungi. Fungi (Figure 8–8) are typically aerobic and **saprophytic,** meaning that they feed on decaying organic matter. They are able to use a wide range of complex organic substances as food sources and are much more tolerant of acidic conditions than most other microorganisms. They reproduce by either sexual or asexual spores.

Molds grow by extending long thread-like structures called **hyphae,** which form a mass called **mycelium** (plural: mycelia). The vegetative mycelium penetrates into the substrate to absorb dissolved nutrients, while the reproductive mycelium forms reproductive structures (spore sacs, spores, etc.).

Yeasts are unicellular, considerably larger than bacteria (1 to 5 μm in width and 5 to 30 μm in length), and generally egg-shaped, spherical, and ellipsoidal cells that are widely distributed in nature. They reproduce asexually by binary fission or by budding. Sexual reproduction is by the formation of ascospores. Unlike molds, yeasts are facultative, that is, they can grow both aerobically and anaerobically. Yeasts are used in a wide variety of fermentative processes (for making wine, beer, and bread) and for synthesis of certain vitamins, fats, and proteins from simple sugars and ammonia nitrogen. Others, like *Candida,* can cause serious human infections.

Mushrooms are highly differentiated forms of fungi. The mycelium is in the soil, and in the right conditions the **basidia** are formed above ground as the structures we call mushrooms.

The three groups of **fungi** are differentiated on the basis of their structure and method of reproduction in the simplified classification scheme shown in Table 8–3.

Protozoa.   Protozoa are the most highly specialized unicellular organisms. Most are nonphotosynthetic, reproduce asexually by binary fission, and lack true cell walls. Most

TABLE 8–3 CLASSIFICATION OF SOIL AND AQUATIC FUNGI

Type	Division	Characteristics/examples
Molds (filamentous)	Phycomycetes	Sexual or asexual spores; *Mucor, Rhizopus*
	Fungi imperfecti	No sexual stage; *Penicillium, Aspergillus*
Yeasts (nonfilamentous)	Ascomycetes	Sexual spores in sacs; *Neurospora, Candida*
Mushrooms (macroscopic)	Basidiomycetes	Sexual stage on basidia; common mushroom

Adapted from Mitchell, 1974

species are motile, and classification can be based on the means of locomotion (see Table 8–4). Their size varies from a few to several hundred microns. Protozoa are widespread in nature and are found in most habitats where moisture is present. They survive adverse conditions by forming cysts with thick walls. Protozoa may be saprophytic (obtain food in dissolved form). They are predators on bacteria and can be found wherever bacteria are prevalent. Some are parasites capable of causing disease in animals and humans.

The sarcodina have a cell membrane that continually changes shape. They move by extending their cytoplasm in search of food. These extensions are called **pseudopodia,** or false feet, and are typical of the **amoebae** (Figure 8–9). Sarcodina are saprophytic. ***Entamoeba histolytica*** is a common pathogen causing amoebic dysentry in humans.

The mastigophora have flagella, and some species are photosynthetic. Photosynthetic organisms (such as ***Euglena***) exhibit some of the characteristics of both protozoa and algae.

TABLE 8–4 CLASSIFICATION SCHEME
FOR COMMON AQUATIC AND SOIL PROTOZOA

I. **Pseudopods** (Sarcodina)
 Motile by pseudopods; flowing amoeboid motion; ***Amoeba, Entamoeba.***

II. **Flagellates** (Mastigophora)
 Motile by flagella; many photosynthetic; ***Euglena, Volvox, Giardia.***

III. **Ciliates** (Ciliophora)
 Free–swimming; motile by many cilia that move in unison; ***Paramecium.***
 Attached; fixed by stalk to a surface; ***Vorticella.***

IV. **Parasitic Protozoa**
 (Suctoria)
 Free-swimming ciliates early in life cycle, then tentacles in later adult stalked stage.
 (Sporozoa)
 Usually nonmotile; rarely free living; parasitic; ***Plasmodium.***

Adapted from Mitchell, 1974

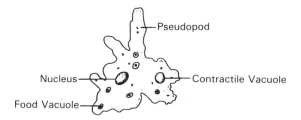

An amoeba, a member of the subphylum Sarcodina.

Figure 8–9 Sketch of *Amoeba*. Source: R. Mitchell, 1974

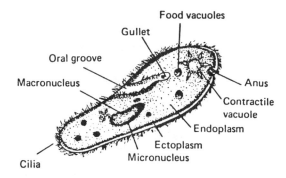

Figure 8–10 Sketch of *Paramecium*

A member of the subphylum Ciliophora, *Paramecium*. The cilia are used to capture food and for cell motility. Source: R. Mitchell, 1974

Some mastigophora are parasitic. ***Trypanosoma***, a blood parasite, causes sleeping sickness in humans.

The ciliophora are characterized by having fine hairs or cilia. In addition to providing motility, cilia aid in the capture of food. Very few of these are parasitic. ***Paramecium*** is a typical ciliate (Figure 8–10).

The parasitic protozoa include suctoria (free–swimming) and sporozoa (nonmotile). Four species of ***Plasmodium***, the cause of malaria in humans, are members of the latter group. The vector (carrier) which conveys these parasites to a human host is the female ***Anopheles*** mosquito.

8.3 APPLIED MICROBIOLOGY

8.3.1 Microbiology of Soil and Solid Wastes

Most land-based living things—plants, animals, and protista—and their associated wastes eventually find their way into the soil. There, microbial activity transforms this material into the substances that constitute soil. Without this activity, nutrient cycles such as the carbon cycle or the nitrogen cycle would not be complete, and life on earth would disappear.

Soils make up a very thin layer of material on the earth's surface. Soil depth and the physical and chemical properties of soil vary with location, but in general there are five major components:

1. *Inorganic mineral particles.* These particles, primarily of aluminum, silica, and lesser amounts of other minerals, range in size from very small clay particles (0.002 mm) to sand grains and pebbles. The proportions of such particles in the soil determine its water-holding capacity, its structure, and the availability of air and nutrients. Inorganic soils (those consisting chiefly of mineral particles) are the most common soils.

2. *Organic Residues.* Plant and animal remains which make up the organic component of the soil through various stages of decomposition to a fairly stable substance known as humus. Organic soils (those formed mostly of organic residue) are found in peat bogs and marshes.

3. *Water.* Water is necessary for microbial activity. The amount of water in soil depends on a number of factors, including precipitation, soil structure, and microbial population. Water is contained in the pore spaces between particles in saturated soils and is adsorbed on particle surfaces in dry soils. Various nutrients are dissolved in the water and are therefore available to microorganisms.

4. *Gases.* Gases, principally nitrogen and oxygen, but also carbon dioxide (particularly where biological activity is occurring), fill the pore spaces not filled by water. In saturated soils, small amounts of gas will be dissolved in the water.

5. *Biological Systems.* Plant root systems, small animals, and microorganisms make up the fifth component of soil. One gram of rich agricultural soil may contain 2.5 billion bacteria, half a million fungi, 50,000 algae, and 30,000 protozoa.

Bacteria and fungi constitute the largest group of microorganisms in soils. Autotrophic and heterotrophic bacteria degrade complex organic and inorganic substances, some under aerobic conditions and others under anaerobic conditions. The fungi decompose cellulose and lignin, the major components of plant tissues, and, as might be expected, are generally found near the surface, where aerobic conditions prevail.

In a fertile soil, the activities of algae are not as important as those of the bacteria and fungi. However, on very barren or inorganic soils and rocks they are the primary producers of organic material. Protozoa are also abundant wherever there are bacteria and aerobic conditions.

The extent and type of microbial growth in soil depend on the same factors that control growth in aquatic environments, namely,

- Whether sufficient nutrients are present
- The availability of moisture and, for aerobic organisms, air
- Suitable temperature and pH.

Under favorable environmental conditions created by landfilling or composting, soil organisms can be used to degrade municipal solid wastes. The most common means of disposing of solid waste is in sanitary landfills (see Chapter 14). The waste material, rich in organic matter, is placed in trenches or pits, compacted, and covered each day with a layer of organic soil which provides a large and diversified population of microorganisms. Microbial activity takes place initially under aerobic and later anaerobic conditions. The facultative or anaerobic microorganisms break down the complex organic substances into simpler organic acids which can be oxidized by the fungi and aerobic bacteria into CO_2 and H_2O. In time the aerobic activity is limited by a lack of oxygen, once the organic material is buried and saturated with water. Carbon dioxide, organic acids, ethanol, ammonia, and other products from the decomposition of organic matter in landfills can cause environmental problems. These are discussed in Chapter 14.

When the decomposition of municipal solid waste is carried out in a controlled aerobic environment, organic degradation is accelerated and the process is called **composting.** The objective of composting is to convert easily degraded organics into humus and produce a nutrient-rich, stable product useful for reclaiming land or improving soil. Whether composting is done naturally in long piles called windows or mechanically in special equipment, the biological activity generates heat which destroys the pathogens in about three days at 60°C. However, temperature must be kept below about 70°C so as not to kill the fungi (*Mucor, Rhizopus, Penicillium,* and *Aspergillus*) and the thermophilic bacteria (the same genera as in other aerobic waste treatment processes) which produce the compost. Further information on composting is presented in Section 14.5.3.

Other applications where biological activity in the soil plays a role include land-based wastewater treatment methods and sludge utilization on agricultural land. These are considered in Sections 12.6.1 and 12.7.3, respectively.

8.3.2 Microbiology of Water, Wastewater, and Indicator Organisms

All water derives from precipitation in the form of rain, snow, hail, or sleet which, as it falls, removes particles of dust from the air. However, after the first few minutes of precipitation, the dust, along with the few microorganisms it contains, is washed out of the air and the rainfall thereafter is relatively free of these contaminants. After reaching the ground, the water not taken up by vegetation either percolates into the ground to become groundwater or runs over the ground into streams, ponds, rivers, and lakes.

Because of the filtering action of the soil, low nutrient levels, low temperature, and the lack of light, groundwaters are normally free of organisms. However, in some rocky areas, especially limestone formations, there can be fairly large underground conduits, and surface water reaching the groundwater system through cracks or tunnels can cause microbial contamination of the groundwater.

Surface water picks up many substances during its travel over agricultural lands and through industrial areas. Agricultural lands contribute nitrates, phosphates, and other nutrients, plus microorganisms from the soil. Organic material like leaves, grass clippings, bird and animal droppings, and wastes from food processing plants, all with their associated microbial population, also have access to surface water. The result, unless toxic contaminants are excessive, is that virtually all surface waters in the world (with two possible exceptions[1]) support a thriving microbial population.

---

[1] Lake Tahoe, straddling the California-Nevada border and Lake Baikal in the U.S.S.R., are so lacking in nutrients that they contain no microbial life.

TABLE 8–5 TYPICAL BACTERIAL COUNTS IN WATER

Source	Bacteria per 100 mL	Coliform Bacteria[1] per 100 mL
Tap water	10	0–1
Clean, natural water	10^3	$0–10^2$
Polluted water	$10^6–10^8$	$10^3–10^5$
Raw sewage	10^8	10^5

[1] Coliform bacteria are present in sewage, but die off with time in natural waters. Their natural habitats are the intestines of warm-blooded mammals and the soil.

Many forms of microbial life can exist in water provided that the appropriate physical and nutritional requirements for growth are met. Dissolved oxygen is necessary for the growth of aerobic bacteria and protozoa. Nitrogen and phosphorus, as well as light, are essential to algae. The number and types of microorganisms present give an indication of water quality. In clean water or water with a low nutrient content, the total number of microorganisms is limited, but a great variety of species can exist. As the nutrient content increases, the number of microorganisms increases, while the number of species is reduced. In a polluted, anaerobic stream, a few species of anaerobic or facultative bacteria will predominate. Typical numbers of bacteria for various waters are presented in Table 8–5.

In addition to the independent behavior of the diverse types of microorganisms that has been described in the preceding sections, microorganisms can interact with each other—in either a cooperative or a competitive way. Such interaction occurs frequently in the environment, and we need to be aware of these relationships in the design of biological waste treatment systems. The following three examples illustrate the phenomenon and are described in more detail in Chapter 12:

1. *Algae-Bacteria.* A close association between algae (which need carbon dioxide and produce oxygen) and aerobic bacteria (which need oxygen and produce carbon dioxide) develops in oxidation ponds, swamps, lakes, and similar environments.

2. *Protozoa-Bacteria.* In the treatment of municipal wastewater by the activated sludge process, bacteria are the primary agents in the conversion of organic wastes to stable end products. At the same time, protozoa consume and limit the bacterial population in a predator-prey relationship, thus maintaining a dynamic balance in the microbial population.

3. *Bacteria-Bacteria.* The anaerobic digestion of organic matter demonstrates the interdependence of two groups of bacteria: the acid-forming bacteria, which convert organic matter to acetic and other organic acids, and the methane formers, which use these acids to produce methane.

Indicator organisms. Water used for drinking and bathing can serve as a vehicle for the transmission of a variety of human enteric pathogens which cause so-called waterborne diseases. A detailed discussion of waterborne diseases is presented in Section 8.4.3. The detection of pathogens in water is difficult, uneconomical, and impractical in routine water analyses. Instead, water is tested using a surrogate that is an indicator of fecal contamination. Since nonpathogenic microorganisms also inhabit the intestines in large numbers and are always present in feces together with any pathogens, these may be used as indicators of fecal contamination.

The main characteristics of a good indicator organism are that (1) its absence implies an absence of any enteric pathogens, (2) the density of the indicator organisms is related to the probability of the presence of pathogens, and (3) in the environment the indicator organisms will survive slightly longer than the pathogens. Obviously, no such ideal indicator organism exists. However, certain bacteria, such as the total coliforms, fecal coliforms, fecal streptococci, and *Clostridium perfringens*, are regarded as evidence of fecal contamination and have been in use for assaying water quality for many years.

Of the many indicator organisms, the total coliform group of bacteria is the one most commonly used. It includes, by definition, "all aerobic and facultative anaerobic, gram-negative, non-spore-forming, rod-shaped bacteria that ferment lactose with gas formation within 48 hours at 35°C" (APHA et al. 1985). The coliform group is comprised of *Escherichia coli*, *Enterobacter aerogenes*, *Citrobacter fruendii*, and other related bacteria.

In drinking water, where no coliforms of any kind should be present, total coliforms are used as an indication of fecal pollution. In the case of polluted streams, sewer outfalls, swimming areas, etc., fecal coliforms are enumerated by using the elevated temperature test (44.5°C $\pm$ 0.5). Differentiation between total and fecal coliforms is based on their ability or inability to grow at 44.5°C $\pm$ 0.5. In temperate climates, *Escherichia coli* is the most frequent and predominant type of coliform found in the human intestine. Other members of the coliform group, usually found in soil and on vegetation, may also be encountered in feces, but in low numbers. In tropical countries, *E. coli* is not the predominant intestinal coliform, so in that case the total, rather than the fecal coliform test is a more useful measure of pollution.

Fecal streptococci, another type of intestinal bacteria, which are more common to animals than to humans, are frequently enumerated in conjunction with the fecal coliforms, and the ratio of fecal coliforms to fecal streptococci (FC/FS ratio) is used to differentiate the source of pollution. With a ratio of 4.0 or more, the pollution is considered to be from human wastes, whereas ratios less than 0.7 indicate pollution from animal wastes. The presence of *Clostridium perfringens* indicates remote fecal pollution.

The enumeration of the bacterial indicators is carried out by two alternative methods, namely, the multiple-tube fermentation technique, also called the most probable number or MPN procedure, and the membrane filter or MF method. Details of these two methods are available elsewhere (APHA et al 1985).

8.3.3 Microbiology of the Atmosphere and Indoor Air

Because of its lack of moisture, the atmosphere is not an environment where microorganisms can live and grow. They can survive, however, in their vegetative state to varying degrees, depending mainly on their resistance to drying and, to a lesser extent, their resistance to ultraviolet radiation. Those bacteria and fungi that form spores can exist for a very long time in the atmosphere. Vegetative cells do not survive more than a few days in the air, whereas spores may remain viable for years. Some protozoa form cysts which, like spores, enable them to survive adverse conditions for long periods. When conditions are favorable, the spores or cysts break apart and vegetative cells develop.

Air is important in microbiology because it provides a mechanism of transfer for microorganisms that is much wider ranging than water. Microorganisms make up a portion of the particulate matter in the atmosphere. Other particles include dust particles and liquid aerosols, or fine droplets. Of course, the smaller the particle, the longer it will remain airborne. The cysts formed by protozoa are relatively heavy, settling out of the air in a matter of minutes. On the other hand, spores of bacteria and fungi are very small and have been found miles above the earth's surface.

Particles in the atmosphere stem from both natural causes and human activities. Examples of natural sources are forest fires, volcanic eruptions, aerosols from ocean spray, and dust picked up by the wind from open fields and vegetation. Human sources are mostly energy related, such as particles from the combustion of fuel for power and transportation as well as dust created by industrial and agricultural processes. The bulk of the particulate matter in the atmosphere is of natural origin, although estimates of society's contribution range from 5 to 45 percent of the total (Perkins, 1974).

Smoke particles, industrial dust, and dust from volcanic eruptions make up the bulk of particulate matter in the air, but few microorganisms are attached to these particles. Wind-generated dusts from fields, oceans, and forests on the other hand are likely to carry spores throughout the lower atmosphere. Figure 8–11 gives an indication of the types and numbers of microorganisms found in the urban environment as a function of height. Air samples taken at heights up to 3,000 m over the North Atlantic ocean have contained spores

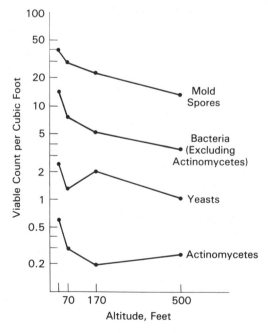

Figure 8–11 Types and numbers of organisms at different elevations in urban air. Source: T. J. Wright, V. W. Greene, and H. J. Paulus, Journal of the Air Pollution Control Association 19 (1969):337

of bacteria and fungi. Knowledge of the concentration and distribution of microorganisms in the atmosphere is fairly limited, but it is quite clear that they are dispersed all over the world.

The microbial content of indoor air is of more immediate significance. It is influenced by the rate and means of ventilation, by the degree of crowding, and by the types of activities occurring in the building. Outdoors, the amount of air available per person is essentially unlimited, and the particulate matter is constantly being diffused and distributed by natural air turbulence. Indoors, the activities of the occupants generate airborne microorganisms which are distributed in a relatively confined space. Their removal depends on the efficiency of the ventilation and filtration systems. Figure 8–12 is a graph showing the bacterial content of a number of indoor environments.

Of particular concern are the pathogenic microorganisms which cause respiratory diseases. The cough or sneeze or even the conversation of an infected person may release

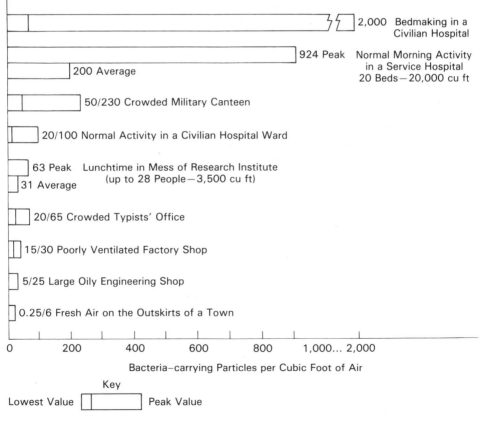

Figure 8–12 Typical bacterial content of indoor air. Source: F. P. Ellis and U. F. Raymond, *Studies in Air Hygiene*, Medical Research Council (GB) Spec. Rep. Ser. 262, 1948. By permission of the controller of H. M. Stationery Office, London

pathogenic organisms into the air and spread the infection. Health effects of microorganisms in the atmosphere are discussed further in Section 8.4.4.

8.4 EPIDEMIOLOGY AND DISEASE

8.4.1 Sanitation and Health

The history of society's concept of, and battle against, epidemic diseases is well illustrated in Winslow (1943). Baker (1981) gives an account more specifically concerned with the history of the purification of public water supplies. Feachem et al. (1983) present a more recent account of the health aspects of excreta and wastewater management. Sartwell (1973) is a valuable reference on the epidemiology of disease. Subsequent sections in this chapter are derived in part from this source.

One of the earliest theories of diseases is called the miasma theory of disease. **Miasma** means bad or unhealthy air. The term "malaria," literally, "bad air," comes from this concept. The etiological or causative agent for disease was considered to be generated and resident within these miasmas. The miasma theory was generally accepted as the explanation for the cause of disease until the tentative establishment of the germ theory of disease in the seventeenth century.

The Dark Ages, with their disastrous epidemics, provided numerous case studies from which scientists could learn about disease. A leprosy epidemic in the sixth century, bubonic plague in 1348, and syphilis in 1500 were the major misfortunes. By the sixteenth century, Fracastoro, after careful observation, had developed a clear concept of contagious disease, but gases rather than microorganisms were thought to be the means of transfer.

Microorganisms were not seen until Leeuwenhoek, a seventeenth-century native of Holland, devised the first microscope with sufficient magnification: the science of microbiology began with his observation of protozoa and bacteria.

The belief in the spread of contagious disease by bad air and the observation that disease was more abundant in the filthy, crowded areas of cities, together, led to the sanitary awakening of the early and middle nineteenth century. The cities of those times contained public squares heaped with decomposing filth, and when these were cleaned up, the incidences of typhoid, cholera, and typhus were markedly reduced. The conclusion that degree of sanitation had something to do with disease was based on empirical observation rather than on any theoretical understanding of contagion.

In 1849, John Snow, a medical doctor, published a pamphlet entitled *On the Mode of Communication of Cholera*. Two cholera epidemics in London served to test his theories. He explained how "minute quantities of the ejections and dejections of cholera patients must be swallowed." He backed this up by his classic study of the Broad Street pump epidemic. He was able to show that of 77 cholera victims, 59 of them had used water from the suspected pump. There was a workhouse in the area almost surrounded by houses in which cholera deaths had occurred, but which remained relatively free of the epidemic. It had its own well. Removing the handle from the Broad Street pump ended the epidemic and also identified the source of pollution as a drain from the house of an infected patient. The drain passed within three feet of the well.

In the latter half of the nineteenth century, Pasteur, Lister, and Koch finally established the germ theory of disease conclusively. They managed to isolate and grow cultures of microorganisms which they were able to show produced specific diseases. The role of the carrier (one who is not clinically ill but still carries the infection) and the role of the insect host in certain diseases were demonstrated later and explained how disease could be transmitted with no apparent contact between infected patients.

The potential of water to spread massive epidemics is well known today. In the early part of this century, attempts to control typhoid fever and various enteric diseases resulted in a number of patents on the use of oxidizing agents and other water purification techniques. The first large-scale application of chlorination was in 1908 at the Boonton Reservoir of the Jersey City Water Works in the United States. Since 1920, enteric diseases have been almost entirely eradicated in most parts of developed countries. However, in the less developed nations, half the population still does not have adequately protected water supplies, and 75 percent lack safe waste disposal systems.

Table 8–6 compares the death rate and average life expectancies at birth between the developed and the developing regions of the world. The table clearly shows the lower death rates and higher life expectancies, attributable to better public health, sanitation, and medicine, in the developed countries. The statistics for Africa and South Asia are similar to those of developed countries over 100 years ago.

TABLE 8–6 ESTIMATED CRUDE DEATH RATES AND LIFE
EXPECTANCIES FOR MAJOR AREAS AND REGIONS
OF THE WORLD, 1965–1970

Major areas and regions	Crude death rate (deaths per 1,000 population)	Expectation of life at birth (years)
World total	14	53
Developing regions	16	50
More developed regions	9	70
Africa	21	43
Asia (excluding the U.S.S.R.)		
East Asia	14	52
Japan	7	71
South Asia	17	49
Europe (excluding the U.S.S.R.)	10	71
Latin America	10	60
North America	9	70
Oceania	10	65
U.S.S.R.	8	70

Source: United Nations, 1973

8.4.2 Pathogens

A **pathogen** is an agent that causes infection in a living host. It acts as a parasite within the host or host cells and disrupts normal physiological activities. This disruption is what causes the symptoms of disease, such as high temperature, an upset in the digestive process, a change in blood chemistry, and so forth.

Infection, by definition, means that a disease-producing agent is growing and multiplying within the host, who may or may not have symptoms of the disease. The presence of the agent stimulates the host to produce antibodies to combat the agent. Depending on the effectiveness of the antibodies, the result can be sickness, recovery, or death, or the antibodies may limit the growth of the agent to the point at which symptoms of infection are not apparent, i.e., the host is infected without appearing to be sick.

The ability of the pathogen to inflict damage on the host is termed its **virulence.** Virulence is a relative concept, comparing the attacking ability of the pathogen to the defensive ability of the host. Virulence can be influenced by factors inherent to the pathogen and the host, as well as by environmental conditions.

An infection is a pathological condition due to the growth of microorganisms in a host.

A pathogen is an agent that causes infection in a living host.

A toxin is a poisonous substance from certain organisms, e.g., bacterial toxins.

Virulence is the capacity of a microorganism to cause disease.

Pathogenic organisms that are virulent enough to infect man under appropriate conditions include some species of bacteria, viruses, algae, and fungi, as well as protozoa and helminthic (parasitic worm) organisms. Table 8–7 indicates, for the various groups of microorganisms, some common diseases and the means by which the pathogens are transmitted.

Virulent pathogens cause epidemics which, as noted earlier, affect an abnormally high number of people in a localized area. The actual numbers of people affected need not be large. For example, a few cases of botulism or food poisoning occurring simultaneously in the same area could be considered an epidemic because the disease is so rare. Yet a sharp increase in head colds among school children in the fall is normal and not regarded as an epidemic. As we learned, an **endemic** disease is one like **schistosomiasis,** a parasitic worm infection, which is prevalent among a particular group, in this case the rural population in parts of Africa.

For an epidemic to occur, three factors must be present: an infected host, a number of non-infected potential hosts, and a mechanism of transfer between the two. An infected host implies the presence of a virulent pathogen. The pathogen eventually brings about its

TABLE 8–7 PATHOGENIC ORGANISMS

Disease	Agents and Vectors
Bacterial	
Anthrax	Contaminated animal hair, wool hides; contaminated undercooked meat; inhalation of airborne spores.
Botulism	A thermolabile toxin produced in nonacid food under anaerobic packaging; organisms from soil and intestinal tract of animals.
Brucellosis or undulant fever	Contact with infected pigs, cattle, goats, and horses; use of raw milk and milk products are the cause of sporadic cases and outbreaks.
Cholera	Feces of cases or carriers contaminate water, milk, food, and flies; initial wave of epidemic cholera is waterborne.
Plague, bubonic and sylvatic	Organism transmitted by fleas from rats and wild rodents; contaminated vomitus of flea enters skin during biting.
Salmonellosis	Feces of animals and infected persons contaminate foods; organisms multiply in unrefrigerated foods to deliver massive doses.
Shigellosis or bacillary dysentery	Four groups of the dysentery bacillus, *Shigella dysenteriae, S. flexneri, S. boydii,* and *S. sonnei,* leave via feces and return to the mouth directly or via water, food, flies, or fecally soiled objects.
Typhoid and paratyphoid fever	Feces and urine of cases and carriers contaminate water, milk, food, and flies.
Endemic (murine)typhus and epidemic typhus	The diseases are caused by rickettsiae. Fleas transmit the rickettsiae from rat to rat and from rats to man; organisms in fleas' feces enter through fresh bites and abrasions.
Viral	
Infectious hepatitis	Outbreaks have been related to contaminated water, milk, and food, including shellfish.
Yellow fever	Urban yellow fever from human cases by *Aedes aegypti*; jungle yellow fever from monkeys and marmosets by forest mosquitoes; presence of *Aedes aegypti* in large areas of Africa and Southeast Asia requires vigilance despite absence of yellow fever.
Protozoan	
Amebiasis or amebic dysentery	Hand-to-mouth transfer, contaminating raw vegetables, flies, soiled hands of food handlers, water.
Malaria	Three plasmodium types are transmitted from man to man by one of about 20 anopheline mosquito species, which are efficient vectors.
Giardiasis	Cysts of *Giardia lamblia* in feces of humans contaminate water and cause diarrhea.
Helminthic	
Ascariasis or roundworm	Soil contaminated with feces of infected persons contains embryonated eggs; ingestion of such soil or raw foods after soil contact is the infection routine.
Hookworm diseases	Penetration of skin by larvae developing in soil contaminated by feces of infested persons.
Schistosomiasis	Eggs of *Schistosoma mansoni* and *S. japonicum* pass with feces from man and *S. hematobium* with urine to cycle in water through specific snail types, to the cercaria form which penetrates human skin; domestic animals and wild rodents host *S. japonicum*.

Adapted from Sartwell, 1973 and Feachem *et al.* 1983

own destruction either by destroying the host or by stimulating the production of antibodies which destroy the pathogen. Therefore, for an epidemic to persist, there must be a continuous supply of new nonimmune hosts for the pathogens to invade.

All epidemics are self-limiting in that there comes a time when the combined factors of a shortage of new hosts, increased distance between infected and uninfected hosts, and perhaps a reduction in virulence lead to very few new outbreaks of the disease. Unfortunately, a tremendous loss of life can occur before the limit is reached, as evidenced by the loss of at least one-third of the population of medieval Europe in the fourteenth century from bubonic plague. In England, about one-third of the population of $4\frac{1}{2}$ million perished in a $2\frac{1}{2}$-year period between 1347 and 1350 (Ziegler, 1969).

Why some microorganisms are more virulent than others is not clear. Most microorganisms are not pathogenic and are **saprophytes,** that is, organisms that live and feed on dead and decaying organic matter. However, included among the saprophytes are a few organisms that are **facultative pathogens** or **opportunists.** These may cause disease under very special circumstances. There are, of course, pathogens that are **obligate parasites,** such as viruses that cannot live outside a host cell. Most pathogens, though, can grow outside a host until a suitable portal of entry for infection is found.

One of the ways pathogens cause disease is through the production of poisonous substances known as **toxins.** Not all pathogens have been shown to have these characteristics, but it is clear that the production of toxins and their potency are major factors in the virulence of some microorganisms. The toxins which are excreted into the surroundings are called **exotoxins.** Those that remain in the microorganism are called **endotoxins** and are not released until the microorganism dies. The amount of exotoxin released at the death or breakup of the cell can also be significant. Exotoxins are generally more potent than endotoxins.

In some cases, enzyme production by pathogens is thought to contribute to their virulence. Enzymes may help spread the disease through the host by increasing the permeability of cell walls, by destroying specific tissues and cells, or by enabling the pathogen to resist attack by antibodies.

The capsular material surrounding some pathogenic bacteria may affect their virulence. The capsules themselves are nontoxic, but they seem to protect pathogenic organisms from attack by antibodies. If this capsular material is removed, the pathogen can lose its pathogenicity or virulence. However, there are many capsulated bacteria that are nonpathogenic, and the virulence of some pathogens is unaffected by the presence or absence of a capsule.

Each pathogen has a specific portal of entry into its host. The enteric pathogens are those that cause disease in the alimentary tract or the digestive system. Organisms producing typhoid, dysentery, gastroenteritis, and cholera are examples of enteric pathogens. They must be ingested to cause infection. Organisms that attack the lungs must be inhaled. Others must enter through abrasions in the skin where they can set up local infections, or enter the circulatory system and spread through the body. Many pathogens grow in the bodies of animals or insects, and the only humans infected are those bitten by these creatures.

The infective agent usually enters and leaves the host by the same route. For example, the enteric pathogens enter and exit via the digestive system. They enter via the mouth,

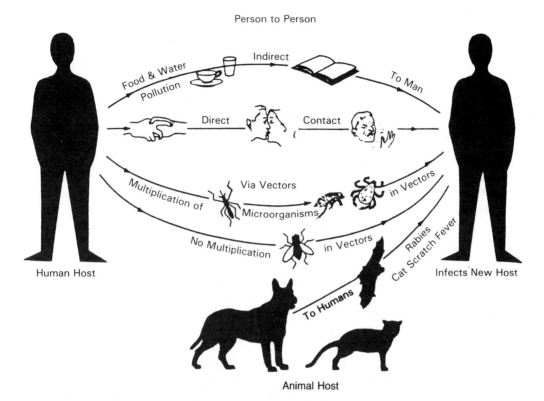

Figure 8–13   Transmission of disease. Source: Pelczar, Reid, and Chan, 1977
Transmission of pathogenic microorganisms from person to person occurs in a variety of ways, ranging from direct contact to fomites and by intermediate vectors. Animals may also transmit pathogens to humans.

continue through the alimentary canal, and exit with the feces. Respiratory pathogens that are inhaled are released in secretions from the nose or throat or by aerosols of these secretions from sneezing and coughing, which carry the microorganisms through the air. Pathogens that infect the skin through cuts and abrasions are discharged in the drainage from skin lesions and abscesses.

After the microorganism leaves its host, it is in an unfavorable environment and must survive until it reaches a new host. Figure 8–13 is a representation of various transfer mechanisms. Most diseases are spread indirectly, either by the pollution of food and water, by contaminated airborne particles, or by bites from infected insects or other intermediate hosts (called vectors). In the remainder of this section we shall examine the waterborne, airborne, and insect- and rodent-borne diseases, since these are of primary interest in the environmental field.

8.4.3 Waterborne Diseases and Water Quality

Diseases transmitted by water are almost all of intestinal origin. Fecal matter from infected hosts or carriers can get into a water supply or swimming area in a number of ways. The most common is by direct discharge of raw sewage into the receiving water. Pit privies located too close to a well or stream can also be a source of contamination. Specific outbreaks of disease have been traced to cross-connections between water and sewer pipes, to breaks in water mains, and to contamination of water supplies during flooding or temporary failure of a sewage treatment facility.

Pathogenic organisms are unable to grow in water, but may survive for several days. Those pathogens capable of forming spores or cysts can exist outside of a host for a much longer time. For example, the spores of *Clostridium tetani*, the pathogen responsible for tetanus infection, can survive for years in nature.

Some of the common types of waterborne disease are salmonellosis, shigellosis, cholera, infectious hepatitis, amebiasis, giardiasis, and schistosomiasis. Epidemics of all these diseases occur periodically in many parts of the world, but are rare in North America because virtually all of the population is served by adequate water supply and waste disposal systems.

Salmonellosis.   Three forms of salmonellosis occur in humans: acute gastro-enteritis, septicemia (blood poisoning), and enteric fevers. These complications in humans are caused by a variety of species of *Salmonella*. The common symptoms in gastroenteritis are diarrhea and abdominal cramps, followed by fever which lasts one to four days. These may be severe, but fatality rates are low. The species most commonly isolated from patients with salmonella gastroenteritis is *Salmonella typhimurium*. During septicemia, some bacteria spread to the spleen, kidneys, heart, and lungs, and lesions may develop on these organs. Typhoid fever (an enteric fever) caused by *Salmonella typhi*, develops only in humans. After being swallowed, the bacteria cause a generalized infection, and following an incubation period of 10 to 14 days, a fever develops (40°C) that lasts several weeks. Accompanying the fever are abdominal pain and bowel disturbances. About three percent of typhoid patients eventually become chronic carriers of the disease. They recover from the symptoms, but they continue to harbor the microorganisms. *Salmonella typhi* generally lives less than one week in nature, but lasts much longer in very cold water or ice. Chloramphenicol is effective in the treatment of salmonellosis.

Shigellosis.   Shigellosis is also called **bacillary dysentery** or acute diarrhea. Like typhoid fever, it is a disease associated with poor sanitation, overcrowding, and unsafe water supplies. A number of species of the genus *Shigella* are pathogenic for humans. Shigellosis is characterized by abdominal cramps, diarrhea, and fever following an incubation period of one to four days. Antibiotics, such as the tetracyclines and chloramphenicol or ampicillin, are effective in the treatment of bacillary dysentery. There has not been as much success in preventing shigella infection as there has been in controlling typhoid fever.

Cholera.   Humans acquire cholera by the ingestion of bacteria known as *Vibrio cholerae* (also *Vibrio comma*), which may be present in polluted water or contaminated food. The ingested bacteria multiply in the small intestine and, after two to five days, cause abdominal cramps, nausea, vomiting, and profuse diarrhea that may lead to dehydration, shock, and death. Cholera is endemic in the Bengal region of India and Bangladesh, where several thousand cases are reported each year. Areas where cholera remains endemic are typically low-lying farmlands subject to periodic flooding and which experience a hot, humid climate and high population densities. Since the turn of the century, cholera has been confined mainly to Southeast Asia, with occasional incursions into neighboring areas. Recent outbreaks of cholera are not positively identifiable as waterborne, since there are many other direct and indirect contact mechanisms. Good sanitation practices play an important role in the control of cholera. Sulfonamides and antibiotics are useful in the treatment of the disease.

Infectious hepatitis.   A viral disease, infectious hepatitis has been shown to be transmitted by water in a number of epidemics. Typical symptoms of the disease are fever, loss of appetite and energy, headache, and back pains. After a few days the fever may subside, and recognizable jaundice (a yellow tint in the skin) may appear. The disease is rarely fatal, and it is suspected that there is probably a large proportion of infected people who do not show clinical symptoms but who can still carry and pass on the disease. There was a large waterborne outbreak of infectious hepatitis in New Delhi, India, in 1955–56 resulting from contamination of the city's water supply by sewage. Surprisingly, there was no increase in enteric bacterial infections, such as typhoid fever, during this same period. This suggests that the chlorine dosage and contact time in the water treatment plant were sufficient to destroy the bacteria but not the viruses. Viruses are small enough to pass through most sand filters and have been shown to resist chlorination (see Chapter 11). Many viruses survive outside of a host for long periods of time. For these reasons, it is most important to prevent contamination of water supplies rather than depend on later purification. Other viral diseases that may be waterborne are epidemic gastroenteritis and poliomyelitis. The problems in isolating and culturing viruses make it difficult to conclusively determine the mechanism of transfer in any given situation.

Amebiasis.   Amebiasis is also called amebic dysentery. Its symptoms are stomach cramps and diarrhea. The causative organism is the protozoan *Entamoeba histolytica*. Its normal habitat is the colon or large intestine. It is relatively small, feeds on bacteria and cellular debris, and produces cysts which pass in the feces to spread the infection. Under conditions not yet understood, the protozoan will penetrate the host tissue, become pathogenic, grow much larger, and continue to reproduce. Other amoebae remain small after division and become encysted. When the cysts are exposed to an external environment and then swallowed, they are capable of returning to a vegetative (active) condition.

Although not as resistant to adverse conditions as fungal spores, the cysts can withstand the normal chlorination of drinking water. However, they are killed by drying and by ultraviolet irradiation from the sun. In water, cysts survive longer at low temperatures. They can survive for over one month in ice, but for only one or two days at 34°C. Although endemic in hot tropical areas, severe outbreaks of amebiasis have occurred in temperate areas as well.

Giardiasis.   Giardiasis is caused by *Giardia lamblia*, a flagellated protozoan of the small intestine. Characteristic symptoms are abnormal cramps, diarrhea, fatigue, anorexia, and nausea. The mean duration of acute illness is often two to three months. An infected individual may excrete more than 10^6 cysts/g of feces. The cysts are ovoid, refractile, and attain a size of 8–14 μm by 6–10 μm. Passed in the feces to the external environment, the cysts may survive for months and contaminate food and water. Most waterborne outbreaks of giardiasis reported in North America are associated with smaller water supplies and recreational areas, where the only treatment was chlorination. *Giardia* cysts are not destroyed by chlorination at dosages and contact times commonly used in water treatment plants, nor are they effectively removed by rapid sand filtration unless it is preceded by coagulation and flocculation (Chapter 11).

Schistosomiasis.   Also known as bilharzia, schistosomiasis is caused by parasitic worm members of the genus *Schistosoma*. It is a chronic disease, endemic to Africa, South America, and parts of Asia, and affects millions of people. Symptoms are enlargement of the liver, diarrhea, and anemia. Schistosoma or blood flukes are not strictly microorganisms, and they are not necessarily transmitted by the fecal-oral route. The life cycle of *Schistosoma* is shown in Figure 8–14. The eggs are laid by the parasite in the infected person's intestine and passed in the feces. The eggs hatch in water into small motile miricidia. These must

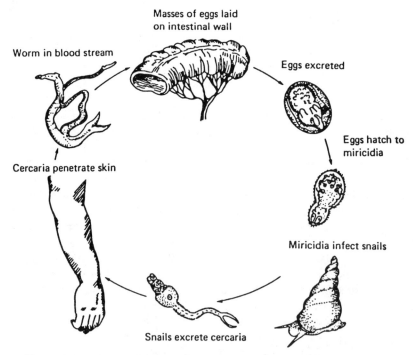

Masses of eggs laid on intestinal wall

Worm in blood stream

Eggs excreted

Eggs hatch to miricidia

Cercaria penetrate skin

Miricidia infect snails

Snails excrete cercaria

Figure 8–14   Life cycle of *Schistosoma*. Source: Mitchell, 1974 (Permission by R. D. Barnes, *Invertebrate Zoology*, 3rd ed., Philadelphia, Pa.: Saunders, 1974)

find a snail host within a few hours, or they will die. After a period of incubation, infected snails excrete cercaria, which can survive for two or three days in water. Cercaria attach to human skin and penetrate to the bloodstream. They mature in the liver into adults. Although the life cycle and control methods are understood, the disease is actually on the increase in endemic areas. Lack of adequate sanitation is the main obstacle to its control. Unfortunately, the increase in irrigation systems and water impoundments associated with hydroelectric projects that provide energy necessary for the development of tropical countries also results in an increase in the numbers and spread of the snail vector. Control of the snail population with poison is only partially effective, because it is often lethal to other aquatic life.

A mild form of schistosomiasis, referred to as swimmers itch, occurs at times in parts of Canada and the United States.

Water quality.   As explained in Section 8.3.2, it is impractical to detect, differentiate, or enumerate the pathogenic organisms which may be present in water and wastewater. Therefore, to monitor water quality, water is tested for indicator organisms which are present when fecal contamination occurs. The coliform group is the most common bacterial indicator of fecal contamination and standards (legal requirements) or guidelines (objectives or goals) have been established by most countries to limit the geometric mean density of total and/or fecal coliform bacteria in water used for different purposes. Representative values are as follows:

	Maximum Allowable Coliforms	
Water use	Total	Fecal
Drinking water	1/100 mL	0
Raw water supply	5,000/100 mL	500/100 mL
Recreational	1,000/100 mL	100/100 mL
Treated wastewater	—	200/100 mL

Water quality standards and guidelines for water supplies and effluent receiving waters are considered in detail in Chapters 11 and 12.

8.4.4 Airborne Diseases

Respiratory diseases such as pulmonary tuberculosis (bacterial), influenza (viral), and pulmonary mycosis (fungal) are transmitted by air. The pathogenic microorganisms from the lungs, sinuses, and bronchioles leave the infected host via the mouth and nose. Coughing, sneezing, and talking produce aerosols or fine droplets which may contain these organisms. Saliva and nasal discharges can also be transferred by hand to intermediate objects, e.g., bedclothing. The aerosol method of transfer is generally termed direct contact, the latter indirect contact. Droplets expelled by coughing or sneezing do not travel very far through the air, so to be infected directly by such droplets, people have to be in close contact with

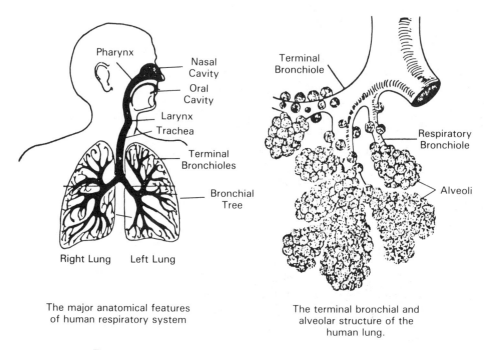

The major anatomical features
of human respiratory system

The terminal bronchial and
alveolar structure of the
human lung.

Figure 8–15 The human respiratory system. Source: Perkins, 1974

an infected person. The droplets that fall on intermediate objects dry out, leaving a dry nucleus to which pathogens may be attached.

Airborne infections are those transmitted by the pathogens carried on the very small droplet nuclei which become resuspended, circulated by the air, carried by dust, or recirculated through inadequate ventilation systems. These particles are taken into the body via the respiratory system (Figure 8–15). The upper respiratory system consists of the nasal cavity, the trachea (or windpipe), and the bronchial tree—a branching of the trachea into successively smaller and smaller tubes in each lung. It functions basically as a passageway for air to get into the lungs. The lower respiratory system consists of the bronchioles—the very small ends of the bronchial tubes about 0.05 cm in diameter—and the alveoli, which are clusters of small air sacs at the end of each bronchiole that are about the size of a pinhead. There are several hundred million alveoli in a healthy lung, providing an interior surface area of about 56 m² (600 ft²).

The only particles that can penetrate to the lower respiratory system are those in the size range 0.1–1.0 μm, which is the size range from viruses through fungal spores to single bacterial cells. Larger particles are removed by the defense mechanisms of the upper respiratory system, and most smaller particles are exhaled. Infections whose portal of entry is the alveoli or bronchioles are necessarily airborne. They could not penetrate that far into the lung unless they were suspended in the airstream. Examples of lower respiratory infections are pulmonary tuberculosis, pulmonary mycosis, and pneumonic forms of plague.

As air is drawn into the lungs, it must negotiate a number of sharp bends in the upper respiratory system. The larger particles cannot make the corners and therefore impact against the lined walls of the sinuses, trachea, and bronchial tree, which are the portal of entry for diseases of the upper respiratory system such as diphtheria, influenza, the acute viral respiratory diseases, and the acute contagious diseases such as measles and mumps. Larger particles can reach their portal of entry more easily than particles responsible for infections of the lower respiratory system and therefore are harder to classify as airborne infections. For example, pathogens deposited in the mouth would reach the upper respiratory system. The airborne mechanism, though, is certainly important.

Pulmonary tuberculosis is the most significant of the airborne diseases. It remains one of the main causes of disability and death throughout the world. There are over a million new cases reported annually. The World Health Organization estimates that this represents only 10 percent of the actual number of cases and that tuberculosis is responsible for an estimated two to three million deaths per year. Like many contagious diseases, it is most prevalent in areas with a low standard of living.

The causative agent, the tubercle bacillus (*Mycobacterium tuberculosis*), is a non-spore-bearing rod-shaped bacterium. Its growth needs are relatively simple, but it grows quite slowly, with a generation time of 18 to 24 hours. The bacilli can be destroyed by exposure to direct sunlight, heat, and disinfectants. They are more resistant to chemical agents and antibacterial agents such as penicillin than are most pathogenic organisms. The length of time they survive in the environment depends on the nature of the secretions in which they are contained. The bacteria are resistant to drying and therefore remain viable on droplet nuclei, as noted earlier. Legionnaire's disease is another respiratory illness caused by a small pneumonic bacteria which can be transmitted through contaminated ventilation systems.

Pulmonary mycosis is a fungal infection of the bronchioles and alveoli. The fungal growth destroys these structures and thus reduces the lungs' capacity. This can result in suffocation, but usually the heart is overtaxed and fails before the suffocation stage is reached.

Other airborne infections of a specialized nature are those of surgical wounds and of laboratory personnel working with pathogenic organisms. The tremendous success in curbing infections by strictly aseptic techniques in surgery indicates that infection during surgery is airborne. However, controlling airborne or suspected airborne diseases by limiting the number of bacteria or dust particles in the air has generally been unsuccessful. In hospital experiments, reductions of 50 to 75 percent in the number of airborne bacteria did not result in any significant decrease in infection. It seems that for effective control, access of the microorganisms to the atmosphere must be prevented. Containing the disease, therefore, depends upon early diagnosis followed by proper medical treatment and complete isolation of the patient or the source of the infection.

8.4.5 Insect- and Rodent-borne Diseases

The bloodstream is the portal of entry and exit for a number of pathogens. Insect bites facilitate the entry of the microorganisms through the host's skin. Insects that have been

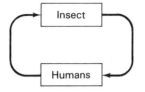

Figure 8–16 Insect–human cycle in insect-borne diseases

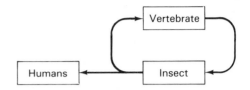

Figure 8–17 Insect–lower vertebrate cycle in rodent-borne diseases

incriminated include mosquitoes, sand flies, tsetse flies, midget ticks, fleas, and lice. A few of these pathogens are vector specific, i.e., a specific species of insect transfers a particular pathogen to a human host.

There are two basic epidemiological cycles in insect-borne diseases. The simplest one is a cycle from insect to humans back to insects, as illustrated in Figure 8–16. The insect must be one that feeds on humans preferentially, and it must have a high susceptibility and infectivity to the disease. The ability to meet all of these conditions is relatively rare, but when the conditions are met, outbreaks of disease can be explosive. Examples are malaria and trypnosomiasis (sleeping sickness),[1] protozoan infections from the female *Anopheles* mosquito and the tsetse fly, respectively.

One of the worst infectious diseases in terms of the annual number of cases is malaria. The causative organism is one of four species of parasitic protozoa (of the genus *Plasmodium*) that destroys the host's red blood cells. Humans are the primary reservoirs in areas of the world where malaria is hyperendemic, i.e., where the adult population has become immune through continuous transmission of the disease and acts merely as a reservoir for the protozoa.

The most common epidemiological cycle is illustrated in Figure 8–17 and involves a cycle of insect to lower vertebrate and back to insect, with humans as an occasional tangent infection. The infection of humans is usually a dead end in the cycle (urban yellow fever being an exception). Examples of diseases affecting humans are yellow fever (viral) and bubonic plague (bacterial). Mammals and birds act as reservoirs for the pathogens.

Yellow fever, an insect-borne viral infection, has been essentially eliminated from many countries in which it was once endemic; however, it still exists in the West Indies and parts of Africa. The basic reservoir cycle includes mosquitoes of the *Aedes* species and monkeys. There is one urban species called *Aedes aegypti* which is anthropophilic (preferentially bites humans). If one or more of these feed on an infected person, an epidemic of urban yellow fever may start.

Plague is actually a disease of rats, with humans infected only tangentially. It is transmitted from rat to rat by fleas. It was often noted (but the significance was not understood) that immediately preceding the plague outbreak there would be an epidemic among the rat population. When an infected flea bites a human, bubonic plague develops. Plague bacilli are often found in the sputum of the infected and can be spread from person to person by direct contact or through airborne infection in crowded areas.

[1] A viral disease (viral encephalitis), common in Manitoba, Canada, is also called sleeping sickness.

Plague epidemics had a profound effect on European society in the fourteenth century. It is estimated that one-third of the population died of the black death, a form of bubonic plague. Plague is uncommon today, but persists in a few areas, such as South Vietnam, where there is still a significant number of cases. Wild rodents such as squirrels, gerbils, marmots, and wild guinea pigs are also reservoirs of plague and could infect the rat population, thereby threatening the human population.

Rickettsia is a general term given to a small group of microorganisms whose size and characteristics lie between those of bacteria and viruses. The group is included here because of its epidemiological importance in causing typhus and typhus-like diseases. Of the rickettsial diseases, the typhus group has been responsible for much sickness and misery in the past. There are two organisms responsible for typhus fevers: ***Rickettsia prowazekii,*** which causes epidemic typhus and under natural conditions infects only humans and the human body louse, and ***Rickettsia typhi,*** common in rats and occasionally transmitted to humans by rat fleas. The rickettsiae are obligate intracellular parasites in fleas, lice, mites, and ticks, and are often pathogenic for humans. These small, non-motile microorganisms appear as spherical forms about 0.3 μm long. Rickettsiae are susceptible to chemical disinfection and are destroyed by heat and dehydration. Antibiotics such as the tetracyclines and chloramphenicol are effective against epidemic typhus in humans.

Control of insect-borne diseases is initiated through control of the vector organism. Spraying the rooms of infected households with pesticides isolates the host from the vector organism. The draining of swamps can affect the ability of mosquitoes to reproduce. If the epidemiological cycle is effectively broken, then the disease can be controlled. Unfortunately, in the developing countries lack of sufficient funds makes eradication of vector organisms difficult. The problem is further aggravated by the tropical climate, which is congenial for effective incubation of microorganisms within the vector organisms. Presently, the strategy of the World Health Organization for developing countries is control of these diseases rather than total eradication. The goal is to slowly increase the general level of public health throughout a region by way of education and comprehensive planning as well as by control techniques.

8.5 NONINFECTIOUS DISEASES

In the developed countries, infectious parasitic diseases have been brought under control and in some areas virtually eradicated. This is due to the sanitary disposal of wastes, the disinfection of water supplies, increased care and cleanliness in food preparation, and the medical advances in drugs, antibiotics, immunization, and the early diagnosis of infected people. In these same countries, however, there has been an increase in the proportion of deaths resulting from degenerative ailments or non-infectious diseases such as cancer, diseases of the heart and the circulatory system, bronchitis, and emphysema.

Part of the reason for the shift to the non-infectious diseases is that a greater proportion of the population is older and the incidence of death due to these diseases increases with age. Table 8–8 gives the percentage distribution of deaths by cause in two model populations from a United Nations study on world mortality trends. Model A is typical of populations

TABLE 8-8   PERCENTAGE DISTRIBUTION OF DEATH BY CAUSE

Cause of Death	Model A Young age structure, life expectancy 50 years	Model B Old age structure, life expectancy 70 years
All causes	100.0	100.0
Infectious, parasitic diseases	34.1	6.5
Cancer	5.6	16.4
Heart and circulatory diseases	18.7	46.5
Violence	4.3	5.2
All others	37.3	25.4

Adapted from United Nations, 1973

in developing countries, with a young age structure (approximately 45 percent less than 15 years old) and an average life expectancy of 50 years. Model B is typical of populations in developed countries, with a relatively old age structure (only 20 percent less than 15 years old) and an average life expectancy at birth of 70 years.

It is also believed, but difficult to prove, that over time the pollution of our air and water with inorganic and organic chemicals contribute to these degenerative diseases. There have been well-documented episodes of gross pollution of water and air which have clearly resulted in an increase of non-infectious diseases. The Minimata Bay mercury poisoning in Japan in the 1950s and the air pollution episodes in London in December 1952 are frightening reminders of the effects pollution can have on humans. In the former, methyl mercury compounds in industrial effluents dumped into Japan's Minamata Bay were concentrated in fish which were, in turn, consumed by the residents. Between 1953 and 1960, 111 cases of mercury poisoning were recorded. In the latter case, a temperature inversion and the resulting static air mass caused a buildup of smoke and SO_2 in London to five times their normal levels for a period of approximately five days. This incidence is considered responsible for between 1,500 and 4,000 deaths, mainly among those already suffering from respiratory diseases.

Generally, it is impossible to attribute a particular health effect to any one pollutant in the water or the atmosphere. The health effects in animals resulting from administering large doses of a particular pollutant may be determined in the laboratory. In the environment, however, pollutants exist in very small concentrations and in combination with too many other factors for particular effects to be attributed to a single cause. It is difficult to be much more specific than to observe that those who live in more polluted urban environments have a greater incidence of noninfectious types of diseases than those living in cleaner rural environments.

Health effects as a result of exposure to certain inorganic and organic chemicals are not known. This is particularly true in the case of long-term exposure to low concentrations of toxic and hazardous contaminants. Epidemiological studies are usually restricted to workers who have been exposed to relatively high levels of specific pollutants in the work place. Taking the results of these studies and extrapolating them to the general population may be of some value in setting allowable limits of exposure. However, more research will be required before the matter of "safe limits" is adequately understood. In the remainder of this chapter, 15 of the most common contaminants affecting health are discussed. They are presented in alphabetical order and grouped as either inorganic or organic. Under each heading that follows, the available knowledge is summarized from Kruss, 1979. Tables 8–9 and 8–10 in Sections 8.5.1 and 8.5.2 respectively provide a summary of these contaminants, their sources, their occurrence in the environment, and their major health effects.

8.5.1 Inorganic Contaminants

Arsenic is a by-product of copper and lead smelting and the roasting of gold, silver, and cobalt ores. It is primarily an airborne pollutant, but it can contaminate bodies of water and accumulate in fish. It is also a component of some agricultural insecticides and fungicides. The main health effect is arsenic poisoning of workers in the gold mining industry and of agricultural workers who handle materials containing arsenic. One symptom is the paralysis of the lower limbs, although acute poisoning includes gastric and intestinal upsets. It is considered to be a potential carcinogen contributing to occupationally related lung cancer.

   Asbestos is used for the production of asbestos-cement floor tiles, brake linings, and gaskets, and the manufacture of fireproof linings and textiles. Those working directly in the mining of the mineral and the manufacture of its products suffer the most adverse effects, but consumers are also regularly exposed to asbestos due to the great variety of asbestos products in daily use. In the form of fibers, asbestos is an airborne pollutant, and inhalation of these fibers over an extended period of time can result in asbestos pneumoconiosis, or asbestosis, which is a severe scarring of the lungs. In addition to forcing the heart to work harder, this scarring can complicate other existing respiratory diseases. Fatalities from asbestosis are often the result of cardiac failure. Asbestos is generally regarded as a carcinogen. Lung cancer among people exposed to asbestos "dust" occurs with a frequency of more than twice that of the general population. This has led to public concern over the presence of fibers in air from products made with the mineral. Asbestos fibers can also be present in water supplies if asbestos products are manufactured nearby, and in processed beverages (soft drinks, beer, and wine) when asbestos filters are used for removing impurities. There is no evidence that ingestion of asbestos fibers is harmful to humans.

   Cadmium, a metal toxic to most species, is released into the environment from industries (electroplaters, battery producers, etc.) in sufficient quantities to warrant its classification as a pollutant. As a result of these industrial effluents, it is commonly found in municipal sewage sludges at concentrations higher than normal background values. The most significant health effects are found in workers subjected to cadmium fumes. Exposure to the fumes, which are suspected to be carcinogenic, can result in degeneration of the joints. Cadmium is taken up at all levels of the food chain, from microorganisms to humans. Human consumption of leafy vegetables, fish, shellfish, and drinking water is the usual method by which cadmium enters the body. Japan recorded over 230 cases of cadmium poisoning between 1962 and 1977 as a result of people eating food contaminated with cadmium. The source of the metal was traced to runoff from mine tailings.

Lead may be present in the food and water we consume and in the air we breathe. The combustion of leaded gasoline is the largest source of lead pollution in the atmosphere and is discussed further in Chapter 13. Raw water supplies can be contaminated by lead from the discharge of sewage treatment plants and from agricultural runoff. Water distribution systems may also contain high concentrations of lead due to the use of lead joints in water mains and lead pipes for water lines inside buildings, which were common at one time. The problem is more severe with soft water, which has a greater tendency to dissolve lead than hard water does. Although most ingested lead is excreted and 60 percent of inhaled lead is exhaled, lead still accumulates slowly in the body. The initial symptoms of lead poisoning are stomach-ache and physical weakness. The final stages may lead to a collapse of the central nervous system. Lead poisoning appears to be most prevalent in children, due partially to their greater capacity to absorb lead and their tendency, at early ages, to chew on cribs, toys, etc., that may have been painted with a lead-base paint.

Mercury poisoning was mentioned earlier in regard to the Minimata Bay disaster in Japan. Inorganic mercury compounds are used in the production of electrical components such as switches, and in the chlor-alkali industry for the production of chlorine and sodium hydroxide. Organic mercury compounds are used as slimicides and fungicides in the pulp and paper industry and in agriculture. Poisoning by methyl mercury (the most common organic compound) is characterized by numbness, speech impairment, and loss of motor coordination progressing to paralysis, deformity, coma, and death. Poisoning by inorganic mercury (particularly vapors) results in damage to the central nervous system and possibly psychotic disorders. Exposure to mercury can be either through the food chain or in the work place.

A steady diet of seafood from a mercury-contaminated source poses a substantial risk to the consumer. Mercury released in past decades is still present in bottom sediments of lakes and rivers, and will continue to be a source of pollution for the forseeable future.

Nitrates and Nitrites derived from the excessive use of fertilizers, can result in significant nitrate pollution of surface water and groundwater. Manure from livestock, food lots, and poultry operations also has a high nitrate content. The human body is capable of reducing nitrates to nitrites in the digestive system. There are two distinct threats to human health from nitrites. First, nitrites can oxidize the hemoglobin (containing Fe^{+2}), to methemoglobin (containing Fe^{+3}), which is incapable of transporting oxygen in the blood stream. This illness, known as methemoglobinemia or blue baby disease, is especially harmful to infants since they are particularly susceptible to asphyxiation by methemoglobinemia. Second, nitrites can combine with various amines in the gastrointestinal tract to form nitrosamines, many of which are known to be carcinogenic. Concern has been expressed about the use of nitrites in cured meats (bacon, prepared meats, hot dogs, etc.) to retard bacterial growth and prevent botulism. Unfortunately, a suitable chemical replacement for nitrites has not been found.

Sulfur dioxide is emitted primarily from the burning of coal and oil having a high sulfur content. Under the right atmospheric conditions (resulting in static air masses for a number of days), SO_2 can build up to deadly concentrations. This happened in London in

TABLE 8–9   COMMON INORGANIC CONTAMINANTS CAUSING NONINFECTIOUS DISEASES

Inorganic contaminant	Major source	Sphere most affected	Primary health effects
Arsenic	—Ore smelting, refining —Pesticides	—Air, water	—Arsenic poisoning (gastrointestinal disorders, lower limb paralysis)
Asbestos	—Heat-flame-resistant applications	—Air	—Asbestosis (scarring of lung) —Carcinogen
Cadmium	—Electroplaters, battery manufacturers	—Air, food, water	—Cadmium fumes, joint pain, lung, kidney disease —Possibly carcinogenic, teratogenic
Lead	—Leaded gasoline, batteries —Solder, radiation shielding	—Air, food, water	—Impairs nervous system, red blood cell synthesis —Depends on exposure

Mercury	—*Inorganic form*
	  Electrical goods
	  Chlor-alkali industry
	—*Organic form*
	  Slimicides
	  Fungicides
		—Water, food	—*Inorganic*: disorders of central nervous system, possible psychoses
			—*Organic*: numbness, impaired speech, paralysis, deformity, death.

Nitrates/Nitrites	—Nitrates: agricultural runoff —Nitrites: meat preservatives	—Food, water	—Nitrates + amines (in body) yield carcinogenic nitrosamines —Nitrates can cause methomoglobinemia in infants.
Sulfur Dioxide	—Combustion of sulfur-containing fuels	—Air	—Irritation of respiratory system —Precursor of acid rain, which is widely destructive.
Particulates	—Smoke from combustion —Dust, pollens, etc.	—Air	—Can lead to cardiac, respiratory ailments (emphysema, bronchitis) —Health effects more noticeable if particulates are in combination with other pollutants (e.g., SO_2)

1952, and in Belgium's Meuse Valley in 1930. Similar incidents have occurred in New York and Pennsylvania. High concentrations of SO_2, especially when accompanied by smoke particles, result in irritation of the respiratory system. Those affected are mainly the elderly with chronic respiratory problems. The contribution of SO_2 emissions to the acid rain problem was discussed in Chapter 5 and will also be dealt with in Chapter 13.

Particulate matter with a specific size of about 0.1 μm (the size of cigarette smoke particles) can penetrate deeply into the lungs and be deposited there. The greater the amount of foreign material in the lungs, the less efficient is the respiration system and the harder the heart has to work. This buildup of particulates is believed to lead to chronic heart and respiratory ailments such as emphysema and bronchitis. To make matters worse, the synergistic effects of particles and gaseous pollutants, such as the combination of SO_2 and smoke particles, are greater than the sum of their individual effects.

8.5.2 Organic Contaminants

Numerous manufactured organic chemicals are considered to be a potential threat to the health of many species, including humans. Of these, DDT, fenitrothion, and Mirex were developed as pesticides. Others, such as PCBs (polychlorinated biphenyls) were developed for quite benign uses, namely, cooling agents in electrical transformers, but through accidental release find their way into the environment. The epidemiological evidence against these chemicals varies, but considering that they are poisonous and have had some link to cancer or other degenerative diseases, public alarm at being unwittingly exposed is not unfounded.

DDT (*d*ichloro*d*iphenyl*t*richloroethane) has been used widely throughout the world. The World Health Organization (WHO) estimates that its effectiveness in controlling the *Anopheles* mosquito freed one billion people from the risk of malaria in the 1950s and 1960s, thereby preventing millions of deaths. However, DDT is a very stable and persistent chemical which remains in the environment for years. It is relatively insoluble in water, but readily soluble in fats and oils. It bioaccumulates in the fatty tissues of many species, including humans. The average concentration of DDT in humans ranges from about 1 ppm in countries with little DDT use to as high as 27 ppm in India, where DDT has been used extensively. Health risks exist as a result of exposure either to spraying or by ingestion. Symptoms of DDT poisoning include nervous disorders and abnormal decreases in white blood cell counts. There have been no reports of human fatalities directly attributable to DDT, but there is concern over the long-term effects of low-level concentrations.

Dioxin, although it represents a family of chemicals, is the common name applied to one of the deadliest chemicals ever manufactured. (Technically, it is more properly referred to as TCDD, or 2, 3, 7, 8-*t*etrachlor*d*ibenzopara-*d*ioxin). Dioxin occurs as an impurity in the manufacture of various chemicals and pesticides having a trichlorophenol base. Small amounts are also released to the atmosphere when plastics are burned; thus the concern when municipal solid wastes are incinerated.

Among the more prominent of the contaminated products are hexachlorophene, a germicide used for acne control, cleansing of newborn infants, and disinfection (it has been banned as a nonprescription drug), and 2, 4, 5-T, or trichlorophenoxyacetic acid, an herbicide, now banned in North America but used extensively in the Vietnam war in a 50:50 ratio with 2, 4-D (creating the defoliant called Agent Orange). Symptoms associated with exposure include changes to most internal organs, chloracne, nervous disorders, and death if exposed to sufficiently high concentrations. Dioxin is a confirmed teratogen (causes birth defects) and a suspected carcinogen. In Seveso, Italy, in 1976, equipment failure resulted in the release of between two and ten pounds of dioxin to the atmosphere. Before evacuation of the town, many animals died and people became ill with chloracne. The residents of the area are undergoing long-term monitoring to determine the permanent effects, if any, of short-term exposure to high concentrations of dioxin.

In the past few years Niagara Falls, New York, has become infamous for dioxin leaking out of old chemical dumps into the Niagara River, creating potential pollution of the water supply of 5 million North Americans. Even the spray from Niagara Falls is a contributor to pollution, releasing volatile organics with small amounts of dioxins into the atmosphere on a continuous basis. (McLachlan, 1987). In 1983, Times Beach, Missouri, a community of 2,000, was declared by the U.S. Environmental Protection Agency to be uninhabitable because of dioxin. The inhabitants were relocated at a cost of more than $30 million. The source of the dioxin was traced to contaminated waste oil that had been used as a dust suppressant on local roads and private properties. Mixed in with the oil were liquid wastes from a factory that had previously produced hexachlorophene.

Fenitrothion is an insecticide used in Eastern Canada for the control of the spruce budworm. It has a fairly rapid rate of decay in the environment and is of low toxicity to mammals. It is the subject of much medical debate concerning its role in the initiation of Reye's syndrome, a disease causing convulsions, brain damage, and possibly death following recovery from a minor viral infection. The cause of Reye's syndrome is unknown, but it is suspected by some researchers that the insecticide could be a contributing factor.

Mirex, another manufactured chlorinated hydrocarbon, has been found at low concentrations in bodies of water, various fish species, and aquatic birds. Mirex was developed as an insecticide (to control fire ants in the southern U.S.) and is also used as a fire retardant in plastics and for generating smoke in military exercises. Although there have been no documented cases of Mirex poisoning in humans, there is concern regarding the health of persons whose diet includes large amounts of Mirex-contaminated fish and other seafood. Mirex is toxic to many organisms, from algae to mammals, and its use has been curtailed due to its toxicity to nontarget species.

PCBs (polychlorinated biphenyls) are chemically inert, soluble in water, and not broken down at normal temperatures. As a result, PCBs have a number of industrial uses, e.g., as a dielectric fluid for industrial capacitors and transformers and as a fluid in hydraulic and heat transfer applications. Their high stability makes them persistent in the environment. They are related to DDT, but are even more stable. PCBs have been shown to cause cancer in rats, and it is feared that cancer could develop in workers exposed to the liquid or to the fumes from transformer fires. It is the cause of skin disorders (such as chloracne), headaches, and viral disturbances. An epidemic of PCB-related effects was discovered in Kyushu, Japan, in 1968 and was traced to a leak in the heating system of a rice-oil manufacturer. Samples of the oil were found to contain 2,000–3,000 ppm of a PCB compound. PCB contamination is usually accidental, and its use has been restricted in many countries.

Trihalomethanes, of which chloroform is the most common example, are produced in water and wastewater treatment plants when natural organic compounds combine with chlorine added for disinfection purposes. **Chloroform** has been used as an anaesthetic for many years, but it is now suspected of being carcinogenic, although at low concentrations its carcinogenic effects have not been established. The concentrations of chloroform found in chlorinated water supplies are normally very low and depend on the total organic content of the water being treated. Considerable research is now being carried out on the significance and control of trihalomethanes.

Ideally, we would like to set "safe" limits on the concentrations of all inorganic and organic contaminants which can cause noninfectious diseases. There is, of course, no such

TABLE 8–10   COMMON ORGANIC CONTAMINANTS CAUSING NONINFECTIOUS DISEASES

Organic contaminant	Major source	Sphere most affected	Primary health effects
DDT	—Application of pesticide throughout world	—Water, food chain	—Bioaccumulates in fatty tissues —Results in nervous disorders —Decreased white blood cell count —Persists in environment
Dioxin (specifically TCDD)	—Impurity of manufacture of trichlorophenols used in various biocides —Released by application or accident	—Water, food chain	—Extremely toxic in concentrated form, damage to kidney, liver and nervous system —Powerful teratogen —Possibly carcinogenic
Fenitrothion	—Insecticide spray on cultivated crops, forested land	—Water, air	—Only toxic to mammals at high dosages —May be partly responsible for initiating Reye's syndrome in children
Mirex	—Insecticides, fire retardant in plastics	—Water, food chain	—Biologically active, persistent —Toxicology varies with species —Bioaccumulates in food chain
PCB	—Dielectric, heat transfer and hydraulic fluid	—Food chain	—Persistent in environment —Probably carcinogenic, exposure results in chloracne, headaches, visual disturbances
Chloroform	—Previously used as anaesthetic —Presently in consumer goods, pharmaceuticals, pesticides —May be produced during chlorination of water supplies.	—Food, water	—Acutely toxic in high concentrations —Liver, heart damage —Carcinogenic to rodents
Trihalomethanes (includes chloroform)	—Accidentally produced in water as a result of certain organics (humic acids, etc.) and chlorination	—Water	—Possibly carcinogenic

thing as a safe level, but only an acceptable level of risk. Even this is difficult and in some cases impossible to determine. Establishing guidelines for contaminant concentration, be it in the air, soil, water, or food, is a challenge for researchers, engineers, and administrators alike. The epidemiology of many noninfectious diseases is extremely complex. The health effects of the elements or compounds previously discussed may be intensified or diminished by the presence of other chemicals. Due to the complexity and the innumerable unanswered questions concerning the effects of these contaminants, regulators must err on the side of caution and set limits as low as is reasonably possible, to minimize the risk to which the general public and the environment are exposed.

PROBLEMS

8.1. Describe and compare the nutritional requirements of autotrophic and heterotrophic bacteria.

8.2. Draw a typical bacterial cell and describe the major components and their functions.

8.3. (a) Draw the growth-death curve for a bacterial culture. Label the axes and all phases of the curve, and briefly explain the diagram.
(b) What comparisons can be made between the growth-death curve of the bacterial culture and a graph showing world human population growth. Can we learn any lessons from this comparison?

8.4. A batch culture of 100 unicellular bacteria has grown from a single bacterium in two hours through exponential growth. Assuming continued exponential growth, how many bacteria will the culture have after one additional hour?

8.5. Why are viruses difficult to remove from a water supply?

8.6. What are the basic differences between algae, fungi, and bacteria?

8.7. What is the rationale for studying domestic water supply and treatment concurrently with wastewater treatment and disposal?

8.8. Cite and describe another example of cooperative behavior between microorganisms similar to the cases noted in Section 8.3.2.

8.9. Why is filtration of water without chlorination partially effective in controlling pathogenic bacteria?

8.10. Outline two methods by which the spread of schistosomiasis in rural Africa might be controlled.

8.11. Compare the relative contributions of a treated water supply and the collection and treatment of human wastes toward the control of epidemic diseases in developing countries. If there are insufficient resources for both systems, should one be given priority over the other? Explain.

8.12. What are the factors affecting the virulence of a particular disease?

8.13. During an epidemic of a contagious disease, why doesn't everybody get infected?

8.14. Name four of the mechanisms of transfer of disease between humans.

8.15. Are you more likely to get sick by drinking water from a polluted stream in the winter or in the summer? Give details of your reasoning.

8.16. Name three waterborne diseases, and note their symptoms and causative organisms.

8.17. What are the requirements for an organism to be an indicator organism?

8.18. What are coliforms? Why is *Escherichia coli* considered an indicator of pollution? Why are the results from the coliform test considered to be presumptive?

8.19. Why are you less likely to contact an airborne infection outdoors than indoors?

8.20. What are the differences between airborne infectious diseases and diseases caused by air pollution?

8.21. Name two airborne infections, their symptoms, and their causative organisms.

8.22. How does an outbreak of bubonic plague occur? How does the outbreak continue to spread? How can it be controlled?

8.23. Describe the necessary conditions under which an insect-borne disease becomes an epidemic.

8.24. What methods are there for controlling insect-borne diseases?

8.25. Should the use of the insecticide DDT be stopped completely? Give reasons for your answer.

8.26. Why is there so much concern and controversy about the amounts of dioxin, Mirex, and other synthetic organic chemicals found in water supplies?

8.27. How do you account for the increase in deaths due to noninfectious diseases in North America over the past century?

REFERENCES

APHA, AWWA, and WPCF *Standard Methods for the Examination of Water and Wastewater*. 16th ed. Washington, D.C.: (American Public Health Association, American Water Works Association, and Water Pollution Control Federation), 1985.

BAKER, M. N. *The Quest For Pure Water*. Vol. 1. New York: American Water Works Association, 1948.

BUCKMAN, O. H., and BRADY, N. C. *The Nature and Properties of Soils*, 6th ed. New York: Macmillan, 1960.

CLARK, J. W., VIESSMAN, W., and HAMMER, M. J. *Water Supply and Pollution Control*. 3rd ed. New York: IEP, 1977.

COMMONER, B. *The Closing Circle*. New York: Knopf, 1971 and Bantam, 1972.

ECKHOLM, E. P. *Down to Earth: Environment and Human Needs*. New York and London: Norton, 1982.

ECKHOLM, E. P. *The Picture of Health: Environmental Sources of Disease*. New York and London: Norton, 1977.

GAUDY, A. F., JR., AND GAUDY, E. T. *Microbiology for Environmental Scientists and Engineers*. New York: McGraw Hill, 1980.

FEACHEM, R. G., BRADLEY, D. J., GARELICK, H., and MARA, D. D. *Sanitation and Disease: Health Aspects of Excreta and Wastewater Management*. Washington, D.C.: Published for The World Bank by Wiley, 1983.

KRUUS, P., and VALERIOTO, I. M. *Controversial Chemicals*. Montreal: Multiscience Publication Ltd., 1979.

MCLACHLAN, M. S. "A Model of the Fate of Organic Chemicals in the Niagara River." M.A.Sc. thesis, University of Toronto, 1987.

MITCHELL, R. *Introduction to Environmental Microbiology*. Englewood Cliffs, N.J.: Prentice-Hall, 1974.

PALMER, C. M. *Algae in Water Supplies: An Illustrated Manual on the Identification, Significance and Control of Algae in Water Supplies*. Public Health Service Publication No. 657. Washington, D.C.: U.S. Department of Health, Education and Welfare, 1959.

PELCZAR, M. J., JR., REID, R. D., and CHAN, E. C. S. *Microbiology*. 4th ed. New York: McGraw Hill, 1977.

PERKINS, H. C. *Air Pollution*. New York: McGraw Hill, 1974.

SARTWELL, P. E. (ed). *Maxy-Rosenau Preventive Medicine and Public Health*. 10th ed. New York: Appleton-Century-Crofts, 1973.

United Nations. *The Determinants and Consequences of Population Trends*. New York: United Nations, 1973.

WINSLOW, C. E. *The Conquest of Epidemic Disease*. Princeton, N.J.: Princeton University Press, 1943.

ZIEGLER, P. *The Black Death*. London: Penguin Books, 1969.

9

Ecology

Thomas C. Hutchinson

9.1 INTRODUCTORY CONCEPTS

The term "ecology" is derived from the Greek "oikos," meaning "house," combined with "logy," meaning "the study of." Thus, literally, ecology is the study of the earth's households.

For our use, **ecology** can be defined as the study of the relationship between organisms and their environment. Here, "environment" is taken to mean both the physical and chemical environment of air, soil, and water, and also the biological environment itself. The range of ecological studies is very broad. Examples include an investigation of the chemistry of the soils to which a particular plant species is restricted, a study of the relationship between the number of eggs a song bird lays and the amount of food available for the chicks to eat, and the changes occurring in a lake or river when untreated sewage is added.

The nonliving or **abiotic** physicochemical and the living, **biotic** assemblages of plants, animals, and microbes comprise interdependent ecological systems or **ecosystems**. Ecosystems can be large or small, consisting of a very large number of, or a few, species. They are frequently defined by their dominant vegetation types and have a certain recognizable unity of their own. Examples of ecosystems which occupy large parts of the earth's surface are the tropical rain forests, boreal coniferous forests, deciduous or hardwood forests, tundra and Alpine ecosystems, prairie grasslands, cactus deserts, salt marshes, coral reefs, swamps and marshes, lakes, marine continental shelf, and open ocean. The boundaries around a defined ecosystem are generally very unclear, and substantial heterogeneity can exist in such large-scale ecosystems as those just listed. Small ecosystems can also exist within a much larger major type. A small pond on top of a mountain amongst the predominantly Alpine terrain would be such an example. Within each ecosystem there is a dependence of one species on other species. Ecosystems are also controlled by, and a consequence of, climate.

The two overriding factors which keep the ecosystem together and functioning as an interdependent whole are the need for *nutrients* and the need for *energy*. While the nutrients within an ecosystem are continuously cycled and recycled through all of its components with quite limited losses from and inputs into the system, the energy is ultimately always supplied from the sun, by incoming solar radiation, and is passed through the system largely unidirectionally.

Ecology is the study of the interrelationships among plants and animals and the interactions between living organisms and their physical environment.

An **ecosystem** is a group of plants or animals, together with that part of the physical environment with which they interact. An ecosystem is defined to be nearly self-contained, so that the matter which flows into and out of it is small compared to the quantities which are internally recycled in a continuous exchange of the essentials of life.

Biota are all the living elements of an ecosystem or a given area.

9.2 ENERGY FLOW IN ECOSYSTEMS

This section draws heavily on the account given by E. J. Kormondy in Chapter 2 of his book, *Concepts of Ecology* (1969). All biological activity is dependent upon green plants successfully utilizing energy that comes ultimately from the sun. In this process, the radiant energy of the sun is transformed first to chemical energy and then to mechanical energy (heat) in cellular metabolism.

The sun can be considered to be a continuously exploding hydrogen bomb, with a temperature and composition such that hydrogen is transmuted to helium, with concomitant release of considerable energy in the form of electromagnetic waves. While these extend from shortwave X- and gamma rays to long-wave radio waves, about 99 percent of the total energy is in the region of wavelengths from 0.2 to 4.0 μm, i.e., the region from ultraviolet to infrared. About 50 percent of this energy is in the region of the visible spectrum (0.38 to 0.77 μm) and is partially utilized in photosynthesis. Because the earth is such a small target in the solar system, only about one fifty-millionth of the sun's tremendous energy output reaches the earth's outer atmosphere (190 km above the earth's surface) and it does so at a constant rate. This constant rate is referred to as the **solar flux**

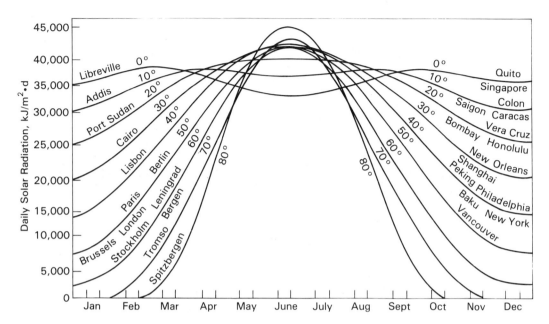

Figure 9–1 Daily totals of solar radiation received on a horizontal surface for different geographical latitudes at different times of the year and based on a solar constant value of 81.2 kJ/m² min. Source: Gates, 1962

Note: Original figure in units of g cal/cm² yr based on solar constant value of 1.94 g cal/cm² yr. Conversion by author.

1 g cal = 4.1855 J; 1 J = 1 W-sec.

or **solar constant,** defined as the amount of radiant energy of all wavelengths that cross a unit area or surface per unit of time. This value is estimated to be approximately 1.4 kJ per m² per second, for a total energy income of 5.5×10^{21} kJ (1.5×10^{18} kW-hrs) (see also Section 7.3). Because of the elliptical orbit of the earth around the sun, the flux at any given location varies seasonally with latitude. Because of the earth's rotation, the flux at a given place also varies diurnally (Figure 9–1).

The process by which chlorophyll-bearing plants use energy from the sun to convert carbon dioxide and water to sugars is called photosynthesis. The generalized equation of photosynthesis is

$$6CO_2 + 12H_2O + 2{,}800 \text{ kJ energy} \xrightarrow[\text{(chlorophyll)}]{} C_6H_{12}O_6 + 6CO_2 + 6H_2O \qquad (9.1)$$

Photosynthetic activity and rates of carbon dioxide fixation into plant carbohydrates can be estimated in a number of ways, including rates of CO_2 removal and rates of O_2 production, as well as rates of accumulation of photosynthetic intermediate compounds.

A few ecological definitions are necessary before proceeding.

An **autotroph*** is an organism that obtains its cell carbon from an inorganic source (CO_2, HCO_3) and its energy *from the sun* (actually a photoautotroph as distinct from a chemoautotroph, which gets its energy from the oxidation of inorganic chemical compounds).

A **heterotroph*** is an organism that obtains both its cell carbon and its energy from organic matter.

A **chemotroph*** is an organism that obtains its energy from the oxidation of simple inorganic compounds, such as FeS and H_2S, and its cell carbon from organic matter. Chemotrophs are relatively insignificant in the energy relations of an ecosystem, but play a significant role in the movement of mineral nutrients in the ecosystem.

The **food chain** is an idealized pattern of flow of energy in a natural ecosystem. In the classical food chain, plants are eaten only by primary consumers, primary consumers are eaten only by secondary consumers, and so forth.

The **food web** is the actual pattern of food consumption in a natural ecosystem. A given organism may obtain nourishment from many different trophic levels, thus giving rise to a complex, interwoven series of energy transfers.

Productivity is the rate of fixation of energy into tissue. Primary productivity is energy fixation by plants; secondary productivity is at higher trophic levels.

Trophic levels are levels of nourishment. A plant that obtains its energy directly from the sun occupies the first trophic level (autotroph). An organism that consumes the tissue of an autotroph occupies the second trophic level (herbivore), and an organism which eats the organism that had eaten autotrophs occupies the third trophic level (carnivore).

Transpiration is the controlled evaporation of water vapor from the surface of leaf tissues.

   * In ecology the energy source is the main basis for the differentiation of organisms, whereas in microbiology (see Section 8.2.3) the carbon source is usually emphasized.

9.2.1 Estimates of Primary Production

The term **primary producer** (engaging in **primary production**) is used to designate any autotrophic organism that is capable of directly utilizing the radiant energy of the sun. This includes organisms capable of photosynthesis. Transeau (1926) calculated primary production in a midwestern U.S.A. cornfield based on an estimated harvest of 10,000 plants per 0.405 hectare, together weighing 6,000 kilograms, and on chemical analysis of

TABLE 9–1   ENERGY BUDGET OF AN ACRE OF CORN (0.405 HECTARE) DURING ONE GROWING SEASON OF 100 DAYS

	Glucose (kg)	Kilojoules (millions)	Solar energy utilized (%)
Incident solar radiation	—	8,550	100.0
Biological utilization:			
Net production (*NP*)	6,687	105.9	1.2
Respiration (*R*)	2,045	32.2	0.4
Gross production (*GP*)	8,732	138.1	1.6
$GP = (NP + R)$			
Energy utilized in transpiration	—	3,808	44.5
Energy not utilized	—	4,604	53.9

Source: Transeau, 1926

this material. He calculated that the corn plants contained 2,674 kg of carbon, all of which had entered as CO_2 through photosynthesis. This is equivalent to 6,687 kg of glucose. But, in addition, the corn plants had to have used up some glucose in respiration to maintain themselves. Transeau estimated this as 2,045 kg of glucose, to give a total or gross production of 8,732 kg. One kg of glucose requires an energy input of 15.7×10^3 kJ. Therefore, 138.1×10^6 kJ have been utilized in gross production, of which 32.2×10^6 kJ has been used in metabolic-respiratory activities. Another energy requirement in plants is that of **transpiration**, i.e., biologically controlled evaporation from the plants by which water and nutrients are taken up from the soil and moved through the plants to the leaves. The water is then evaporated through small pores of the leaves which can be opened or closed so as to control water loss. It is perhaps surprising to learn that the efficiency of energy utilization by the cornfield is only 1.6 percent (Table 9–1), i.e., only 1.6 percent of the total energy available is incorporated into carbohydrate through photosynthesis. Mathematically,

$$\frac{\text{Gross production}}{\text{Solar radiation}} \times 100 = \frac{138 \times 10^6 \text{ kJ}}{8.55 \times 10^9 \text{ kJ}} \times 100 = 1.6\% \qquad (9.2)$$

Most natural ecosystems operate with an overall efficiency between 0.1 to 2 percent in nature, while the very best agricultural system can achieve values as high as 3 percent. Open ocean systems, which cover the majority of the earth's surface, are closer to the 0.1 percent conversion rate, although recent information suggests that ocean productivity may have been strongly underestimated.

Another interesting factor in energy conversion is the efficiency with which the autotrophs use the energy that they have incorporated. This utilization is really the difference between gross production and net production expressed as a percentage. For Transeau's example, it is

$$\frac{\text{Energy of respiration}}{\text{Energy of gross production}} \times 100 = \frac{32.2 \times 10^6 \text{ kJ}}{138.1 \times 10^6 \text{ kJ}} \times 100 = 23.4\% \qquad (9.3)$$

TABLE 9-2 ANNUAL ENERGY BUDGET OF (A) LAKE MENDOTA, WISCONSIN, AND (B) CEDAR BOG LAKE, MINNESOTA

	kJ/m²-yr	Solar energy utilized (%)
A. Lake Mendota, Wisconsin		
Incident solar radiation	4,975,390	100.0
Plant utilization		
Phytoplankton		
Net production (NP)	12,515	
Respiration (R)	4,185	
Gross production (GP)	16,700	
Bottom-living plants		
Net production (NP)	920	
Respiration (R)	290	
Gross production (GP)	1,210	
Gross production by autotrophs	17,910	0.36
B. Cedar Bog Lake, Minnesota		
Incident solar radiation	4,975,390	100.0
Plant utilization		
Net production (NP)	3,680	
Respiration (R)	970	
Gross production (GP)	4,660	0.10

Source: (a) Lake Mendota: Juday, 1940. (b) Cedar Bog Lake: Lindeman, 1942.

In other words, although only 1.6% of the total energy was utilized in carbohydrate production, the corn plants are quite efficient in converting the captured energy to biomass, utilizing 76.6 percent of it ($105.9/138.1 \times 100\%$).

In aquatic systems, energy capture is considerably less efficient than in terrestrial systems. Data for two freshwater lakes are given in Table 9–2. Juday (1940) found that only 0.36 percent of the solar flux for Lake Mendota, Wisconsin, was incorporated in gross production at the autotroph level. Over 90 percent of this incorporated energy was used by the phytoplankton, while less than 10 percent was used by the plants growing attached to the mud at the bottom of the pond.

The acidic Cedar Bog Lake in Minnesota, with its brown humic-stained waters, is only one-quarter as efficient as Lake Mendota, because its colored waters do not transmit light as well as the clearer Lake Mendota waters. Gross production, including losses due to grazing and decomposition, was found by Lindeman (1940) to be at an efficiency of 0.10 percent. Respiratory maintenance used 21 percent of this. This difference in primary energy capture is in large measure due to the decrease in light penetration in water to the depth at which the plants are growing. Since light is measured at the water surface, 0.10 percent (Cedar Bog Lake) is the overall photosynthetic energy conversion efficiency. If measurement of radiation were taken at plant depths in the water, the efficiency would rise to about 1 and 3 percent, respectively, for the two examples in the table. These are closer to the values of the terrestrial systems.

9.2.2 Comparison of Primary Productivity in Different World Ecosystems

Typical productivity values in different world ecosystems are given in Table 9–3. Gross production is controlled by a number of factors such as respiration and nutrient supply (mainly nitrogen and phosphorus), and a number of key climatic variables, notably light supply, length of growing season, temperature, and water supply.

Some of these major constraints emerge out of a consideration of Table 9–3. It is useful to have a rather careful look at this table, as it tells us a great deal about world food production potential and the major differences among ecosystems.

From the table, the most productive ecosystem in terms of rate of annual production is the tropical rain forest. Perhaps surprisingly, swamps and marshes are equally productive on average, though the area they occupy is much less. This high marsh productivity is due to their high nutrient status, while the high tropical rain forest productivity is due to high temperatures, continuous growth (no seasonal dormancy), and high rainfalls. The major constraint in the tropics is nutrient supply. Clearing such forests is fraught with dangers, as experienced in many parts of the world, since soil erosion in such high-rainfall areas can be devastating and much of the ecosystem's nutrients are, in fact, in the forest trees themselves and are removed with the cut timber.

Temperature is the major constraint in the reduction in productivity per year in moving

TABLE 9–3 THE PRIMARY PRODUCTION OF THE EARTH

Ecosystem type	Area 10^6 km^2	Net primary productivity per unit area, normal range dry g/m^2/yr	Mean	World net primary production 10^9 dry t/yr
Tropical rain forest	24.5	1,000–3,500	2,000	49.4
Temperate forest	12.0	600–2,500	1,250	14.9
Boreal forest	12.0	400–2,000	800	9.6
Woodland and shrub land	8.5	250–1,200	700	6.0
Savanna	15.0	200–2,000	900	13.5
Temperate grassland	9.0	200–1,500	600	5.4
Tundra and alpine	8.0	10–400	140	1.1
Desert and semidesert	42.0	0–250	40	1.7
Cultivated land	14.0	100–3,500	650	9.1
Swamp and marsh	2.0	800–3,500	2,000	4.0
Lake and stream	2.0	100–1,500	250	0.5
Total continental	149.0		773	115
Open ocean	332.0	2–400	125	41.5
Continental shelf, upwelling	27.0	200–1,000	360	9.8
Algal beds, reefs, estuaries	2.0	500–4,000	1,800	3.7
Total marine	361.0		152	55
World total	510.0		333	170

Source: Westlake, 1963, as modified and expanded from Whittaker, 1961

from tropical to temperate to arctic (tundra and Alpine) regions. Light intensity and the length of the growing season are also involved. The effect of water supply can be seen by comparing temperate grassland productivity (600 dry g/m² yr) with desert (40 dry g/m² yr). Higher productivity in estuaries (1,800 dry g/m² yr) is due to the high nutritional status of these areas (e.g., the deltas of the Mississippi, Nile, and other rivers).

In looking at world net primary productivity, the key factors are the area of the earth's surface occupied by the different ecosystems and the net primary productivity per unit area. Enormous areas are occupied by deserts and semideserts with their low productivity. Thus, their 42 million km² produce an annual total of only 1.7×10^9 dry tons, while the 24.5 million km² of tropical rain forests yield approximately 49.4×10^9 dry tons. World agriculture produces about 9.1×10^9 tons, much of this, of course, being in nonedible portions of crops, and much of it being lost to diseases, to pests, in storage, and in spoilage. The total land yield is about 115×10^9 tons of dry matter per year, while the oceans, with two-and-one-half times the surface, yield only 55×10^9 tons of dry matter per year. It is thus unrealistic to consider the ocean as providing a huge untapped food supply for the world's predicted population growth. In fact, most fisheries are already overexploited and are in need of replenishment. The Antarctic is presently an exception, but krill, an amazingly abundant crustacean, are being caught in unrestricted amounts by Japanese and U.S.S.R. trawlers, so the surplus there may not last for long.

One other point is worthy of note. Since oxygen is released in stoichiometric proportions to carbon dioxide utilization in photosynthesis, it follows that the land, with almost twice the productivity of the oceans, produces about two times as much O_2 as do the oceans. A high proportion of the oxygen from vegetation is produced in tropical regions. Thus, destruction of green plants in the oceans and in tropical land areas can be expected to have long-term consequences on atmospheric oxygen levels, just as fossil fuel burning and forest destruction are causing a buildup of CO_2 in the atmosphere (see Chapter 5). However, an as yet unknown amount of oxygen replenishment occurs as a result of iron oxide reduction by bacteria (Stumm and Morgan, 1981).

9.2.3 Energy Flow in Ecosystems beyond Primary Producers

We need to see how the initial conversion of incoming radiant energy by primary producers in fact sustains all the organisms in the ecosystem, not just the green plants. In the example of Cedar Bog Lake studied by Lindeman, gross production was 4,660 kJ/m² yr, while 970 kJ/m² yr were consumed in the metabolic activities needed to sustain the primary producers and allow them to reproduce. Figure 9–2 indicates that 630 kJ/m² yr of the 4,660 kJ/m² yr are consumed by the herbivores. This is 17 percent used by primary consumers of net autotroph production. Only 125 kJ/m² yr out of 440 kJ/m² yr, or 28.4 percent, of this available herbivore energy is actually used by the carnivores. Although this is a more efficient utilization of resources than occurs at the autotroph-herbivore transfer level, it still leaves room for greater exploitation.

At the level of the carnivore, about 60 percent of the energy intake is consumed in metabolic activity, while the unconsumed remainder becomes part of the sediment. Thus,

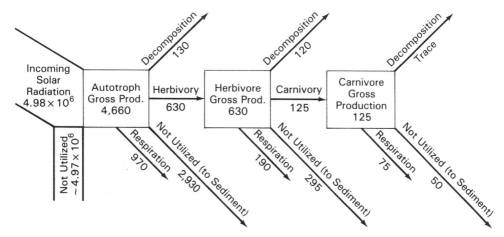

Figure 9–2 Fate of energy incorporated by carnivores in Cedar Bog Lake, Minnesota in Kilojoules per square meter per year. Source: Lindeman, 1942

Note: Conversion from original g cal/cm² yr to kJ/m² yr by author.

(g cal/cm² yr × 41.855 = kJ/m² yr).

the percentage of available energy used for metabolic activity increases through the *trophic series*, autotroph → herbivore → carnivore, at Cedar Bog Lake from 17 to 28 to 60 percent. This is typical of many food chains. One of the main reasons is that the herbivores have to move around in search of the green plants on which they feed, while a still greater expenditure of energy is needed by the carnivores as they search for and capture the herbivores.

9.3 FOOD CHAIN AND TROPHIC LEVELS

The sequence of consumption from the autotrophs through the carnivores represents the sequence of the food chains, in which each link is dependent for its food (energy) supply on the immediately preceding link. These positions along food chains are called **trophic levels.** Often, the boundaries of the levels are not sharp. Many animals find food that is suitable in size range and other characteristics at several different trophic levels. In order to describe the network of various trophic levels, showing their interconnections, the term **food web** is often used.

In considering the flow and utilization of energy in the food chain, it is clear that the energy movement in the ecosystem is one way, or unidirectional. As it moves progressively through the various trophic levels, it is no longer available to the previous level. The relationships between the various trophic levels can be expressed as shown in Figure 9–3. These are known as the **productivity pyramids.** Pyramids can also be used to represent several of the other relationships in an ecosystem, e.g., biomass (total organic matter) and numbers of organisms (see figure).

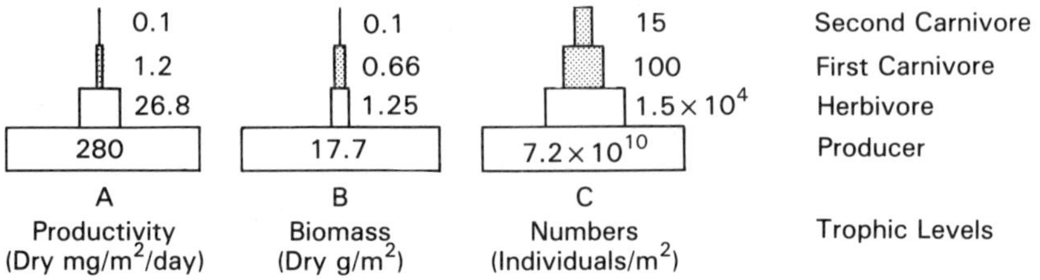

0.1	0.1	15	Second Carnivore
1.2	0.66	100	First Carnivore
26.8	1.25	1.5×10^4	Herbivore
280	17.7	7.2×10^{10}	Producer
A	B	C	
Productivity (Dry mg/m^2/day)	Biomass (Dry g/m^2)	Numbers (Individuals/m^2)	Trophic Levels

Figure 9–3 Community pyramids for an experimental pond. Source: Whittaker, 1961
Productivity was estimated from the rate of phosphorus uptake in a shallow pond of low
nutrient content. The fourth trophic level was estimated as a fraction of carnivores feeding
on both the second and third levels. The widths of the steps for numbers of organisms are
on a logarithmic scale.

TABLE 9–4 EXAMPLES OF CONCENTRATION FACTORS DUE TO FOOD CHAIN
ACCUMULATION OF POLLUTANTS OF VARIOUS KINDS*

**A. C.F. in Biota for Various Levels of Cadmium (Cd) in Water
(Based on Dry Weight)**

Cadmium Level in water (mg/L)	0.01**	0.0025	0.05	0.010
Floating fern	21,000	25,000	31,000	31,000
Duckweed	120,000	155,000	140,000	48,000
Water hyacinth				
roots	42,000	50,000	27,000	17,000
leaves	7,000	7,000	7,000	4,000
Zooplankton	—	2,600	16,600	6,000
Snails				
tissue	22,000	22,000	6,000	19,000
shell	800	2,000	400	1,000
Fish	600	600	2,600	700
Sediment	10	36	11	25

B. Food Chain Accumulation of DDT in a California Lake

	DDT (ppm)	Concentration factor (C.F.)
Water	0.00005	1
Plankton	0.04	800
Silverside minnow	0.094	18,800
Pickerel (predatory fish)	1.33	26,600
Tern (feeds on small fish and animals)	3.91	78,200
Herring Gull (scavenger)	6.00	120,000
Merganser (fish-eating duck)	22.8	460,000

Source: Hunt, and Bischoff, 1960
* Numbers shown are concentration factors (C.F.) at various levels of concentration of pollutants.
** The leftmost column (headed 0.01) is from one study, the other three from a second study.

Because, in general, each carnivore has to consume large numbers of herbivores to maintain itself, and each of these herbivores has to consume many times its own biomass per year of autotrophs to maintain itself, substances which are nonbiodegradable by nature, when they enter a food chain, are biomagnified at each succeeding trophic level. This has led to many of the well-known environmental problems of the past 20–30 years. The best known example is the biomagnification of organic pesticides such as the chlorinated hydrocarbons, which include DDT (Table 9–4). The concentration of DDT can be increased many thousandfold in the fatty tissues of carnivores such as owls, peregrine falcons, ospreys, pike, muskie, and bass, as well as in fish-eating birds such as brown pelicans, loons, and gannets. DDT can enter bodies of water by drainage from agricultural land, from aerial drift during spraying, and from deliberate insecticidal additions to such bodies to kill gnats and mosquitoes. Drastic declines in the number of many birds of prey that survive on aquatic life have resulted from this bioaccumulation, since one effect of DDT (and chlorinated hydrocarbons in general) is interference with calcium metabolism. This leads to the production of thin-walled eggshells, which are easily broken. Other examples of adverse effects of food chain biomagnification have come from the use of mercury as a fungal-killing seed dressing, which is then picked up by grain-feeding birds, such as pheasants and grouse, and in a number of instances has accumulated to lethal levels.

People, being at the ends of food chains on which they depend, can also be recipients of accumulations of toxic chemicals in the food they eat. Food processing and quality control are essential to safeguard our health and to avoid such situations. One example of a safeguarding action was the banning of fishing in the Great Lakes and other lakes in the area in the early 1970s because of excessive mercury in the fish.

9.4 NUTRIENT CYCLES

The supply of nutrients other than CO_2 to an ecosystem comes principally from the soil, but also to a smaller extent from the air, in rain and snow and as dust. The supply of many nutrients is quite limited because they are in short supply in the soil and in other sources. Nutrients are cycled in such a way that they are either incorporated into plants and animals, or else they are made available for plant uptake by the decomposition of dead plant and animal remains. The pathways from sources to sinks and back to sources are termed **elemental cycles,** and they differ among the various elements. We shall consider briefly the three most important nutrient cycles, those of carbon, nitrogen, and phosphorus.

9.4.1 Carbon Cycle

Carbon is required in large amounts as a basic building block for all organic matter. The ultimate source of carbon for organic matter is carbon dioxide, converted to organic matter in photosynthesis. In nature, the movement of carbon is from the atmospheric reservoir of carbon dioxide to green plants and on to consumers, and from both of these groups on to the microbial decomposer organisms. Algae and autotrophic bacteria also incorporate or "fix" carbon from atmospheric CO_2, producing carbohydrates and other complex organic substances. These are distributed through the food chain and make up the tissues of living matter. Fossil fuels, carbonate rocks, and carbon dioxide dissolved in the oceans are major additional reservoirs of carbon, although the first two of these are not naturally accessible to plants and animals. These "bound" sources of carbon become available when CO_2 is released during the burning of fossil fuels and through the action of CO_2 (from microbial decomposition) in converting insoluble carbonate to soluble bicarbonate.

The return of carbon dioxide to the atmospheric reservoir is achieved in a number of ways. Perhaps the best known is through the respiratory processes of humans and animals. However, by far the greatest quantities of carbon dioxide are returned to the atmosphere through the activities of groups of bacteria and fungi, which use dead organic matter as their food source. They thus oxidize the dead material, either directly or in a number of stages, with CO_2 and H_2O as end products, completing the cycle. Other sources returning CO_2 to the atmosphere are forest fires and the combustion of fossil fuels and other organic matter. The burning of dried-out peat, coal, or oil is an example of utilizing an ancient photosynthetic biomass for a source of heat energy. The carbon is oxidized to carbon dioxide in each case.

The geological component of the carbon cycle involves (a) the accumulation, slow decomposition, and compaction of plant material, forming peat, coal, and oil, and (b) the accumulation and compaction of animal shells and microscopic diatom skeletons, forming carbonate rocks. Calcium carbonate can also be precipitated in fresh waters when algae remove CO_2 from the water, thus decreasing its pH. When mixed with clay, these deposits form calcareous marls, which in time become compacted as limestone. Huge coal deposits and much of the limestone were laid down during the Carboniferous period when shallow water and warm climate predominated over the earth. As discussed in Chapter 6, carbon dioxide also diffuses into and out of water, in which it dissolves to form carbonic acid (H_2CO_3).* This dissociates in a series of reactions to form a hydrogen ion (H^+) and a bicarbonate ion (HCO_3^-), the latter of which in turn dissociates to form another hydrogen ion (H^+) and a carbonate ion (CO_3^{-2}). All reactions are reversible and depend upon diffusion gradients and pH.

The discharge of domestic sewage and organic industrial wastes can contribute large quantities of carbon to receiving waters. The need to reduce organic matter in these wastewaters is one of the principal reasons for wastewater treatment (see Chapter 12).

An overall picture of the carbon cycle is shown in Figure 9–4. The most important points to bear in mind are that (1) all green land plants obtain their carbon from gaseous carbon dioxide, (2) water plants obtain their carbon from bicarbonate, and (3) the carbon complexes formed in (1) and (2) are returned to their original forms by microbial decomposition.

9.4.2 Nitrogen Cycle

Another important nutrient cycle is that of nitrogen, shown schematically in Figure 9–5. Nitrogen is a critically important element for all life. Proteins, which are constituents of

---

* H_2CO_3 has been defined by Stumm and Morgan (1981) to be $H_2CO_3 = CO_2 \cdot H_2O + CO_2(aq) + H_2CO_3$, with the $CO_2 \cdot H_2O$ being truly carbonic acid in its predominant form.

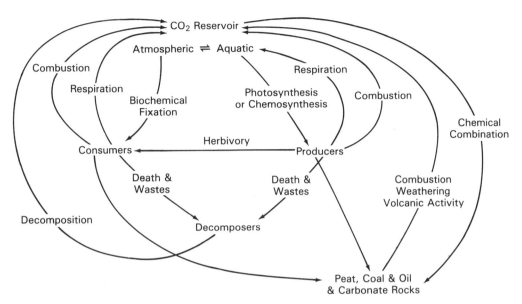

Figure 9–4 The carbon cycle. Source: Kormondy, 1969

all living cells, contain an average of 16 percent nitrogen by weight. Other complex nitrogenous substances important to life are nucleic acids and amino sugars. Without a continuous supply of nitrogen, life on earth would cease.

The nitrogen cycle is somewhat like the carbon cycle, but with a number of critical differences. Even though 79 percent of the earth's atmosphere is composed of elemental nitrogen (N_2), this inert gas is entirely unavailable for uptake by most plants and animals. This is in stark contrast to the small amount of CO_2 (0.03 percent, or 330 ppmV) in the atmosphere, which is readily available for plant uptake. A relatively few microbes are capable of "fixing" atmospheric nitrogen from the inorganic to the organic form. Such microbiological fixation averages 140–700 mg/m² yr. In very fertile agricultural areas it can exceed 20,000 mg/m² yr.

A number of bacteria, fungi, and blue-green algae are known to be able to fix nitrogen. Nitrogen fixation involves the direct incorporation of atmospheric nitrogen into the organic "body" of the fixing organism. The nitrogen fixers constitute only a very small portion of these groups overall. They can be divided into (a) **symbiotic nitrogen fixers,** which are largely bacteria and which are associated with the roots of legumes (members of the pea and bean family) and some other flowering plants, and (b) **free-living nitrogen fixers.** The genus **Rhizobium** includes those bacteria which inhabit the nodules which develop on the roots of members of the pea and bean family. They are present in soil and infect the fine roots as seedlings grow. The root produces a special nodule which houses the rhizobia, in which the bacteria convert atmospheric nitrogen (N_2) into the organic nitrogen constituents of their own cells. Since bacterial cells die very rapidly, this nitrogen becomes available to the higher plants. Crops of clover and beans actually add nitrogen to the soils

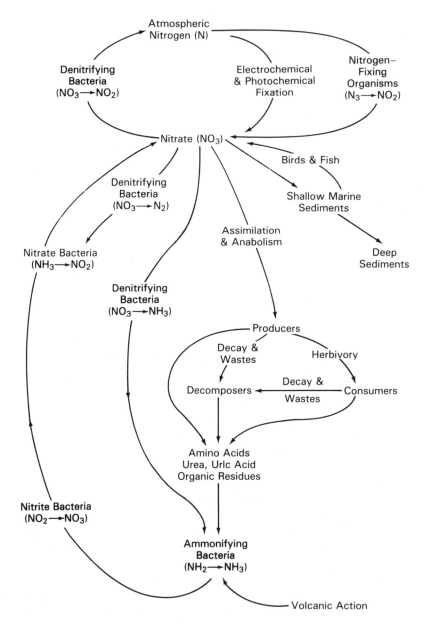

Figure 9–5 The nitrogen cycle. Source: Kormondy 1969

in which they grow and eliminate the need for expensive fertilizers. A large scientific effort is under way in many countries to find bacteria which can form a similar association with the cereal grain crops. The symbiotic* nitrogen fixers seem to be confined to terrestrial ecosystems and have not been found in aquatic habitats, the one exception being a marine worm that attacks submerged wood.

Among the nonsymbiotic nitrogen fixers are both aerobic and anaerobic free-living bacteria, as well as cyanobacteria. These occur in soils and in both marine and fresh waters and can add substantially to the nitrogen content of these environments. An additional but generally minor source of atmospheric nitrogen to soils and waters are lightning storms in which electrochemical nitrogen conversions take place.

Nitrogen enters the producer-consumer food chain when plants take it up from the soil solution either as nitrates (NO_3^-) or as the ammonium ion (NH_4^+). Nitrate can also be converted to ammonia by **denitrifying** bacteria in the soil, especially by bacteria and fungi in waterlogged soils. Such conversion also occurs in low-oxygen conditions in lakes. The process is called **denitrification.** The so-called **nitrifying** bacteria, in turn, can use ammonia nitrogen (NH_3) as a source of energy to synthesize their own protoplasm. This process occurs only slowly, if at all, under acidic conditions. First the ammonia is converted to nitrite (NO_2^-) by the bacterial genus **Nitrosomonas,** and the nitrite is then converted to nitrate (NO_3^-) by another genus, **Nitrobacter** (Figure 9–5). This two-step process is called **nitrification.** Both bacterial groups obtain their energy from this oxidation process and then utilize some of the energy to convert CO_2 to cellular carbon. Finally, after nitrate (NO_3^-) has been taken up and converted by higher plants and microbes into protein and nucleic acids, it is metabolized and returned to the major part of the cycle as waste products of that metabolism, i.e., as inanimate organic nitrogen. Many heterotrophic bacteria and fungi in both soil and water utilize this organic nitrogen-rich material, converting it and releasing it as inorganic ammonia in a process called **ammonification.** Other parts of the cycle involve the release of gaseous nitrogen and nitric oxides back into the atmosphere, although these are of limited significance (Figure 9–5).

As noted earlier, nitrogen is introduced into the aquatic environment through the discharge of domestic sewage and organic industrial wastes. Organic nitrogen (proteins) and ammonia are the main constituents. These may be partially oxidized in the treatment process to nitrites and nitrates. The discharge of excessive quantities of nitrogenous compounds to rivers and lakes can result in excessive nuisance growth of algae and macrophytic plants (see Section 9.6).

9.4.3 Phosphorus Cycle

The phosphorus cycle, particularly in the aquatic system, is of special interest to environmental scientists and engineers. Phosphorus, an essential element for growth, is very frequently found to be in limited supply in rivers and lakes, whereas carbon and nitrogen are more readily available. Therefore, excessive growth of algae and aquatic weeds in rivers

---

* In a symbiotic relationship, two dissimilar organisms live together with advantages for each.

and lakes can often be reduced or prevented by limiting the supply of phosphorus alone. Phosphorus is thus a **limiting factor.**

Phosphorus occurs in soils and rocks as calcium phosphate ($Ca_3(PO_4)_2$) and as hydroxy apatite ($Ca_5(PO_4)_3(OH)$). Since phosphate rock is only slightly soluble, quite small amounts of phosphorus are leached into solution, resulting in concentrations as low as 1 ppb! Since phosphorus is required for all life processes, its concentration in natural waters is further reduced by the biological system. Because of seasonal changes in plant and animal production, and because of increased phosphorus input to natural waters from spring runoff, the concentration of phosphorus in water will vary markedly over the year.

The input of phosphorus from human activity can be far greater than from natural sources. Domestic sewage contains phosphorus in feces and from commercial detergents, in which phosphates are used (as wetting agents), although the latter contribution has been greatly reduced in many places, following legislation. Runoff from agricultural areas which have received fertilizers (normally containing nitrogen, phosphorus, and potassium) can be another important source of phosphorus. Therefore, in many polluted waters, soluble phosphorus can reach much higher concentrations than in nonpolluted waters. This readily available phosphorus can often lead to the growth of nuisance organisms such as filamentous algae, which can cause taste and odor problems in water supplies and clog filters in water treatment plants. The general problem of eutrophication of lakes is discussed in Section 9.6.

Phosphorus is a constituent of nucleic acids, phospholipids, and numerous phosphorylated compounds. It has been noted that the ratio of phosphorus to other elements in organisms tends to be considerably greater than the ratio of phosphorus to other elements in the external sources such as soil or water, thus indicating that the supply of phosphorus is critical to biological growth in lakes.

For their nutrition, plants (and bacteria) require phosphorus in the phosphate (dissolved) form, generally as orthophosphate (PO_4). They assimilate it directly, converting the PO_4 to the organic (insoluble) form in their protoplasm. Decay of these organisms dissolves and releases (mineralizes) the phosphorus for reuse. However, in lakes, much of the phosphate is removed from the water by the sediment, which eliminates it from the seasonal water circulation.

A simplified phosphorus cycle is shown in Figure 9–6. The solid arrows represent major pathways of flow, and the dashed arrows are flows of "much less importance."

Particulate organic phosphorus is contained within dead and living cells, and part of the dissolved inorganic phosphorus in the water is derived from this organic material by excretion and decomposition. Rigler (1964) found that two patterns of phosphorus change predominated: (1) inorganic phosphorus was low for most of the year, but increased from December to April (because of limited biological activity): (2) particulate organic phosphorus showed no consistent seasonal patterns, but a minor increase occurred in winter. The inorganic phosphorus is taken up very actively in the spring by phytoplankton in the lake waters, so that a rapid drop in dissolved phosphorus concentration occurs. This is illustrated in Figure 9–7 for four of the lakes studied. The three forms of phosphorus indicated are particulate (organic), dissolved (PO_4), and inorganic (polyphosphates), which would eventually break down (hydrolyze) in the water to PO_4.

Phosphorus becomes bound to sediments under oxidizing conditions, (as occur in

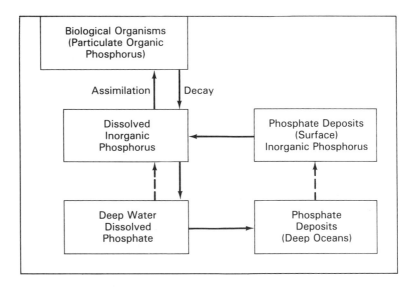

Figure 9–6   The phosphorus cycle. Source: Adapted from Odum, 1971, *Fundamentals of Ecology*, 3rd ed., with permission of W. B. Saunders, Co.

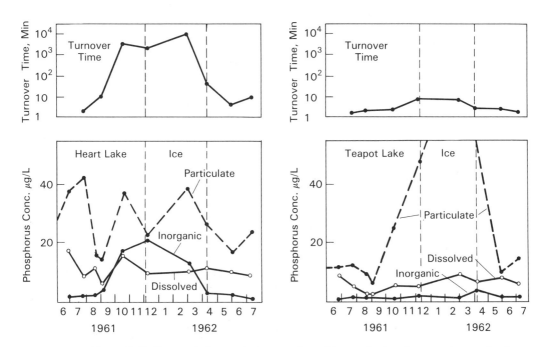

Figure 9–7   Seasonal changes in turnover time of inorganic phosphorus and amounts of the three forms of phosphorus in two lakes in Ontario. Source: Rigler, 1964

the fall during lake mixing), and is released into the water column again under winter anoxic (low-oxygen) conditions of stagnation (see Section 9.5). The summer distribution of these components appeared to be static, yet plant and animal cells were continuously dying and being replaced by new ones. Using the radioactive isotope of phosphorus, ^{32}P, Rigler was able to show that ^{32}P turned over, or was reused, with remarkable speed and efficiency. During the summer, average turnover time ranged from 0.9 to 7.5 minutes. As the season advanced into fall and winter, there was a striking lengthening of the turnover time until the maximum which occurred under ice and snow cover was reached. At the two extremes, Rigler found turnover in winter to be 7 minutes in one lake—only 3.5 times longer than in summer—and 10,000 minutes, i.e., 7 days, in another, which was, in fact, more typical of the rest of the lakes studied.

It is this rapid and continuous turnover of phosphorus which enables phytoplankton populations to expand rapidly in the spring, although the actual scarcity of phosphorus in many waters means that it is a limiting nutrient, additions of which can be rapidly assimilated by algae, which then grow profusely.

9.5 ELEMENTS OF LIMNOLOGY

The study of the physical, chemical, and biological characteristics of rivers and lakes (i.e., fresh water) is called **limnology.** Its counterpart, dealing with the geographically much larger oceans, is called **oceanography.** It is important that those dealing with the use and protection of water resources, including such related activities as irrigation, waste disposal, and shore erosion, understand how freshwater systems work. Excellent information on these practical aspects of limnology is available from Ruttner (1965) and Wetzel (1975).

Some important limnologically related definitions are as follows:

A **benthic organism** is a plant or animal that lives at or near the bottom of a lake, river, stream, or ocean.

The **epilimnion** is the upper waters of a lake.

The **euphotic zone** is that surface volume of water in the ocean or a deep lake which receives sufficient light to support photosynthesis.

The **hypolimnion** is the lower levels of water in a lake or pond which will remain at a constant temperature during the summer months.

Limnology is the study of the physical, chemical, and biological characteristics of rivers and lakes (i.e., fresh water).

Plankton are any small free-floating organisms living in a body of water; *phytoplankton* refers to the plant species (algae), and *zooplankton* to the animal species (crustaceans, rotifers, protozoa) feeding on other forms of plankton.

The **metalimnion** is the middle waters of a lake, where the thermocline occurs (temperature and oxygen content fall off rapidly with depth).

9.5.1 Quantity and Quality of Water

As an ecosystem, a lake or river by itself is a rather artificial unit in that many of its characteristics are determined by the nature, size, and shape of the land surrounding it, and by the drainage waters which enter it. The ecological unit, then, is the lake or river together with its **drainage basin.** The latter is also known as the **catchment area** or the **watershed.** The amount of water entering a lake or river is determined by the amount of precipitation falling in the drainage basin—both rain and snow (in the mountains, fog and intercepted stratus clouds can be an important additional source of precipitation)—by the size of the drainage basin, and by the nature of the vegetation and soil which surround the body of water. Deep organic peat soils with bog vegetation on them may act like a sponge, so that much of the rainfall over a short time period is retained in the watershed, whereas a similar amount of rain on a hard, granitic rock basin, with sparse vegetation and shallow soil cover, might mostly run off and enter the lake or river within a few hours of the occurrence of the rainfall.

Surface water quality will be affected by the atmosphere through which the rain falls, by the nature of the soil and vegetation over which the surface water runs off, and by the extent of human activity in the basin. The composition of waters entering rivers and lakes may be changed by industrial gases drifting in from distant sources and dissolving in the rainwater falling on the catchment (see Chapter 5). Major bird migrations can have a substantial effect on the quality of water in lakes which lie along their paths; for example, waterfowl may bring in salts and nutrients in their excreta. Runoff from heavily forested areas will be rich in organic matter and of brownish-yellow color due to humic acids. Runoff from barren, deforested land may be very turbid because of a heavy load of silt from soil erosion. Land-use activities by people in the watershed may also be of great significance. Agricultural activities can substantially influence the levels of nitrogen, phosphorus, organic matter and bacteria entering a body of water. Mining will increase the concentration of metals in water, as well as the acidity of water (through acid mine drainage). The effect of sewage and industrial waste discharges on water quality may be severe, particularly with regard to organic content.

9.5.2 Biotic Communities

Organisms which live suspended in the water column are called **plankton.** In swiftly moving waters they do not have time to develop significant populations, but as the flow rate slows on the lower, gentler slopes of the catchment, and as the volume and depth of water increase, they begin to build up distinctive plant and animal planktonic populations, termed **phytoplankton** and **zooplankton** respectively. The phytoplankton are a diverse group of microscopic green algae, from a dozen different groups. The predominant groups are the single-celled green algae. The desmids and the blue-green algae can form enormous populations under favorable conditions, causing so called **algal blooms,** in which they color the water, and can produce distinctive odors and tastes difficult to remove in water purification plants. Cell numbers can reach 8×10^6 per mL. Other groups include the beautiful sculptured cells of the diatoms with their silicon skeletons, the yellow-green algae, the euglenoids, and the dinoflagellates (see Section 8.2.4). In marine waters the brown

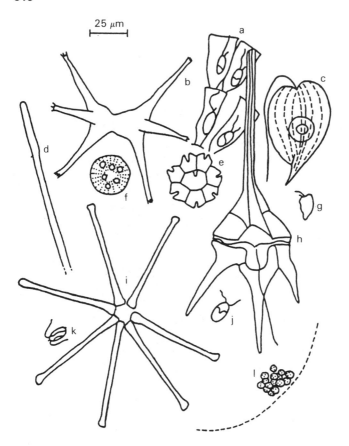

Cyanophyta (blue–green algae)—(d) *Oscillatoria,* (l) *Microcystis;*
Chrysophyta (yellow–green or golden algae)—(a) *Dinobryon;*
Chlorophyta (green algae)—(e) *Pediastrum,* (b) *Staurastrum* (a member of a group called the
desmids strictly now placed in the Charophyta), (j) *Chlamydomonas;*
Bacillariophyta (diatoms)—(f) *Cyclotella,* (i) *Asterionella;*
Euglenophyta (euglenoids)—(c) *Phacus;*
Cryptophyta (cryptomonads)—(g) *Rhodomonas;*
Pyrrophyta (dinoflagellates)—(h) *Ceratium;*
Haptophyta—(k) *Prymnesium.*
Microcystis (l) is a very large alga, of which a diagram of the entire colonly could occupy as
much as this page. Only a few cells are shown.

Figure 9–8   Some typical freshwater phytoplankton algae, drawn to scale. Source:
Moss, 1980 (See also Table 8–2 for a simplified classification system)

and red algae are of great importance, but in fresh waters they have very few representatives.
Examples of some of the freshwater phytoplankton are shown in Figure 9–8.

In fresh water, the zooplankton include mostly small crustaceans (the crab and shrimp
family) and rotifers. Many of these filter large volumes of water each day, from which they
extract phytoplankton and smaller zooplankton (protozoa), as well as bacteria and dead

organic matter. Others are more actively carnivorous, seeking out prey and grabbing, biting, and tearing them. Zooplankton vary in size from about 70 to 3,000 μm. Some of them have specific buoyancy mechanisms, involving oil droplets or air bladders. Most of them are quite mobile, possessing groups of cilia (hair-like structures) or flagella (small whip-like structures), or, in the larger forms, swimming legs. They can thus pursue their prey and also change position in the water column to maximize feeding grounds or to avoid predators. Some show remarkable diurnal migrations of several meters per day, feeding at night in the upper waters and sinking to the darker depths by day. Some insect larvae and also fish fry form part of this floating world, feeding on phytoplankton and on zooplankton. In turn, fish-eating fish, reptiles, birds, and mammals may join the ecosystem.

As the flow rate of the water drops in the lower parts of the catchment area, sediment may deposit on the bottom, providing a rooting medium for larger aquatic plants **(macrophytes)** and a habitat for mud-living benthic (bottom-living) invertebrates (Figure 9–9). The latter burrow in the rich deposit of silt, clay, and organic material, and feed on the new supply of food at the surface of the mud. The aquatic oligochaete worms (earthworm family), chironomids (midge larvae), and bivalve molluscs, such as freshwater mussels and clams, live in such benthic muds. Reed beds emerge from the water and, together with the submerged macrophytes, provide a physical habitat for a great diversity of invertebrates and fish, and a food supply for waterfowl, algae, and bacteria.

In the case of a river, as it increases in size and the flow rate drops further, it increasingly comes to resemble a shallow lake in its biota. Sand, silt, and clay particles accumulate, with the clays and silts especially sinking very slowly through the water. Dead algal and zooplankton cells also accumulate on a seasonal basis, together with pollen from

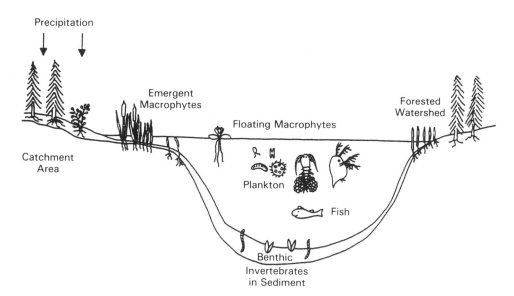

Figure 9–9 Diagram of a lake (or large river), with biota in the shallow waters, in the benthic muds, and in the water column (not drawn to scale)

the spring pollen rain and the feces of the myriads of zooplankton and fish of the upper waters. Low oxygen levels reduce rates of breakdown of this organic material. Precipitates of iron, manganese, and phosphates can occur under such anoxic conditions and are incorporated into the sediments. The nutrients being added to the upper layers of sediment can be released back into the water column and provide part of the spring flush of available phosphorus, as discussed in Section 9.4. Once the nutrients are buried beneath a few millimeters of sediment, they are sealed off and effectively lost to the water column and its biota.

9.5.3 Light in Lakes

The amount of light available at different water depths in lakes (or in very large rivers) is important to the ecology of the lake. Visible light is absorbed by the water itself, by dissolved substances, and by particulate matter. The highest and lowest wavelengths (the reds and blues) are absorbed best, so that below a few meters depth light quality is predominantly of the green and yellow wavelengths. Organic materials very effectively absorb the red and the blue light. In each successive increment of water depth, light of a given wavelength is reduced by a fixed proportion (Figure 9–10). Theoretically, light is never totally extinguished, but before it reaches visually undetectable levels it falls to about one percent of the surface intensity. This has a conventional significance, since it approximately describes the level where algal photosynthesis is reduced to the point that it only just matches respiration. This is called the **compensation point.** Below it, algal growth

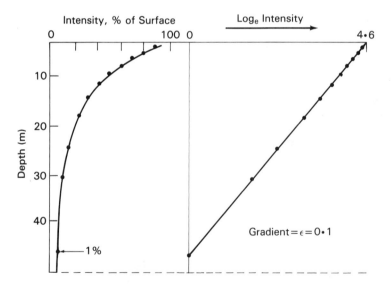

Figure 9–10 Absorption of light in a uniform water column. Source: Moss, 1980
The left-hand panel shows light intensity as a percentage of that found at the surface (after reflection losses have been allowed for). The right-hand panel shows the same data with the light intensities as natural logarithms. The gradient is the extinction coefficient.

cannot occur; above it is the **euphotic** zone, in which phytoplankton cells can thrive. Sometimes the bottom of a lake is in the euphotic zone, in which case rooted water weeds cover it. In very clear lakes this can be quite deep (20–50 meters), with a theoretical maximum of around 200 m. In most of the world's bodies of water, primary productivity is confined to less than half of the mass of water, and in some deeper lakes the euphotic zone may be only a thin surface skin below which lies a huge dark world. This world is not devoid of life, however, as zooplankton, bacteria, fungi, fish, and invertebrates can live there permanently or enter for short periods.

9.5.4 Temperature and Vertical Stratification of Lakes

In turbulently flowing streams, in which the waters are continuously mixed, no temperature gradients develop in the summer months. In slow-flowing deep rivers, and especially in lakes, the surface water in summer tends to heat up more rapidly than the water below it. When winds and currents are insufficient to mix the lake waters from top to bottom, a common occurrence in deeper bodies of water, a steep gradient of temperature develops. Heat radiation is very largely absorbed in the top one or two meters of water. This upper warmer layer is called the **epilimnion,** and the lower denser cold layer down to the bottom of the lake is called the **hypolimnion** (Figure 9–11). Between the two is a transitional layer called the **metalimnion.** In this transitional zone, the temperature changes very rapidly over a short change in depth. The occurrence of this rapid vertical change in temperature with development of stratification is called the **thermocline.**

The productivity of a lake is directly affected by thermal stratification and seasonal mixing. Lakes in the temperate zone thermally stratify during winter and summer (Figure 9–11) and circulate each spring and autumn. In the winter, ice near 0°C covers the surface, while the more dense water at 4°C (more precisely, 3.94°C, the temperature at which the density of water is a maximum) sinks to the bottom. The cold temperature and reduced light penetration inhibit biological productivity. During summer, as the surface waters warm rapidly and spring winds subside, a less dense surface layer of water forms. This epilimnion mixes continuously during the summer and supports the growth of phytoplankton.

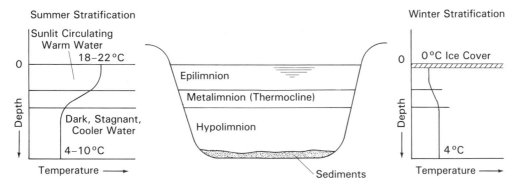

Figure 9–11 Thermal stratification of a deep lake

Thermal stratification also influences water quality. The epilimnion supports abundant algal growth, while the hypolimnion in eutrophic lakes decreases in dissolved oxygen content. Hydrogen sulfide, odorous organic compounds, and reduced iron can be released from bottom sediments as a result of anaerobiosis. During thermal stratification, water supply of highest quality is usually found just below the thermocline.

9.6 EUTROPHICATION

9.6.1 The Problem

The word **eutrophication** comes from two Greek words; *eu*, meaning "good" or "well," and *trophos*, meaning "food"; thus, **eutrophic** can be translated as "nutrient (food) rich." All lakes undergo a natural enrichment over time. Sediments are carried in from the surrounding watershed, and soluble nutrients are leached from them. This **natural eutrophication** is a slow process from a human point of view, frequently taking place over periods of thousands of years. The discharge of untreated sewage and agricultural or industrial wastes into a lake hastens the process greatly. This accelerated process is sometimes referred to as **cultural eutrophication.** Lakes in which the nutrient level is particularly high, which are characterized by abundant littoral (shore-dwelling) vegetation, frequent summer stagnation with algal booms, and absence of cold-water fish species, are said to be **eutrophic;** Lake Erie is such a lake. Lakes with low nutrient levels are called **oligotrophic** ("oligo" meaning "small" or "deficient in"); Lake Superior is such a lake. Lakes with intermediate nutrient levels are called **mesotrophic.**

Eutrophication is the natural process of nutrient enrichment that occurs, over time, in a body of water. The resulting biological growth, mainly algae, in the epilimnion dies and settles to the hypolimnion, where it decays and depletes the oxygen from the water.

Eutrophication is one of the most significant and worldwide water quality problems. The most important problems created by excessive eutrophication are:

- The detrimental effect on the commercial and sportfishing industry due to changes in the species of fish found in lakes, caused principally by the low levels of oxygen found in the lower waters.
- The effect on recreation and tourism through excessive growth of algae and other aquatic plants, rendering the water and beaches unfit for recreational purposes. The filamentous algae are washed onto the beaches during storms, leaving stinking, rotting piles of organic matter.
- The abundant algal blooms, which create an unpleasant taste and odor in water supplies and plug intakes and filters in water treatment plants.

Thus, a biologically poor lake is preferable to a fertile one from the standpoint of water use and recreation. This appears to be a paradox since in some parts of the biosphere we are doing everything possible to increase fertility in order to produce more food, whereas in other parts we are trying everything possible to prevent fertility. In the case of land, more food from increased fertility does not harm the land. However, overfertilization of lakes and the resulting algae can, in its early stages, impair the quality of the fish we are trying to produce, and eventually, in the extreme, destroy all aquatic life.

9.6.2 Physical–Chemical and Biological Changes

The importance of phosphorus and nitrogen in the eutrophication process was not recognized widely until the early 1960s. Therefore, there is relatively little long-term information on lakes available for these two elements. Changes in the chemical composition of lakes over a long period have been documented, but not necessarily for these most critical elements. Figure 9–12 gives information on a number of chemical changes for Lake Erie over a 70-year period.

Pioneering work was done on several Wisconsin lakes to establish the levels of

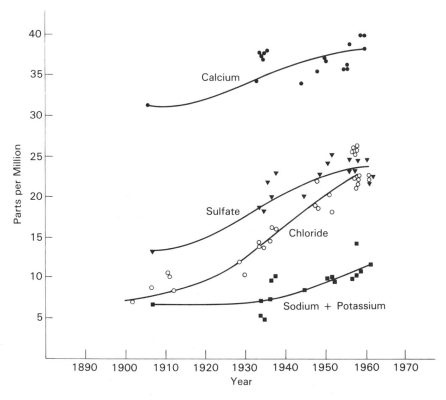

Figure 9–12   Chemical changes in Lake Erie, 1890–1960. Source: Beeton, 1965

phosphorus and nitrogen concentrations below which nuisance growth will not likely occur. Upper concentration limits of 0.3 mg/L of inorganic nitrogen and 0.02 mg/L of ortho-phosphate phosphorus at the time of spring turnover of the lake were thought to be suitable then, but values one-third to one-half of these have now been found to trigger algal blooms when all other nutrients are available (see Section 12.3.1).

Rising nutrient levels in a lake change the numbers and types of biota present there. Oligotrophic lakes normally have clear water and contain little biomass. Annual productivity is likewise low. At the same time, they have a great diversity of species of plants and animals. As nutrient enrichment occurs, the trend is to more biomass, a higher rate of production, and less or changed species diversity. Large increases in algal production occur, and in turn benthic algae and zooplankton increase. Plankton may change in composition from the initially sparse numbers of diatoms to a dense growth of desmids, which results in decreased light penetration. Finally, blue-green algae come to dominate the water column. During such algal blooms the cell density may reach 10,000–100,000 cells per milliliter.

As some of the algae die, either through exhaustion of some essential nutrients or for other reasons, they become food for bacteria. Bacterial decomposition of the dead algae and bacteria then consumes the dissolved oxygen, leaving the water beneath the slimy surface oxygen deficient and therefore unable to support the species that people value most. Trout and bass are replaced by coarse fish such as suckers, carp, sunfish, yellow perch, and smelt. The latter group is much less valuable for commercial and sportfishing, thus causing an economic loss.

9.6.3 Control of Eutrophication

It has been demonstrated, at Lake Washington in Seattle, that cultural eutrophication can be reversed if the inflow of nutrients is substantially reduced. Before taking corrective measures, a quantitative survey of nitrogen and phosphorus sources and limnological studies of the lake are essential to establish its trophic levels and to determine whether the majority of its nutrients are from point or diffuse sources. Point sources, such as municipal wastewater discharges, can be controlled by alternative disposal on land, diversion around the lake, or removing the nutrients from the wastewater prior to discharge to surface waters. Nonpoint source loads, such as agricultural runoff, can be reduced through land management techniques that prevent soil erosion and avoid excessive use of fertilizers. Since conventional treatment methods reduce phosphorus by less than 20 percent and nitrogen by about 30 percent, supplementary nutrient removal in the plant may be required to protect the receiving waters. Typical biologically treated wastewater contains approximately 8 mg/L of phosphorus and 30 mg/L of nitrogen (see Table 12–1). Emphasis is normally placed on phosphorus removal, since phosphate is believed to be the controlling factor in the enrichment of most lakes.

Several temporary controls can be used to arrest or reduce nuisance effects in eutrophic lakes and reservoirs. Chemical control through application of copper sulfate, an algicide, is effective for a short time. Harvesting of aquatic weeds can be practiced in near-shore areas. (Examples are the Trent Canal and Kawartha Lakes system in Ontario.) Where oxygen depletion is a serious concern, mechanical aeration and mixing may help. These methods

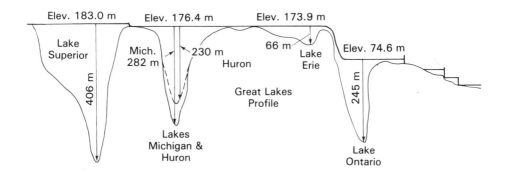

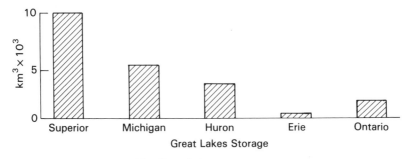

Figure 9–13 The Great Lakes. Source: U.S. Department of the Interior, 1968

can only be considered stop-gap measures, since they do not control input but merely ameliorate its consequences.

9.6.4 A Case Study: The Great Lakes

The Great Lakes system (Figure 9–13) is a good example of international cooperation between the United States and Canada to prevent further deterioration of water quality and, if possible, reverse the trend of eutrophication. The serious deterioration of the very large lower Great Lakes is a sharp reminder of the inadvertent damage a large industrialized population can do to water quality in a short period of time, and the large and expensive effort needed to reverse this trend. The Great Lakes is the greatest reservoir of fresh water on the earth's surface, with a shore line of 10,500 km, an area of 245,000 km², and a volume of 25,000 km³. In its basin and catchments live about one-eighth of the population of North America, discharging its wastes and industrial by-products to the Great Lakes. The relative effects on the lakes are determined to a very substantial extent by the human population along the shoreline and the size of the lake. Details of some of the morphometric data are given in Table 9–5.

Lake Superior is by far the largest and least polluted of the Great Lakes and has the smallest population density in its watershed. Also, it is estimated to take more than 190 years for the water in it to be totally replaced. Lake Erie, in contrast, is a much smaller, shallower lake, with an estimated population of over 15 million in its drainage basin. Water replacement is much faster than in the other Great Lakes, so benefits from a reduction in incoming pollution would be evident much sooner. It is important to recognize, however, that if the pollution problem is associated with the sediments rather than with the water column itself, then it is a much more permanent feature. In Lakes Erie, Michigan, and Ontario, the occurrence of mercury, PCBs and dioxins are substantially related to sediments immediately downstream of particular industrial discharges.

Agricultural drainage into the lakes affects water quality in Lakes Michigan, Erie, and Ontario, and has been a major factor in the algal problems of Lakes Michigan and Erie, where both phosphorus and nitrogen have been entering the lake waters from farms in the watershed. Problems of increased concentrations of such salts as chlorides from road salting in the winter (e.g., in Lakes Erie and Ontario) and sulfates from major industries are of growing concern.

TABLE 9–5 MORPHOMETRIC DATA FOR THE ST. LAWRENCE GREAT LAKES

Lake	Area (km²)	Maximum depth (m)	Mean depth (m)	Volume (km³)	Shoreline (km)	Hydraulic residence time (yrs)
Superior	83,300	397	145	12,000	3,000	190
Huron	59,510	223	76	4,600	2,700	40
Michigan	57,850	265	99	5,760	2,210	36.5
Erie	28,280	60	21	540	1,200	3
Ontario	18,760	225	91	1,720	1,380	8

Data derived from Hutchinson, 1957

Use of the lakes for swimming, boating, fishing, and other water sports as well as for cottage development is of great importance to the public. Unfortunately, many beaches are grossly polluted, especially around Cleveland and Detroit, and Toronto beaches have been posted as unsafe for swimming since 1983, causing a considerable public outcry. The closing of beaches is always associated with excessive fecal coliform bacterial counts, which indicate potentially dangerous levels of pathogenic bacteria.

The Great Lakes provide vital navigational routes to and from the major trading centers. This has been a key factor in the development of cities and towns around the Great Lakes, and it spurred development of the Welland Canal and later the St. Lawrence Seaway system, both of which have stimulated international trade.

The Great Lakes and the rivers draining into them have often been used as dumping grounds for waste from industry and from urban populations. Demand for water for both domestic and industrial needs has greatly increased. Power stations, especially nuclear power stations, make huge demands for a constant supply of cooling water. Unwanted materials dumped in the past have actually created fire and navigational hazards. In fact,

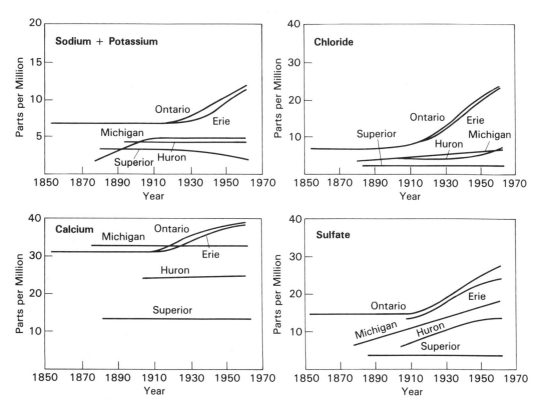

Figure 9–14   Changes in the chemical characteristics of Great Lakes waters. Data for Lake Erie, 1958; Lake Huron, 1956; Lake Michigan, 1954, 1955. 1966; Lake Ontario, 1961; Lake Superior, 1952, 1953, 1961, 1962 from the Ann Arbor Biological Laboratory, U.S. Bureau of Commercial Fisheries. All lakes, *Great Lakes Water Quality Board Report*, 1982

some of the rivers draining into Lake Erie are officially declared fire hazards. The Cuyahoga River caught fire in 1969, and the Buffalo River has had at least three fires due to hydrocarbon dumping since then. Figure 9–14 shows the increases that have occurred in a wide range of elements in each of the five Great Lakes since about 1900 (Beeton, 1965).

The decline of the commercial fishing industry over the past 30 years has been quite dramatic in all of the lower Great Lakes. When the Welland Canal opened up the Great Lakes to shipping by providing a bypass around Niagara Falls, it also opened up a path for the sea lamprey to enter the Great Lakes from the Atlantic. The lamprey is a parasitic fish which attaches itself to the body of lake trout and to other fish, including salmon, whitefish, chub, blue pike, and suckers. It rasps holes in the unfortunate victims and sucks their body fluids, eventually killing them. Its entry into the Great Lakes has been a disaster, and one which no one really could have foreseen. In Lakes Superior, Michigan, and Huron, it has almost eliminated lake trout populations (Figure 9–15). The alewife is a small fish which

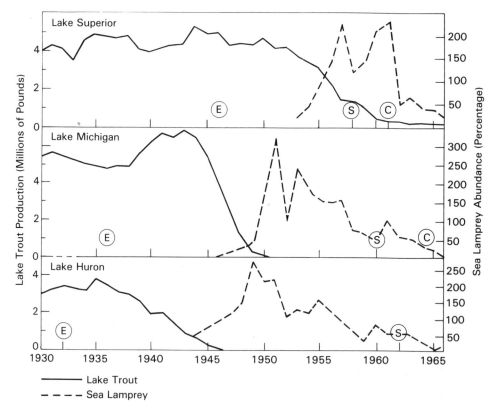

Figure 9–15 Production of lake trout and abundance of the sea lamprey in Lakes Superior, Michigan, and Huron. Source: Smith 1971. Reproduced with the permission of the Journal of the Fisheries Research Board of Canada, 1968, vol 25(4) pp 667–693

(E) marks the first sea lamprey record, (S) is initiation of chemical control, and (C) is completion of the initial series of chemical treatment.

has also invaded through the canal and become a problem. It feeds on other fishes' eggs and competes with their young for food. The alewife was not a problem until the sea lamprey eliminated the larger fish which naturally preyed upon them. Beeton (1965) gives a detailed account of the many associated changes in biota which have taken place, especially the changes in the phytoplankton, the spread of nuisance algae such as *Cladophora*, and the decline of many insect larvae which depend upon clean, well-oxygenated waters for their survival.

While the emphasis in what has just been said has been on land and water as sources of nutrients, it is important to realize that atmospheric deposition is a significant contributor of both phosphorus and nitrogen to the Great Lakes. Over 30 percent of the phosphorus loading to Lake Michigan, for example, comes from the atmosphere.

Finally, it needs to be stressed that it was largely because of the public outcry in the Great Lakes region about the polluted, eutrophic state of Lakes Erie, Michigan, and Ontario that the phosphorus ban on detergents was introduced in Canada in the early 1970s. This, together with mandatory treatment of wastewater, has brought about significant improvement in the water quality of the Lower Great Lakes. Lake Erie in particular is being restored to health.

PROBLEMS

9.1. Define the following terms, using examples where necessary:
   (a) autotroph and heterotroph
   (b) primary producer
   (c) trophic level
   (d) food chain biomagnification
   (e) denitrification

9.2. The rates of productivity on an annual basis differ very widely among different ecosystems. Comment on this and speculate as to the major climatic factors controlling these differences.

9.3. The flow of energy within an ecosystem is unidirectional and dependent upon constant solar input, whereas the flow of nutrients is cyclical. Comment on this and explain how it creates constraints.

9.4. It has been said that the relative efficiency of energy utilization in animals is higher than in plants. Is this true, and what are the key factors underlying the situation?

9.5. Respiratory use of energy at different trophic levels in different ecosystems varies a great deal. Using data given for Cedar Bog Lake, as well as an old field and salt marsh, discuss this statement.

9.6. Describe the cycling of nitrogen in ecosystems and the role of atmospheric nitrogen in the supply of nitrogen to higher plants.

9.7. What is meant by the following?
   (a) drainage basin
   (b) macrophytes
   (c) zooplankton
   (d) algal blooms

(e) benthic organisms

(f) euphotic zone

(g) oligotrophic

(h) thermocline

(i) hypolimnion

9.8. Describe the development of a thermocline in lakes of different latitude, and how the thermocline varies in persistence.

9.9. Why is the temperature of 3.94°C of great importance in understanding seasonal changes in lakes?

9.10. Phosphorus has been shown to have a very rapid turnover in lakes. What is the relevance of this to potential problems regarding eutrophication and algal blooms?

9.11. Eutrophication has become a very serious problem in many lakes in Europe and North America. What is it, why is it a problem, and how can it be solved?

9.12. The lower Great Lakes have seen profound changes in their fisheries in the past 60 years. What are some of these changes, and why have they occurred?

REFERENCES

BEETON, A. M. "Eutrophication of the St. Lawrence Great Lakes." *Limnology and Oceanography* 10 (1965): 240–254.

GATES, D. M. *Energy Exchange in the Biosphere*. New York: Harper and Row, 1962.

Great Lakes Water Quality Board. *Report to the International Joint Commission: 1982 Report on Great Lakes Water Quality*. Windsor, Ont.: International Joint Commission, 1982.

HUNT, E. G., and BISCHOFF, A. I., "Inimical Effects on Wildlife of Periodic DDT Application to Clear Lake." *California Fish and Game* 46, 1960: 91–106.

HUTCHINSON, G. E. *A Treatise on Limnology*. Vol. 1. New York: Wiley, 1957.

JUDAY, C. "The Annual Energy Budget of an Inland Lake." *Ecology* 21 (1940): 438–450.

KORMONDY, E. J. *Concepts of Ecology*. Englewood Cliffs, N.J.: Prentice Hall, 1969.

LINDEMAN, R. L. "The Trophic-Dynamic Aspect of Ecology." *Ecology* 23 (1942): 399–418.

MACAN, T. T. *A Guide to Freshwater Invertebrate Animals*. London: Langman, 1959.

MOSS, B. *Ecology of Freshwaters*. Oxford: Blackwell Scientific Publications, 1980.

ODUM, E. P. *Fundamentals of Ecology*. 3rd ed. Philadelphia: W. B. Saunders Co., 1971.

OVINGTON, J. D. *Advances in Ecological Research*. Vol. 1. New York: Academic Press, 1962.

RIGLER, F. H. "The Phosphorus Fractions and the Turnover Time of Inorganic Phosphorus in Different Types of Lakes." *Limnology and Oceanography* 9 (1964): 511–518.

RUTTNER, F. *Fundamentals of Limnology*. 3d ed. Translated by D. G. FREY and F. E. J. FRY. Toronto: University of Toronto Press, 1965.

SCHINDLER, D. W., WELCH, H. E., KALY, J., BRUNSKILL, G. J., and KRITSCH, N. "Physical and Chemical Limnology of Char Lake, Cornwallis Island 75°N Lat." *Journal Fisheries Research Board of Canada* 31 (1974): 585–607.

SMITH, S. H. "Species Succession and Fishery Exploitation in the Great Lakes." In *Man's Impact on the Environment*, edited by T. R. DETWYLER. New York: McGraw-Hill, 1971.

STUMM, W., and MORGAN, J. J. *Aquatic Chemistry*. 2nd ed. New York: Wiley-Interscience, 1981.

TRANSEAU, E. N. "The Accumulation of Energy by Plants." *Ohio Journal of Science* 26 (1926): 1–10.

U.S. Department of the Interior, Federal Pollution Control Administration. *Proceedings, Progress Evaluation Meeting, Pollution of Lake Erie and Its Tributaries*, Washington, D.C.: U.S. Department of the Interior, 1968.

WESTLAKE, D. F. "Comparisons of Plant Productivity." *Biological Reviews* 38 (1963): 385–425.

WETZEL, R. G. *Limnology*. Philadelphia: Saunders, 1975.

WHITTAKER, R. H. "Experiments with Radiophosphorus Tracers in Aquarium Microcosms." *Ecological Monographs* 31 (1961): 157–175.

PART 3
TECHNOLOGY AND CONTROL

10

Water Resources

J. Glynn Henry

10.1 INTRODUCTION

Water resources management in North America has evolved from its original goal of simply supplying water at minimum cost to promote development to the contemporary approach in which a wide spectrum of objectives is examined. Benefits, from flood control to the aesthetic enjoyment of the environment, are now evaluated in terms of human needs and activities. The application of this broadened concept must be based on an understanding of the many factors which influence decisions. These factors and their influence on water resources management are described in this chapter.

Technical background information on the overall quantities of water available and the requirements for various uses precedes a discussion of management alternatives. The need for adequate data and the role of systems analysis in assessing the environmental consequences of these alternatives are explained.

Elements of successful planning including flexibility, methods for measuring benefits (especially noneconomic ones), and utilization of public involvement are outlined. The importance of these requirements and suggestions for implementing them are also discussed.

Subsequent sections of the chapter deal with legislation affecting water management, with emphasis on the agencies involved and their areas of jurisdiction. Economic and political considerations are other major concerns dealt with.

Techniques to improve the management of our water resources advanced greatly during the seventies. The challenge today is to apply these new methods to practical situations. Two case studies are used to illustrate this approach. The first example is the controversial Peripheral Canal in California. In addition to its unusual engineering and economic considerations, this project demonstrates, on a grand scale, the unavoidable and often conflicting interrelationships among levels of government, business interests, the general public, and environmental protection. The second case study deals with problems of pollution in the St. John river system in New Brunswick, for which a specialized planning agency created by Canadian federal and provincial agreement was formed. Further examples of problems in water resource management are provided at the end of the chapter.

10.2 WATER RESOURCES MANAGEMENT

10.2.1 Importance of Water

Water resources have been critical to human society since people discovered that food could be produced by cultivating plants. The cities and towns which arose from Egypt east to Mesopotamia (modern-day Iraq) following the agricultural revolution about 3500 B.C. required a ready supply of water for domestic as well as agricultural needs. Eventually, running water drove machines which cut wood, milled grain, and provided motive power for many industrial processes. Water's abundance made it ideal as a universal solvent for cleaning and flushing away all manner of waste from human activities. Until recently the approach to providing water, for whatever purpose, was simple: either locate close to water, as many cities did, or store and transport the water to wherever it was required. After use, water was generally discharged to the nearest body of water, often the same source from which it came. The low-cost supply of large quantities of water was one of the foundations of modern society.

Exponentially growing population and industrial expansion primed the need for increased water supply and distribution. This need was met by constructing dams, reservoirs, river diversions, pipelines, and aqueducts to bring water from more distant, unpolluted sources. The widespread application of modern technology to the supply of abundant water for unrestricted municipal, industrial, and agricultural uses, with no incentive for reuse or conservation, has greatly increased the competition for limited sources of easily accessible water. Activities—such as huge withdrawals of water for mining or agricultural purposes—that formerly did not affect other water users, now sometimes impinge directly on municipal water supplies for cities hundreds of miles distant. In addition to the technical problems of meeting water needs, there are growing environmental concerns to be satisfied. Concerns about the long-term effects of water use and the loss of water for aesthetic and recreational purposes are often in conflict with the objective of providing and maintaining a low-cost water supply.

10.2.2 Need for Control

The obvious effect that various water users have on each other is to create a shortage of water for themselves and for other users. The less obvious, indirect effects of water use include those caused by pollution from waste disposal and surface runoff, changes in aquatic life, and increasing stream salinity. For example, the runoff of herbicides, pesticides, and fertilizers from cultivated lands may affect the aquatic food chain sufficiently to cause the loss of local sportfishing, or encourage the explosive growth of unsightly algae, which in turn may foul water supplies for municipal and industrial users. The agricultural practices needed to halt this negative impact would require changes in land-use practice which may be difficult and costly to enforce. This network, or system of interactions among water users, characterizes the complexity of the problems encountered in attempting to reconcile many uses for the same supply of water. With the diversity of needs and the interrelated effects of water use, it is not hard to see how difficult questions concerning the legal right to water of a certain quality can arise.

In the past, **riparian rights,** that is, the rights of those "on the river bank," dealt with the quantity of water that a person could rightfully claim because of private ownership of abutting land. The emerging awareness of water as a common resource to which others besides riparian owners have rights has highlighted the inadequacies of these laws regarding water use. Failure to resolve the difficult legal, economic, and social issues raised by the interactions of multiple water use can lead to serious consequences. Those having the rights to large volumes of water at unreasonably low or subsidized prices will tend to use excessive amounts, with no incentive to conserve water or to restore its quality after use. This wastefulness penalizes those having to pay a higher price for possibly inferior-quality water, places an additional burden on taxpayers, and limits the availability of water for future development. The geographic locations of watersheds often cross national and jurisdictional boundaries, complicating the application of policies that try to regulate water use.

Politicians and other policymakers must recognize the limitations posed by intuitive human understanding of the complex physical, biological, and social processes involved in managing water resources. Engineers, biologists, sociologists, geographers, and many other specialists are all intimately involved in researching and predicting all aspects of water resource management. Each approaches the problem with a different viewpoint. A mathematical description or model for understanding the various relationships of a water resource system on a quantitative basis is normally necessary, one that allows for the complexity of interaction between parts of the system without requiring an impractical amount of precise data. Once this is done, proposed projects such as dams or aqueducts can be judged on the basis of their predicted benefits and costs. To accomplish this, we need somehow to quantify the direct and indirect noneconomic benefits derived from water use by various parts of society. That is, we need to express such things as aesthetic enjoyment, recreational use, and quality of life in economic or at least common quantifiable terms in order to compare these benefits with the cost of providing them, in the same way that we do with economic benefits. We should also be aware that the average price for water charged by public utilities is probably set by precedent from historically low charges and may not represent its true market value. True market value would recognize not just the cost of delivering the water, but also the subsequent cost of recovering it, restoring its quality, and returning it for reuse.

The various options possible in the development of water resources may benefit different groups of water users unequally. Major projects like dams and reservoirs—whether they are part of a water supply, flood control, power generation, or recreation scheme—can serve many purposes, not all of which are compatible. Because of the large expenditures and long-term environmental impacts involved, government support is usually needed. Consequently, many of the key decisions in such projects are of a political nature. Sound policies for such water resource developments require the participation not only of proponents of the development and specialists, but of an informed public as well.

10.2.3 Objectives in Water Resources Management

Water resources in nature seldom exist when and where they are needed. Erosion, flooding, and drought also affect the availability and quality of water for use and result in loss of property each year and in the case of flood and drought, loss of human life as well. Yet properly harnessed to correct these shortcomings and reduce these fluctuations, water resources can attract regional industry and provide recreational facilities, along with a myriad of direct and indirect benefits that result. Sound water resources management requires not only control of the flow of water, but also an understanding of the need for coexistence of all types of water users within a particular watershed. The general objective of water resources management is therefore to maximize the benefits obtained from the utilization and control of water resources. Projects may have several objectives, and the relative importance of each must be established. This evaluation will be influenced by the amount of water to be supplied or controlled, the need for protection or improvement of its quality, and the cost of providing the potential benefits to the various users.

One of the earliest examples of this comprehensive style of water resource management is the network of hydroelectric dams and water control structures built in the Tennessee River valley in the United States before World War II. In addition to the fact that the entire country was in the midst of an economic depression, the people of this region had suffered greatly from floods that displaced thousands of residents and eroded arable land that had been stripped of vegetation by uncontrolled logging and strip mining practices. The Tennessee Valley Authority (TVA) was created to oversee river development in a fashion that would benefit public, as opposed to private, interests. Reformed land practices and improved methods of irrigation were also initiated to prevent the silting up of dam structures. The availability of inexpensive electric power encouraged industry to locate nearby. The ensuing dramatic increase in the quality of life in the region demonstrated the validity of the concept of unified river planning and remains today a prime example of progressive water resource management.

Proper water resource planning techniques depend on adequate data. Also important is an understanding of the agencies involved, their areas of jurisdiction, and the legislation that governs them. Economic and political considerations are as significant as engineering decisions in successful water resources management. Accordingly, each of these factors is a part of the decision-making process.

10.3 TECHNOLOGICAL CONSIDERATIONS

10.3.1 Properties of Water

Water is the most abundant chemical component within the biosphere. It is also perhaps the most important. Almost all life on earth, including man, uses water as the basic medium of metabolic functioning. The removal and dilution of most natural and human-made wastes are also accomplished almost entirely by water. In addition, water possesses several unique physical properties that are directly responsible for the evolution of our environment and the life that functions within it. Its ability to conduct (thermal conductivity) and store (heat capacity) heat is unmatched by any other substance. Water also has an extremely high heat of evaporation: while it takes only 0.239 J (1 calorie) to raise the temperature of 1 gram of liquid water 1°C (or 1 BTU to raise the temperature of 1 pound of water 1°F), it takes 540 times as much energy to evaporate it. Freezing of water releases 335 kJ per kg (144 BTU/lb). Every day the sun's energy removes roughly 1,230 km³ (300 mi³) of water from the seas, lakes, rivers, and soil through evaporation and from plants by transpiration (Miller, 1975). The vast storage of energy in the sun warms bodies of water, and atmospheric water vapor is responsible for driving the global weather engine which redistributes this solar energy and moderates our climate, as was discussed in detail in Chapter 7.

Example 10.1

Calculate, in J and energy-equivalent barrels of oil, the amount of solar energy used daily to evaporate water from the surface of the earth. One barrel of oil (42 gal or 159 L) yields approximately 6.7×10^9 J (6.4×10^6 BTU).

Solution
Amount of water evaporated daily

$$= 1,250 \text{ km}^3 \ (300 \text{ mi}^3)$$
$$= 1,250 \text{ km}^3 \times \frac{10^9 \text{ m}^3}{\text{km}^3} \times \frac{1,000 \text{ L}}{\text{m}^3}$$
$$= 1,250 \times 10^{12} \text{ L} \quad (3.3 \times 10^{14} \text{ gal})$$

Solar energy used daily to evaporate water

$$= 1,250 \times 10^{12} \text{ kg} \times 540 \times \frac{239 \text{ J}}{\text{kg}}$$
$$= 1.61 \times 10^{20} \text{ J} \ (1.52 \times 10^{17} \text{ BTU})$$

Daily solar energy input in equivalent barrels of oil to evaporate water

$$= 1.61 \times 10^{20} \text{ J}/6.7 \times 10^9 \text{ J per barrel}$$
$$= 24 \text{ billion barrels of oil!}$$

Comment

Using the information in Chapter 3, we can show that the solar energy used to evaporate water from the global water surface is over 4,000 times greater than the energy consumed by the daily world use of oil.

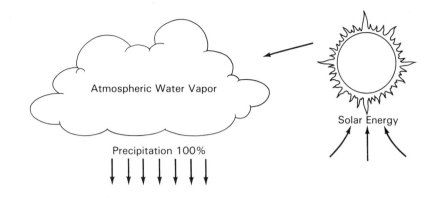

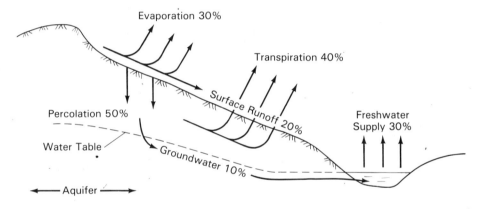

Figure 10–1   Hydrologic cycle. Source: McGauhey, 1968.

The water environment of aquatic life is protected from sudden temperature changes by the fact that it takes a great deal of heat to raise the temperature of water. Water is one of only two substances—the other being mercury—that is more dense as a liquid than as a solid. If the reverse were true, lakes and rivers would freeze from the bottom up, killing most aquatic life within them.

Solar energy drives vast amounts of water through the ecosphere in a closed system known as the **hydrologic cycle.** This cycle has been discussed before in Chapter 6 (Figure 6–10) in connection with mass balances, and again in Chapter 7 (Figure 7–10) as it relates to global climate. The latter figure depicts, in a simplified way, how surface and subsurface sources of water are derived. Rain falling on land fills soil pores in much the same way as water saturates a sponge. If the rate of rainfall exceeds the speed at which water can percolate downward through the soil, the water forms puddles and rivulets which eventually contribute to the surface runoff of streams and rivers, shaping our topography by erosion. Figure 10–1 shows how water eventually starts flowing horizontally as soil pores and rock

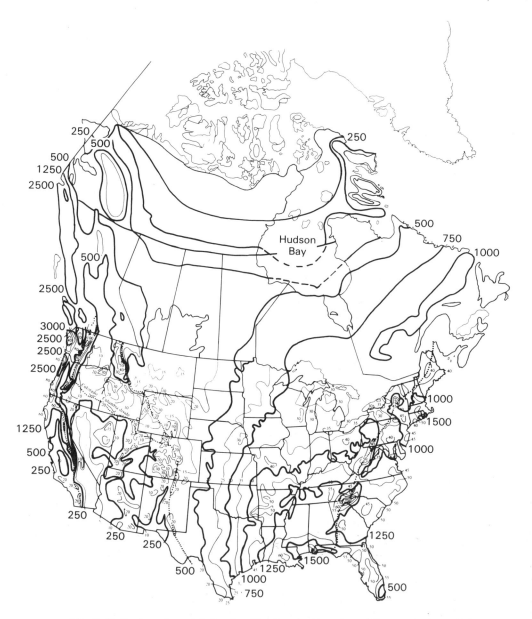

Figure 10–2 Annual precipitation in millimeters for Canada and the conterminous United States. Adapted from Canada, Dept. of Fisheries and the Environment, Atmospheric Environment Service (1941–1970); U.S. Dept. of Agriculture, Soil Conservation Service (1899–1938).

cracks are filled. The boundary formed is called the **water table** and may be found just below the ground surface in areas of heavier rainfall to hundreds of meters down in dry areas. These combinations of water, soil, and rock structures are called **aquifers,** which, through well drilling, form a major source of water for municipalities and rural areas where surface supplies would be too costly to develop.

10.3.2 Annual Precipitation

Although we refer to rivers and wells as sources of water, these, in fact, depend on precipitation in the form of rain, sleet, or snow for their replenishment. The amount of precipitation and the quantity of water available are therefore closely related.

Figure 10–2 shows the mean annual distribution of precipitation in Canada and the conterminous United States. Locations of equal annual precipitation are joined by lines called **isopleths.** Isopleths indicate lows of less than 250 mm (10 in) of precipitation per year in the arid southwest United States and northern Canada to highs from 1,500 mm (60 in) on the east coast of the continent to 2,500 mm (100 in) in the humid regions of the west coast.

Values for annual precipitation are of little direct use in estimating water quantities, but they do indicate what regions are likely to be short of water and therefore arid, and what areas probably have more abundant water supplies. Table 10–1 provides information on annual precipitation for some selected locations in the world. The uneven distribution of rainfall throughout the world is evident from these figures. Arid locations like Los Angeles and Las Vegas in the United States and Cairo in the United Arab Republic are at one extreme, while wet areas such as Buena Vista in Colombia and Cherrapunji in India are at the other.

10.3.3 Quantity of Water Available

As indicated in Figure 10–3, water in all forms constitutes a fixed supply of about 1.36 $\times$ 10^{18} m^3 (360 billion billion gallons) (Todd, 1970). This astronomical sum makes it hard to understand why shortages exist. However, when we consider the water actually available for use, the amount is reduced drastically. Approximately 97.2 percent of the global water supply is found in the oceans. The remaining 2.8 percent is fresh water, but over 75 percent of this is locked up in the polar ice packs, soil and rock formations, and the atmosphere, leaving less than 25 percent available as surface water and groundwater. Unfortunately, over 99 percent of this surface water and groundwater is not very accessible, and we rely on the approximately 0.6 percent available (about 0.004 percent of the original quantity) for our water supplies. Figure 10–3 shows these relationships diagramatically. To appreciate the relative amount of water available for use, suppose the earth's total water supply is represented by a 4-liter (about 1 gal) container. Then total amount of groundwater would be less than 40 mL (1$\frac{1}{2}$ oz). After removing that water which is too deep underground, or too far away, or too polluted, we would have only one drop left. This drop will still represent about 10 million liters (about 2.6 million gallons) per person for a world population of 5 billion. The rate at which this seemingly abundant supply of fresh water can be used is

TABLE 10–1 ANNUAL PRECIPITATION FOR SOME SELECTED LOCATIONS IN THE WORLD

Country	Location	$\dfrac{mm}{year}$	$\dfrac{in}{year}$
Canada	Vancouver, B.C.	1,460	57.4
	Calgary, Alta.	425	16.7
	Toronto, Ont.	820	32.2
	Montreal, P.Q.	1,035	40.8
	Halifax, N.S.	1,415	55.7
United States	Seattle, WA	990	39.0
	Los Angeles, CA	325	12.8
	Las Vegas, NE	95	3.8
	Chicago, IL	845	33.2
	New Orleans, LA	1,365	53.7
	New York, NY	1,075	42.3
	Miami, FL	1,520	59.9
Mexico	Mexico City	585	23.0
Costa Rica	San Jose	1,800	70.8
Bahamas	Nassau	1,180	46.4
Argentina	Buenos Aires	950	37.4
Brazil	Rio de Janiero	1,080	42.6
Chile	Santiago	360	14.2
Colombia	Buena Vista	8,690	342.0
Czechoslovakia	Prague	490	19.3
Denmark	Copenhagen	590	23.3
France	Paris	565	22.3
Germany	Berlin	585	23.1
Greece	Athens	400	15.8
Italy	Rome	750	29.5
Poland	Warsaw	560	22.0
Sweden	Stockholm	570	22.4
England	London	580	22.9
Russia	Moscow	630	24.8
Ethiopia	Addis Ababa	1,235	48.7
Kenya	Nairobi	960	37.7
Morocco	Casablanca	405	15.9
Nigeria	Lagos	1,835	72.3
South Africa	Capetown	510	20.0
Sudan	Khartoum	160	6.2
Egypt	Cairo	30	1.1
China	Shanghai	1,145	45.0
Japan	Tokyo	1,565	61.6
Indonesia	Jakarta	1,800	70.8
Singapore	Singapore	2,415	95.0
India	Cherrapunji	10,800	425.1
India	New Delhi	640	25.2
Saudi Arabia	Riyadh	80	3.2
Australia	Sidney	1,180	46.5

Source: Todd, 1970

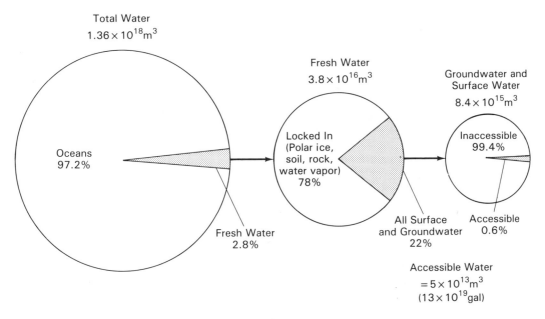

Figure 10–3 Water sources as a percentage of total supply. Source: Todd, 1970.

limited by the rate at which water moves through various portions of the hydrologic cycle. The time to replenish (i.e., completely replace) water varies from about two weeks in the atmosphere to 10–100 years in lakes, depending upon their depth (Miller, 1982). It is difficult to estimate how much of the total water budget is available on a continuous basis. If we consider only the water participating annually in the hydrologic cycle, this precipitation (and an equal amount of evaporation) is estimated at 420,000 km³ per year, of which 25 percent falls on the land (Todd, 1970). If 30 percent of this total amount (see Figure 10–1) were available to the world population of 5 billion, then the supply of fresh water in liters per capita per day (Lpcd) would be

$$25\% \text{ of } 30\% \times \frac{420{,}000 \text{ km}^3/\text{yr} \times 10^{12} \text{L/km}^3}{365 \text{ days/yr} \times 5 \times 10^9 \text{ people}} = 17{,}300 \text{ Lpcd}$$

Even this amount represents an unrealistic figure for many areas because of unequal distribution of accessible water, rapidly rising demand, and the pollution of water supplies close to urban areas. These three factors combine to limit present world per capita use to about 35 Lpcd (see Chapter 11). Of course, only part of this amount comes directly from precipitation; much of it is previously used water, pure enough for reuse without going through the hydrological cycle.

In many areas of the world today, more water is being withdrawn for use than is being replaced by rainfall. These are primarily regions with low rainfall and relatively high densities of urban population. Countries around the Mediterranean and parts of Africa, Iran, Australia, New Zealand, norther Mexico, and the southwestern United States are all

experiencing water shortages. By the year 2000 many other parts of the United States and
Mexico, the Soviet Union, Europe, the Caribbean, and parts of South America will be
suffering from chronic water shortages as well (Miller, 1982). The seriousness of this
problem, particularly in light of present efforts by developing countries to improve their
standard of living, was indicated by the United Nations declaration that the 1980s would
be the International Water Supply and Sanitation Decade. Although water shortages are a
problem in both developed and developing countries, the reasons for these shortages are
fundamentally different. Developed countries have the technology and water management
organization to support a higher standard of living based on an extremely high rate of
water use. Their problem is that water withdrawals are becoming so high, that even these
facilities cannot keep up with the demand. Less developed countries, on the other hand,
lack the facilities to properly treat and distribute the water resources within their reach.
This fact, coupled with burgeoning population, has led to severe water shortages in many
developing countries.

Question: Why is it misleading to say; "By the year 2000 we will run out of water?"

10.3.4 Water use

It is important to distinguish between consumptive and non-consumptive water use. Consumptive water use is that use which renders water unavailable for further use, either because
of evaporation, extreme pollution, or seepage underground, until the hydrologic cycle returns
it as rain. The nonconsumptive use of water leaves the water available (after treatment if
necessary) for reuse without going through the hydrologic cycle. On this basis, agriculture,
because of evaporation and percolation of the water used on crops, is responsible for about
80 percent of the water unavailable for reuse. Figure 10–4 shows the relative magnitudes
of various water uses in the United States. Water conducted many miles in open channels
and spread over large field areas is very susceptible to losses by evaporation and seepage
before reaching crop roots. Water percolating through irrigated fields may only be reused
a few times, because of the increasing load of dissolved soil salts it picks up from its
passage through the soil. Nonconsumptive water uses, on the other hand, leave water clean
enough after purification, by natural or mechanical processes, to be used again. Industrial
and thermoelectric power uses account for about 58 percent of water withdrawals. However,
97 percent of this use is nonconsumptive because the water is used only once to cool
machinery before being discharged. The remaining volume of industrial water is either
consumed by evaporator-type coolers or too contaminated by industrial chemicals to be
restored. Water for domestic use makes up less than 7 percent of total withdrawals. The
pollutants in municipal wastewater account for less than 0.5 percent of its mass, and therefore
purification for reuse is technically possible. The end result of our total water use is that
almost 80 percent of all water withdrawn is returned to the surface water portion of the
hydrologic cycle where, unless it is too disruptive of natural systems, it is partly rejuvenated
by natural processes.

Water shown as being returned in Figure 10–4 may not be fit for direct reuse for
recreation or municipal water supply and irrigation without some form of treatment.

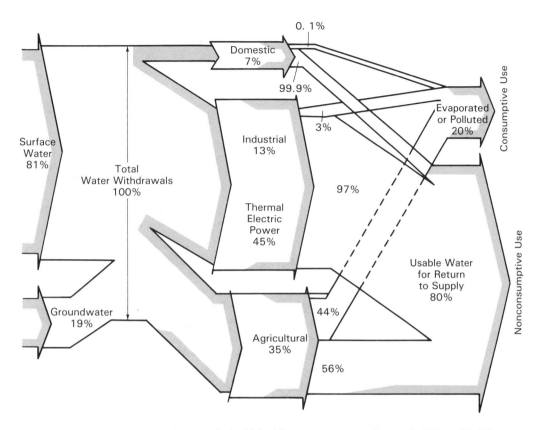

Figure 10–4 Water use in the United States. Adapted from Clark et al, 1977 pp. 105–106.

Example 10.2

California has approximately 8.7 million acres of land under irrigation; however, only 38% of the water supplied is taken up by the crops, the balance being lost by evaporation and seepage. To alleviate this problem, researchers are experimenting with a chemical that reduces evaporation losses in irrigation reservoirs and channels by forming a monomolecular layer over open bodies of water.

(a) Assuming that each acre of irrigated land requires 4 acre-feet of water per year (1 acre-foot = 1 acre covered to a depth of 1 foot), calculate the number of acre-feet that would be saved in one year if these chemical methods reduced consumptive losses by 1%.

(b) If an average family of four persons requires a maximum of 1 Ac-ft/yr, calculate the size of residential city that could be served with the water saved.

Solution

(a) Current consumptive loss = 62% of water used. *Water saved* by a 1% reduction in consumptive losses is

$$1\% \times \frac{8.7 \times 10^6 \text{ Ac} \times 4 \text{ Ac-ft}}{\text{yr·Ac}} \times 0.62 = 2.16 \times 10^5 \text{ Ac-ft/yr.}$$

(b) *Population* which could be saved by reduction in losses is

$$\frac{2.16 \times 10^5 \text{ Ac-ft/yr}}{1 \text{ Ac-ft/4 people · yr}} = 864,000 \text{ people}$$

As intimated earlier, the reusable character of water resources makes estimates of national or regional averages of water use misleading. Averages erroneously suggest that individual rates of water use can be added and compared to total rainfall or river flow figures. Also, total water use can exceed the total water budget when used water is returned and serves as a source of supply. Thus, in estimating total water use in a drainage basin or region, the water available for reuse and the extent of recycling must be considered.

Problem: Use of the word "consumptive" in describing water use can be vague and confusing. Suggest one or more terms which could be used to distinguish more precisely between the different types of water use.

10.3.5 Options for Meeting Water Demands

The growing demand for water has caused many countries, including even the United States and Canada, which together contain about 30 percent of the world's freshwater supplies, to examine ways in which essential water can be provided while at the same time preserving supplies for future use. Two major approaches can be identified, examples of which are shown in Photo 10-1(a) and (b). The first consists of using large engineering projects to obtain more water from various freshwater systems before they discharge to the ocean. This is a **supply-type** of solution. The second is based on increased water recycling, using both constructed and natural purification systems before the water is lost by evaporation or returned to the ocean reservoir. This is called a **reuse-type** of approach, which, in effect, recirculates water as a subcycle of the global hydrologic cycle shown in Figure 10–1. The latter approach will become more essential as freshwater supplies become more inaccessible.

Supply options

1. Dams and reservoirs are the oldest means of controlling water flow. Their benefits are equalization and control of stream flow, power generation, flood and drought control, and recreation. Problems, however, include silting up of reservoirs over time, greater evaporation losses due to large reservoir areas, and lowering of river delta flows in coastal areas, which allows the intrusion of saltwater.

2. Large-scale water diversions from one area to another have come into greater use, notably in California, since the major diversion of the Colorado River in 1931 to serve Los Angeles. The benefits of supplying abundant water for domestic and industrial development are obvious. The disadvantages of these major projects are their cost, evaporative losses, and the tendency to cause salt buildup and soil deterioration through improper drainage of irrigation projects.

(a)

(b)

Photographs 10–1   Options for meeting water demands

(a) Dams (a supply option) that store water for various purposes including water supply can create problems such as excessive algae growth as shown here (Photo courtesy B. J. Adams) (b) Water reclamation plants (a reuse option) like the San Jose Creek Plant in Los Angeles County, restore polluted water for such uses as groundwater recharge and watering of golf courses. (Photo courtesy J. G. Henry)

3. Groundwater contains 97 percent of all the freshwater in the United States and supplies about 20 percent of the country's needs. It is usually of higher quality than surface water and can be withdrawn for use in areas far from municipal distribution networks. Groundwater withdrawal must be limited to the rate at which the aquifer is recharged, otherwise the groundwater table will drop in level, reducing the amount available and increasing the cost of extracting it. Recharging relatively shallow aquifers can be done fairly easily, but deep reservoirs in dry areas may take hundreds of years to recharge and are thus, in practical terms, not renewable.

4. Desalination is receiving more attention as arid countries report success in some applications. Reverse osmosis (RO), forcing water through a semi-permeable membrane which passes water but traps dissolved salts, is the most practical of several desalination methods, including conventional distillation. RO units are expensive and relatively energy intensive, but they will become more economical for water purification as their use grows.

5. The use of icebergs as a water supply for dry coastal cities receives attention from time to time. Unresolved problems include the environmental effects, melting during transit, and methods for melting the ice and moving the water ashore.

6. Relocation of the population away from areas that are short of water or are already supporting as many water users as possible is one obvious plan. This option will receive greater attention as the cost for water increases and recycling and conservation have been implemented.

Reuse options

1. Better treatment to permit more reuse of waters that have become polluted will be a vital element of future water resource policies. When no new sources can be tapped, increasing the number of times that water can be reused before its return to the hydrologic cycle will be the only way to meet water demand in the long term, since the total amount of available water is fixed.

2. Reducing evaporation from water surfaces has the potential of significantly lowering water consumption in agriculture, the largest single user of water resources.

3. Water conservation techniques could be immediately effective in extending freshwater resources. Even relatively simple measures, such as installing special faucet and shower fittings, can save a great deal of water. Most industrial equipment that uses water was designed with abundant water supplies in mind. Efficient design could therefore drastically reduce industrial water needs. However, not all conservation techniques are technical. Changes in social and economic attitudes regarding freshwater supply and distribution can also play an important role in conserving water.

10.3.6 Quantifying Ecological and Social Effects

Water unites physical systems, such as the atmosphere, soil, and rock, with living systems. It is also an important factor in human society and affects the way individuals relate to each other by an intricate web of laws, rights, services, and activities. The use of limited supplies of water by any party in society affects other people and other living organisms. This relationship between water users is sometimes obvious. For example, a fisherman's catch will depend on the degree to which proper wastewater treatment methods are used in a nearby pulp and paper mill that discharges huge quantities of its effluent. Other situations, such as the dependence of an estuarine environment on river flow, may not be as obvious. The delta area of San Francisco Bay in California has a marine ecology that supports a large salmon population because of the character of the food chain resulting from the delicate balance between the flushing action of freshwater from the Sacramento River and the landward flow of an underlying layer of sea water. Reducing the seaward flow of fresh water by diversion projects upstream could upset this natural balance enough to endanger the six-million-pound annual catch of delta area salmon (Seckler, 1971).

Projects that alter river flow patterns in order to control and supply badly needed water resources must be planned and executed with a view toward their ecological and social consequences. This is not to say that water resource projects should not be undertaken until all such conflicts are eliminated, just that all factors should be considered before choosing a course of action.

Before a water control structure, such as a dam, can be designed, precise data must be collected and numerous decisions made. How much water will the dam have to retain? At what rate will settlement of silt reduce reservoir capacity? When should water be released for flood control, stream augmentation, or recreational use? Will undesirable plant growth be stimulated by the impounded water? What is the expected benefit of recreational facilities, and how will having land submerged by reservoirs affect local residents? These and many other questions were seldom examined closely until recently, because it was assumed that the capacity of the environment to absorb these changes was sufficient for any projects of a scale that man could execute. However, from projects like the Aswan Dam in Egypt, which destroyed the country's sardine industry, created downstream erosion, and promoted the spread of disease, we now know otherwise. There is a need to integrate data from various studies in a manner that will allow more rational decisions to be made. Computer models that represent, if only approximately, the relationship of many natural and economic activities, are presently receiving much attention for three reasons. The first is that these models, although they have limitations, can be run quickly to assess the possible cumulative effects that long- and short-term variations in river characteristics can have on the anticipated benefits of any water control proposal. Natural and social subsystems that affect and are affected by proposed projects can, to the extent that their relationships are quantifiable, be modeled as a study tool as well. The second reason for using modeling involves probability. Very often key data for design purposes can only be estimated within certain bounds of probability. River flow and rainfall records may not be adequate to determine the size of facility needed to lower the probability of flooding to socially acceptable levels. The complexities of balancing flooding losses against construction costs, while taking the probability of flooding occurrences into consideration, is a task requiring the use of computer models. The third, and perhaps decisive, reason is that the many diverse considerations and interactions involved in planning long-term water resource projects are becoming so numerous, and the effects of poor design so costly, that intuition, unaided by any computing

Photograph 10–2 The Stewartville generating station, dam and reservoir on the Madawaska River in Ontario, Canada. Source: Ontario Hydro

assistance in project evaluation, is too unreliable. Computer models, however, require careful scrutiny and interpretation to be beneficial in the decision-making process. Sound data are the foundation of mathematical models, and although collecting and interpreting this information may be time consuming, it is essential that this be done if decisions supported by numerical results are to withstand the criticism of concerned and informed opponents.

    Question: The dam pictured in Photo 10-2 was built across a river to provide hydroelectric power. Using a dam for which information is available as an example, outline the beneficial changes (physical, social, and economic) that the dam has created. What harmful effects has it caused? In each case your answers should be quantified.

10.4 PLANNING REQUIREMENTS

10.4.1 Purpose of Planning

Planning, the process preceding the implementation of a project, establishes objectives, critically examines them, and evaluates the means and results of implementing the plan. Today, its primary purpose is to inform those making a decision as to the consequences of their actions.

    Until recently, most water resources planning was oriented toward providing hydroelectric power and water for industrial, urban, and agricultural expansion because of the economic benefits involved. However, these benefits, as the public has become well aware, are often accompanied by environmental degradation and the loss of water for recreational or aesthetic enjoyment.

    Conflicts between those competing for water have become more visible, partly due to the power of the media to inform and provoke response from concerned citizens, and partly because water quality has become a critical factor inseparable from problems of water quantity. The goals of many strong special interest groups, each with claims to limited water supplies, are frequently in conflict. For example, manufacturers discharging "used" water without treatment may impair water quality, which has to be maintained for recreational uses that support a local tourist industry. Pollution abatement, flood control, land reclamation, and conservation should be considered simultaneously in water management planning. The challenge to the planner is to find an acceptable compromise between competing needs, while at the same time using water resources in the most efficient manner.

Confusion as to the purpose of planning has caused some resistance to wider application of its techniques. Planning is not the design and implementation of a preconceived project, nor does it imply state control of human activities. What planning should do is provide insights into the problems and alternatives so that all parties concerned, particularly elected representatives, can make intelligent decisions. Many people believe that planning can be applied as needed to resolve a specific problem. It is presumed, that if enough expertise is employed and sufficient money spent, then the proper course of action will emerge. This is unlikely to happen. Water resources planning is an educated look into the future of a dynamic system. Economic, political, social, and technological factors play a part in both the creation and solution of the problems. Uncertainty is unavoidable in determining such important parameters as the level of industry to be served, the timing and magnitude of water flows, and the political commitment to carry out the plan. Planning studies may be carried out over several years, during which time the situation must be constantly reassessed if appropriate responses to changing conditions are to be made.

10.4.2 Stages in the Planning Process

Although each planning situation is unique—depending on the project, the region, who makes the decisions, and the means and consequences of making them—there are enough similarities among successful plans that a general structure for the planning process can be identified. Figure 10–5 shows the stages in the planning process, suggested by Environment Canada (1975), that are common to many planning situations.

In the following discussion of this planning model, it is important to realize that planning is a social process. To treat the interaction between people as an analytical exercise will produce no meaningful results. Planning is, in essence, a rational process for determining the most appropriate course of action under a given set of circumstances. For the discussion following, the stages in the planning process have been grouped into three categories: formulation, evaluation, and adoption.

10.4.3 Formulation of the Study

In forming a plan, stages 1(a) through 1(d), as outlined in Figure 10–5, must be considered. The need for planning should be evident from Section 10.4.1. Awareness of the need for planning is the first step in the planning process. It is frequently difficult, particularly with natural systems such as rivers, to appreciate how changes in some components of a system

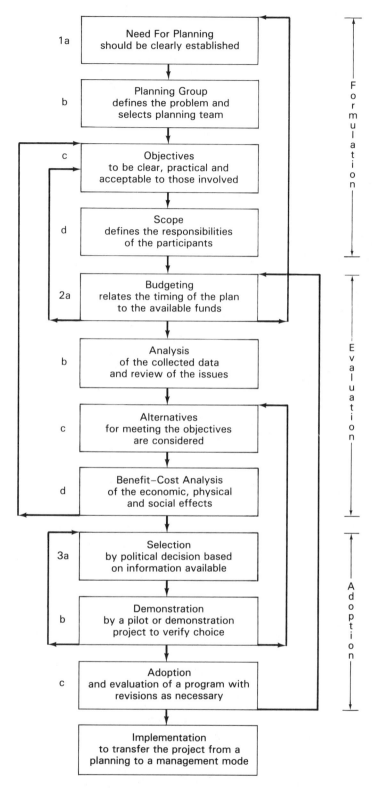

Figure 10–5 Outline of the planning process. Adapted from: Environment Canada, 1975.

can affect other components. A slight drop in the dissolved oxygen content of the river water, a change in temperature, an alteration in flow, or small additions of toxic material may set off a chain reaction affecting many aquatic species. Unless such interdependence is recognized, the full benefits of planning cannot be achieved.

The second step, establishment of a planning group, is important because it is at this point that key assumptions about the effect of the project are made in order to select the team of experts who will produce the report on the project. If, for example, a problem is perceived as requiring only engineering specialists when in fact land management, social behavior, and transportation patterns are influenced, then these vital elements will be neglected in the plan. The prime consultant must be able to provide the diversity of expertise necessary to deal with all those participating in or affected by the plan, including the approving authority, sponsoring agency, beneficiaries, landowners, other consultants, and public interest groups.

The objectives set for the study should be clearly understood by all parties so that subsequent work is directed toward accomplishing only what the study sets out to do. Setting objectives is not an easy task. The programs of government agencies with related interests should be examined to ensure that problems of jurisdiction and duplication of effort do not occur later. One problem with comprehensive water resource planning is that the logical geographic and hydrologic unit of study—the drainage basin—often crosses jurisdictional boundaries of all types, from the local to the international level. Planners must also be aware of the tendency for conflicts between several reasonable points of view. Attempting to accommodate the wishes of all water users can cause objectives to be set too high to be practical or too vague to be of use in evaluating proposals at a later stage. The only method of ensuring that practical goals are maintained is to constantly review them in the light of public and private consultation as the planning process advances. In many instances, negative public reaction to planning may be understood and overcome by recognizing the natural apprehension that occurs when the views of those affected do not seem to be represented. The effectiveness of all planning efforts will be improved dramatically if each party, whether public, administrative, or technical, can identify with the goals of the project and why these goals were selected.

Establishing the scope of a planning study is similar to a contract in that it clearly outlines what is expected of each participant. This step is also a key element in the planning process because it requires extensive dialogue among all parties, including the public, to determine the depth of study required by each component discipline to achieve the stated objectives. At this time, various agencies and advisors may be called upon to present specific viewpoints that will set the overall scope of the endeavor.

10.4.4 Evaluation of Alternatives and Their Effects

The evaluation of alternatives covers the four stages from budgeting to analysis, through alternative approaches, to benefit-cost analysis.

Budgeting, that is, setting schedules and costing the planning effort, can be a sobering experience when the idealistic goals set earlier are found to be prohibitively expensive. This is not unusual, because the need for data almost always exceeds available funds.

Compromise and ingenuity at data synthesis are usually required. Scheduling is important because it will also govern and be governed by the time needed to gather even rudimentary field data. Some data can only be obtained at certain times of the year, thus requiring a project schedule to be a detailed series of time slots in which certain activities must take precedence or else delay other stages, and thereby increase costs. It is also likely that obstacles will be uncovered during scheduling that cannot be overcome by new proposals or goals. In this case the proposal must be reviewed with respect to the time needed for each study, and new priorities and schedules must be drawn up. Realistic contracts for study can then be awarded by the prime consultant as required by the planning mandate.

Analysis of the problems requires the collection of pertinent chemical, physical, social, biological, and economic data. Early data will indicate where additional work must be done to gather and refine information so that the problems can be further clarified and discussed with those involved. Varying perceptions as to what the issues actually are will often occur even when data collection is in its later stages. Planners must not ignore, or appear to ignore, any concerns that may be voiced by affected parties. Disregarding earnest objections, as has been common, may lead to such strong opposition, with media support, that the objectives of the original plan are no longer politically palatable. Many studies have failed to be implemented because of this lack of flexibility. One Canadian example is the proposed transfer of water from northern to southern Alberta to ease the shortage of irrigation water in the south. The project has not been implemented because of environmental concerns, financial limitations, questionable benefits, and negative political reaction (Smith, 1981).

The generation of alternatives follows the analysis of the data. The issues which have now been defined are investigated, with a special emphasis on meeting the requirements of the participating regulatory agencies. For example, river flow regulation and waste discharge schedules would be related because of the pollution problems associated with dumping waste into rivers with low flows. Alternatives may apply to any of the human, physical, and biological systems that govern water use. The engineering approach which looks at permutations of dams, diversions, or other works to effect a certain goal is only one possibility. Water management options, for instance, may focus on better control of flooding through regulation of floodplain lands. Institutional alternatives may involve the creation or modification of agencies designed to monitor and respond to the abuse of water resources by one party at the expense of another. Phasing, that is, implementing a particular scheme in small segments over time, is another approach which may merit study. The main point is that all the present and future approaches should be studied so that no one alternative is favored for lack of information or consideration of another.

Benefit-cost analysis, first used in the United States in the 1930s (Phelps et al., 1978), was quite successful in terms of quantifying the tangible, that is the economic benefits and costs, in dollar terms, and choosing, within budgetary limits, the best proposal as the one with the highest benefit-to-cost ratio. However, planners now know the importance of evaluating competing proposals in terms of multiple objectives that attach sufficient importance to intangibles such as social betterment and natural environmental quality as well as to tangible economic factors. Benefit-cost analysis has been adapted to include intangible (noneconomic) benefits by using a computer to simulate the technical, biological, and social relationships involved in a project. In this way, input (costs) and output (benefits) can be compared to arrive at a ratio similar to the simple economic benefit-cost ratio. For this procedure computer simulation has become an increasingly effective and essential tool for finding the optimal solution of a multiobjective proposal under specified financial and physical constraints.

Digital computers are ideally suited to modeling the dynamic nature of water resource systems because of their ability to perform quickly the thousands of calculations needed to represent the state of the model during each increment of time. The most severe limitation of this evaluation method is that all inputs and outputs must be expressed in economic terms in order to permit comparisons. Economists and social scientists have been seeking a means of measuring natural environmental quality and social betterment ever since the advent of public concern for the inclusion of these intangibles in traditional evaluation procedures. Monetary techniques have sought to place a dollar value on water-resource-related benefits in terms of the amount consumers would be willing to pay for them in a hypothetical free-market economy. The foremost problem here has been the inability to objectively assess these community benefits in an economic system that depends on individual ownership to set market values. Also troublesome is the fact that in past projects many recreational and aesthetic facilities have been provided at no cost, thus depressing the market "value" that can be realistically assigned to these benefits.

Among the various ways of estimating the value of intangible benefits, ranking is a possible technique. In this method, impartial observers are asked to rate, on an appropriate scale, their opinion of the values of different individual and community assets, in much the same way that students fill out course evaluations in college. Based on the relative ranking, monetary values are assigned to intangible items by relating these to the benefits with known costs.

Evaluation techniques now at an early stage will undoubtedly become more sophisticated in the future. However, if we are to cope effectively with the complexities of water resource management, greater citizen understanding and acceptance of new approaches and more public involvement in the process will be necessary.

The following examples illustrate in a simplified way how the comparison of two alternatives by cost-benefit analysis can differ, depending upon whether recreational benefits are included or not.

Example 10.3

A planning authority for a small river basin has proposed two alternatives for a flood-control dam, each with an expected life of 40 years. Calculate the benefit-cost ratio for each alternative, using the following data:

Annual budget for dams (with no recreational benefits)	Alternative A	B
Yearly payment on construction cost	$10,487	$41,950
Expected average yearly decrease in flood damage claims	$18,000	$60,000
Yearly maintenance costs	$ 2,500	$ 5,000

Solution

	Alternative	
	A	B
Total benefits from dams	$18,000	$60,000
Total costs of dams	$12,987	$46,950
Benefit/cost	1.37	1.28

Alternative A, which returns $1.37 in benefits for every dollar spent, seems to be the better alternative.

Example 10.4

Objections have been raised concerning the fact that recreation was not included in the analysis of Example 10.3. One way of putting a dollar value on the benefits expected from this option is to look at prices charged at a similar project which operates for 100 days at 70% capacity:

Additional annual budget (for recreational facilities)	Alternative	
	A	B
Additional yearly payment on construction cost	$3,200	$ 7,100
Additional annual maintenance cost	$6,000	$10,000
Estimated daily park fee/person	$1.00	$ 2.00
Park capacity (people) per day	60	200

Solution

Annual recreational benefits for alternative A = 70 × 60 × $1.00 = 4,200
Annual recreational benefits for alternative B = 70 × 200 × $2.00 = 28,000

	Alternative	
	A	B
Total Annual Project Benefits		
From dams (see Example 10.3)	$18,000	$60,000
From recreational facility (see above)	$ 4,200	$28,000
	$22,200	$88,000
Total Annual Project Costs		
For dams (see Example 10.3)	$12,987	$46,950
For recreational facility:		
Capital	$ 3,200	$ 7,100
Maintenance	$ 6,000	$10,000
	$22,187	$64,050
Benefit/cost	1.00	1.37

Alternative B, which returns $1.37 in benefits for every dollar spent, now seems to be the better alternative.

Problem: Benefit predictions like that made for recreational facilities in Example 10.4 are usually subject to wide variation. As a result, most benefit-cost calculations include a "sensitivity" analysis to see what effect an error in predicted benefits will have on the final benefit-cost ratio. Calculate the percent variation in the benefit-cost ratio caused by a 40% variation in the original prediction of the daily fee required to encourage full use of park facilities for each alternative. Which alternative is more sensitive to changes in the benefit-cost ratio caused by the variation in fees?

10.4.5 Adoption of a Plan

Progress through the steps outlined in Figure 10–5 will require that many diverse opinions be reconciled before an acceptable solution is found. The selection of the "best" plan and eventual adoption of a program comprise a political decision in which many factors— economic conditions, the level of unemployment, other priorities, etc.—must be considered. The planner should not choose an alternative; it is the prerogative of the parties that commissioned the study to select from the choices that have been presented. Nonetheless, the planner, who perhaps better than anyone recognizes the risk of forecasting the response of a dynamic system, must predict the consequences of whatever decision is made.

Figure 10–5 clearly shows the need for iteration in the planning process. Unless planning retains the flexibility to examine and redefine its actions as new information is uncovered, its conclusions may become invalid and alienate those whose support is needed for the plan to succeed. Usually, one or more cycles of the complete planning process are required to provide planners and the public alike with the insights necessary to make sound decisions.

Planning does not end with the selection of an alternative. Rather, the choice may be tested by pilot studies leading to possible reevaluation of the program, or the planning process may move to the implementation mode. It is during the implementation stage that any problems or consequences of changes brought about by the project itself or any new developments within the drainage basin will be dealt with. Ongoing management of water resources by constant monitoring and follow-up action will be necessary to maintain public confidence in the plan.

10.5 LEGISLATIVE CONTROLS

The primary control of water resources is accomplished in North America and Europe by institutions and agencies created by governments, both federal and state (or provincial), under their mandate to serve the public interest. Federal institutions, like the U.S. Power and Resources Service (formerly the Bureau of Reclamation) or the Canadian Inland Waters Directorate, and state or provincial bodies like the Departments of Water Resources or Natural Resources, are empowered by government to plan and control water use. In specific situations these institutions often delegate authority to other agencies, such as water management commissions, planning boards, and conservation authorities. The problem of jurisdictional confusion, as it relates to the pervasive nature of water resources, has already been mentioned. To this should be added the natural reluctance of many water authorities and planning boards to relinquish their control over programs into which they may have invested considerable time and effort. Situations like this are inevitable because legislative mechanisms now exist for creating new planning authorities which, while more closely representing hydrologic boundaries, often give rise to conflict with existing water management bodies.

Considerable diplomacy, especially when dealing with international waters, is a prerequisite for any successful water planner. There is no substitute for knowing how to frame the planning process within the context of the many agencies which will eventually be involved. This can only be achieved by a thorough understanding of the existing authorities, the legislative tools they use to fulfill their purpose, and the historical reasons for their existence. To describe the various jurisdictional levels is beyond the scope of this text; however, a brief outline of the basic institutions and their relationship to one another will give some perspective on their importance in water resource management. Figure 10–6 indicates the major elements of the North American systems.

In Canada, the division of authority over water resources between the three levels of government—federal, provincial, and municipal—was specified by the British North American Act (BNA) now incorporated into the Canadian Constitution of 1982. The federal government has legislative rights over navigable waters, fisheries, waters in national parks and federal lands, and waters involving international boundaries. It also shares jurisdiction over irrigation and any other water resource undertakings mutually agreed upon with the provinces. The provinces are constitutionally designated as the sole proprietors of water resources within their boundaries, with the exceptions just noted. This grants to the provinces the right to manage, develop, license, and regulate water resources for any public purpose by using legislative powers also provided for in the constitution. Municipalities come under provincial supervision and are generally given responsibility for public water supply and wastewater treatment. Ontario has had success in organizing municipalities into regional conservation authorities for implementing comprehensive water resource management plans.

The Canada Water Act of 1970 was passed to allow the federal government to work more closely with provincial water resource authorities to facilitate a more flexible form of planning and funding of interprovincial and long-term river basin water projects. Under this act, river basin planning boards may be formed when federal and provincial authorities agree that they are needed. These boards are empowered to engage planners to advise on means for using water resources more efficiently. An extensive program of monitoring and reporting is encouraged so as to keep all parties informed of work being done. In most of these projects, water use regulation is stated in terms of water quality rather than quantity (Environment Canada, 1975).

The situation in the United States differs from that in Canada in a number of ways. In the U.S., federal powers cover a much broader range of water uses. Defense, treaty negotiations, taxes, and any projects that advance the good of the country are all legitimate federal activities in water resources. The federal government, under the U.S. Water Resources Planning Act (1965) can claim three major roles in water resource development within the country. First, it may, through the Water Resources Council, formulate and enforce standards and procedures for use by all federal agencies in preparing and evaluating water resource projects. Second, it funds state and interstate river basin water management programs. Because large sums of money are allocated annually for this purpose, the federal government

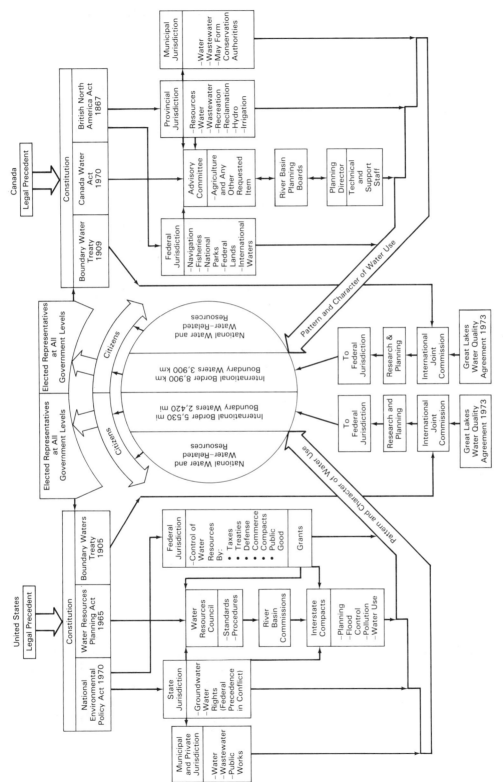

Figure 10–6 Legislative control of water resources in the United States and Canada. Adapted from Environment Canada, 1975.

can exert considerable influence in determining what projects will proceed under what conditions. Third, similarly to the function of the Canadian federal government, the U.S. federal government provides for the creation of river basin commissions to coordinate planning for water resource projects.

In addition to these three arrangements permitted under the Water Resources Act, another type of agreement, called a **compact,** is often used when the management requirements of an interstate water system go beyond the financial and jurisdictional limitations of individual states. Under the compact, in the event of overlapping federal and state interests in an area of water resource jurisdiction, the federal needs take precedence. The activities of the state in water resource management are primarily arbitration of water rights and distribution of water resources, both surface water and groundwater, through their power over individual proprietary rights. Municipalities, dependent on the states for their authority, generally manage water supply, wastewater treatment, and local public works.

Britain and France are both highly oriented toward comprehensive water management techniques. As a result, each relies on river basin authorities to serve as planning and implementation units that operate in a manner that is consistent with government water resource policy. In Britain, the river authorities regulate water withdrawal and discharge, through the use of fees. In France, the management of all aspects of water resources is aided by the extensive powers that river basin authorities are given. User charges are also a key aspect of their regulatory system (Environment Canada, 1975).

Internationally shared water resources have proven to be difficult to manage under the usual framework of water resource jurisdictions. Institutions created especially to deal with this problem have had considerably more success in this regard. Of the 8,900 km (5,530 miles) of border between Canada and the United States, fully 3,900 km (2,420 miles) are over water. This fact led to early recognition of the need to work together. The Boundary Waters Treaty of 1909 authorized the creation of the International Joint Commission. Including members from both countries, this advisory body was empowered to coordinate and negotiate all water resource programs for waters shared by the two nations. Later ratification of the Canada–U.S. Great Lakes Water Quality Agreement of 1973 recognized the immediate importance that both countries, with large populations centered around the lower Great Lakes, attached to problems of boundary water pollution.

10.6 POLITICAL INFLUENCES

10.6.1 Pressure Groups

Although thorough planning and comprehensive legislation are essential ingredients for orderly water resource development, other influences are often more important in determining what will ultimately be built. Figure 10–6 illustrates the extent to which planning is controlled by other agencies and government bodies. Equally important in the decision-making process are the special interest groups such as environmentalists, taxpayers, business organizations, and public action groups. These groups, organized to attract attention to their particular concerns, solicit political support to oppose or promote resource development projects.

Few water resource developments now proceed without lively debate among those whose often conflicting interests are involved. Such exchanges help to clarify the issues, but not necessarily to everyone's satisfaction.

The political process, however, is based on more than public debate and enumeration of preferences. Large business interests are well aware of the increased development opportunities associated with water development projects. Their views, backed by intense political lobbying and contributions to those who support them, are forcefully presented. On the other side, environmentalists pressure politicians to consider non-economic concerns that are often in conflict with public or private needs for low-cost water. The public, too, can exert considerable pressure through elected representatives. Conflicts are generally envisaged as being between big business and environmentalists or between the public and big government. This notion is fostered by the tendency of the media to simplify complex issues. In fact, it frequently happens that segments of similar groups find themselves on opposite sides in contentious issues. This was certainly the case in California in the debate over the Peripheral Canal. This project, designed to divert water southward from the Sacramento and San Joaquin rivers before they flow into the San Francisco Bay delta area, is discussed in Section 10.8.

Proponents of water resource projects must explore all possible avenues for resolving objections to proposals. The last resort, court action, leads to lengthy and costly delays, which may of course be precisely what those opposed are trying to accomplish.

10.6.2 Management Policies

Of the factors determining the scope of the work to be done, cost is probably the most significant. The fact that there is difficulty in applying traditional economic analysis to benefits such as flood control and environmental quality does not mean that economic forces are not at work. It just means that economic relationships are not understood sufficiently to predict the economic effects of these projects. The true cost of providing, storing, controlling, and treating water is seldom charged. Table 10–2 gives some idea of the prices charged for water delivered by public utilities and water authorities in North America. Subsidization of water resource facilities is the chief means by which government ensures that water will be available when and where it is needed. Social and economic policy is therefore reflected by the degree to which water benefits are subsidized. Table 10–3 illustrates

TABLE 10–2 TYPICAL CHARGES (1980) FOR WATER IN NORTH AMERICA

Use	Typical charge (U.S. dollars)		
	per acre-foot	per 1,000 gal	per 1,000 m³
Domestic	$133.00–267.00	$0.40–0.80	$106.00–212.00
Industrial	$ 67.00–133.00	$0.20–0.40	$ 53.00–106.00
Irrigation	$ 6.70–13.30	$0.02–0.04	$ 5.30–10.60

Adapted from Miller, 1975; Phelps et al., 1978; and Baker, 1980

TABLE 10–3 ALLOCATION OF WATER RESOURCE BENEFITS AND COSTS: CENTRAL VALLEY PROJECT, CALIFORNIA

Participants	Project capacity % use	Unsubsidized cost % share	Project costs % share	Benefit-cost ratio[1]
Residential, commercial industrial	3	10	8.7	0.34
Hydroelectric power	34	73	63.5	0.54
Irrigated farms	63	17	14.8	4.26
Federal government	0		13.0	0
TOTALS	100	100	100	

Adapted from Taylor, 1971, p. 121

[1] Benefit/cost = % use of project capacity/% share of project costs.

how hydroelectric, municipal, and irrigation water rates are related to actual project costs attributable to these uses in the Central Valley Project in California. The benefit-to-cost ratios indicate that water for municipal and hydroelectric use costs, respectively, about three times and two times the average cost of producing the water. The effect of this cost allocation is to subsidize farms which pay less than one-quarter of the average cost.

Example 10.5

Using average values for the typical water charges shown in Table 10–2 and the percent use of project capacity shown in Table 10–3, determine (a) what the average charge to users would be if all users paid the same rate (neglect the 13% government subsidy) and (b) the subsidy, in percent, that is provided for irrigation water by the other users.

Solution Assume a convenient water-use quantity of, say, 100,000 units (m³ or gallons). Then we have the following table:

Water use	% Water use	Unit costs per 1,000 m³	per 1,000 gal	Cost for share of 100,000 m³	100,000 gal
Municipal	3	$160.00	$0.60	$ 480.00	$ 1.80
Hydroelectric	34	$ 80.00	$0.30	$ 272.00	$10.20
Irrigation	63	$ 8.00	$0.03	$ 504.00	$ 1.90
TOTAL	100	—	—	$3,704.00	$13.90

(a) The average cost of the water to users is

$$37.04 \text{ per } 1,000 \text{ m}^3 = \$0.14 \text{ per } 1,000 \text{ gal}$$

(b) The percent subsidy for irrigation water is

$$\frac{\$37.04 - \$8.00}{\$37.04} \times 100 = 78.5\%$$

Two problems exist with California's subsidy program: it has not encouraged the efficient use of water necessary in a state that is approaching its limit in water resources, and it has allowed individuals with large holdings of marginal land to make high profits from the use of subsidized irrigation water to raise land values and produce valuable crops. This transformation of public resources into the private wealth of a few large agribusinesses was foreseen and restricted by the National Reclamation Act of 1902. The act limited the use of subsidized state water resources to owners who resided on 65 hectares (160 acres) or less of land. Large farming interests have argued against this law ever since, because it limits the amount of subsidy that they can realize. Even so, the amounts available are significant. When property taxes are used to finance part of the cost of a water project, the subsidy to irrigation water users is even greater. This is because areas with high assessment and a small proportion of the total water use (urban regions) then pay a disproportionately large share of the costs compared to farms with low property taxes and high water use. The following example illustrates the sums of money involved.

Example 10.6

In California, water is bought from the federal and state water projects by local water "districts" for subsequent distribution to individual water users. If 60% of project costs are charged to general assessment and other sources, with the remainder collected directly from the users, estimate the effective total public subsidy per acre of irrigated land if 1 acre requires 4 acre-feet (326,000 gallons) per year. Assume that the direct charges are proportioned as shown in Table 10–2 and that the district charges the irrigator $10.00/Ac-ft ($8.00/1,000 m³).

What increase in subsidy would an agribusiness with 160 acres gain if the land ownership limit specified in the 1902 Reclamation Act were doubled?

Solution   The yearly charge per acre to farmers for state water used in irrigation is

$$\frac{4 \text{ acre-feet}}{\text{acre-yr}} \times \frac{\$10}{\text{acre-feet}} = \$40.00 \text{ per acre/year}$$

The yearly cost to the state to supply irrigation water if users pay 14.8% (Table 10–3) is

$$\frac{\$40.00}{\text{ac}} \times \frac{1}{0.148} \times \frac{1}{0.40} = \$675.00 \text{ per acre/year}$$

So the subsidy is

$$\$675.00 - \$40.00 \doteq \$635.00 \text{ per acre/year}$$

The increase in annual subsidy to an agribusiness allowed to double its land holdings would be

$$160 \text{ Ac} \times \$635.00/\text{Ac} = \$102,000$$

Note: For simplicity, the estimated charges and subsidies to irrigators, as calculated, neglected property taxes. The cost of water to irrigators, therefore, would be increased by any water payments based on assessment, and the subsidies shown would be reduced by a similar amount. The changes would be relatively small.

Governments and their agencies are the only participants in the development of water resources that can enforce a long-term outlook on water quality and quantity. The concern of private water users is limited to those expenditures that promise a quick return. Such a view is incompatible with the decades often needed to formulate and implement water-use policies. The fact that government regulatory agencies are staffed by civil servants rather than elected officials enables long-term policies to be established and decisions based on political expendiency for short-term gains to be avoided.

10.7 FUTURE CHALLENGES

Fresh water is a limited renewable resource. The rising demand for clean, safe water by the public, industry, and agriculture has made us aware that past practices, which assumed an infinite supply of inexpensive water, can no longer continue. Human activity has affected the quantity and quality of every body of water on earth. The future beneficial use of water will depend on our determination to employ new social, technical, economic, and political methods of dealing with water resource management. The reason these techniques are so necessary now is that we have reached a point in many areas where water demand exceeds the readily available supply. The opportunity of contributing to the prosperity and well-being of many millions of people by developing better management skills is an exciting challenge for those in water resource development.

Future challenges in the technical aspects of water resource management are many. Our comprehension of the hydrologic cycle is limited to an understanding of the major pathways of water movement, and we are still unable to predict with any accuracy the quality and quantity of water at different points in the cycle. Hydrologists are only beginning to examine the complex relationship between rainfall and water runoff patterns. From extensive field data, geologists are studying the movement of water through porous soil and rock so that this information about groundwater behavior can be incorporated into water planning schemes. Meteorologists, limnologists, biologists, and many other scientists are contributing to our knowledge about the natural systems that relate human beings to their environment. A better understanding of natural phenomena will lead to more intelligent decisions in water resource management. As water demands grow, the fixed amount of fresh water available for the hydrologic cycle will eventually compel greater efforts at recycling and reuse of municipal and industrial wastewaters. Conservation is another way to utilize limited water resources more efficiently. For example, reductions in the excessive evaporative losses from reservoirs and aqueducts by means of surface films and membrane covers is sometimes practical. Irrigation by sprinkling systems reduces the large seepage losses associated with the ridge-and-furrow irrigation method which has been in use for 5,000 years. The "drip" method developed in Israel in the 1950s uses perforated piping installed on or below the soil surface. By delivering water directly to the root zone, water losses are reduced from the 50–60 percent lost by conventional methods to perhaps 15–25 percent. By 1985, an estimated 1 million acres (about 1% of the total irrigated area) were under drip irrigation in the U.S. Urban water use can also be curtailed by various means, including metering of water services, higher charges for water, increasing unit costs with increasing quantity, mandatory use of water-saving fixtures, and public education.

Conservation by all possible means should be part of water management policy if efficient use of water is important.

Desalination is one of the technological options in water resources planning that is becoming more and more an economic reality. Originally developed to extract pure water from seawater, its potential use is now much broader. Traditional desalination methods employed distillation, in which boiling and subsequent condensation of the stream left the salt and other impurities behind. Large energy requirements for heating made the process extremely expensive (about 10 times the cost of most municipal water) and limited its usefulness. Recent developments in reverse osmosis (R.O.), a less expensive process in which pressure forces pure water through a permeable membrane, leaving organic and inorganic impurities behind, have reduced the energy requirement by half. This has opened up the possibility of using R.O. to purify many types of polluted water, such as groundwater supplies that have become brackish due to saltwater intrusion or industrial wastewaters that are needed to supplement inadequate freshwater supplies. The world's largest R.O. plant, 72.4 mgd (million gallons per day) near Yuma, Arizona, was built to reduce the salinity of drainage water from 3,000 mg/L to 285 mg/L so that the water can be returned to the Colorado River for reuse in Mexico (Applegate, 1986). Desalination may help to rectify the imbalance between the uneven geographical distribution of freshwater resources and the desire to develop more land in water-scarce areas.

Controversy surrounding water-use developments has affected the planning process greatly. The exhaustive investigation and seemingly endless consultations now demanded for water resource projects were unheard of a generation ago. The greatest challenge to water resource planners is to develop an acceptable method for evaluating alternative proposals, considering the many economic and noneconomic functions these projects must now fulfill. The methods must allow a broad interpretation of river basin planning that includes all aspects of social betterment, economic growth, and natural environmental quality. Balancing these multiple objectives to maximize overall social enjoyment may, for example, require that one benefit not be optimized, or else other benefits will be lost. Education of the public in specific choices, trade-offs, and limitations of the project is essential if public preferences are to be accommodated in the planning. Planners themselves have been guilty of producing recommendations and policies that are far too general to be applied by municipalities and industries having limited funds available. Planning that is not implemented, for whatever reason, has failed its purpose. Benefit-cost analysis must reduce all factors involved in a project to dollar terms. However, inflation and the fluctuating cost of money are making the assessment of long-term projects difficult. One way that planning can alleviate this uncertainty is by careful phasing of the work in stages so that maximum flexibility for modifying the project in the future is maintained.

The procedures employed by various agencies for evaluating water resource projects are still evolving. Whether a standardized approach will develop or is even desirable is uncertain. There is no doubt that the methods will become more complicated.

Legislative controls on water use have evolved from constitutional and legal precedent. These are sometimes inadequate to deal with contemporary water resource problems that involve water diversions, control of water pollution, jurisdictional conflicts, and other similar difficulties. The formation of a single authority responsible for an entire drainage basin has resolved some, but not all, of the issues. When an agency's jurisdiction crosses political boundaries, whether at the local, national, or international level, controversy can be expected. The success of large water resource projects of the future will depend on close cooperation between the sponsoring agency, elected representatives, the media, the public, and other participants. Fostering this harmony is the responsibility of the planners, who will have to be completely familiar with the legal and political structures in which they must operate. Failure to generate a cooperative attitude may not just cripple support for the project—it may even generate strong opposition to its implementation.

Economists have had difficulty in determining equitable charges for water resources. Unlike most commodities, whose value is based on what people are willing to pay, water, in developed countries at least, has been looked upon as a free commodity and, because of its abundance, of little value. Governments have reinforced this attitude by their policy of subsidizing water supply as a device for encouraging development. In southeast Asia the problem of financing water projects is even more difficult, because of the people's belief that water is a gift from God and free to all. In Egypt in 1985, it cost about ten cents to purify and deliver 250 gallons of Nile water; the charge to the homeowner was about one cent.

Determining what people should or would pay for water and its associated benefits based on what they have paid in the past is not realistic. Perhaps social scientists involved in water resource problems will be able to suggest a logical basis for levying costs. Until then, rates will probably continue to be set in conformity with the previous inadequate charges.

Better water management techniques are urgently needed in developing countries to replace present inefficient and hazardous practices. Why they should do this when the developed countries, until recently, pursued the same environmentally damaging course is not hard to explain. Certainly, the needs of the two worlds are different. In developing countries, where an estimated 80 percent of all illness is attributed to an inadequate water supply, the justification for proper water management should be based not on aesthetic enjoyment and recreational benefits, but on necessary improvements in water supply and sanitation to reduce disease and protect public health, while at the same time, where possible, stimulating food production and industrial development. Clearly, because of their predicament, developing countries have a greater need for effective water management than developed nations. Modifying the technological and institutional aspects of modern water resource management techniques to suit differing priorities in developing countries is necessary for these methods to be useful. "Appropriate technology" is the name given to this type of approach.

As the number of occurrences of contaminated water supplies, water shortages, and flood damage multiply inexorably in the future, the need for water resource management will become more evident. Like energy, adequate supplies of clean water are essential to the modern way of life. Our response in the coming years will, to a large extent, govern the quality of life for future generations. Success in this area will depend not so much on technological advances as on improvements in our social, political, and economic institutions, which have lagged far behind scientific progress.

10.8 CASE STUDIES

Problems in water resource management are seldom as simple as those described in Examples 10.1 through 10.4. Most are complex and site-specific, and involve special interest groups with conflicting objectives. Two case studies have been selected to illustrate some of the difficulties in implementing water resource projects. The first, the Peripheral Canal, typifies the social conflicts that can arise over water use as a result of trying to increase supply by redirecting surface water flows. The information presented is based on reports by Seckler (1971), Phelps et al. (1978), and Baker (1980). The second case describes an unsuccessful attempt to reduce water pollution in the Saint John River basin in Eastern Canada through the use of a specially created federal-provincial planning board to coordinate research and implementation (Environment Canada, 1975).

10.8.1 The Peripheral Canal

California covers an area of 411,000 km^2 (100 million acres), stretching about 1,500 km (900 miles) northward along the Pacific coast from Mexico to Oregon. Its gross product exceeds that of all but seven nations in the world. California leads the U.S. in agriculture, manufacturing, and population (23,670,000 in 1982). One in every ten Americans lives in California, largely in cities surrounding the coastal metropolises of Los Angeles, San Diego, and San Francisco. The state is bordered on the east by the Sierra Nevada mountain range and on the west by the Pacific Ocean and Coastal Mountains. Most rainfall occurs north of the latitude of San Francisco in the winter and spring. However, 75 percent of water use is in the lower two-thirds of the state in the summer and fall. One of the major reasons why the Los Angeles region has become the hub of California's industrial activity and one of the largest metropolitan areas of the nation is because of the supply of water it receives from the statewide network of reservoirs and aqueducts that store and transport water from the humid north to the arid south.

Of the 600 mm (2 ft) of annual precipitations that falls on California, approximately 20 percent, or 40 million acre feet, is used by the state (Phelps et al., 1978). About 55 percent of this comes from surface supplies, with 45 percent obtained from groundwater.

Agriculture, which accounts for 85 percent of water use, has always been an important industry in California. Early settlers found that the soil was very fertile, capable of growing all types of fruits and vegetables when irrigated. Because of the importance of water, institutions and laws governing its use have often been the subject of public debate. During the 1920s and 1930s, farmers in the San Joaquin valley, following the lead of San Francisco and Los Angeles, decided that waters from the Sacramento and San Joaquin rivers could be stored and moved southward by a series of canals and pumping stations. Although the state legislature approved the financing, the Depression intervened and forced the federal government to take over the Central Valley project (CVP). The Shasta Dam on the Sacramento River, the Friont Dam on the San Joaquin River, and several aqueducts were then built. These works served delta-area water users as well as Central Valley farmers (see Figure 10–7).

The United States Bureau of Reclamation (now the United States Water and Power

Figure 10–7 State and federal water projects in California

Resources Service) was responsible for administering the CVP, including the enforcement of the 1902 federal Reclamation Act, which limited the size of farms receiving subsidized water to 160 acres. The demand for more water by the burgeoning postwar population of Los Angeles and by large agribusinesses seeking to avoid the 1902 act prompted the state to offer to buy the CVP from the federal government. After rejecting the price asked by the federal authorities, California proposed damming the Feather River as a major storage facility for another series of aqueducts stretching southward. This scheme, called the State Water Project (SWP), was approved by a narrow margin in 1960 over the objections of people in the north, who were becoming uneasy about the large reservoirs being constructed for the benefit of southern farmers and cities. The California Department of Water Resources (DWR) was created to plan the nonprofit SWP operation. Today, water releases from the Oroville Reservoir travel down the Feather River to the Sacramento–San Joaquin delta, wash through the delta, serving the City of San Francisco and delta area farmers, and are then pumped south over 400 miles to an elevation of 1 kilometer (3,280 feet) higher than the delta. In all, 23 dams and reservoirs, 9 aqueducts, 22 pumping stations, and 8 power plants make up the 23-billion-dollar project. Ninety percent of water rights in California are gained through legal, or appropriative, means. Appropriative water use is governed by legislation requiring that such use be "reasonable and beneficial." Riparian uses, although comprising only ten percent of all water use in the state, generally take precedence over appropriative uses. Nonprofit regional water distribution agencies known as **water districts,** of which there are approximately 1,000 scattered throughout the state, exercise their appropriative water rights by contracting for water delivery with the SWP or the federal CVP. District income for this water is typically 55 percent by water tolls, 25 percent by property taxation, and 20 percent from other sources, although some districts finance water projects wholly by water charges while others use only property taxes (Phelps et al., 1978).

The purpose of the proposed 400-foot-wide, 43-mile-long Peripheral Canal was to allow water from the north to bypass the delta on the east and south, thus increasing the rate of water available for pumping into the federal CVP and the California SWP. First proposed in the mid-1960s, this project has generated heated controversy ever since. Political careers were made and lost over the project, alliances were formed on either side, and lobbying was intense.

Proponents of the scheme included the California DWR, the San Joaquin valley farmers, the Metropolitan Water district (MWD) of Southern California, known also as the "Met," and southern corporate interests. The DWR insisted that in order for the SWP to meet its future contract obligations, it must increase water deliveries from 2.9 million acre-feet to 4.2 million acre-feet, 700,000 acre-feet of which were expected to be contributed by the Peripheral Canal. Although 31 water districts have claim to this state water, two of them—the Met, serving 12 million people in 131 communities throughout Southern California, and a water district in the San Joaquin valley known as the Kern County Water Agency—make up 75 percent of the demand. Both of these organizations vigorously supported the canal project also. They contended, with the aid of economic and engineering studies, that California's continued development and way of life were dependent on increased water flows from the wilderness area of the northern part of the state.

Critics of the plan disagreed. Water use in California, they argued, was very inefficient under the present system, which automatically linked water rights to land use. There was no incentive for a farmer to use less water under a state water policy that set prices for new water by averaging in the cost of earlier, less expensive projects. They recommended a more realistic pricing schedule called **marginal pricing,** to reflect the actual cost of providing more water by the construction of facilities much costlier than earlier projects (Phelps et al., 1978). Efficient water use would also be encouraged by having water tolls increased so that property taxes, paid by all landowners whether they use water or not, would not be used to subsidize water use. Another proposal was to allow holders of water rights to sell water to other users willing to pay more for it, thus creating a water "market" which would automatically benefit the most efficient water user. Other people, not necessarily advocates of the Peripheral Canal, rejected this idea, saying that the huge amount of private capital generated by this use of low-cost public water would result in an unacceptable redistribution of wealth within society that would outweigh the benefits of increased efficiency. The efficiency of unregulated groundwater use also became a topic of debate. Landowners have traditionally viewed groundwater as a resource for their own unrestricted use. However, depletion of readily available groundwater has necessitated recharge of aquifers from state-owned surface water supplies. The cost of this is borne by many taxpayers, who, in effect, subsidize excessive groundwater withdrawals. Consequently, pressure for better management of groundwater use and recharge has mounted steadily.

Regional differences created unusual alliances between various water users. Some large agribusinesses in the San Joaquin area joined forces with delta area farmers in opposing the canal. They were fearful that the canal would lower freshwater flows through the delta, allowing salt water to intrude farther up the channels in the summer and fall, when river flows were lower. These lower flows might also be unable to dilute municipal wastewaters and salt-laden irrigation runoff enough to avoid environmental problems. Delta-area farmers, who in the past refused to pay for high-quality water brought into the delta by the SWP, saying that they did not request the project, now sought water contracts and water quality legislation to minimize the effect of the Peripheral Canal. The U.S. Fish and Wildlife Service and the Marine Fisheries Service joined ecologists in opposing the plan. They claimed that the sensitive estuarine environment of the delta, described by Goldman (1971), would be harmed by lowered flows of fresh water that presently dilute pollution and carry wastes farther out to sea. A degree of eutrophication, similar to what happened in Lake Erie, was feared for the delta and San Francisco Bay. Ecologists predicted a reduction in the variety of life that supported, among many other activities, a valuable fishing industry.

Residents in the north of the state were concerned that the Peripheral Canal would make possible diversion of more northern rivers, thus avoiding water restrictions and encouraging further waste. The statewide controversy caused the U.S. Water and Power Resources Service to reduce its initial high estimate of project benefits, and federal participation in the project, once assured, seemed uncertain.

As it turned out, the referendum to build the Peripheral Canal was soundly defeated (Proposition 9, June 1982), as was a later vote (Proposition 13, November 1982) that would have required water districts to develop a conservation program by 1985. In further attempts to get water around the Sacramento–San Joaquin Delta to the farms and cities in the south, new water bills continue to be proposed (and vigorously opposed) in the state legislature.

It is evident that the controversy over water resources in California is far from over. Water laws and institutions are unlikely to remain unaffected. The broad economic and social implications raised by the debate on the Peripheral Canal will no doubt be part of all future water development projects in California.

10.8.2 St. John River Basin

Information for this case study has come largely from a treatise on comprehensive river basin planning prepared by Environment Canada (1975) for the NATO Committee on Challenges for Modern Society. The treatise demonstrates the range of social, political, and biological problems which arise in a river basin as population and development increase.

The St. John River, located on the east coast of Canada, flows 676 km (420 miles) from its headwaters in the state of Maine (U.S.) to its mouth on the Bay of Fundy at St. John, New Brunswick (see Figure 10–8). The river basin covers 55,200 square km (21,300 square miles) in the provinces of New Brunswick and Quebec in Canada and in Maine in the United States. Most of the region is covered by secondary forest growth, two-thirds of which is coniferous and supports a pulp-and-paper industry. Seventy percent of the basin's estimated 445,000 inhabitants are rural, living mainly in the river valley. Fredericton, the provincial capital, and St. John, an industrial center and maritime port, hold most of the urban population. The major uses of river water are for wood processing, waste discharge, food processing (mainly potatoes), power generation, and cooling. Groundwater from river valley aquifers is used for 60 percent of the public supplies because of the aquifers' higher quality water.

The river flow varies considerably. Mean flow is approximately 736 m³/s (26,000 cfs), with a daily low of 28 m³/s (1,000 cfs) and a maximum of 11,330 m³/s (40,000 cfs). A pulp-and-paper mill located at Edmunston is the largest point source of pollution on the river. The organic wastes from the mill have been estimated to be equivalent to the sewage from a city of 2.4 million people. A thick matting of wood fibers covers the river bottom below Edmunston, and will constitute a continuing source of pollution even after waste discharges at the mill are reduced. Potatoes are the major crop between Grand Falls and Woodstock, where a hydroelectric power dam and potato processing plant are located. Mactaquac Dam, located 24 km (15 miles) above Fredericton, has a salmon hatchery from which fish must be trucked to other rivers in the basin because the waters of the St. John River are too toxic at this point. Water quality improves downstream until St. John, where it is affected by tidal action which traps municipal and industrial wastes in the harbor.

Despite its natural resources, the region is relatively poor by North American standards. Unemployment is high, and much of the work that is available is seasonal in nature. Any attempts to expand the economic base of the region must involve pulp and paper, food processing, agriculture, and tourism. One of the major environmental problems in the area was the conflict between the use of the river for the discharge of industrial wastes and its use for tourism and fishing. The chemicals used for agriculture and for the control of a tree parasite that threatens forest production were other pollution problems in the basin.

In the 1960s, water resources were administered on a province-wide basis by the New Brunswick Water Authority (NBWA). The small staff was greatly increased during

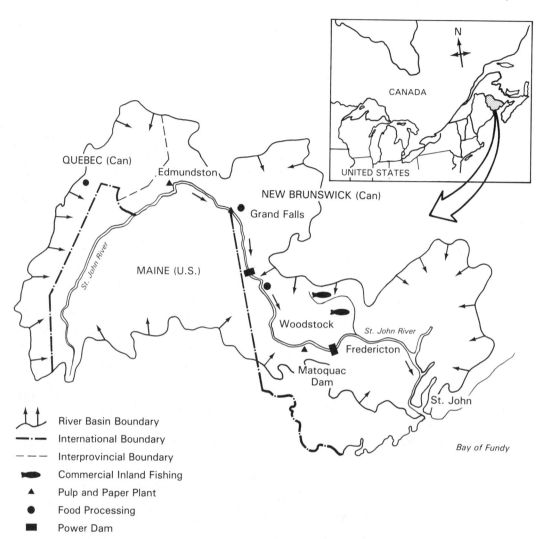

Figure 10–8 Saint John river basin. Adapted from Environment Canada, 1975.

the early 1970s to include the New Brunswick Environment Agency and the New Brunswick Department of Fisheries. Passage of the Canada Water Act of 1970 prompted the federal and provincial governments to form the St. John River Basin Board in an attempt to explore techniques and concepts for river basin, as opposed to provincial, water quality management. This board, consisting of three federal and three provincial members, was to prepare a comprehensive basin plan optimizing social, economic, and ecological aspects of water use. A liaison with the Northern Maine Regional Planning Commission was arranged to ensure that policies being developed on both sides of the Canada—U.S. border in the basin would be compatible. A public participation group was set up to explain planning alternatives to the people and to inform the board of public reaction to the plan.

Between 1972 and 1976, 30 specialist reports prepared for the Board were published. In 1975 (after expiry of the Board's mandate), a summary of the previously published three volumes of the Comprehensive Plan was released. Later a group commissioned by the federal and provincial governments found that by 1978 some of the Board's recommendations had been implemented, but that many were either too confusing or too theoretical to form a basis for action. Other suggestions by the Board seemed to be general statements of fact rather than definite recommendations. Consequently, the Board's plan was only slowly being implemented as part of the existing programs within several departments. Obviously, as far as implementation was concerned, the plan had been a failure.

Water management authorities within the basin have offered several possible reasons why the plan did not work. Water pollution, the main reason for the study, stemmed mainly from the obvious industrial sources. Before the Board was formed, both levels of government had made it clear that these industries would have to do a better job of treating their effluents. In response, the companies had threatened to close down rather than commit funds for new treatment equipment. Municipalities also were pressed to complete sewage treatment facilities already under way. The Board's work was helpful in two respects. First, through its public participation program, it helped bring citizen pressure to bear in convincing the polluting industries to clean up their discharges. And second, the ecological studies helped planners to understand how the capacity of the river to absorb pollution could be utilized. Unfortunately, the Board's research focused, not on these problems, but on the whole range of living systems within the basin. Although this produced good research, it diffused the focus of the recommendations, making them too general, particularly in terms of providing economic justification for the actions specified.

Another reason the Board was not as effective as it might have been was that the administration of the study was dispersed among federal and provincial agencies. The result was poor management of expenditures and confusion over responsibility for project control. For example, better control over the timing of technical report submissions would have helped the public participation group, which lacked the material needed to discuss specifics with the public, once public interest in the Board's work had been aroused. Although there was wide interest, most people seemed satisfied that the federal and provincial authorities were working in the public's best interests, and they did not actively participate in the study.

The work of the St. John River Basin Board was successful in that it allowed federal and provincial authorities to gain some experience in working together under the terms of the Canada Water Act. However, if the Board had been more aware of the needs of the agencies already at work and the water management programs under way at the time it commissioned various studies, a plan of more immediate use would have resulted. It is also evident, from hindsight, that chances for the success of the plan would have been improved if a single rather than shared government authority had placed more emphasis on solving the major problem of water pollution and less on extraneous matters. Although in some ways the operation can be deemed a success, the patient, the river in this case, has not recovered.

PROBLEMS

10.1. Approximately 378 m³ (100,000 gallons) of water are used in the production of one automobile. How much would the cost of water for this purpose have to increase in order to produce a 1% change in the $6,000 cost of producing a new car? Compare this to the increase in industrial water rates needed to produce the same increase in the 10¢ cost of producing a 25¢ newspaper. Assume that the production of a newspaper requires 1 m³ of water and that the cost of industrial water is $50.00 per 1000 m³.

10.2. Desalination seems to be the key to unlocking the vast supply of ocean water for our use. Explain why it is unlikely that desalination—even if the costs can be brought down to those of most municipal water supplies—will ever be widely used in irrigation, the world's most common use of water. Note that seawater contains 18,000 to 35,000 mg/L of salt.

10.3. Because of improvements in flood control, municipal development is taking place on lands that were formerly prone to flooding. Explain why annual flood damage costs have been rising steadily despite the billions of dollars that have been spent in the last 25 years on these flood-control projects. Assume that meteorological conditions have remained constant.

10.4. Recycling of municipal wastewater is technically, and sometimes economically, feasible for most municipalities. Why is it not more commonly used?

10.5. There is little doubt that we will need to increase the efficiency of water use. Some people think that that will only come about when a "free market," as opposed to a government-subsidized, approach to water supply and distribution is adopted. List the major points for and against this "water market" concept in which the rights to use water would be sold to the highest bidder just as mineral or logging rights are sold.

10.6. In most cases, individuals withdrawing water from aquifers pay no water charges except the expenses involved in pumping. However, high rates of water extraction can lower the water table enough to significantly increase pumping costs for other groundwater users in the area. For example, suppose a farmer with 65 hectares estimates that increasing his pumping by enough to raise annual gross revenues by $200 per hectare will cause the water table to drop 30 cm per year. If the cost of this higher rate of pumping for his farm is $30 per meter of additional lift, calculate for how long his pumping benefit-cost ratio will remain greater than unity. How long will pumping be profitable if he also has to pay a surcharge covering half of the extra pumping costs, which are increasing at a rate of $2.75 per year per hectare in 20 neighboring farms, each 65 hectares in size?

10.7. What would municipal and agricultural water charges to users in California be if they were equalized and set to cover 40% of the water projects' costs without the 13% federal subsidy? Translate this into a percentage increase or decrease from present water charges. Use average figures from Table 10–2 for present water costs in each category.

10.8. Suppose that the authority for the water "district" in Example 10.6 decides that the irrigators should pay 100%, as opposed to 40%, of the cost of water from federal projects.
(a) How much more would a 65-hectare (160-acre) farm have to pay for water yearly?

(b) What would the extra cost be if the total cost of the project were charged to users at the same rate and the federal subsidy of 13% was not available?

REFERENCES

American Water Works Association. ''42 Mile Canal Splits California Electorate.'' *Mainstream* 16–5, May 1982: 3–5.

APPLEGATE, R. ''World's Largest RO Desalting Facility.'' *Waterworld News* 2–3, May–June 1986: 17–19.

BAKER, L. B. ''U.S. Aides Tread Water on Peripheral Canal.'' *Sacramento Bee*, March 16, 1980.

BOLSTAD, W., MATHUR, B., and MACKNIGHT, H. ''A Case Study of Rural Opposition toward River Valley Planning.'' *Canadian Water Resources Journal* 6 (1981): 141–156.

CARDY, W. F. G. ''River Basins and Water Management in New Brunswick.'' *Canadian Water Resources Journal* 6 (1981): 66–79.

CLARK, J. W., VIESSMAN, V. L., JR., and HAMNER, M. J. *Water Supply and Pollution Control.* 3d ed. New York: Harper and Rowe, 1977. 4th ed., 1985.

Environment Canada. *Monograph on Comprehensive River Basin Planning.* Ottawa: Information Canada, 1975.

GOLDMAN, C. R. ''Ecological Implications of Reduced Freshwater Flows on the San Francisco Bay–Delta System.'' *California Water*, edited by D. SECKLER, p. 121. Berkeley: University of California Press, 1971.

MCGAUHEY, P. H. *Engineering Management of Water Quality.* New York: McGraw-Hill, 1968.

MILLER, G. T. *Living in the Environment.* Belmont, Calif.: Wadsworth Publishing Company, 1975. 2nd ed. 1979; 3d., 1982.

PHELPS, C. E., GRAUBARD, M. H., JAQUETTE, D. L., LIPSON, A. J., Moore, N. Y., SHISKO, R., and WETZEL, B. *Effective Water Use in California: Executive Summary.* Santa Monica, Calif.: Rand Corporation, 1978.

SECKLER, D., ed. *California Water.* Berkeley: University of California Press, 1971.

SMITH, D. ''Alberta Dream That Won't Die.'' *Toronto Star*, December 26, 1981.

TAYLOR, P. S. ''The 160 Acre Law.'' *California Water*, edited by D. SECKLER, p. 121. Berkeley: University of California Press, 1971.

TODD, D. K. *The Water Encyclopedia.* Port Washington, N.Y.: Water Information Center, Inc., 1970.

11

Water Supply

Gary W. Heinke

11.1 INTRODUCTION

This chapter deals with issues related to providing the quantity and quality of water required for society's various needs: the selection of alternative sources of water; the means of upgrading the quality of raw water through treatment methods; and the transportation and distribution of water, with particular emphasis on public water supplies. Water for irrigation, public water supplies, and industrial uses must be withdrawn from the source. Uses of water that do not entail withdrawal from the source include transportation, recreation, and fishing. Each of these uses places different constraints on the quality of water.

Irrigation, by far the largest withdrawal use of water, makes agriculture possible in many areas that could not otherwise support crops.

Public water supply refers to safe, clean water for use in homes, schools, hospitals, work places, commercial and some industrial activities, street cleaning, and fire protection. Water for drinking, personal hygiene, and sanitary purposes is of paramount importance to the health and well-being of the society.

Industry relies heavily on adequate supplies of water to be used either as a product component, as in beverages, or indirectly in controlling the process of production, as in the cooling of heat-generating machines.

Transportation by boat has been a practical and convenient means of moving people and products since ancient times. Water transport remains the most economical form of transportation even in this age of airplanes, railroads, and automobiles. Surface water pollution caused by shipping has become a significant problem, and regulations have been introduced for its prevention.

Recreation occupies a high priority in terms of the benefits that society realizes from an unpolluted water source. Swimming and bathing in particular are dependent on clean water. Propagation of fish and other aquatic flora and fauna is directly affected by pollution of the body of water.

11.2 WATER QUANTITY REQUIREMENTS

11.2.1 Water Demand

Total water demand on a municipal water supply system is the sum of all the individual demands (from toilet flushing, lawn watering, industrial cooling, street washing, etc.) during a stated period. Demand is not constant, but varies during the day and with the season. Variations decrease as the period over which we measure the demand increases— from hourly, to daily, to monthly, to yearly. Consequently, water demand in a particular community is normally specified in terms of **average daily demand,** defined as follows:

$$\text{Average daily demand (in community)} = \frac{\text{Total water use in one year (volume)}}{365 \text{ days (time)}} \quad (11.1)$$

Units are m³/d or million m³/d, or gallons per day (gpd), or million gallons per day (mgd).

It is often convenient to express the rate of demand per person:

$$\text{Average daily demand (per person)} = \frac{\text{Average daily demand in community}}{\text{midyear population in community}} \quad (11.2)$$

The units here may be liters per person (capita) per day (Lpcd) or gallons per person (capita) per day (gpcd).

Table 11–1 provides information on average daily per capita water consumption for various uses in North American cities. The data represent an average of actual water-use rates in several cities and from different references. Wide variations from these average figures occur, depending mainly on the extent of industrial and commercial activity, and on the climate of the city. Water consumption has increased at a rate of about one-half to one percent per year in the past two decades. Under the category of "Other" are included such uses as fire fighting, street flushing, and water lost through leakage from pipe joints.

Within the home, toilet flushing and bathing are the two single largest uses of water, accounting for almost 80 percent of total domestic use. Drinking water and kitchen use account for about 10 percent, and the remaining 10 percent is for clothes washing, house and car cleaning, and garden watering.

Water consumption in other developed countries is generally lower than in North America. In underdeveloped countries water consumption may be much lower. The amount of water used depends on the existence of a public water system; on the capability of that system to deliver water, whether the water is piped, trucked, or hand carried; on the extent,

TABLE 11-1 WATER USE IN NORTH AMERICAN CITIES

Use	Average daily consumption per person*		Percentage of total use
	Lpcd	gpcd	%
Domestic	300	79	45
Commercial	100	26	15
Industrial	160	44	25
Other	100	26	15
TOTAL	660	175	100

* Consumption in small residential communities and large industrialized cities can vary from -50% to $+50\%$, respectively, from the above quantities.

Adapted from Steel and McGhee, 1979.

if any, of plumbing in the home; on the existence of industrial users; on climate; and on social and economic conditions in general.

The results of a survey by the World Bank in 1976 on water use in rural areas of the developing world are provided in Table 11–2. In cities of the developing world, particularly in the business and wealthier residential areas, complete water systems are usually installed, and water consumption in these areas would be closer to North American figures.

Table 11–3 provides water consumption figures for a few selected industries. Industries requiring large quantities of water often develop their own water supply and do not use process water from the public system.

Water consumption in a particular community will vary because of several factors. For example, climatic conditions influence activities such as lawn sprinkling, bathing, and

TABLE 11-2 WATER USE IN RURAL AREAS OF THE DEVELOPING WORLD

Region	Average daily water consumption per person Lpcd	
	Minimum	Maximum
Africa	15	35
Southeast Asia	30	70
Western Pacific	30	95
Eastern Mediterranean	40	85
Latin America and the Caribbean	70	190
Normal range	35	90

Source: World Bank, *Village Water Supply*, Sector Policy Paper, Washington, D.C.: World Bank, 1976

TABLE 11–3 INDUSTRIAL WATER USE

	Water use	
Industry/product	Liters/unit	Gallons/unit
Oil Refining	18,000/tonne	770/bbl
Paper	160,000/tonne	39,000/ton
Steel	150,000/tonne	35,000/ton
Thermoelectricity	300/kWh	80/kWh
Woolens	580,000/tonne	140,000/ton

Source: Linsley and Franzini, 1972; SI conversion by authors

air conditioning. Also, water use tends to increase in direct proportion to the economic status and the standard of living of the people served. The extent and type of industrial activity can significantly increase water requirements as well, and price may also be a factor in water consumption, particularly where water supply is scarce and therefore expensive. Many other factors, such as the presence or absence of sewers, water quality, the pressure in the mains, and control of leakage, likewise affect water use.

11.2.2 Fluctuations in Water Use

The demands on a water system vary not only from year to year and from season to season, but also from day to day and hour to hour. An example of short-term variation in residential water demand during summer and winter is shown in Figure 11–1. Note that during the early hours of summer evenings, a substantial increase in water consumption may result due to lawn watering. It is common practice to express demand fluctuations as a fraction of the average daily demand. Records of water demand in similar areas can then be analyzed statistically to yield ratios such as those given in Table 11–4.

Most community fire departments obtain water for fire fighting from the nearest fire hydrant connected to the local water distribution system. If there are no fire hydrants, tank trucks or portable pumps and hoses must bring water from the nearest water source. A water distribution system is designed to provide the larger of the maximum hourly demand or the maximum daily demand, plus the fire demand to any group of hydrants in the system. This fire demand is often the governing requirement in establishing pipe sizes, pumping capacity, and reservoir capacity for cities under 200,000 people. The flow required to put out or at least contain a fire in an individual group of buildings can be estimated from an empirical formula recommended by the Insurance Services Office (1974), viz.,

$$F = 224\, C\sqrt{A} \qquad (11.3)$$

where F = the required fire flow in liters per minute
C = a coefficient which takes into account the type of construction, existence of automatic sprinklers, and building exposures. Its value is 1.5 for wood frame

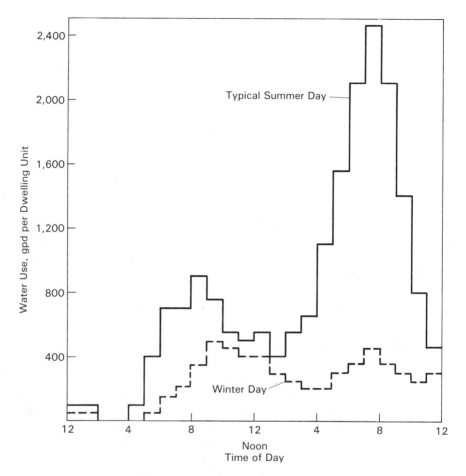

Figure 11–1   Residential water-use fluctuations. Adapted from Viessman and Hammer, 1985.

construction, 1.0 for ordinary construction, 0.8 for noncombustible construction, and 0.6 for fire-resistive construction.

A = the total building area or floor space, in square meters.

The equivalent formula in the American Engineering System (AES) is

$$F = 18\,C\sqrt{A}$$

where F is in gpm and A is in ft^2.

For residential areas with single- or double-family housing, required fire flows vary from a minimum of 1,800 Lpm (about 500 gpm) when the separation distance between buildings is over 30 m (about 100 ft) to 9,600 Lpm (about 2,500 gpm) for contiguous

TABLE 11–4 DEMAND VARIATIONS

Average Daily Rate	yearly 1.0
	summer 1.25
	winter 0.80
Maximum Daily Rate	1.5 (range, 1.2–2.0)
Maximum Hourly Rate	2.5 (range, 1.5–12)

Adapted from Viessman and Hammer, 1985

buildings. For the normal case of separation distances of 3 to 9 m (about 10–30 ft), the required fire flow would be between 3,600 and 5,700 Lpm (about 950 to 1,500 gpm).

When a fire occurs, the public water supply system must be able to deliver the required fire flow for 2 to 10 hours. Therefore, sufficient water has to be stored in a reservoir and additional pumping capacity has to be available to accomplish this, even during power failure. The nature of the buildings to be protected will determine the rate and duration of the required flow. The recommended duration varies from a minimum of 2 hrs for fire flows of 9,600 Lpm (2,500 gpm) or less to a maximum of 10 hrs for major fires.

Example 11.1

Calculate the water consumption (average daily rate, maximum daily rate, maximum hourly rate, and fire flow) of a North American mixed industrial-commercial-residential city of 100,000 people. The total floor area of the largest office building complex downtown is 25,000 m² (269,100 ft²). Assume that the coefficient C is 1.0 for this building, and determine the required capacity of the pipe distribution system.

Solution   From Table 11–1, assume average daily consumption of water is 660 Lpcd. Then

(a) $$\text{Average daily rate} = 660 \times 100,000 = 66.0 \times 10^6 \text{ Lpd}$$

From Table 11–4, assume that

$$\text{Maximum daily rate} = 1.5 \times \text{average daily rate}$$
$$\text{Maximum hourly rate} = 2.5 \times \text{average hourly rate}$$

(b) $$\text{Maximum daily rate} = 1.5 \times 66 \times 10^6 = 99.0 \times 10^6 \text{ Lpd}$$

(c) $$\text{Maximum hourly rate} = 2.5 \times 66 \times 10^6 = 165 \times 10^6 \text{ Lpd} = 6.87 \times 10^6 \text{ Lph}$$

(d) The required fire flow is determined using Equation 11.3:

$$F = 224C\sqrt{A} = 224(1.0)\sqrt{25,000} = 35.4 \times 10^3 \text{ Lpm}$$

(e) The design flow for the pipe distribution network is the greater of (1) the sum of the maximum daily demand and the fire flow, or (2) the maximum hourly rate. We have:

(1) $$99.0 \times 10^6 \frac{\text{L}}{\text{day}} \times \frac{1 \text{ day}}{1,440 \text{ min}} + 35.4 \times 10^3 \text{ Lpm} = 68,800 + 35,400$$
$$= 104,200 \text{ Lpm} = 6.25 \times 10^6 \text{ Lph}$$

(2) From part (c), maximum hourly rate = 6.87×10^6 Lph. Therefore, the pipe capacity must be 6.87×10^6 Lph.

Example 11.2

A steel mill producing 1,000 tons of steel per day is going to be built near the city of 100,000 people, discussed in Example 11.1. The site is located adjacent to a large river, from which the city also obtains its water supply. Calculate the amount of process water required by the steel mill daily, and compare this to the city's requirement.

Solution   From Table 11–3, one ton of steel requires 35,000 gal of water. Therefore, the daily water demand of the steel mill will be 35,000 × 1,000 = 35 mgd. Since the average daily demand of the city was calculated to be 66 × 10⁶ Lpd, or 17.4 mgd, the steel mill will require about twice as much process water as the entire city! It will obviously construct its own water system.

11.3 WATER QUALITY REQUIREMENTS

11.3.1 Water Quality Standards

Water contains a variety of chemical, physical, and biological substances which are either dissolved or suspended in it. From the moment it condenses as rain, water dissolves the chemical components of its surroundings as it falls through the atmosphere, runs over ground surfaces, and percolates through the soil. Water also contains living organisms which react with its physical and chemical elements. For these reasons, water must often be treated before it is suitable for use. Water containing certain chemicals or microscopic organisms may be harmful to some industrial processes while being perfectly adequate for others. Disease-causing (pathogenic) microorganisms in water can render it dangerous for human consumption. Groundwater from limestone areas may be very high in calcium carbonates, making it hard and thus requiring softening before use.

Water quality requirements are established in accordance with the intended use of the water. Quality is usually judged as the degree to which water conforms to physical, chemical, and biological standards set by the user. It is not as easy to measure as water quantity because of the many tests needed to verify that these standards are being met. Knowing the water quality requirements of each water use is important in order to determine whether treatment of the raw water is required and, if so, what processes are to be used to achieve the desired quality. Water quality standards are also essential in monitoring treatment processes.

Water is evaluated for quality in terms of its physical, chemical, and microbiological properties. It is necessary that the tests, used to analyze the water as regards each of these properties, produce consistent results and have universal acceptance, so that meaningful comparison with water quality standards can be made. *Standard Methods for the Examination of Water and Wastewater* (APHA, et al 1985) is a compendium of analytical methods followed in the United States and Canada to assess water quality. Table 11–5 lists the allowable limits set by the U.S., Canada, and the World Health Organization for various contaminants in drinking water. Chemicals listed under the Aesthetics heading are so limited because they cause undesirable taste, odor, or color, and (unless in excess) are seldom a threat to health. The limits suggested may be exceeded in some areas where treatment is

TABLE 11–5 DRINKING WATER STANDARDS

Contaminant		United States EPA	Canada NHW	International WHO
PRIMARY STANDARDS (Health) MCL				
Total Coliforms (Memb Filter)		Avg 1/100 mL	2/100 mL	0
		Max 4/100 mL	3/100 mL	—
Turbidity		1–5 TU	1–5 TU	<1
Inorganic Chemicals (mg/L)				
Arsenic	(As)	0.05	0.05	0.05
Barium	(Ba)	1.0	1.0	—
Cadmium	(Cd)	0.010	0.005	0.005
Chromium	(Cr)	0.05	0.05	0.05
Fluoride	(F)	1–2 (15°C)	1.5	1.5
Lead	(Pb)	0.05	0.05	0.05
Mercury	(Hg)	0.002	0.001	0.001
Nitrate	(N)	10.0	10.0	10.0
Selenium	(Se)	0.01	0.01	0.01
Silver	(Ag)	0.05	0.05	—
Organic Chemicals (mg/L)				
Endrin		0.0002	0.0002	—
Lindane		0.004	0.004	0.003
Methoxychlor		0.1	0.01	0.030
Toxaphene		0.005	0.005	—
2-4-D		0.1	0.1	0.1
2,4,5 TP		0.01	0.01	—
Trihalomethanes		0.10	0.35	—
SECONDARY STANDARDS (Aesthetics) RCL				
Chloride	(Cl)	250 mg/L	250 mg/L	250 mg/L
Color		15 color units	15 color units	15 color units
Copper	(Cu)	1.0 mg/L	1.0 mg/L	1.0 mg/L
Iron	(Fe)	0.3 mg/L	0.3 mg/L	0.3 mg/L
Manganese	(Mn)	0.05 mg/L	0.05 mg/L	0.1 mg/L
Odor		3 (Threshold)	Nil	Nil
pH		7.5 ± 1	7.5 ± 1	7.5 ± 1
Sulfate	(SO₄)	250 mg/L	500 mg/L	400 mg/L
Total Diss. Solids		500 mg/L	500 mg/L	1000 mg/L
Zinc	(Zn)	5.0 mg/L	5.0 mg/L	5.0 mg/L

Where two values are noted (turbidity, fluorides), the lower one indicates the recommended contaminant level (RCL), the higher one the maximum contaminant level (MCL) acceptable.

Sources: U.S. Environmental Protection Agency (EPA), *National Interim Primary Drinking Water Regulations*, EPA 570/9–76–003, 1976; Dept. of National Health and Welfare Canada (NWH), *Guidelines for Canadian Drinking Water Quality*, 1978; World Health Organization, (WHO) *International Standards for Drinking Water*, 1983

difficult and water users have become used to a particular taste or odor. Characteristics under the Health category are known to affect humans adversely; exceeding their specified limits can constitute grounds for rejection of the water supply.

The interim regulations of the U.S. EPA are continuously under review, and revised standards, when promulgated, may include any of the following:

- The use of a standard plate count (SPC) to monitor water quality
- Lowering of the turbidity guidelines for filtered water (to 1 TU) to protect against *Giardia lamblia*, Legionella, and viruses
- Separate MCLs for the different valences or forms of As, Cr, and Hg
- Changes to MCLs for Pb (lower), Ba, and Se (higher)
- The addition of presently unregulated contaminants.

In 1987, under this last category the EPA adopted final drinking water standards for eight volatile organic chemicals to be enforced after Jan. 9, 1989. Twenty-five other organic chemicals are to be regulated by 1991. The regulations are aimed primarily at protecting well supplies and are not likely to be a major concern as regards surface water sources.

11.3.2 Physical Characteristics

Tastes, odors, color, and turbidity are controlled in public water supplies partly because they make drinking water unpalatable, but also because of the use of water in beverages, food processing, and textiles. Tastes and odors are caused by the presence of volatile chemicals and decomposing organic matter. Measurements for these are conducted on the basis of the dilution needed to reduce them to a level barely detectable by human observation. Color in water is caused by minerals such as iron and manganese, organic material, and colored wastes from industries. Color in domestic water may stain fixtures and dull clothes. Testing is done by comparison with a standard set of concentrations of a chemical that produces a color similar to that found in water. Turbidity, as well as being aesthetically objectionable, is a health concern because the particles involved could harbor pathogens. Water with enough suspended clay particles (10 turbidity units) will be visually turbid. Surface water sources may range in turbidity from 10–1,000 units; however, it is possible for very turbid rivers to have 10,000 units of turbidity. Turbidity measurements are based on the optical properties of the suspension that cause light to be scattered or absorbed rather than transmitted in straight lines through the sample. Results are then compared to those from a standard suspension.

11.3.3 Chemical Characteristics

The many chemical compounds dissolved in water may be of natural or industrial origin and may be beneficial or harmful depending on their composition and concentration. For example, small amounts of iron and manganese may not just cause color; they can also be oxidized to form deposits of ferric hydroxide and manganese oxide in water mains and industrial equipment. These deposits reduce the capacity of pipes and are expensive to remove.

Hard waters are generally considered to be those waters that require considerable amounts of soap to produce a foam or lather and that also produce scale in hot water pipes, heaters, boilers, and other units in which the temperature of water is increased materially. Water hardness is expressed as equivalent milligrams per liter of calcium carbonate. The bicarbonates of calcium and magnesium precipitate as insoluble carbonates when carbon dioxide is driven off by boiling. This "temporary" hardness, called **carbonate hardness,** should be limited where it causes scale formation in boilers and industrial equipment. Sulfates, chlorides, and nitrates of calcium and magnesium are not removed by boiling. These salts cause **noncarbonate hardness,** sometimes called "permanent" hardness.

Synthetic organic compounds, like DDT, which are products or byproducts of chemicals used in agriculture and industry, can build up to toxic levels in water and living organisms. Measurement techniques have advanced much further than our ability to establish the relationship between synthetic organic compounds now in use and human health. Most governments have set arbitrary limits on the more dangerous of these chemicals until more complete knowledge in this area is available.

The microbiological characteristics of water are discussed in detail in Chapter 8.

11.4 SOURCES OF WATER

The quality and quantity of water from surface water and groundwater, the two main sources, are influenced by geography, climate, and human activities. Groundwater can normally be used with little or no treatment. Surface water, on the other hand, often needs extensive treatment, particularly if it is polluted. In arid regions of the world, the lack of groundwater or surface water may make the desalination of seawater and the reclamation of treated wastewater necessary. Such treatment is costly, but water of adequate quality for any purpose can be produced.

11.4.1 Groundwater

Groundwater is water that has percolated downward from the ground surface through the soil pores. Formations of soil and rock which have become saturated with water are known as groundwater reservoirs, or **aquifers.** Water is normally withdrawn from these reservoirs by wells. Soil pore size, water viscosity, and other factors combine to limit the speed at which water can move through soil to replenish the well. This flux (velocity) may vary from 1 m/day to 1 m/yr. A groundwater reservoir can only support a water withdrawal rate as high as is continually supplied by infiltration. Once this flow is exceeded, the water table will begin to drop, causing existing wells to run dry and requiring expensive deep drilling to locate new wells. There is growing concern that vast stretches of productive farms may lose irrigation water as wells go dry. Figure 11–2 shows how each category of water use is apportioned between groundwater and surface water in the United States. Note that public and rural water supplies make up only a small fraction of total water withdrawls, with irrigation and industrial water use each being one order of magnitude larger. Most rural water users rely on groundwater because it can be tapped and used directly right where it is needed, eliminating the need for expensive pipelines and purification. Figure 11–3 shows the total water use by source.

Groundwater is not as susceptible to pollution as surface water, but once polluted, its restoration, if even possible, is difficult and long term. Most pathogenic organisms and

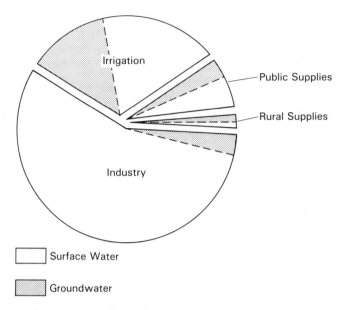

Figure 11–2 Water withdrawals for public supplies, rural supplies, irrigation, and self-supplied industry, 1980. Adapted from: Solley, Chase, and Mann, 1983.

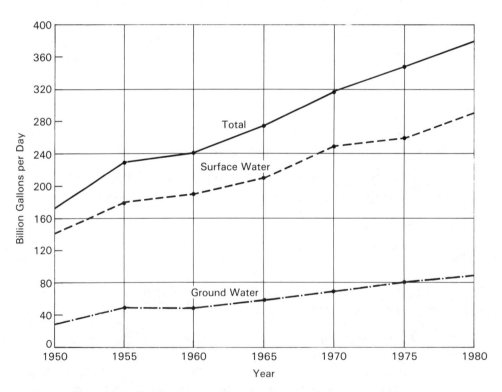

Figure 11–3 Total water use in the United States by source, 1950–1980. Source: Solley, Chase, and Mann, 1983.

many undesirable substances are removed by the filtering action of soil particles. This is why municipalities, even those located close to surface waters, prefer wells for a municipal water supply. Much less treatment and therefore expense is needed to bring groundwater to drinking water standards. Well waters, though of limited quantity, are usually of uniform quality and free of turbidity, but may require softening. Groundwater quality is difficult to monitor when large numbers of wells are in use. Proper well construction is essential to prevent contamination of the well water and hence the water table. Siting of septic tanks in relation to wells is critical if pollution of the water source is to be avoided.

Rainwater trickling through industrial and sanitary landfill sites can dissolve substances that pose a serious hazard to local groundwater quality. This can be prevented with a proper leachate management system as discussed in Chapter 14.

11.4.2 Surface Water

Surface waters from rivers and lakes are important sources of public water supplies because of the high withdrawal rates they can normally sustain. One disadvantage of using surface water is that it is open to pollution of all kinds. Contaminants are contributed to lakes and rivers from diverse and intermittent sources, such as industrial and municipal wastes, runoff from urban and agricultural areas, and erosion of soil. Water with variable turbidity, and a variety of substances that contribute to the taste, odor, and color of the water, can necessitate extensive treatment.

The problems with algae as related to water treatment were mentioned in Chapter 9. These problems, together with the additional costs (1) to control algae at the water supply source with copper sulfate, (2) for more frequent backwashing of the filters, and (3) for the extra chlorine or other disinfectant consumed by the algal organic matter are the reasons why eutrophication of lakes is of concern from a water treatment standpoint. It should be evident that the removal of phosphorus and occasionally nitrogen in our wastewater treatment plants, the limiting of phosphorus in detergents, and the restrictions on nutrients in agricultural runoff are not just for aesthetic reasons but have their economic motivations as well.

Direct use of rainfall is a limited but important water source in a few areas that are remote from fresh water but that receive regular precipitation. In Bermuda, for example, rainwater is collected on roofs and stored in cisterns for later use.

11.4.3 Seawater

Seawater, available in almost unlimited quantities, can be converted into fresh water by a number of processes. However, conversion costs (not including the costs for disposal of the masses of salt residue generated) are perhaps two to five times higher than those of treating fresh water.

Desalination is the general term used for the removal of dissolved salts from water. Distillation, the oldest desalination technique, depends upon the evaporation and condensation of water. The process is energy intensive, but using solar energy to evaporate water may make it practical in countries with plentiful sunshine. Another method, freezing, lowers the water temperature until ice crystals free of salt can be separated from the brine. Electrodialysis involves forced migration of charged ions through cation-permeable or anion-permeable membranes by applying an electric potential across a cell containing mineralized water. Reverse osmosis uses membranes that are permeable only to water; however, the driving force in this case is pressure provided by pumps. This process seems promising because energy costs are well below those for other technologies.

Currently, desalination plants for municipal water supply have found wide use in the Middle East. Future use will occur in areas of extreme freshwater shortage, particularly for industrial water uses.

11.4.4 Reclaimed Wastewater

Reclaimed wastewater refers to water that has been treated sufficiently for direct reuse in industry and agriculture, and for limited municipal applications. Such recycling or closed-loop operations may offer the only alternative in areas that cannot obtain enough fresh water. Suspended solids, biodegradable organics, and bacteria can all be removed or degraded by normal wastewater treatment processes, but color, the inorganic salts of magnesium, sodium, and calcium, synthetic organics like pesticides, and other toxic substances must be removed by advanced techniques similar to those used for desalination. Activated carbon is effective in removing many organic pollutants because of its extremely large surface ($\approx$ 1,000 m^2/g) that can trap and adsorb water impurities. Allowing water to cleanse itself by percolating through soil is another technique that removes impurities from water and has wide application in recharging groundwater supplies.

Currently, the use of reclaimed wastewater as a water source is practiced mainly in the Middle East, South Africa, and arid parts of the United States.

11.5 WATER TREATMENT PROCESSES

11.5.1 The Water Treatment Plant

One of the great achievements of modern technology has been to drastically reduce the incidence of waterborne diseases such as cholera and typhoid fever. These diseases are no longer the great risks to public health that they once were. The key to this advance was the recognition that contamination of public water supplies by human wastes was the main source of infection, and that it could be eliminated by more effective water treatment and better waste disposal. Filtration of drinking water was used as early as 1802 in Paisley, Scotland, and by water vendors in London, England, in 1828. In the United States, filtration of drinking water was first practiced in 1872 by the city of Poughkeepsie, New York. By the beginning of this century, improvements in the technology of making water safe for public use had become widespread throughout Europe and North America.

Today's water treatment plants are designed to provide water continuously that meets drinking water standards at the tap. There are four main considerations involved in accomplishing this: source selection, protection of water quality, treatment methods to be used, and prevention of recontamination. Common precautions to prevent groundwater and surface water pollution include prohibiting the discharge of sanitary and storm sewers close to the water reservoir, installing fences to prevent pollution from recreational uses of water, and restrictions on the application of fertilizers and pesticides in areas that drain to the reservoir. Instituting regulations that comprehensively deal with protection of the source can be difficult because several jurisdictions, from local to federal, may be involved in one project. Considerable political cooperation is therefore a prerequisite to the safe development of many large-scale water supplies.

Screening, coagulation/flocculation, sedimentation, filtration, and disinfection are the main unit operations involved in the treatment of surface water. Water treatment operations fulfill one or more of three key tasks: removal of **particulate** substances such as sand and clay, organic matter, bacteria, and algae; removal of **dissolved** substances such as those causing color and hardness; and removal or destruction of **pathogenic bacteria** and **viruses.** The actual selection of treatment processes depends on the type of water source and the desired water quality. Figure 11–4 shows a schematic outline of (a) a typical surface water treatment plant, and (b) a groundwater treatment plant. In the former, water flows by gravity through an intake structure and pipe, screens remove larger items, such as fish, sticks, and leaves, and low-lift pumps raise incoming water to the level of the treatment plant. From this point on, water moves through the plant by gravity.

Occasionally, raw water with low turbidity can be treated by plain sedimentation (no chemicals) to remove larger particles and then filtration to remove the few particles that failed to settle out. Usually, however, particles in the raw water are too small to be removed in a reasonably short time through sedimentation and simple filtration alone. To remedy

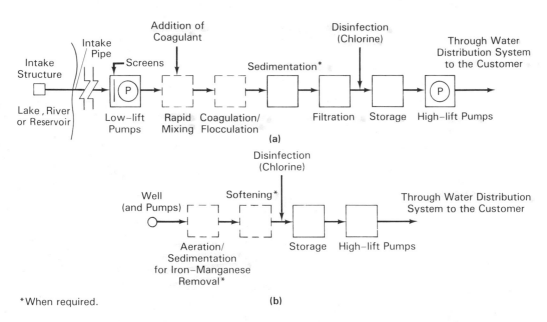

Figure 11–4 Schematic of water treatment plant (a) using a surface water source, (b) using a groundwater source.

this, a chemical is added to coagulate/flocculate the small particles, called **colloids,** into larger ones, which can then be settled out in sedimentation tanks or removed directly in filters. Where sedimentation precedes filtration, filters can operate for longer periods, or at higher rates, before they have to be backwashed. Clarified water drawn off the top of the sedimentation tanks is conveyed to the filters. Any remaining suspended particles are removed by straining, settling, and adhering to the sand or other filtering material as the water flows through the small pore openings of the filter bed. Filtration of chemically coagulated/flocculated water with *no* prior sedimentation (called direct filtration) is effective for waters with low to moderate turbidity (5 to 20 turbidity units) and is in fact the practice in many of the newer water treatment plants. Following filtration and before it flows into the storage reservoir, the water is disinfected, usually with chlorine. Fluoride may also be added because of its ability to retard tooth decay. Treated water is then pumped by the high-lift pumps into the distribution system to serve customers and to maintain water levels in storage reservoirs if required.

The rate at which water can be processed in a water treatment plant is usually based on the maximum daily demand rather than the average daily demand thus reducing the need for large storage capacity and allowing for shutdown of parts of the plant for maintenance during off-peak hours.

It is important to recognize that water treatment still remains somewhat of an art, despite many scientific advances in understanding the physical and chemical principles involved. There is a need for more research in order to meet the increasing demands being made on limited water resources. The rest of this section examines closely each unit operation grouped under the main functions of a water treatment plant.

11.5.2 Removal of Particulate Matter

The unit operations employed for the removal of particulate matter from water include screening, sedimentation, coagulation/flocculation, and filtration.

Screening to remove large solids such as logs, branches, rags, and small fish is the first stage in the treatment of water. Allowing such debris into the treatment plant could damage pumps and clog pipes and channels. For the same reasons, water intakes are located below the surface of the lake or river in order to exclude floating objects and minimize physical damage from ice. In a lake this intake is located far enough offshore to minimize the pollution effects of shore vegetation or waste discharges; in rivers it is located in a protected area. Coarse screens consisting of vertical bars spaced approximtely 25 mm (1 in) or more apart are placed at the intake point to exclude larger objects. Water then flows by gravity through the intake pipe to the low-lift pumping station at a velocity sufficient to prevent settling of particles in the pipe. Mechanically cleaned bar screens and fine screens (6 mm, or $\frac{1}{4}$ in spacing) are placed just ahead of the low-lift pumps that raise the water to the plant level. These fine screens are also used at the base of groundwater wells to exclude larger soil particles which may damage pumps and clog piping.

Sedimentation, the oldest and most widely used form of water and wastewater treatment, uses gravity settling to remove particles from water (see Chapter 6). It is relatively simple and inexpensive and can be implemented in basins that are round, square, or rectangular. As noted earlier, sedimentation may follow coagulation and flocculation (for highly turbid water) or be omitted entirely (with moderately turbid water). Particulates suspended in surface water can range in size from 10^{-1} to 10^{-7} mm in diameter, the size of fine sand and small clay particles, respectively. Turbidity or cloudiness in water is caused by those particles larger than 10^{-4} mm, while particles smaller than 10^{-4} mm contribute to the water's color and taste. Such very small particles may be considered, for treatment purposes, to be dissolved rather than particulate.

Water containing particulate matter flows slowly through a sedimentation tank and is thus detained long enough for the larger particles to settle to the bottom before the clarified water leaves the tank over a weir at the outlet end. Particles that have settled to the bottom of the tank are removed manually or by mechanical scrapers to be discharged to a sewer, returned to the water source if permitted, or stored on the site pending their treatment and/ or removal. Progressively smaller particles are settled out as the detention time is increased by making the tanks larger. The removal of very small particles using plain sedimentation would be impractical because of the high cost of making a sedimentation tank large enough to provide the needed settling time. Detention time is typically three hours in tanks 3–5 m (10–15 ft) deep. Particles too small to be settled out in this time must be removed by filtration or other methods.

Coagulation/flocculation is a chemical—physical procedure whereby particles too small for practical removal by plain sedimentation are destabilized and clustered together for faster settling. A significant percentage of particulates suspended in water are so small that settling to the bottom of a tank would take days or weeks. These colloidal particles would never settle by plain sedimentation.

Coagulation is a chemical process used to destabilize colloidal particles. The exact mechanism is not well understood, but the general idea is to add a chemical which has positively charged colloids to water containing negatively charged colloids. This will neutralize the negative change on the colloids and thus reduce the tendency for the colloids to repel each other. Rapid mixing for a few seconds is required to disperse the coagulant. Gentle mixing, called **flocculation,** of the suspension is then undertaken to promote particle contact. This is achieved by mechanical mixing through the use of slowly rotating paddles inside the coagulation/flocculation tank, or by hydraulic mixing which occurs when flow is directed over and around baffles in the tank. Detention time in the coagulation/flocculation tank is usually between 20–40 minutes in tanks 3–4 m (10–13 ft) deep. Through the combined chemical/physical process of coagulation/flocculation, the colloidal particles which would not settle out by plain sedimentation are agglomerated to form larger solids called floc. These appear as fluffy growths of irregular shape that are able to entrap small noncoagulated particles when settling downward. Aluminum sulfate (alum) is the most common coagulant but organic polymers may also be used alone or in combination with the alum to improve flocculation. The floc suspension is gently transferred from the coagulation/ flocculation tanks to settling tanks, or directly to filters where the flocs are removed. A cross section of a coagulation/flocculation and settling tank is shown in Figure 11–5.

The chemistry of coagulation is complex, but simplified equations can illustrate the

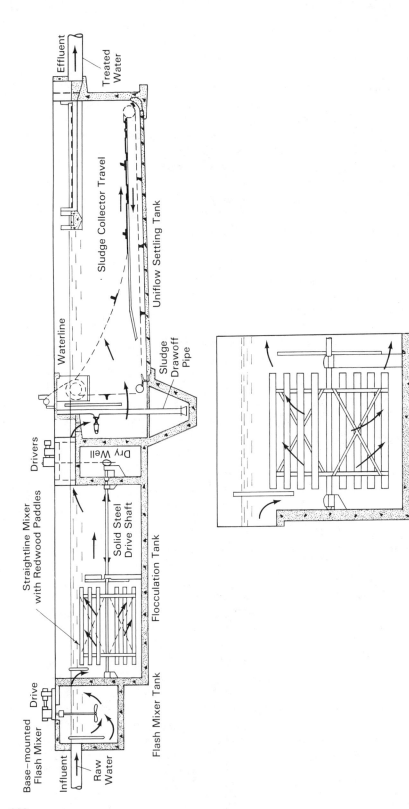

Figure 11–5 Cross-section of a rapid mixing, flocculation, and sedimentation tank. Courtesy Materials Handling Systems Div, FMC Corp.

388

process. The positively charged cations needed for coagulation of the negatively charged colloids can be provided by metallic salts, aluminum and iron salts being the most common.

The coagulation process using filter alum, $Al_2(SO_4)_3$ $14.3H_2O$, the standard coagulant in water treatment, is thought to proceed in the following three stages:

1. The alum ionizes in the water, producing Al^{+3} and SO_4^{-2} ions. Some of the Al^{+3} ions neutralize the negative charge on the colloids, but . . .

2. Most of the Al^{+3} ions combine with OH^- ions (from the water) to form colloidal $Al(OH)_3$, which adsorbs positive ions from solution

$$Al_2(SO_4)_3 + 6H_2O \rightarrow 2Al(OH)_3 \downarrow + 6H^+ + 3SO_4^{-2} \qquad (11.4)$$

3. The positively charged $Al(OH)_3$ sol then helps to neutralize the negative colloids, and the excess is neutralized by the SO_4^{-2} to produce a precipitate of $Al(OH)_3$ and adsorbed sulfates.

Note that the excess H^+ ions formed in step 2 tend to depress the pH, which would stop the formation of the $Al(OH)_3$ since it is pH dependent. Normally, the excess H^+ ions are removed by the alkalinity (HCO_3^-) present in the water according to the equation

$$6H^+ + 3SO_4^{-2} + 3Ca(HCO_3)_2 \rightarrow 3CaSO_4 + 6CO_2 \uparrow + 6H_2O \qquad (11.5)$$

The overall reaction, combining equations 11.4 and 11.5, is

$$Al_2SO_4 \ 14.3H_2O + 3Ca(HCO_3)_2 \rightarrow 2Al(OH)_3 + 3CaSO_4 + 6CO_2 + 14.3H_2O \quad (11.6)$$

which reveals that 600 parts of filter alum have used up three hundred parts of alkalinity (expressed as $CaCO_3$, as noted in Section 6.3.3).

The overall chemical effect will be a decrease in the pH of the water, a conversion of some calcium hardness $(Ca(HCO_3)_2)$ to sulfate hardness $(CaSO_4)$, and the production of CO_2. If insufficient alkalinity is present in the water for this reaction to occur, then the pH must be raised by adding lime, $(Ca(OH)_2)$, soda ash, (Na_2CO_3), or lye $(NaOH)$. The optimum pH for coagulation with alum is about 6. Coagulation does not require much additional pH control, because the introduction of alum lowers the pH of normally neutral surface waters to an acceptable value.

It is not normally possible to achieve adequate clarity of water with either plain sedimentation or the combination of coagulant/flocculation and sedimentation. **Filtration** therefore follows these unit processes in virtually all surface water treatment plants. Filtration is a process in which water passes through a filter bed made up originally of fine sand over a layer of supporting gravel. Other filter media and support systems are now common. Mechanisms involved in filtration include straining of the particles larger than the pore openings; flocculation, which occurs when the particles are brought into closer contact within the filter; and sedimentation of the particles in the pores of the filter. In time, the pore openings in the filter, particularly those at the surface, become clogged, and the filter must be cleaned by backwashing.

Two basic types of filter have been used: the slow sand filter and the rapid sand filter.

Slow sand filters were first used in Britain in the nineteenth century. They can process water at a rate of about 3–4 Lpm/m² (less than 0.1 gpm/ft²). River or lake water is pumped into large open-air slow sand filters, with or without plain sedimentation preceding, depending on the raw water quality. The filter bed thickness is approximately 0.6 to 1.2 m (2–4 ft), with water drains to carry the filtered water to storage. When the pore openings in the filter become too clogged, it is necessary to stop the application of water and remove the upper layers of sand manually for cleaning. Slow sand filters require large areas of land and are labor intensive because of the frequent cleaning needed to produce sufficient quantities of water. Although slow sand filters became outmoded by rapid sand filters at the turn of the century in Europe and North America, they still offer a practical means of water filtration for smaller municipalities and communities in developing countries, particularly where the climate is more favorable than in the northern United States and Canada. Compared to rapid sand filters, they are cheaper to build, simpler to operate, and do a better job of removing bacteria—an important consideration if other means of disinfection are unreliable.

Rapid sand filters can process water at a rate of 80–160 Lpm/m² (2–4 gpm/ft²) or higher, which is about 40 times faster than slow sand filters. The filter medium is still a layer of fine sand or anthracite and other materials supported over a layer of gravel or other supporting structure. Figure 11–6(a) shows a cross section of a sand filter bed, and Figure 11–6(b) shows a sand-anthracite filter bed. Figure 11–7 is a cross section of a typical rapid sand filter showing the filter box, bed, and appurtenances. These filters are normally housed in a building because of their small area, and to protect the water from possible sources of pollution. Clarified water from the settling tanks or flocculation tanks flows into the filter box and moves by gravity through the filter bed to the underdrains, which lead to storage reservoirs for treated water. The rate at which water passes through a filter slowly decreases as particulates build up on the filter grains and the pore openings become smaller.

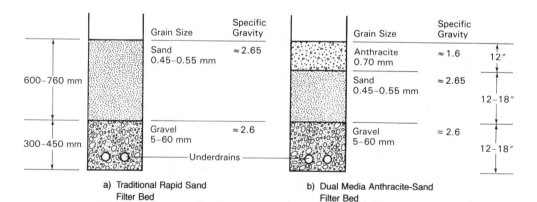

a) Traditional Rapid Sand Filter Bed

b) Dual Media Anthracite-Sand Filter Bed

Note: The difficulty with sand filters is that clogging occurs on the top layers of fine sand. Anthracite (coal), having a greater particle diameter and being lighter than sand, remains on top of the sand and makes more of the filter bed effective in the removal of suspended solids.

Figure 11–6 Construction of a filter bed.

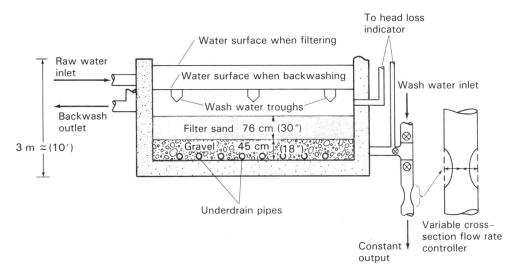

Figure 11-7   Cross-section of rapid sand filter. Adapted from Linsley and Franzini, 1972.

To provide a uniform flowrate, an external rate controller—a form of adjustable restriction in the outlet pipe—is used to keep the total loss of head through a filter, and hence the flow, approximately constant. The 2.5–3.0 m (8–10 ft) depth of the filter box limits the head available to force water through the filter bed. When the limit for head loss is exceeded, the filter is cleaned by an operation known as backwashing. Water under pressure is pumped through the pipes and underdrains and upwards through the filter. This reverse flow expands the filter bed by up to 50 percent and allows the lighter dirt particles to be removed with the washwater that overflows into the washwater troughs and out to the sewer. Where no sewer is available, the washwater is treated on the site, and, if necessary, the solids are hauled away for disposal.

The rate of backwash must be controlled to ensure that the sand or anthracite grains are not swept out with the washwater. Backwashing takes about 10–15 minutes and is normally carried out once a day, or more frequently if required. The water required for backwashing is generally about 4 percent of the water produced. When the backwashing operation is stopped, the filter medium will settle in place as it was before backwashing, since according to Stokes' law (see Equation 6.7), the larger (or denser) particles will settle faster than the smaller (or lighter) particles.

For small municipal installations, industrial applications, and swimming pool systems, pressure filters are often used. These are closed, usually cylindrical vessels that contain filter material through which water is forced under pressure, rather than by gravity as in the case of rapid sand or dual-media filters.

The relative effectiveness of the treatment operations covered thus far is roughly as follows. Turbid lake water at up to 100 TU (Turbidity Units) is reduced to approximately 10 TU by coagulation/flocculation and sedimentation. Filtration further decreases turbidity to less than 1 TU. A general rule of thumb is that turbidity is lowered by an order of magnitude by each process. Very turbid river waters (1,000 TU) will require presedimentation before undergoing the processes described. On the other hand, lake water withdrawn in winter can have turbidity below 10 TU, and coagulation/flocculation may not be necessary.

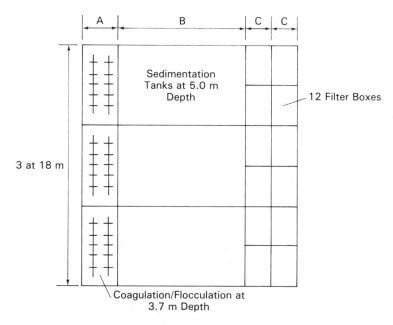

Example 11.3

The preceding sketch is a plan of a proposed water treatment plant for the city of 100,000 people in Example 11.1. The detention time for coagulation/flocculation (A) is 25 minutes, and the tank is 3.7 m deep. The detention time for the sedimentation tank (B) is 2 hr, and the tank is 5.0 m deep. The filtration rate (C) is 110 Lpm/m². Select appropriate dimensions for the units. Three parallel sets of tanks are used to provide flexibility of operation.

Solution   The required processing rate is the maximum daily rate for the city in Example 11.2, or 99 × 10⁶ Lpd. Each tank will handle one-third of this flow, or 33 × 10⁶ Lpd (22,916 Lpm). Accordingly, the required capacity of the coagulation/flocculation tank is

$$25 \text{ min} \times \frac{22,916 \text{ L}}{\text{min}} = 572.9 \times 10^3 \text{ L}$$

$$= 572.9 \text{ m}^3$$

So the length B of the coagulation/flocculation tank is

$$\frac{572.9}{18 \times 3.7} = 8.6 \text{ m}$$

The required capacity of the sedimentation tank is

$$120 \text{ min} \times \frac{22,916 \text{ L}}{\text{min}} = 2,749.9 \times 10^3 \text{ L} = 2,749.9 \text{ m}^3$$

So the length B of the sedimentation tank is

$$\frac{2,749.9}{18 \times 5.0} = 30.6 \text{ m}$$

Each filter will handle one-twelfth of the total flow, or 5,729 Lpm. Thus, the required area of each filter is

$$\frac{5,729}{110.0} = 52.1 \text{ m}^2$$

and the length C of each filter box is

$$\frac{52.1}{9.0} = 5.9 \text{ m}$$

Comment

In practice, it is customary to provide for maximum daily demand with any single unit out of service. In this case a fourth set of parallel tanks would probably be built. Alternatively, each of the three sets of tanks could have been designed to accommodate 50% of the maximum daily demand. The arrangement of the units is made compact and symmetrical for several reasons, viz.,

- To ensure uniform flow with as few changes in direction and thus as little turbulence as possible
- To permit economical common-wall construction and simplify enclosing the plant in a building
- To permit easy shutdown of one parallel stream for maintenance while the other units supply the demand
- To facilitate future expansion of the plant.

11.5.3 Disinfection

To ensure that water is free of harmful bacteria it is necessary to **disinfect** it. **Chlorination** is the most common method of disinfecting public water supplies. Sufficient quantities of chlorine from chlorine gas or hypochlorites are added to treated water to kill pathogenic bacteria. Chlorination is a reliable, relatively inexpensive, and easy disinfection method to use. Other disinfectants include chloramines, chlorine dioxide, other halogens, ozone, ultraviolet light, and high temperature. Ozonation, which has been used extensively in France, is now gaining acceptance in North America, especially as an alternative to pre-chlorination where natural organics are present. Although effective, ozone does not leave a lasting residual for long-term disinfection. The $2,000 \times 10^6$ Lpd (600 mgd) Los Angeles filtration plant contains one of the largest municipal ozone disinfection systems in the world, but the water is still chlorinated prior to distribution.

Chlorine gas hydrolizes in water almost completely to form hypochlorous acid:

$$Cl_2 + H_2O \rightleftarrows HOCl + H^+ + Cl^- \tag{11.7}$$

The hypochlorous acid, HOCl, dissociates into hydrogen (H^+) ions and hypochlorite (OCl^-) ions in the reversible reaction

$$HOCl \rightleftarrows H^+ + OCl^-$$ (11.8)

Chlorine lowers the pH of the water because of the hydrogen ions released in the preceding reactions. The pH of the water is quite important in determining how far the hypochlorous acid can dissociate to produce hypochlorite ions. Hypochlorous acid, the prime disinfecting agent, is dominant at a pH of less than 7.5 and is approximately 80 times as effective as the hypochlorite ion that dominates above pH 7.5. HOCl and OCl$^-$ together are called **free available chlorine,** meaning available for disinfection. The disinfecting qualities of hypochlorous acid are greatly increased at lower levels of pH because of the greater proportion of HOCl present.

When added to water, chlorine, a very reactive element, will oxidize organic and inorganic matter alike. Therefore, not all of the chlorine added to water results in the production of free available chlorine. The amount of chlorine that reacts with inorganics (Fe^{+2}, Mn^{+2}, NO$_2^-$, and NH$_3$) and organic impurities, is known as **chlorine demand** and must be satisfied before free available chlorine is formed. The application of chlorine to water to the point where free residual chlorine is available is called **break-point chlorination.**

The reaction of chlorine with nitrogenous impurities such as ammonia (NH$_3$) is of special interest because chloramines are produced. Chloramines are effective for disinfection, but to a lesser degree than free available chlorine. However, they persist longer in the treated water than free available chlorine does and are useful in guarding against possible contamination in the distribution system, caused by improper construction and maintenance. The addition of ammonia to chlorinated water to produce chloramines is called **chloramination.** Combined available chlorine is that residual existing in chemical combination with ammonia (chloramines) or organic nitrogen compounds. In some cases it is necessary to use chlorine to remove tastes and odors in water. This requires the addition of much larger quantities of chlorine in a process called **superchlorination.** To get rid of the excess chlorine, it is necessary to dechlorinate with sulfur dioxide, sulfite, or sodium metabisulfite. The exact mechanism by which chlorine attacks organisms in water is unknown, but what is known is that the water must be relatively free of organic matter for disinfection to be complete. Consequently, chlorination cannot be used as a substitute for poor water treatment practices.

The two basic parameters for effective chlorination are dosage and contact time. The rate of bacterial kill for a particular dosage does not follow a first-order reaction, so empirical equations are used to relate dosage and contact time for the desired percent destruction. Sufficient chlorine must be added to the water to both satisfy the chlorine demand and produce a concentration of free available chlorine of 0.2 mg/L after 10 minutes of contact at a pH of 7. The equivalent minimum combined available chlorine residual is 1.5 mg/L after 60 minutes of contact time at a pH of 7. Exceeding proper dosage levels is undesirable because this causes an unpalatable chlorine taste in water. Tests must be carried out frequently to determine the proper dosage of chlorine. To ensure sufficient time for chlorine to kill the bacteria, under varying pH and temperature, it is necessary to provide at least 30 minutes of contact time. Many authorities stipulate 2 hours at design flow.

If chlorination is the only form of treatment required, as is often the case with a groundwater source, it is applied at the distribution system pump well. In surface water treatment plants chlorination is normally performed as the last stage of treatment just before the water flows into the storage reservoir.

Chlorine, a gas under normal pressure and temperature, can be compressed to a liquid and stored in cylinders or containers. Because chlorine gas is poisonous, it is dissolved in water under vacuum, and this concentrated solution is applied to the water being treated. For small plants, cylinders of about 70 kg (150 lb) are used; for medium to large plants, ton containers are common; and for very large plants, chlorine is delivered by railway tank cars. Chlorine is also available in granular or powdered form as calcium hypochlorite $Ca(OCl)^2$, or in liquid form as sodium hypochlorite ($NaOCl$), (bleach).

One of the problems with chlorine is that it combines with natural organic substances which may be present in water (from decaying vegetation) to form trihalomethanes (THM), including chloroform, which is a carcinogen. Since THMs are not removed by conventional treatment methods, water to be chlorinated should be free of natural organics, or an alternative disinfectant should be used.

Example 11.4

Calculate (a) the number of kg of chlorine needed per day, and (b) the capacity of the contact tank in a water treatment plant supplying the city of 100,000 people given in example 11.2. The chlorine demand is 1 mg/L.

Solution

(a) We know that at least 1.2 mg of chlorine must be added to every liter to overcome the chlorine demand of 1 mg/L and produce a free available chlorine concentration of 0.2 mg/L. Since the treatment plant must be capable of operating at the maximum daily flowrate, we can make the following calculation to determine the amount of chlorine needed:

$$\frac{\text{kg of chlorine}}{\text{day}} = \frac{\text{liters of water treated}}{\text{at maximum daily flowrate}} \times \frac{1.2 \text{ mg of chlorine}}{\text{liter}} \times \frac{\text{kg}}{1 \times 10^6 \text{ mg}}$$

$$= 99.0 \times 10^6 \text{ Lpd} \times 1.2 \text{ mg/L} \times \frac{1}{10^6} \text{ kg/mg}$$

$$= 118.8 \text{ kg of chlorine added daily (at maximum production)}$$

(b) If we assume a minimum contact time of 30 minutes, then:

$$\text{Required capacity of contact tank} = \text{flowrate} \times \text{contact time}$$

$$= 99.0 \times 10^6 \frac{\text{liters}}{\text{day}} \times \frac{1 \text{ day}}{1,440 \text{ min}} \times 30 \text{ min}$$

$$= 2.063 \times 10^6 \text{ L}$$

$$= 2063 \text{ m}^3$$

The contact time in water treatment plants is usually provided by a large storage chamber called a clear well or storage reservoir. The primary function of the clear well is to isolate the plant from the hourly fluctuation of municipal water demand, but it also serves the purpose of allowing sufficient contact time so that the concentration of free available chlorine will stabilize at the proper value before being pumped to users.

Ozonation is the disinfection of water by adding ozone (O_3), which is a powerful oxidant of inorganic and organic impurities. Its advantages over chlorine are that it leaves no tastes or odors, and unlike chlorine, it apparently does not react with natural organics to form compounds hazardous to humans. Ozonation is widely used in Europe, particularly in France where many municipalities use it to disinfect public drinking water. In North America, except for the cities of Montreal and Los Angeles, disinfection with ozone is limited to a few smaller plants. This will probably change in the future. The disadvantages of ozone are that (1) it cannot be transported easily and therefore must be generated on site, (2) it does not provide a combined residual like chloramines to guard against distribution system infection, and (3) it is still quite costly.

11.5.4 Removal of Dissolved Substances

Several of the unit operations discussed so far are partially effective in removing objectionable dissolved substances. For example, color in water caused by colloidal or dissolved matter is reduced by coagulation/flocculation. Generally, though, conventional processes are not intended to remove dissolved substances or gases. If these are a problem, several other unit operations are available.

Aeration is used to remove excessive amounts of iron and manganese from groundwater. These substances cause taste and color problems, interfere with laundering, stain plumbing fixtures, and promote the growth of iron bacteria in water mains. By bubbling air through water, or by creating contact between air and water by spraying, dissolved iron or manganese (Fe^{+2}, Mn^{+2}) is oxidized to a less soluble form (Fe^{+3}, Mn^{+4}), which precipitates out and can be removed in a settling tank or filter. Aeration also removes odors caused by hydrogen sulfide (H_2S) gas.

Softening of water is a process that removes hardness, caused by the presence of divalent metallic ions, principally Ca^{+2} and Mg^{+2}. Hardness in water is the result of contact with soil and rock, particularly limestone, in the presence of CO_2. As noted in Section 11.3.2, the concentrations of both carbonate and noncarbonate hardness in water are expressed as $CaCO_3$.

Softening is seldom necessary for surface waters (where hardness above 200 mg/L is unusual), but it is occasionally desirable for groundwaters (where hardness above 1,500 mg/L is not uncommon). Hard water is acceptable for human consumption, but may be unsuitable for industrial use because of the scaling problems it causes in boilers. Lime-soda softening and ion exchange are two of the methods available for softening hard water. In **lime-soda softening,** lime (CaO) removes carbonate hardness by converting soluble bicarbonate (HCO_3^-) to insoluble carbonate (CO_3^{-2}), and soda ash (Na_2CO_3) removes noncarbonate hardness. With **ion exchange,** hard water is forced through an ion exchange resin such as Zeolite, which preferentially removes Ca^{+2} and Mg^{+2} ions from the water and releases Na^+ ions which form soluble salts.

Activated carbon is an extremely adsorbent material used in water treatment to remove organic contaminants. Activated carbon is produced in a two-stage process. First, a suitable base material such as wood, peat, vegetable matter, or bone is carbonized by heating the material in the absence of air. Then the carbonized material is activated by heating it in the presence of air, CO_2, or steam to burn off any tars it has and to increase its pore size. Adsorption of gases, liquids, and solids by activated carbon is influenced by the temperature and pH of the water as well as the complexity of the organics being removed. Powdered activated carbon can be added to water just after the low-lift pumps or at any point ahead of the filters. It has been used mainly for the removal of organics which cause tastes and odors. However, as concern grows over the presence of toxic organics in our water supplies, the role of granular activated carbon (made from anthracite) will increase.

In **reverse osmosis** (RO), fresh water is forced through a semipermeable membrane in the direction opposite to that occurring in natural osmosis. Because the membrane removes dissolved salts, the main application for RO has been in desalination. However, the process also removes organic materials, bacteria, and viruses, and its application in water treatment is growing.

11.6 TRANSMISSION, DISTRIBUTION, AND STORAGE OF WATER

11.6.1 Transmission

The conveyance of large quantities of water for a relatively long distance, between the points of supply and distribution is called **transmission.** A small city of 30,000 people in North America typically requires about 15,000 m³/day (4 mgd) of water (almost 15,000 tonnes daily!). The transmission line would be sized for at least twice this amount to allow for the maximum daily flowrate at the end of the design period plus allowances necessary for increases in population and per capita water use during that time. Gravity flow, whenever possible, is the preferred method for transporting these large quantities. Water at an elevation above that of its destination has potential energy that can be converted to the kinetic energy of moving water by the slope of an aqueduct. The steeper the slope, the faster the water velocity (within limits) and the smaller the aqueduct can be. Since frictional losses are directly proportional to the square of the water velocity, there is an optimum aqueduct slope to move water at a desired flow rate while minimizing losses due to friction. The most economical route for gravity flow, in which the effects of aqueduct size and slope on excavation (or tunneling) costs are considered, must be compared on an annual cost basis with a pressurized system where the energy costs for pumping may be offset by a smaller, shallower conduit.

There are three basic types of aqueducts. **Open channels** operate at atmospheric pressure and are called flumes if they are supported at or above ground level. They are usually chosen if topographic conditions are favorable for gravity flow with a minimum of excavation. Lining open channels with impervious materials may be required if the local soil is too porous and significant seepage losses occur. Evaporation and pollution concerns may also necessitate covering the channel. Materials such as concrete, butyl rubber, and synthetic fabrics may be used to line open channels. **Pipelines** are built when topographic conditions rule out the use of open channels. Placed above or below ground, these conduits often work under high operating pressures, so they are built of reinforced concrete, steel, cement-lined steel, or cast iron pipe. Reliable operation requires the installation of a system of check valves, surge control equipment, expansion joints, inspection ports, pumps, and many other appurtenances. Massive increases in pressure caused by sudden changes in flow are called hydraulic surges and must be minimized and controlled to prevent costly damage to the pipelines. **Tunnels** are used when open trenching for a pipeline is impractical.

11.6.2 Distribution

A water distribution system must be able to deliver either the maximum hourly flow or the maximum daily demand plus the fire requirements (whichever is greater) to any point in the municipality. Mains at least 150 mm (6 in) in diameter are needed to do this in residential areas. The pattern of distribution mains, street layout, topography, and pipe sizes all affect the cost and reliability of the system. Figure 11–8 is an example of a grid distribution system that will continue to serve most water users by at least one other route in the event of pipe failure. Shutoff valves at grid junctions can isolate any pipe segment for maintenance or repair without interrupting service to other parts of the grid. This is an important feature for system reliability, particularly in case of fire. Municipal fire insurance rates are based largely on the availability of a minimum pressure and flow at fire hydrants while the system is meeting the needs of regular users.

Water pressures in the distribution network range from 130–260 kPa (20–40 psi) in residential areas with buildings not over four stories in height to 400–500 kPa (60–75 psi) in areas with taller commercial or residential buildings. It is impractical to install costly additional pumps at the plant or reservoir to increase system pressures enough to supply the upper floors of very tall buildings with adequate water. To solve this problem, booster pumps in the buildings pump water to rooftop reservoirs which serve the upper floors and provide water for fire fighting.

Watermains are located within municipal road allowances so as to be accessible for maintenance. For temperate climates in the northern hemisphere, installation on the north

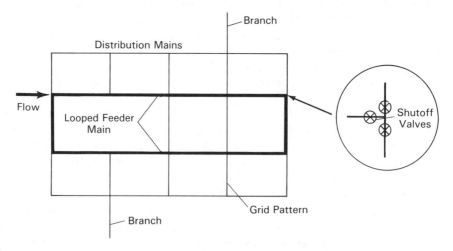

Figure 11–8 Distribution system configuration

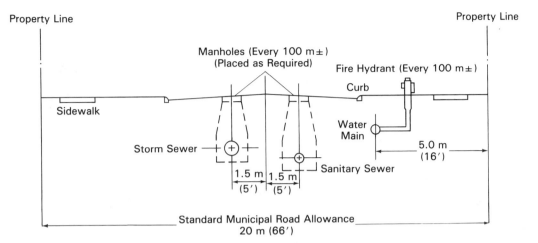

Figure 11–9 Typical arrangement of water and sewer services on a residential street

and east sides of the street (the warmer sides) is preferred at sufficient depth to be both safe from traffic loads and below frost level (1 to 3 m). Figure 11–9 shows a typical arrangement for water (and sewer) services.

Topography is another factor in distribution system design. Extreme ranges in elevation over an area can cause excessive pressure on watermains in low-lying areas and insufficient pressure at higher elevations. The high pipe pressure increases water leakage and may damage hot water tanks; the low pressure is not just inconvenient, but can result in contamination of the mains and inadequate fire protection. The solution is to divide the distribution system into separate zones, with a reservoir and pumping station in each zone fed directly by high-pressure feeder mains from the water plant or main reservoir.

Designing a pipe network involves selecting a system of pipes that vary in size and that will provide the desired flows and pressures for any reasonable combination of demands at different locations.

11.6.3 Storage

Storage is necessary in any municipal water supply system to meet variable water demand, to provide fire protection, and for emergency needs. Three types of reservoirs are used: surface reservoirs, standpipes, and elevated tanks. Surface reservoirs are located where they will provide sufficient water pressure, either by natural elevation on a hill or through the use of pumps. They are usually covered to avoid contamination. Standpipes are basically tall cylindrical tanks whose upper portion constitutes the useful storage to produce the necessary pressure head and whose lower portion serves to support the structure. Standpipes over about 15 m (50 ft) in height are uneconomical, and above this height elevated storage tanks become the preferred choice.

Residential water demand varies according to reasonably predictable patterns over the day. High-lift pumps at the treatment plant are not normally designed to meet these changes in demand. Instead, the usual practice is to pump water into the distribution system

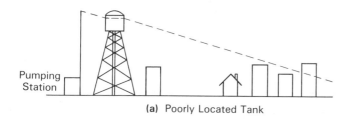

(a) Poorly Located Tank

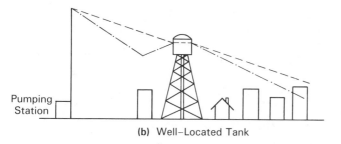

(b) Well–Located Tank

———— Pressure during
        Periods of Low Demand

—·——·—— Pressure during
        Periods of High Demand

Figure 11–10 Effect of water storage
reservoir location on pressure
distribution. Adapted from Linsley and
Franzini, 1972.

at a fixed rate for a given period and allow a reservoir to either supply extra water if demand exceeds this rate or receive water if demand is less than the pumping rate. Reservoirs that operate this way are known as floating reservoirs, that is, they accomplish their regulating function by hydrostatic pressure alone. In large cities, reservoirs may be located in the center of several distribution areas. Figure 11–10 illustrates how the location of a reservoir affects its ability to equalize operating pressures throughout a distribution system. Note how high water use and the accompanying friction losses increase the slope of the pressure profile so that water starts to flow from the reservoir to the surrounding area. Once the demand decreases, the slope of the hydraulic profile from the pump to the tank also decreases, allowing water to enter the tank and replenish the storage. In recent years elevated tanks have become less popular, partly because of their increased cost and partly because of the availability of relatively inexpensive variable-speed pumps and controls which make it possible to adjust pumping rates to varying demand. Photos 11–1 (a) and (b) show examples of elevated water storage tanks.

In addition to selecting the type and location of storage, storage size must be determined. This requirement depends on the population (water demand) and the purpose of the storage. Volumes for the three purposes—i.e., flow equalization, fire protection, and emergency needs—are calculated separately according to the time period over which they may be needed.

Equalizing storage, also called operating storage, is used to meet variable water demands while maintaining adequate pressure on the system. Where information on water demand is available, storage volume can be calculated or found graphically (from a mass

Elevated water tanks, either steel or concrete, are commonly used to provide equalizing storage for water distribution systems.

(a) The City of Welland, (population 50,000) in Canada handles the variable water demand in part of the city with a 5700 m³ (1.5 × 10⁶ gal) steel tank 39 m (129 ft) high, supported on 12 adjustable legs that provide for initial differential settlement. (Photo courtesy of R. V. Anderson Associates Limited)

(a)

(b) The City of Al Kharji (population 100,000) in Saudi Arabia stores 800 m³ (2.1 × 10⁶ gal) of well water in a concrete tank, 116 m (380 ft) high overall. The structure includes a 400 seat revolving restaurant. (Photo courtesy of R. V. Anderson Associates Limited)

Photographs 11–1 Elevated water storage tanks.

(b)

TABLE 11–6   DURATION OF REQUIRED FIRE FLOW

Required fire flow		
Million gallons per day	Liters per second	Duration (hours)
3.60 or less	160 or less	2
4.32	190	3
5.04	220	3
5.76	250	4
6.48	280	4
7.20	320	5
7.92	350	5
8.64	380	6
10.08	440	7
11.52	500	8
12.96	570	9
14.40 or more	630 or more	10

Source: National Fire Protection Association, *Fire Protection Handbook*, 14th ed., 1976. SI units by authors.

diagram, for example). When no information is available, operating storage is taken to be 15–25 percent of maximum daily consumption.

Fire storage is calculated by taking the product of fire flow and fire duration. Fire-flow duration times suggested by the National Fire Protection Association (NFPA) are given in Table 11–6. Fire-flow capacity may be raised or lowered depending on the reliability of the water supply source. For example, a municipality may increase its flow storage capacity if a water source such as a single well is used.

Emergency storage of up to five times the maximum daily demand is suggested by the Insurance Advisory Organization, in order to provide water during shutdowns for maintenance or repair to the system. This is seldom done in practice, and emergency storage is usually estimated to be one-quarter to one-third the sum of the operating and fire capacity requirements.

The sum of the three volumes for equalization, fire, and emergency is the storage capacity provided in a municipal water supply system. It will normally be about one day's average consumption.

Example 11.5

Calculate the required storage capacity for the mixed industrial/commercial/residential city of 100,000 people used in Example 11.1.

Solution   From Example 11.1, the maximum daily consumption is 99.0×10^6 Lpd. The fire flow is 35,400 Lpm (589 L/s), and therefore, from Table 11–6, the recommended flow duration is 9 hr. We thus have:

Operating storage $= 99 \times 10^6 \text{ L} \times 0.20 = 19.8 \times 10^6 \text{ L}$

$$\text{Fire-flow storage} \quad = 35,400 \frac{L}{min} \times \frac{min}{hr} \times 9 \text{ hr} = 19.1 \times 10^6 \text{ L}$$

$$\text{Emergency storage} \quad = \tfrac{1}{3}(\text{operating storage plus fire-flow storage})$$

$$= \tfrac{1}{3}(19.8 \times 10^6 + 19.1 \times 10^6) = 13.70 \times 10^6 \text{ L}$$

$$\text{Total storage required} = (19.8 + 19.1 + 13.0) \times 10^6 \text{ L} = 51,900 \text{ m}^3$$

11.7 FUTURE NEEDS AND DEVELOPMENT

Despite the many achievements of water supply engineering in contributing to human health and welfare, there still remain three formidable obstacles to the establishment of a balance between our need for clean water and the proper functioning of ecological systems. First, at least one-half of the world's people do not enjoy an adequate supply of clean water. We in the developed nations have grown up with the financial and technical infrastructures of water supply systems and institutions firmly in place. Less developed countries often do not have this advantage at the very time population growth and increased water needs make it most imperative. The demand for clean water for both rural and urban populations in the less developed countries was identified by the United Nations as the single most pressing challenge of the 1980s, which were designated as the International Drinking Water Supply and Sanitation Decade. A 1987 UNICEF progress report on the program indicated that by 1990 half of the original objective might be reached. Second, the spread through the biosphere of an increasing number of chemical compounds used by industry has created some doubt as to the effectiveness of present water treatment methods in preventing potential long-term health hazards associated with their presence in drinking water. And third, the general quality of freshwater sources is now deteriorating because of the increasingly intensive use of such sources by our growing industrial societies. The difficulties in meeting these three challenges are formidable.

An adequate supply of clean water is an absolute prerequisite to the provision of proper health care, nutrition, and industrialization. The reasons why advances in the technology of water treatment and supply have not been applied in less developed countries are both financial and institutional. Present water supply systems in developed countries evolved slowly in an environment where the capital resources for their installation and upkeep were not a major problem. This approach is not possible in poorer countries that must quickly install capital-intensive water treatment and distribution systems to meet the demand for clean water by an exponentially growing population. Although the methods covered in this text are applicable to many urban areas in less developed countries, other appropriate technologies for water supply must be developed for remote rural areas. Appropriate technology is a concept that recognizes low cost as a crucial factor. Simple solutions tailored to local needs in the developing countries are now receiving more attention.

The institutional barrier to implementing water supply technology, although not as obvious, is far more insidious. Adequate funding does not guarantee success. Large-scale modern water supply facilities have often gone unused because of a lack of skilled personnel to maintain them. Lowry (1980) notes that methods must be used which can be quickly taught to local technicians, implemented properly by unskilled and semi-skilled labor, and accepted culturally by the inhabitants.

The second challenge for water supply engineers is the control of many new chemicals for which conventional water treatment plants have not been designed. Over 1,000 new chemicals a year are being added to the large inventory of chemicals that appear as products or by-products of industrial processes. Synthetic organic chemicals like PCBs (polychlorinated biphenyls), trihalomethanes, Mirex, and dioxin do not break down in natural ecosystems. The biggest problem facing health scientists is identifying the relationships between various chemicals and human ailments. This information is needed if meaningful drinking water standards are to be developed.

The third problem, namely, the decreasing quality of water sources, is forcing municipalities to consider alternatives. One is to look farther afield for unexploited water supplies. Large water transmission projects such as those described in Chapter 10 are undertaken if population growth simply makes other alternatives too expensive. Environmentalists have voiced their concern over the effect that massive withdrawals of water from remote wilderness reservoirs might have on wildlife. The fear of not having sufficient fresh water for urban centers is demonstrated by the vast sums that governments spend to construct these water transmission projects.

Another alternative is greater reuse of extensively treated municipal wastewaters for certain water needs. It has been much more difficult to overcome the negative psychological reaction of people to the reuse of municipal wastewaters than to develop the treatment methods. Perhaps public objection will moderate as water recycling systems begin to prove themselves in areas that have very little choice but to apply these techniques.

PROBLEMS

11.1. Why are distribution pipes not sized according to maximum hourly demand plus fire flow instead of maximum daily demand plus fire flow? (See Example 11.2.)

11.2. Calculate the water consumption (average daily rate, maximum daily rate, maximum hourly rate, and fire-flow rate) for a New England town of 10,000 people. The only industry in town, a wool and textile mill with a production of 100 tons per month, has its own water supply. If there are any data missing, make your own assumptions and give your reasons for them.

11.3. A small community of population 1,000 located in the Canadian arctic has a trucked water supply system, drawing from a neaby lake (3 km from the village) as a water source. There are 200 houses, one hotel, one hospital, one school, one nursing station, and two general stores in the community. The total road system in town is 2 km long. Each house is equipped with a water storage tank of 1,000 L capacity, with bigger tanks in the other establishments. The average water consumption is 40 Lpcd. It is quite common for winter storms to prevent the trucks from traveling to the lake for up to three days. Based on this information, determine (a) the size of the storage reservoir in the village, and (b) the number of trucks required, if each truck has a 4,000 L. tank. The trucks also serve the purpose of providing fire protection.

Make any assumptions you feel are necessary to complete the assignment, giving reasons for each assignment.

11.4. A well will be used to supply a village of 5,000 people with water. The following chemical analysis on a water sample is available:

Constituent	mg/L
Sulfates (SO_4)	50
$CaCO_3$	220
Chlorides (Cl)	200
Iron (Fe)	1.8
Lead (Pb)	0.01
Manganese (Mn)	0.1
Nitrate nitrogen (as N)	4.0

Recommend a treatment process for the well supply, and draw a schematic diagram along the lines of Figure 11–4(b).

11.5. Presedimentation can reduce the suspended solids concentration of raw river water from 500 mg/L to 200 mg/L. How many pounds of dry solids does this represent per million gallons? If the specific gravity of these solids is 2.60 and the sludge removed has a solids concentration of 2%, what volume of sludge must be removed every week at a design flow of 4 mgd?

11.6. The water treatment plant for the city of 100,000 studied in Example 11.3 is supplied by a river intake. Normally, the turbidity of the water is 50 TU. Three or four times a year heavy rains cause the turbidity to increase to 500 TU. In addition, the sewage treatment plant of a nearby upstream town is known to release partially treated sewage at these times because of the high rate of storm-water runoff flowing through the combined stormwater and sewage lines feeding the plant. In its present configuration the plant cannot provide the required quality of drinking water during these periods, and several prominent citizens have begun to complain. Discuss the strategies open to you as the waterworks engineer for the town.

11.7. Fluoride is often added to municipal drinking water as an aid in preventing tooth decay. Too much fluoride in drinking water can cause fluorosis (a mottling of the teeth). If the proper dosage in the community is 0.8 mg/L calculate the amount of fluoride needed weekly for the city in Example 11.2.

11.8. Figure 11–6(a) shows the construction of a rapid sand filter bed. One might wonder why a layer of larger gravel particles is not placed on top to filter the larger suspended particles first, before the finer sand particles are used to filter small suspended particles. In other words, a filter more like the one in Figure 11–6(b) might be used. Explain why this is not done.

11.9. Figure 11–6(b) shows a anthracite-sand filter bed. The anthracite particles are larger than the sand particles. How is it possible for the filter to return to the proper grading scheme after backwashing?

11.10. The nominal detention time of the sedimentation tanks in Example 11.3 is 2 hr. Effluent from the middle tank is more turbid than that from each of the other two tanks. You are asked to investigate what the problem may be and suggest a remedy if you can. How will you proceed?

11.11. Chlorination is the usual method for disinfecting water in North America.
  (a) Name the two parameters that control the extent of disinfection.
  (b) Why is it necessary to guard against an overdose of chlorine?

(c) Why does the presence of ammonia in water reduce the bactericidal efficiency of chlorine? Why, then, do some plants add ammonia to chlorinated water?

(d) Why does coagulation with alum ahead of chlorination increase the efficiency of disinfection of water by chlorine?

(e) Assume that disinfection with chlorine follows a first-order reaction. In a chlorinated water sample containing 1.0 mg/L chlorine, the initial concentration of viable bacteria is 100,000 per mL. At the end of a 5-min contact time the number of viable bacteria has decreased to 10 per mL. What effect would a contact time of 10 min have on the bacterial count?

11.12. A new residential community in the midwestern U.S.A., served by private wells, is considering a groundwater source of supply. If the population is expected to reach a maximum of 5,000, determine the following (make necessary assumptions and state your reasoning):

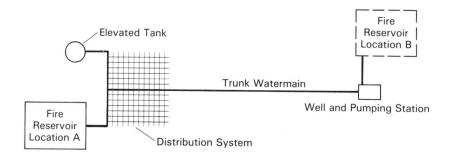

(a) What is the minimum capacity of the well that is required to adequately supply water to the town (in gpm and Lpm)?

(b) Fire insurance regulations require that there be sufficient storage for fire-fighting purposes. Calculate the total volume necessary for a ground-level, fire reservoir at either location A or B (in gal and L).

(c) What should be the capacity in gpm and Lpm, of the trunk watermain if the fire reservoir is in Location A? If it is in Location B?

(d) Assuming that the town will use an elevated tank to provide all the water required during the night (10P.M. to 6A.M.), what should be the capacity of the elevated tank (in gal and L)? Assume that during this period the hourly water consumption is one-third of the average hourly consumption.

11.13. You have been lucky and obtained a summer job with the World Health Organization. Your assignment is to act as an engineering student advisor to the mayor of a small village in Central America, located in the tropical coastal flatland. The village has 200 families, or about 1,500 people. The majority of the families carry their water in buckets from a nearby stream. The average carry is 500 m. About 25% of the families purchase their water from a local vendor at 4¢ per gallon. Five of the wealthier families have private wells on their premises. Toilet facilities throughout the village are outdoor privies. The mayor informs you that there have been incidents of various kinds of sicknesses, which have been caused by the local sanitation situation. He asks you for suggestions to improve the situation. (Problem courtesy of Professor W. M. McLellon, Florida Technological University.)

(a) What factors of costs and benefits must you consider to determine whether a community water supply project should be undertaken?

(b) Suggest two (or more) systems of different levels of sophistication (and cost).

(c) What resources and services must the village mobilize and sustain to make a community water supply a successful venture?

11.14. A water supply, providing a maximum of 1000 m³/d to a town, contains 100μg/L of toxic organics which conventional treatment processes cannot remove. Fortunately, these organics are extremely soluble in liquid Zorbitol (SG = 1.5), resulting in a 100-fold difference in the concentration of toxic organics in Zorbitol compared to water.

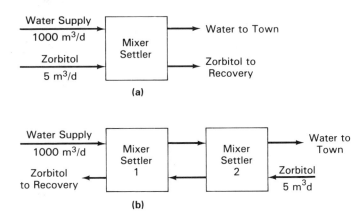

Using 5m³/d of Zorbital, what concentration of toxic organics in the town water would be achieved in:

(a) a single-stage mixer-settler

(b) a two-stage counter-current, mixer-settler system
     assuming complete mixing in all cases.

REFERENCES

APHA, AWWA, and WPCF. *Standard Methods for the Examination of Water and Wastewater*. 16th ed. (Washington, D.C.: American Public Health Association, American Water Works Association, and Water Pollution Control Federation). 1985.

HAMMER, M. J. *Water and Waste Water Technology*. New York: Wiley, 1977.

Insurance Services Office. *Guide for Determination of Required Fire Flow*. New York: Insurance Services Office, 1974.

LINSLEY, R. K., and FRANZINI, J. B. *Water Resources Engineering*. New York: McGraw-Hill, 1972.

LOWRY, E. F. "Breaking the Cost Barrier to Household Water Service." *Journal of the American Water Works Association*, December 1980, pp. 672, 677.

NEMEROW, N. L. *Industrial Water Pollution*. Reading, Mass.: Addison-Wesley, 1978.

RICH, L. G. *Unit Operations of Sanitary Engineering*. New York: Wiley, 1961.

ROBERTSON, J. A., and CROWE, C. T. *Engineering Fluid Mechanics*. Boston: Houghton Mifflin, 1980.

SOLLEY, W. B., CHASE, E. B., and MANN, W. B. IV. *Estimated Use of Water in the United States in 1980*. U.S. Geological Survey Circular 1001. Washington, D.C.: U.S. Government Printing Office, 1983.

STEEL, E. W., and MCGHEE, T. J. *Water Supply and Sewerage*. New York: McGraw-Hill, 1979.

VIESSMAN, JR., W., and HAMMER, M. J. *Water Supply and Pollution Control*. 4th ed. New York: Harper and Row, 1985.

12

Water Pollution

J. Glynn Henry

12.1 INTRODUCTION

As noted in Chapter 8, the relationship between polluted water and disease was firmly established with the cholera epidemic of 1854 in London, England. Protection of public health, the original purpose of pollution control, continues to be the primary objective in many areas. However, preservation of water resources, protection of fishing areas, and maintenance of recreational waters are additional concerns today. Water pollution problems intensified following World War II when dramatic increases in urban density and industrialization occurred. Concern over water pollution reached a peak in the mid-seventies. In the United States, where national control is exercised by the federal government, Public Law 92-500 (1972) was the official recognition of this concern. In Canada, where pollution control is a provincial responsibility, Ontario, through the Water Resources Act (1970), was the most active province in prodding municipalities into action. The situation was similar in Great Britain, Europe, Japan, and other industrialized countries where increasing urbanization and industrialization were accompanied by serious water pollution problems. In less developed regions, wastes from burgeoning populations are a threat to public health and endanger the continued use of often scarce water supplies.

Water pollution is an imprecise term that reveals nothing about either the type of polluting material or its source. The way we deal with the waste problem depends upon whether the contaminants are oxygen demanding, algae promoting, infectious, toxic, or simply unsightly. Pollution of our water resources can occur directly from sewer outfalls or industrial discharges (point sources) or indirectly from air pollution or agricultural or urban runoff (nonpoint sources).

This chapter deals primarily with point sources. It provides information on the origins, quantities, and characteristics of wastewater and the effects of pollutants on the water environment. The use of stream standards and water quality objectives to control pollution are reviewed, and the features of combined and separate sewers evaluated. The principles of wastewater treatment and the methods available for both large and small installations are explained. Systems covered range from large municipal facilities employing combinations of physical, biological, and chemical methods to units suitable for single-family needs.

Legal and economic controls are other measures used to control water pollution. Fines, surcharges, financial incentives, subdivision agreements, and sewer-use bylaws are some of the tools available. Principles covered in the chapter are illustrated by examples.

12.2 WASTEWATER

Municipal wastewater, also called sewage, is a complex mixture containing water (usually over 99 percent) together with organic and inorganic contaminants, both suspended and dissolved. The concentration of these contaminants is normally very low and is expressed in mg/L, that is, milligrams of contaminant per liter of the mixture. This is a weight-to-volume ratio used to indicate concentrations of constituents in water, wastewater, industrial wastes, and other dilute solutions. Since the specific gravity (SG) of these dilute solutions is similar to that of water, the concentrations can also be considered to be weight-to-weight ratios such as mg/kg or ppm (parts per million). However, where the SG of the mixture is not 1.0, mg/L and ppm are not interchangeable terms.

12.2.1 Constituents

Microorganisms.   Wherever there is suitable food, sufficient moisture, and an appropriate temperature, microorganisms will thrive. Sewage provides an ideal environment for a vast array of microbes, primarily bacteria, plus some viruses and protozoa. As mentioned in Chapter 8, most of these microorganisms in wastewater are harmless and can be employed in biological processes to convert organic matter to stable end products. However, sewage may also contain pathogens (disease-causing organisms) from the excreta of people with infectious diseases that can be transmitted by contaminated water. Waterborne bacterial diseases such as cholera, typhoid, and tuberculosis, viral diseases such as infectious hepatitis, and the protozoan-caused dysentery, while seldom a problem now in developed countries, are still a threat where properly treated water is not available for public use. Tests for the few pathogens that might be present are difficult and time consuming, and standard practice is to test for other more plentiful organisms that are always present (in the billions) in the intestines of warm-blooded animals, including humans. These tests were described in Chapters 8 and 11.

TABLE 12–1   SOLIDS IN WASTEWATER

Sample	Total solids (residue 103°C)	Inorganic (residue 550°C)	Organic (loss 550°C)
Unfiltered (Suspended + Dissolved)	Total Solids	Total Fixed Solids	Total Volatile Solids
Filtered (Dissolved)	Total Dissolved Solids	Fixed Dissolved Solids	Volatile Dissolved Solids
By Difference	Suspended Solids		Volatile Suspended Solids

Solids.   The total solids (organic plus inorganic) in wastewater are, by definition, the residues after the liquid portion has been evaporated and the remainder dried to a constant weight at 103°C. Differentiation between dissolved solids and undissolved, that is, suspended, solids is accomplished by evaporating filtered and unfiltered wastewater samples. The difference in weight between the two dried samples indicates the suspended solids content. To further categorize the residues, they are held at 550°C for 15 minutes. The ash remaining is considered to represent inorganic solids and the loss of volatile matter to be a measure of the organic content. Table 12–1 shows the categories of solids in wastewater.

Of the categories listed, suspended solids (SS) and volatile suspended solids (VSS) are the most useful. SS and BOD (biochemical oxygen demand), which will be discussed in Section 12.2.2, are used as measures of wastewater strength and process performance. VSS can be an indicator of the organic content of raw wastes and can also provide a measure of the active microbial population in biological processes.

Inorganic constituents.   The common inorganic constituents of wastewater include:

1. *Chlorides and sulphates.* Normally present in water and in wastes from humans.
2. *Nitrogen and phosphorus.* In their various forms (organic and inorganic) in wastes from humans, with additional phosphorus from detergents.
3. *Carbonates and bicarbonates.* Normally present in water and wastes as calcium and magnesium salts.
4. *Toxic substances.* Arsenic, cyanide, and heavy metals such as Cd, Cr, Cu, Hg, Pb, and Zn are toxic inorganics which may be found in industrial wastes.

In addition to these chemical constituents, the concentration of dissolved gases, especially oxygen, and the hydrogen ion concentration expressed as pH are other parameters of interest in wastewater.

Organic matter.   Proteins and carbohydrates constitute 90 percent of the organic matter in domestic sewage. The sources of these biodegradable contaminants include excreta and urine from humans; food wastes from sinks; soil and dirt from bathing, washing, and laundering; plus various soaps, detergents, and other cleaning products.

Various parameters are used as a measure of the organic strength of wastewater. One

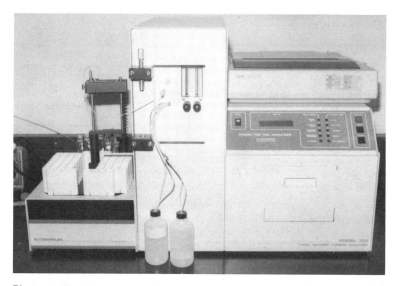

Photograph 12–1   Total organic carbon analyzer
Current model of an automatic analyzer in which chemical oxidation is used to determine the TOC in water and wastewater.

method is based on the amount of organic carbon (total organic carbon, or TOC) present in the waste. TOC is determined by measuring the amount of CO_2 produced when the organic carbon in the sample is oxidized by a strong oxidizer and comparing it with the amount in a standard of known TOC. An instrument which measures TOC is shown in Photo 12–1.

Most of the other common methods are based on the amount of oxygen required to convert the oxidizable material to stable end products. Since the oxygen used is proportional to the oxidizable material present, it serves as a relative measure of wastewater strength. The two methods used most frequently to determine the oxygen requirements of wastewater are the COD and BOD tests. The **COD,** or chemical oxygen demand, of the wastewater is the measured amount of oxygen needed to chemically oxidize the organics present; the **BOD,** or biochemical oxygen demand, is the measured amount of oxygen required by acclimated microorganisms to biologically degrade the organic matter in the wastewater.

BOD is the most important parameter in water pollution control. It is used as a measure of organic pollution, as a basis for estimating the oxygen needed for biological processes, and as an indicator of process performance.

12.2.2 BOD Measurement

The amount of organic matter in water or wastewater can be measured directly (as TOC, for example), but this doesn't tell us whether the organics are biodegradable or not. To measure the amount of biodegradable organics, we use an indirect method in which we measure the amount of oxygen used by a growing microbial population to convert (oxidize)

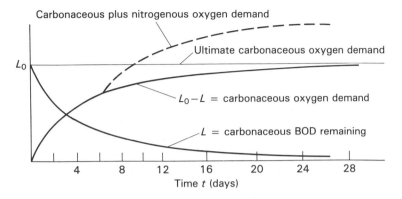

Figure 12–1   BOD curves at 20°C

organic matter to CO_2 and H_2O in a closed system. The oxygen consumed, or biochemical oxygen demand (BOD), is proportional to the organic matter converted, and therefore BOD is a relative measure of the biologically degradable organic matter present in the system. Because biological oxidation continues indefinitely, the test for ultimate BOD has been arbitrarily limited to 20 days, when perhaps 95 percent or more of the oxygen requirement has been met. Even this period, however, is too long to make measurement of BOD useful, so a five-day test, BOD_5, carried out at 20°C, has become standard. The rate of the BOD reaction depends on the type of waste present and the temperature and is assumed to vary directly with the amount of organic matter (organic carbon) present (a first-order reaction).

A plot of L. The carbonaceous BOD remaining versus time t is shown in Figure 12–1. Since the amount of oxygen used versus time represents the amount of organic matter oxidized, the curve of $L_0 - L$ shows the carbonaceous organic matter oxidized at time t. The equations for L and $L_0 - L$, shown as solid lines in the figure, are

$$L = L_0(10^{-kt}) \tag{12.1}$$

and

$$L_0 - L = L_0(1 - 10^{-kt}) \tag{12.2}$$

where L = carbonaceous BOD remaining at time $t = t$
   (O_2 needed to oxidize carbonaceous organic matter remaining)
   L_0 = ultimate carbonaceous oxygen demand i.e. ultimate BOD
   (O_2 needed to oxidize carbonaceous organic matter initially present)
   $L_0 - L$ = carbonaceous oxygen demand i.e. BOD satisfied
   (O_2 used to oxidize carbonaceous organic matter at time $t = t$)
   t = time (days)
   k = rate constant (base 10) (day^{-1})

The curves L and $L_0 - L$ indicate the oxidation of carbon (carbonaceous BOD) in organic matter to CO_2 and water. However, after five to ten days, nitrogenous compounds

TABLE 12–2   BOD REMOVAL RATE CONSTANTS

Source of organic contamination	Rate of change k in BOD per day (base 10)
River water	0.10
Domestic sewage	0.17
Glucose solution	0.25

Source: Sawyer, C. N., and McCarty, P. L., *Chemistry for Environmental Engineers*, 3d Ed., New York: McGraw Hill, 1978

will begin to be oxidized. The dotted line in the figure indicates the effect on oxygen demand when nitrogen present in the waste is oxidized in the conversion (nitrification) of ammonia to nitrates. This second-stage exertion of BOD can be inhibited in the BOD test with the addition of suitable chemical agents. Typical values for the rate constant k (base 10) at 20°C for carbonaceous oxidation are presented in Table 12–2.

The rate constant k in the table applies at 20°C. For other temperatures, a simplified version of the van't Hoff-Arrhenius expression can be used to modify k:

$$\frac{k_T}{k_{20}} = \theta^{(T-20)} \tag{12.3}$$

where $\theta = 1.047$
T = Temperature, °C

The value of the dimensionless temperature coefficient Θ is also temperature dependent, but for temperatures between 10 and 25°C, the change in Θ is slight and can be neglected.

12.2.3 Municipal Wastewater

The excreted waste from humans is called **sanitary sewage,** and wastewater from residential areas is referred to as **domestic sewage** and includes kitchen, bath, laundry, and floor drain wastes. These, together with the liquid wastes from commercial and industrial establishments, are termed **municipal wastewater.** This wastewater is normally collected in a public sewerage system (sewers, manholes, pumping stations, etc.) and directed to treatment facilities for safe disposal.

Quantities of municipal wastewater are commonly determined from water use. Because water is consumed by humans, utilized in industrial products, used for cooling, and required for activities such as lawn watering and street washing, only 70 to 90 percent of the water supplied reaches the sewers. However, the assumption is frequently made that the water loss is balanced by infiltration (groundwater leakage into the sewer system through poor joints) or storm water, which enters the sanitary sewer system through illicit connections (roof downspouts and road catchbasins) or through manhole openings.

In North America, municipal water use and the resulting wastewater flow range from about 280 liters (75 gallons) per capita per day for small, mainly residential municipalities to over 900 Lpcd (240 gpcd) for large industrialized cities (see also Table 11–1). These daily averages are based on annual amounts. However, flows vary from day to day. From

TABLE 12–3   CHARACTERISTICS OF DOMESTIC SEWAGE BASED ON TYPICAL PER CAPITA
CONTRIBUTIONS TO WASTEWATER

Parameter	BOD_5	SS	DS	COD	TOC	P_1	P_2	N_1	N_2
Per Capita									
g/day	76	90	180	128	54	1.6	4.0	16	0
lb/day	0.167	0.20	0.40	0.28	0.12	0.0035	0.009	0.035	0
Concentration									
mg/L	190	225	450	320	135	4.0	10	40	0

Values based, in part, on Metcalf and Eddy (1979)

mg/L	based on 400 Lpcd wastewater flow
P_1	Phosphorus from human wastes as P
P_2	Total phosphorus (with PO_4-based detergents) as P
N_1	Kjeldahl nitrogen (organic + NH_3) as N
N_2	Nitrate nitrogen as N
COD	Estimated assuming $BOD_5/COD = 0.6$
TOC	Estimated assuming $BOD_5/TOC = 1.4$.

larger (500,000 people) to smaller (10,000 people) municipalities, flows as a percentage
of the annual daily average might range from maximums of 150 to 200 percent on a daily
basis and 200 to 300 percent on an hourly basis and minimums of 70 to 50 percent on a
daily basis and 50 to 30 percent on an hourly basis, with the extremes in each case (the
second value) applying to the smaller municipality. The quality of municipal wastewater
varies with the proportion of residential, commercial, and industrial contributors and the
nature of the industrial wastes which the system receives.

The concentration of contaminants in sewage from residential areas can be estimated
from the daily per capita contribution if water use (and hence wastewater flow) is known.
Typical values are indicated in Table 12–3.

The total dry weight of the wastewater constituents from residential areas will be
relatively constant, but their concentration will vary with the amount of water used. As
the community grows larger and more diversified, the addition of contaminants from
commercial and industrial establishments will change the characteristics of the wastewater.

12.2.4 Industrial Wastewater

Wastewaters from industries include employees' sanitary wastes, process wastes from manufacturing, wash waters, and relatively uncontaminated water from heating and cooling
operations. The wastewaters from processing are the major concern. They vary widely with
the type of industry. In some cases, pretreatment to remove certain contaminants or equalization to reduce hydraulic load may be mandatory before the wastewater can be accepted
into the municipal system. In contrast with the relatively consistent characteristics of
domestic sewage, industrial wastewaters, often have quite different characteristics, even
for similar industries. For this reason, extensive studies may be necessary to assess pretreatment requirements and the effect of the wastewater on biological processes.

Wastes are specific for each industry and can range from strong (high BOD_5) biodegradable wastes like those from meat packing, through wastes such as those from plating shops and textile mills, which may be inorganic and toxic and require on-site physical-chemical treatment before discharge to the municipal system. The volume or strength of industrial wastewaters is often compared to that of domestic sewage in terms of a **population equivalent** (PE) based on typical per capita contributions (Table 12–3). The basis using BOD_5 is illustrated in Example 12.2(c).

TABLE 12–4 REPRESENTATIVE VALUES OF CONTAMINANTS IN WASTEWATER

Waste	Municipal (R + C + I)			Industrial (Process)				Stormwater (annual runoff)
Parameter	Small	Medium	Large	Food	Meat	Plating	Textile	Small/medium/large
Volume (L)								
/capita/day	400	500	600	—	—	—	—	—
/tonne prod.	—	—	—	10,000	12,000	—	100,000	—
% runoff	—	—	—	—	—	—	—	30/35/45
MPN (10^6/100/mL)	100	80	70	0	—	0	0	0.008
BOD_5	190	240	300	1,200	640	0	400	14
COD	320	400	500	—	—	—	—	100
TOC	135	170	215	—	—	—	—	—
Susp. Solids	225	300	350	700	300	0	100	170
Diss. Solids	450	600	700	—	200	—	1,900	170
Total N	40	30	25	0	3	0	0	3.5
Total P	10	8	7	0	—	0	0	0.35
pH	7.0	7.0	7.0	—	7.0	4 or 10	10	—
Copper	0.14	0.17	0.21	0.29	0.09	6	0.31	0.46
Cadmium	0.003	0.010	0.016	0.006	0.011	1	0.03	0.025
Chromium	0.04	0.08	0.16	0.15	0.15	11	0.82	0.16
Nickel	0.01	0.06	0.11	0.11	0.07	12	0.25	0.15
Lead	0.05	0.1	0.2	—	—	—	—	—
Zinc	0.19	0.29	0.38	1.08	0.43	9	0.47	1.6

Note: Concentrations of constituents are in mg/L. Values for all parameters may vary widely from those noted.

Small	Small residential (R) community
Medium	Medium-size diversified municipality, residential (R), commercial (C), and industrial (I) areas with separate sewers
Large	Large industrialized city (R + C + I) with combined sewers
Food	Food waste (canning factory: pickles, beets, tomatoes, pears)
Meat	Meat processing (poultry plant with no manure or blood recovery)
Plating	Plating shop (wastes are acidic with chromate baths, alkaline with cyanide baths, and are less than 2,000 m³/d for most plants)
Textile	Textile mill (spun cotton yarn processed into cotton goods, sized with starch)

Adapted in part from Collins, P. G., and Ridgway, J. W., Journal, Environmental Engineering Division, American Society of Civil Engineers, Vol. 106, Reading, Mass: EE-1, 1980; Klein, L. A., et al., Journal, Water Pollution Control Federation, Vol. 46:12, 1974; Nemerow, N. L., *Industrial Water Pollution*, Addison-Wesley, 1978; Polls, I., and Lanyon, R., Journal, Environmental Engineering Division, American Society of Civil Engineers, Vol. 106, EE-1, 1980; and Waller, D. H., and Novak, Z., Journal, Water Pollution Control Federation, Vol. 53:3, 1981.

12.2.5 Stormwater

The runoff from rainfall, snowmelt, and street washing is less contaminated than municipal wastewater. It therefore receives little or no treatment before being discharged into storm sewers (for direct disposal into receiving waters) or before being combined with the municipal wastewater for delivery to the wastewater treatment plant.

The quantity of stormwater which runs off from a municipality varies widely with the time of year, the type of terrain, and the intensity and duration of the storms which occur. As indicated in Chapter 10, North America, excluding desert areas, has a range of 250 to 2,000 mm (10 to 80 inches) of annual precipitation across the continent. In temperate areas with, for example, 750 to 900 mm (30 to 36 inches) of rainfall per year, stormwater runoff would amount to about 25 percent of the total annual municipal wastewater volume. However, during a storm the rate of stormwater runoff can often be several times, and on occasion can be as high as 100 times, the normal flow to sanitary sewers. This explains why, in order to minimize the possibility of sewer backup and basement flooding, the admission of stormwater into a separate sanitary sewer system should be prohibited.

Of the total global annual rainfall, two-thirds may be lost through evaporation and transpiration, with the balance going to surface water and groundwater storage. Even during storms, not all rainfall becomes runoff. The proportion which does varies from about 20 percent for parks and lawns up to 100 percent for roofs and paved areas. An overall average value for a municipality might range between 30 and 50 percent during fairly intense storms. Example 12.1 compares the quantity of stormwater to the volume of municipal wastewater from a predominantly residential community.

Stormwater runoff, particularly in cities, contains dust and other particulates from roads, leaves from trees, grass cuttings from lawns and parks, and fallout from air pollution. The concentration of these contaminants is highest when they are first flushed into the sewer system during the early stages of runoff and then decreases as the rain continues.

Table 12–4 lists representative values of contaminants in the three different types of wastewater with which a municipality must deal. Any one of the three types may be the major concern most of the time or may be only an occasional problem, depending upon the relative volumes and the characteristics of each type of wastewater. Example 12.2 illustrates how the type of information contained in Table 12–4 may be used to estimate wastewater characteristics when field sampling is not practical.

Example 12.1

A municipality with a population of 10,000 in an area of 200 ha (494 ac) uses water at an average rate of 400 Lpcd (105 gpcd). Annual rainfall is 900 mm (36 in) per year, and the maximum intensity of rainfall during a heavy storm which occurs, on average, about once every two years is 25 mm (1 in) per hour, when the entire area is contributing to runoff. Compare the stormwater runoff with the average sanitary sewage flow on an annual basis and during a heavy storm.

Solution   Annual basis:

Sanitary flow (assumed to be equal to water use) is

$$Q = 10,000 \times \frac{400}{10^3} \times 365 = 1.46 \times 10^6 \text{m}^3/\text{year} (386 \times 10^6 \text{ gal/year})$$

Storm runoff is assumed to be 20% of rainfall for this case. Thus,

$$Q = 200 \text{ ha} \times 10^4 \frac{\text{m}^2}{\text{ha}} \times \frac{900 \text{ mm/yr}}{10^3 \text{ mm/m}} \times 0.2 = 0.36 \times 10^6 \text{ m}^3/\text{yr} (95.1 \times 10^6 \text{ gal/yr})$$

Therefore,

$$\frac{\text{Storm runoff}}{\text{Sanitary flow}} = \frac{0.36}{1.46} \times \frac{10^6}{10^6} \times 100 = 25\% \text{ on an annual basis}$$

During storm:

Sanitary flow (assumed to be equal to water use) is

$$Q = \frac{10,000 \times 400}{24 \times 60} = 2,777 \text{ L/min (734 gpm)}$$

Storm runoff is assumed to be $\frac{1}{3}$ of the rainfall. Thus,

$$Q = 200 \text{ ha} \times \frac{10^4 m^2}{\text{ha}} \times \frac{25 \text{ mm/hr}}{60 \text{ min/hr}} \times \frac{1}{3} = 277,777 \text{ L/min (73,400 gpm)}$$

Therefore,

$$\frac{\text{Storm Runoff}}{\text{Sanitary Flow}} = \frac{277,777}{2,777} = \frac{100}{1}$$

during a storm with an intensity of 25 mm (1 in) per hour.

Example 12.2

A small municipality of 9,000 people has two industries: a cannery producing 4,000 tonnes of whole tomatoes and other canned goods over a seven-month season, and a textile mill which produces 2,000 kg of cotton goods per day. Estimate the BOD_5 and SS content of the municipal wastewater (a) with and (b) without these industries being served by the municipal system, and (c) determine the population equivalent (PE) of the cannery in terms of BOD_5.

Solution   From the information given, and from Table 12–4, we can set up the following table:

Parameter	Residential	Cannery	Textile	Total
Quantity	9,000 people	$\frac{4,000 \text{ t}}{210 \text{ d}}$	$\frac{2 \text{ t}}{\text{d}}$	—
Unit Vol.	400 Lpcd	10,000 L/t	100,000 L/t	—
Volume	3,600 m³/d	200 m³/d	200 m³/d	4,000 m³/d
BOD_5	190 mg/L	1,200 mg/L	400 mg/L	$\frac{(190 \times 3.6) + (1,200 \times 0.2) + (400 \times 0.2)}{3.6 + 0.2 + 0.2} = 250$ mg/L
SS	225 mg/L	700 mg/L	100 mg/L	$\frac{(225 \times 3.6) + (700 \times 0.2) + (100 \times 0.2)}{3.6 + 0.2 + 0.2} = 240$ mg/L

Then we have:

(a) 3,600 m³/day, BOD$_5$ = 190 mg/L, SS = 225 mg/L.

(b) 4,000 m³/day, BOD$_5$ = 250 mg/L, SS = 240 mg/L.

(c) $$PE = \frac{(200 \text{ m}^3/\text{d} \times 1{,}200 \text{ mg BOD}_5/\text{L})}{76 \text{ g BOD}_5/\text{capita} \cdot \text{d}} \times \frac{10^3 \text{L}}{\text{m}^3} \frac{10^{-3}/\text{g}}{\text{mg}} = 3{,}160 \text{ people}$$

12.3 POLLUTION OF RECEIVING WATERS

12.3.1 Effects of Pollutants

Water pollution occurs when the discharge of wastes impairs water quality or disturbs the natural ecological balance. The contaminants which cause problems include disease-causing organisms (pathogens), organic matter, solids, nutrients, toxic substances, color, foam, heat, and radioactive materials. Discharge of specific pollutants is not the only cause of water pollution. The construction of dams, reservoirs, and river diversions can also seriously degrade water quality. These latter influences were considered in Chapter 10.

Pathogens.   Concern for public health arises when sewage that may contain pathogens is discharged into receiving waters used for water supplies or recreation. Although limitations on the density of "indicator" organisms control the degree of pollution from human wastes, they do not assure the absolute safety of the water.

Organic matter (BOD).   The higher the BOD, that is, the more organic matter, the greater is the problem created by the decomposition of the organic matter. Microbial activity by bacteria requiring oxygen may reduce the normal dissolved oxygen (DO) content in a stream or lake to less than 4 mg/L, below which most fish cannot survive. When all DO disappears, anaerobic conditions occur and objectionable odors ensue. Because the amount of dissolved oxygen (DO) in water decreases with increasing temperature, the amount of oxygen in streams is more critical to aquatic life in summer (when flows are low and temperatures are high) than it is in winter.

Solids.   Organic and inorganic particulates in wastewater consist of settleable, floating, and suspended solids that can result in unsightly deposits, odorous sludge banks, and reduced penetration of sunlight through the water.

Nutrients.   Nitrates and phosphates, derived from municipal wastewaters, are inorganic nutrients which promote plant and algal growth. The amounts necessary to trigger algal blooms are not well established, but concentrations as low as 0.01 mg/L for phosphorus and 0.1 mg/L for nitrogen may be sufficient for eutrophication when other elements are in excess (see Chapter 9). In addition to having a detrimental aesthetic effect on lakes (odor, appearance), algae can be toxic to cattle, spoil the taste of the water, plug filtration units, and increase chemical requirements in water treatment.

Toxic and hazardous substances.   Low concentrations of acids, caustics, cyanide, arsenic, many heavy metals, and numerous chemicals are toxic to living organisms, including humans and to the microbial population utilized in wastewater treatment processes. Two of the more harmful metals are mercury and cadmium, which are biologically accumulative.

Radioactive materials, which are also harmful to biological life and bioaccumulative, are more carefully managed than industrial wastes, and cases where water supplies have been rendered unsuitable because of radioactivity are rare.

Other pollutants.   Color, foam, and heat are other pollutants that cause problems. Color (from textile dyeing for example) and foam (pulp and paper mill wastes) are not just aesthetically objectionable; they also limit light penetration and can reduce DO levels, both of which upset the natural ecological balance in the water. Thermal discharges, primarily the cooling water from power plants, offer the potential for heat recovery. They also cause an increase in the rate of oxygen utilization, because at the higher temperature there is more rapid growth of aquatic life and accelerated organic decomposition. At the same time, less DO is available in the water at the higher temperature. Acclimatization of lower classes of fish and promotion of the growth of troublesome blue-green algae can result from a permanent temperature increase.

12.3.2 Water Quality Requirements

In the past, when water was plentiful and development sparse, common law determined water rights. The rights of owners living on the banks (nonsalable riparian rights) allowed ''reasonable'' use of the water, provided downstream users were not adversely affected. Later, as development occurred, some authorities allocated salable water rights to industries, mining companies, and others for as long as they continued to use the water and whether they were riparian owners or not.

The simple approaches of the past, based on common law, are not adequate for highly developed societies with limited water resources that need to be protected from pollution. Control of water quality may be under federal jurisdiction, as in the United States and most other countries, or under a regional authority (for example, a state, province, or county), as in Canada. Whatever the basis, it is essential that standards (rigid limits enforceable by law) or objectives (guidelines or criteria for meeting desirable goals) be established. The quality of either the receiving water or the effluent may be the controlling requirement. If, as seems reasonable in theory, the receiving water governs, then administration and enforcement of water quality standards or criteria are necessary. The quality of municipal and industrial discharges must be related to the stream or lake standards and enforced, difficult though that may be. On the other hand, if the requirements regulate effluent quality, monitoring and control are much simpler in practice. Offsetting this advantage is the problem that treatment requirements are not related to the water quality of the stream or lake. The result may be more costly treatment than is warranted where ample dilution is available, but insufficient treatment where it is not. A combination of stream and effluent requirements may be the best compromise. Assimilation studies could establish allowable loadings to a stream or lake, and this capacity could be allocated to users on a priority basis. Anyone discharging wastewater to these waters would have to meet these requirements or minimum applicable effluent standards, whichever was least detrimental to the receiving water.

Under the Clean Water Act as amended in 1977, the United States has required since July 1, 1988, that all discharges receive the "best conventional pollution control technology" regardless of the assimilation capacity of the stream. This requirement, which implies secondary treatment, may, like the original plan for zero discharge of pollutants, be modified where ample dilution is available, the location is remote, essential industries are involved, needed employment is provided, or other politically justified reasons make it expedient to do so. Several thousand plants, in fact, missed the 1988 deadline.

Most developed countries have effluent and ambient water quality requirements that vary depending upon the characteristics of the wastewater and whether the receiving waters

TABLE 12–5   EFFLUENT AND WATER QUALITY REQUIREMENTS

Parameter	Wastewater effluent	Stream quality
Canadian Objectives (Ontario)[1]		
BOD$_5$	15 mg/L max.	4 mg/L max.
SS	15 mg/L max.	—
DO	2 mg/L min.	4 mg/L min.
Total Coliforms	—	5,000/100 mL max. (water supply)
		1,000/100 mL max. (swimming area)
Fecal Coliforms	200/100 mL max.	500/100 mL max. (water supply)
United States Standards (Typical)		
BOD$_5$	30 mg/L max.	4 mg/L max.
SS	30 mg/L max.	—
DO	—	4 mg/L min.
Total Coliforms	—	5,000/100 mL max. (water supply)
		1,000/100 mL max. (swimming area)
Fecal Coliforms	200/100 mL max.	500/100 mL max. (water supply)
Japanese Standards[2]		
BOD$_5$	20 mg/L max.	2 mg/L max.
SS	70 mg/L max.	25 mg/L max.
DO	—	7.5 mg/L min.
Total Coliforms	—	5,000/100 mL max. (water supply)
		1,000/100 mL max. (swimming area)
Fecal Coliforms	30/100 mL max.	

[1] Ontario Ministry of the Environment (1978).

[2] Hayashi, T., "Water Pollution Control in Japan," *Journal of the Water Pollution Control Federation* 52 (1980): 855; Tamaki, T., "Wastewater Treatment Works in Japan," *Journal of the Water Pollution Control Federation* 52 (1980): 864.

will be used for water supply, recreation, irrigation, or industrial purposes. Examples of effluent and ambient water quality requirements for Ontario, the United States, and Japan are given in Table 12–5. The effluent requirements apply generally to municipal wastewater where at least 85 percent removal of organic matter is provided. The criteria for stream quality pertain to receiving waters that are either suitable (with filtration) as a raw water supply or acceptable for swimming. The relationship between effluent and ambient water quality requirements is demonstrated in Example 12.4.

Example 12.4

The average daily flow in a small river during the driest month is 100 L/s. If a wastewater treatment plant could consistently produce an effluent with a BOD of 20 mg/L or less, what population could be served if the BOD in the river, after dilution, must not exceed 4 mg/L? Assume that there is no upstream pollution and that the municipal water supply does not come from the river.

Solution Assuming complete mixing and choosing 1 second as a convenient time interval write a material balance on BOD_5.

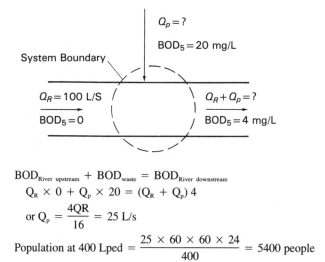

$$BOD_{River\ upstream} + BOD_{waste} = BOD_{River\ downstream}$$
$$Q_R \times 0 + Q_p \times 20 = (Q_R + Q_p)\,4$$
$$\text{or } Q_p = \frac{4QR}{16} = 25\ L/s$$
$$\text{Population at 400 Lped} = \frac{25 \times 60 \times 60 \times 24}{400} = 5400 \text{ people}$$

12.3.3 Need for Pollution Control

It has often been said, and only partly in jest, that the solution to pollution is dilution. There is logic in the statement. Where small quantities of sewage discharge into relatively large rivers or water bodies, incidents of contaminated water supplies or hazards to public health are rare. This is because of the dilution of the contaminants and the natural purification that takes place. With increasing population, these mitigating factors become less effective, and eventually, some form of wastewater treatment is warranted. Initially, partial (primary) treatment, consisting normally of screening and settling, may suffice to prevent the more obvious evidences of pollution, but with continuing urbanization, additional (secondary) treatment by biological methods may be required. The necessary treatment efficiency can be related to the assimilative capacity of the receiving water, that is, its ability to accept organic matter, nitrogen, and other pollutants without creating problems. This is determined from field studies.

The need for wastewater treatment arises in all countries. In the less developed regions, the treatment of domestic wastes for the protection of public health is still the principal concern. For example, in vast areas of India, Africa, and South America, untreated wastes enter receiving waters that are used directly by large populations for washing, bathing, and drinking. In most developed countries, the need is shifting from purely public health considerations to the control of eutrophication, protection of aquatic life, and concern over toxic substances in the environment.

Treatment methods in a country or region vary with the population density and the state of technological development. Sparsely settled rural communities can employ simple treatment processes to reduce the concentration of BOD, SS, and pathogens in domestic sewage. However, in urban centers, as municipal and industrial waste becomes more complex and the protection of receiving waters more necessary, wastewater treatment methods must become more sophisticated and more efficient.

12.4 WASTEWATER COLLECTION

12.4.1 Early Systems

The remnants of drains from the Minoan civilization of 3000 B.C. can still be seen on the Greek island of Crete. These sewers carried runoff from rainfall, water from the baths, and perhaps other wastes from the palace. Despite these early beginnings, widespread use of sewers did not occur until Roman times, when pipes were installed primarily to carry away stormwater from the streets since few houses or buildings had drains. Practically no further advances were made until the middle of the nineteenth century, when the discharge of human wastes into sewers began. The reasons for using ''storm drains'' this way were partly for convenience and aesthetics, but more importantly because people had finally become aware that sewage and household waste, if not removed, could contaminate their water and cause disease. Construction of this type of combined system, i.e., sewers carrying both stormwater and human wastes, continued well into the twentieth century. During this period, as sewer systems became more extensive, the concentration of wastes at the outlets resulted in unsightly conditions, deposition of solids, and odor, thereby forcing varying degrees of wastewater treatment to be provided. Over the years, problems with combined sewers (deposition of solids, odor, basement flooding, and others) led to the concept of separate sewers: one system for stormwater and one for sanitary wastes. This is the preferred practice today, although it can be argued, with some justification, that all drainage requires treatment and that this can be best accomplished with a single combined system.

12.4.2 Present Systems

Sewage collection systems today normally consist of separate storm and sanitary sewers in the newer areas and combined sewers in the older sections of cities.

Sanitary sewers. Sanitary sewers carry domestic sewage, liquid commercial and industrial wastes, and undesirable contributions from infiltration and stormwater. Because sewage discharges from homes, buildings, and factories can occur simultaneously, sanitary sewers must be designed to handle the peak rate of flow. The Harmon formula is one of several formulae, based on population, that are used to estimate the ratio of peak to average flowrate. It is given by

$$\frac{\text{Peak flowrate}}{\text{Average flowrate}} = 1 + \frac{14}{4 + \sqrt{P}} \tag{12.4}$$

where P is the population in thousands. Typical ranges for flow variation were noted in Section 12.2.3.

In some municipalities, sink disposal units or "garbarators" may be installed in sinks to macerate household organic (food) wastes so that the wastes can be flushed to the sanitary sewer rather than collected with the municipal garbage. The convenience to the homeowners and the possibility of less frequent garbage pickup are offset by the cost of the unit and the increased cost for treating the additional SS and organic load discharged to the sewers and municipal treatment facilities. Where sink disposal units are used, increases in the SS and BOD of the sewage have been estimated at 30 and 60 percent, respectively (APWA, 1970).

Storm sewers. Storm sewers receive stormwater runoff from roads, roofs, lawns, and other surfaces. Various methods for estimating the rates or volumes of runoff are available. All start with a storm of specified frequency (from 2 to 100 years) and information on an actual (or representative) storm, or with curves relating rainfall intensity to storm duration. Typical storm frequencies used for design are 2 to 5 years for residential areas, 10 to 25 years for high-value and commercial districts, and 50 to 100 years for major drainage systems.

Storm systems may be partial systems with relatively small pipes, providing only road drainage and not connected to building drains. Because these sewers are not connected to buildings, they can surcharge (overflow) without flooding basements, causing only minor inconvenience. On the other hand, complete storm systems that provide both road drainage and storm connections to buildings must have larger pipes designed with less likelihood of surcharging if backup of stormwater around basement footings is to be avoided.

Buildings with basements are protected against high groundwater levels by the installation of perforated or open-jointed drainage pipes laid in a gravel trench around the basement footings. The uncontaminated water collected by these foundation drains may then discharge to (a) a storm or combined sewer if sewer backup is unlikely (a common practice in larger cities), (b) a sanitary sewer or basement sump (typical of smaller municipalities), or (c) a separate foundation drain collector (limited to occasional installations where a third street sewer with a free outlet can be installed).

As has been mentioned previously (see also Table 12–4), runoff can contain high concentrations of pollutants. Increasing public awareness of the pollution caused by storm-water, forced improved methods for collecting and treating runoff to be considered. Sed-imentation, filtration, and chlorination were the early approaches, but more recently, the emphasis has shifted from treatment to the reduction and detention of stormwater runoff by ponding, flow regulation, and porous pavement.

Combined sewers.   Combined sewers perform the functions of sanitary and storm sewers and are common in the older sections of most large municipalities. Because these sewers carry sanitary wastes and are connected to basement floor drains, any surcharging could cause a backup of untreated sanitary sewage into basements. Consequently, such systems must be designed to accommodate large storms (recurring at 25-to-50-year average intervals) without surcharging, while still receiving sanitary flow. Since the volume of sanitary sewage, in comparison with the peak amount of stormwater flow, is insignificant, it can normally be neglected in design.

12.4.3 Pollution from Combined Sewer Overflows

During dry weather all flow in combined sewers goes to the treatment plant. During rainfall, the plant may be able to accept $1\frac{1}{2}$ to 3 times the dry weather flow (DWF), but the excess stormwater, now combined with municipal wastewater, must be discharged to the receiving waters without proper treatment. In a city where all sanitary sewage is collected for secondary treatment (90 percent removal of BOD and SS), the pollution (in terms of BOD and SS) from untreated combined sewer overflows can exceed that from the treated sanitary sewage. Most solutions to the problem are expensive, and none are completely satisfactory. Two basic approaches have been used: either some degree of separation or storage of some proportion of the excess flow for subsequent treatment.

Complete separation of stormwater from municipal wastewater has generally been considered the best long-term solution. Unfortunately for many cities, this solution is economically impractical. Washington, D.C., is one of the few large cities to have adopted the method.

Partial separation, a less costly alternative, is another approach. Toronto is imple-menting this scheme in some older sections of the city at about half the cost of full separation. The new sewers provide road drainage and create reserve storage capacity in the existing combined sewers. This reserve capacity can be used to reduce the frequency of overflows from the combined sewer system to less than half the number of previous occurrences. It can also be used to allow redevelopment of low-density housing with higher density apartment development, which of course creates increased sanitary sewage flows on account of the much higher population.

Stormwater retention during peak runoff periods is used frequently to reduce com-bined sewer overflows. Storage can be provided by unused sewer capacity, tanks, reservoirs, or even the roofs of buildings. It may be in-line or off-line storage, that is, part of, or separate from, the sewer system, and it may be at ground level as well as above or below it.

Whether separation of stormwater from sanitary sewage or the storage (and subsequent treatment) of combined sewer overflows is a better way to control pollution is not obvious. There are those who contend that complete separation of sanitary sewage and stormwater should be the objective, because then the risk of sanitary wastes backing up into basements or overflowing into receiving waters is minimized. On the other side are those who argue that stormwater is so contaminated anyway that it requires treatment and that this can be accomplished most effectively by controlling and treating the overflow from the combined sewers that already exist in the older cities. Obviously, neither approach is the answer in all cases. From a theoretical standpoint, separate storm and sanitary systems with control and treatment of the flows from both, if necessary, would be the best solution. However, practical considerations, including political and economic factors, greatly influence the type of sewage collection system that is used. Determination of the best solution in each location requires a site-specific investigation.

Photograph 12–2   Stormwater holding reservoir. Source: R. V. Anderson Associates, Ltd.

To prevent pollution of the Welland Canal from combined sewer overflows, a 45,000 m³ (12 × 10⁶ gal) stormwater holding reservoir was constructed as a more economical alternative than installing separate sewers (Photo 12–2). The two interconnected cells, one concrete lined to facilitate cleaning and the other sodded, accommodate a storm with a 2-year frequency. Following the storm, the reservoir contents are drained by remote control to the wastewater treatment plant for processing.

12.5 PRINCIPLES OF WASTEWATER TREATMENT

12.5.1 Effluent Requirements

The primary objective of wastewater treatment is to remove or modify those contaminants detrimental to human health or the water, land, and air environment. Land disposal, evaporation from ponds, and deep-well injection are occasional options, but usually the only practical outlets for the disposal of treated (or untreated) wastewater are streams, rivers, lakes, and oceans. To protect these water resources, the discharge of pollutants into them must be controlled. This is done in North America by setting effluent requirements for BOD, SS, and fecal coliforms. Secondary treatment is normally necessary to meet these requirements.

In the United States, for example (see Table 12–5), BOD_5 or SS values in the effluent must not exceed an average of 30 mg/L. The maximum limit for fecal coliforms is 200/100 mL (geometric mean). A further stipulation is that not less than 85 percent removal of BOD and SS must be provided. Canadian requirements are similar, but some provinces use 15 mg/L as the maximum for effluent BOD_5 and SS. This implies a greater than 90 percent removal efficiency in the treatment of "normal" municipal wastewater with a BOD_5 of 200 mg/L.

Where receiving waters have limited assimilation capacity, contain excessive nutrients, provide essential water use, or support valuable aquatic life, then more stringent effluent requirements are warranted. Such situations entail detailed investigations to evaluate the need for additional wastewater treatment.

12.5.2 Treatment Processes

The suspended, colloidal, and dissolved contaminants (both organic and inorganic) in wastewater may be removed physically, converted biologically, or changed chemically. The basic components that accomplish all of these in a municipal wastewater treatment plant are shown in Figure 12–2, but before assessing the processes in detail, an overview of a conventional plant and the principles employed is helpful.

Contaminants are generally removed from wastewater in order of increasing difficulty. First, rags, sticks and a miscellaneous array of large objects are retained on coarse screens (100–150 mm spaces) when necessary to protect small pumps. Then grit, the material that wears out equipment, takes up space and settles according to Stoke's Law, is removed in grit tanks or chambers. At this point most of the small solids are still in suspension and

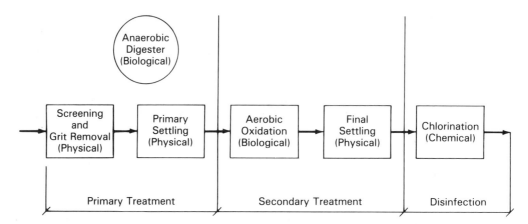

Figure 12–2 Municipal wastewater treatment plant components

the settleable portion of these (about 50%) can be removed and concentrated in the primary gravity settling tanks. The concentrated solids, called raw sludge, are pumped to an anaerobic digester for biological decomposition and the clarified primary tank effluent flows to the secondary treatment units. These consist of (a) a biological oxidation section where the dissolved and colloidal matter in the wastewater provides food for microorganisms which then convert the organics to CO_2 and H_2O and (b) a final gravity settling tank where the microorganisms are settled out. Part of this concentrated biological sludge is returned to "reseed" the oxidation section but most, after further thickening, goes to the anaerobic digester. The last stage in treatment of wastewater is disinfection of the plant effluent before discharge to the receiving water.

Many of the processes used in water treatment are also used in wastewater treatment, suitably modified for the removal of greater amounts of pollutants. Therefore, much of the material in Chapter 11 is relevant here and will not be repeated.

Physical processes.   Gravity settling is the most common physical process for removing suspended solids from wastewater. It is employed in:

- Removing grit (defined as sand particles of 0.2 mm diameter or greater)
- Clarifying raw sewage and concentrating the settled solids (called raw or primary sludge)
- Clarifying biological suspensions and concentrating the settled floc (called biological, activated, or secondary sludge)
- Gravity thickening of primary or secondary sludges.

Ideal Settling.   Settling of discrete particles as previously described for water treatment also occurs with the removal of grit from wastewater. Stokes' Law for small spherical particles under laminar flow conditions is applicable here. In an ideal settling basin, as represented by the schematic, continuous-flow, rectangular tank shown in Figure 12–3, discrete particles settle at a constant velocity u_t.

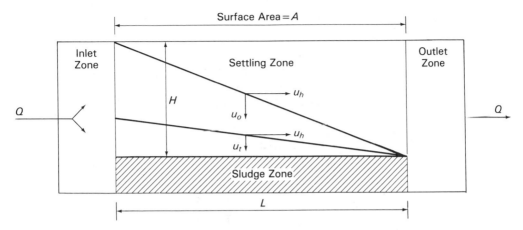

Figure 12–3 Settling of discrete particles in an ideal rectangular settling basin. Source: Clark, Viessman, and Hammer, 1977.

In this "ideal" tank the solids distribution at each cross-section is uniform, the flow, undisturbed by eddies, is horizontal and steady, and when a particle reaches the sludge zone it is removed and stays removed. These conditions in which no intermixing occurs, result in "plug" flow as described in section 7.4.1. In plug flow, the detention time "t_0" for all particles $= V/Q$ and

$$t_0 = \frac{V}{Q} = \frac{AH}{Q}$$

For complete removal of particles of a given size, the settling velocity u_t must be equal to or greater than H/t_0, in other words, sufficient time to settle to the bottom of the tank within the detention time.

Therefore

$$u_t = \frac{H}{t_0} = \frac{H}{AH/Q} = \frac{Q}{A} \tag{12.12}$$

Q/A is called the "overflow rate" and sometimes surface loading rate or clarification rate. It is expressed as

$$\frac{\text{Flowrate}}{\text{Surface area}} \left(\text{i.e., } \frac{Q}{A} \right) \text{ or } \frac{\text{distance}}{\text{time}} \left(\text{i.e., } \frac{H}{t_0} \right)$$

and is usually reported in m^3/m^2d (gpd/ft^2). It is in fact the settling velocity of those particles which enter the tank at the surface and just reach the bottom of the tank before their theoretical time for discharge. This concept provides a useful basis for designing settling basins.

Theoretically, all particles having settling velocities u_t greater than or equal to a specified settling velocity u_0 ($= Q/A$) will be completely removed, while those particles having $u_t < u_0$ will be removed in proportion to the ratio of the particle settling velocity to the selected velocity. Thus,

$$\text{Proportion removed} = \frac{u_t}{u_0} = \frac{u_t}{Q/A} \qquad (12.6)$$

Equation 12.6 reveals that, perhaps contrary to intuition, the proportion of particles removed in an ideal settling tank is independent of the depth of the tank: for a given flow Q, it varies only with the surface area A of the tank. Of course, from a practical standpoint, the depth must be sufficient to provide for sludge collection and removal and to limit the horizontal scouring velocity of the liquid.

Example 12.5

What proportion of spherical sand particles 0.01 mm in diameter with specific gravity (SG) = 2.65 will theoretically be removed in an ideal, horizontal-flow, settling basin 3 m deep with a surface area of 900 m² if the flow is 80,000 m³/d and the water temperature is 20°C? ($\mu = 1.0 \times 10^{-3}$ kg/m·s and $\rho = 1000$ kg/m³).

Solution   From Stokes' Law (Equation 6.7),

$$u_t = \frac{g}{18} \frac{(\rho_s - \rho)}{\mu} d_p^2 = \frac{9.81}{18} \frac{(2,650 - 1,000)}{(1 \times 10^{-5})} (1 \times 10^{-5})^2 = 90 \times 10^{-6} \text{ m/s}$$

The overflow rate is

$$u_0 = \frac{Q}{A} = \frac{80,000 \text{ m}^3/\text{d}}{900 \text{ m}^2 \times (60 \times 60 \times 24) \text{ s/d}} = 102.9 \times 10^{-6} \text{ m/s}$$

The removal is

$$\frac{u_t}{u_0} = \frac{90 \times 10^{-6}}{102.9 \times 10^{-6}} \times 100 = 87.5\%$$

Note that in this ideal settling basin, the depth of the basin was of no significance. Consequently, grit removal tanks based on gravity settling are constructed as very shallow basins.

Nonideal Settling.   In the previous development of settling tanks, we assumed uniform flow (plug flow) undisturbed by eddy currents or wind, and sludge that stayed settled. In fact, because of turbulence—particularly at the tank inlet and outlet—short circuiting of the flow, dead spots in the tank, and the movement of sludge collectors, ideal settling does not exist in practice. Furthermore, unlike the particulates (turbidity) found in water treatment processes and the grit described earlier, most of the suspended solids to be removed from wastewater are usually not discrete particles which settle at a constant rate. In sewage the particulates tend to be flocculent, and when larger, faster settling particles overtake slower ones, they coalesce into larger clumps during settling. In cases where ideal settling doesn't apply, settling column tests with the wastewater can be used to establish design criteria. Where settling tests are not practical, various criteria for overflow rates for designing settling tanks are used, depending upon the type of waste being treated.

Applications of Physical Processes.   Retention of raw sewage for 5 to 15 minutes in grit tanks or channels removes about 99 percent of the material classified as grit.

Sedimentation of raw sewage will remove 40 to 60 percent of the SS from the incoming sewage and 30 percent of the BOD_5 (contained in the settled sludge) if at least one hour detention is provided for maximum flow and the overflow rate does not exceed 100 m^3/m^2d (2,500 gpd/ft²). More conservative criteria (40 m^3/m^2d) (1,000 gpd/ft²) must be used for the settling of lighter biological floc (Metcalf and Eddy, 1979).

Sedimentation tanks, also called clarifiers, may be circular or rectangular, and occasionally square. Round tanks (10–50 m in diameter) have fewer submerged moving parts, and therefore lower maintenance costs should result. Rectangular units (up to 50 m in length) like that shown in Figure 12–4 permit more economical, common-wall construction and are said to be more hydraulically efficient. The advantage of one shape over another has not been conclusively demonstrated, so either type may precede biological treatment. Both types have scraper mechanisms to remove the solids that float (scum) and those that settle (primary or raw sludge).

Other physical unit operations in wastewater treatment include

- Screening to remove large objects and perhaps grinding to reduce particle size. These operations precede sedimentation
- Granular media filtration (by gravity or pressure) to remove effluent SS
- Flotation or centrifugation to thicken sludges.

(ii) Biological Processes.   Most of the organic constituents in wastewater can serve as food (substrate) to provide energy for microbial growth. This is the principle used in biological waste treatment, where organic substrate is converted by microorganisms, mainly bacteria (with the help of protozoa), to carbon dioxide, water and more new cells. The microorganisms may be aerobic (requiring free oxygen), anaerobic (not requiring free oxygen), or facultative (growing with or without oxygen). Processes in which microorganisms use bound oxygen (from NO_3 for denitrification, for example) are often called anoxic rather than anaerobic. The microbial population may be maintained in the liquid as

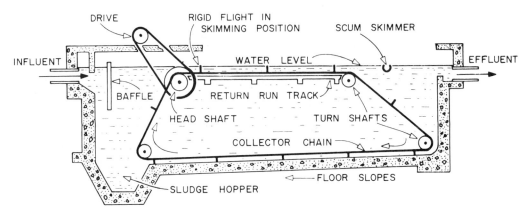

Figure 12–4 Rectangular settling tank for primary clarification. Source: Water Pollution Control Federation Manual of Practice, No. 11, 1976

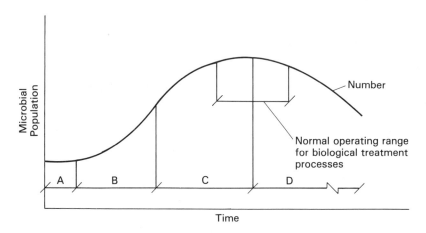

Key: A Lag phase
       B Log growth phase
       C Declining growth phase
       D Endogenous phase

Figure 12–5   Microbial growth in a batch culture. Adapted from Clark, Viessman, and Hammer, 1977.

suspended growth, referred to as mixed liquor suspended solids or volatile suspended solids (MLSS or MLVSS), or it may be attached to some medium in a fixed-film process. These two possibilities are described later in Section 12.7 in more detail. The rate of microbial growth varies directly with the amount of available substrate. In a batch culture when food is not limiting, the microbial population, after an initial lag period, grows rapidly at a logarithmic rate. As food decreases, growth slows until, at some point, growth stops and the number of new cells produced is balanced by the number of old cells which are dying. When the substrate is exhausted, the number of microorganisms declines as old cells decompose (lyse) releasing their nutrients for use by new microorganisms. These four phases referred to as the lag (A) log growth (B) declining growth (C) and endogenous (auto-oxidation) phase (D) are shown schematically in Figure 12.5.

In a continuous biological process, the system normally operates at some point on the growth curve toward the end of the declining growth phase, or into the endogenous phase where cells utilize their own protoplasm to obtain energy. Utilization of the logarithmic growth phase in wastewater treatment has been impractical, because substrate removal is incomplete and no economical way has been found to separate the microbial population from the liquid.

Aerobic/Anoxic Processes.   In aerobic processes (i.e. molecular oxygen is present), heterotrophic bacteria (those obtaining carbon from organic compounds) oxidize about one-third of the colloidal and dissolved organic matter to stable end products (CO_2 + H_2O)

and convert the remaining two-thirds into new microbial cells that can be removed from the wastewater by settling. The overall biological conversion proceeds sequentially, with oxidation of carbonaceous material as the first step, i.e.,

$$\text{Organic matter} + O_2 \rightarrow CO_2 + H_2O + \text{new cells} \tag{12.7}$$

Under continuing aerobic conditions, autotrophic bacteria (those obtaining carbon from inorganic compounds) then convert the nitrogen in organic compounds to nitrates according to the simplified equations

$$\text{Organic N} \rightarrow NH_3 \text{ (decomposition)} \tag{12.8}$$

and

$$NH_3 + O_2 \xrightarrow[\text{bacteria}]{\text{nitrifying}} NO_2^- \rightarrow NO_3^- \text{ (nitrification)} \tag{12.9}$$

No further changes in the nitrates take place unless the process becomes anoxic (i.e. only bound oxygen is present). Under such conditions, heterotrophic bacteria convert the nitrates to odorless nitrogen gas:

$$NO_3^- \xrightarrow[\text{bacteria}]{\text{denitrifying}} NO_2^- \rightarrow N_2^- \text{ (denitrification)} \tag{12.10}$$

Under continuing anoxic conditions, any sulfates present are reduced to odorous hydrogen sulfide gas according to the descriptive equation

$$SO_4^{-2} \xrightarrow[\text{bacteria}]{\text{sulfate reducing}} H_2S \tag{12.11}$$

The preceding naturally occurring reactions are used in various biological processes for the treatment of wastewater. In every case, nutrients essential for biological growth must be present in the waste or must be added. Unlike municipal wastewater, which contains the necessary ingredients, many industrial wastes, including those from the pulp and paper, cannery, and meat processing industries, need nitrogen and/or phosphorus added (as ammonium or phosphate salts) for biological growth to occur.

Ever since the importance of wastewater treatment became recognized, municipalities and industries have relied almost exclusively on aerobic rather than anaerobic biological processes for treating their liquid organic wastes. Aerobic treatment has predominated because of its simplicity, stability, efficient and rapid conversion of organic contaminants to microbial cells, and relatively odor-free operation.

Although all aerobic biological oxidation processes use microorganisms to convert organic contaminants, the methods for accomplishing this conversion are varied and numerous. They are discussed in Section 12.7.

Anaerobic Processes.   In anaerobic biological processes (i.e. no oxygen is present), two groups of heterotrophic bacteria, in a two-step liquefaction/gasification process, convert over 90 percent of the organic matter present, initially to intermediates (partially stabilized end products including organic acids and alcohols) and then to methane and carbon dioxide gas:

$$\text{Organic matter} \xrightarrow[\text{bacteria}]{\text{acid-forming}} \text{intermediates} + CO_2 + H_2S + H_2O \quad (12.12)$$

$$\text{Organic acids} \xrightarrow[\text{bacteria}]{\text{methane}} CH_4 + CO_2 \quad (12.13)$$

The process is universally used in heated anaerobic digesters, where primary and biological sludges are retained for approximately 30 days at 35°C to reduce their volume (by about 30 percent) and their putrescibility, and thus simplify their disposal, usually on agricultural land.

Two major advantages of anaerobic processes over aerobic ones are, that they provide useful energy in the form of methane and that sludge production is only about 10 percent of that from aerobic processes for converting the same amount of organic matter. This is advantageous in the treatment of high-strength wastes, where the handling of large volumes of sludge would be a problem.

Chemical Processes.   Many chemical processes, including oxidation, reduction, precipitation, and neutralization, are commonly used for industrial wastewater treatment. For municipal wastewater, precipitation and disinfection are the only processes having wide application.

Chemical treatment alone or with other processes is frequently necessary for industrial wastes that are not amenable to treatment by biological means. The oxidation of toxic cyanide to manageable cyanate (with SO_2) or of hexavalent chromium to the nontoxic trivalent form in the disposal of plating wastes are examples. Chemical processes are also useful in municipal waste treatment, where phosphorus concentrations are reduced and removal of solids increased by precipitation of these contaminants with metallic salts.

Disinfection of the effluent from wastewater treatment plants, generally by chlorination, is desirable where there is a potential health hazard. However, the uncertainty as to when a hazard exists has resulted in wide variation in practices. Some authorities in the United States and Canada require chlorination of wastewater effluent all year round. A minimum residual of 0.5 mg/L after a contact time of 15 to 30 minutes is often specified. In some areas, this empirical definition of chlorination has been replaced by a requirement that fecal coliforms must not exceed 200 per 100 mL in the final effluent. In most municipal wastewater treatment plants, an average chlorine dosage of 10 mg/L to the secondary effluent and a contact time of at least 30 minutes at peak flow will accomplish these objectives and provide adequate destruction of pathogens. Chlorination of wastewater is a North American practice. In other jurisdictions, including Britain and all western European countries, disinfection of wastewater is rare. It is felt that the cost and possible disadvantages of disinfection are not worth the uncertain protection of public health.

Chlorine is the least expensive and most often used chemical for wastewater disinfection, but unfortunately it produces some undesirable side effects. Organic matter present will combine with the chlorine to form chlorinated organics, some of which are known or suspected **carcinogens** (capable of causing cancer). The fear is that these chlorinated organics are a potential hazard to water supplies. Another concern is the toxicity of chlorine residuals to aquatic life. Residuals as low as 0.05 mg/L are toxic to various species of freshwater fish (Environment Canada, 1978). Where disinfection is necessary for the protection of public health, but toxicity from chlorine residuals is unacceptable, either the effluent must be dechlorinated or alternatives to chlorination must be considered. Unfortunately, experience with alternatives such as chlorine dioxide, other halogens, ozone, or ultraviolet light is extremely limited. Costs for chlorination and dechlorination or alternative disinfection methods are two to three times the cost of chlorination alone. Also, substitutes for chlorine have other disadvantages and unknown long-term effects. There is no ideal disinfectant.

Not all of the available physical, biological, and chemical processes have been described, nor are all of the processes mentioned required in every wastewater facility. The basic units in a typical municipal plant, as shown in Figure 12–2, might consist of primary

TABLE 12–6   BASIC PROCESSES IN A TYPICAL WASTEWATER TREATMENT PLANT

Process	Purpose	Common methods	Typical reductions	
1. *Physical*				
Primary Settling (at peak inflow)	Remove SS from raw sewage, and concentrate to 4 to 6%* $\left(\dfrac{VSS}{SS} = 0.6–0.8\right)$	Detention $\ll$ 1 hr $Q/A \gg 100$ m³/m²·d (primary sludge only) $Q/A \gg 60$/m³/m²·d (primary + activated)	SS BOD	50% 30%
Final Settling (at peak outflow)	Remove SS from MLSS, and concentrate to 0.5 to 2% $\left(\dfrac{VSS}{SS} = 0.7–0.9\right)$	Detention $\ll$ 2 hr $Q/A \gg 40$ m³/m²·d $\dfrac{\text{Peak}}{\text{Avg.}} = 1.5–2.0$	SS BOD N P	90% 90% 40% 20%
2. *Biological*				
Aerobic Oxidation	Convert colloidal and soluble organics to microorganisms	Aeration of wastewater and microorganisms for 6 hours	BOD (conversion)	90%
Anaerobic Digestion	Stabilize primary and secondary sludges, and concentrate to 6–8%* $\left(\dfrac{VSS}{SS} = 0.3–0.5\right)$	Anaerobic decomposition for 30 days at 35°C	VSS Volume	50% 30%
3. *Chemical*				
Disinfection	Destroy pathogens	Detention $\ll$ 15 min at peak flow with 0.5 mg/L chlorine residual	Coliforms 99%	

Adapted from Metcalf and Eddy, 1979
* Decrease values by 25% if waste-activated sludge is returned to the primary tank.

settling tanks, preceded by screening and grit removal; secondary-treatment units for oxidation and settling; anaerobic sludge digesters; and chlorination facilities. Information about these is provided in Table 12–6.

Example 12.6

A municipality with diversified industrial and commercial areas must provide primary clarification for an average wastewater flow of 20,000 m^3/d.

(a)   If circular primary tanks 3 m deep are used, what total volume is needed? How many tanks, of what diameter, should be provided?

(b)   What volume of primary sludge would be produced each day?

Solution   (a) In the absence of data from settling tests, the size of clarifiers can be based on detention time and overflow rate. If $Q_{average}$ = 20,000 m^3/d, then the population equivalent (for a medium-size municipality, use 500 Lpcd, from Table 12–4), is given by

$$PE = \frac{20,000 \times 1,000}{500} = 40,000 \text{ people}$$

and

$$\frac{\text{Peak flow}}{\text{Average flow}} = 1 + \frac{14}{4 + \sqrt{40}} = 2.9$$

Therefore, $Q_{max.}$ = 20,000 × 2.9 = 58,000 m^3/d.

For a detention of 1 hour (Table 12–6), we have

$$\text{Volume} = Qt = 58,000 \frac{m^3}{d} = \frac{1 \text{ hr} \cdot d}{24 \text{ hr}} = 2,420 \text{ m}^3$$

$$\text{Surface area} = \frac{\text{volume}}{\text{depth}} = \frac{2,420}{3} = 807 \text{ m}^2$$

For an overflow rate of 100 m^3/m^2 d (Table 12–6), we have

$$100 = \frac{58,000}{A}$$

so that

$$A = 580 \text{ m}^2$$

Thus, with the larger area governing and the design based on detention time, the possible choices of tanks are as follows:

Use 4 tanks 16 m in diam. × 3 m deep (A = 804 m^2)

or 3 tanks 19 m in diam. × 3 m deep (A = 850 m^2)

or 2 tanks 23 m in diam. × 3 m deep. (A = 831 m^2)

Economics, space limitations, and flexibility would influence the final choice. At least two tanks are necessary to permit one tank to be taken out of service. An odd number of tanks is usually avoided because of difficulties in dividing and proportioning flows. Here, four tanks would provide maximum flexibility and, depending on their cost, might be the preferred choice if no plant expansion is likely for many years.

(b) For an average dry weather flow = 20,000 m^3/d, assume SS = 300 mg/L (Table 12–4). Then the weight of SS in the plant inflow is

$$Q \times SS = \frac{20 \times 10^6 \text{ L/d} \times 300 \text{ mg/L}}{10^6 \text{ mg/kg}} = 6,000 \text{ kg/day}$$

Assume a 50% SS removal in the primary = 3,000 kg/d and sludge concentration = 5% = 50,000 mg/L (From Table 12–6). Then

$$\text{Sludge volume} = \frac{3,000 \times 10^6 \text{ mg/d}}{50,000 \text{ mg/L}} = 60,000 \text{ L/d} = 60 \text{ m}^3/\text{d}$$

Since primary settling removes only about one-third of the organic matter (BOD) from municipal wastewater, it is seldom suitable alone. For BOD removals up to 80 or 90 percent, secondary treatment, which is biological oxidation with final settling, must be added. The options for secondary treatment are described, with examples, in Section 12.7.

Where the basic units in a secondary treatment plant are inadequate to meet effluent and water quality objectives or standards, the treatment must be upgraded. Improvements might be required to reduce effluent BOD and SS concentrations, provide nutrient removal, and nitrify, denitrify, or detoxify the effluent. These additional treatment methods for improving secondary facilities are described generally as tertiary or advanced wastewater treatment processes. They are not considered here.

12.5.3 Selection of Treatment Method

Wastes must satisfy effluent and water quality objectives or standards if they are to be discharged without creating a nuisance. Treatment facilities for accomplishing this may range from relatively simple land-based treatment systems to complex automated wastewater treatment plants. In specific situations, several treatment methods may be equally suitable. As a result, the final choice is based only partly on engineering analysis. Many intangibles, such as local preference, the experience of the consultant, and the track record of the process, influence the decision. The next three sections outline features of the most common treatment methods.

12.6 LAND-BASED TREATMENT METHODS

Land-intensive waste treatment methods include land (and wetland) application systems in which algae, or green plants and soil play a major role in the treatment process, oxidation ponds which utilize algae, and lagoons where supplementary aeration may be used to accelerate biological treatment.

Land-based methods require large areas for the storage and treatment of wastewater, for the disposal of effluent when no suitable outlet is available, and for isolation of the site from neighboring land uses. Hydraulic and organic loading rates for these systems are commonly based on land area requirements, and this provides a convenient way for comparing the various processes. Their advantages over wastewater treatment plants are their simplicity and, in most cases, their lower capital and operating costs. In these relatively uncontrolled systems, physical, biological, and chemical processes are not separated. They may, in fact, occur simultaneously. Sedimentation or filtration may physically remove particulates while the oxidation of organic matter by bacteria (in the soil, attached to plants, or in ponds) is proceeding, together with the assimilation of nitrogen and phosphorus by plant and microbial growth. Percolation may be facilitating the adsorption of inorganic ions such as phosphorus and many of the heavy metals (Cd, Cu, Ni, Pb, Zn) by the soil. Other inorganic contaminants may be removed from the wastewater by ion exchange in the soil. Reduction of pathogens takes place because of natural die-off and the effect of ultraviolet rays in sunlight.

12.6.1 Land Application

Land application of sewage is not new. Raw sewage was applied to agricultural land outside Edinburgh, Scotland, 250 years ago—not for disposal, since no health hazard was recognized, but because of the benefit to crops. Since 1893, Melbourne, Australia, has discharged its wastes to a sewage farm for treatment and disposal. Despite its long and successful history, land treatment of raw or primary treated sewage has not been accepted in North America, except for a few cases. The method has been rejected, and in many areas prohibited, because of concern about the pathogens that the raw waste might contain. The approach in the United States has been to encourage the application of secondary effluent to the land following some form of biological treatment. In fact, in 1977, the United States Environmental Protection Agency (US EPA) decreed that all communities seeking construction grants must employ land treatment and disposal unless it can be demonstrated that it is not cost effective. The objective was to reduce waste discharges to surface waters while utilizing the nutrient content for crops, and encourage the use of effluent for irrigation and groundwater recharge. A discussion of several land and wetland applications follows.

Spray irrigation and infiltration.   Wastewater that has received secondary treatment, generally in an oxidation pond or aerated lagoon, is spread by flooding or channeling, or sprayed from nozzles, on pervious soil, where filtering, biological decomposition, and adsorption by the soil remove the contaminants. For year-round disposal in moderately porous soil, about 1 ha is needed for every 240 people (1 ac/100 people). For organic industrial wastes, the corresponding rate is 18 kg BOD_5/ha (16.7 lb/ac). When disposal on land is only possible for part of the year, area requirements must be increased proportionately. Land for storage, secondary treatment, and a buffer zone is added to this. In northern areas with severe winters, land disposal of effluent is limited to small communities, resorts, and institutional use. Spray irrigation is the most popular land treatment method, but its use should be limited to secondary effluent and industrial wastes such as cannery, tannery, and food wastes that contain no pathogens.

Overland flow.   Settled sewage or secondary effluent is applied to fairly impervious land having a slope of two to eight percent so that substantial runoff (as laminar flow) occurs. Sedimentation and the bacterial population in the cover crop are the major treatment mechanisms. The method has been used for some time for settled sewage in Melbourne, Australia, but in the United States, except for some trial installations, it has been limited to industrial waste applications (Metcalf and Eddy, 1979). Collection and disposal of runoff amounting to perhaps 50 percent of the wastewater applied is a major consideration.

Wetlands.   Natural or artificial wetlands such as marshes and swamps provide the environment for physical sedimentation and bacterial activity in the treatment of preaerated wastewater or secondary effluent. Floating plants such as water hyacinth, or emergent ones like cattails, serve as mechanical filters and a support structure for the bacteria (Stowell et al., 1980). Where wetlands also provide useful crops or aquatic species, the process is called **aquaculture.** Growing fish in waste ponds has been practiced since Roman times, particularly by the Japanese, who have made wide use of this waste to fish to human to waste recycle system. Commercial applications of the concept are under investigation in North America.

12.6.2 Oxidation Ponds

An oxidation or stabilization pond is a shallow basin 1 to 2 m (3 to 6 ft) deep receiving raw sewage. Conditions in such ponds can range from aerobic through facultative to anaerobic, depending upon the depth of pond and the degree of natural or induced mixing. Detained settleable solids undergo anaerobic decomposition at the bottom of the pond. In the upper, aerobic part of the pond, soluble organic wastes are converted to CO_2 and water by bacteria working in combination with algae, which provide oxygen for the bacteria (a symbiotic relationship). Unfortunately, the algae, which grow in excess, are difficult to remove, and escape with the effluent. The algae may in fact be responsible for more BOD in the effluent than the organic contaminants cause in the influent. Other problems associated with oxidation ponds include the considerable reduction of treatment efficiency in cold weather under ice cover, and odors occurring after the breakup of ice.

12.6.3 Lagoons

Lagoons are ponds 3 to 5 m deep, which may be anaerobic throughout, facultative (part aerobic, part anaerobic), or completely aerobic when sufficient mixing and oxygenation are provided. Anaerobic lagoons provide only an intermediate degree of treatment and are a continual source of odors. They are worth considering in remote locations, especially where high-strength wastes (BOD > 1,000 mg/L) are encountered, because the quantity of sludge created from biological growth is much less than it would be under aerobic conditions.

Aerated lagoons are a logical development from oxidation ponds. Instead of oxygen from algae or natural addition from the atmosphere, air is introduced into the lagoon by mechanical means such as diffused air tubing, mechanical mixers, or air distributors. The continuous supply of air permits aerated lagoons to have a smaller surface area and be deeper than an oxidation pond. Strictly speaking, aerated lagoons, particularly aerobic ones where sludge recycling is practiced, are more akin to wastewater treatment plants than to land systems. However, the light loadings per unit volume and the resulting large land areas required by the process are the reasons for their inclusion here.

TABLE 12–7 COMPARATIVE DESIGN PARAMETERS FOR LAND-BASED TREATMENT METHODS

Process	$\dfrac{D^{1}}{m}$	Loading[2] kg BOD/ha.d	Detention[3] days	Comments
Spray Irrigation	—	10–20 (0.5–1 cm/d)	60–200 (storage)	Pervious soil required
Overland Flow	—	20–40 (1–2 cm/d)	60–150 (storage)	Impervious soil, 2–8% slope
Wetlands Treatment	0.5	10–200 (0.5–10 cm/d)	12 (in system)	No process control
Anaerobic Lagoon	4.5	2,000–3,000	30 (in lagoon)	Partial treatment of high-strength wastes
Oxidation Pond (Photosynthetic)	2.0	10–50	100 (in pond)	Variable BOD, excess algae in effluent
Aerated Lagoon (facultative)	3.0	50–200	20 (in lagoon)	Variable SS in effluent
Aerated Lagoon (Aerobic)	3.0	250–350	4 (in lagoon)	High cost of operation

[1] Depth: D (typical), used to calculate surface loading.

[2] Loading: Wastewater with BOD_5 of 200 mg/L is assumed.

[3] Detention: Variations of 50% are common.

Adapted in part from:

Clark, J. W., Viessman, W., Jr., and Hammer, M. J., *Water Supply and Pollution Control*, 3d ed., New York: Dun-Donnelley, 1977.

Metcalf and Eddy, Inc., *Wastewater Engineering: Treatment Disposal Reuse*, 2d ed., New York: McGraw Hill, 1979.

U.S. Environmental Protection Agency, *Aquaculture Systems for Wastewater Treatment, An Engineering Assessment*, Washington, D.C.: EPA Publication No. MCD 68, June 1980.

U.S. Environmental Protection Agency, *Process Design Manual for Land Treatment of Municipal Wastewater*, Washington, D.C.: Technology Transfer Report, EPA 625/1-77-008, 1977, p. 506.

Because treatment proceeds at a slow rate in land-based systems, applications are restricted to installations where small flows are generated or large land areas are available. Industries and small municipalities often have such conditions. Favorable climate, suitable soil, and appropriate topography are additional prerequisites. Table 12–7 lists comparative loadings and detention periods for various land-based treatment systems. A typical application for an oxidation pond is described in Example 12.7.

Example 12.7

A 400-room resort hotel proposed for a tropical area with moderate rainfall (500 mm/year) must provide its own treatment system until municipal facilities are available in 15 to 20 years. A staff of 300 will run the hotel during the peak season of 6 months, and a skeleton staff of 100 will serve for the remainder of the year, when the hotel is at 20 to 40% capacity. A flat, 30-ha, square-shaped site with pervious soil adjacent to a large river is available 1 km from the hotel. What land-based treatment methods might be considered, assuming that a 200-m buffer zone must be maintained between any treatment works and the property boundary? Assume a flow of 200 Lpcd for all hotel guests and staff (including all hotel services).

Solution The location of the treatment area (x^2) within the 30 ha parcel is shown below.

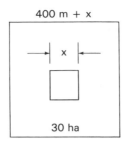

400 m + x

x

30 ha

Let the treatment area $= x^2$. Then the area of the site $= (400 + x)^2 = 300,000$ m^2 and $x = 548 - 400 = 148$ m. Also, $x^2 = 21,830$ m^2 $= 2.2$ ha. At full occupancy, the hotel has 800 guests and 300 staff, for a total of 1,100 people. So the waste volume per day $=$ 1,100 x 200 $= 220,000$ L/d. And the BOD$_5$ per day at 76 gm per capita (Table 12–3) $=$ 1,100 $\times$ 76 $= 83,600$ gm/d $= 83.6$ kg/d. Thus, the BOD loading, using the available treatment area, $= 83.6/2.2 = 40$ kg BOD$_5$/ha.d.

Comment

An oxidation pond could probably handle this loading (see Table 12–7) more economically than the other processes noted. Spray irrigation without adequate (secondary) pretreatment should not be used for wastes which might contain pathogens. Furthermore, the available land area is too small and storage for days when spraying is impractical would have to be provided.

Overland flow and wetland treatment are possibilities, but in both cases, since natural conditions (sloping ground, impervious soil, or a marsh) for the processes do not exist, the cost of creating these conditions would likely make them uncompetitive with an oxidation pond.

Of the other alternatives, anaerobic lagoons can be ruled out because of the intermediate (partial) treatment provided and the problems with odors. Aerated lagoons are not necessary in this situation, but may be a consideration in the future if the oxidation pond becomes

Photograph 12–3 Oxidation pond, Ontario Hospital, Goderich, Ontario, **Canada.** Source: Northway Map Technology Ltd., Don Mills, Canada

overloaded. A septic tank and tile system (discussed later) might be a consideration, but for this large an installation it is likely to be more costly than the pond.

An oxidation pond with two cells could be considered. One cell would be enough for the off-peak season. With a 4-m berm between cells, the dimensions of each cell would be 72 m × 148 m × 2 m depth.

Photo 12–3 is an aerial view of a small 2.5-ha (6-ac) oxidation pond serving a hospital.

12.7 WASTEWATER TREATMENT PLANTS

If simple systems are impractical or inadequate, treatment must be carried out in a wastewater treatment plant. By mechanization and control, the plant can meet effluent or water requirements in a smaller space, in less time, and perhaps, where land is expensive, at less cost than equivalent land-based treatment systems. Physical, biological, and chemical processes are utilized, but in a plant they are more likely to occur in separate units. The basic components of a conventional plant as outlined in Section 12.5.2 provide primary clarification, secondary (biological) treatment, disinfection, and sludge processing. Primary clarification and, in North America, disinfection embodying the principles described earlier are common to most wastewater treatment plants. Further elaboration of these processes, which do not vary significantly from one plant to another, is unnecessary. It is in the other two components that wide variations in process design can occur.

In the secondary phase, where the remaining solids and BOD are removed, biological treatment takes place in either a suspended-growth system or a fixed-film process. In the former, the microorganisms are kept in suspension in the wastewater by mixing or aerating devices. Such devices must satisfy the requirements for both mixing and oxygen transfer. With attached growth systems, the active microorganisms grow on the surface of rock, plastic, or other medium with which the waste is brought into contact.

12.7.1 Suspended-Growth Systems

Conventional activated sludge. The most widely used application of suspended growth is the conventional activated sludge process. Microorganisms are kept in suspension for four to eight hours in an aeration tank (see Photo 12–4a) by mechanical mixers and/or diffused air, and their concentration in the tank is maintained by the continuous return of the settled biological floc from a secondary settling tank to the aeration tank. The flocculent suspension which is settled and returned to "reseed" the aeration tank was given the name "activated sludge" by Ardern and Lockett in 1914, presumably because they found that by returning these solids, substrate oxidation was accelerated or activated (Arden and Lockett, 1914). The contents of the aeration tank are referred to as mixed liquor, and the solids are called mixed liquor suspended solids (MLSS). The latter include inert material as well as living and dead microbial cells. Determination of the living or active proportion is not easy, and it is usual to assume that the volatile portion of the solids (MLVSS), which accounts for about 80 percent of the MLSS, represents the active mass.

(a)

(b)

Photographs 12–4 (a and b)
Suspended growth systems

Aeration and final settling tanks, the two major components of suspended growth systems, are shown with the tanks dewatered.

(a) Aeration may be provided by various proprietary mechanical aerators such as the draft tube type shown here (Milton, Ontario) or by a variety of diffused air systems, ranging from those providing course bubbles (to minimize diffuser plugging) to those generating fine bubbles (to maximize oxygen transfer efficiency). (Courtesy J. G. Henry)

(b) Gravity settling of the biological growth developed in the aeration tanks takes place in the final settling tanks which may be round, as shown here (Truckee, California), rectangular, or square.
(Courtesy J. G. Henry)

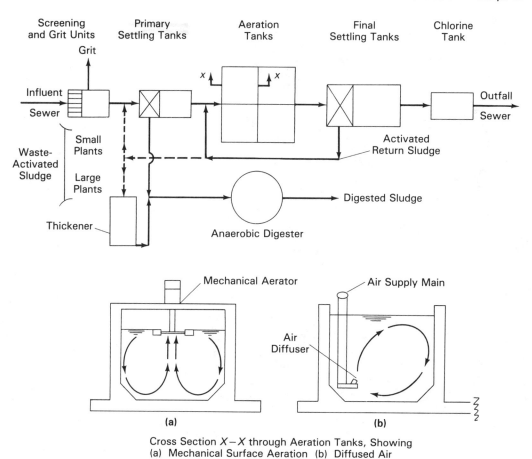

Cross Section $X-X$ through Aeration Tanks, Showing
(a) Mechanical Surface Aeration (b) Diffused Air

Figure 12–6   Flow diagram of conventional activated sludge process

A flow diagram for the conventional activated sludge (CAS) process is shown in Figure 12–6.

Like primary tanks, final tanks may be rectangular or circular, and occasionally square, but they provide longer detention (2 hours) and lower overflow rates (30–50 m^3/m^2d) at peak flow rates (see Photo 12–4b). The solids loading (MLSS) coming from the aeration tank is an additional design consideration, and values of 6–9 kg of MLSS per m^2 of settling tank surface area per hour (1.25–1.875 lb/ft²/hr) at the peak tank outflow rate are common (Metcalf and Eddy, 1979).

Of the biological floc which settles in the final tanks (also called **secondary clarifiers**), about 25 to 40 percent is returned to the aeration tank. The rest, called **waste activated sludge,** must receive further treatment. In small and medium-size plants, this waste activated sludge is returned to the primary tanks to settle with the primary sludge (resulting in a 40 percent decrease in the allowable primary tank overflow rate). In larger plants, the waste activated sludge from the final tanks may be thickened by gravity settling or dissolved air flotation before digestion with the primary sludge.

Many variations of the activated sludge process have been developed. Differences in aeration time, MLSS concentration, solids retention time, and loading are some of the distinguishing characteristics of the processes, making them more suitable for one application than another.

Extended aeration.   The flow diagram for the extended aeration process is similar to that for conventional activated sludge (CAS), but with no primary tanks. Extended aeration provides 24-hour aeration (vs. 6 for CAS), which results in process stability but costly operation. Detention time in final tanks is about twice that of a conventional plant. The oxidation ditch is a variation of the extended aeration process in which aeration and settling are combined in a shallow (1-m-deep) channel through which the waste is circulated by rotating paddles or brushes. Simplicity of operation has resulted in wide use of the extended aeration process for small installations and in package plants.

Contact stabilization.   In the contact stabilization process, also called **biosorption,** the biological sludge, rather than the mixed liquor, undergoes long aeration. Figure 12–7 shows the arrangement.

The process is suitable for wastes having a high proportion of the organic contaminants in a particulate form, since it depends on the adsorption of these particulates by the stabilized sludge during a short (20–40 min) contact period. Assimilation of the adsorbed organics takes place in the stabilization (sludge reaeration) tank over a 4-hour period. Since there are no primary tanks, no primary sludge is produced, and stabilization of the aerobic waste sludge has generally been carried out by aerobic digestion. High energy costs for aeration limit this approach to smaller plants (less than 4,000 m³/d).

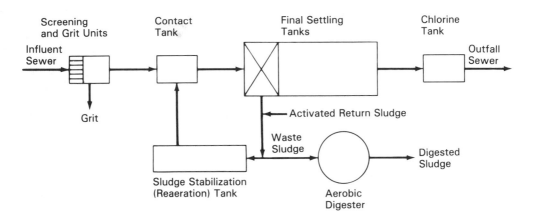

Figure 12–7   Flow diagram of contact stabilization process

Oxygen-activated sludge.   The oxygen-activated sludge process, in which pure oxygen, rather than air, is used in covered tanks, was developed by Union Carbide in the late 1960s. It has some advantages over air-activated sludge. It can operate at higher loadings, thereby reducing aeration tank size, and can accept wider fluctuations in load and stronger wastes than can air-activated sludge (Parker and Merrill, 1976). Process control of oxygen requirements by pressure sensors is simple and reliable. Where a high rate of oxygen transfer is necessary, as with high-strength wastes, the oxygen system has an advantage over the air-activated sludge with its limited transfer capability. Another situation where the oxygen system is more likely to have an economic advantage is where odor control is important and covering of the aeration tanks is intended. Also, control of odors will be much simpler since the volume of gas to be vented is only about 1 percent of that for air-activated sludge. However, the high cost of providing oxygen and the skills necessary to operate the process make it uncompetitive with CAS except for large plants.

Reactor design.   The chemical reaction kinetics introduced in Chapter 6 can be applied in modified form to the biological reactions in suspended growth systems. These relationships, coupled with mass balances around the aeration system, provide a means for process evaluation and design.

Type of Reactor.   Aeration tanks can be designed to operate as plug-flow or complete-mix systems. Reactor theory (Chapter 6) indicates that for a first-order reaction, plug flow, because of more efficient conversion of substrate, would be preferable to a complete-mix reactor. In large plants, a series of long, narrow aeration tanks with limited longitudinal mixing approaches plug-flow conditions. On the other hand, it is easier to approach complete mixing in small plants with square tanks operated in parallel. Even these two extremes lie somewhere between an ideal PFTR and an ideal CSTR. As noted in Section 6.6.4, the difference in performance between the two ideal reactors decreases for reactions that are less than first order, and at zero order there is no difference. Also, in practice no significant difference in performance between the two types of reactors is evident, and whether this can be attributed to the reactor being less than ideal, the reactions being less than first order, the generous tank volumes provided, or any combination of these, is not clear.

Process Kinetics.   The parameters applicable to suspended growth systems are the substrate removal rate q, specific growth rate μ, mean cell residence time Θ_c, and organic loading rate F/M. Let us examine them and consider some examples of their application to process design.

Process Kinetics: Substrate Removal Rate (q).   In chemical reactions (Chapter 6), the rate r of substrate conversion or removal is a function of the chemical substrate concentration A. Mathematically,

$$r = -k_r[A]^n \text{ (mg/L} \cdot \text{t)} \tag{12.14}$$

where, when $n = 0$ (zero order), k_r is in mg/L $\cdot$ t
  when $n = 1$ (first order), k_r is in 1/t
 and when $n = 2$ (second order), k_r is in L/mg $\cdot$ t

In biological processes, the rate q of substrate (organic matter) conversion is also a function of $[X]$, the concentration of biological solids present, as well as the organic substrate concentrations (Brackets indicating concentration are omitted). Mathematically,

$$q = k_q S^n \left(\frac{\text{mg substrate converted/t}}{\text{mg solids present}} \right) \qquad (12.15)$$

where S = concentration of organic substrate present
q = specific substrate conversion rate = $r/[X]$
$[X]$ = concentration of biological solids present as MLVSS
and where, when n = 0 (zero order), k_q is in 1/t
when n = 1 (first order), k_q is in L/mg $\cdot$ t
and when n = 2 (second order), k_q is in $(\text{L/mg})^2/\text{t}$.

It should be evident that to avoid confusion, the units for substrate removal coefficients need to be clearly indicated.

In order to apply the integrated kinetic equations in Table 6–6 to suspended growth systems, the substrate removal coefficient k_r needs to be multiplied by the biological solids concentration $[X]$. For example, in a first-order reaction in a CSTR, the equation becomes

$$[X]\, k_q t = \frac{[A_0] - [A]}{[A]} \qquad (12.16)$$

Example 12.8

For a first-order reaction and a substrate removal rate coefficient k_q = 0.025 L/mg·d, what hydraulic detention time is needed in a CSTR for 90% BOD removal from domestic sewage?

Solution   Assume MLVSS = $[X]$ = 2,000 mg/L (Range 1,000–3,000). Then, from $k_q [X] = \dfrac{[A_0]}{[A]} - 1$ (Table 6–6),

$$0.025t \times 2,000 = \frac{100}{10} - 1$$

$$t = 0.167\text{ d} = 4\text{ hours}$$

Another expression for q can be developed based on the definition of

$$q = \frac{\text{mg substrate converted/t}}{\text{mg solids present}}$$

from which we can write

$$q = \frac{(S_0 - S)}{[X]\, V}\, Q \qquad (12.17)$$

where $[X]$ = solids concentration in the aeration tank as MLVSS
V = volume of aeration tank
Q = flow into aeration tank

and since V/Q is the hydraulic detention time t in the aeration tank, it follows that

$$q = \frac{S_0 - S}{[X]\,t} \qquad (12.18)$$

Example 12.9

If an aeration tank with 4 hours detention and an MLVSS concentration of 2,000 mg/L reduces BOD_5 from 200 to 20 mg/L, what solids concentration $[X]$ would be necessary for 95% BOD_5 reduction? What is the specific substrate removal rate in each case?

Solution Assuming a first-order reaction, then from

$$k_q t\,[X] = \frac{S_0 - S}{S}$$

we obtain

$$k_q = \frac{200 - 20}{20} \times \frac{24}{4 \times 2,000} = 0.027 \, \frac{L}{mg \cdot d}$$

So from Equation 12.15

$$q = k_q S = 0.027 \times 20 = 0.54/d$$

or from Equation 12.18

$$q = \frac{S_0 - S}{[X]t} = \frac{180}{2,000} \times \frac{4}{24} = 0.54/d$$

To find the solids concentration $[X]$ for 95% BOD_5 reduction, we use

$$k_q t\,[X] = \frac{200 - 10}{10}$$

so that

$$[X] = \frac{19}{0.027} \times \frac{24}{4} = 4,222 \text{ mg/L}$$

and

$$q = k_q S = 0.027 \times 10 = 0.27/d$$

Process Kinetics: Specific Growth Rate μ. The rate dX/dt at which biological solids are produced is proportional to the rate dS/dt at which substrate is utilized, and we can write

$$\frac{dX}{dt} = Y \frac{dS}{dt}$$

or

$$\mu = Yq \qquad (12.19)$$

where μ = specific growth rate per day

$\quad\quad q$ = specific substrate removal rate per day

$\quad\quad Y = \dfrac{\text{mg of biological solids produced}}{\text{mg of substrate converted}}$

Example 12.10

For a specific growth rate of 1.0 per hour (base e), what mass of microorganisms will be present after one hour if 100 mg were present initially?

Solution   Assuming a first-order reaction, we can write

$$\frac{dX}{dt} = \mu X$$

where X is the total number of cells present. Solving for dX/X and integrating, we obtain

$$\int_{100}^{x} \frac{dX}{X} = \int_{0}^{t} \mu \, dt$$

so that

$$\ln x - \ln 100 = \mu t$$

$$x = e^{5.61} = 272 \text{ mg}$$

Comment

To make use of a specific growth rate, we must know or assume an order of reaction and we must know whether the exponential rate is based on the natural (base e), common (base 10), or other logarithm. Any of these could be used in plotting the logarithm of the quantity versus time. Each line would have the same constant slope, but the change in quantity with time would depend on what logarithm had been plotted to determine μ.

In this example, the specific growth rate μ was expressed in day^{-1} and the rate for this first-order reaction is actually an increase of 1 per hour in the value of the natural logarithm of the quantities involved. The fractional rate of increase is

$$\frac{X_2 - X_1}{X_1} = \frac{272 - 100}{100} = \frac{1.72 \text{ mg cells produced/d}}{\text{mg cells present}}$$

From pure culture studies, the specific growth rate μ for microorganisms has been found to follow a saturation-type reaction of the form

$$\mu = \mu_m \frac{S}{k_s + S} \tag{12.20}$$

where μ = specific growth rate of microorganisms per day

$\quad\quad \mu_m$ = constant = maximum growth rate of microorganisms per day when S does not limit growth

$\quad\quad S$ = concentration (mg/L) of growth-limiting substrate

$\quad\quad k_s$ = constant = S when $\mu = \mu_m/2$.

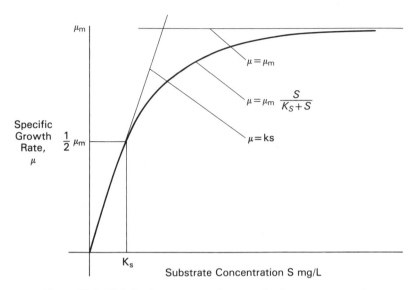

Specific Growth Rate, μ

μ_m

$\frac{1}{2}\mu_m$

K_s

$\mu = \mu_m$

$\mu = \mu_m \dfrac{S}{K_S + S}$

$\mu = ks$

Substrate Concentration S mg/L

Figure 12–8   Relation between growth rate and substrate concentration

This relationship, shown graphically in Figure 12–8, applies whether the substrate is at a limiting concentration or not, since

$$\text{when } S \gg k_s,\; \mu = \mu_m \text{ (a zero-order reaction)}$$

and

$$\text{when } S \ll k_s,\; \mu = k_s S \text{ (a first-order reaction)}$$

It should be evident that if the specific growth rate μ of the microorganisms follows a saturation reaction with respect to the substrate concentration, then, since $\mu = Yq$, the specific substrate removal rate q must be similarly related to the limiting substrate concentration, i.e.,

$$q = \frac{1}{Y}\mu_m \frac{S}{k_s + S} \tag{12.21}$$

Because of the low suspended solids concentration and the relatively high k_s in suspended growth systems, μ and q can be approximated by first-order rather than by saturation-type reactions. This approach will normally be satisfactory, but the simplifying assumption on which it is based should be kept in mind.

If, on the other hand, we do assume that the growth of microorganisms in a suspended growth system follows a saturation-type reaction, then application of the specific growth rate μ to process design requires information regarding the values of k_s, μ_m, and Y. Although these are usually unknown, they can be estimated from laboratory tests if the specific waste to be treated is available and if the results from the laboratory model represent full-scale performance. Taking account of these qualifications, this approach is more useful in evaluating an existing plant than in designing a new one. For a new plant, solids retention time and organic loading as described subsequently are the parameters normally used for process design.

Process Kinetics: Mean Cell Residence Time Θ_c.   The mean cell residence time Θ_c, also referred to as the solids retention time or sludge age, is the time in days that the biological solids remain in the system, i.e.,

$$\Theta_c = \frac{\text{Mass of solids in system}}{\text{Mass of cells produced/day}} \tag{12.22}$$

At steady-state operation, the mass of cells produced equals the mass of cells wasted. If we neglect the small amount of biological solids lost in the effluent, then the mass of cells wasted is $Q_w [X]_r$, where

Q_w = rate at which sludge (biological solids) is returned to the aeration tanks

$[X]_r$ = concentration of biological solids as SS or VSS in the return flow.

Also,

$$\Theta_c = \frac{X}{Q_w [X]_r} \tag{12.23}$$

where X = total biological solids as SS or VSS in the aeration tank

And since

$$\mu = \frac{\text{mass of cells produced/day}}{\text{mass of cells in system}} = \frac{Q_w [X]_r}{X}$$

it follows that

$$\Theta_c = \frac{1}{\mu} \tag{12.24}$$

Thus, from

$$\mu = Yq \tag{12.19}$$

we obtain

$$\Theta_c = \frac{1}{Yq} \tag{12.25}$$

Process Kinetics: Organic Loading Rate F/M.   At one time, aeration tanks were designed on the basis of hydraulic retention times ranging from less than 4 hours to over 24 hours and organic loadings (as BOD_5) of 0.5 to 1.6 kg BOD_5 per day per m³ of aeration capacity (30–100 lb/d per 1,000 ft³), depending upon the type of process selected. The weakness in these procedures was that the effect of the biological solids concentration in the aeration tanks was not considered. To remedy this deficiency, loading expressed as a food-to-microorganism (F/M) ratio was introduced. It is the ratio of BOD_5 applied per day

TABLE 12–8 CHARACTERISTICS OF SUSPENDED GROWTH SYSTEMS

Process	F/M [1] d^{-1}	MLVSS [2] mg/L	Det hr	Comments
Conventional Activated Sludge	0.3	2,000	6	Upset by shock loads good for "normal" wastes
Extended Aeration	0.1	4,000	24	Stable, simple, costly operation, good for small plants
Contact Stabilization [3]	0.4 $(a + b)$	(a) 2,000 (b) 6,000	(a) $\frac{1}{2}$ (b) 4	Stable, reliable system for wastes high in particulates
Oxygen-Activated Sludge	0.8	6,000	2	Complex, good odor control, for high-strength wastes

[1] $\dfrac{F}{M} = \dfrac{\text{kg BOD}_5 \text{ applied/d}}{\text{kg MLVSS in system}}$ and $\dfrac{\text{MLVSS}}{\text{MLSS}} = 0.7 \text{ to } 0.9$

[2] Variations of $\pm 50\%$ in the MLVSS are common.

[3] Contact stabilization: (a) = Contact tank
(b) = Stabilization tank

to the aeration tanks divided by the MLVSS under aeration. The MLVSS are assumed to represent the active mass of microorganisms present.

Summary. The major characteristics, including F/M ratio, for the preceding four suspended growth systems, are summarized in Table 12–8. Use of this information is illustrated in Examples 12.11 and 12.12.

Example 12.11

A conventional activated sludge plant treats 100,000 m³/d of municipal wastewater with SS and BOD of 225 and 200 mg/L, respectively. If the MLSS concentration in the 24,000-m³ capacity aeration tanks is 1,800 mg/L, is the plant organically overloaded? If so, how might the situation be rectified?

Solution The influent BOD_5 is

$$\frac{100 \times 10^6 \times 200}{10^6} = 20,000 \text{ kg/d}$$

Assume 30% BOD removal in the primary tank (Table 12–6). Then the rate of BOD_5 to aeration is 14,000 kg/d, and the total MLSS under aeration is

$$\frac{24,000 \times 10^3 \times 1,800}{10^6} = 43,200 \text{ kg}$$

Assume that the MLSS are 80% volatile (see footnote, Table 12–8). Then the total MLVSS under aeration is $0.8 \times 43,200 = 34,560$ kg, and

$$\frac{F}{M} = \frac{14,000}{34,560} = 0.4 \text{ kg BOD}_5/\text{kg MLVSS·d}$$

Comment

According to Table 12–8, the organic load on the aeration tanks is higher than desirable. This could be remedied by one of the following options, which are listed in order of increasing difficulty and cost:

- Increase the MLSS (to 2,300 mg/L) in the aeration tanks by returning more biological floc.
- Increase the BOD removal in the primary tanks by adding chemical precipitation (not covered in this text).
- Modify the plant to utilize higher rate processes such as contact stabilization.
- Expand the plant.

Example 12.12

A 1,000-m³/d activated sludge plant was designed for an F/M of 0.3 and a BOD_5 reduction from 150 to 15 mg/L through the aeration system. If the concentration of MLVSS of 2,000 mg/L can be increased to 1% in the final settling tank and is recycled to the aeration tanks at an average rate of 258 m³/d, determine:
(a) The aeration tank volume V and hydraulic detention time t
(b) The mass of activated sludge to be removed from the system each day
(c) μ, q, Y, and Θ.

Solution   As in most problems of this nature, the first step is to make a sketch showing all the relevant information, given and required:

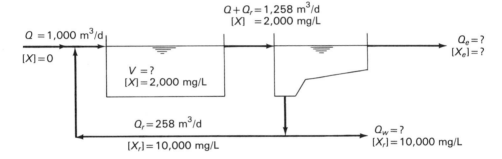

(a) From

$$F/M = 0.3 = \frac{150 \text{ mg/L} \times 1,000 \text{ m}^3/\text{d}}{2,000 \text{ mg/L} \times V \text{ m}^3}$$

we find that $V = 250$ m³, and since $t = V/Q$, it follows that

$$t = \frac{250 \text{ m}^3}{1,000 \text{ m}^3/\text{d}} \times \frac{24 \text{ h}}{\text{d}} = 6\text{h}$$

(b) From a mass balance on the microorganisms (solids) in and out of the aeration tank, we obtain

$$\text{IN} = Q_r [X_r] = 258 \frac{\text{m}^3}{\text{d}} \times \frac{10^3 \text{ L}}{\text{m}^3} \times 10,000 \frac{\text{mg}}{\text{L}} \times \frac{\text{kg}}{10^6 \text{ mg}} = 2,580 \frac{\text{kg}}{\text{d}}$$

$$\text{OUT} = (Q + Q_r) [X] + \mu [X]V \text{ (solids produced)}$$

$$= 1,258 \frac{\text{m}^3}{\text{d}} \times 2,000 \frac{\text{mg}}{\text{L}} \times \frac{10^3 \text{ L}}{\text{m}^3} \times \frac{\text{kg}}{10^6 \text{ mg}} + \mu [X] V$$

$$= 2,516 \text{ kg} + \mu [X] V$$

The mass of activated sludge to be removed from the system each day is

$$\mu \, [X] \, V = 2,580 - 2,516 = 64 \text{ kg/d}$$

(c) $\mu = \dfrac{\text{mass of volatile solids produced per day}}{\text{mass of volatile solids in system}}$

$$= \frac{64 \times 10^6 \text{ mg/d}}{2,000 \text{ mg/L} \times 250 \times 10^3 \text{ L}} = 0.128\text{/d}$$

$q = \dfrac{\text{organic substrate removed per day}}{\text{volatile solids in system}}$

$$= \frac{(150 - 15) \text{ mg/L} \times 1,000 \text{ m}^3\text{/d}}{2,000 \text{ mg/L} \times 250 \text{ m}^3} = 0.27\text{/d}$$

$Y = \text{yield}$

$$= \frac{\mu}{q} = \frac{0.128}{0.27} = 0.47 \, \frac{\text{g VSS produced}}{\text{g BOD}_s \text{ satisfied}}$$

$\Theta_c = \text{sludge age}$

$$= \frac{1}{\mu} = \frac{1}{0.128} = 7.8 \text{ days}$$

Comment

In this example, the values selected provide reasonable results. However, operating plants do not behave so predictably. Their performance is influenced by many parameters which we have not even considered. For example, the value of 0.47 calculated for yield represents the total mass of cells produced, but in fact, this quantity could be reduced by about 50% because of the decomposition of microbial cells with time.

Temperature effects, changes in waste strength, fluctuations in F/M, and variations in MLSS concentration are among other factors which make it difficult to model suspended growth systems. Nevertheless, the integrated kinetic chemical equations, modified to allow for organic substrate and biological solids, are useful tools. When combined with mass balances of materials, they provide a means for solving many problems in reactor evaluation and design.

12.7.2 Fixed Film Processes

In the fixed film, or attached growth, method of secondary treatment, wastewater is brought into contact with microorganisms attached to a solid medium such as rock, plastic, or sand. Trickling filters and rotating biological contactors, two of the more common processes, fall under this category.

Trickling filter.   Trickling filters, first used in 1893, do not actually filter wastewater. They act as ''contact beds'' where settled sewage is spread by a rotary distributor over 2 m (6 ft) deep, circular beds with diameters of up to 50 m containing 50 to 100 mm (2 to 4 in) diameter stones. In passing through the bed, the colloidal and soluble organics in the wastewater support the growth of microbial slimes on the medium. In a typical high-rate trickling filter plant, the flow pattern is similar to that of conventional activated sludge, and the system requires primary settling tanks, the trickling filter, final settling tanks, and a recirculation system to maintain a minimum flow through the filter in order to operate the distributor. Trickling filters are more resistant to shock from high organic and hydraulic loads and cheaper to operate than activated sludge plants, but they cost more to build, are more temperature sensitive, and provide 10 percent lower BOD removals. Deeper beds, up to 12 m (40 ft) deep, called **biological towers,** have been developed with high-void-ratio plastic media to reduce bed plugging and improve treatment efficiency.

Rotating biological contactor (RBC).   First conceived in 1900, the RBC concept became popular for small systems in the 1960s in Europe, when inexpensive polystyrene became available. The first commercial unit in the United States, installed in 1969, has been followed by hundreds of installations. Present models of large-diameter polyethylene media with about 40 percent of their surface area submerged in the waste rotate slowly on a horizontal shaft in a concrete tank. A biomass film 1–4 mm thick grows on the surface and shears off periodically into the tank below, to be removed by final sedimentation.

The process may have oxygen transfer limitations with high-strength wastes; however, for most wastes, this simple, stable system, which has the lowest power requirements (50–60 percent of those for activated sludge) of any competitive biological process, seems to offer advantages over conventional treatment methods. Like the trickling filter, it is temperature sensitive, but since the units are normally covered anyway to reduce shaft stress from wind, the added cost of insulating the housings may not be significant. A unique combination of RBCs and the activated sludge process is used in Philadelphia's 800 × 10^6-L/d (210-mgd) North East Water Pollution Control Plant, where the RBCs are mounted in the aeration tanks and rotated by the bubbles from the aeration system.

Comparative parameters for the three fixed, or attached, film processes just described are shown in Table 12–9, and the information is used in Example 12.13.

TABLE 12–9   COMPARATIVE PARAMETERS FOR FIXED FILM PROCESSES

Process	Depth or Det[1]	Loading BOD[2] g/m³ · d	Loading Hyd[3] m/d	Comments
High-Rate Trickling Filter (Rock)	2 m	560	20	Simple stable process with low efficiency
Biological Tower (Plastic)	6 m	1,600	70	Higher void ratio (92% vs. 42%) gives less plugging
RBC at 13°C (Polyethylene)	3 hr	NA	0.1 (no recirc)	Reliable, stable process with low energy requirements

Adapted in part from Clark et al., 1977; Metcalf and Eddy, 1979.

[1] Det: Detention time.

[2] BOD: BOD_5 loading applied (g/m³d), not including recirculation.

[3] Hyd: Hydraulic loading (m³/m²d), with 100% recirculation.

Note: 1 m³/m²d = m/d = 24.57 USG/ft²/d; 1 kg/m³d = 62.4 lb/1,000 ft³ d.

Example 12.13

A town with a sewer system and a population of 12,000 is planning to construct two high-rate trickling filters, each 20 m in diameter and 2 m in depth, to treat its wastewater. Will these be large enough to satisfy normal loading parameters?

Solution Assume that the per capita wastewater flow is 400 Lpcd (Table 12–4). Then the average municipal wastewater flow is

$$\frac{12,000 \times 400}{10^3} = 4,800 \text{ m}^3/\text{d}$$

So the flow to each filter is 4,800 m³/d, including 100% recirculation. Assume that the influent BOD$_5$ is 190 mg/L (Table 12–4). Then the influent BOD$_5$ loading to the plant is

$$\frac{4,800 \times 10^3 \times 190}{10^6} = 912 \text{ kg/d}$$

Assume 30% BOD removal in the primaries (Table 12–6). Then the BOD loading to each filter is $\frac{1}{2} \times 0.70 \times 912 = 319$ kg/d.

(a) *Filter Size Based on BOD Loading*

Assume that the design BOD loading is 0.560 kg/m³d (Table 12–9). Then we have

$$\text{Volume of filter} = \text{area} \times \text{depth} = \frac{\pi d^2}{4} \times 2 \text{ m}$$

and

$$\text{Unit BOD loading} = \frac{319 \text{ kg/d}}{\left(\dfrac{\pi d^2}{4} \times 2\right) \text{m}^3} = 0.560 \text{ kg/m}^3\text{d}$$

from which it follows that $d = 19$ m.

(b) *Filter Size Based on Hydraulic Loading*

Assume a design hydraulic load of 20 m³/m²d (Table 12–9). Then we have

$$\text{Area of filter} = \frac{\pi d^2}{4}$$

$$\text{Unit hydraulic loading} = \frac{4,800 \text{ m}^3/\text{d}}{\left(\dfrac{\pi d^2}{4}\right) \text{m}^2} = 20 \text{ m}^3/\text{m}^2\text{d}$$

from which it follows that $d = 17.5$ m.

Comment

Thus, the two 20-m-diameter filters should be large enough for normal BOD and hydraulic loadings. In practice we should also consider what happens when loadings higher than normal occur or when one filter is out of service. Oversizing of the two filters, providing a third or allowing a temporary overload are possibilities.

12.7.3 Sludge Processing

Sludge processing, the most neglected area of wastewater treatment, deals with less than 1 percent of the total waste volume, but accounts for 30 to 40 percent of the capital and operating costs. Raw sludge from the primary tanks, together with biological sludge from the final tanks, must be concentrated and stabilized before disposal on the land. Aerobic or anaerobic digestion is the usual method for accomplishing this, and it may be sufficient, with no further dewatering or oxidation required before ultimate disposal.

The quantities of sludge to be processed vary with the type of wastewater process employed. The volume of sludge produced by gravity settling can be determined if we know the percent removal of SS and the concentration of the settled material.

The quantities produced by aerobic biological processes depend on the BOD loading. For fixed film systems, such as trickling filters, excess biological sludge production is usually between 0.2 to 0.5 g VSS/g BOD_5 applied, with the lower value for light BOD loadings and the larger applicable to higher ones. With suspended growth processes like activated sludge, the excess sludge produced varies with the *F/M* ratio employed. The net VSS yield ranges from about 0.2 g VSS/g BOD_5 applied in the case of extended aeration plants (low *F/M*) to approximately 0.4 for activated sludge plants operating with a high *F/M* ratio (Clark et al., 1977).

In anaerobic digestion, microbial growth is limited by the small amount of suitable substrate available and is about one-tenth of that for aerobic processes.

Anaerobic digestion.   Escalating energy costs and the need for energy conservation have discouraged the use of aerobic digestion for plants processing over 4,000 m^3/day and have renewed interest in the anaerobic process. As mentioned in Section 12.5.2, anaerobic digestion occurs in two phases. In the first stage, hydrolysis (liquefaction) of the organics and their biological conversion to organic acids proceeds rapidly. In the second stage, slow-growing, environmentally sensitive methane bacteria utilize the organic acids to produce gas that is about two-thirds methane and one-third CO_2, with traces of H_2S. Any shock to the methane bacteria from excess acid, oxygen, or toxic substances, or from extremes in pH (7 $\pm$ 0.3) or temperature (33°C $\pm$ 2°), upset the process, causing it to fail and turn acid or ''sour.''

With a single digestion tank, mixing to promote digestion, settling for sludge separation and thickening, and storage for digested sludge must all take place together. This inefficient arrangement, called **conventional or single-stage digestion,** limits digester loadings (based on VSS) to about 0.5 kg/m^3 (30 lb/1,000 ft^3) per day. By separating the mixing from the thickening and storage functions, loadings can be 1.6 kg VSS/m^3 (100 lb/1,000 ft^3) of primary digester capacity per day, and the term high-rate digestion is then applied. The reduction in VSS would be about 50 percent in both cases.

Land application of sludge   The application of digested sludges to agricultural lands is becoming more restricted as we learn more about the environmental consequences. The contemporary approach is to consider sludge utilization rather than sludge disposal. The nitrogen assimilation capacity of the soil and cover crop is frequently the basis for determining the amount of digested sludge that may be applied to the land. However, limitations are also set for various heavy metals, with cadmium being one of particular concern where leafy vegetables are to be grown. Application rates allowed by Ontario's Ministry of the Environment are based ostensibly on the available nitrogen content

(NH_4 + NO_3). However, the sludge is rejected as unsuitable for land disposal if its nitrogen-to-metal ratio is less than stipulated values.

The simplest method of collecting and applying the sludge is by means of tank trucks (4 to 8 m^3 capacity) equipped with a piping manifold to spread the liquid sludge on the agricultural land. Rates of application vary widely with the nitrogen and heavy metal content of the waste and with the characteristics of the site (climate, soil, cover crop). Typical application rates for digested sludges from municipal activated sludge plants are 2 to 8 m^3/ha·d (Clark et al., 1977). Simplified calculations for estimating the approximate quantities of sludge resulting from the treatment of municipal wastewater are outlined in Example 12.14.

Example 12.14

A conventional activated sludge plant treating 2,000 m^3/d of municipal wastewater disposes of its anaerobically digested sludge on relatively impervious farm land. (a) How many tank trucks (3.8 m^3 capacity) are required each week to haul the liquid digested sludge? (b) What land area (excluding buffer zones) is required for disposal of this liquid sludge?

Solution

ASSUMPTIONS

1. Raw sewage   SS = 225 mg/L (70% volatile, Table 12–6)
   (Table 12–4)   BOD = 190 mg/L
                   (Excess activated sludge returned to primary)

2. Primary settling   SS = 50% removal
   (Table 12–6)   BOD = 30% removal

3. Excess activated sludge   = 0.4 g VSS produced per g BOD_5 applied
   (Section 12.7.3)   (80% volatile; see Table 12–6, "Final Settling,"
                   or footnote, Table 12–8)

4. Anaerobic digester   VSS reduced 50%
   (Table 12–6)   Digested sludge concentration = 6%
                   Sludge S.G. = 1.0

5. Field application   2 m^3/ha · d (range 2 to 8 m^3/ha · d based on the
   (Section 12.7.3)   annual amount of nitrogen allowed on the soil).

CALCULATIONS

1. Primary Sludge = $\dfrac{0.5 \times 2,000 \times 225}{10^8}$ = 225 kg/d SS

                              = 225 × 0.70 = 160 kg/d VSS

2. Excess activated sludge = $\dfrac{0.4 \times 0.7 \times 2,000 \times 190}{10^3}$ = 106 kg/d VSS

3. Total VSS to be anaerobically digested = 160 + 106 = 266 kg/d

4. VSS remaining after digestion (50%) = 133 kg/d

5. Fixed solids remaining after digestion

- From primary sludge $= 0.3 \times 225 = 68$ kg/d
- From excess activated sludge $= 0.2 \times 106 = 21$ kg/d

6. Total solids remaining after digestion $= 133 + 68 + 21 = 222$ kg/d

7. Volume of digested sludge $= \dfrac{222}{10^3} \times \dfrac{100}{6} = 3.8$ m³/d

8. Tank trucks per week $= \dfrac{3.7 \times 7}{3.8} = 6.82$, or 7 tank trucks per week

9. Area required $= \dfrac{3.7 \text{ m}^3/\text{d}}{2 \text{ m}^3/\text{ha} \cdot \text{d}} = 1.9$ ha (4.6 ac)

Example 12.15

For the plant described in Example 12.14, determine the sizes of
(a) Two rectangular primary tanks with length-to-width ratios of 5:1 and depths of 3 m.
(b) Two final settling tanks of similar proportions, but 3.5 m deep and handling a peak flow twice the plant DWF (excluding recirculation).
(c) A single-stage, circular, heated anaerobic digester with a 7-m average water depth. Assume that digested sludge can be removed on a regular weekly basis all year and that no extra sludge storage capacity need be provided.

Solution

$$Q_{\text{average}} = 2,000 \text{ m}^3/\text{d}$$

$$\text{Population (based on 400 Lpcd)} = \frac{2,000,000}{400} = 5,000 \text{ people}$$

$$\frac{\text{Peak flow}}{\text{Average flow}} = 1 + \frac{14}{4 + \sqrt{P}} = 3.25$$

$$Q_{\text{max}} = 2,000 \times 3.25 = 6,500 \text{ m}^3/\text{d}$$

(a) *Primary Tanks*

For 1-hour detention (Table 12–6), the volume is $6,500 \times 1/24 = 270$ m³, and the area is $270/3 = 90$ m². For an overflow rate ≤ 60 m³/m²·d (Table 12–6), we have

$$60 = \frac{6500}{A}$$

so that the area is 108 m². (The larger area governs.)
For two rectangular tanks with $l/w = 5{:}1$,

$$\text{Area} = \frac{108}{2} = lw = 5w^2$$

and

$$w^2 = 10.8$$

Therefore, the width is 3.3 m, and the length 16.4 m.

(b) *Final Tanks*

For $Q_{\text{average}} = 2{,}000$ m³/d,

$$Q_{\text{peak}} = 2{,}000 \times 2.0 = 4{,}000 \text{ m}^3/\text{d}$$

Based on a detention time of 2 hr (Table 12–6), $V = 333$ m³.

For a depth of 3.5 m, $A = 95.1$ m²; also, based on $Q/A \leq 40$ m³/m² · d (Table 12–6), $A = 100$ m². Thus, the latter governs, and we select two tanks 3.2 m × 16 m × 3.5 m deep.

(c) *Digester*

The total VSS to be anaerobically digested is 266 kg/day (from Example 12.10). The volume, based on a loading of 0.5 kg VSS/m³·d (Section 12.7.3), is

$$\frac{266}{0.5} = 532 \text{ m}^2 = \frac{\pi d^2}{4} \times 7$$

so the diameter is

$$\frac{532 \times 4}{\pi \times 7} = 9.8 \text{ m}$$

Sludge dewatering and incineration. With the shift in emphasis from land disposal to land utilization of sludge, many municipalities currently using land disposal are being forced into on-site incineration of those sludges that cannot meet low metal content requirements. For incineration to be practical, costs for supplementary fuel must be minimized.

In the United States, the trend to incineration, which had grown rapidly during the 1970s, ended with rising fuel prices. In fact, some states have shut down sludge incinerators to conserve fuel. To incinerate sludge that is too wet to burn alone, without supplementary fuel, is costly. The problem arises because a cake with over 30 percent solids is necessary for self-sustaining (autogenous) burning, and vacuum filters and centrifuges, the two most widely used dewatering devices, can only dewater sludge to 15–25 percent solids. The obvious answer to the dilemma when land disposal is not possible and self-sustaining incineration is necessary is either to add combustible material to the sludge, or to use more efficient dewatering methods.

Fortunately, more efficient dewatering devices are now available from more than a dozen manufacturers and suppliers of belt pressure filters and plate and frame pressure filters. Belt filters compress the sludge between moving porous belts, forcing out the water. Plate and frame filters use extremely high pressure to force the water through a fixed porous medium. Both methods are more efficient than the traditional vacuum filters or centrifuges. In the United States, belt filters are rapidly replacing vacuum filters or are being installed to further dewater vacuum filter cake. Energy requirements for the units are one-sixth those for vacuum filters. Studies have shown that cake solids from belt filters range from 25 to 30 percent, which is 5–12 percent higher than those from vacuum filters operating at the same plant on similar sludge (Villiers and Farrell, 1977).

Example 12.16

If the concentration of digested sludge solids could be increased from 6% to 12% by thickening, what percent reduction in volume would be accomplished? Assume that the digested sludge contains 60% fixed solids and 40% volatile solids (on a dry weight basis), having specific gravities of 2.4 and 1.0, respectively.

Solution   Consider a weight of 100 kg of wet sludge before thickening. We can set up a mass balance between the quantity of sludge and its components at each stage, keeping in mind that the amount of solids remains constant and only the water content is reduced. We can write

$$\text{Sludge} = \text{solids} + \text{liquid}$$

before and after thickening. Calculations for the before stage are as follows:

Before Thickening

Quantity	Sludge	Solids		Liquid
By Weight		6 kg (6%)		
		Fixed 60%	Volatile 40%	
	100 kg	3.6 kg	2.4 kg	94 kg
By Volume $\left(\dfrac{\text{Weight}}{\text{SG}}\right)$	$\dfrac{100}{\text{SG}}$ L	$\dfrac{3.6}{2.4}$ L	$\dfrac{2.4}{1.0}$ L	$\dfrac{94}{1.0}$ L

From the table, the volume of the unthickened wet sludge is

$$\frac{100}{\text{SG}} = \frac{3.6}{2.4} + \frac{2.4}{1.0} + \frac{94}{1.0} = 98.0 \text{ L}$$

and the specific gravity (SG) is

$$\frac{100}{98} = 1.02$$

After thickening, the quantity of solids in the sludge remains unchanged at 6 kg, but now represents 12% of the sludge by weight. The liquid therefore accounts for the remaining 88%, or

$$\frac{88}{12} \times 6 = 44 \text{ kg}$$

and the sludge (= solids + liquid) is

$$6 \text{ kg} + 44 \text{ kg} = 50 \text{ kg}$$

Calculations of the after stage are then:

After Thickening

Quantity	Sludge	Solids		Liquid
By Weight		6 kg (12%)		
		Fixed 60%	Volatile 40%	
	6 + 44 = 50 kg	3.6 kg	2.4 kg	$\dfrac{88}{12} \times 6 = 44$ kg
By Volume $\left(\dfrac{\text{Weight}}{\text{SG}}\right)$	$\dfrac{50}{\text{SG}}$ L	$\dfrac{3.6}{2.4}$ L	$\dfrac{2.4}{1.0}$ L	$\dfrac{44}{1.0}$ L

From the table, the volume of the thickened sludge is

$$\frac{50}{\text{SG}} = \frac{3.6}{2.4} + \frac{2.4}{1.0} + \frac{44}{1.0} = 47.9 \text{ L}$$

and the specific gravity (SG) is

$$\frac{50}{47.9} = 1.04$$

Therefore, the reduction in volume is

$$\frac{98.0 - 47.9}{98} \times 100 = 51\%$$

Comment

The specific gravity of the sludge could be assumed to be 1.0 before and after thickening, in which case the reduction in sludge volume would be

$$\frac{12 - 6}{12} \times 100 = 50\%$$

which is accurate enough for most sludge volume calculations.

12.7.4 Odor Problems

Virtually every process treating domestic and industrial wastes will exude unpleasant odors at some time, and even well-operated plants cannot escape the problem completely. In most cases the odors are confined to the immediate vicinity of the process units, and operators need be concerned only with those odors escaping beyond the plant site. Often, this is a function of the topography and the prevailing winds.

Probably the most common cause of odors is the development of anaerobic conditions. With the exception of the anaerobic digestion process and the anaerobic zones required for denitrification, anaerobic conditions should be avoided. Sludge deposits and slime growths, rich in anaerobic sulfate-reducing bacteria, are the major cause of H_2S production in sewers.

Similar deposits, if permitted to occur, will create odors in any tank, including aeration tanks, where anaerobic conditions may result because of poor mixing. The long detention time of interceptor sewers, especially those at flat grades, allows anaerobic degradation of organic matter to begin before the waste reaches the treatment plant; hence the typical smell of "septic" wastewater.

Local authorities are under increasing pressure to reduce or eliminate odors from their treatment plants. The solutions are seldom obvious. They may range from simple operating and process changes or chemical additions to expensive odor control devices, such as air scrubbers and adsorbers. Each situation is unique, and in many cases the best solution cannot be determined until after the problem arises.

12.8 ON-SITE TREATMENT FACILITIES

At one time, before sewer systems were in wide use, treatment and disposal of wastewater at its source was necessary. Even today for many small communities, on-site systems may be preferable to a municipal facility from the standpoint of overall cost and pollution control. Currently, about 25 percent of all housing units in the United States dispose of their wastewater using on-site wastewater treatment and disposal systems (U.S. EPA, 1980). This percentage has remained relatively constant for the past 30 years, despite extensive sewer construction in the 1960s and 1970s. Resorts, service centers, camps, and isolated institutions frequently have no access to municipal sewers and must rely on small on-site installations. Options for small systems range from pit privies and toilets not requiring water through septic tanks to package units. The features of these options are described in the rest of this section. Much of the information has been summarized from a report by Ross et al. (1980).

12.8.1 Waterless Systems

Where no pressurized water is available or soil conditions are unsuitable for effluent disposal, the choices for on-site treatment are limited to privies or waterless toilets. Provided that surface or groundwater contamination is not a problem, then a privy, which is an outhouse over an earthen pit, is the simplest and cheapest solution. When the pit is full, the privy is relocated. If contamination of the water supply is a potential problem, then impervious pits are used and the collected wastes pumped out on a regular basis. Both types of privies are widely used for unserviced campgrounds, parks, and recreation areas.

Various types of waterless toilets, namely, compost, chemical, and incinerator units, are a less primitive approach than privies for small installations and single-family use. Compost toilets are a recent development in which wastes, dried to a critical moisture content by evaporation, are composted aerobically at a controlled temperature and forced ventilation. Chemical toilets stabilize organic matter and destroy bacteria with strong caustic solutions which must be removed and replaced on a regular basis. Incinerator toilets use electricity, oil, or gas to completely oxidize organics to a dry inert ash that must be removed

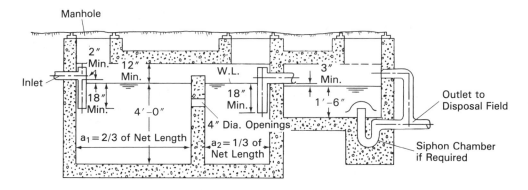

Figure 12-9 Two-compartment septic tank with dosing chamber. Source: Septic Tank Systems, Ontario Dept. of Health, 1965.

periodically. Several patented versions of these three types of unit are available commercially for small installations.

12.8.2 Septic Tanks

A septic tank system, the most common method of on-site wastewater treatment and disposal, can be designed for a small family or a large institution. The concrete, fiberglass, or coated steel tank (Figure 12-9) is usually located below, but is accessible from, ground level. It provides at least 24-hour retention of the wastes in one or two compartments, thus allowing heavy solids (sludge) to settle to the bottom of the tank and oil and grease (scum) to rise to the surface. These solids, which undergo only slight compaction and anaerobic decomposition, accumulate and should be removed every three to five years to maintain tank storage capacity and effluent quality. In most municipalities, plans and design criteria, which the homeowner must follow, are provided by the Medical Officer of Health.

The liquid effluent, with a BOD and SS concentration of about 200 mg/L, is then discharged, preferably through an intermittent dosing siphon, to a subsurface disposal system. This system may consist of open tile drains laid in trenches filled with gravel or crushed stone that permit downward percolation and upward evaporation and transpiration.

The length of trench required (0.5 m wide) varies from about 10 to 50 m/m³ of waste applied per day for soils of low to high permeability. For example, for disposal of 2,500 L/d of septic tank effluent in moderately permeable soil, a total trench length of 50 m would be necessary based on 20 m of trench/m³ of waste per day. The length of each trench should not exceed 30 to 35 m.

Leaching cesspools (seepage pits) are an alternative type of subsurface disposal, but are less satisfactory than tile beds. In this system, the septic tank effluent leaches through the open-jointed walls of the pit into the surrounding pervious soil. The cellpool tends to be anaerobic and therefore odorous, and is subject to plugging. In rock or clay, or where permeable soil is not available, tile fields or cesspools should not be used for effluent disposal.

12.8.3 Package Plants

Where treatment by septic tank would be inadequate, aerobic package plants can produce an effluent with a BOD and SS concentration 30 to 50 percent of that from a septic tank. Units from several manufacturers are available in sizes ranging from those for single families to larger systems suitable for a small community. Ensuring proper maintenance of package plants for private use is difficult, and authorities may require the owner to have an ongoing maintenance contract with an approved contractor.

The simplest package unit is a single-compartment tank in which the sewage is aerated for a fixed period and then settled, before the supernatant is pumped to subsurface disposal. More elaborate units that provide a higher degree of treatment may contain separate aeration and settling compartments and the means for sludge recycling. Like septic tanks, package plants require periodic removal of sludge and effluent discharge to a subsurface disposal system. A small sand filter may be located between the package plant and the disposal bed in order to upgrade effluent quality and protect the bed from plugging.

TABLE 12–10   FEATURES OF ON-SITE WASTEWATER TREATMENT AND DISPOSAL SYSTEMS

System	Application	Need W[1]	Need P[2]	Advantages	Disadvantages
WATERLESS SYSTEMS [used where effluent disposal is undesirable]					
Privy	Campgrounds, recreation areas	No	No	Low cost, little maintenance	May endanger water supply (unless wastes are collected)
Compost Toilet	Single-family (SF) residences	No	Yes	Avoids groundwater contamination	Upset by shock loads (moisture content of waste is critical)
Chemical Toilet	Single- or multiple-family residences	No	No	Avoids groundwater contamination	Chemicals are costly, wastes must be hauled
Incinerator Toilet	Single-family residences	No	Yes	Residue is inert ash	Odors, high power costs
EFFLUENT-PRODUCING SYSTEMS [used where effluent disposal to subsurface soil systems (or receiving streams) is possible]					
Septic Tank	SF residence to institution	Yes	No	Simple, reliable system	Solids removal is necessary
Package Plant	SF residence to institution	Yes	Yes	High-quality effluent	Solids removal is necessary, maintenance is costly

[1] W: Water supply (pressurized) is needed
[2] P: Power supply is needed.
Adapted in part from Ross et al., 1980

For large installations, package plants may become uneconomical, and the various processes employed in municipal waste treatment should be examined. Oxidation ponds, extended aeration, rotating biological contractors, and other methods are adaptable to small installations.

The features of on-site wastewater treatment and disposal systems are summarized in Table 12–10.

12.9 GOVERNMENT/PUBLIC ROLE IN POLLUTION CONTROL

Despite the convenience and the increased property values resulting from municipal sewerage systems, their implementation is often opposed by the taxpayers. Even when the benefits, or the needs, are obvious, authorities must generally provide some legal or economic inducement to get the project under way and to control it once it is in operation. Incentive payments, direct regulation, and municipal bylaws are some of the tools available for improving environmental quality. Public awareness of environmental problems can be another effective means for stimulating corrective measures. The following is a brief description of the means available to municipalities for pollution control.

12.9.1 Government Subsidies

Municipalities have been the recipients of government subsidies in the form of low-cost, long-term loans and outright grants. In Canada, the Canada Mortgage and Housing Corporation (CMHC) supplied funds during the sixties and seventies, covering two-thirds of the cost of pollution control projects. Of this amount, 25 percent was forgiven, with the balance repaid over 25 years at a low fixed interest rate. In the United States, federal loans covering up to 75 percent of the cost of wastewater treatment plants have been available to municipalities since 1973. Repayment for these projects may be from general taxation or by revenue bonds repaid from sewer rental fees based on each user's water bill.

12.9.2 Direct Regulation

Many regulations to control water pollution have been enacted. In Ontario, for example, there are 25 federal statutes, 40 provincial statutes, and countless bylaws to control environmental pollution (Estrin and Swaigen, 1974). These regulations are in addition to the federal, state, and provincial requirements for the acceptability of wastewater effluents or the quality of receiving waters. Direct regulation, enforced by fines and penalties, has been the predominant practice for the past 20 years or more. However, control by regulation has limitations: it is politically unpalatable, fines are usually a fixed amount that is too low to encourage compliance, and in some cases the rigid standards are unwarranted. Fines proportional to the degree of violation provide a better approach, but unless fines are greater than the cost of correcting the situation, little improvement can be expected.

12.9.3 Municipal Bylaws

At the municipal level, considerable control over water pollution is exercised through bylaws which control subdivision development, sewer use, and industrial waste discharges. These bylaws may stipulate that developers provide a certain level of sewage collection and storm drainage services in new subdivisions. The capital costs of these services are borne by those who purchase the lots, the operating and maintenance costs by the municipality. Sewer-use bylaws and industrial waste bylaws establish the types of wastes that are accepted by the municipality and may include a system of surcharges which are assessed against those whose wastes are stronger than normal domestic sewage. Certain industries may have to pretreat their wastes in order to make them acceptable to the municipality.

12.9.4 Public Involvement

Public awareness of the contamination of our water supplies, the consequences of eutrophication, and the loss of recreational waters have contributed significantly to the progress in wastewater technology. As measures to control water pollution from municipal systems have been implemented, the emphasis has been shifting from this relatively narrow concern to the broader implications associated with the treatment and disposal of solid and hazardous wastes. These are considered in Chapters 14 and 15.

PROBLEMS

12.1. If the BOD test follows a first-order reaction in which the rate of removal is proportional to the organic matter remaining, develop the expression

$$L_0 - L = L_0(1 - 10^{-kt})$$

for the carbonaceous portion of the BOD curve, where

$$L_0 - L = \text{BOD satisfied (mg/L) at time } t \text{ (days)}$$
$$L_0 = \text{Ultimate carbonaceous BOD}$$
$$k = \text{a rate constant (base 10) per day}$$

Sketch the curve of BOD satisfied vs. time, showing the effect if a substance toxic to the microbial population were added to the mixture on the third day after starting the BOD test.

12.2. If 8,000 m^3/d of wastewater from an industry has a BOD_5 of 190 mg/L and $k = 0.17$ per day (base 10),
   (a) How much oxygen (kg/day) is required to satisfy the BOD demand of this waste, assuming that 1 kg of oxygen is to be supplied per kg of ultimate BOD in the waste?
   (b) What is the population equivalent of the waste (based on BOD_5)?

12.3. In a BOD test, 5 ml of wastewater (DO = 0.0) is mixed with 295 ml of diluting water (DO = 9 mg/L). After 5 days' incubation of the mixture, DO = 5.2 mg/L. What is the BOD_5 of the wastewater?

12.4. A 40-ha residential subdivision with a density of 10 houses per ha has separate storm and sanitary sewers. Unfortunately, downspouts from the roofs (140 m² average area) of 5% of these houses have been inadvertently connected with the sanitary system. If the sanitary sewer outlet from the subdivision has 20% spare capacity at peak sanitary flow, is surcharging of this sewer likely to occur when a rainfall with an intensity of 60 mm/hr occurs?

12.5. For the municipality described in Example 12.2, estimate the number of kilograms of copper that would be discharged annually by a stormwater outlet serving the town.

12.6. A small municipality with only separate sanitary sewers has a population of 5,000 and a poultry plant (with no manure or blood recovery) which processes 3,000 chickens per hour during two 8-hour shifts per day. Sink disposal units are allowed and estimated to be in 10% of the homes. Estimate the volume of municipal wastewater and its BOD and SS, as well as its total nitrogen content.

12.7. A 5,000-m³/day wastewater treatment plant discharges to a river with a minimum drought flow of 137.7 m³/min (immediately upstream of the outfall).
   (a) What total allowable BOD_5 (kg/day) could be discharged by the plant if the BOD_5 in the river after initial dilution must not exceed 1 mg/L and no upstream pollution is occurring?
   (b) What degree of treatment (% BOD satisfied) does this represent, assuming the plant influent to have a BOD_5 of 250 mg/L?

12.8. During recent winters there has been considerable criticism of the practice of dumping salt-laden snow from the streets into rivers or lakes for disposal. What problems are caused by this? Should the practice be prohibited? What alternatives are available? How would these resolve the problems?

12.9. Investigate the policy in your municipality regarding (a) the disposal of water from foundation drains, and (b) the use of residential sink disposal units. Discuss the actual (or potential) problems created by this policy.

12.10. What size of rectangular sedimentation tank, not over 3.7 m deep and with a length-to-width ratio of about 5 to 1, would be required to provide at least 4 hours detention and an overflow rate less than 30 m³/m² day of surface area for a flowrate of 4,250 m³/day? (Assume an ideal tank.)

12.11. If the cannery in Example 12.2 were to treat one-third of its own waste, how many kg of nitrogen and phosphorus salts as NH_4Cl (32% N by weight) and KH_2PO_4 (23% P by weight), respectively, would have to be added each day to the cannery waste in order to achieve a BOD/N/P ratio of 100/5/1?

12.12. A lake 75 ha by 30 m deep is fed by a river with an annual mean flow of 0.2 m³/s. It is proposed that a metal refinery be established on the lakeshore near the river inlet. The refinery would operate for 30 years, drawing 4,000 m³/day of process water from the river and discharging an equal amount of wastewater into the lake. The company guarantees to limit the concentration of Pb in its waste to 0.5 mg/L, 70% of which will be insoluble (will settle). The community draws its water supply from the opposite (outlet) end of the lake. Assuming that the waste does not short-circuit and stays in the lake for its theoretical detention time (plug flow), when will steady-state conditions be reached? Show the results on a graph. As the city engineer, advise the council as to whether the proposed refinery will impair the water supply to unacceptable levels (0.05 mg/L Pb) or cause other problems. State any assumptions.

12.13. As an alternative to the high-rate trickling filters selected in Example 12.13, two biological towers have been proposed. Based on the design parameters in Table 12–9, what size would be suitable? If construction costs per unit volume were the same for the towers as for the filters, which system would be preferable? Why? In order to provide for loadings above normal, what options might be considered? What factors would influence the choice?

12.14. A primary treatment plant for 5,000 people is to have two circular primary clarifiers and a single circular, heated anaerobic digester with sludge disposal by tank truck.

(a) Using typical design parameters, determine suitable diameters for primary tanks 3 m deep and a digester with a side water depth of 6 m. Allow for capacity in the digester for the 4 months of the year when removal of digested sludge is not possible.

(b) If the digester reduces volatile suspended solids to 45%, how many cubic meters of digested sludge would have to be hauled away each month during the hauling season?

(c) What area (ha) should be provided if the sludge is to be utilized as a soil conditioner on agricultural land and if a 1,000-m buffer zone is required?

12.15. A 4,000-m^3/day conventional activated sludge plant is needed for a small municipality. Determine the tank volumes required for:

(a) Two rectangular primary tanks 3.5 m deep and having an overflow rate not exceeding 60 m^3/m^2d and a detention time of at least 1 hour (at peak flow).

(b) Two aeration tanks operating at an F/M of 0.25 kg BOD_5/kg MLVSS per day and an MLSS concentration of 2,000 mg/L.

12.16. Replacement of vacuum filters by belt filters in a wastewater treatment plant results in an increase in the sludge solids concentration from 15 to 25% on a dry weight basis. Assuming the sludge (again, on a dry weight basis) contains 20% fixed solids and 80% volatile solids, having specific gravities of 2.5 and 1.0, respectively, what reduction in sludge volume will be achieved by the change in the method of filtration?

12.17. Suppose secondary treatment is added to the plant in Problem 12.14. Determine:

(a) The capacity of two aeration tanks if detention must be at least 4 hours, F/M must not exceed 0.3, and an MLSS concentration of 1,800 mg/L will be maintained.

(b) The diameter of two circular final tanks 4 m deep and handling a peak flow of 1.75 times DWF (excluding recirculation).

(c) The loading (kg $VSS/m^3 \cdot d$) on the original digester if sludge storage need only be provided for 3 months instead of 4.

REFERENCES

American Public Works Association. *Municipal Refuse Disposal.* 3d ed. Chicago: Public Administration Service, 1970.

ARDEN, E., and LOCKETT, W. T. "Experiments on the Oxidation of Sewage Without the Aid of Filters." *Journal Society of the Chemical Industry.* 33 (1914): 1122.

CLARK, J. W., VIESSMAN, W., JR., and HAMMER, M. J. *Water Supply and Pollution Control.* 3d ed. New York: Dun-Donnelley, 1977.

DESILVA, P. F. "Operational Experience with Zimpro Process." *Proceedings, Seminar on Current Approaches in Wastewater Treatment.* Toronto: Pollution Control Association of Ontario and Ontario Ministry of the Environment, 1978.

Environment Canada. *Wastewater Disinfection in Canada.* Ottawa: Water Pollution Control Directorate, Environmental Protection Service Report, EPS 3-WP-78-4, May 1978.

Estrin, E., and Swaigen, J. *Environment on Trial: A Citizen's Guide to Ontario Environment Law.* Toronto: Canadian Environmental Law Association, 1974.

Metcalf and Eddy, Inc. *Wastewater Engineering: Treatment, Disposal, Reuse.* 2d ed. New York: McGraw Hill, 1979.

Ontario Ministry of the Environment. *Water Management: Goals, Policies, Objectives, and Implementation Procedures of the Ministry of the Environment.* Toronto: Ontario Ministry of the Environment, 1978.

Parker, D. S., and Merrill, M. S. "Oxygen and Air Activated Sludge: Another View." *Journal of the Water Pollution Control Federation* 48 (1976): 2511.

Polls, I., and Lanyon, R. "Pollutant Concentrations from Homogenous Land Uses." *Proceedings, American Society of Civil Engineers* 106 (1980): 69.

Prasad, D., Henry, J. G., and Kovacko, R. "Pollution Potential of Autumn Leaves in Urban Runoff," *Proceedings, International Symposium on Urban Storm Runoff.* Lexington: University of Kentucky, 1980, p. 197.

Ross, S. A., Guo, P. H. M., and Jank, B. E. *Design and Selection of Small Wastewater Treatment Systems.* Ottawa: Economic and Technical Review Report, EPS-3-WP-80-3, Environmental Protection Service, Environment Canada, March, 1980.

Stowell, R., Colt, J., and Tchobanoglous, G. *Toward the Rational Design of Aquatic Treatment Systems.* Paper presented at the National Conference on Environmental Engineering, American Society of Civil Engineers, Seattle, Wash., March, 1980.

Sutton, P. M. "Fixed Film Biological Processes." *Proceedings, Seminar on Current Approaches in Wastewater Treatment,* Toronto: Pollution Control Association of Ontario and Ontario Ministry of the Environment, 1978.

U.S. EPA (U.S. Environmental Protection Agency). *Design Manual, On-site Wastewater Treatment and Disposal Systems.* Washington, D.C.: Technology Transfer Report, EPA 625/1-80-102, October, 1980.

Villiers, R. V., and Farrell, J. B. "A Look at Newer Methods for Dewatering Sewage Sludge." *Civil Engineering* 47 (1977): 66.

13

Air Pollution

William J. Moroz

13.1 AIR POLLUTION IN PERSPECTIVE

13.1.1 Introduction

Air pollutants are substances which, when present in the atmosphere, adversely affect the health of humans, animals, plants, or microbial life; damage materials, or interfere with the enjoyment of life and the use of property. In this chapter, we shall be concerned primarily with urban and industrial sources of atmospheric pollutants, their effects, and methods of controlling them. In some cases we do not control emissions adequately. We must then rely on dispersion and subsequent natural atmospheric cleansing processes to avoid exceeding pollutant concentrations which would result in undesirable effects.

Throughout the world, emphasis has been on controlling the ambient atmospheric concentrations of pollutants to levels at which no health effects will be observed. In the United States the levels established to protect human health are referred to as primary air quality standards. Secondary standards are established on the basis of effects on plants and animals or damage to materials.

Control of air pollution is not always easy, for it is impractical to eliminate all emissions of a specific pollutant. On the other hand, it is reasonable to expect control of emissions to the lowest possible level consistent with available technology and within reasonable cost. In practice, emission control limits or standards are frequently established rather than ambient air quality standards, because they are far easier for a control agency to enforce, although it is the latter which are really desired.

13.1.2 Air Pollution Episodes

The science and technology of air pollution control is only a few decades old, and our knowledge is still developing very rapidly. For example, most of the instruments which are used today to measure air quality were developed in the last decade. Despite changes in pollutant characteristics, we can benefit from a brief historical review of air pollution and control actions. We list only a few of the recorded air pollution episodes, emphasizing those which have had the most influence on control activities and what was learned from each of the incidents. The history of industrial air pollution is naturally longest in England, where an industrial society developed earliest. In reading this list, try to recall the way people lived at the time of the incident, how many and what kinds of industries existed, how large each industrial establishment might have been, and what transportation and energy systems were in use.

852:            London, England. Complaints of foul air due to the burning of coal inefficiently on open grates for heating.

1100–1200:   London, England. Several Acts of Parliament were passed relating to air pollution.

1661:   London, England. A scholar named John Evelyn wrote a lengthy report on air pollution in London. He proposed zoning of domestic, commercial, and industrial activities, with green belts between them and location of the strongest sources downwind in the direction of the prevailing winds. He was far ahead of his time, and his analysis and recommendations went unheeded.

1864:   The first air-pollution-related lawsuit was initiated in St. Louis, Missouri, U.S.

1873:   English medical records indicate ''excess deaths'' associated with periods of heavy smog.

1880:   London, England. A 27 percent increase in mortality was reported for a two-week period.

1891:   London, England. 1,484 excess deaths were attributed to air pollution.

1922:   London, England. An 11.8 percent increase in fatalities was reported.

1939:   London, England. 1300 excess deaths were reported during a four-day period of heavy fog.

1926:   The U.S. Public Health Service made observations of particle loading in seven major cities (Buffalo, New Orleans, Baltimore, Detroit, Los Angeles, San Francisco, and Washington). They identified these as Class I cities with a ''fallout'' of 3,600 micrograms (μg) of particles per cubic meter of air. Today, the US EPA primary air quality standard for particles is 75 μg/m^3 for the annual geometric mean. Instrumentation used in the 1926 study was quite different from that used today, so air quality observations are not directly comparable but the reduction in particle loading is noteworthy.

1930:   An "epidemic" was reported in the Meuse Valley, Belgium. Sixty-three deaths were reported, and about 8,000 people became ill during a period of intense smog.

1948:   Donora, U.S.A. A severe air pollution episode occurred from October 25 through 31 in a Pennsylvania community of about 14,000 people. Heavy smoke and fog did not dissipate during the day. Winds were light, and horizontal pollutant transport was restricted by the mountains. An inversion restricted vertical dispersion of the pollutants in the valley as well. Primary industries included coke ovens, blast furnaces, steel mills, sulfuric acid plants, a zinc smelter, a fertilizer plant, and electrical generating stations. There was heavy coal-fueled train and boat traffic.

The Donora episode was extensively investigated after the event. Conclusions reached include the following:

1. About 43 percent of the population was affected. Twenty persons died, 1,440 persons were severely affected, 2,322 moderately affected, and 2,148 mildly affected. Health effects were classified as severe if a person could not breathe lying down; moderate if breathing was labored or chest constriction or a productive cough, vomiting, or diarrhea was reported; and mild if only smarting eyes, a nasal discharge, sore throat, dry cough, headache, or dizziness was reported.

2. All sexes, races, and occupational statuses were affected, including housewives, students, and children.

3. There were marked variations in the percentage of people affected in each age group. About 60 percent of persons over 65 and 50 percent of all adults, but only 16 percent of children less than 6 years old, were affected.

4. About 90 percent of the persons affected reported respiratory symptoms, and 34 percent reported gastrointestinal (stomach) symptoms.

5. Persons with respiratory or cardiac histories were most severely affected.

6. Reconstruction of events and estimates of air quality indicated that no single pollutant caused the observed effects.

1952:   London, England. 4,000 excess deaths were attributed to air pollution during a four-day period. The maximum sulfur dioxide concentration measured was 1.34 ppm (3,510 μg/m^3), and concentrations as high as 1 ppm (2,620 μg/m^3) were observed frequently. By comparison, primary air quality standards for SO_2 in the United States today are 365 μg/m^3

for a 24-hour averaging period and 80 $\mu g/m^3$ as an annual arithmetic mean.

1956:   The British Clean Air Act was passed.

1963:   The U.S. Clean Air Act was passed.

1971:   The Canadian Clean Air Act was passed.

Despite the implementation of air pollution control legislation, further serious air pollution episodes have occurred in New York, Pittsburgh, Birmingham, Los Angeles, and San Francisco in the United States, Toronto in Canada, several cities in Great Britain and Europe, Tokyo, Japan, and other heavily industrialized areas of the world. The Los Angeles air pollution problems were of a different type and are so important that they are considered separately in the next section.

All these episodes have had significant health effects. In addition, there have been incidents of severe crop, forest, and materials damage, and the costs have been substantial. Based on the experience gained in the episodes described in the preceding list, early emphasis was placed almost exclusively on controlling sulfur dioxide, particle, and nitrogen oxide concentrations in the atmosphere.

13.1.3 The Los Angeles Smog

During the 1940s, air pollution in Los Angeles became so objectionable that the citizens demanded action to clean up the atmosphere. It was observed that cracks formed in rubber tires after about one year and some synthetic fabrics aged and discolored very quickly. The most apparent symptom was a severe decrease of visibility starting before noon and continuing throughout the day despite steady breezes off the ocean. Laws were passed to limit emissions of sulfur dioxide, particles, and nitrogen oxides and were stringently enforced. While some improvement in air quality was noted, the most objectionable symptoms, such as irritation of the eyes, nose, and throat, did not change significantly. These symptoms were different from those experienced at other locations, and it was apparent that other pollutants must be active.

In the latter part of the decade a photochemist named Haagen-Smit postulated that the pollutants primarily responsible for the Los Angeles smog were not emitted directly from any source (Haagen-Smit, 1952). He speculated that secondary pollutants were formed in the atmosphere as a result of chemical reactions involving the primary pollutants.* Continuing investigations confirmed that nitrogen dioxide (NO_2) dissociated when irradiated by intense, short-wavelength, radiant energy from the sun and a series of photochemical reactions led to the formation of the powerful oxidant, ozone (O_3). The sequence of chemical reactions is simplistically described as follows (the molecular oxygen, O_2, is present in the atmosphere):

$$2NO + O_2 \rightleftarrows 2NO_2 \tag{13.1}$$

---

* The terms "primary pollutants" (those emitted by an identifiable source) and "secondary pollutants" (those formed in the atmosphere by chemical reactions) are common in the air pollution literature today.

$$NO_2 + h \rightleftarrows NO + O \tag{13.2}$$

$$O + O_2 + M \rightleftarrows O_3 + M \tag{13.3}$$

In these equations, h represents an input of short-wavelength radiation and M represents a molecule which acts roughly as a catalyst, although it may also react with O and O_3 to form new compounds, many of which are in the form of solid particles. Certain reactive hydrocarbon species are particularily effective in this role. Peroxyacetyl nitrate (PAN), which causes severe plant damage at very low concentrations, is another secondary pollutant formed.

Despite the lack of a complete understanding of what actually happens, we can identify the automobile as a significant source of pollutants leading to the formation of Los Angeles smog. It is the largest source of nitrogen oxides (NO and NO_2) in both the United States and Canada, and it is by far the largest source of reactive hydrocarbons as well. (We frequently use the symbol HC to represent compounds of hydrogen and carbon.)

Example 13.1

In Toronto, Canada, there are about 750,000 automobiles registered. The average nitrogen oxides emission rate from the cars is 3.1 g per vehicle mile (VM), and the HC emission rate is 1.6 g/VM. Each car travels about 30 miles round trip per working day. Calculate the amount (volume) of NO_x (NO and NO_2) and the amount of HC put into the city atmosphere each working day.

Solution The NO_x^* produced is $750,000 \times 30 \times 3.1 = 69.8 \times 10^6$ g/day $NO_x = 69.8$ tonnes/day NO_x. The HC** produced is $750,000 \times 30 \times 1.6 = 36 \times 10^6$ g/day HC $= 36$ tonnes/day HC. Since the molecular weight of $NO_2 = 46$ g/mol volume $= 46$ g/22.4 L, it follows that the volume of NO_2 produced is

$$\frac{69.8 \times 10^6 \times (22.4 \times 10^{-3})}{46} = 3.4 \times 10^4 \text{ m}^3/\text{day.}$$

The molecular weight of $CH_4 = 16$ g/mol volume $= 16$ g/22.4 L. So the volume of CH_4 produced is

$$\frac{36 \times 10^6 \times (22.4 \times 10^{-3})}{16} = 5.0 \times 10^4 \text{ m}^3/\text{day.}$$

Thus, the total volume of NO_2 and HC produced is 8.4×10^4 m³/day.

---

* The nitrogen oxides emitted by high-temperature combustion operations are almost all NO. The NO oxidizes to NO_2 in the atmosphere (Equation 13.1) in minutes to hours, so that the nitrogen oxides observed at street level away from the source outlet are mostly NO_2. However, there is always some NO present in the ambient atmosphere along with the NO_2, so we measure nitrogen oxides (NO_x) as NO_2 for convenience. We shall assume they are reported as NO_2 in this example.

** There are many different hydrocarbons emitted from internal combustion engines. The average molecular weight of all of the hydrocarbons is roughly equal to that of methane, and emissions are frequently reported as methane (CH_4) for convenience in converting from mass to volume. We shall assume that they are reported as CH_4 in this case. Reporting emissions as CH_4 is in contrast to the reporting of hydrocarbon concentrations in the atmosphere (air quality) as nonmethane hydrocarbons only because methane is relatively nonreactive in generating photochemical smog.

Comment: If this large volume of gas is not continuously dispersed vertically and transported out of the area, the air at street level can become unfit to breathe.

13.1.4 Global and Regional Pollutants

The anthropogenic (arising out of human activities) pollutants identified thus far result in local or regional impacts. More recently, we have recognized certain pollutants as having continental and global effects. These pollutants include acid rain; carbon dioxide; chlorofluorocarbons (previously called chlorofluoromethanes); radioactive pollutants from nuclear tests, natural, and industrial sources; and conventional pollutant emissions from large-scale natural sources, including volcanoes and forests.

Acid rain is presently regarded as a regional or continental rather than a global problem. The most significant precursors of acid rain are considered to be SO_2 and NO, which oxidize and hydrolyze to form H_2SO_4 and HNO_3. Carbon dioxide (CO_2) also contributes to the acidity of rainfall, but more importantly, CO_2 is transparent to short-wavelength radiation from the sun but opaque to longer wavelengths radiated back to space from the earth. Thus, increased concentrations of CO_2 may result in a heating of the earth's atmosphere and global warming. These phenomena are described more fully in Chapter 5.

The chlorofluorocarbons (CFCs), and in particular the freons, are very stable chemically, i.e., they persist for decades in the atmosphere. As a result, these substances mix fairly uniformly through a very deep layer of the atmosphere, finally reaching the ozone layer aloft between 25 and 50 km above the earth's surface. The ozone layer aloft is formed by photodissociation of oxygen molecules, an entirely different process than that leading to ozone formation at the earth's surface in photochemical smogs. It protects us from very short-wavelength ultraviolet radiation from the sun, which causes skin cancer. The CFCs react photochemically at high altitudes to diminish ozone concentrations, an unacceptable effect. Legislation has been passed in many countries to restrict the production of these chemicals.

13.2 EFFECTS OF AIR POLLUTION

Only a few examples of air pollution effects can be considered in the sections to follow. All of the examples identify local, observable, or measurable impacts, because it is very difficult to develop direct relationships between specific pollutants and effects for exposures over the longer term or at great distances. A more complete description could fill several books. One of the most obvious local effects of particles in the atmosphere is a reduction in visibility. Only a few decades ago, soiling by soot and smoke was clearly apparent in almost every urban center. Reduction in visibility results in a social cost due to slowdown of air traffic and the need for instrument-guided landing systems.

In the air pollution episodes cited, the local presence of gases was objectionable because of odor, taste, or obvious corrosive or chemical effects. Today these gross sensory insults are rarely encountered. However, subtle health effects persist, such as eye or nose irritation or difficult breathing. In the extreme, health effects extend to the brain (CO) and stomach (several pollutants alone or in combination). Damage to vegetation due to chronic exposure to atmospheric pollutants may be one of the more apparent precursor symptoms leading to identification of chronic air pollution.

13.2.1 Health Effects

Health effects were the dominant considerations in early air pollution episodes for obvious reasons. While the specific pollutant or groups of pollutants generating the observed effects frequently could not be identified, there was sufficient information to implicate certain pollutants as significant contributors. Early research to relate pollutant concentrations and effects was concentrated on these clearly identifiable pollutants.

As discussed in Chapter 8, the human respiratory system is quite efficient in filtering the larger particles out of the air we breathe. Particles smaller than about 5 μm, however, can penetrate to the lungs and be deposited in the alveoli. Figure 13–1 shows the fraction of particles of various sizes deposited in the respiratory system, established using an aerodynamic model. For example, cigarette smoke particles are smaller than 1 μm, and they enter and are deposited in the alveoli.

Some particles are particularly damaging because they adsorb gases which cause

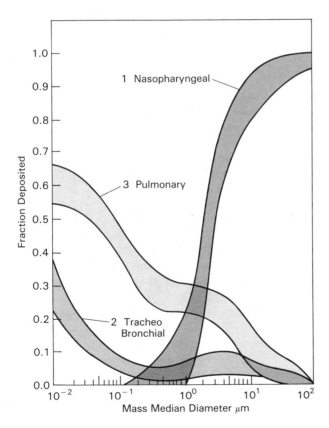

Figure 13–1 Aerodynamic deposition of particles by size in the respiratory tract (nasopharyngeal: nose and throat regions; tracheobronchial: tubes leading to lungs; pulmonary: lung regions).
Source: *Air Quality Criteria for Particulate Matter,* National Air Pollution Control Administration, Publication AP 49, 1960

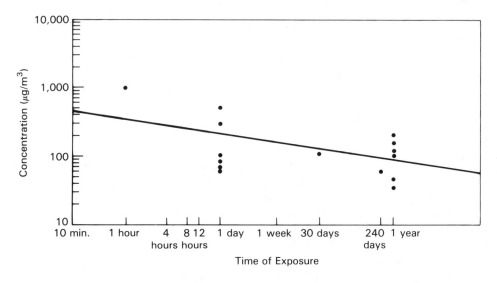

• Health Effects

Figure 13–2 Health effects of suspended particles. Source: W. J. Moroz (compiled from several sources)

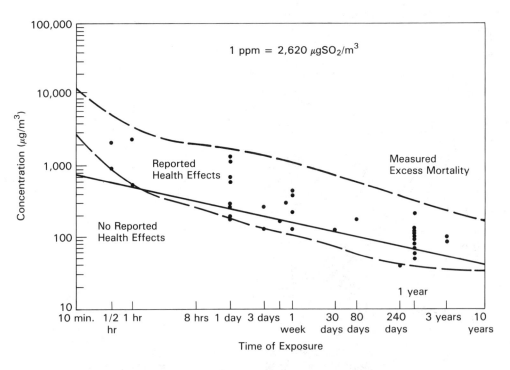

• Health Effect in Man

Figure 13–3 Health effects of SO₂. Source: W. J. Moroz (compiled from several sources)

more intense irritation locally. Gases also penetrate into the deepest lung pockets. Both particles and gases entering the body through the respiratory system can affect the gastrointestinal system, as was reported in the Donora incident. Some chemicals, such as lead, can enter the human bloodstream either from the digestive system (ingestion) or by passing through the lung membranes (the respiratory system), and airborne tritium and a few other chemicals can enter the bloodstream through the skin.

Each pollutant affects the human body differently, and records of effects have been assembled relating the intensity to the period of exposure for various pollutants. Figure 13–2 shows the concentration vs. time, or dosage, at which effects of suspended particulate matter on humans have been observed as deduced from the reports of several investigators. Figures 13–3 and 13–4 show the effects on humans of exposure to gaseous sulfur dioxide (SO_2) and fluorine (F). Health effects of carbon monoxide are clearly detectable at 4 percent carboxyhemoglobin (COHb) in the blood, and possibly as low as 2.5 percent COHb.

13.2.2 Effects on Plants and Animals

The detrimental impacts of air pollution are not limited to those involving human health. Plants and animals are also susceptible. For example, fluorine is emitted from aluminum, glass, phosphate, fertilizer, and some clay-baking operations in significant quantities. Figure 13–4 also shows the exposures at which phytotoxicological effects (effects on plants) of fluorine are observed. Frequently, the plant damage is observed on the fruit or on the flowers, either of which significantly lowers the value of the crop. Fluorine affects plants at concentrations several orders of magnitude below that at which human health is affected.

Fluorine also has an effect at even lower concentrations when it is taken up by shrubs, trees, or grass which are subsequently eaten by cattle or other animals. The animals may develop fluorosis, although the plants may not show signs of damage. The animals act as concentrators of the fluorine, resulting in poor animal health and associated lower animal value or survival capability. Some heavy metals, such as mercury and lead, and most radionuclides also become concentrated in plants and animals, frequently in specific organs.

Different plants and animals have different susceptibilities to pollutants, and figures similar to Figure 13–4 could be drawn for each plant. For example, sugar maple can tolerate relatively high concentrations of sulfur dioxide alone, but it is susceptible to damage from exposure to SO_2 and O_3 together. White pine, on the other hand, is very sensitive to damage from either pollutant alone.

13.2.3 Effects on Materials

Sulfur and nitrogen oxides react in the atmosphere to form acidic compounds which attack metal surfaces, a problem which has been particularly acute for the communications, switchgear, and computer industries. Fluorine is particularly reactive, and at high atmospheric concentrations etching of glass has been observed. These impacts are taken into consideration when sensitive components are designed, and the required protective measures or design modifications add to the cost of the item being produced. Hydrogen sulfide in the ambient atmosphere reacts with lead oxide in white paint to form lead sulfide, so that

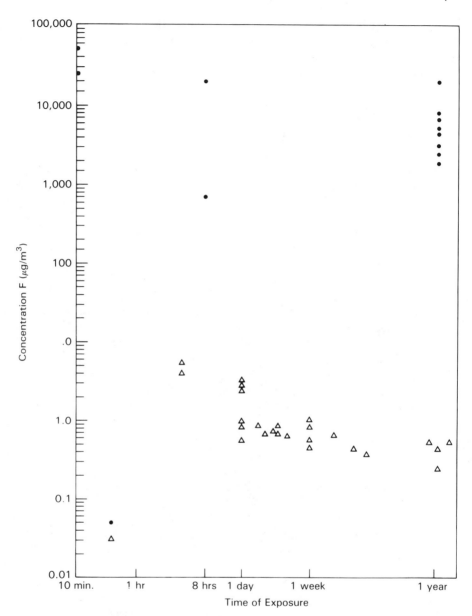

Figure 13–4 Health and phytotoxicological effects of fluorine. Source: W. J. Moroz (compiled from several sources)

white houses have been observed to take on a brownish tint overnight. Accelerated aging of synthetics and rubber due to exposure to atmospheric oxidants has already been noted. (Both the strength and color of the materials are affected.)

It is extremely difficult to estimate the dollar value of accelerated deterioration of materials and aesthetic items such as buildings, statues, or horticultural plantings. Halvorsen and Ruby (1982) provide a thorough discussion of these costs, and it is generally agreed that the damage from atmospheric pollution is measured in the billions of dollars.

13.2.4 Ambient Air Quality Standards

Curves similar to those shown in Figures 13–2 through 13–4 have been prepared for many other pollutants. Using these curves, we can establish ambient air quality standards at which pollutant effects will not be detectable. For example, Figure 13–4 shows that health effects of fluorine have been observed after about an 8-hour exposure at 800 μg/m³, but that plant effects are observed after an 8-hour exposure at less than 1 μg/m³. These data have led to the establishment of primary standards to protect health only and secondary standards to protect against effects to other systems. Standards are established for various sample averaging periods, because effects are dependent upon pollutant concentration and exposure time.

Jurisdictional control of air pollution lies with different levels of government in different

TABLE 13–1   AMBIENT AIR QUALITY STANDARDS* FOR SELECTED POLLUTANTS (1982)

Pollutant	Jurisdiction		
	US EPA**		Ontario, Canada
	Primary	Secondary	
Suspended particulate matter	75—AGM 260—24h	60—AGM 150—24h	60—AGM 120—24h
Sulfur dioxide	80—AAM 365—24h	60—AAM 260—24h 1300—3h	55—AAM 275—24h 690—1h
Nitrogen oxides (as NO₂)	100—AAM		200—24h 400—1h
Hydrocarbons	160 between 0600h and 0900h as nonmethane HC		No standard
Oxidants (as O₃)	160—1h		165—1h
Fluorides (Total)			0.138—30d 3.44—24h

* All concentrations given are in μg/m³ for the averaging period indicated. AAM = annual arithmetic mean; AGM = annual geometric mean; 1h = 1 hour; 30d = 30 days, etc.

** Primary standards define air quality levels which protect the public health. Secondary standards protect the public welfare. In the United States and most other countries, emission standards are enforced to achieve air quality standards. In Ontario, the air quality standards themselves rather than the emissions are enforced.

TABLE 13–2   REPRESENTATIVE AMBIENT AIR QUALITY OBSERVATIONS IN CANADA

Pollutant	Averaging Period, hr	Heavily Industrialized Region Concentration, $\mu g/m^3$		Remote Forested Northern Site Concentration, $\mu g/m^3$	
		Average	Maximum	Average	Maximum
Suspended particulate matter	24	104	255	11	44
Sulfur dioxide	1	26	416	13	78
	24	26	208		
Carbon monoxide	1	12	115	NA*	NA*
	8	11	69		
	24	10	46		
Nitrogen dioxide	1	60	282	4	75
	24	55	206		
Ozone	1	20	220	NA	232**
	24	35	120		

* Not available at this site.

** The maximum was probably associated with a thunderstorm and lightning.

Source: W. J. Moroz (compiled from several sources)

countries. For example, in the United States the federal government sets air pollution standards and the individual States can adopt standards which are equivalent to or more stringent than the federal standards. In Canada, the individual provinces set the standards and the federal government recommends guidelines. In both cases responsibility for enforcement lies with the individual states or provinces, except in the cases of interstate or international pollutant transport and disputes.

Table 13–1 gives ambient air quality standards enforced by the United States Environmental Protection Agency and by the Province of Ontario in Canada for several atmospheric pollutants. Table 13–2 presents air quality observations downwind from a heavily industrialized city in Canada and at a remote site in Northern Ontario.

Where observed air quality is worse than the adopted air quality standards, vigorous air pollution control programs are underway to reduce pollutant emissions. Control programs may include regulations to control fuel quality, enforcement of emission control standards, paving roads to reduce dust reentrainment into the atmosphere, and many other techniques.

13.3 SOURCES OF AIR POLLUTION

13.3.1 The Principal Atmospheric Pollutants

The early air pollution episodes were clearly detectable by the senses without special aids or instrumentation. Particles in urban atmospheres reduced visibility and were aesthetically dirty. Sulfur dioxide smelled, caused silvered surfaces to turn black, caused plant damage, and, in extreme situations, made breathing difficult. Ozone caused rubber and synthetic materials to deteriorate very quickly, and photochemical smog containing high concentrations of ozone caused lacrimation (formation of tears). Nitrogen oxides, NO and NO_2, and hydrocarbons, of which there are several thousand different species, were found to be precursors of photochemically formed ozone and PAN in a shallow layer of the atmosphere at the earth's surface. It was also well recognized that carbon monoxide represented a severe health hazard at extremely low concentrations. As the popularity and use of the automobile grew, it became apparent that carbon monoxide concentrations at street level in heavily congested areas could be sufficiently high to affect the performance of persons exposed for long periods, such as traffic officers and parking lot and garage attendants.

The subjective identification of air pollution effects led to the identification of the following principal pollutants: particles; sulfur dioxide; carbon monoxide; nitrogen oxides, usually measured as NO_2; hydrocarbons, measured in the atmosphere as nonmethane hydrocarbons because methane is relatively nonreactive in photochemical smog formation; and ozone, (a surrogate for other oxidants, PAN, and possibly other compounds). These pollutants are the most common and pervasive ones in urban centers, where populations are concentrated. We know today that when an air pollution effect is observed, it is necessary to consider many other pollutants in addition to the principal pollutants.

13.3.2 Identifying Air Pollutants

By definition, pollutants have an observable or detectable effect. Effects are not always easily identifiable, however, and frequently, observed effects cannot be related directly to specific pollutants. For example, nitrogen oxides and hydrocarbons were not identified as principal primary pollutants until it was understood that these were the precursors of ozone and PAN in photochemical smog.

In the past few decades, methods of identification of air pollution sources have changed from simple sensory techniques based on sight, smell, taste, and odor to objective techniques which permit quantification or measurement of air quality. Today, with few exceptions, if the pollutants are detectable by the senses, or if direct effects can be observed, the sources are considered to be contributing to ''gross insult'' in the air environment.

Gross insult was most easily detectable from industrial sources, usually related to the combustion of fossil fuels in the early days of air pollution control, because the pollutants were released through a chimney or stack at an elevated level where they could be easily seen and identified. As identification and measurement techniques improved, it became clear that emissions from industrial sources of air pollution were frequently exceeded by emissions from domestic, commercial, agricultural, and transportation-related, sources. For example, there are few strong industrial sources of air pollution in the heart of Denver, Los Angeles, or Toronto, but in general air quality deteriorates towards the city core.

In the sections that follow, a few of the generic sources of pollutants are identified. Tables 13–3 and 13–4 present total nationwide emissions of the principal pollutants from selected source groups in the United States and Canada, respectively. Note that the distribution of the emissions from similar source categories in the two countries is quite different. The difference in total emissions is a result of differences in industrial activities, size and distribution of population, forested areas, energy sources, and other factors. A

TABLE 13–3   NATIONWIDE EMISSIONS OF PRIMARY AIR POLLUTANTS, UNITED STATES, 1980
(10^6 tonnes/yr)

Source Category	Particulate matter t/yr	Particulate matter %	Sulphur oxides* t/yr	Sulphur oxides* %	Nitrogen oxides** t/yr	Nitrogen oxides** %	Hydrocarbons t/yr	Hydrocarbons %	Carbon monoxide t/yr	Carbon monoxide %
Transportation										
Highway	1.1	12.6	0.4	1.7	6.7	33.8	6.8	29.6	65.3	68.7
Aircraft	0.1	1.1	—	—	0.1	0.5	0.2	0.9	1.0	1.1
Railroads	0.1	1.1	0.1	0.4	0.8	4.0	0.2	0.9	0.3	0.3
Vessels	—	—	0.3	1.3	0.1	0.5	0.5	2.2	1.5	1.6
Other	0.1	1.1	0.1	0.4	1.0	5.1	0.5	2.2	4.7	4.9
		15.9		3.8		43.9		35.8		76.6
Stationary Sources—Fuel										
Electric utilities	1.0	11.5	15.5	66.2	6.4	32.3	—	—	0.3	0.3
Industrial	0.4	4.6	2.4	10.3	3.0	15.2	0.1	0.4	0.6	0.6
Commercial	0.1	1.3	0.7	3.0	0.3	1.5	—	—	0.1	0.1
Residential	0.6	6.9	0.2	0.9	0.4	2.0	0.8	3.5	5.2	5.5
		24.3		80.4		51.0		3.9		6.5
Industrial Processes	3.7	42.5	3.7	15.8	0.7	3.5	10.7	46.5	6.2	6.5
Solid Waste Disposal										
Incineration	0.2	2.3	—	—	—	—	0.3	1.3	1.2	1.3
Open burning	0.2	2.3	—	—	0.1	0.5	0.3	1.3	1.0	1.1
		4.6								
Miscellaneous										
Forest fires	1.0	11.5	—	—	0.2	1.1	0.9	3.8	6.9	7.3
Other burning	0.1	1.2	—	—	—	—	0.1	0.4	0.7	0.7
Organic solvents	—	—	—	—	—	—	1.6	7.0	—	—
		12.7								
Total	8.7	100.0	23.4	100.0	19.8	100.0	23.0	100.0	95.0	100.0

Source: *National Air Pollution Emission Estimates, 1970–1981*, US EPA, Report No. EPA-450/4-82-012, 1982
* Expressed as SO_2.
** Expressed as NO_2.

careful study of the relative contributions by each of the different sources to each of the principal pollutants is highly informative. Air pollution analysts must identify all of the sources in a region and prepare a list of pollutants and their order of significance to develop an air pollution control program for the region or area in question.

13.3.3 Natural Sources

Natural pollutant emissions vary from one location to another, with seasonal, geological, and meteorological conditions, and with the type of vegetation. Human activities may even contribute to conditions which promote pollutants from natural sources. For example, the "dust bowl" conditions in the southwestern United States during the 1930s were partly a result of overzealous cultivation for agricultural purposes; spores from the ragweed plant,

TABLE 13–4   NATIONWIDE EMISSIONS OF PRIMARY POLLUTANTS, CANADA, 1978
(10⁶ tonnes/yr)

Source Category	Particulate matter		Sulfur oxides*		Nitrogen oxides**		Hydrocarbons		Carbon monoxide	
	t/yr	%	t/yr	%	t/yr	%	t/yr	%	t/yr	%
Transportation	0.1	4.4	0.1	2.3	1.1	61.1	0.9	40.9	7.2	73.4
Stationary sources—Fuel										
Electric utilities	0.2	8.6	0.7	15.9	0.2	11.1	(a)		(a)	
Industrial, commercial, residential	0.1	4.4	0.8	18.2	0.4	22.2	0.1	4.55	0.3	3.1
Industrial processes	1.3	56.5	2.8	63.6	0.1	5.6	0.6	27.3	1.1	11.2
Solid waste disposal	(a)		(a)		(a)		(a)		0.4	4.1
Miscellaneous										
Forest fires	0.5	21.7			(a)		0.1	4.55	0.5	5.1
Other	0.1	4.4	(a)		(a)		0.5	22.7	0.3	3.1
Total	2.3	100.0	4.4	100.0	1.8	100.0	2.2	100.0	9.8	100.0

(a) Less than 0.1
* Expressed as SO_2.
** Expressed as NO_2.
Source: Report EPS 3-EP-83-10, Environment Protection Programs Directorate; Environment Canada, 1983

which are an important contributor to hay fever, thrive on unplanted earth banks at construction sites and along roadways.

Volcanic eruptions present a concentrated, localized natural source of all types of gases and particles. As an example, the Mount St. Helens eruption on May 18, 1980, ejected an estimated 4 km³ (approximately 10 billion tonnes) of solids alone into the atmosphere. This is several orders of magnitude greater than the total mass of particles discharged annually from human-related activities in North America. The particles vary in size from boulders to very small particles (0.001 μm). The energy of volcanic eruptions is frequently sufficient to force the gases and small particles through the lower atmospheric layers into the stratosphere (see Figure 7–1), where natural removal processes are very slow. Thus, these pollutants can remain suspended in the atmosphere for very long periods.

Chemical analysis of dust (ash) samples taken as far as 650 km downstream from Mount St. Helens (Fruchter et al., 1980) showed the ash to be 60 to 70 percent silicon dioxide (SiO_2) and 16 to 18 percent aluminum oxide (Al_2O_3). The chemical composition and the particle size distribution of the ash varied with distance from the source, indicating that certain chemicals are concentrated in specific particle size ranges. This occurs because some elements, such as lead, zinc, and arsenic, have relatively low vaporization temperatures and are concentrated in very small particles formed by sublimation after release.

Example 13.2

Estimate the travel distance of a 70-μm particle of density 1600 kg/m³ ejected at 3,000 m from Mount St. Helens to a final height of 10,000 m (i.e., the plume rise for 70-μm particles is 7,000 m) into the atmosphere when mean wind speed through the atmospheric layer is 15 ms^{-1}.

Solution Suppose that the particle falls to the earth's surface at mean sea level (0 m) and that the properties of air at atmospheric pressure and 20°C can be used in the solution as a first approximation. Then, from Equation 6.7, the Stokes' terminal settling velocity for the particle is

$$u_t = \frac{gd_p^2(\rho_p - \rho)}{18\mu}$$

where ρ for air at 20°C and atmospheric pressure is 1.2 kg/m^3 and μ for air at 20°C and atmospheric pressure is 1.81×10^{-5} N · s/m^2 = 1.81×10^{-5} kg/m · s. Therefore,

$$u_t = \frac{9.81 \times (70 \times 10^{-6})^2 \times (1600 - 1.2)}{18 \times 1.81 \times 10^{-5}}$$
$$= 0.237 \text{ m s}^{-1}$$

and the fall time for the particle is

$$t = \frac{10^4}{0.237} = 4.22 \times 10^4 \text{ s}$$

During this fall, the particle will travel horizontally

$$4.22 \times 10^4 \times 15 = 6.33 \times 10^5 \text{ m}$$
$$= 633 \text{ km}$$

Note that for particles falling in a gas, the density of the gas can be neglected in the determination of Stokes' fall velocity. In the actual Mount St. Helens eruption, the plume centerline was at a height of about 12 miles (19.3 km), and some of the small particles in the plume actually circled the globe.

Natural background radioactivity in the atmosphere results from the bombardment of gaseous molecules in the upper layers of the atmosphere by ionizing cosmic radiation from the sun and emissions of radon and thoron from the earth's crust. Cosmic radiation is most intense north and south of about 50° latitude, due to deflection of solar particles by the earth's magnetic field. Seepage of radioactive gases is slower through a snow or

TABLE 13–5 RADIATION EXPOSURE (MILLIREMS/YEAR) AT SELECTED LOCATIONS

Radiation source	Pennsylvania	Texas	Colorado[1]
Cosmic	45	45	120
Terrestrial			
Internal	55[2]	30	105
External	25	25	25
Total	125[2]	100	250[2]

[1] These data are for exposure at about 5,000 ft (1,525 m) altitude.
[2] These values vary at least by a factor of two at a specific location.
Source: W. J. Moroz (combined from several sources)

ice cover. Radiation from surface and upper atmospheric sources is distributed throughout the atmosphere by global circulation, but atmospheric radioactivity increases upward and near the more powerful geological sources. Table 13–5 indicates the relative exposure of a person to cosmic and terrestrial radiation at selected locations.

Dust and sandstorm particles reentrained during windy periods can be transported great distances and result in very high particle concentrations for short time periods at remote locations. It is of particular interest that forest fires are the largest collective source of particles in the Canadian National Inventory of Emissions. Forest fires are also a prodigious source of CO and CO_2. Living forests are powerful sources of hydrocarbon and other organic emissions. This source is so strong that it is detectable by smell and sight, and releases occur over very broad areas. These releases are thought by many to be the cause of the blue haze observed over heavily forested regions.

13.3.4 Domestic Sources

In residential areas, domestic activities are the major causes of pollutant emissions. A few of the activities and the types of pollutants released are as follows:

Activity	Pollutants released
Space heating	CO, CO_2, NO_x, SO_x, soot, smoke (if fossil fuels are burned at the residence)
Cooking	Fats (as solids, liquids, and vapors), particles, odors
Cleaning	Solvent vapors, dust, lint, spray can propellants.
Gardening	Pesticides, fertilizers (some of which may be highly toxic)
Painting	Principally solvent vapors
Washing	Detergent particles, soap particles, lint.

Prior to the industrial revolution, domestic fires were the most important pollutant source in London, England. The problem was amplified by the fact that emissions were from low chimneys and fireplaces are notoriously inefficient combustion and heating systems. Modern domestic furnaces are much better, but pollutants are still emitted through low chimneys. Improvement is possible when district heating systems and electric heating transfer emissions to a different site, where tall stacks and emission control systems can be installed.

Example 13.3 indicates the importance of emissions from painting and cleaning activities, much of which occur in the household.

Yet another source of air pollution from domestic and commercial activities is associated with the disposal of solid wastes. Formerly the burning of leaves, backyard incinerators, and open dumps were common sources of particle and gaseous releases. Today these activities or sources are banned in most communities, and solid waste is disposed of in sanitary landfill operations or by burning in large, efficient municipal incinerators. At old sanitary landfills the decay of refuse over many years releases methane gas (a hydrocarbon) to the atmosphere, and hydrogen sulfide released from decaying organic and other materials is also frequently detectable by smell over old landfill sites.

13.3.5 Commercial Sources

Commercial sources of air pollution include the public service industries. As an example, consider the dry cleaning of clothes; almost all of the solvent used in the process evaporates to the atmosphere. Emissions from dry cleaning range from 15.9 kg of solvent lost per 100 kg of clothes cleaned in small uncontrolled cleaning machines to less than 10 kg of solvent lost per 100 kg of uniforms, towels, etc., cleaned in large industrial machines. The solvent used in most small machines for domestic and commercial cleaning is perchlorethylene, a chlorinated hydrocarbon; in larger industrial machines simple hydrocarbons are used because of their lower cost. There are well over 50,000 cleaning centers in the United States today.

Other commercial establishments or activities releasing pollutants to the atmosphere include restaurants, hotels, schools, printing, and painting. Preparation of meals results in the disposal of $\frac{1}{2}$ to 1 kg of solid waste per meal. There are 3 to 5 kg of solid waste per hospital bed per day and 4 kg of waste per school class. The amount of plastic to be disposed of from these activities is increasing rapidly today. Many of these plastics are chlorinated hydrocarbons, and on combustion they release chlorine, which rapidly hydrolizes in the atmosphere to hydrochloric acid, a highly corrosive pollutant which also causes damage to sensitive vegetation at very low concentrations and is a contributor to acid rain.

Example 13.3

Assume that the annual average urban consumption of oil- (HC) based paint and coatings is about $\frac{1}{2}$ gal per capita for all purposes (domestic, commercial, and industrial consumption). Each gallon of paint contains 6 lb of hydrocarbon carrier for the pigment. The U.S. national annual average consumption of solvents for dry cleaning (assume they are all hydrocarbons) is about 2 lb per capita.

Estimate the total hydrocarbons discharged into an urban region of about 2.3 million people (the population of Metropolitan Toronto, Canada) from the evaporation of paint and dry-cleaning solvents, assuming that the averages are representative and released locally. Compare this number with the amount of HC discharged from controlled automobiles (Example 13.1).

Solution   The HC evaporation from paint and coating materials used is $2.3 \times 10^6 \times 0.5 \times 6 = 6.9 \times 10^6$ lb/yr $= 3.1 \times 10^6$ kg/yr $= 8.6 \times 10^3$ kg/day $= 8.6$ tonnes/day. The HC evaporation from cleaning solvents used is $2 \cdot 3 \times 10^6 \times 2 = 4.6 \times 10^6$ lb/yr $= 5.7$ tonnes/day. From Example 13.1, or a city of 2.3 million, HC emission from cars is 36 tonnes/day.

Substantial quantities of HCs are released from paints and solvents, and there is currently considerable emphasis on developing paints, painting techniques, and cleaning solvents which will reduce the amount of hydrocarbons emitted.

13.3.6 Agricultural Sources

Agricultural sources which have been affected directly by air pollution control legislation include slaughterhouses and animal feedlot operations. For example, chicken production for meat has been concentrated in very large operations, frequently housing several hundred

TABLE 13-6 EMISSION RATES AND EMISSION FACTORS FOR TYPICAL HARVESTING OPERATIONS

Operation	Emission rates lb/hr (kg/hr)		Emission factors lb/mi^2 (kg/km^2)	
	Wheat	Sorghum	Wheat	Sorghum
Harvest machine (combine)	0.027 (0.012)	0.18 (0.083)	0.96 (0.17)	6.5 (1.1)
Truck loading	0.014 (0.007)	0.014 (0.006)	0.007 (0.012)	0.13 (0.022)
Field trucking	0.37 (0.17)	0.37 (0.17)	0.65 (0.11)	1.2 (0.2)

Adapted from Supplement 10, AP42, US EPA Compilation of Emission Factors, 3d Edition, Feb. 1980

thousand birds at one location. Observations of particles discharged through the ventilation system at a small "house" at the Pennsylvania State University indicated that more than 40 percent of the particles were smaller than 5 microns in aerodynamic diameter. These are well within the respiratory size range.

Another example of gross insult agricultural air pollution is the release of cotton particles during harvesting and processing in sufficient quantities that they have been reported to be the cause of respiratory problems in residential areas near the processing centers. On the farm itself, exposure to particles from crop harvesting operations, to pesticides and insecticides, and to ammonia used for fertilizer represent chronic health hazards. Many farmers die each year as a result of exposure to "silo gas," which is NO_2.

Table 13–6 represents typical particle emission rates and factors for combining, truck loading, and transporting wheat and sorghum in the field. Note that the emissions per unit area are highest for harvesting because of the number of hours required for this operation.

13.3.7 Industrial Sources

Industrial sources of air pollution are the most noticeable, because emissions are usually discharged through a single stack or duct. Where a particular industrial contaminant is the major objectionable pollutant in a community, it may be traced to its origin by a knowledge of the industrial processes being used. In these cases, corrective measures are frequently easy to implement.

The following paragraphs briefly consider some of the industrial pollutant sources most frequently encountered in air pollution problems.

Nitrogen oxides (NO_x) are produced by any high-temperature combustion operation. Process sources include plants for manufacturing fertilizer and explosives.

Sulfur oxides (SO_x) are emitted primarily as SO_2 from the combustion of fuel oil and coal at stationary sources. Sulfur is usually removed from natural gas at the well so that the gas can be used in domestic applications. A very small amount of SO_x is emitted from the combustion of gasoline and diesel fuels. Combustion sources also emit small quantities of SO_3. The refining of sulfide ores generates very large quantities of SO_2. Oil refineries are also important sources of SO_2. The sulfur oxides react with atmospheric water to form H_2SO_4 within hours.

Hydrogen sulfide may be emitted in large quantities from paper plants, natural gas cleaning and processing plants, oil refineries, and some plants manufacturing synthetic fibers—for example, rayon. The H_2S reacts in hours in the atmosphere, oxidizing to SO_2 and H_2O. Most of these emissions are controllable today.

Carbon monoxide may be released at high concentrations in the production of cast iron and other metallurgical processes where it is desirable to minimize the presence of oxygen. It is released at extremely low concentrations at stationary fuel-burning installations, but the quantities generated are still quite substantial because of the amount of fuel burned. Industrial emissions of CO are exceeded by emissions from automobiles in both Canada and the United States.

Hydrocarbons are released in large quantities from a host of industrial processes usually related to the petroleum and natural gas industries or industries which use their products. They may be in the form of vapor, liquid, or tarlike particles, as in the case of asphalt paving operations. The hydrocarbon releases from the paint, roofing, and cleaning industries have already been noted. Hydrocarbons are also released at plants manufacturing plastics, most of which use petroleum or natural gas derivatives as base products, and at rubber or synthetic manufacturing or processing facilities.

Particles may be liquid or solid. The public is most aware of particle emissions, because they are visible to the naked eye. The chemical and physical nature of the particles is extremely important in assessing the significance of the emissions. Metallic oxides from spray painting and the coating industries; catalyst dusts from refineries; asbestos fibers from the insulation, cloth, and pipe industries; and special chemical releases such as barium, beryllium, boron, and arsenic from the metals processing or manufacturing industries are designated as hazardous particles because they are highly toxic or carcinogenic and are in the respirable size range. The largest industrial particle emissions are ash from combustion of coal, oil, and refuse; carbon particles from the combustion and processing of fossil fuels, including natural gas; and particles from quarrying and mining and their associated industries.

It is informative to consider a few specific industries as typical examples and the types of emissions which they discharge.

Iron and steel production facilities emit large quantities of small particles to the atmosphere. Most of the particles are oxides of iron, carbonate fluxing materials, or oxides of metals used to produce special alloys. Most are smaller than 2 μm, i.e., they are in the respirable size range. The shift to basic oxygen furnaces in the last two decades has resulted in a shift of particle emissions to even smaller sizes and in greater quantities. Table 13–7 gives typical emission factors for iron- and steel-producing operations.

Cast iron furnaces operate with an oxygen-deficient (reducing) atmosphere by reacting almost all of the oxygen in the air used with the fuel. The resulting gas discharge from a cast iron cupola during the heat contains about 12 percent CO_2, 11 percent CO, less than 1 percent O_2, and the balance N_2 with much smaller percentages of sulfur and nitrogen oxides and hydrocarbons. The CO, SO_x, NO_x, hydrocarbons, and small particles are all important pollutants from these sources. Table 13–7 also gives typical emission factors for gray iron-melting operations.

Feed, grain, and cereal mills discharge large quantities of particles during drying, dehusking, screening, grinding, and processing of grain. More than 50 percent of the

TABLE 13–7 EMISSION FACTORS FOR TYPICAL UNCONTROLLED STEEL AND IRON
PRODUCTION OPERATIONS

| | Emissions, lb/ton (kg/tonne) | | | |
| | | | Fluorides | |
Process	Particles	CO	Gaseous	Particles
Blast Furnace				
Pig iron production	150 (75)	1,750 (875)	—	—
Sintering	22 (11)	44 (22)	—	—
Open-Hearth Furnace Steel Production				
No O_2 lancing	8.3 (4.15)	—	0.1 (0.05)	0.03 (0.015)
With O_2 lancing	17.4 (8.7)	—	0.1 (0.05)	0.03 (0.015)
Basic Oxygen Furnace	51 (25.5)	139 (69.5)	Negligible	0.20 (0.10)
Electric Arc Furnace				
No O_2 lancing	9.2 (4.6)	18 (9)	0.012 (0.006)	0.238 (0.12)
Gray Iron Foundry				
Cupola	17 (18)	145 (72.5)	—	—
Reverberatory furnace	2 (1)	—	—	—
Electric induction furnace	1.5 (0.75)	—	—	—

Source: Part B, AP42, *US EPA Compilation of Emission Factors*, 3d Edition, August 1977

particles from these sources are larger than 250 μm and are easily collected. Particles in the atmosphere in the mills present a serious explosion hazard, so "housekeeping" is usually good. The internal air in the buildings is controlled by exhausting the uncollected particles to the atmosphere.

Woodworking operations such as cutting, sanding, and shaping release large particles which, although aesthetically objectionable, are above the respirable size ranges.

As a specific example of the nonferrous metals industry, consider primary lead smelting. Lead (Pb) usually occurs as the sulfide in ore, containing small amounts of copper, iron, zinc, and other trace elements. It is normally concentrated at the mine from ore of 3 to 8 percent Pb to ore concentrate containing 55 to 70 percent Pb and 13 to 19 percent free uncombined sulfur (S) by weight. Processing involves three major steps, starting with the concentrate:

1. In the sintering step, concentrated Pb and S are oxidized to produce PbO as sinter and SO_2 as a gas in an autogenous operation at about 1,000°C. Sintering machines can be updraft or downdraft, with the updraft preferred because it makes cleaner, harder sinter and because the SO_2 gas stream can be maintained at higher concentrations for subsequent flue gas desulfurization. Particles are also created which may be emitted.

2. In the reduction step, the PbO is reduced (oxygen is removed) to form molten Pb (bullion) in a blast furnace. The chemical reactions are

$$PbO + CO \rightleftarrows Pb + CO_2 \tag{13.4}$$

$$C + O_2 \rightleftarrows CO_2 \tag{13.5}$$

$$C + CO_2 \rightleftarrows 2CO \text{ (i.e., combustion with a deficiency of } O_2 \text{ in a reducing atmosphere)} \tag{13.6}$$

$$2PbO + PbS \rightleftarrows 3Pb + SO_2 \tag{13.7}$$

and

$$PbSO_4 + PbS \rightleftarrows 2Pb + 2SO_2 \tag{13.8}$$

The gaseous pollutants generated are thus CO_2, CO, and SO_2. Small particles are also created in this step.

The bullion is further processed by cooling to about 370–430°C and treating to remove impurities in a reverberatory furnace.

3. Refining takes place in cast iron kettles at an elevated temperature to remove antimony, tin, arsenic, precious metals, zinc, bismuth, and other trace metals. The end product is Pb pigs of 99.99 to 99.999 percent purity.

Typical emission factors and the size distribution of particles from an updraft sintering machine for lead-processing operations without controls are given in Table 13–8.

TABLE 13–8 TYPICAL EMISSION FACTORS AND PARTICLE SIZE DISTRIBUTIONS FOR UNCONTROLLED LEAD-PROCESSING OPERATIONS

	Emission factors	
Process	Particles lb/ton (kg/tonne)	Sulfur dioxide[1] lb/ton (kg/tonne)
Ore crushing	2 (1)	—
Sintering[2]	213 (106.5)	550 (275)
Blast furnace	361 (180.5)	45 (22.5)
Dross reverberatory furnace	20 (10)	—
Materials handling	5 (2.5)	—

Notes: 1. Only about half the sulfur in the concentrate is emitted as SO_2, the remainder going into the slag.

2. The particle size distribution from updraft sintering machines is as follows:

Size (μm)	% by weight
20–40	15–45
10–20	9–30
5–10	4–19
5	1–10

Source: Part B, AP42, *US EPA Compilation of Emission Factors*, 3d Edition, August, 1977

Particles from the blast furnace are mostly submicron size (92% < 4 μm). They readily agglomerate, and the agglomerates are cohesive but difficult to wet. They bridge and arch in hoppers and are difficult to handle. Some NO_x is generated in all of the high-temperature processes, of course.

Lead smelting is actually a chemical process. Moreover, if we maintained a material balance on all of the trace elements in the original ore, a host of other chemical releases would have been identified which are far too varied to discuss here.

Example 13.4

A lead sintering machine processing 100 tons per day of sinter at 6 percent S content removes 80 percent of the S in the ore as SO_2. The exhaust gases from the machine are processed through a single-stage sulfuric acid absorption plant where 96 percent of the SO_2 is converted to concentrated sulfuric acid (H_2SO_4). The carryover of H_2SO_4 as sulfuric acid mist is about 2 percent. An electrostatic precipitator (ESP) following the acid plant removes 99.5 percent of the mist. The combined acid plant (acting as a scrubber) and the ESP remove 99.9 percent of the solid particles emitted from the sintering machine. Estimate the mass of SO_2, H_2SO_4, and solid particles emitted to the atmosphere per day. The molecular weights are as follows: S = 32, SO_2 = 64, H_2SO_4 = 98.

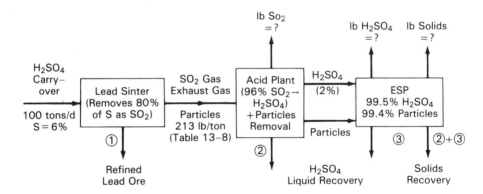

Solution   The daily rate of S oxidized to SO_2 is $0.06 \times 100 \times 2000 \times 0.8 = 9,600$ lb/day. The rate of SO_2 produced at the sintering plant is $9,600 \times (64/32) = 19,200$ lb/day, and the rate of H_2SO_4 produced is $0.96 \times 19,200 \times (98/64) = 28,224$ lb/day. The rate of SO_2 emitted per day is $0.04 \times 19,200 = 768$ lb/day, and the carryover of H_2SO_4 mist to ESP is $0.02 \times 28,224 = 565$ lb/day. The rate of H_2SO_4 discharged to the atmosphere is $0.005 \times 565 = 2.82$ lb/day ≈ 3 lb/day.

From Table 13–8, typical particle emissions from the sintering machines are 213 lb per ton of ore processed. So the total mass of the particles entering the acid plant is $213 \times 100 = 21,300$ lb/day, and the mass of particles emitted to the atmosphere is $0.001 \times 21,300 = 21.3$ lb/day ≈ 21 lb/day.

The combined sintering plant, sulfuric acid plant, and particle removal device emits to the atmosphere 768 lb/day of SO_2 and 24 lb/day of particles composed of 3 lb of H_2SO_4 mist and 21 lb of solids. It also recovers 28,220 lb/day of H_2SO_4 and has to dispose of 21,280 lb/day of ash.

13.3.8 Transportation-related Sources

The spark ignition engine which powers our automobiles releases more carbon monoxide, reactive hydrocarbons, and nitrogen oxides to the North American atmosphere than all other urban and industrial sources combined. The CO and HC compounds are products of inefficient combustion, and they would be eliminated by burning to CO_2 and H_2O in the engine to produce power if possible.

Most of the hydrocarbon emissions are from the tailpipe. These are controlled using catalytic reactors and by injecting air at the exhaust ports of the engine to burn emitted hydrocarbons in this high-temperature zone. Neither process recovers useful energy, so efforts to modify engine design have been intense, particularly since 1973, when the oil price spiral started. More than 20 percent of the uncontrolled automobile engine hydrocarbon emissions are from the crankcase vent (blowby and evaporated lubricating oil) and from the carburetor vent to the atmosphere. These emissions are controlled using a crankcase vent pipe to the carburetor (requiring a pollution control valve, or PCV) and a "carbon canister" adsorption unit, respectively.

Researchers have found that it was necessary to reduce emissions of nitrogen oxides to reduce photochemical smog and acid rain precursor emissions. (Pitts et al., 1983). Production of nitrogen oxides increases very rapidly with combustion temperature, so engines were redesigned to lower the cylinder temperature primarily by reducing engine compression ratios, modifying the ignition (spark) timing, introducing exhaust gas recirculation, and using two-stage combustion. Figure 13–5 shows the effect of exhaust gas recirculation on the production of nitrogen oxides for a spark ignition engine.

Concern also developed that lead emissions were approaching the limits of acceptability in the atmosphere for health reasons. However, lead-free gasoline was actually mandated

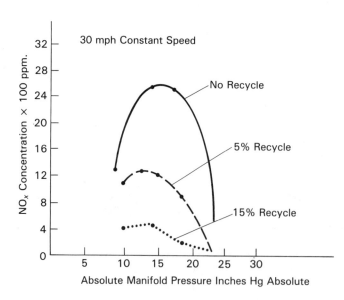

Figure 13–5 Reduction of nitrogen oxides in automobile exhaust gases as a result of exhaust gas recirculation. Source: Benson, J. D.: Reductions of Nitrogen Oxides in Automobile Exhaust. S.A.E. Int. Auto. Eng. Congress, Detroit, Mich., Jan 13–17, 1969.

TABLE 13–9 (A) TYPICAL CONCENTRATION OF CO, NO$_x$, AND HYDROCARBONS
IN THE EXHAUST FROM AN UNCONTROLLED SPARK IGNITION 300 CID, 4-STROKE ENGINE

Mode (operation)	Air/fuel ratio[1]	exhaust emissions			
		gas flow cfm	NO$_x$[2] ppm	HC[2] ppm	CO[2] ppm
Idle	12	6.5–7.0	30	350	17,000
Cruise					
City	13.5–14	40	1,000	⎰ 200–	6,000
Highway	13.5–14	35	1,000	⎱ 400	5,000
Acceleration					
Part throttle	13.3	45	1,700–2,500	⎰ 350–	7,000
Maximum	12.7	100–125	700–1,200	⎱ 400	
Deceleration					
Free	12	6.8	60	1,200	
Braking	12	6.8	60	1,200	18,000

[1] Mass Basis

[2] Volume Basis

TABLE 13–9 (B) U.S. FEDERAL EMISSION STANDARDS FOR LIGHT-DUTY
(AUTOMOBILE) GASOLINE-POWERED VEHICLES IN 1982, g/mi (g/km)

exhaust emissions	NO$_x$	HC	CO
emission standard, g/mi (g/km)	2.0 (1.24)	0.41 (0.25)	3.4 (2.1)

because the lead (about 3 g per gallon in high-octane leaded gasoline) "poisons" the catalysts in the reactors, significantly reducing their effectiveness. Table 13–9(A) shows that the emissions from an engine are also heavily dependent upon the way the engine is being operated, the so-called driving mode.

A review of Table 13–9(A) clearly shows that controlling the concentration of each pollutant in the automobile engine exhaust does not tell the whole story. Different engines have different displacements, and the gasoline flow through the engine varies with engine rpm (revolutions per minute) and the throttle opening. Consideration of these factors resulted in a change in the method of regulating automobile emissions to restrict the mass emission of pollutants per vehicle mile traveled, starting in 1970. Subsequently it was recognized that the vehicle driving cycle would have to be specified in emissions control legislation. Today, legislation limits emissions on the basis of grams per vehicle mile, using a specified driving cycle which includes each of the driving modes in a ratio typical of a normal trip to or from work in an urban environment (Table 13–9(B)).

Figure 13–6 is a schematic diagram showing the components of an uncontrolled spark ignition engine, the points at which pollutants are emitted, and the types and relative magnitudes of pollutant discharge at each point. Comparative emission data for well-tuned diesel and gasoline engines are given in Table 13–10.

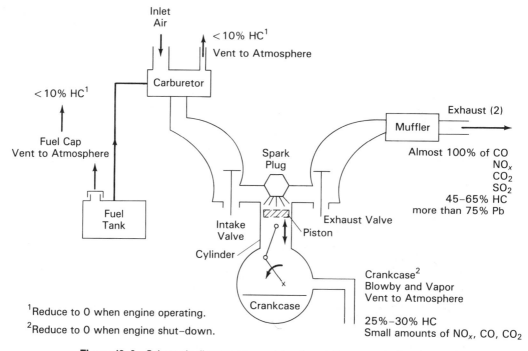

Figure 13–6 Schematic diagram summarizing the relative amounts, location, and types of emissions from an uncontrolled spark ignition engine

Small engines on boats and garden tools can also be substantial contributors to the overall air pollution problem. Table 13–11 shows the relative emissions from comparable four-stroke and two-stroke engines (usually oil and gas are mixed for two-stroke engines) at various loads. The dramatically higher hydrocarbon emissions from small two-stroke engines as compared to four-stroke engines is obvious.

TABLE 13–10 COMPARATIVE EMISSIONS[1] OF CO, NO_x, AND HYDROCARBONS FROM HEAVY-DUTY (TRUCK) GASOLINE AND DIESEL ENGINES, g/mi (g/km)

	Gasoline powered	Diesel
CO	117.0 (73)	28.7 (17.8)
HC	6.0 (3.7)	4.6 (2.9)
NO_x*	11.4 (7.1)	18.1 (11.2)

[1] For 1978 model year, assuming warmed up vehicle operation at an average of 18 mi/hr in urban traffic.

* As NO_2.

Source: Part B, AP42 U.S. EPA *Compilation of Emissions Factors*, 3d ed., August 1977.

TABLE 13–11 AVERAGE EMISSIONS FROM SMALL UTILITY TWO-STROKE AND FOUR-STROKE ENGINES

Load condition[1]	Exhaust emissions		g/Hp-h
	HC	CO	NO_x[2]
Summary of 27 four-stroke engines, rated 2 to 20.5 horsepower			
Full Load	8	180	5
Partial Load	12	230	5
Summary of 6 two-stroke engines, rated 3 to 6 horsepower			
Full Load	140	250	2
Partial Load	150	300	1
Summary of 1 two-stroke engine, rated 22 horsepower			
Full Load	79	89	10
Partial Load	43	201	1

[1] Adjusted for "lean-best power" air/fuel mixture setting.

[2] As NO_2.

Source: Eccleston, B. H., and Hurn, R. W., "Exhaust Emissions from Small Utility Internal Combustion Engines," Society of Automotive Engineers Paper No. 720197, Automotive Engineering Congress, Detroit, Mich., January 10–14, 1972.

13.4 CONTROL OF AIR POLLUTION

13.4.1 Natural Cleansing of the Atmosphere

Some typical particle sizes and removal mechanisms were identified in Section 6.2. The physical characteristics of particles influence natural removal mechanisms. Very small particles bounce around in random motion like gas molecules, and if they collide with other particles they grow by coagulation and fall out as large particles. Particles carrying an electrical charge grow or coagulate by attracting particles of opposite charge. Small particles, acting as nuclei, may fall within a raindrop. Alternatively, a raindrop may collide with and collect particles as it falls. These processes are called rainout and washout, respectively. Particles as large as 100 microns may impact on and stick to surfaces such as buildings, plants, and automobiles, to be subsequently washed off by rain. Some materials are hygroscopic and grow by collecting water molecules from the atmosphere to become liquid droplets. They continue to behave like and are still called particles.

Gases may be washed out of the atmosphere by precipitation (absorption), or they may be adsorbed (deposited) on solid particles and be removed by gravity. Gases also react chemically with other gases or particles in the atmosphere, forming new compounds which may be solid, liquid, or gaseous. For example, a series of elementary reactions leading to removal of hydrogen sulfide as particulate matter might be the following:

$$2H_2S + 3O_2 \rightleftarrows 2H_2O + 2SO_2 \text{ (both gaseous)} \qquad (13.9)$$

$$SO_2 + H_2O \rightleftarrows H_2SO_3 \text{ (sulfurous acid vapor,} \atop \text{or possibly a liquid droplet)} \qquad (13.10)$$

$$H_2SO_3 + PbO \rightleftarrows PbSO_4 + H_2 \text{ (lead sulfate as}$$
$$\text{a solid and hydrogen as gas)} \tag{13.11}$$

The reaction noted in Equation 13.9 might take hours in the atmosphere, and that of Equation 13.10 might occur in hours or days depending upon the availability of photochemical energy and other compounds which might act as catalysts. The last reaction could take place in minutes in an urban atmosphere where there was a relatively high concentration of PbO from combustion of leaded gasoline in cars.

In a more remote and cleaner atmosphere, Equation 13.11 might be replaced by

$$2H_2SO_3 + O_2 \rightleftarrows 2H_2SO_4 \tag{13.12}$$

The H_2SO_4 might be present as dispersed vapor molecules (i.e., in very low concentrations) which are readily absorbed in raindrops to fall as acid rain, or as solid particles which coalesce at relative humidities as low as 25 percent to form liquid droplets. Clearly, many secondary reactions between pollutants can occur—so many, in fact, that we are still identifying significant new pollutants.

13.4.2 Air Quality Control

The objective in air pollution control is to maintain an atmosphere in which pollutants have no negative impact on human activities. Obviously, the best way to control air pollution is not to produce the pollutants. For example, lead emissions from automobiles are eliminated by burning nonleaded fuels, and nitrogen oxide emissions have been significantly reduced by redesigning engines. We do not know how to eliminate nitrogen oxide emissions completely, but changing our means of transportation might do so. A possible alternative is to shift the location of the nitrogen oxides emissions—for example, from the automobile tailpipe to the stack of an electric generating station, by using electric or hydrogen-fueled cars. Legislating the quantity of ash and sulfur in fuels is a way of reducing emissions of these materials or their end products.

Other solutions include reducing emissions by using "add-on" devices. In the case of the automobile, carbon canisters are used to adsorb hydrocarbon vapors emitted from the carburetor and the gas tank. The vapors are subsequently returned to the engine for burning. In the automobile exhaust system, catalytic converters chemically reduce emissions of hydrocarbons. The energy in these hydrocarbons is lost as heat. In industrial applications, scrubbers (absorbers) may be used to remove pollutants from gas streams and possibly to induce them to react chemically to form more stable substances for collection and storage.

Planned dispersion may be used to control local air quality where emissions are not controllable by other techniques. Emissions from tall stacks have more time to disperse in the atmosphere before reaching the ground where they impact on humans, materials, and vegetation. For example, electric heating shifts emissions from short chimneys on houses in residential areas to tall stacks at remote locations.

The next section deals with add-on devices routinely used for air pollution control and then, as a less desirable alternative, atmospheric dispersion processes.

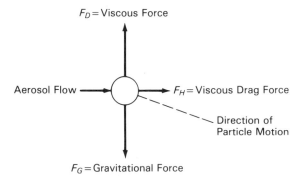

Figure 13–7 Forces acting on and the direction of motion of a particle in a moving air stream

13.4.3 Particle Emission Control

Designers of particle emission control equipment must deal with solid and liquid particles ranging from smaller than 1 μm to larger than 100 μm in diameter. The smaller particles are far more difficult to collect. Collectors are broadly categorized according to the physics of the collecting mechanism.

Gravitational settling chambers.   Gravitational settling chambers are simple, inexpensive collectors in which gravitational forces dominate vertical particle motions. They are essentially simple expansions in a duct in which the horizontal velocity of the particles is reduced to allow time for the particles to settle out by gravity. The forces acting on the particle are shown in Figure 13–7. In the figure, $F_D = F_G$ and the particle is assumed to fall at u_t, its terminal velocity. The horizontal component of the viscous force is negligible because the particle moves at the velocity of the gas stream.

Figure 13–8 is a sketch of a very simple gravitational settling chamber. A theoretical expression for the efficiency of this collector is

$$\eta_g = 1 - \exp\left\{-\frac{u_t L}{uH}\right\} \tag{13.13}$$

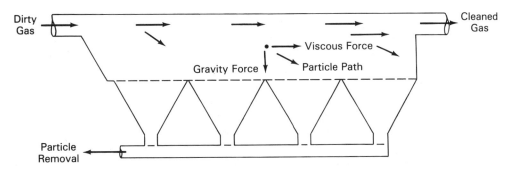

Figure 13–8 Gravitational settling chamber

or

$$\eta_g = 1 - \exp\left\{-\frac{g\,d_p^2\,\rho_p L}{18\,\mu u H}\right\} \tag{13.14}$$

where $\exp\{A\}$ is used to represent the exponential e^A, e being the base for natural logarithms, and where

η_g = efficiency of removal, as a fraction
L = length of the collector in m
H = depth of the collector in m
u = horizontal velocity of the gas and particles through the collector in m sec^{-1}

and other symbols are as previously defined (see Section 6.2.4).

Example 13.5

Calculate the 50 percent cutoff diameter for particles of CaO suspended in an airstream at 100°C and at atmospheric pressure for a gravitational settling chamber 3 m long and 1 m high when the gas velocity in the collector is 1 m/sec. The 50 percent cutoff diameter is defined as the particle diameter at which $\eta_g = 50$ percent, i.e., 50 percent of the particles of this diameter are collected and 50 percent are lost.

Solution   Using Equation 13.14, we obtain

$$\eta_g = 0.5 = 1 - \exp\left\{-\frac{g\,d_p^2\,\rho_p L}{18\,\mu u H}\right\}$$

ρ_p for CaO = 3310 kg/m³, and μ for air = 2.17×10^{-5} N · s/m² = 2.17×10^{-5} kg/m · s (at 100°C). Therefore,

$$0.5 = \exp\left\{-\frac{9.81 \times 3310 \times 3 \times d_p^2}{18 \times 2.17 \times 10^{-5}}\right\}$$

so that

$$d_p^2 = \ln{(0.5)} \times \left\{-\frac{18 \times 2.17 \times 10^{-5}}{9.81 \times 3310 \times 3}\right\}$$

$$= (-0.693) \times (-40 \times 10^{-10}) = 28 \times 10^{-10} \text{ m}^2$$

Therefore, the 50 percent cutoff particle diameter $d_p/50 = 5.3 \times 10^{-5}$ m = 53 μm

Note from this example that the cutoff diameter is not sharp, and in practice particles smaller and larger than $d_p/50$ are collected but the frequency distribution for collection particles is centered on $d_p/50$. Further, we might find that the performance of the collector varied considerably from design predictions because of turbulence and variations of flow in the collector. We compensate for the deviations by introducing empirical coefficients dependent upon the physical characteristics of the equipment and particles.

Inertial collectors.   A very simple particle skimmer is sketched in Figure 13-9(a), and a common particle separator called a cyclone is sketched in Figure 13-9(b).

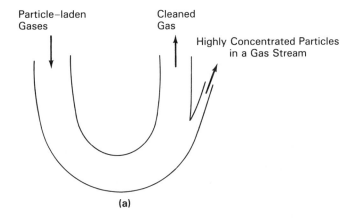

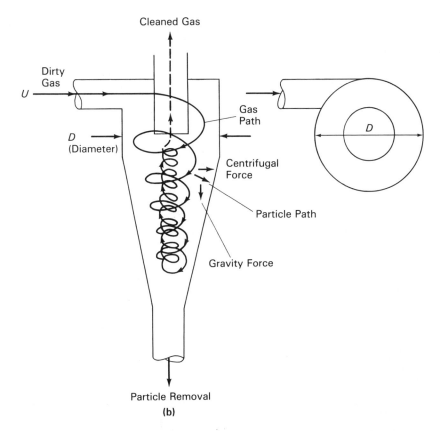

Figure 13–9 Simple inertial particle collectors. (a) A very simple centrifugal particle skimmer. (b) Cyclone collector.

Both collectors rely on centrifugal forces to separate the heavier particles from the lighter gas molecules. The skimmer shown in Figure 13–9(a) simply increases the particle concentration in a separated gas stream, which might then be passed through a gravitational collector (Figure 13–8) or possibly through a cyclone like the one in Figure 13–9(b).

As shown in Figure 13–9(b), particle-laden gases enter the cyclone tangentially at the top and, theoretically, spiral downward along the casing in solid body rotation at the entrance velocity u. Particles migrate to the outside of the spiral, where they slide down the casing to the hopper bottom. The only exit for gases from the cyclone is vertically upward through the central pipe, and to exit, the spiral must contract to a smaller diameter. The decrease in the radius of the particles' trajectory results in increased centrifugal force as the particles move toward the inner spiral. In practice, the gas velocity is found to vary roughly as $1/r^n$, where n varies from 0.5 to 0.7.

The magnitude of the centrifugal force F_c is

$$F_c = m_p \frac{u_T^2}{r} \tag{13.15}$$

or

$$F_c = \frac{\pi d_p^3}{6} \rho_p \frac{u_T^2}{r} \tag{13.16}$$

where u_T is the tangential velocity of the particle, and r is the radius of curvature of the particle trajectory.

Examination of Equation 13.16 indicates that the removal efficiency of centrifugal collectors increases

- Directly as the particle diameter cubed, i.e., very rapidly with d_p
- Directly as the density of the particle (ρ_p)
- Directly as the square of the tangential velocity u_t of the particle in the collector
- Inversely as the radius r of the particle trajectory.

This dependence of cyclone collection efficiency on the radius r has led to an arbitrary terminology in the trade which identifies large-diameter units as conventional cyclones and units having a diameter D less than about 15 cm as high-efficiency cyclones. A typical collection efficiency for a cyclone 1 m in diameter might be 50 percent for 20 μm particles, and a high-efficiency cyclone might have a collection efficiency $\eta_c = 80$ percent for $d_p > 10$ μm. The typical pressure drop through a conventional cyclone is 5 to 15 cm of water, and through a high-efficiency cyclone is 10 to 30 cm of water.

Gravitational settling chambers and simple inertial separators contain no moving parts. They may be fabricated using metals that can withstand high temperatures and resist corrosive attack by particles or gases. They are equally effective for solid or liquid particles.

Wet collectors. Wet collectors, or scrubbers, are designed to increase particle sizes using water or slurry droplets, because larger particles are easier to collect. There

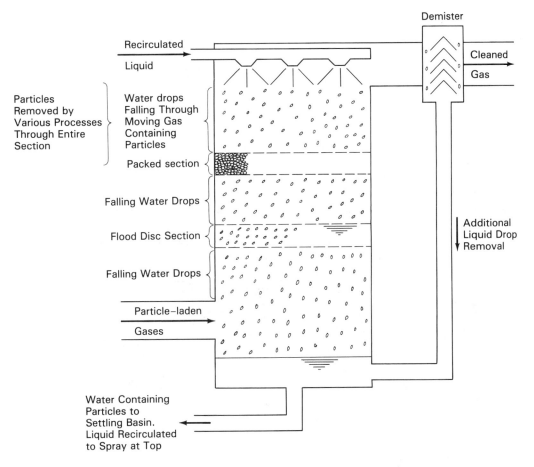

Figure 13–10   Sketch of an absorber or scrubber

are many different scrubber designs, but we shall consider only two types here: conventional and venturi scrubbers.

Figure 13–10 is a sketch of a scrubber showing several modes of particle collection. In the upper part of the tower, falling water drops collide with and collect particles from the upward-flowing gases. In the packed section, special shapes are added to increase the area of contact between the liquid and the aerosol (gas plus particles). Despite this use of special shapes, plugging remains a problem in this section. Below the packed section is a flooded perforated disc which may support several centimeters of water, allowing contact between the bubbles containing the particles and the liquid. The liquid drains through the perforations to develop another falling-drop collector section.

Not every collison of a water droplet and a particle results in collection, because of the surface tension of the droplets and particle wettability characteristics. Chemicals are sometimes introduced to reduce the surface tension of the droplet or to improve the ability of the droplets to absorb gases selectively in addition to collecting particles.

The liquid containing the particles is collected at the bottom of the tower and pumped to a settling basin or filtration device, where the particles are removed (see Chapter 11). The liquid may be recirculated with or without chemical treatment, resulting in a zero-discharge system and reduced makeup water requirements.

The demister at the exit of the scrubber is actually a particle collector. It is designed to remove drops of liquid carried over in the gas stream that leaves the scrubber (called an absorber when the scrubber is designed primarily to control gaseous emissions rather than particle emissions).

The size of the water drops is critical in determining the performance of a scrubber. If the water drops are very large relative to the particles, aerodynamic forces (drag) will displace the particles out of the path of the falling drops, and the number of collisions will decrease significantly. If the drops are the same size as the particles, collisions are also decreased because drag causes the collecting liquid droplets to move with the gas stream and particles. To optimize scrubber performance, the water drops should be a little larger than or smaller than the particles to be collected. The scrubber must be maintained to ensure that the desired drop distribution persists. Scrubber performance is also highly dependent upon the physical and chemical characteristics of the particles, the collecting liquid, and the final droplet (particle) collector.

In venturi scrubbers (Figure 13–11), gases and particles accelerate in the throat followed by rapid deceleration as expansion of the gas stream occurs. Liquid is injected into the venturi throat, usually in a solid stream perpendicular to the gas flow. The high relative velocity between the gas and the liquid streams results in the liquid stream being torn apart, and drops are formed. Particles moving with the gas stream collide with the

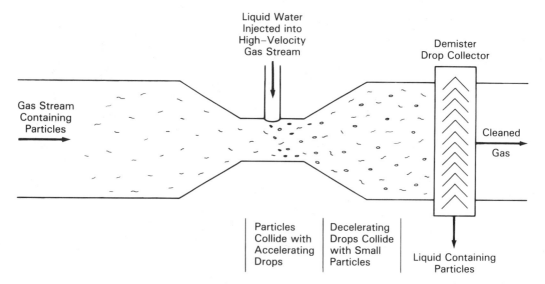

Figure 13–11 Venturi scrubber

liquid as the drops are being formed and become entrained in the drops. Drops which are larger than the particles decelerate more slowly than the particles in the expanding section of the venturi, and additional particle collection occurs through this section. The drops containing the particles are subsequently removed as large particles.

The performance of venturi scrubbers is critically dependent upon the gas stream velocity and the physical and chemical properties of the liquid and particles. Thus, the venturi scrubber must be operated at a constant gas flowrate if performance is to be maintained for a given particle size distribution and concentration in the gas stream. To overcome this limitation, venturis have been designed in which the throat area can be changed while the device is operating.

The pressure loss Δp through conventional scrubbers ranges from 15 to 40 cm of water. Collector efficiency increases with the pressure loss and may be as high as 95 percent for $d_p > 5$ μm. The pressure drop through venturi scrubbers ranges from about 50 to 200 cm of water. At high pressure drops, venturi scrubbers can collect particles as small as 1 μm at efficiencies approaching 99 percent, although this performance is extremely difficult to maintain continuously.

Scrubbers collect solid or liquid particles. They can be designed to resist corrosion, and they can be operated at relatively high temperatures as long as the liquid used does not boil and excessive evaporation losses can be prevented. For these reasons, venturi scrubbers are frequently used to collect the small particles generated in steel-making or smelting operations. Operating costs are relatively high for high-pressure loss scrubbers, but capital cost is low compared to other collectors of equivalent performance.

Fabric and fibrous mat collectors.   Fabric or baghouse collectors are similar to a vacuum cleaner on a grand scale. They are used to remove dry particles from dry, low-temperature (0 to 275°C) gas streams. Cloth socks 15 to 30 cm in diameter and up to 10 m long are suspended in a chamber, and air forced through the sock, discharges through the fabric. The fabric may be woven or made of felt, but woven cloth is by far the most common. Fabric materials include cotton, synthetics, and fiberglass, each having different adaptability to gas and particle temperature and physical and chemical characteristics, as indicated in Table 13–12.

A schematic of an industrial baghouse is shown in Figure 13–12. Use of multiple cells allows maintenance of the baghouse in individual cell blocks while the unit is in operation.

The cloth from which the bags or socks are made may have holes exceeding 100 μm across, but when properly operated, the collector performs with an efficiency greater than 99 percent for particles of $d_p > 1$ μm. Small particles are collected using the filter cake on the cloth surface as the filtration medium. As the thickness of the filter cake builds up, the pressure loss through the baghouse, and hence power costs, increase. If the porous filter cake becomes too thick, the pressure on the upstream side of the cake may collapse the cake into a more compact mass, and the pressure loss through the cake will then increase dramatically. If the pores in the filter become filled with liquid, a similar action occurs; hence, baghouses are limited to dry particle collection, and special precautions must be taken to prevent excessive condensation from the gas stream.

TABLE 13–12 SUMMARY OF DATA ON THE COMMON FILTER MEDIA USED IN INDUSTRIAL BAGHOUSES

Fiber	Operating exposure °F		Supports combustion	Air permeability* cfm/ft²	Composition	Resistance[+]				Cost[++] rank
	Long	Short				Abrasion	Mineral acids	Organic acids	Alkali	
Cotton	180	225	yes	10–20	Cellulose	G	P	G	G	1
Wool	200	250	no	20–60	Protein	G	F	F	P	7
Nylon	200	250	yes	15–30	Polyamide	E	P	F	G	2
Orlon	240	275	yes	20–45	Polyacrylonitrile	G	G	G	F	3
Dacron	275	325	yes	10–60	Polyester	E	G	G	G	4
Polypropylene	200	250	yes	7–30	Olefin	E	E	E	E	6
Nomex	425	500	no	25–54	Polyamide	E	F	E	E	8
Fiberglass	550	600	yes	10–70	Glass	P–F	E	E	G	5
Teflon	450	500	no	15–65	Polyfluoroethylene	F	E	E	E	9

* cfm/ft² at 0.5 in. of water.

+ P = Poor, F = Fair, G = Good, E = Excellent.

++ Cost rank: 1 = lowest cost, 9 = highest cost.

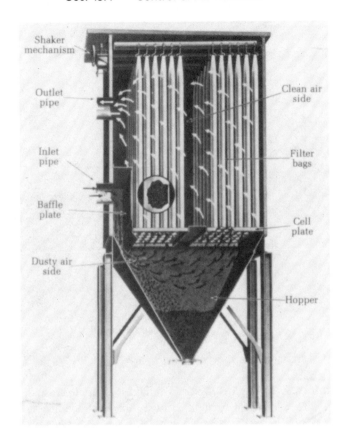

Figure 13–12 Typical simple baghouse with mechanical shaking (Courtesy, Wheelabrator Canada Inc.)

The filter cake is removed from small collectors by simply shaking the bag so that the cake falls off. Large industrial collectors may be cleaned more gently by passing a ring jet of air along the length of the bag or by momentarily reversing the flow through the bag. Some particle reentrainment does occur into the gas stream during the cleaning process, but most of the reentrained particles are simply collected again and removed in the next cleaning cycle. To avoid the necessity for very frequent bag shaking while maintaining a reasonable filter cake thickness for efficient particle collection without excessive pressure loss, the volume flowrate through the cloth is usually restricted to 0.5 to 2 m^3 sec^{-1} per m^2 of cloth surface.

Typical pressure drops across a baghouse range from 5 to 40 cm of water for shaking periods ranging from 4 or 5 times per hour to once in several hours. A typical bag life might be 2 to 3 years.

Fiber mat particle collectors operate at very low pressure drops and are frequently disposable, although many may be washed and reused several times. Fiber mat filters are used extensively in air conditioning and hot-air domestic heating systems, and for filtration of air entering internal combustion engines.

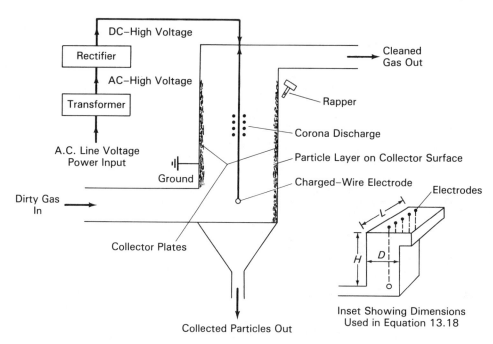

Figure 13–13 An elementary electrostatic precipitator

Electrostatic precipitators.   Figure 13–13 shows a model of an electrostatic precipitator (ESP). The voltage difference (field strength) between the electrode and the collector plates is maintained at as high a level as possible, but below the field strength at which spark-over occurs. Electrons are released at the electrode in a corona discharge and attach themselves to particles, thus charging the particles. The charged particles or molecules (ions) of the same polarity as the electrode migrate toward the grounded surfaces due to electrostatic forces. Positive or negatively charged electrodes may be used. The negative corona generates a slightly greater quantity of O_3 and is slightly more effective for industrial operation.

Migrating ions collide with liquid or solid particles in the gas stream, giving the particles a charge which results in particle motion toward the collector plates. The force acting on the particle is

$$F_E = q_p E_c \tag{13.17}$$

where F_E = electrostatic force acting on the particle

q_p = charge on the particle (a typical value is 0.3×10^{-15} coulomb (C)

E_c = electric field strength (a typical value is 10,000 volt cm^{-1}).

When particles touch the plates, they stick there. In time, a layer of particles which acts as an insulating blanket will collect on the plates, and the blanket surface charge may actually approach that of the electrode. This blanket must continually be removed by rapping the vertical plates so that the particle layer slides downward, by flooding the plates by washing them with liquid, or, if the particles collected are liquid, by having them run down the plate surfaces much like condensation on a window. The particles falling off the bottom of the plate are collected in hoppers for disposal.

The Deutsch equation for the efficiency of a plate-type electrostatic precipitator is

$$\eta_E = 1 - \exp\left\{-\frac{2u_e HL}{HDu}\right\} \tag{13.18}$$

where u_e = terminal migration velocity of the particle, m s^{-1} (typical values are 0.03 to 0.30 m s^{-1})

H = height of the collector plate, m

L = length of the collector plate, m

D = distance between the collector plates, m

u = velocity of the gas and particles through the precipitator, m s^{-1} (typically 0.50 to 2.50 m s^{-1})

and the factor 2 accounts for collection on both sides of the plates. Note that $HDu = Q$, the volume flowrate of gas through the collector in m^3 s^{-1}.

Gas and particle resistivity are important precipitator design variables because they determine the rapidity with which particles can be charged and the field strength in the collector sections. Resistivities vary with temperature and chemical composition. Conditioning agents, i.e., chemicals that significantly change resistivity, such as SO_3 and NH_3, are frequently used to improve collector performance.

Precipitator collection efficiencies are as high as 99 percent for particles larger than 2 μm at pressure losses of 5 cm of water or less. The units can be built entirely of metal. They are used almost exclusively on processes discharging corrosive gases at elevated temperatures in very large volumes containing a high percentage of particles larger than 1 μm.

Fire and explosion are always hazards when collecting combustible particles in dry collectors. They are particularly hazardous in electrostatic precipitators because of the danger of ignition by spark-overs.

Example 13.6

A large, uncontrolled cast iron production process emits 10 tonnes of particles per day. Two types of collectors are being considered for the process, an electrostatic precipitator (ESP) and a high-efficiency cyclone. The particle size distribution for the process and the collector efficiency for each size category are as follows:

Particle size range (μm)	0–10	10–20	20–44	>44
Average for class (μm)	5	15	32	say, 50
% by weight in class	20	35	30	15
% ESP efficiency for class	90	97	99.5	100
% cyclone efficiency for class	55	78	90	99

(a) Calculate the overall efficiency for the ESP and the high-efficiency cyclone.
(b) What is the efficiency of each collector for respirable particles?

(c) How many respirable particles of mean diameter 5 μm are released each day? Compare this with the number of particles 20 to 44 μm in diameter released each day. Assume that the particle density is 2,300 kg/m³.

Solution Sample calculations for 0 to 10 μm particles of nominal mean diameter 5 μm follow. The mass of 0 to 10 μm particles generated $= 10 \times 10^3 \times 0.2 = 2 \times 10^3$ kg/day. The mass collected by the ESP $= 2 \times 10^3 \times 0.9 = 1.8 \times 10^3$ kg/day, and that collected by the cyclone $= 2 \times 10^3 \times 0.55 = 1.1 \times 10^3$ kg/day. Also, the mass of particles emitted each day by the ESP $= (2 - 1.8) \times 10^3 = 2 \times 10^2$ kg, and that by the cyclone $= (2 - 1.1) \times 10^3 = 9 \times 10^2$ kg.

(a) The overall collector removal is given in the following table:

					Total
Particle size range, μm	0–10	10–20	20–44	>44	
Mass of particles in class, kg/d	2×10^3	3.5×10^3	3×10^3	1.5×10^3	10×10^3
Mass of particles collected, kg/d					
ESP	1.8×10^3	3.4×10^3	3×10^3	1.5×10^3	9.7×10^3
Cyclone	1.1×10^3	2.7×10^3	2.7×10^3	1.5×10^3	8.0×10^3

Thus, the overall collection efficiency for the ESP is

$$\frac{9.7 \times 10^3}{10 \times 10^3} \times 100 = 97\%$$

and for the cyclone is

$$\frac{8 \times 10^3}{10 \times 10^3} \times 100 = 80\%$$

(b) Only the particles smaller than about 10 μm are respirable. Therefore, the collector efficiency of the ESP for respirable particles is 90% by weight, and that for the cyclone is 55% by weight.

(c) The mass of a particle of mean diameter 5 μm is

$$\rho_p \frac{\pi d_p^3}{6} = 2,300 \times \frac{\pi}{6} \times (5 \times 10^{-6})^3 = 1.5 \times 10^{-13} \text{ kg}$$

The number of respirable particles of mean diameter 5 μm released each day from the ESP is

$$\frac{(2 - 1.8) \times 10^3}{1.5 \times 10^{-13}} = 13 \times 10^{14}$$

while that for the cyclone is

$$\frac{(2 - 1.1) \times 10^3}{1.5 \times 10^{-13}} = 60 \times 10^{14}$$

The number of particles of mean diameter 32 microns released each day from the ESP is

$$\frac{(3 - 2.985) \times 10^3}{3.9 \times 10^{-8}} = 1.2 \times 10^9$$

and that for the cyclone is

$$\frac{(3 - 2.7) \times 10^3}{3.9 \times 10^{-8}} = 7.7 \times 10^9$$

Note that there are roughly 100,000 times as many respirable particles as particles 20 to 44 μm in diameter released to the atmosphere each day from each collector despite the fact that the mass of particles in this size range is only 1.5 times as great!

13.4.4 Gas Emission Control

There are four fundamental ways to reduce emission of undesirable gases:

1. Reduce or eliminate the production of the undesirable gases.
2. Induce the gases to react after production in chemical processes to produce different, less objectionable emissions.
3. Selectively remove the undesirable product from a gas stream by *absorption*, which is the transfer of gas molecules into a liquid.
4. Selectively remove the undesirable gas by *adsorption*, which is the deposition of gas molecules on a solid surface.

Where the gas is recovered from the absorbing liquid or the adsorbing solid, the process is called regenerative because the liquid or solid is used repeatedly in the same process. In these cases the gas is most frequently processed further, making it a salable by-product from which part or all of the collection costs can be recovered.

Absorption processes.   Absorbers or spray towers are designed to selectively remove a specific gas from a mixture of gases and are similar to low-pressure-drop scrubbers (Figure 13–10). Where emissions of the gas must be reduced to very low concentrations, packed towers are frequently used. A typical absorption equilibrium curve describing the relationship between the partial pressure of a gas over a liquid and the concentration of gas in the liquid is shown in Figure 13–14. In the figure, C^* is the equilibrium concentration of gas molecules in the absorbing liquid corresponding to a partial pressure p of a gas above the liquid, and p^* is the equilibrium partial pressure in the bulk gas corresponding to a concentration C of the gas molecules in the liquid. In industrial scrubbers, flowing bulk liquid contacts the gas mixtures in which the partial pressure of the gas to be removed is p. The driving gradient of pressure or concentration moving the gas into the liquid is then $p - p^*$ or $C^* - C$, respectively. By analogy with heat transfer, an expression describing the absorption rate into the liquid for the gas is

$$\frac{dN}{dt} = K_G(p - p^*)A = K_L(C^* - C)A \tag{13.19}$$

where K_G and K_L are empirical coefficients whose values depend on the gas-liquid combination, flow patterns and turbulence in the scrubber, temperature, and other factors; N

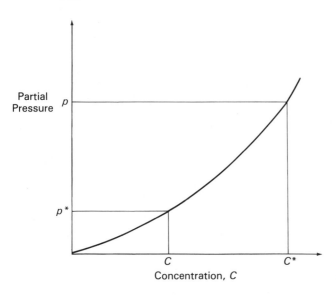

Figure 13–14 A typical absorption equilibrium curve

is the number of molecules of the gas being transferred, and A is the area of contact between the liquid and the gas. Henry's law is a special case of this equation where the equilibrium concentration curve is linear.

Equilibrium values of the percent by weight of HCl in water at various partial pressures are shown in Table 13–13 for several temperatures. The table clearly shows that the concentration of gas molecules in the liquid is much higher for a given gas pressure at lower temperatures. This is a typical absorption characteristic for gases in liquids. (Beer and soda water are kept refrigerated to maximize the CO_2 absorption in the liquid.) Absorption cannot easily be attained in industrial gas streams at elevated temperatures, so heat exchangers are frequently an integral component of gas scrubbers.

Adsorption processes.   Selective adsorption of gases is achieved in beds of solid adsorbent through which the gases are passed. The adsorbent is selected for its ability to collect the desired gas. For example, the small bags packed with delicate instruments or cameras contain silica gel, an adsorbent for water vapor. Adsorption is an extremely

TABLE 13–13 EQUILIBRIUM WEIGHT PERCENT OF HCl IN WATER AT VARIOUS TEMPERATURES AND PARTIAL PRESSURES

Weight of HCl in H₂O %	Partial pressures of HCl in mm of Hg				
	10°C	30°C	50°C	80°C	110°C
78.6	840	—	—	—	—
47.0	11.8	44.5	141	623	—
25.0	0.084	0.48	2.21	15.6	83
2.04	1.2×10^{-5}	1.5×10^{-4}	1.4×10^{-3}	2.5×10^{-2}	0.28

Source: Strauss, W., *Industrial Gas Cleaning*, London: Pergamon Press, 1966

TABLE 13–14 ADSORPTIVE CAPACITY AND RETENTIVITY OF ACTIVATED CHARCOAL
FOR THREE COMMON VAPORS

Substance	Adsorptive capacity weight %	Retention after removal weight %
Carbon Tetrachloride	80–110	27–30*
Gasoline	10–20	2–3*
Methanol	50	1.2**

* Regenerated by passing air at 25°C through the bed for 6 hours.
** Regenerated using steam at 150°C for 1 hour.
Source: Strauss, W., *Industrial Gas Cleaning*, London: Pergamon Press, 1966

complex thermo-chemical process which is not fully understood. As a gas deposits on the adsorbent surface, the heat of adsorption is released, resulting in heating in the solid. In special cases this heating can result in ignition of a carbon bed. For these reasons, the design of adsorbers is still an empirical process, and each manufacturer has developed proprietary coefficients for a given configuration, gas mixture and quality of gas recovered (or processed for disposal), adsorbent, operating temperature, gas-to-adsorbent volume ratios, bed depth, and gas flowrate.

The amount of adsorbate which the solid can take up is a function of the chemical and physical properties of the solid, particularly the surface area of the pores and fissures in the solid particles within which the gas molecules are deposited. In industrial processes, the adsorbent is frequently regenerated by passing hot steam through the bed, allowing the water molecules to displace the gas molecules at elevated temperatures. The concentrated gas may be recovered, dried, and reprocessed to yield a salable by-product, and the adsorbent is recycled.

Activated carbon and activated alumina are excellent adsorbents for several gases. Silica gel, as noted, is a good adsorbent for water vapor and other selected gases. The surface area of the adsorbents varies from 500 to 1,500 m² per g for activated carbon to 175 m² per g for silica gel. Table 13–14 shows the adsorptive capacity of activated charcoal for three common vapors.

13.4.5 Flow Diagrams for Typical Recovery Processes

The significance of SO_2 and H_2S as pollutants was recognized very early in the history of air pollution because of gross insult effects on plants and health in the case of SO_2 and because of the foul odor of H_2S at very low concentrations. Extensive work has been done to reduce SO_2 and H_2S in industrial gas streams over a wide range of concentrations, so we shall consider SO_2 and H_2S recovery processes as examples for further description.

Selective removal of SO_2 at high concentration in smelter gases. The dimethylanaline (DMA) process (Figure 13–15) is cost effective for removal of SO_2 at concentrations higher than 3.5 percent in the exhaust gases from the smelting of sulfide ores. The SO_2 is recovered as a liquid and is used in the production of fertilizer (ammonium

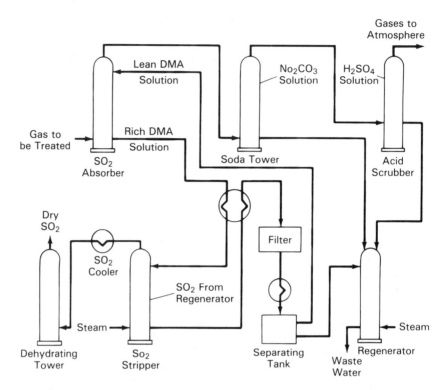

Figure 13–15 Flow diagram for the dimethylaniline process for SO$_2$ absorption and recovery. Source: Magill, P. L.; Holden, E. R., and Ackley, A. C., *Air Pollution Handbook*, McGraw-Hill, New York, 1956

sulfate), paper, and sulfuric acid (H$_2$SO$_4$). The DMA used in the process is regenerated by heating and is then cleaned and recirculated. The soda and acid scrubbers are secondary cleanup devices to recover any DMA and SO$_2$ that might be discharged from the primary scrubbers. The gases entering the primary scrubber must be cleaned of solid particles using an appropriate particle removal device to avoid contamination of the DMA solution. Demisters are required at the exit of each of the scrubbers to minimize particle (droplet) carryover to the next stage or to the atmosphere.

Selective removal of H$_2$S at high concentration from sour natural gas. Hydrogen sulfide, present at concentrations exceeding one percent by weight in sour natural gas streams, must be removed to make the gas acceptable as a domestic or engine fuel. Figure 13–16 shows a flow diagram of an H$_2$S scrubber used to remove H$_2$S and trace quantities of CO$_2$ from raw natural gas by means of a triethanolamine solution absorber following particle removal. Triethanolamine or diethanolamine solutions may be used in the scrubbers. The latter has a high absorptive capacity for H$_2$S, but also has a higher vapor pressure at scrubber operating temperatures and therefore must be cleaned

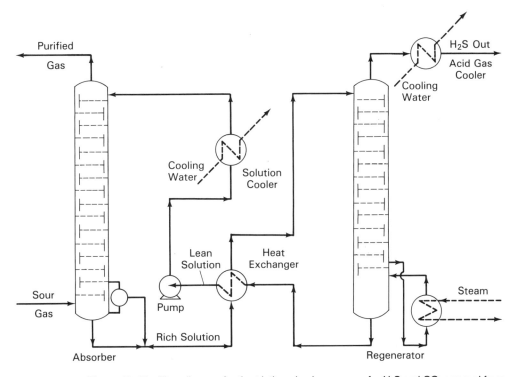

Figure 13–16   Flow diagram for the triethanolamine process for H_2S and CO_2 removal from sour gas streams. Source: Strauss, W., *Industrial Gas Cleaning,* London: Pergamon Press, 1966

from the exhaust gas stream in a secondary scrubber. The absorbent is regenerated in this example, and the H_2S is recovered for subsequent conversion to elemental sulfur or to H_2SO_4.

Selective removal of SO_2 at low concentration from gas streams.

Removal of SO_2 from coal or oil combustion gas streams has presented a particularly difficult problem, because the SO_2 is present at concentrations of less than 0.5 percent in a complex mixture of other gases and particles at elevated temperatures. Although the concentration of SO_2 is low, the quantity of gases discharged from combustion processes is very large.

Figure 13–17 is a sketch of a typical lime or limestone slurry scrubber and supplementary sludge processing components for removal of SO_2 from combustion exhaust gases. In the scrubber, the SO_2 reacts with the lime or limestone and forms sulfite or sulfate solids that remain in solution or suspension in the slurry. The solids are separated from the slurry in conventional settling tanks, and the liquid containing some suspended solids is enriched by adding more lime or limestone and is then recirculated. The sulfite may be stored in basins or ponds, or it may be converted to sulfate (sludge stabilization) by oxidizing in aeration tanks. The sulfite is thixotropic, i.e., it settles as a gel rather than as a solid. The

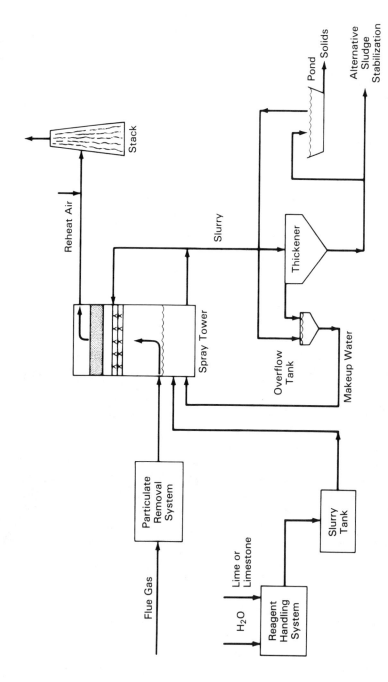

Figure 13-17 Schematic of a lime/limestone slurry scrubber system for removal of SO_2 selectively from fossil fuel combustion gases

sulfate is a stable solid (gypsum) which can be used in the building trades if it is not contaminated by other compounds or substances.

Common slurry reactants for removal of SO_2 include lime (CaO), slaked lime ($Ca(OH)_2$ = (CaO + H_2O)), limestone ($CaCO_3$ + $MgCO_3$; $CaCO_3$ > 50 percent of mass), and dolomite ($MgCO_3$ + $CaCO_3$; $MgCO_3$ > 50 percent of mass). Reactions (dry at elevated temperatures in the furnace, or wet in a scrubber and sludge tank) include the following:

$$CaO + SO_2 \rightleftarrows CaSO_3 \text{ (dry)} \qquad (13.20)$$

$$CaCO_3 + SO_2 + H_2O \rightleftarrows CaSO_3 + CO_2 + H_2O^* \text{ (wet)} \qquad (13.21)$$

$$2CaSO_3 + O_2 \rightleftarrows 2CaSO_4 \qquad (13.22)$$

$$2\ CaCO_3 + 2SO_2 + O_2 + 2H_2O \rightleftarrows 2CaSO_4 + 2CO_2 + 2H_2O^* \text{ (wet)} \quad (13.23)$$

More recently, the concept of dry scrubbing to remove SO_2 has gained wide acceptance and prototype dry scrubbers are being tested. In dry scrubbers, liquid slurry drops containing very little water are sprayed into a hot gas stream. The liquid evaporates as chemical reactions and absorption of SO_2 take place in similar reactions to those in wet scrubbers. The solid particles can be collected with fly ash in the precipitator or baghouse. The advantages of dry scrubbers are that a dry powder is produced, avoiding problems associated with disposal of a liquid slurry or wet sludge, much less water is required, liquid slurry pumping costs and equipment are reduced, and corrosion and plugging of scrubber sections where a slurry is being recirculated are almost eliminated.

It should be noted that the absorbers (scrubbers) described for removal of SO_2 from combustion gases are "chemisorbers". That is, SO_2 is removed by chemical reaction with scrubber liquid additives rather than being retained as the original gas in solution in the liquid. These processes are not regenerative and thus discharge material which must be disposed of. Some $CaSO_4$ can be sold as gypsum for construction and agricultural use if an acceptably pure product is produced; other materials are disposed of in storage basins.

Example 13.7

Coal containing 2.65% S and 10% ash with a heat value of 12,000 BTU per lb is burned in a 1,000-MW (1 MW = 3.413 × 10⁶ BTU/hr) generating station which operates at a 60% annual capacity factor. The station thermal efficiency is 33.3%. It is proposed that a limestone scrubber should be installed at the station to remove 90% of the SO_2 from the flue gases. The scrubber would precede the existing ESP, which currently operates at 99.5% removal efficiency. When the scrubber is installed, the efficiency of the precipitator will drop to 98% as a result of removal of the SO_2, which significantly affects the resistivity of both the ash particles and the flue gas. However, the scrubber will remove 60% of the fly ash from the exhaust gases in addition to the SO_2. Estimate on an hourly basis

    1. The amounts of coal burned and fly ash and SO_2 discharged to the atmosphere before installation of the scrubber.

    2. The amount of limestone required if 10% excess limestone (i.e., an amount greater than the stoichiometric requirement for 100% of SO_2 removal) is required to attain 90% SO_2 removal in the scrubber.

---

* The H_2O included in these equations represents the scrubber liquid or carrier for the chemical reactant.

3. The mass of dry sludge to be discharged from the scrubber and the mass of particles removed by the ESP after installation of the scrubber.

4. The volume of dewatered solids discharged to the ash storage area each year, assuming a specific gravity of 1.83 for the dewatered sludge (as $CaSO_4$) and 0.8 for the dry ash collected by the precipitator.

5. The mass of SO_2 and particles discharged to the atmosphere after installation of the scrubber.

A flow diagram of the process and Table 13–15 listing emission factors are provided for the problem: (Molecular weights: $CaCO_3$ = 100, $CaSO_4$ = 136, $CaSO_3$ = 120, S = 32)

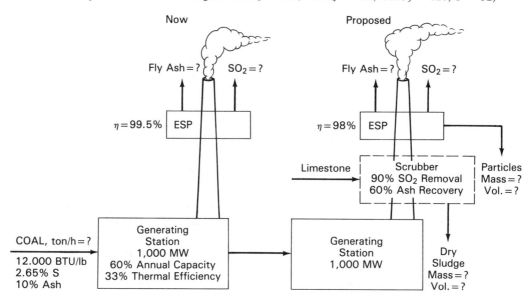

Solution

1. Coal required $= \dfrac{1{,}000 \times (3.413 \times 10^6) \times 0.60}{12{,}000 \times 0.333 \times 2.000}$ = 256 ton/h averaged over the year
                                                                            = 232 tonne/h

Fly ash generated = 8.5* × 10 × 232 = 19,700 kg/h = 19.7 tonne/h
Fly ash emitted from ESP = 19,700 × 0.005 = 98.6 kg/h
Sulfur generated = 19* × 2.65 × 232 = 11,700 kg/h S
                                    = 11.7 tonne/h S
Sulfur emitted = 11.7 tonne/h S (before scrubber installation)
SO_2 emitted = $11.7 \dfrac{64}{32}$ = 23.4 tonne/h SO_2

2. The rate at which limestone is required is obtained as follows. Assume that the limestone is $CaCO_3$ and contains 5% nonreactive materials (inerts). The chemical reactions are

$$SO_2 + CaCO_3 + \text{inerts} \rightarrow CaSO_3 + CO_2 + \text{inerts}$$

* See Table 13–15 for dry bottom, pulverized coal firing furnace.

TABLE 13–15 EMISSION FACTORS FOR BITUMINOUS COAL COMBUSTION WITHOUT CONTROL EQUIPMENT. EMISSION FACTOR RATING: A

Furnace size,[a] 10⁶ Btu/hr heat input	Particulates[b]		Sulfur oxides[c]		Carbon monoxide		Hydrocarbons[d]		Nitrogen oxides		Aldehydes	
	lb/ton coal burned	kg/t coal burned	lb/ton coal burned	kg/t coal burned	lb/ton coal burned	kg/t coal burned	lb/ton coal burned	kg/t coal burned	lb/ton coal burned	kg/t coal burned	lb/ton coal burned	kg/t coal burned
• Greater than 100 (Utility and large industrial boilers)												
—Pulverized												
General	16A	8A	38S	19S	1	0.5	0.3	0.15	18	9	0.005	0.0025
Wet bottom	13A[e]	6.5A	38S	19S	1	0.5	0.3	0.15	30	15	0.005	0.0025
Dry bottom	17A	8.5A	38S	19S	1	0.5	0.3	0.15	18	9	0.005	0.0025
—Cyclone	2A	1A	38S	19S	1	0.5	0.3	0.15	55	27.5	0.005	0.0025
• 10 to 100 (large commercial and general industrial boilers)												
—Spreader stoker[f]	13A[g]	6.5A	38S	19S	2	1	1	0.5	15	7.5	0.005	0.0025
• Less than 10 (commercial and domestic furnaces)												
—Underfeed stoker	2A	1A	38S	19S	10	5	3	1.5	6	3	0.005	0.0025
• Hand-fired units	20	10	38S	19S	90	45	20	10	3	1.5	0.005	0.0025

[a] 1 Btu/hr = 0.252 kcal/hr.

[b] The letter A on all units other than hand-fired equipment indicates that the weight percentage of ash in the coal should be multiplied by the value given. For example, if the factor is 16 and the ash content is 10 percent, the particulate emissions before the control equipment would be 10 times 16, or 160 pounds of particulate per ton of coal (10 times 8, or 80 kg of particulates per t of coal).

[c] S equals the sulfur content (see footnote b above).

[d] Expressed as methane.

[e] Without fly-ash reinjection.

[f] For all other stokers, use 5A for particulate emission factor.

[g] Without fly-ash reinjection. With fly-ash reinjection, use 20A. This value is not an emission factor, but represents loading reaching the control equipment.

Source: U.S. EPA, *Air Pollution Emission Factors*, AP42, 3d ed., 1980

and

$$2CaSO_3 + O_2 + inerts \rightarrow 2CaSO_4 + inerts$$

Thus, 1 mol of SO_2 reacts with 1 mol of $CaCO_3$ to form 1 mol of $CaSO_4$ (gypsum). Therefore, the limestone requirement, based on 5% inerts and 10% excess, is

$$11,700 \times \frac{100}{32} \times \frac{1.10}{0.95} \times \frac{1}{1,000} = 42.3 \text{ tonne/h } CaCO_3$$

3. The mass of dry sludge to be removed from the scrubber is obtained as follows:
   (a) The rate of fly ash removal as scrubber sludge is $0.60 \times 19,700 = 11,830$ kg/h.
   (b) The rate of $CaSO_4$ removal as scrubber sludge is

$$0.90 \times 11,700 \times \frac{136}{32} = 44,750 \text{ kg/h}$$

   (c) The rate of unreacted limestone usage is 42,300 kg/h, and the rate at which limestone reacts to $CaSO_3$ (MW = 120) is

$$0.90 \times 11,700 \times \frac{120}{32} = 39,500 \text{ kg/h}$$

Therefore, the unreacted limestone, including inerts, amasses at the rate of $42,300 - 39,500 = 2,800$ kg/h, and the rate at which dewatered scrubber sludge mass is removed is $11,830 + 44,750 + 2,800 = 59,300$ kg/h = 59.3 tonne/h.

   The rate at which particles are removed in the ESP following scrubbing is $(1 - 0.60) \times 19,700 \times 0.98 = 7,730$ kg/h = 7.7 tonne/h. Therefore, the rate at which dewatered solids are removed from the scrubber and the ESP = *67.0 tonne/h*. This compares with 19.7 tonne/h of fly ash removed before installation of the scrubber.

4. The volume of dewatered solids discharged per hour to the storage area from the scrubber/ESP system is

$$59.3 \text{ tonne/h} \times \frac{1}{1.83} + 7.7 \times \frac{1}{0.8} = 32.4 + 9.6 = 42.0 \text{ m}^3/\text{h}$$

This compares with $19.7/0.8 = 24.6$ m³/h before installation of the scrubber. Thus, the storage area must be increased by a factor of $42.0/24.6 = 1.7$.

5. The mass of SO_2 discharged to the atmosphere after installation of the scrubber is $23.4 \times 0.1 = 2.3$ tonne/h, and the mass of fly ash discharged to the atmosphere after installation of the scrubber is $19.7 - 11.8 - 7.7 = 0.2$ tonne/h.

13.4.6 Nitrogen Oxide Emission Control

Control of nitrogen oxide emissions from combustion processes provides a good example of air pollution control by reducing the amount of pollutant produced. Chemical kinetics reveals that nitric oxide is produced much more rapidly as the temperature of the reaction increases. Table 13–16 gives the theoretical reaction times for NO formation at several

TABLE 13–16 TIME FOR NO FORMATION IN A GAS
CONTAINING 75 PERCENT NITROGEN
AND 3 PERCENT OXYGEN

Temperature °F	Time to 500 ppm NO sec	NO at equilibrium ppm
2,400	1,370	550
2,800	16.2	1,380
3,200	1.10	2,600
3,600	0.117	4,150

Source: U.S. EPA, *Control Techniques for Nitrogen Oxides Emissions
from Stationary Sources*, AP 67, 1970

elevated temperatures in a gas mixture containing 3 percent O_2 and 75 percent N_2, the approximate concentrations in combustion gases from boilers. The table shows that if the reaction temperature can be reduced, there will be much less NO produced for the same residence time.

In combustion processes, lower flame or combustion chamber temperatures can be achieved by burning the fuel more slowly and using multistage combustion. Recirculation of exhaust to dilute the fuel-air mixture in the combustion chamber of internal combustion or stationary combustion sources has a similar effect (See Figure 13–5). Research indicates that there are many supplementary factors which affect the reaction rates of N_2 and O_2. The presence of other substances in solid, liquid, or gaseous forms, the rate and extent of quench or gas cooling after formation of the NO, and the concentration of the participating gases are important controlling factors. Design for minimum NO_x production is, therefore, largely based on empirical factors.

13.4.7 Ambient Air Quality Control by Dilution

For some operations, the technology to control emissions is still in a developmental state. In these cases superimposing an air pollution control process on a production process may result in unacceptable downtime, excessive costs for maintenance and operation, and high capital and interest charges. According to our original definition of air pollution, a substance is not a pollutant unless it causes an effect. In the vicinity of the source (called the near field), dilution is frequently used to achieve acceptable air quality. Dilution using tall stacks may be a more economical way to attain an air quality standard than installation of removal systems.

During the sixties and seventies, when many control processes were in the developmental state, dilution was used extensively for air pollution (air quality) control. As additional information developed, it was recognized that dilution is not always an acceptable solution even though near-field effects can be eliminated. For example, consider sulfur and nitrogen oxides and the chlorine compounds in the atmosphere. On the continental or global scale, it is recognized that a reduction in acid rain impacts can be achieved only by reducing emissions at the source or by instituting mitigation measures such as lake liming in the field. This has resulted in reconsideration of the value of dilution to control air quality and renewed and increased emphasis on source emission control processes.

13.5 PREDICTING AIR POLLUTANT CONCENTRATIONS

The design of industrial complexes, community planning, identification of significant sources, and prediction of pollutant concentrations at selected receptors are usually done using mathematical models. Important inputs to air pollution models include the type, character, and distribution of the sources; the pollutants emitted; meteorological variables which determine the transport and dispersion of pollutants; and the chemical reactions of pollutants in the atmosphere. In the near field (less than 20 km from the source), except for selected pollutants such as fluorine, H_2S, and photochemical oxidants, we can usually neglect atmospheric chemical reactions and removal processes. In the far field (greater than 100 km from the source), chemical reactions and removal processes become increasingly important. Because models are greatly simplified representations of actual processes, single-valued predictions should be regarded as being within a factor of two at best. We might consider this accuracy analogous to the safety factor used in the civil and mechanical design of structures and machine components.

13.5.1 Air Pollution Meteorology

The key factors in air pollutant transport, i.e., winds, turbulence, and temperature in the atmosphere, were discussed in detail in Chapter 7. Pollutants are transported at the speed and in the direction of the wind. At the same time, pollutants released as a plume from a continuous source or as a puff from an instantaneous source disperse under the action of turbulence. Wind speed, direction, and turbulence in the layer of air from the earth's surface to about 1 km above the surface, where most pollutants are released, are all strongly influenced by the vertical atmospheric temperature structure referred to as the lapse rate of temperature.

    If a parcel of air is moved vertically in the atmosphere, and no entrainment of outside air, condensation, or evaporation occurs, the temperature inside the parcel changes at the dry adiabatic lapse rate Γ_D, equal to 9.8 K/km of rise. If the air temperature in the atmosphere outside the parcel decreases by $\gamma = 5$ K/km of rise (see Section 7.4), then after rising 1 km, the temperature of the air in the parcel will be 4.8 K lower than that of the surrounding air. (γ is called the actual lapse rate.) Since the pressure both inside and outside of the parcel will be the same at all levels (pressure variations are transmitted at the speed of sound, but heat transfer is very, very much slower), the air in the parcel will have a higher density than the surrounding air, given by

$$\rho_p = \frac{p_p}{RT_p} \qquad \text{and} \qquad \rho_A = \frac{p_A}{RT_A} \qquad (13.24)$$

where the subscripts p and A refer to the parcel and the ambient atmosphere, respectively, and R is the gas constant for air. Note that $p_p = p_A$ and if $T_p < T_A$, then $\rho_p > \rho_A$.

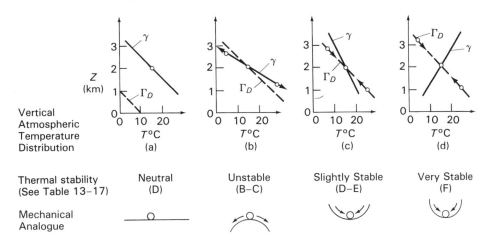

Figure 13–18 Atmospheric lapse rates and stability

It follows that the parcel of air at higher density will fall through the surrounding air until it reaches its original position, where the temperatures inside and outside of the parcel are the same. If the parcel is moved downward 1 km, the opposite occurs, and the buoyant parcel at higher temperature returns to its original position. Because the parcel always has a tendency to return to its original altitude in this atmosphere, we say that the atmosphere is thermally stable. In a thermally stable atmosphere, both vertical and horizontal atmospheric turbulence and the vertical transfer of momentum are suppressed.

The effect of atmospheric temperature distributions on vertical motions is depicted in Figure 13–18. In (a), the curve γ denoting the temperature change with height in the environment is the same as the dry adiabatic lapse rate, denoted by the curve Γ_D. A parcel of air moved vertically will remain where it is placed or will continue to move vertically if its motion is not stopped by external forces. We describe this atmosphere as being thermally neutral (see also Equation 7.7 and the accompanying discussion).

In Figure 13–18(b), $\gamma > \Gamma_D$, and a parcel moved vertically will accelerate as it moves further from its original height. Vertical motions are thus strongly enhanced, and we say that the atmosphere is unstable. Part (c) of the figure is the case described in detail for $\gamma < \Gamma_D$ and is considered to be slightly stable. Part (d) shows an inversion, where the temperature in the atmosphere increases with height instead of decreasing. In an inversion, vertical motions and turbulence are very strongly suppressed and the atmosphere is classified as being very stable.

At this point, you may well ask how representative our air parcel analysis is when it is applied to an actual situation. Observations indicate that the analysis is in fact remarkably good for vertical motions and turbulence in the real atmosphere.

In Figure 13–19, an instantaneous and time-averaged plume outline is shown corresponding to each of the stability regimes in Figure 13–18. Clearly, when vertical motions are enhanced, the time-averaged plume cross section is much larger and we say dispersion is greater and concentrations in the plume are lower. In the stable atmosphere represented

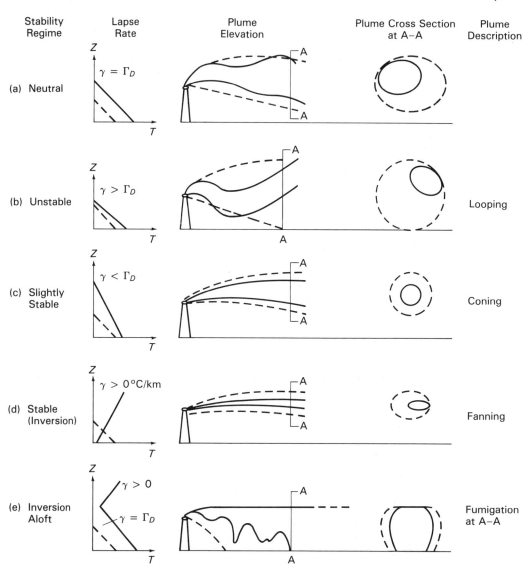

Figure 13–19 Plume descriptions for selected atmospheric stability regimes. Solid and dashed plume lines represent, respectively, instantaneous and time averaged (over several minutes or hours) plume boundaries.

by an inversion, dispersion is very slow and if the plume approaches vegetation, a building, or a surface, these may be exposed to very high pollutant concentrations.

In Figure 13–19(e), a two-layer atmospheric lapse rate is depicted with strong dispersion near the ground, but with an inversion and limited vertical diffusion aloft. This lapse rate configuration might develop inland when cold air blows off a lake over warmer (heated) land during summer. The inversion acts almost like a lid for vertical diffusion, but the strong lapse rate and dispersion below the inversion bring pollutants to the surface at high concentrations close to the source in a "fumigation."

In a surface inversion, pollutants remain in a shallow layer and winds are always light, so high concentrations of pollutants are observed in pollutant releases near the ground. Surface inversions are common on windless, very clear nights when the earth and the air in contact with it cool by radiating to space (called a **radiation inversion**), when warm air passes over a cold surface such as a lake or snow-covered land (called an **advection inversion**), and when air in contact with the ground along the slopes of a valley cools at night and drains down the slopes into the valley to create a pool of cool air overlain by warmer air aloft (called a **drainage inversion**).

Inversions aloft form when a deep layer of cold air passes over a warmer surface such as heated land or an open lake in winter when land is snow covered. In this case, heating at the surface creates an unstable layer below the inversion, as in Figure 13–19(e). Inversions aloft may also occur over a city because of heat release in the "heat island" (see Chapter 7). Large-scale inversions aloft also develop by complicated dynamic and thermodynamic processes. The persistent Los Angeles smog develops under a large-scale inversion aloft that restricts vertical dispersion. Other important factors in the L.A. smog are the low coastal mountain range to the east, prevailing westerly winds, and more rapid atmospheric chemical reactions when pollutant buildup occurs.

Observations indicate that wind speed varies with height above the ground as a function of atmospheric stability and surface roughness. The variation of wind speed with height is frequently described using the power law

$$u_z = u_{10} \left(\frac{Z}{Z_{10}} \right)^p \tag{13.25}$$

where u_z = wind speed at height Z above the ground

$\quad u_{10}$ = wind speed at 10 m (measurement height specified by World Meteorological Organization for meteorological stations)

$\quad p$ = exponent depending upon atmospheric stability and the character of the underlying surface (varies from about 0.1 to 0.4).

Other important meteorological variables for air pollution analysis include precipitation, cloud cover, and radiation to or from the earth's surface. Precipitation removes pollutants from the atmosphere by the physical processes of rainout and washout and by the chemical processes of oxidation and hydrolization. Cloud cover and radiation are excellent indicators of atmospheric stability when considered with other variables such as wind speed, time of day, and characteristics of the underlying surface.

13.5.2 Pollution Dispersion Models

By making many simplifying assumptions, Gaussian diffusion equations can be developed to describe the atmospheric dispersion of a puff in three dimensions or a steady-state plume from a continuous source in two dimensions (Slade, 1968; Randerson, 1981). For the simplest model, we assume that a plume traveling horizontally (in the x direction) at a mean speed $\bar{u}$ disperses horizontally (y) and vertically (z) so that the concentration of a pollutant at any cross section of the plume follows the normal (Gaussian) probability distribution. If also, for any point (x, y, z) in the plume, the concentration C of pollutant at that point is such that

$$C_{(x,y,z)} \propto \frac{1}{\bar{u}} \quad (\bar{u} = \text{average wind speed})$$

$$C_{(x,y,z)} \propto Q \quad (Q = \text{source strength})$$

and

$$C_{(x,y,z)} \propto G \quad (G = \text{normalized Gaussian curve in the } y \text{ and } z \text{ directions})$$

then

$$C_{(x,y,z)} = \frac{Q}{\bar{u}} G_y G_z$$

The expression for the Gaussian function G_y normalized so that the area under the curve is unity (Perkins 1974) is

$$G_y = \frac{1}{\sqrt{2\pi}\,\sigma_y} \exp\left[-\frac{1}{2}\left(\frac{y}{\sigma_y}\right)^2 \right]$$

and similarly for G_z so that

$$C_{(x,y,z)} = \frac{Q}{\bar{u}} \left\{ \frac{1}{\sqrt{2\pi}\sigma_y} \exp\left[-\frac{1}{2}\left(\frac{y}{\sigma_y}\right)^2 \right] \right\} \left\{ \frac{1}{\sqrt{2\pi}\sigma_z} \exp\left[-\frac{1}{2}\left(\frac{z}{\sigma_z}\right)^2 \right] \right\}$$

$$= \frac{Q}{2\pi\sigma_y\sigma_z\bar{u}} \exp\left[-\frac{1}{2}\left(\frac{y}{\sigma_y}\right)^2 \right] \exp\left[-\frac{1}{2}\left(\frac{z}{\sigma_z}\right)^2 \right] \tag{13.26}$$

where σ_y and σ_z are the standard deviations of the dispersion in the y and z directions, respectively, $x = 0$ at the source and y and z are zero on the plume center line.

In order to relate this expression to the ground level rather than to the centerline of the plume, we can make the height of any point C in the plume a distance Z above the ground. In this case, the vertical height of point C above the centerline of the plume becomes $Z - H$ (see Figure 13–20) and the equation becomes

$$C_{(x,y,z)} = \frac{Q}{2\pi\sigma_y\sigma_z\bar{u}} \exp\left[-\frac{1}{2}\left(\frac{y}{\sigma_y}\right)^2 \right] \exp\left[-\frac{1}{2}\left(\frac{Z-H}{\sigma_z}\right)^2 \right] \tag{13.27}$$

Provided that the plume does not impinge on the ground (and that our previous assumptions are valid), this model should apply. However, because the ground tends to reflect rather than remove pollutants, a technique assuming 100 percent reflection of pollutants is used to account for the increased pollutant concentration at ground level (see Figure 13–20 again). A mirror image of the plume is envisaged, and the concentration of pollutant at an imaginary

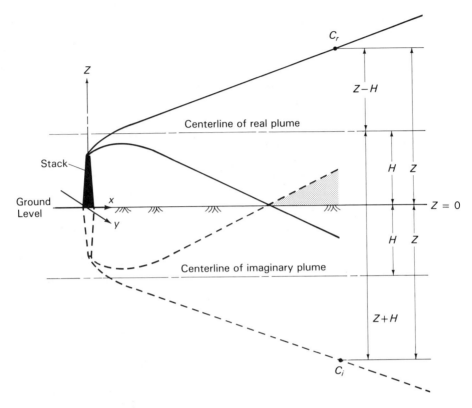

Figure 13–20 Definition sketch showing plume dispersion in the vertical direction and the reflection of pollutants at ground level

C_r and C_i = concentration due to real and imaginary sources, respectively.

point (at a location $Z + H$) is added to the concentration in the real plume. The plume diffusion equation, in its most common form, then becomes

$$C = \frac{Q}{2\pi\sigma_y\sigma_z\bar{u}} \exp\left[-\frac{1}{2}\left(\frac{y}{\sigma_y}\right)^2\right]\left\{\exp\left[-\frac{1}{2}\left(\frac{Z-H}{\sigma_z}\right)^2\right] + \exp\left[-\frac{1}{2}\left(\frac{Z+H}{\sigma_z}\right)^2\right]\right\}$$

(13.28)

where,

- C is the pollutant concentration (kg/m³) at a receptor located at (x, y, z)
- σ_y and σ_z are diffusion coefficients in the y and z directions, respectively (m), and are functions of the downwind distance x from the source;
- $\bar{u}$ is the mean wind speed through the layer in which diffusion takes place (m/s);

- x, y, and z are spatial coordinates of the receptor (m) relative to the source (the x-axis is oriented in the direction of the mean wind, y is at right angles to x in the horizontal plane, z is in the vertical plane, and Z is the vertical coordinate relative to ground level);
- H is the effective height of the pollutant release (m); and
- Q is the source emission rate (kg/s)

Some of the assumptions made to develop this equation are as follows:

1. All of the pollutants are emitted from a point source of infinite strength.
2. The wind is uniform through the layer in which dispersion occurs, and an average or mean wind can be used in the equation. In practice, the wind used in the equation is taken to be the wind at the top of a stack for an elevated source, estimated using Equation 13.25.
3. The concentration distribution across the width and depth of the plume is Gaussian.
4. The edges of the plume are where the concentration of pollutant has decreased to one-tenth of the plume centerline value.
5. The pollutant under consideration is not lost by decay, chemical reaction, or deposition; i.e., it is conservative. The method of images is used to assure that pollutants are not lost to the ground. (It is supposed that all of the pollutant which impinges at the earth's surface is fully "reflected.")
6. The equation is to be used over relatively flat, homogeneous terrain. It should not be used routinely in coastal or mountainous areas, in any area where building profiles are highly irregular, or where the plume travels over warm bare soil and then over colder snow or ice-covered surfaces.
7. The equation represents a steady state solution ($\delta Q/\delta t = 0$) over the averaging period.
8. The pollutants have the same density as the air surrounding them. This assumption is remarkably close for the case of stack gases from fossil-fuel-fired combustion processes. It is satisfactory for small particles, but not for particles which have a finite and significant fall velocity.

Note how the equation reduces to a simpler form for concentrations at specific ground-level locations, such as at a distance y from the centerline ($Z = 0$) or on the centerline of the plume ($y = 0$, $Z = 0$).

The values of σ_y and σ_z have been determined empirically and are conveniently graphed as functions of x and atmospheric turbulence or stability categories in Figure 13–21. Table 13–17 describes the method for determining the stability categories based on wind speed, time of day (radiation), and cloud cover. Category A corresponds to an extremely unstable atmosphere, F to a very stable atmosphere, and D to a near-neutral atmosphere. The curves in Figure 13–21 are for continuous point-source plumes over averaging periods of 10 minutes or so. They should not be used to describe the diffusion of a puff in three dimensions. Using data from Figure 13–21, Equation 13.28 gives the 10-minute average

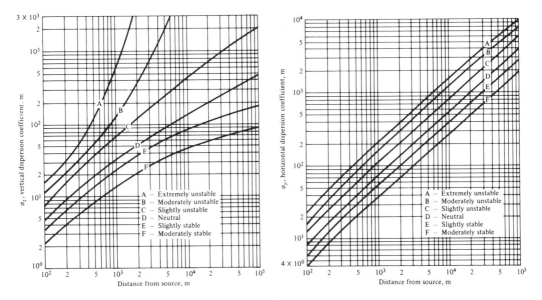

Figure 13–21   Plume dispersion coefficients as a function of downwind distance from the source. Source: Davis 1973

concentration in a plume having dimensions described by the dashed lines in Figure 13–19.

13.5.3 Plume Rise Models

The final variable required in the application of Equation 13.28 is the plume height H. Observation of a plume emitted from a stack at a temperature T_s above the ambient air

TABLE 13–17   KEY TO STABILITY CATEGORIES

Surface wind speed (at 10 m), m sec^{-1}	Day			Night	
	Incoming solar radiation			Amount of overcast	
	Strong	Moderate	Slight	$\geq$ 4/8 Low cloud	$\leq$ 3/8 Low cloud
<2	A	A–B	B		
2–3	A–B	B	C	E	F
3–5	B	B–C	C	D	E
5–6	C	C–D	D	D	D
>6	C	D	D	D	D

Note: Categories A, B, C range from extremely unstable to slightly unstable. Category D is neutral and should be assumed for overcast conditions during day or night. Categories E, F indicate slightly stable and moderately stable conditions respectively. Adapted from Turner, 1970.

temperature T_A shows that the plume rises above the top of the stack due to its discharge momentum and its thermal buoyancy. For plumes from combustion sources, the momentum rise is so small relative to the buoyancy rise (due to the high temperature of the plume) that it can be neglected. The final plume height H is the sum of the stack height H_s and the combined momentum and buoyancy plume rise ΔH, i.e. $H = H_s + \Delta H$.

Many plume rise equations have been proposed, but those developed by Briggs (1969) using dimensional analysis are the most widely used today. Briggs postulated that plume rise occurs simultaneously with a relatively rapid plume expansion (diffusion) as a result of entrainment of ambient air into the plume. Therefore, the plume rise must also be a function of the stability of the atmosphere. Briggs proposed the following equations to describe the buoyant rise of a warm plume:

1. For stable and near-neutral conditions,

$$\Delta H = 2.6 \left(\frac{F}{\bar{u}S}\right)^{1/3} \tag{13.29}$$

where F is the initial buoyancy flux of the emitted plume defined by

$$F = \frac{g(T_s - T_A)W}{T_s}\left(\frac{D}{2}\right)^2 \tag{13.30}$$

where, in turn, g = acceleration due to gravity (m/s²)
T_s and T_A = temperatures of the emitted gas and the environment, respectively, at the point of emission (K)
W = exit velocity of the plume (m/s)
D = diameter of the stack at the top (m)
$\bar{u}$ = mean wind speed through the atmospheric layer of depth H, frequently taken to correspond to u at the height H_s of the stack
S = a stability parameter defined by

$$S = \frac{g}{T_A}\left[\frac{\Delta T_A}{\Delta z} + 0.01°C/m\right] \tag{13.31}$$

The coefficient 2.6 in Equation 13.29 was determined empirically, and the value of $\Delta T_A/\Delta z$ through the layer of plume rise should be used to determine S.

2. For unstable atmospheres where the plume theoretically would never stop rising as a result of ambient air entrainment,

$$\Delta H = 1.6 \frac{F^{1/3}x^{2/3}}{\bar{u}} \tag{13.32}$$

For unstable atmospheres, there is no general agreement on where the plume rise should be terminated, but it is reasonable to consider the rise terminated by the time the plume has traveled 10 stack heights or so downstream. (i.e. $u = 10H_s$)

Example 13.8

A proposed paper processing mill is expected to emit $\frac{1}{2}$ tonne of H_2S per day from a single stack. The nearest receptor is a small town 1,700 m northeast of the mill site, and southwest winds are expected to occur 15% of the time. The stack at the mill must be sufficiently high that the H_2S concentration in the town will not exceed 20 ppb by volume (28 $\mu g/m^3$ on a mass basis) at the ground. The physical characteristics of the emissions and the ambient atmosphere are as follows:

Gas exit velocity W	= 20 m/s
Gas exit temperature T_s	= 122°C
Stack diameter D at the top	= 2.5 m
Ambient air temperature T_A	= 17°C
Wind velocity u assumed for conservative analysis	= 2 m/s
Temperature lapse rate γ	= 6°C/km (assumed)

Estimate the required stack height at the mill.

Solution   Using Equation 13.28, the maximum ground-level ($Z = 0$) concentration will occur on the horizontal centerline ($y = 0$). At $y = 0$,

$$\exp\left[-\frac{y^2}{2\sigma_y^2}\right] = 1$$

and at $Z = 0$,

$$\left\{\exp\left[-\frac{(Z-H)^2}{2\sigma_z^2}\right] + \exp\left[-\frac{(Z+H)^2}{2\sigma_z^2}\right]\right\} = 2\exp\left[-\frac{H^2}{2\sigma_z^2}\right]$$

Thus, Equation 13.28 reduces to the simple form

$$C_{0,0,x} = \frac{Q}{\pi\sigma_y\sigma_z\bar{u}}\left\{\exp\left[-\frac{H^2}{2\sigma_z^2}\right]\right\}$$

The source strength is

$$Q = \frac{500 \text{ kg/d}}{86,400 \text{ s/d}} = 5.79 \times 10^{-3} \text{ kg/d}$$

A temperature lapse rate of 6°C/km represents a slightly stable atmosphere, say, stability category E.

At $x = 1,700$ m, for stability E, from Figure 13–21 (a) and (b); we have $\sigma_y = 80$ m and $\sigma_x = 30$ m. Solving for H, we obtain

$$28 \times 10^{-9} = \frac{5.79 \times 10^{-3}}{\pi \times 80 \times 30 \times 2}\left\{\exp\left[-\frac{H^2}{2 \times (30)^2}\right]\right\}$$

$$\exp\left[-\frac{H^2}{1,800}\right] = \frac{28 \times 10^{-9} \times 3.14 \times 80 \times 30 \times 2}{5.79 \times 10^{-3}} = 0.729$$

$$H^2 = 1,800 \ln(0.729) = 569 \text{ m}^2$$

$$H = 23.9\text{m}$$

The plume rise ΔH is estimated using Equation 13.29 for a slightly stable atmosphere. For the conditions specified,

$$F = \frac{g(T_s - T_A)W}{T_s}\left(\frac{D}{2}\right)^2 = \frac{9.81 \times (395 - 290) \times 20}{395}\left(\frac{2.5}{2}\right)^2 = 81.4 \text{ m}^3/\text{s}^3$$

and from Equation 13-31

$$S = \frac{9.81}{290}[-0.006 + 0.01] = 1.35 \times 10^{-4}$$

Therefore, from Equation 13.29

$$\Delta H = 2.6\left(\frac{81.4}{2 \times 1.35 \times 10^{-4}}\right)^{1/3} = 17.4 \text{ m}$$

Thus, the stack height H_s must be $H - \Delta H = 23.9 - 17.4 = 6.5$ m high.

Note that the prediction of stack height in Example 13.8 is a minimum height for flat terrain and is not sufficiently precise to warrant three-figure accuracy. The design stack height would thus be 7 m for the conditions specified. The actual stack height would be selected after repeating the calculation many times for various meteorological conditions and after consideration of other factors such as damage to vegetation at various distances downstream.

13.5.4 Other Pollutant Dispersion Models and the Accuracy of Predictions

Air pollution control engineers use many other types of models to predict the effects and concentrations of air pollutants and to identify specific sources of pollutants. Long-range transport models incorporating chemical reactions and wet and dry deposition processes have undergone intensive development in the last decade for prediction of acid deposition from distant sources. Elementary models which incorporate pollutant decay terms have been used to predict radionuclide concentrations in plumes and puffs for many years. These models have been used to quantify the source terms for accidental releases of radionuclides by back-calculation; that is, if one knows the downwind concentration at the point (x, y, z) and the travel time, decay rates, and meteorological conditions, the source strength Q can be estimated. Receptor models using "fingerprinted" emissions are under intensive development today to identify pollutant sources. For example, gasoline contains a very specific ratio of bromine to lead, so these elements can be used to identify the concentration of hydrocarbons contributed by automobiles at a given receptor site (x, y, z). The reader is referred to Yanskey, Markea, and Richter (1966); Slade (1968); Turner (1970); and Hanna, Briggs, and Hosker (1982) for excellent and detailed discussions of some of the more common models, together with their applications and uses.

PROBLEMS

13.1. Plank's radiation formula for the emissive power of a black body at wavelength λ is

$$E_\lambda = \frac{c_1 \lambda^{-5}}{\exp\left(\dfrac{c_2}{\lambda T}\right) - 1}$$

where $c_1 = 3.74 \times 10^{-12}$ watt cm^2
$\quad\quad c_2 = 1.44$ cm K
$\quad\quad \lambda$ is measured in micrometers $= 10^{-6}$ m.

The black-body radiation temperature of the sun is estimated to be 6,000 K and that of the earth together with its atmosphere is 255 K.
   (a) Plot the radiation power curve as a function of wavelength on linear scales for the sun and the earth on separate sheets.
   (b) Atmospheric photochemical reactions are stimulated by ultraviolet radiation. What is the dominant source of this short-wavelength radiation?
   (c) Using the curves developed, explain the CO_2 greenhouse effect.

13.2. For air at atmospheric pressure, $\rho = 1.2$ kg/m³ @ 20°C and $\mu = 1.81 \times 10^{-5}$ N · s/m².
   (a) Plot the fall velocity, in cm s^{-1}, of a water droplet in air at 20°C as a function of particle diameter (in 10^{-6} m) on log-log paper.
   (b) The highest concentration of dust from the El Chichon volcanic eruption on April 3, 1982, was observed at 24 km above the earth's surface (Robock and Matson, 1983). How long would a 10 μm particle of density 2100 kg/m³ remain suspended in the atmosphere?
   (c) How far would the particle travel horizontally during this period if the average wind speed through the atmospheric layer between 24 km and the earth's surface were 20 km/h?

13.3. Determine (a) the lead oxide (SG = 8.0) particle emission rate (kg/s) from a process and (b) the overall collection efficiency of a particle control device given the following particle statistics:

Average diameter for class (μm)	Number sec^{-1}	Number %	Cumulative %	Precipitator removal efficiency %
1.0	48×10^9	15	15	40
5.0	80×10^9	25	40	40
10.0	112×10^9	35	75	80
20.0	48×10^9	15	90	90
40.0	32×10^9	10	100	99

13.4. Estimate the daily emissions of sulfur dioxide, nitrogen oxides as NO_2, and particulate matter from an industrial steam-generating plant equipped with a dry-bottom pulverized-coal firing system. The plant burns 500 kg of coal per hour, and the coal contains 12% ash and 2.6% sulfur. The plant is equipped with an electrostatic precipitator which is 95% efficient.

13.5. Briefly describe the difference between an oxidizing smog and a reducing smog. Use chemical equations if that will *shorten* your explanation.

13.6. An expression for light attenuation in the atmosphere due to scattering and absorption is

$$I = I_0 e^{-kx}$$

where I is the intensity of electromagnetic energy received at a receptor from a source of intensity I_0 after the light has passed through a distance x in an atmosphere having an extinction coefficient k.

(a) If the NO_2 extinction is 1 ppm mile^{-1} for light at 0.4 μm, what fraction of 0.4 μm energy remains after a light beam has passed through five miles of air containing 0.2 ppm NO_2?

(b) Observation indicates that visibility, defined as the greatest distance that a black object can be seen and identified, occurs when $I/I_0 = 0.02$ to 0.05. What is the visibility in the atmosphere described in part (a)?

13.7. On a summer day when the wind speed is 3 m/s, a tank truck loaded with liquid chlorine is involved in an accident which results in a small split in the tank and a leak of 30 kg/min at the top of the tank. (The chlorine escaping is all in vapor form.) The accident occurs on the throughway in a city at a location where there are three very large apartment buildings about 300 m downstream. The threshold limit value (TLV) for chlorine is 3 mg/m^3. (Hint: In your solution, assume that in an emergency estimate, for all practical purposes, gas density differences can be neglected and the leak is at ground level.

(a) Would you give the order to evacuate the apartment buildings?

(b) How far downstream would you establish barricades to keep people away from the accident at street level?

(c) The fire department proposes to approach the truck from the upstream side and heavily spray the truck with water to prevent an explosion. Will this help to reduce the chlorine concentration downstream?

13.8. For a gravitational settling chamber, determine the length required to collect 50-μm-diameter particles if the flowrate is 1,000 m^3/min, the chamber width is 5 m, the chamber height is 2 m, and the particle specific gravity is 2.5. Assume the air temperature is 20°C.

Good settling chamber design practice calls for a bulk velocity of less than 3 m/sec. Does this design meet that requirement?

13.9. Compare the gravitational forces acting on a particle of mercury falling in air and the centrifugal forces acting on the same particle moving at a tangential velocity of 300 m/s at a radius of 1 m.

(a) Suppose the particle is 100 μm.

(b) Suppose the particle is 1 μm.

(c) Does the ratio of forces suggest that cyclones would be relatively better than settling chambers for particle removal?

(d) If so, why are settling chambers used?

(e) Could the same cyclone collect mercury and oil droplets of the same size simultaneously?

(f) Could the same cyclone collect mercury droplets and particles of uranium simultaneously?

(g) Would you recommend this cyclone for collection of the 1-μm droplets at 85% efficiency?

13.10. A new fossil-fuel-fired thermal generating station is proposed consisting of four 500 MW units for a total capacity of 2,000 MW. The design load factor is 60%; i.e., the station is expected to operate at 60% of the maximum design capacity averaged over a year. Coal, residual oil, and natural gas are all considered as potential fuels.

Sulfur dioxide emissions are to be controlled to a maximum of 1.2 lb $SO_2/10^6$ BTU in the fuel for coal firing or 0.8 lb $SO_2/10^6$ BTU for oil firing, assuming that all of the sulfur in the fuel is converted to SO_2 in the exhaust gases. Particle emissions are to be controlled using a device which is 99.5% efficient by weight for either oil or coal firing.

(a) What devices or methods can be used to control:
   (i) Particle emissions to the level required?
   (ii) Sulfur dioxide emissions to the level required?
   (iii) Nitrogen oxide emissions?
(b) Which of the devices or methods would you consider to be most acceptable, and why?
(c) Why would you control the emission of each of these pollutants?
(d) Estimate the hourly emission rate for each of the pollutants for each fuel.
(e) Annual owning, operating, maintenance, and fixed costs (primarily interest) for an SO_2 scrubber are about $70/kW of installed capacity for a coal-fired plant (1987 dollars) and about $50/kW for an oil-fired plant. Corresponding costs for an ESP are about $24/kW of installed capacity for a coal-fired plant and $16/kW for an oil-fired plant. Compare the combined costs of fuel and air pollution control for the proposed station, considering the above constraints. (Note that these costs do not include the costs of ash and/or sludge handling and disposal; for simplicity, do not try to include them here.)

13.11. It has been said that atmospheric particle loadings have decreased severalfold during the last 40 years, but the health impacts of particles have not changed significantly during the same period and may, in fact, be greater. How can this be so?

13.12. What are the principal pollutants produced by the automobile engine? Why would you control each of these pollutants? Is their effect significant locally (say, within 40 km), regionally (say, within 800 km), internationally, or globally?

13.13. A baghouse (fabric filter) and a venturi scrubber are being considered for particle removal from the exhaust gases of a basic oxygen furnace at a steel plant. Discuss the relative merits of each type of control equipment, preferably in tabular form, under the following headings:
   (i) Particle size
   (ii) Temperature
   (iii) Particle abrasion characteristics
   (iv) Acid or alkali particles
   (v) Efficiency
   (vi) Condensation (in exhaust gas or on particles)
   (vii) Particle disposal methods
   (viii) Operating costs
   (ix) Capital (first) cost
   (x) Variation in exhaust gas flowrate
   (xi) Variation in particle loading of exhaust gases
   (xii) Chemical composition of exhaust gases (acid, alkali, reaction with or in control device).

13.14. A 100 MW electric generating station burns 278 tonnes of residual fuel oil per hour. The fuel contains 2.6% S and 1.2% ash. The station is equipped with particle, sulfur dioxide, and nitrogen oxide emission control systems. The particle collector must operate at efficiencies exceeding 98% because the ash is valuable for its vanadium content. The sulfur dioxide removal device operates at a nominal efficiency of 80%.
(a) Estimate the hourly SO_2 and particle emissions from the station.
(b) What type of particle collector would you use at the station? Give three reasons for choosing this type of collector.

(c) What type of SO_2 removal device before (in front of) or after (following) the particle collector would you use?

(d) How would you control the nitrogen oxide emissions?

13.15. A lead sintering furnace discharges particles and SO_2 through a 150-m stack. The hot gases and particles rise an additional 100 m above the top of the stack when the wind speed is 5 m/sec and the lapse rate of temperature is 6°C/km. How far downwind will a 10-μm particle travel before falling to the ground? Assume flat terrain and Stokes' fall velocity for the particle.

13.16. (a) What are the major pollutants emitted from the tailpipe of an automobile?

(b) Why is each pollutant important?

(c) Name one control method or device used on present-day automobiles to control each pollutant.

REFERENCES

BRIGGS, G. A. *Plume Rise*. Washington, D.C.: U.S. Atomic Energy Commission, 1969.

DAVIS, M. L. *Air Resources Management Primer*. New York: ASCE, 1973.

FRUCHTER, J. S., et al, "Mount St. Helens Ash from the May 18, 1980 Eruption: Chemical, Physical, Mineralogical and Biological Properties." *Science*, Sept. 5, 1980, pp. 1116–1124.

HAAGEN-SMIT, A. J. "Chemistry and Physiology of Los Angeles Smog." *Industrial Engineering Chemistry*, 44 (1952): 1342–1346.

HALVORSEN, R., and RUBY, M. G. *Benefit-Cost Analysis of Air Pollution Control*. Lexington, Mass.: Lipington Books, 1982.

HANNA, S. R., BRIGGS, G. A. and HOSKER, R. P., JR. *Handbook on Atmospheric Diffusion*, Washington, D.C.: U.S. Dept. of Energy, 1982.

NOLL, K. E., and PATEL, M. "Evaluation of Performance Data from Fabric Filter Collectors on Coal Fired Boilers." *Filtration and Separation*, May–June, 1979.

PITTS, J. N., JR., et al, "Comment on 'Effect of Nitrogen Oxide Emission Rates on Smog Formation in the California South Coast Air Basin' and 'Effect of Hydrocarbon and NO_x on Photochemical Smog Formation under Simulated Transport Conditions.' " Environmental Science and Technology 17 (1983): 54–63.

RANDERSON, D., ed. *Atmospheric Science and Power Production*. Washington, D.C.: U.S. Dept. of Energy, 1981.

ROBOCK, A., and MATSON, M. *Circumglobal Transport of the El Chichon Volcanic Dust Cloud*. *Science*, July 1983, pp. 195–197.

SLADE, D. H., ed. *Meteorology and Atomic Energy*. Springfield: U.S. Dept. of Commerce, 1968.

TURNER, D. B. *Workbook of Atmospheric Diffusion Estimates*. Springfield: U.S. Dept. of Commerce, 1970.

YANSKEY, G. R., MARKEA, E. H., and RICHTER, A. P. *Climatology of the National Reactor Testing Station*. Springfield: U.S. Dept. of Commerce, 1966.

U.S. Public Health Service. *Air Pollution in Donora, Pa*. Cincinnati: Bulletin No. 306, 1949.

Supplementary Reading

BUTCHER, S., and CHARLSON, J. *An Introduction to Air Chemistry*. New York: Academic Press, 1972.

CRAWFORD, M. *Air Pollution Control Theory*. New York: McGraw-Hill, 1976.

FRIEDLANDER, S. K. *Smoke, Dust, and Haze*. New York: Wiley, 1977.

GREEN, HENRY L. *Particulate Clouds: Dust, Smokes, and Mists*. 2d ed. Princeton, N.J.: Van Nostrand, 1964.

MASON, B. J. *The Physics of Clouds*. 2d ed. Oxford: Clarendon Press, 1971.

PERKINS, H. C. *Air Pollution*. New York: McGraw-Hill, 1974.

SCHWARTZ, E. *Trace Atmospheric Constituents: Properties, Transformations and Fates*. New York: Wiley-Interscience, 1983.

SEINFELD, J. H. *Air Pollution: Physical and Chemical Fundamentals*. New York: McGraw-Hill, 1975.

SPRINGER, G. S., and PATTERSON, D. J. *Engine Emissions: Pollutant Formation and Measurement*. New York: Plenum Press, 1973.

STARKMAN, E. S. *Combustion-Generated Air Pollution*. New York: Plenum Press, 1971.

STERN, A. C. *Air Pollution*. 3d ed. New York: Academic Press, 1976.

SUGDEN, T. M., ed. *Pathways of Pollutants in the Atmosphere*. London: The Royal Society 1978.

WARK, K., and WARNER, C. F. *Air Pollution: Its Origin and Control*. New York: IEP-Dun-Connelly, 1976.

WILLIAMSON, J. *Fundamentals of Air Pollution*. Reading, Mass.: Addison-Wesley, 1973.

14

Solid Wastes

J. Glynn Henry

14.1 INTRODUCTION

Solid waste in the broadest sense includes all the discarded solid materials from municipal, industrial, and agricultural activities. However, for the discussion to follow, solid waste will refer only to those solid wastes which are the responsibility of, and usually collected by, a municipality. Residential and commercial areas, together with some industrial operations, are the sources of these "nonhazardous" municipal wastes.

Municipal solid waste is a material difficult to characterize and generally uneconomical to utilize. The objectives of solid waste management are to control, collect, process, utilize, and dispose of solid wastes in the most economical way consistent with the protection of public health and the wishes of those served by the system. Information on the amounts and characteristics of solid wastes helps to explain why this "refuse" is now a major concern and why conservation of these "resources" is receiving more attention.

In this chapter, various aspects of collection systems are discussed. Landfilling, incineration, composting, and other methods for the disposal of solid waste are compared from the standpoint of performance and economics. Design criteria for landfills and the problems encountered in their operation, including the control of leachate and gas, are covered in detail.

14.2 CHARACTERIZATION OF SOLID WASTES

14.2.1 What Is Solid Waste?

In general terms, solid waste (sometimes called refuse) can be defined as waste not transported by water, that has been rejected for further use. For municipal solid wastes, more specific terms are applied to the putrescible (biodegradable) food wastes, called garbage, and the nonputrescible solid wastes referred to as rubbish. Rubbish can include a variety of materials that may be combustible (paper, plastic, textiles, etc.) or noncombustible (glass, metal, masonry, etc.). Most of these kinds of waste are discarded on a regular basis from specific locations. However, there are wastes—sometimes called ''special wastes''—such as construction debris, leaves and street litter, abandoned automobiles, and old appliances, that are collected at sporadic intervals from different places.

Not included in the components of municipal waste as just described are many other solid wastes that are not normally a municipal responsibility. Such things as ashes from coal-fired generating stations, sludges from water and wastewater treatment plants, wastes from animal feedlots, mine tailings, and other industrial solid wastes are in this category and require separate arrangements for their disposal. The estimated quantities of solid wastes generated from all sources in the United States in 1976 are given in Table 14–1. The municipal portion, representing only five percent of the total, receives the most attention because of the effect its improper disposal can have on public health and on water supplies from both surface-water and groundwater sources.

14.2.2 Changes in Municipal Solid Waste

Until the late forties, the bulk of municipal solid waste consisted of ashes from coal-burning furnaces and food wastes. The few scrap materials, such as metals and rags, that were recoverable were collected on a casual basis by scavengers. With the shift of the burgeoning population of the 1950s to cities, urban population densities increased, oil and natural gas heating grew in popularity, and society became increasingly industrialized. The two root causes for the increasing urgency of solid waste problems are urbanization and industrialization. Urbanization, i.e., the influx of people to metropolitan areas, affects living habits

TABLE 14–1 SOLID WASTE QUANTITIES (1976) GENERATED IN THE UNITED STATES

Solid waste source	Generated (10^6 tons/year)	%
Municipal	230	5.2
Industrial	140	3.1
Agricultural	640	14.5
Mining	1,700	38.6
Animal	1,700	38.6
Total	4,410	100.0

Source: Tchobanoglous et al., 1977.

and consequently waste characteristics. Also, with more people, the areas requiring solid waste collection have expanded and sites for waste disposal are further away. Industrialization, because it generates inexpensive, labor-saving products, has created a "throwaway" society. During the 1960s and 1970s, new products appeared in abundance. Cans, bottles,

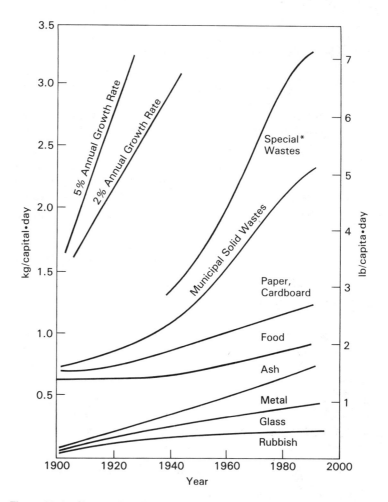

Figure 14–1   Changes in solid waste generations in the united states since 1900

Note: Estimates for the United States have been adapted from information available for Edinborough (1900), London (1925), New York (1940), and the U.S. (1965, 1980).
Source:
1. Handbook of Solid Waste Management, D. G. Wilson Ed., Van Nostrand Reinhold, New York, 1977.
    (a) Niessen, W. R., Estimation of Solid-Waste-Production Rates, pp. 544–574.
    (b) Wilson, D. G., History of Solid Waste Management, pp. 1–9.
   * Special Wastes include construction debris, street cleanings, yard wastes, large discards, etc.

plastic containers, appliances, tires, and many other items were considered to be cheaper to throw away than to reclaim. Recovery of material has become more difficult because of the use of numerous synthetic materials, bonded plastics, and nonferrous alloys. Packaging for convenience foods, hardware, household items, and other goods has created a vast array of material that is easy to discard. Solid waste increased significantly in quantity and complexity with the advent of the "throwaway" society and the growth of the packaged-and-processed food industry. Figure 14–1 indicates how solid waste has changed over the years. Of course individual regions may show quite different trends, proportions, and quantities, and therefore studies of solid waste management problems should be based on site-specific surveys.

Today, food wastes are generated more by processing plants than by home or farm. While such wastes are a problem because of their large volume, high strength, seasonal nature, and rural location, the change has enabled better control through an industry-wide approach to waste disposal with the costs being borne by the users of the products. The growth of the food processing industry does not seem to have changed the amount of food wastes from urban residents, but the increase in packaging associated with convenience foods is certainly part of the reason for the growing per capita waste production.

Example 14.1

In 1920, five trucks were required to collect 5,800 t (6,400 tons) per year of residential and commercial waste from a municipality of 20,000 people. By 1980, when the population had reached 100,000,

(a) How would the annual quantity of refuse collected have changed if "special" wastes were not included in the annual amounts?

(b) Neglecting lost time because of vehicle breakdown and maintenance, how many trucks of 4.5 t (5 tons) capacity, operating 5 days per week, were required for twice weekly collections if trucks averaged 2 loads/day at 80% capacity?

(c) How did the quantity of paper change during the 60-year period?

Solution

(a) 1920 quantity = 6,400 ton/year

$$= \frac{6,400 \times 2,000}{20,000 \times 365} = 1.75 \text{ lb/capita} \cdot \text{day (versus 1.85}$$
$$\text{from Figure 14–1)}$$

1980 quantity = 4.5 lb/capita · day (from Figure 14.1)

$$= \frac{4.5 \times 100,000}{2,000} \times 365 = 82,125 \text{ ton/year (an increase of}$$
$$\text{about 13 times the 1920 quantity)}$$

(b) Annual capacity of one truck $= \dfrac{5 \text{ tons}}{\text{load}} \times \dfrac{2 \text{ loads}}{\text{day}} \times \dfrac{5 \text{ days}}{\text{week}} \times \dfrac{52 \text{ weeks}}{\text{year}} \times 80\%$

capacity = 2,080 ton/year

Trucks required $= \dfrac{82,125}{2,080} = 39.5 = 40$ (8 times the number previously used)

(c) In 1920, paper = 14% = 0.14 × 6,400 = 896 tons
   In 1980, paper = 45% = 0.45 × 82,125 = 37,000 tons (a more than forty fold increase in paper)

14.2.3 Quantities

Typical municipal and industrial solid waste generation rates for Canada and the United States are indicated in Table 14–2. The amounts of residential and commercial wastes collected in the United Kingdom and other European countries (France, Germany, Sweden) are about 50 percent of these values.

In wet areas, because of moisture absorbed by the solid waste, the amount collected may exceed the amount generated, which is usually reported on a dry basis. On the other hand, with the use of home grinders, on-site storage of recyclable materials, and other conservation measures, the amount collected may be less than that generated. In this chapter, differences between the quantities generated and collected are ignored because variations due to other factors are much more significant.

During the seventies, the quantity of solid waste generated was increasing at a rate above 5 percent per year, a greater rate than the annual increase in the Gross National Product (GNP) at that time. However, in the 1980s, the per capita waste generation leveled off, and an annual increase of 2 to 3 percent now seems more realistic.

As a rough guide to municipal solid waste generation in North America, a figure of 1 tonne per capita per year (2.7 kg or 6 lb per capita per day) is commonly applied. It should be clear, however, that this average value is subject to wide variation from one municipality to another and at different times of the year.

TABLE 14–2 URBAN SOLID WASTE GENERATION IN NORTH AMERICA

Solid waste source	kg/capita per day	lb/capita per day
Residential	1.1	2.5
Commercial	0.9	2.0
Special*	0.9	2.0
Total Municipal	2.9	6.5
Industrial**	1.4	3.0
Total Municipal & Industrial	4.3	9.5

Adapted from Niessen, 1977, and Tchobanoglous et al., 1977

Quantities noted are representative of 1980 generation rates for a "typical" municipality. Actual amounts for a particular community may vary significantly from these values.

* Special waste includes construction debris, leaves and street litter, and large discards.

** Industrial solid wastes (including water and wastewater treatment plant sludges) not collected with municipal solid wastes.

14.2.4 Characteristics

Composition.   In addition to the variations in quantity, wide differences in waste composition can also occur. Factors influencing the composition of municipal solid waste include such things as

- *Climate*: In wet areas such as São Paulo, Brazil, the moisture content of solid waste is typically 50 percent.
- *Frequency of collection*: More frequent collections tend to increase the annual amount collected. Since the amount of organics is relatively constant, perhaps with more pickups there is a tendency for residents to discard more paper and rubbish.
- *Prevalence of home garbage grinders*: Grinders reduce, but don't eliminate, food wastes.
- *Social customs*: Some ethnic areas use few convenience foods, so less paper and more raw food wastes result.
- *Per capita income*: Low-income areas produce less total waste, but with a higher food content.
- *Acceptability of packaged and convenience foods*: In North America, wide use of packaging has increased the paper content of solid waste.
- *Degree of urbanization and industrialization of the area*: Because of the composting, recycling, and recovery possible in rural areas and areas with single-family dwellings, solid wastes from such sources may be less in quantity and of different constituents than that from industrialized metropolitan areas with multiple-family housing.

Table 14–3 shows the composition of municipal solid waste in the United States and Canada, and in four other industrialized countries. The percentages noted are for the wastes ''as collected,'' and not the dry weight of the refuse. The moisture content of municipal solid waste varies from 15 to 40 percent, depending on the composition of the wastes and

TABLE 14–3. COMPOSITION OF MUNICIPAL SOLID WASTE (% BY WEIGHT)

Component	U.S. & Canada	Sweden	France	Israel	Japan
Paper	45	55	30	23	25
Organics	15	12	24	71	37
Metal	10	6	4	1	3
Glass	10	15	4	1	3
Ash	10	0	24	2	19
Miscellaneous	10	12	14	2	13
	100	100	100	100	100

Source: American Public Works Association, *Solid Waste Collection Practice*, 4th ed., Chicago: American Public Works Association, 1975

Note: The composition of the waste for specific locations may differ by up to 40% from the above annual average percentages. Seasonal variations may be even greater.

TABLE 14–4 MOISTURE CONTENT OF
SOLID WASTES AS COLLECTED*

| Component | Moisture (% by weight) | |
	Range	Typical
Paper	4 to 10	7
Food wastes	50 to 80	70
Yard trimmings	30 to 80	60
Metal	2 to 6	3
Glass	1 to 4	2
Ashes, dirt	6 to 12	8
Other rubbish†	5 to 30	20
Municipal	15 to 40	20

* Adapted from Tchobanoglous et al., 1977

† Includes plastic, textile, rubber, leather, and
wood.

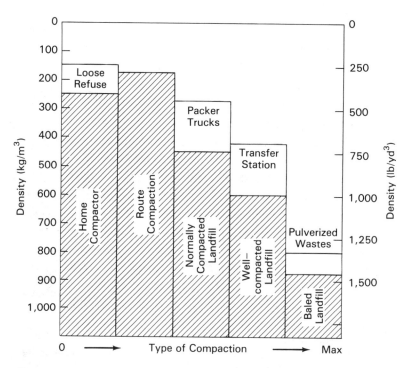

Figure 14–2 Typical densities of municipal solid wastes with different types of
compaction (Adapted in part from Tchobanoglous et al, 1977)

TABLE 14–5 TYPICAL UNCOMPACTED DENSITIES
FOR MUNICIPAL WASTE COMPONENTS*

Component	Density* kg/m^3	Density* lb/yd^3
Paper and cardboard	80	135
Food wastes	300	500
Yard trimmings, rubbish†	160	270
Ashes, dirt, brick, ferrous metal	<u>480</u>	<u>800</u>
Municipal waste	150	250

Adapted in part from Niessen, W. R., "Properties of
Waste," pp. 10–62 in *Handbook of Solid Waste Manage-
ment*, edited by D. G. Wilson, New York: Van Nostrand
Reinhold, 1977

* Actual density may vary up to 50% from the typical values
shown, depending on the nature of the constituents and their
moisture content.

† Including glass, wood, rubber, leather, and nonferrous
metals, but excluding dirt, masonry, and scrap iron.

the weather (temperature, humidity, precipitation). The moisture content of individual
components is shown in Table 14–4.

Density.   The density of municipal solid waste varies with waste composition and
the degree of compaction. Typical values, shown in Figure 14–2, range from an uncompacted
density of 150 kg/m^3 to 800 kg/m^3 for landfilled, pulverized refuse. Uncompacted densities
for the various components (as collected) are noted in Table 14–5.

Example 14.2

Using typical values for individual components, estimate the moisture content and density of
the municipal solid waste whose composition is given in the following table:

Component	Data provided 100-kg Sample kg	Moisture[a] %	Moisture[a] kg	Dry solids %	Dry solids kg	Density[b] kg/m^3	Volume[c] m^3
Paper	45	7	3.2	93	41.8	80	0.56
Organics	20	70	14.0	30	6.0	300	0.07
Metal (Fe)	7	3	0.2	97	6.8	480	0.01
Glass	10	2	0.2	98	9.8	160	0.06
Ashes	3	8	0.2	92	2.8	480	0.01
Miscellaneous	<u>15</u>	20	<u>3.0</u>	80	<u>12.0</u>	160	<u>0.10</u>
Solid Waste	100		20.8		79.2		0.81

[a] Percentage moisture for waste as collected (see Table 14–4).

[b] Density of component in 100-kg sample (see Table 14–5).

[c] Volume of component = weight of component/density of component.

Solution   (Paper component only)

Moisture: Weight of moisture in component = 7% in 45 kg of paper = 3.2 kg

Dry Solids: Weight of dry solids in component = 93% in 45 kg of paper = 41.8 kg

Volume: Volume of component in 100 kg sample $= \dfrac{45 \text{ kg of paper}/100 \text{ kg sample}}{80 \text{ kg of paper}/\text{m}^3 \text{ paper}}$

$$= 0.56 \text{ m}^3 \text{ of paper}/100 \text{ kg sample}$$

Moisture Content of Solid Waste: $= \Sigma$ weights of moisture in components
$$= 20.8 \text{ kg in } 100 \text{ kg sample} = 20.8\%$$

Density of Solid Waste:

$$\text{Density} = \frac{\text{Weight of sample}}{\text{Volume of sample}}$$

$$= \frac{\text{Weight of sample}}{\Sigma \text{ volume of components}}$$

$$= \frac{100 \text{ kg}}{0.81 \text{ m}^3} = 124 \text{ kg/m}^3 \ (206 \text{ lb/yd}^3)$$

Energy content.   Municipal solid waste contains about 50 percent volatile (combustible) matter, the balance being roughly equal proportions of moisture and inert solids.

TABLE 14–6 TYPICAL ENERGY CONTENT FOR COMBUSTIBLE MATERIALS

Material	Typical energy content	
	kJ/kg	BTU/lb
Municipal Solid Waste		
• Per unit weight of refuse	10,500	4,500
• Per unit weight of combustibles	23,200	10,000
• Per unit weight of paper	16,300	7,000
• Per unit weight of organics	5,800	2,500
Primary Sewage Sludge		
• Per unit weight of dry solids	17,700	7,600
Digested Sewage Sludge		
• Per unit weight of dry solids	9,100	3,900
Fuels		
• Per unit weight of #6 fuel oil*	46,500	20,000
• Per unit weight of anthracite	28,000	12,000
• Per unit weight of methane**	49,000	21,000

Source: Sarofim, 1977; Tchobanoglous et al., 1977
Note: BTU/lb × 2.3241 = kJ/kg
* Energy content of fuel oil = 37.3 × 10⁶ kJ/m³ (1 × 10⁶ BTU/ft³)
** Energy content of methane or natural gas = 37,300 kJ/m³ (1,000 BTU/ft³)

Because of the volatile content, the waste is often burned as a means of disposal and occasionally utilized as a source of energy. Table 14–6 indicates typical energy contents for various combustible materials, including uncompacted solid waste, which may have an energy content from 9,300 to 12,900 kJ/kg. The energy content shown for each material is its heat of combustion. In comparing solid wastes with other fuels, the energy required for shredding and classification of the refuse should be considered, as should the difference in operating efficiency between the incinerator and another type of furnace.

Example 14.3

Based on the energy contents of the components of municipal solid waste as collected (Table 14–6), determine the energy content in refuse consisting of 50% paper and 20% metal, glass, and ash, with the balance being food and other organic wastes.

Solution   An outline of the procedure for estimating the energy content of solid waste is given in the following table:

Data provided		Values estimated	
Solid waste		Energy, kJ/kg (BTU/lb)	
Component	*%*	*Component*	*Solid Waste*
Paper	50	16,300 (7,000)	8,150 (3,505)
Organics	30	5,800 (2,500)	1,740 (750)
Noncombustibles	20	Neglected	
Solid Waste	100	—	9,890 (4,255)

The energy content of the solid waste as collected is 9,890 kJ/kg (4,255 BTU/lb).

14.3 CONSIDERATIONS IN SOLID WASTE MANAGEMENT

As with every engineering decision, economic considerations are a major concern in solid waste management. However, there are other factors which must not be neglected. Among these are the public health, which is paramount, waste separation (recycling), and energy recovery.

14.3.1 Protection of Public Health

Under warm, moist conditions, organic wastes become ideal breeding places for disease-causing organisms. Pathogens, even if absent initially, have easy access to the waste via vectors. With solid wastes, the usual vectors (carriers) for disease transmission, i.e., water, air, and food, are not important; flies, rodents, and mosquitoes are the primary vectors. The relationship between solid waste and disease has not been well documented, but it is known that about 50 different diseases are borne by flies, rodents, and mosquitoes, so that protection of public health requires constant vigil. Preventative measures suggested by Wilson (1977) include:

- The use of tightly closed containers for organic wastes.
- Compaction of waste to at least 600 kg/m³ (1,000 lb/yd³) to reduce insect breeding places and rodent access.
- Processing within two days (since fly larvae become flies in a few days).
- Shredding of waste to promote aerobic decomposition, which is a heat-producing process and therefore unattractive to insects and vermin.

The generation of harmful organisms and their transmission is not the only health-related concern. Many potentially harmful materials, such as solvent and pesticide containers, medical wastes, and asbestos debris, even though prohibited, may already be present in the waste when it is collected. Air pollution, caused by the particulates and gaseous pollutants from landfill sites and municipal incinerators, is an additional environmental problem related to solid waste disposal.

Landfilling of solid wastes or the residues from incineration can endanger the quality of groundwater or surface waste supplies. Proper design and careful operation of landfills (as described in Section 14.8) are needed to minimize the risk associated with the contaminated drainage (called leachate) coming from the decomposing refuse.

14.3.2 Waste Separation

Recovery and recycling of the resources in solid wastes are appealing in theory, but difficult in practice. Over the past 250 years, the proportion of domestic wastes which has been recycled has declined steadily from over 90 percent to less than 1 percent in 1977 (Wilson, 1977). Since then, interest in recycling has been revived, and in 1987 New Jersey passed the first statewide mandatory recycling legislation. Under this law, residents in the state's 567 communities are required to recycle three kinds of materials, including aluminum, and all towns are required to compost leaves (Bell, 1987).

Costly materials such as metals are routinely recycled by industry because doing so is cheaper than buying virgin metals. Low-cost items in municipal wastes may be recycled when they become either of value (e.g., newspapers for cellulose insulation) or too costly to throw away (e.g., returnable bottles and cans with a high refundable deposit). Metals such as copper, aluminum, and lead from building demolition are usually salvaged, but their recovery, as well as that of other metals from municipal waste, is rare. Economics is the reason. For example, it takes more energy to reuse old glass than to process new materials. In fact, according to Wilson (1977), when crushed glass is used to replace crushed stone in road pavement, 60 times as much energy is used.

Separation at the source.   Regulations in many areas of the United States require separation of wastes at the source, by the resident, into such components as food wastes, paper, ashes, and glass. The practice is perhaps a carry-over from the days when salvaging was common. Whether it is worthwhile to separate these items today is debatable.

Trial source-separation programs in residential communities have generally received enthusiastic support during the early stages. However, the concept seems to have application only to single-family areas, and even there it has been found that participation by the residents drops from perhaps 90 percent in the beginning to around 50 percent after three or four months.

Unless markets for reclaimed materials exist, separation at the source is seldom warranted. However, under special circumstances, it may be advantageous to collect some types of waste or separated waste components. For example, a separate collection might be considered for such items as

- *Newspapers and cardboard*, where landfill space is limited or costly, incineration of solid wastes is possible, or a use for the material exists.
- *Aluminum cans* in areas where they are in common use.
- *Mixed glass* (called cullet) if a ready market exists (as in California) for cullet to make green glass for wine bottles.
- *Food wastes* from restaurants if the wastes can be utilized for animal feed.

Newspaper recycling is a particularly difficult market to establish: prices for recycled paper can fluctuate drastically from zero to over $100 per ton in a single year, the material is bulky and expensive to handle, and it must be deinked before reuse as paper stock. Without deinking, the paper is limited to reuse in roofing felt, paperboard, and similar paper products.

Despite the lack of success in the past, the principle of separation at the source by residents is receiving increasing attention. The pressure of decreasing landfill capacity, environmental concerns, improving markets, economic incentives, and political support are contributing to the trend.

Central separation.   Central separation of mixed municipal refuse for reclamation is another approach in waste management. It has the most potential in dense metropolitan areas where separation at the source is difficult, i.e., in multiple-family, high-rise, and mixed residential and commercial areas. At any central station, the unsanitary but unavoidable hand sorting is necessary to at least remove oversize and unshreddable objects and potentially explosive material. After this, the amount of hand sorting necessary will depend on the capabilities of the subsequent separation units. According to Wilson (1977), there is no material which repays the cost of hand separation, and mechanical separation is essential if labor costs are to be reduced.

Various unit processes, many of them proprietary, are available for bulk sorting. Machine sorting of refuse, item by item, is at an early stage of development, and the usual approach has been to pulverize and grind the mixed refuse to prepare it for the type of recovery intended. Methods for the separation of dry, pulverized mixed waste are based on the diversity, size, inertia, conductivity, or other characteristics of the ground refuse and include (Wilson, 1977).

- Air (or water) classification for lightweight components such as paper and plastic.
- Magnetic separation for ferrous metals.
- Screening for removal of nonferrous material.
- Optical differentiation of color for separating clear from colored glass.
- Inertial classifiers for separating organic from inorganic particles, or heavy, resilient particles from light, inelastic ones (i.e., for removing contaminants from compost.)

Central bulk sorting and source separation need not be mutually exclusive in a community. The extent to which each is used will depend on such things as the size and type of municipality, its economic status, the extent of separation expected, and media support. Estimates on the effect of source separation of newsprint and beverage containers suggest a loss of 3 to 10 percent of the fuel value of the refuse, assuming 20 to 25 percent maximum recovery in most municipalities and a 10 percent loss in the value of other recoverable materials, such as ferrous metals, aluminum, and glass.

14.3.3 Recovery of Energy

The two principal ways to use the energy contained in municipal solid wastes are to use the material as fuel and to recover material for reuse, thus saving the energy needed for processing and transporting virgin material. Using the U.S. Bureau of Mines estimates for the quantities of organic wastes produced in 1972, Wilson (1977) has determined that urban refuse could supply only 1.5 to 3.0 percent of total U.S. energy requirements.

Fuels from solid wastes.   Solid wastes may be burned directly in incinerators (a process called mass-burning) or converted to more efficient "refuse-derived fuel" (RDF). The controlled burning of either municipal solid waste or RDF can produce hot water or steam for heating, or steam for driving turbines to generate power.

Refuse-derived fuel (RDF) is prepared from the paper and plastic removed from (usually) previously shredded municipal solid waste by standard air classifiers. It may be pelletized into solid fuel particles (1.0–1.5 cm) or burned directly in coal-burning utility boilers up to a proportion of 15 percent of the total heat input. Without pelletizing or other preservative measures, storage of RDF is a problem because of changes in moisture content, mildew, and decomposition (Vesilind, 1981). Other methods for deriving fuels from solid wastes include pyrolysis and anaerobic decomposition of the organic material in solid wastes. Processes for recovering the fuel value of solid wastes are covered in more detail in Section 14.5.

Reuse of materials.   Reuse of recovered materials is the other principal means by which energy conservation can be applied to solid waste disposal. As noted earlier, reuse, at least from an energy standpoint, can be readily justified for ferrous metals, aluminum, and other nonferrous metals, because the mining of virgin material is so energy intensive. With glass, however, energy savings can be realized only by the reuse of the container.

14.4 COLLECTION SYSTEMS

From the early 1900s, when infrequent pickup of ashes and other household wastes in open, horse-drawn carts was the practice, solid waste handling has evolved, gradually at first and then rapidly following World War II. Many of the changes in the past 20 years have resulted in greater convenience to the public and reduced costs for the municipality. Since collection costs are typically 80 percent of solid waste budgets, this is the most effective area for cost reductions.

14.4.1 Ease and Frequency of Pickup

Wide public acceptance of plastic garbage bags, introduced in 1963, has to some extent transferred part of the collection cost from the municipality to the residents. The replacement of metal cans with plastic bags permits easier waste handling and faster, more efficient service by collection crews.

Because home garbage grinders reduce the putrescible waste to be collected, their general use may permit less frequent pickup of garbage. As noted in Chapter 12, organic and solids loading to wastewater treatment facilities could rise (by 30 and 60 percent, respectively) if all residences had these sink disposal units, but the additional treatment costs would be offset by the lower cost for solid waste collection.

Even though home compactors (that reduce volume by one-third to one-half) have not become popular, commercial storage containers for refuse from schools and apartments and stationary compactors for supermarkets, hospitals, and institutions are in widespread use. In addition to having aesthetic advantages over using small cans, the use of site containers and compactors has reduced the frequency of collection and the volume of the waste to be transported.

The frequency of solid waste collection has been declining since the 1950s, when two pickups per week were the most common practice. By the 1970s, about half the U.S. population was served by a weekly collection and the trend to fewer pickups has continued into the 1980s, although at a reduced rate (Chanlett, 1979). The reasons for the pickups being fewer are:

- Wastes become less objectionable as the proportion of putrescibles decreases.
- Better designed packer trucks and trailer units keep odors and flies under control.
- Costs for service rise with higher labor costs and longer hauls to disposal.
- Better management enables wastes to be moved from pickup to transfer station to landfill and burial within hours.

Solid wastes should of course be collected at least once per week and probably more frequently for high-density districts and for wastes with high putrescible content, particularly during warm weather.

14.4.2 Collection Equipment

Packer trucks.   The usual vehicle for residential areas is the manually rear- or side-loaded compaction truck operating with a crew of two or three, including the driver (see Photograph 14–1). Each truck of 14 to 18 m³ (15 to 20 yd³) capacity can carry 4 to 5 tonnes of waste to a disposal site or transfer station. Two loads per day per truck, possibly three, are typical for most operations.

Larger self-loading compactor vehicles with one driver-operator are also available which can automatically unload full-storage containers at apartments, shopping centers, etc., into the vehicle, replace the empty containers for reuse, and deliver the compacted contents along with the contents of several other containers to disposal.

Example 14.4

A residential area of about 40 ha (100 Ac) contains 300 single-family residences and 8 ha (20 Ac) with multiple-family units housing 400 people. With two curb-side pick ups per week, how many trips on each collection day would one packer truck need to make in order to serve this area? Assume 4 residents/single-family unit.

Solution

Population Served:

Single family at 4 residents per unit   = 1,200 people
Multiple family at 50 residents per ha =   400 people
                                     Total = 1,600 people

Waste Quantity:

Assume the per capita waste generation is 1.1 kg/d. Then the amount each collection day is

$$1,600 \times 1.1 \times \frac{7}{2} = 6,160 \, \text{kg} = 6.2 \, \text{tonnes}$$

Thus, on a normal collection day, when half the weekly waste is collected, one packer truck with a 4- to 5-tonne capacity would have to make two trips to serve the area in question.

Container trucks.   The other common type of collection vehicle is one that delivers a large empty storage container to an institution or commercial operation and picks up a full one which is then hauled to disposal. One driver can perform all the unloading and loading of containers unless regulations require that the driver have a helper. The containers may be carried by a hoist truck (see Photograph 14–2(a)) which lifts relatively small containers (with loose, bulky rubbish) into place, or by a tilt-frame truck (see Photograph 14–2(b)) which can handle larger containers called drop boxes containing loose or compacted refuse.

Photograph 14–1 Residential packer truck Source: American Public Works Association, *APWA Solid Waste Collection Practice*, Chicago: APWA 1975

(a)

(b)

Photographs 14–2 (a and b) Container trucks. (a) hoist truck; (b) tilt-frame truck
Source: Stone 1977 (Heil Corporation) and American Public Works Association, APWA Solid Waste Collection Practice, Chicago: APWA 1975

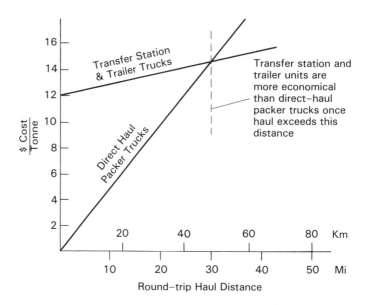

Figure 14–3 Hypothetical comparison of costs for hauling refuse to disposal
The illustration is based on a cost of $0.30/t.Km ($0.30/ton.mi) for packer trucks carrying 5 tonnes and a cost of $0.06/t.Km ($0.10/ton.mi) for trailer units carrying 10 tonnes. The transfer station is assumed for simplification to have a fixed annual cost of $12.00/t. for capital and operating.

With the advent of self-loading packer trucks and the tilt-frame hauled container system which can carry much larger containers, the hoist truck has been replaced for all but a few limited applications.

14.4.3 Transfer Stations

As nearby landfill sites become scarce and the hauling distance increases, the time spent by the crew of the packer truck in unproductive travel to the site becomes excessive. It may then be more economical to transfer the waste from small local collection vehicles to units that can travel longer distances. This operation is carried out at a transfer station to which the three-person packer trucks deliver 4 to 5 tonnes of refuse for transfer (usually with additional compaction) to one-person trailer units which have a capacity of 27 to 46 m³ (35 to 60 yd³) and can carry 15 to 20 tonnes. An economic analysis of alternative hauling arrangements, as illustrated in Figure 14–3, can be used to establish whether a transfer station would be advantageous. For the example given, it would be more economical to use a transfer station with long-haul trailer units instead of direct-haul packer trucks if the average round-trip haul distance is more than about 50 km (30 miles).

14.4.4 Rail Haul

When large quantities of solid waste must be hauled long distances for disposal, rail haul should be examined. The conditions proposed by Stone (1977) for rail haul to be competitive with trailer units are that:

- Round trip haul distance exceed 100 miles (160 km).
- At least 1,000 tons (900 tonnes) are transported.
- Wastes are baled at the transfer station.
- Railroad cars are used exclusively for solid waste.

14.4.5 Route Selection

Determining the most economical route for collecting solid waste and hauling it to disposal points can be difficult. Interrelated variables such as labor costs, crew size, union restrictions, collection frequency, distance (travel time) to disposal, and the performance and annual costs of various types of waste-handling equipment will influence the choice.

Alley or curb pickup are the two basic types of local collection. Alley pickup, where possible, has advantages: a homeowner doesn't have to set out cans, scheduled service is not required, there is no interference with street traffic, and both sides of the alley can be served at the same time with minimum walking.

In some municipalities, setting out and/or returning the cans to the front or backyard is provided, but most residents will forgo these benefits because of the additional cost. Systems analysis can be applied to many waste routing problems, but for local collection a trial-and-error approach is commonly used.

When wastes from different collection districts can be directed to several possible points, it may be difficult to decide which wastes should go to which location for the most economical solution. Waste allocation problems of this type are normally solved by linear programming. Where a program or a computer is not available, approximate methods can be used for finding a solution close to optimal.

14.5 DISPOSAL

14.5.1 Landfilling

Except for the disposal of municipal solid wastes at sea, which is not permitted by most developed countries, solid wastes, or their residues in some form, must go to the land. Landfilling, the most economical and consequently the most common method of solid waste disposal, is used for 90 percent of the municipal solid wastes in the United Kingdom and North America. Even in European countries like West Germany and Switzerland, with massive investments in incineration and composting plants, over 60 percent of domestic and commercial waste is landfilled (Holmes, 1981). Incineration cannot, of course, eliminate landfilling. In fact, it creates a more concentrated residue that may be more hazardous to water supplies than unburned solid wastes. The area needed for landfilling of solid wastes is about 1 ha per year for every 25,000 people (1 Ac/10,000 people). This is illustrated in Example 14.5.

Example 14.5

> For a population of 25,000, estimate the annual area requirements (excluding the buffer zone) for a normally compacted landfill having a refuse depth of 4 m excluding cover material.

Solution   Assuming that per capita waste generation is 2.0 kg/d (Table 14–2) and that the density of a well-compacted landfill is 450 kg/m³ (Figure 14.1), the annual area required is

$$\frac{25,000 \times 2.0 \text{ kg/d} \times 365 \text{ d/yr}}{450 \text{ kg/m}^3 \times 4 \text{ m} \times 10,000 \text{ m}^2\text{/ha}} = 1.0 \text{ ha}$$

Area requirements for landfilling can vary considerably with the type of waste and the degree of compaction. Details of the design and operation of sanitary landfills are discussed in Section 14.6.

The balance of this section reviews some of the processes that may be used prior to land disposal to reduce waste volume and/or utilize waste components, thus reducing landfill needs.

14.5.2 Incineration

Volume reduction.   Large numbers of batch-fed incinerators built during the 1930 and 1940s to reduce waste volume were major contributors to air pollution, performed poorly, and were costly to maintain. Some of these were upgraded, but most were shut down and replaced by land disposal of refuse whenever possible. However, as landfill capacity decreased, volume reduction became more important. At the same time, the fuel value of refuse had been rising steadily. As a result, incineration for reducing waste volume (by about 90 percent) and weight (by 75 percent) with the possibility of energy recovery, became a very popular processing option during the 1970s (see Table 14–7).

The newer municipal incinerators are usually the continuously burning type, and many have "waterwall" construction in the combustion chamber in place of the older, more common refractory lining. The waterwall consists of joined vertical boiler tubes containing water. The tubes absorb the heat to provide hot water for steam, and they also control the furnace temperature. With waterwall units, costly refractory maintenance is eliminated, pollution control requirements are reduced (because of the reduction in quench water and gas volumes requiring treatment), and heat recovery is simpler. Unfortunately, judging by European experience, corrosion of waterwall units may be a serious problem (Sarofim, 1977). The components of a waterwall incinerator are shown in Figure 14–4.

The combustion temperatures of conventional incinerators fueled only by wastes are about 760°C (1,400°F) in the furnace proper (insufficient to burn or even melt glass) and in excess of 870°C (1,600°F) in the secondary combustion chamber. These temperatures are needed to avoid odor from incomplete combustion. Temperatures up to 1,650°C (3,000°F), which would reduce volume by 97 percent and convert metal and glass to ash, are possible with supplementary fuels. Although the first high-temperature pilot installation was built in 1966, application to full-scale units has not followed, presumably because of the high costs involved.

Energy recovery.   Mass-burning of solid wastes to produce steam for heating or for use in power generation has been common in Western Europe and Japan for many years. However, until rising fuel prices through the 1970s and early 1980s made the economics of energy recovery attractive, the practice was rare in North America.

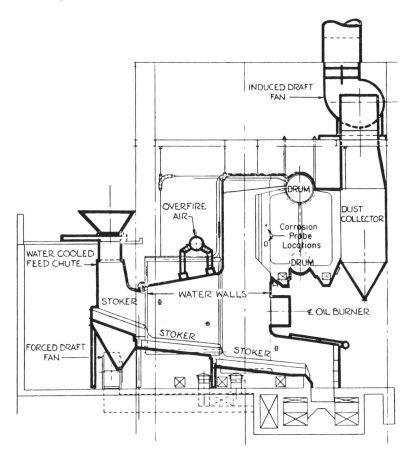

Figure 14–4 Section through a waterwall incinerator. Source: P. E. Miller, U.S. EPA Report, SW-72-3-3, 1972 as cited by Sarofim 1977, p. 174

Wastes burned solely for volume reduction do not need any auxiliary fuel except for start-up. On the other hand, when the objective is steam production, supplementary fuel (usually gas) must be used with the pulverized refuse, because of the variable energy content of the waste or in the event the quantity of waste available is insufficient. Ferrous metals are normally recovered from the ash.

Markets for steam must be close to the waste-burning incinerators for these combustion systems to be competitive with other heating sources. Wilson (1977) has suggested a maximum distance of 1 mile (1.6 km), but even this may be too far. The city of Chicago (Northwest Plant) and the city of Hamilton, Ontario (SWARU Plant), are two locations where no market for the steam from incineration was available during the first 10 years of operation (APWA, 1979). The incinerator in the city of Montreal, Canada, had no market for the steam from its mass-burning waterwall incinerators until 1983 (about 15 years after it was built).

TABLE 14–7 EXAMPLES OF STEAM-PRODUCING, MASS-BURNING MUNICIPAL INCINERATORS IN THE UNITED STATES AND CANADA

Location & start-up date	Design size[1] ton/d	Cost at start-up millions	Recover	Customer	Cost[2] per 1000 ton/d capacity millions
Braintree Mass., 1971	250	$ 2.8	—	Art & Leather Co.	$ 63
Chicago, Ill., 1971	1,600	$25.0	Fe	Candy Co. (1981)	$ 88
Nashville, Tenn., 1974	720	$29.0	—	Downtown Heating	$120
Quebec City PQ, 1974	1,000	$25.0	Fe	Pulp & Paper Mill	$ 88
Harrisburg, PA 1971	720	$ 8.3	Fe	Pennsylvania Power Co.	$ 64
Hamilton, Ont., 1971	500	$ 8.5	Fe	Ontario Hydro (1982)	$ 92
Saugus, Mass., 1976	1,500	$50.0	Fe	General Electric Power Plant	$ 83

Average Cost (1987) $ millions per 1000 ton/d capacity = $ 85

Adapted from APWA, 1979; NCRR, 1980; Schwegler and Hickman, 1981

[1] Design Size is reported in short tons (2,000 lb) per day.

[2] Capital costs in $1,000,000 (U.S.) per 1000 ton (2,000 lb) of daily plant capacity have been updated to a common year (1987) without the extra costs for the more stringent air pollution controls now required.

Some of the municipalities that have been able to sell their steam for heating or power generation are noted in Table 14–7 for comparative purposes.

Of the problems associated with incineration, air pollution control, especially the removal of the fine particulates and toxic gases (including dioxin), are the most difficult. The emission of combustible, carbon-containing pollutants can be controlled by optimizing the combustion process. Oxides of nitrogen and sulfur and other gaseous pollutants have not been a problem because of their relatively small concentration. Other concerns related to incineration include the disposal of the liquid wastes from floor drainage, quench water, and scrubber effluent, and the problem of ash disposal in landfills because of heavy metal residues. Public opposition to incinerators is another serious obstacle to their use. Capital costs (1987 prices) of about $120 million (U.S.) per 1000 ton of daily capacity, and operating costs of $15 to $30 per ton, apply to cities with over 300,000 population. Unit costs for smaller centers are much greater, and this tends to limit the use of incinerators to large cities. Even there, however, because of public concern (and considerable evidence) about the toxic gases generated by the burning of solid wastes, incineration is seldom proposed now unless lack of landfill sites leaves no better alternative. This was the situation in Detroit in 1987, where, despite vigorous protests from environmentalists and the neighboring Canadian city of Windsor, Ontario, a $500 million, 4,000-ton-per-day waste-to-energy plant, the largest in the U.S., was approved.

Example 14.6

A mass-burning incinerator with heat recovery operates on 400 t/d (450 tons/d) of municipal solid waste with natural gas as a supplementary fuel. A plan for residential source separation is expected to reduce the amount of paper and cardboard collected by 20%. For the incinerator to maintain steam production, the heating value of the lost combustibles will have to be replaced by natural gas at an average cost of $0.20/m³ ($0.56/100 ft³).

Neglecting changes in collecting costs, what price per tonne would need to be received for the paper for the municipality to break even? Note: The higher efficiency of incineration of natural gas compared to refuse can be neglected.

Solution   Our assumptions are as follows:

1. Original paper content of solid waste (Table 14–3) = 45%
2. Energy content of paper (Table 14–6) = 16,300 kJ/kg
3. Energy content of natural gas (Table 14–6) = 37,300 kJ/m³

For the original operation, we have:

- Total quantity of refuse burned = 400 t/d
- Quantity of paper burned (45%) = 180 t/d
- Energy content of paper burned = 180 t/d × 16.300 kJ/t = 2.93 × 10⁹

After separation, we obtain:

- Energy content of paper lost (20%) = 5.86 × 10⁸ kJ/d
- Additional volume of gas required:

$$\frac{5.86 \times 10^8 \text{ kJ/d}}{37,300 \text{ kJ/m}^3} = 15,700 \text{ m}^3/\text{d}$$

- Cost of additional gas = 15,700 × $0.20 m³ = $3,140/d

Thus, since the quantity of paper recovered is 36 t/d (40 ton/d), the municipality would have to receive

$$\frac{\$3,140/\text{d}}{36 \text{ t/d}} = \$87.00/\text{t} \ (\$79.00/\text{ton})$$

for separation to be worthwhile. Any additional expense attributable to separation, collection, storage, handling, and delivery of the paper should be added to this cost. Any offsetting savings because of the reduced quantity of refuse to be collected and the smaller amount to be prepared for burning would probably be small.

14.5.3 Other Conversion Processes

Chemical processes (such as fluidized bed incineration, pyrolysis, and wet oxidation) and biological processes (composting and anaerobic digestion) are other potential methods for reducing municipal waste volumes and/or converting the waste to useful products. Information on many of these processes has been provided in the *Handbook of Solid Waste Management* (Wilson, 1977) and by other investigators (Schwegler and Hickman, 1981;

Vesilind, 1980; Dasgupta et al., 1981). However, of all the chemical and biological conversion processes, only incineration with heat recovery and composting have become widely accepted.

Composting is the aerobic decomposition of organic matter by microorganisms, primarily bacteria and fungi. The reactions generate heat, raising compost temperatures during the composting period. Waste volume is reduced by about 30 percent for wastes with a high proportion of newsprint to perhaps 60 percent for garden debris (Golueke, 1977). Section 8.3.1 provides additional information on composting.

Composting may take place naturally under controlled conditions or in mechanized composting plants. In natural systems, ground garbage, preferably with glass and metals removed, is mixed with a nutrient source (sewage sludge, animal manure, night soil) and a filler (wood chips, ground corn cobs) which permits air to enter the pile. The mixture, maintained at about 50 percent moisture content, is placed in windrows, 2 to 3 m wide, and turned over once or twice a week. In four to six weeks, when the color darkens, the temperature drops, and a musty odor develops, the process is complete. The filler may then be removed and the remaining ''humus'' used as soil conditioner. With mechanical plants, continual aeration and mixing enable composting time to be reduced by about 50 percent. A short period usually follows the mechanical process to allow the composting material to ''mature.''

There is a limited market for soil conditioners in North America. Of the 20 or 30 solid waste composting plants built in the U.S. since the first one in Altoona, Pa., in 1951, all but perhaps two or three are closed. Newer composting plants like the Delaware Reclamation Project and others in the northeastern United States combine composting of sewage sludge with municipal solid waste and may indicate a trend toward high-rate composting as one solution to the problems of sludge disposal and solid waste management. By 1985, about 60 composting facilities, primarily for sewage sludge, were operating in 30 states, with the one in Denver, Colo., which opened in 1987, being the largest aerated windrow system in the U.S. The situation is quite different in Western Europe, Israel, Japan, and other advanced countries committed to land reclamation, where many successful solid waste composting plants have been operating for many years. Rotterdam, in Holland, already the location of one of Europe's largest heat recovery incinerators, has a major composting plant to complement its waste management program. The same interest in composting exists in Third-World countries, but in these areas windrow systems are the preferred method.

14.5.4 Resource Recovery Plants

Recovery of materials and energy, discussed in Section 14.3, has also been mentioned in connection with the incineration of municipal wastes when steam is produced and ferrous metals (and occasionally other materials) are recovered. However, in some municipalities, resource recovery, rather than being part of other processes, may be a separate operation to reclaim reusable material such as cardboard, paper, metal, and glass from municipal refuse.

Because of the large and increasing paper component in municipal solid waste, refuse-derived fuel (RDF) can be a useful supplementary fuel for combustion processes and utility

Municipal Solid Waste

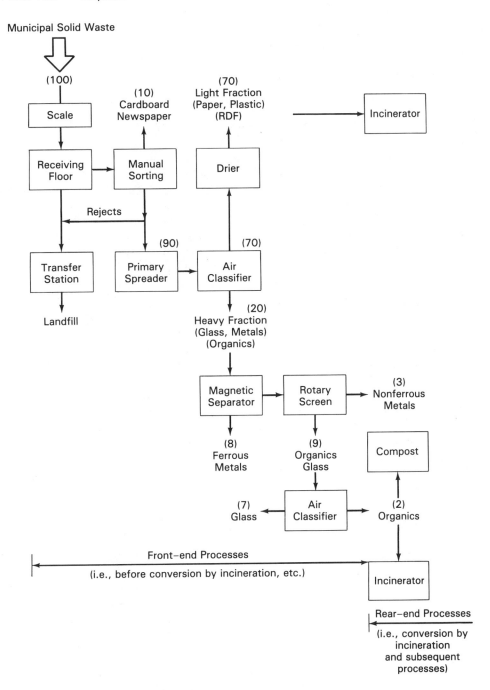

Figure 14–5   Flow diagram for a typical resource recovery plant Note: Figures shown in parenthesis are approximate percentages handled (dry weight).

boilers (see Section 14.3.3). The first full-scale use of RDF with coal in a utility boiler was in 1972 with Union Electric's Meramac Station in St. Louis, Missouri, where RDF provided up to 15 percent of the energy content in combination with the pulverized coal (Wilson, 1977). No serious problems with corrosion, erosion, or air requirements were encountered. Although the 270-t/d (300-ton/d) demonstration project was deemed a success, the full-scale 3,600-t/d (4,000-ton/d) plant was never built, initially because of public opposition, and later because of the questionable economics and the difficulty in ensuring that sufficient municipal refuse could be collected (Vesilind, 1980).

The enthusiasm of the 1970s for recycling and the possibility of realizing "cash from trash" led to the construction of various types of resource recovery plants. The primary objective of most of these was to produce RDF, with reclamation of other material being incidental to this operation. One of the earliest plants employing mechanical separation was the 180-t/d (200-ton/d) resource recovery plant in Ames, Iowa. Since start-up in 1975, this facility has provided RDF for the city-owned power plant and ferrous metal which is sold. The aluminum recovery system has had little use because of operational problems (APWA, 1979). A simplified flow diagram typical of many resource recovery facilities is shown in Figure 14–5.

By 1980, when the recycling movement was at its peak, there were over 40 facilities in the United States and Canada recovering energy and/or resources from municipal wastes. Since then, many of these resource recovery plants have been shut down. The few new ones, like the 2,000-ton/d resource recovery plant in Indianapolis, Ind., usually handle sewage sludge along with the solid waste. None of these plants has been profitable, and none may ever be economically justifiable. The unavoidable hand separation, the inefficiency of mechanical separation, the frequent equipment breakdowns, and the recovery of only 10 percent of the solid wastes were a few of the problems. Perhaps it would be fairer to assess these plants only for their role in relieving a disposal problem and not as a profit-making enterprise—in other words, the same way we consider incineration, wastewater treatment, and other necessary municipal services.

14.6 SANITARY LANDFILL DESIGN AND OPERATION

From the earliest times, disposal of solid wastes into open dumps was standard practice for municipalities. The town dump was usually in a low-lying area near a watercourse. Fires, water pollution, odors, rats, flies, and blowing papers were the visible results. Burial of the waste reduced these problems, but the greatest improvement was obtained by compacting the waste in layers and covering it with earth at the end of each day's operation. This method, called sanitary landfill, presumably to distinguish it from the typical unsanitary open dump, was first used in California in 1934 to reclaim land. Compacting and covering are still the basic operations today. Better compaction, reduced cover, and more recently leachate collection and site monitoring, together with more care in site selection, are some of the improvements that have taken place.

14.6.1 Criteria for Sanitary Landfills

Criteria for access, buffer distance, fencing, ditching, slopes, leachate handling, monitoring, operating procedures, etc., are normally set by the body responsible for site approval. Specific requirements vary greatly between authorities, but Tchobanoglous (1977) has suggested some general guidelines. Ideally, a sanitary landfill site should be on inexpensive land within economical hauling distance, have year-round access, and be at least 1,500 m downwind from residential and commercial neighbors. The area should be reasonably clear, level, and well drained, with capacity for not less than about three years' use until its future role as ''open'' space is realized. Soil of low permeability, well above the groundwater table, is also desirable for protection of underground water supplies and as cover material. Final choice of the site should not be made without a detailed hydrogeological investigation.

Preparation of the site involves fencing, grading, stockpiling of cover material, construction of berms, landscaping, and the installation of leachate collection and monitoring systems. Wells for gas collection may also be provided.

Mixed wastes with varying degrees of compaction are delivered to the site in packer trucks and/or trailer units. Some hand sorting of incoming wastes will be necessary, and pulverizing or high-pressure compaction and baling for volume reduction may precede placement. Loose material is placed in the lower part of the prepared pit or trench and then spread and compacted by machine in layers of about 0.5 m thickness. When the depth reaches 2 to 3 m, and at the end of each day's operation, the refuse is covered with 150 to 300 mm (6 to 12 inches) of earth. This consolidated solid waste enclosed by earth is called a cell and normally contains one day's waste. The cross section of the sanitary landfill in Figure 14–6 shows its salient design features.

Landfilling mixed refuse of low density (300 kg/m³) may be an uneconomical use of a site with limited capacity. Reduction of waste volume can not only extend the life of

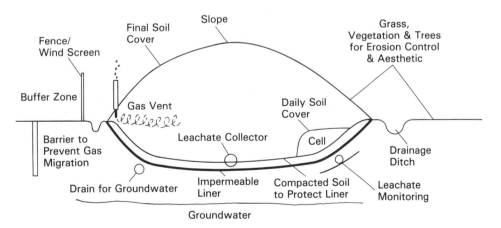

Figure 14–6   Cross section through a sanitary landfill

the landfill, but can contribute other benefits as well. Volume reduction by incineration is too costly, except for larger cities, so for small communities physical methods are the only practical alternative. Milling or pulverizing the refuse by a hammer mill is the most common method. In addition to a volume reduction of 50 percent and the reduction or perhaps elimination of the cover material, problems with odors, blowing papers, rodents, insects, fires, settlement, and mired vehicles are all greatly reduced (Reinhardt, 1969).

High-pressure compaction of municipal wastes into solid bales or blocks (about 1 m^3) weighing 850 to 950 kg is another way to reduce waste volume. The advantages are the same as those for pulverizing, but in addition, the need for on-site sorting and field compaction is eliminated. Furthermore, the resulting stable bales are resistant to infiltration of rainwater and, if necessary, suitable for rail haul (Wolf and Sosnovsky, 1977). Because of the saving in landfill capacity and other advantages, greater use of pulverizing or high-pressure compaction for waste volume reduction seems likely in the future.

14.6.2 Problems with Landfilling

Aesthetic considerations. Many of the shortcomings of a poorly operated landfill are evident: odors and blowing papers carried by the wind; vermin, insects, and scavenger birds attracted by the organic refuse; and dust and noise from trucks and compacting operations. Continuous field compaction of the loose refuse and covering all material with earth at the end of each day alleviate these problems. Volume reduction by pulverizing or high-pressure compaction provides even greater assurance of an aesthetically acceptable operation.

Economic loss. Property devoted to landfilling is no longer available as productive farm land or as taxable property. Even after closure of the site, future use of the area must be restricted to some type of open development such as a park, recreational area, or ski hill, and the construction of buildings must be rigidly controlled.

Environmental effects. The potential effects of solid waste on public health and the environment were noted in Section 14.3. The following discussion deals with the environmental effects of leachate and gas from sanitary landfills.

When municipal solid wastes are landfilled, the organic matter present decomposes aerobically during the first few weeks (in wet areas) or the first year (in dry areas) and then degrades anaerobically when oxygen is no longer present. While the wastes are decomposing, liquid from the waste, seepage from groundwater, and water from precipitation and surface runoff percolate through the refuse, producing a contaminated liquid called leachate. Contamination of the groundwater by leachate high in organics, dissolved solids, and other constituents can be a serious problem where nearby wells are used for water supply. The hazard stems mainly from soluble salts, since biodegradable organics and pathogenic microorganisms are usually removed by the soil before the leachate has traveled very far.

The gases, principally methane (CH_4) and carbon dioxide (CO_2) generated by the anaerobic decomposition of organics in the landfill are also a concern. Depending on the stage of decomposition reached, methane may constitute up to 60 percent of the gaseous components generated by a sanitary landfill. Methane is an odorless, combustible gas that is heavier than air and explosive when its concentration in air is between 5 and 15 percent. It is therefore a hazardous gas and should not be ignored. Methods for controlling methane are considered later. Carbon dioxide in combination with water creates an acidic environment in which minerals such as calcium, magnesium, iron, cadmium, lead, and zinc that are present in the refuse (or in the soil) tend to dissolve and move toward the groundwater table. Calcium and magnesium only add hardness to groundwater, but the toxic heavy metals are a more serious problem because they can make the water unfit for human consumption.

Inappropriate regulations.   Regulations applying to sanitary landfills are often arbitrary or based on studies conducted elsewhere. Having pioneered in the development of landfills, California has served as a guide for many other localities. For example, California regulations require that sanitary landfills be located a minimum distance above the groundwater table, in soil of low permeability, and a minimum distance from the nearest point of water use. This makes sense because, in California, where evaporation is greater than rainfall, keeping the landfill a minimum distance above the groundwater table helps to protect the groundwater against intrusion by leachate. However, in humid areas, around the Great Lakes, for example, where precipitation exceeds evaporation, such a requirement offers no protection for groundwater against the downward movement of leachate. Furthermore, in areas where the depth to groundwater is the greatest, more permeable soils are usually found. Consequently, leachate percolates rapidly through the soil, making protection of groundwater difficult.

On the other hand, soil of low permeability is considered desirable because it restricts the flow of leachate. Whether a completely impermeable soil is better is questionable, however, depending to some extent on site topography and climate. In humid areas, completely impermeable soil could be troublesome since it might allow leachate to collect in the landfill and eventually overflow, causing a local nuisance or polluting surface water. Permeability, therefore, is not a factor to be considered in isolation from other requirements.

Finally, regulation by distance is also not meaningful unless it is related to soil permeability. Tests have shown that over 99 percent of the reduction in dissolved solids occurs in 17 feet (5 m) in silty clay, but in silty sand 650 ft (200 m) is needed for just a 90 percent reduction (Hughes and Cartwright, 1972).

Accordingly, general rules governing landfills may be inappropriate. Requirements for a particular landfill should be based on extensive hydrogeological investigation of the actual site. This has seldom been done in the past, but it is essential if we hope to ensure adequate protection of our surface-water and groundwater supplies in the future. Aerial photography provides the best base for preliminary mapping. A site survey can then furnish additional details on topography, drainage, soil stratigraphy, and groundwater characteristics. Data from local wells and site boreholes can supply information on groundwater depth and movement.

TABLE 14–8 CHARACTERISTICS OF LEACHATE FROM
SANITARY LANDFILLS (mg/L)*

Constituent	Range*	Typical value
Organic strength, COD	1,000–30,000	10,000
BOD$_5$	200–20,000	6,000
Total solids	2,000– 5,000	3,000
Total nitrogen	20– 1,000	200
Alkalinity (as CaCO$_3$)	200– 5,000	400
Soluble Salts (Cl, SO$_4$)	200– 3,000	500
Iron	50– 800	100
Lead	1– 10	2
Zinc	25– 250	50
pH	5– 8	6

Adapted from Chian and DeWalle, 1977, Tchobanoglous et al., 1977,
and Vesilind and Rimer, 1981
* Except for pH

14.6.3 Leachate Generation and Control

Characteristics of leachate.   Leachate, the contaminated liquid draining from
a sanitary landfill, varies widely in its composition depending on the age of the landfill
and the type of waste it contains. Typical concentrations of the constituents and representative
ranges are noted in Table 14–8.

Quantity of leachate.   Vesilind and Rimer (1981) have described a water balance
technique developed by Tenn et al. (1975) based on annual average values, for estimating
leachate quantity. According to this method, leachate production is equal to the precipitation
falling on the landfill less the amount of this precipitation lost from the landfill. Until the
landfill becomes saturated, water entering the landfill is also reduced by the amount of

TABLE 14–9 QUANTITY OF LEACHATE FROM CONTRIBUTING SOURCES

	Source of leachate	Reference	Typical value	Normal range	Governing factor
1	Precipitation	Records	900 mm	400–1,200 mm	Location (climate)
2	Losses from:				
	(a) Evapotranspiration	Records	70%*	40%–90%	Humidity
	(b) Surface runoff	Figure 14.7	20%	10%–45%	Soil type
3	Water retained: (% by volume)				
	(a) Saturated soil	Figure 14.8	30%	10%–40%	Soil type
	(b) Refuse in place	Field test	30%	20%–35%	Refuse

Adapted from Tenn et al., 1975, and Vesilind and Rimer, 1981.

* The loss of 70% of precipitation by evapotranspiration applies to open land with vegetation. In a landfill, the loss
might only be one-half to two-thirds this value, or, in this case, 35 to 45% of the total precipitation. (Studies by
Pohland in Georgia indicated that about 35% of the incident rainfall was evaporated from an open cell).

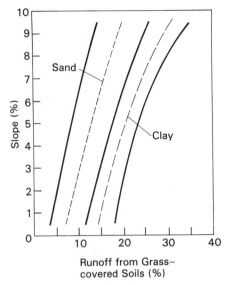

Figure 14–7 Runoff from grass-covered slopes (Adapted from Vesilind and Rimer 1981 and Tenn et al 1975)

Runoff from Grass-covered Soils (%)

moisture the landfill retains in the soil and refuse. The basis for estimating the quantities from these contributing sources is indicated in Table 14–9. Figure 14–7, showing the percent runoff from grass-covered soils at various slopes, and Figure 14–8, which relates the maximum field capacity of soils to their permeability coefficient or particle diameters, supplement the table. Example 14.7 illustrates the use of this information.

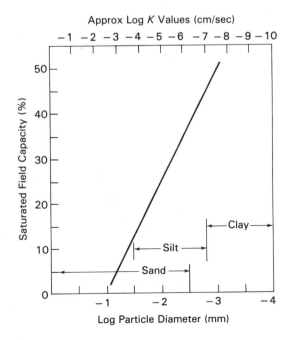

Log Particle Diameter (mm)

Figure 14–8 Approximate field capacity of soils (% Vol. Water/Vol. Soil) Adapted from Vesilind and Rimer, 1981 and American Iron and Steel Institute, Modern Sewer Design, 1980

Example 14.7

Municipal solid wastes from packer trucks are placed and well compacted in a sanitary landfill in three lifts, each 2 m (6.5 ft) deep separated by a 250-mm (10-in) clay layer and topped with a 1-m- (40-in-) thick clay cap having a 4% slope. If annual precipitation in the area is 900 mm (36 in), of which 67% is lost through evapotranspiration, estimate (a) the quantity of leachate that will be generated, and (b) the time until the refuse is saturated and the leachate flows from the landfill. (Note: Neglect the 250-mm (10-in) separating clay layers for estimating the leachate percolation rate through the refuse.

Solution　The assumptions for the problem are as follows:

1. Density of waste as delivered (Figure 14–2) = 300 kg/m³ (500 lb/yd³)
2. Average moisture content as delivered (15–40%) = 25% by weight*
3. Density of well-compacted landfill (Figure 14–2) = 600 kg/m³ (1,000 lb/yd³)
4. Maximum moisture content of compacted refuse (20–35%) = 30% by volume*

(a) Annual precipitation (from local meteorological records) = 900 mm (36 in)

Less loss by evapotranspiration (67% from records) × ½
(Table 14–9)　　　　　　　　　　　　　　　　　　　　= 300 mm (12 in)

Less loss by surface runoff = 17% (Figure 14–7)　　 = 150 mm (6 in)

Annual Leachate Production　　　　　　　　　　　　= 450 mm (18 in)

(b) WATER RETAINED

In clay soil = 45% by volume for the least impervious clay (Figure 14–8)

Soil is assumed to be saturated when placed and able to retain no additional water.

IN REFUSE AS DELIVERED

- As a basis for calculation, assume weight of refuse delivered = 1,000 kg
- Weight of water/unit weight of refuse (Assumption 2) = 25%
- Weight of water in 1,000 kg of refuse = 250 kg
- Volume of water = $\dfrac{\text{weight of water}}{\text{density of water}} = \dfrac{250 \text{ kg}}{1{,}000 \text{ kg/m}^3} = 0.25 \text{ m}^3$

IN REFUSE AS COMPACTED

$$\text{Volume of compacted refuse} = \frac{\text{weight of refuse}}{\text{density of refuse}}$$

$$= \frac{1{,}000}{600} \frac{\text{kg}}{\text{kg/m}^3} = 1.667 \text{ m}^3$$

$$\text{Volume of water/unit volume of compacted refuse} = \frac{0.25 \text{ m}^3}{1.667 \text{ m}^3} \times 100 = 15\%$$

Note: The preceding calculations can be reduced to

$$M_v = M_w \frac{\rho_c}{\rho_w} = 25 \times \frac{600}{1{,}000} = 15\%$$

* Should be found by field tests.

where M_v and M_w = percent water in the compacted refuse (by volume) and in the refuse as collected (by weight), respectively

ρ_c and ρ_w = density of refuse as compacted and density of water, respectively.

Thus, the available capacity is as follows:

$$\begin{array}{ll} \text{Maximum available capacity (Assumption 4)} & = 30\% \\ \text{Less water originally present in refuse} & = 15\% \\ \text{Available water-retaining capacity of refuse} & = 15\% \end{array}$$

Consequently, with leachate produced at the rate of 450 mm/year, percolating through 6 m of refuse which has 15% water-retaining capacity, it will be

$$\frac{15}{100} \times \frac{6\text{ m}}{0.450\text{ m/yr}} = 2 \text{ years}$$

before flow from the landfill occurs.

Example 14.7 illustrates how a mass balance on water can be used to estimate leachate quantity and generation time. In practice, during the working life of the landfill, material is not always in place or surface runoff may drain into the landfill (rather than to a collecting sump), thus creating an immediate leachate control problem. The water balance approach has many flaws, but it does enable a comparison to be made on the relative severity of the leachate problem under different climatic conditions. Leachate may be negligible in dry climates, or it may even exceed 100 percent of precipitation with improperly designed landfills in wet areas.

Control of leachate.   The available liquid-retaining capacity of refuse is sometimes the justification for mixing liquid wastes such as sewage sludge or industrial effluents with solid waste. If refuse were able to accept 15 percent of its volume as liquid (as in Example 14.7), then 15 m³ of liquid per 100 m³ of refuse could be absorbed. This amounts to 450 L/tonne of refuse, for a refuse density of 333 kg/m³ delivered to the landfill, a ratio which is used in the United Kingdom where codisposal of municipal solid waste and liquid industrial waste is practiced.

Once the capacity of the refuse to absorb liquid has been reached (or even before), leachate migrates through the underlying soil toward the groundwater table. The maximum rate of percolation through the soil occurs when its field capacity is reached (Figure 14–8), and at that point the soil no longer absorbs water. Under this condition, the rate of liquid travel through a uniform soil is proportional to the hydraulic gradient that causes the flow. This relationship, known as Darcy's Law, can be expressed as

$$Q = KSA \tag{14.1}$$

where Q = quantity of liquid flowing through area A per unit time = nvA

K = coefficient of hydraulic conductivity (dependent on soil type)

S = hydraulic gradient, i.e., the change in elevation of the "free" water surface between the two points being considered divided by the distance through which the liquid must travel

A = gross cross-sectional area through which the flow passes

TABLE 14-10 VALUES OF POROSITY AND HYDRAULIC CONDUCTIVITY

Soil type	Porosity (%)	Hydraulic conductivity	
		Description	K(cm/sec)*
Gravel	25–40	High permeability	Over 1×10^{-1}
Sand to fine sand	25–50	Medium permeability	1×10^{-1} to 1×10^{-3}
Silty sand to dirty sand	30–50	Low permeability	1×10^{-3} to 1×10^{-5}
Silt	35–50	Very low permeability	1×10^{-5} to 1×10^{-7}
Clay	40–70	Practically impervious	Less than 1×10^{-7}

* Because of the widespread use of cm/sec for soil systems, these units have been retained here. Conversion factors are as follows:

$$\text{cm/sec} \times 1.97 \quad = \text{feet/min}$$
$$\text{cm/sec} \times 2880 \quad = \text{feet/day}$$
$$\text{cm/sec} \times 21{,}200 = \text{qpd/ft}^2$$

Source: For porosity values, Freeze, R. A., and Cherry, J. A., *Groundwater*, Englewood Cliffs, N.J.: Prentice Hall, 1979, pp. 36–37.

For hydraulic conductivity values, Theil, P. "Subsurface Disposal of Storm Water," in *Modern Sewer Design*, Canadian Edition, ed. Committee of Sheet Steel Producers, Washington, D.C.: American Iron and Steel Institute, 1980, pp. 175–193.

v = velocity at which the liquid travels through the soil

n = soil porosity i.e., void volume divided by total volume of the soil mass

Approximate values of porosity and hydraulic conductivity are listed in Table 14–10. The application of Darcy's Law is illustrated in Example 14.8.

Example 14.8

If the sanitary landfill described in Example 14.7 were set in clay having 50% porosity, and if $K = 1 \times 10^{-7}$ cm/s,

(a) How long would it take the leachate to percolate from the bottom of the landfill through the underlying soil to the groundwater table 1.5 m below (assuming that the leachate is not allowed to build up in the landfill and the underlying soil is saturated)?

(b) How long would it take if the soil had a hydraulic conductivity of 1×10^{-6} cm/s?

(c) If the sanitary landfill became filled to overflowing, what effect would this have on the liquid time of travel determined in parts (a) and (b)?

Solution

(a) From Equation 14.1 and the relationship $v = Q/nA$, we obtain

$$v = \frac{KSA}{nA} = \frac{KS}{n}$$

Therefore,

$K = 1 \times 10^{-7}$ cm/s $\times$ 86,400 $\times$ 365 s/yr = 3.15 cm/yr

H = difference in elevation between the water surface at the bottom of the landfill and at the top of the groundwater table = 1.5 m

d = distance over which the liquid differential H is acting = 1.5 m

$$S = \frac{H}{d} = 1.5 \text{ m}/1.5 \text{ m} = 1.0$$

$$v = \frac{KS}{n} = \frac{3.15 \times 1.0}{0.5} = 6.30 \text{ cm/yr}$$

$$t = \text{time of travel} = \frac{d}{v} = \frac{1.5 \text{ m}}{6.30 \times 10^{-2} \text{ m/yr}} = 24 \text{ years}$$

(b) If the coefficient of hydraulic conductivity were only 1 x 10^{-6} cm/s (31.5 cm/yr), then, assuming that the porosity n is still 0.5, the leachate would reach the groundwater table in

$$t = \frac{1.5 \times 0.5}{31.5 \times 10^{-2} \text{ m/yr}} = 2.4 \text{ years}$$

(c) If the sanitary landfill were filled with liquid to overflowing, the hydraulic driving force H becomes the difference in elevation between the water surface at the top of the landfill and the top of the groundwater table, namely, 9 m. The distance d through the soil over which this 9-m liquid differential is driving the leachate is still 1.5 m. Therefore, S becomes 9 m/1.5 m = 6, and $v = 6K/n$, so that for $K = 1 \times 10^{-7}$ cm/s (i.e., 3.15 cm/yr),

$$t = \frac{dn}{KS} = \frac{1.5 \text{ m} \times 0.5}{3.15 \times 10^{-2} \text{ m/yr} \times 6} = 4 \text{ years}$$

and for $K = 1 \times 10^{-6}$ cm/s = 31.5 cm/yr,

$$t = \frac{dn}{KS} = \frac{1.5 \text{ m} \times 0.5}{31.5 \times 10^{-2} \text{ m/yr} \times 6} = 0.4 \text{ year}$$

Comment: Calculations have assumed that the soil through which the leachate passes is saturated. For unsaturated soil, calculations would be much more complex and the time of travel could be even shorter.

Example 14.8 demonstrates the importance of locating landfills in impermeable soils well above the groundwater table and of not allowing the leachate to build up within the landfill. Such ideal conditions are not always available, and additional precautions are frequently necessary to protect groundwater supplies from contamination. Supplementary measures include clay and/or membrane covers and liners for the landfill, a leachate collection, removal, and treatment facility, and a groundwater monitoring system. These procedures are intended to ensure that the buried refuse remains as dry as possible and that leachate is prevented from reaching groundwater supplies. Such measures cannot ensure that no seepage will occur. However, if leakage does occur, attenuation of contaminants as the liquid passes through the soil serves as an additional barrier to groundwater contamination. Because of filtration, adsorption, biological activity, and precipitation, soil is an extremely effective medium for removing many organic materials, heavy metals, and other inorganic ions.

Instead of attempting to maintain dry conditions within the landfill, some designers contend that the refuse should be kept wet. This will accelerate anaerobic digestion and is in fact necessary if recovery of gas is contemplated. Collection and recirculation of leachate through the landfill has been advocated by Pohland (1972) in order to hasten decomposition, thereby reducing the concentration of the contaminants and shortening the time until final stabilization of the site is achieved. The decision as to whether sanitary landfills should be kept wet or dry is not always obvious. For each location, designers must consider to what extent climatic conditions are advantageous or detrimental, whether inflow can be controlled or utilized, to what extent the groundwater supply is important, whether recovery of gas is economical, and whether there is a need for leachate treatment. Because the significance of these factors can vary widely with location, landfill design suitable for one part of the country may be unsatisfactory in another. Consequently, no single design approach is applicable.

14.6.4 Gas Production

The gas produced in sanitary landfills by the anaerobic digestion of organic refuse is usually vented to the atmosphere through gravel-packed seams or wells and causes no problems. In some landfills, waste gas burners are installed at the top of the vents to burn off the escaping gas. If proper venting is not provided, lateral movement of the gas may occur under the landfill cover particularly when the ground surface is frozen. This can be dangerous if the gas should migrate to nearby buildings.

Increasing fuel prices in the 1970s provoked interest in the possibility of gas recovery from sanitary landfills. In the United States, by 1983 over 20 projects, most of them in California, were under way to recover and purify landfill gas for on-site use for heat and power, or for off-site use as fuel (Weddle et al., 1983). Methane (CH_4), which constitutes 40 to 60 percent of the landfill gas, has a heat content of about 37,000 kJ/m^3 (991 BTU/ft^3), or 20,000 kJ/m^3 for landfill gas containing 55 percent methane. Because of the dilution of the gas with air during recovery, 16,800 kJ/m^3 (450 BTU/ft^3) is perhaps a more realistic value (Emcon Associates, 1980). Theoretically, the total amount of gas produced is 200 to 270 L of CH_4 per kg of refuse, depending on the characteristics of the solid wastes and the basis of the determination (Vesilind, 1980). Of the amount generated, an estimated 15 to 35 percent can be recovered.

Landfill stabilization and, hence, gas generation take place over a long time. Thirty years is a commonly mentioned period, but this could be shortened under continuously wet conditions or prolonged if the refuse remained dry. For a stabilization period of 25 to 30 years, one-third to two-thirds of the gas might be generated within the first five years (Tchobanoglous et al., 1977). Methane production rates of 2.5 to 3.7 L per kg of refuse per year have been reported by Emcon Associates (1980) for refuse that has been in place a few years.

Example 14.9

Over a three-year period, wastes from a population of 100,000 have been placed in a sanitary landfill with a gas recovery system. This practice is to continue into the foreseeable future, so a steady supply of gas with 55% methane (CH_4) is expected. A nearby armed forces base has 50 oil-heated detached homes for married personnel. The homes use an average of 100 $\times$ 10^6 kJ of heat energy during the year and have a peak demand during the coldest month of $2\frac{1}{2}$ times the average. Will there be enough landfill gas available to heat these homes?

Solution   The assumptions for the problem are as follows:

1. Quantity of municipal solid waste delivered to site = 1 tonne/capita per year.
2. Total gas production = 200 L CH_4 per kg of refuse over 30 years.
3. Gas recovery will be at least 15% of that generated, i.e., 30 L CH_4 per kg of refuse.
4. At least one-third of this gas will be generated within the first 5 years.
5. Conversion efficiency of the gas furnaces is 75%.
6. The first year that the homes are connected will be the most critical, since the gas available will be dependent on the three-year supply of waste already in place. This amount of waste will therefore be used as the basis for the calculations.

The annual quantity of solid waste produced = 100,000 t/yr, so that the total quantity in 3 years = 300,000 t. The total amount of CH_4 produced over 30 years from the original three-year supply of waste is

$$\frac{200 \text{ L} \times 1 \text{ m}^3}{1,000 \text{ L}} \times 300,000 \text{ t} \times \frac{1,000 \text{ kg}}{\text{t}} = 6 \times 10^7 \text{ m}^3$$

The CH_4 recovery is 15% = 9×10^6 m³, so the CH_4 recovery rate during the first 5 years is

$$\frac{1}{3} \times \frac{9 \times 10^6 \text{ m}^3}{5 \text{ yr}} = 600,000 \text{ m}^3/\text{yr}$$

(Note that 600,000 m³/yr = 2 L of CH_4 per kg of refuse, a very conservative value compared to the 2.5 to 3.7 L/kg reported (Emcon Associates, 1980.)

The energy available for heating at 75% efficiency = 450,000 m³/yr, so the energy content of the fuel (55% CH_4) is

$$\frac{450,000 \text{ m}^3/\text{yr}}{0.55} \times 16,800 \text{ kJ/m}^3 = 13.8 \times 10^9 \text{ kJ/yr}$$

Since the heat requirement per home during the peak period = 250×10^6 kJ/yr, the number of homes that can be served is

$$\frac{13.8 \times 10^9 \text{ kJ/yr}}{250 \times 10^6 \text{ kJ/yr}} = 55 \text{ homes}$$

Consequently, there will be sufficient landfill gas to heat the homes, and with 100,000 t of additional refuse being deposited annually, the amount of recoverable gas would increase for many years. Note, however, that the question as to whether it would be economical to recover, clean, and store the gas and operate the facility for 50 homes has not been answered. What is your opinion?

14.7 FUTURE OPPORTUNITIES

The unfavorable economics of resource recovery and the health concerns about incineration made these options, once so popular in the 1970s, of limited interest in the 1980s. At the same time, vociferous public opposition made approval of new landfill sites time consuming and difficult, if not impossible. Many municipalities, caught in this dilemma, have turned to separation at the source as a way of recycling material and relieving the solid waste problem, essentially a throwback to the early days. There is, of course, no one best solution with universal application. Solid waste management is a relatively new field, and many years of experimentation lie ahead. As the urgency of the problems increases through the nineties, there will be new opportunities for dealing with solid wastes and for reexamining old practices, from the control of waste sources through collection and processing to final disposal of residues. Waste management, if it is to be effective, must integrate all these aspects of a solid waste system for the maximum benefit of those served by the system. The components are all interrelated, and a change in one can affect all the others. Consequently, none of the activities can be considered in isolation.

14.7.1 Source Control

Control of solid wastes should begin at the manufacturing and marketing level before disposition of the refuse by residents. Product design, packaging, and consumer consumption habits are areas where significant improvements in waste reduction are possible, although difficult to achieve. Prohibition of throwaway beverage containers is an example of the type of legislation that can be applied to conserve resources and relieve the municipality of the onus of dealing with these discarded containers.

14.7.2 Collection

For most municipalities, current collection practices with packer trucks and two- or three-person crews will probably continue. However, under special circumstances, newer, more innovative approaches are possible. A few of these are the following:

- One-person collection vehicles that automatically pick up, unload, and return containers are efficient under ideal conditions. However, parked cars, low trees, and containers of a nonuniform size create problems (Stone, 1977).
- Pneumatic solid waste collection systems with 450 mm minimum-diameter pipe are common for short-distance transport from hospitals, large apartment complexes, etc. (Stone, 1977). They are occasionally used for longer distances up to about 3 km (2 mi). Such systems have been operating successfully in Sweden and Japan and at Disney World in Florida since the late 1960s and could have applications for municipal use in North America (Zandi, 1977).
- Liquid slurry pipelines have potential for transporting solid wastes over longer distances than pneumatic systems. Pipes could connect directly to buildings or supplement a pneumatic system (Zandi, 1977). In either case, wastes would be ground and mixed with water before discharge into the pipeline. High initial costs for the system and for grinding of solids have discouraged its use. However, the emergence of low-cost

grinder pumps for wastewater collection systems in the 1970s may encourage the use of simpler systems for solid waste. A common system for handling both solid wastes and wastewater would also seem to have great potential.

Venice, Italy, has one of the more unusual collection systems, as shown in Photograph 14–3(a). Eventual disposal of the solid waste at sea, although simple, is environmentally

(a)

(b)

Photographs 14–3 (a and b) Collection and disposal of municipal solid wastes
Current municipal solid waste collection and disposal practices vary from simple to sophisticated, from undesirable to acceptable.

(a) In Venice, Italy, population 350,000, solid waste is collected by thirty-two motorized gondolas like the one shown here. The collected waste is transferred to a scow and towed 30 km into the Adriatic Sea where it is dumped. This approach, fortunately, does not seem to have wide application. (Photo courtesy J. G. Henry)

(b) In Vancouver, Canada, trailer units, loaded at a transfer station, are quickly unloaded at the Burn's Bog landfill site by the tipping unit shown here. The unit reduces labor requirements at the site and the unit itself is operated by the truck driver (who does not remain in the truck). (Photo courtesy J. G. Henry)

undesirable. A more acceptable disposal method in a modern sanitary landfill is shown in Photograph 14–3(b).

14.7.3 Processing

The first opportunity for processing solid waste occurs at a transfer station. This central collection point would seem to be the best place for materials recovery (as markets develop) and the preparation of RDF for heat recovery incinerators, utility boilers, or institutional heating systems. Reduction of waste volume by compression and baling or pulverizing and compaction could also be economically accomplished here, with resultant savings in hauling the reduced waste volume to disposal.

14.7.4 Disposal

Landfilling is the common element of all disposal methods, receiving solid wastes as collected or their residues left after processing. As long as cost is the determining factor in solid waste disposal, landfilling of unprocessed wastes will continue to be the predominant method. Even in highly developed countries where incineration is common and land is scarce, there are potential sites for landfilling. In the United Kingdom, for example, 25 percent of the space left from surface mining operations in one year could accommodate all residential and commercial wastes leaving 75 percent for inert industrial wastes and construction industry debris (Holmes, 1981).

    Although much has been learned about landfilling since the days of the open dump, there are still a great many questions to be answered, for example,

- Under what conditions should a landfill be designed to be kept dry, and when would additional (or recirculation of) liquid be beneficial?
- What are the consequences of codisposal of municipal solid wastes with sewage sludges or industrial wastes?
- When should landfills be designed for complete containment, and when should soil attenuation and leachate migration be considered?
- What circumstances should prevail to make recovery of gas from landfills practical? (In Italy, recovery of gas from sanitary landfills is required by law.)

    Answers to these and numerous other questions are needed if future landfills are to be designed on a rational basis. Solutions will be influenced by the characteristics of the site and the climate, of course, but also by the types of processing, collection, and source control that precede landfilling and by the size of the area being served. Control of these latter system-interdependent factors (among others) is the function of solid waste management. Guidance for management is set out in a waste management plan prepared in accordance with the objectives of the local, regional, or national authority involved. These objectives and their order of priority will vary from one location to another. In one region, protection of the groundwater supply might be the major requirement, while in another minimal use of landfilling or maximum energy recovery will be predominant. If the objectives are clearly established by a comprehensive plan, decisions at the political level and those taken by management are more likely to be in harmony and therefore acceptable to the public.

PROBLEMS

14.1. A mass-burning incinerator burns 800 t/d (880 ton/d) of municipal solid waste 7 days a week.
   (a) What area of storage is needed to store waste for 3 days if the average depth of refuse is to be limited to 1.5 m (5 ft)? Waste is delivered by packer trucks.
   (b) How many packer trucks arrive at the incinerator on an average day if collections are carried out over 5 days?
   (c) If a newspaper strike reduced the quantity of paper collected by 50%, what percentage reduction would this make in the weight of refuse delivered each day?

14.2. The solid wastes from a summer camp with 100 children and a staff of 25 are to be collected once per week. If bottles and cans (representing 20% of the weight) are removed, paper wastes (40%) are burned in the camp incinerator, and only the wastes from the kitchen (30%) and miscellaneous wastes from the cabins (10%) are collected, what volume will be picked up?

14.3. The composition of solid waste from a residential community is as follows:

COMBUSTIBLES	% BY WEIGHT
Paper and cardboard	35
Food wastes	15
Garden trimmings	10
Others (textiles, rubber, leather, wood, plastic)	10

NONCOMBUSTIBLES	
Metal	10
Glass	10
Ash	10

Estimate (a) the moisture content, (b) the density, and (c) the energy content of this waste based on typical values for the components (Tables 14–4, 14–5, 14–6).

14.4. The solid waste proportions shown in Table 14–3 and the per capita generation rates for 1980 indicated in Figure 14.1 are averages for the United States. Neglecting "special" wastes, indicate quantitatively how the proportions and the per capita generation rates (kg/d) might compare with the national average in each of the following cases:
   (a) The first collection in a residential area following the Christmas holiday if Christmas trees are picked up separately.
   (b) Collection in the same neighborhood as in (a) during the two-month summer holiday season.

(c) Collection in a suburban residential community where many head offices, government agencies, and high-technology industries are also located.

(d) Collection in a town of 20,000 located in a farming region. A private, aggressive source-separation plan for papers, bottles, and cans is in operation.

14.5. Based on the following hypothetical costs, determine the average haul distance to a landfill site before a transfer station and one-person trailer units (capacity 20 tonnes) would be more economical than direct haul by three-person packer trucks (capacity 4 tonnes).

Labor costs (average)	$12.00/hr
Capital and operating costs	
Transfer station	$14.00/t ($12.75/ton)
Packer trucks	$ 0.62/km ($1.00/mi)
Trailer units	$ 1.25/km ($2.00/mi)

Assume that the average speed while hauling is 30 km/hr (20 mph), including delays at the site.

14.6 The Regional Municipality of Urbanville (see sketch) has an industrialized city core A and two centralized cities B and C. Two thousand tons per day of municipal solid wastes from the entire area are currently conveyed directly to a suitable permanent sanitary landfill site D.

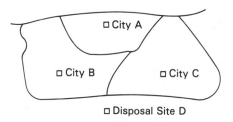

□ Disposal Site D

Consideration is being given to (1) providing transfer stations at A, B, and C, with final disposal at D, or (2) adding central resource recovery processing at A, with transfer stations at B and C and final disposal of residues at D.

(a) Describe briefly or illustrate diagramatically two other alternative systems which should be evaluated.

(b) Why would mathematical programming techniques be useful in evaluating alternative systems?

(c) Is it reasonable to evaluate these options on the basis of costs alone? What other factors might be important? Be specific.

14.7. As part of a program to reduce twice weekly collection to a once-per-week pickup, an American municipality of 300,000 is promoting the installation of sink disposal units in homes and implementing a plan for separate newspaper pickup. An oil-fired, mass-burning waterwall incinerator takes the municipal refuse having a composition as shown in Table 14–3 and Figure 14–1 for 1980 and sells steam to an adjacent power-generating station. If sink units that can reduce collectable food wastes by 50% are used in 20% of the dwellings, then, assuming that the paper content of the municipal solid waste will be 85% of the amount previously collected,

(a) By what percentage will the output of steam to the utility be reduced if no supplementary fuel is provided and the food and paper content represent all the combustibles? Neglect the difference in combustion efficiency between solid waste and oil.

(b) With oil at $0.35/L, what would be the additional daily cost of oil to maintain the previous output?

14.8. An experimental 200-t/d resource recovery plant, salvaging the proportions noted in Figure 14–5, has markets for RDF, ferrous metals, and glass, each at $20/t, and savings from the reduced amount to be landfilled (at $20/t). If capital and operating costs for the plant amount to $50/t, what annual subsidy must be provided to keep the plant operating?

14.9. If the landfill described in Example 14.7 had a compacted density of 600 kg/m³, a refuse depth of 9 m (29.5 ft), a moisture content of 20% by volume, and a 1-m- (3.25-ft-) thick clay cover with a 2% slope, estimate (a) the quantity of leachate that would be generated each year, and (b) how long it will be before each year's deposit of refuse is saturated and leachate flows from that section. Use rainfall and evapotranspiration values appropriate for your area.

14.10. Refuse from an industrialized city with a population of 200,000 has the following composition by weight:

paper	50% (includes cardboard)
organics	20% (food wastes)
glass	10% (colored, noncolored)
metal	10% (50% ferrous)
inerts	10% (ashes, dirt)

The refuse also has at-source bulk densities as noted in Table 14–5. Average waste generation is 2 kg/capita/day (4.5 lb/capita/day).

The city owns agricultural land 1,000 m × 600 m (1,600 ft × 1,000 ft) 15 km (24 mi) from the city core. The site is level in the North-South direction (1,000 m), has a 2% fall in the West-East direction (600 m), and consists of sandy clay with $K = 1 \times 10^{-6}$ cm/s and $n = 0.5$. The maximum groundwater level is 3 m (10 ft) below the lowest ground elevation and level across the site. Information for designing a sanitary landfill with an unlined bottom and no leachate collection system is as follows:

Bottom of landfill	2 m (6.5 ft) minimum above groundwater table
Daily earth cover	8% of cell volume (avg.)
Side earth slopes	2 horizontal to 1 vertical
Final cover	1 m (3.25 ft) soil cap brought up to match existing grade

In-place density of the refuse in the landfill is expected to be three times the uncompacted (at-source) density calculated from the bulk densities of the components noted.

(a) Estimate the maximum life of the site as a sanitary landfill for this municipality, assuming that a 100-m (160-ft) buffer is required between the property boundary and the landfill.

(b) Will the material excavated be sufficient for the daily and final covers for the life of the landfill and for the restoration of a nearby abandoned quarry requiring 1×10^6 m³ of earth fill?

(c) Estimate the effect on the life of the landfill if an at-source recovery program were implemented and resulted in a 20% reduction in the amount of waste paper collected.

(d) If a clay liner 1 m (3.25 ft) thick for which $K = 1 \times 10^{-7}$ cm/s were provided over the horizontal bottom of the landfill, at what velocity would leachate enter the subsoil once this clay liner became saturated? How long would it take for the leachate to pass from the top to the bottom of the liner? Assume that leachate has not built up in the landfill above the liner.

14.11. Consideration is being given to the installation of a gas recovery system for the landfill in Problem 14.10 to supply low-BTU gas to a nearby industry. The industry uses 1×10^7 MJ (1×10^{10} BTU) of energy for heating from coal each year and would enter into a contract with the municipality if this amount could be supplied by landfill gas for 10 years at a price competitive with coal at $190/t.

(a) For what price would the gas have to be sold per m³?

(b) Would sufficient gas be available for the industry if deposition of refuse stopped after four years, i.e., the time at which the industry would start using the gas?

REFERENCES

APWA (American Public Works Association). *Solid Waste Facts Review Update*. Chicago: Institute for Solid Wastes, 1979.

BELL, J. M., ed. *Environmental Engineering News*. Lafayette, Ind.: Purdue University, April 1987.

CHANLETT, E. T. *Environmental Protection*. 2d ed. New York: McGraw Hill, 1979.

CHIAN, E. S. K., and DEWALLE, F. B. *Evaluation of Leachate Treatment. Characterization of Leachate*, vol. 1. Washington: U.S. EPA, 1977.

DASGUPTA, A., NEMEROW, H. L., FAROOQ, S., DALY E. F., JR., SENGUPTA, S., GARRISH, H. P. and WONG, K. F. "Anaerobic Digestion of Municipal Solid Waste." *Bio Cycle* 32, No. 2 (1981): 34–38.

Emcon Associates. *Methane Generation and Recovery from Landfills*. Ann Arbor, Mich.: Ann Arbor Science Publishers, Inc., 1980.

GOLUEKE, C. G. "Biological Processing: Composting and Hydrolysis." In *Handbook of Solid Waste Management*, edited by D. G. Wilson, pp. 197–225. New York: Van Nostrand Reinhold, 1977.

HOLMES, J. R. *Refuse Recycling and Recovery*. Toronto: Wiley, 1981.

HUGHES, G. M., and CARTWRIGHT, K. "Scientific and Administrative Criteria for Shallow Waste Disposal." *Civil Engineering* 42, March, 1972: 70–73.

JOHNSON, W. J. "Shredding of Solid Wastes." In *Handbook of Solid Waste Management*, edited by D. G. Wilson, pp. 150–165. New York: Van Nostrand Reinhold, 1977.

MILLER, S. S., ed. *Environmental Science and Technology, Solid Wastes—II*. Washington, D.C.: American Chemical Society, 1973.

NCRR (National Center for Resource Recovery). "National Resource Recovery Update." *Waste Age* 11 November, 1980: 54–62 and *Solid Waste Management* 23 November, 1980: 57–65.

NIESSEN, W. R. "Estimation of Solid-Waste-Production Rates." In *Handbook of Solid Waste Management*, edited by D. G. Wilson, pp. 544–574. New York: Van Nostrand Reinhold, 1977.

POHLAND, F. G. *Landfill Stabilization with Leachate Recycle* EP00658. Washington, D.C.: U.S. EPA, 1972.

REINHARDT, J. J. "A Report of Milled Refuse and the Use of Milled Refuse in Landfill." City of Madison, Wisconsin, January 20, 1969.

SAROFIM, A. F. "Thermal Processing: Incineration and Pyrolysis." In *Handbook of Solid Waste Management*, edited by D. G. Wilson, pp. 166–196. New York: Van Nostrand Reinhold, 1977.

SCHWEGLER, R. D., and HICKMAN, H. L., JR. "Waste to Energy Products in North America." *GRCDA Progress Report No. 1*. Washington, D.C.: Government Refuse Collection and Disposal Association, 1981.

STONE, R. "Collection and Transportation." In *Handbook of Solid Waste Management*, edited by D. G. Wilson, pp. 95–115. New York: Van Nostrand Reinhold, 1977.

TCHOBANOGLOUS, G., THEISEN, H., and ELIASSEN, R. *Solid Wastes*. New York: McGraw-Hill, 1977.

TENN, D. G., HANEY, K. J., and DEGEARE, T. V. *Use of the Water Balance Method for Predicting Leachate Generation from Solid Waste Disposal Sites*. OSWMP, SW 168, Washington, D.C.: US EPA, 1975.

VESILIND, P. A., ed. *Energy from Waste*. Design and Management for Resource Recovery, vol. 1, and *High Technology—A Failure Analysis*. Design and Management for Resource Recovery, vol. 2. Ann Arbor, Mich.: Ann Arbor Science Publications, Inc., 1980.

VESILIND, P. A., and RIMER, A. E. *Unit Operations in Resource Recovery Engineering*. Englewood Cliffs, N.J.: Prentice Hall, 1981.

WEDDLE, C. L., MCDONALD, H. S., and HOWARD, W. R. "Landfill Gas Utilization at a Wastewater Treatment Plant." *Journal Water Pollution Control Federation* 55 (1983): 1325–1330.

WILSON, D. G. "Resource and Energy Recovery." In *Handbook of Solid Waste Management*, edited by D. G. Wilson, pp. 275–431. New York: Van Nostrand Reinhold, 1977.

WOLF, K. W., and SOSNOVSKY, C. H. "High Pressure Compaction and Baling of Solid Waste." In *Handbook of Solid Waste Management*, edited by D. G. Wilson, pp. 136–149. New York: Van Nostrand Reinhold, 1977.

ZANDI, I. "Pipeline Transportation of Solid Wastes." In *Handbook of Solid Waste Management*, edited by D. G. Wilson, pp. 115–125. New York: Van Nostrand Reinhold, 1977.

15

Hazardous Wastes

J. Glynn Henry and O. J. C. Runnalls

15.1 INTRODUCTION

Hazards in the environment may arise from natural occurrences like floods and hurricanes (Chapter 4), from human environmental disturbances like CO_2 buildup and acid rain (Chapter 5), and from the improper treatment and disposal of the toxic and hazardous wastes (such as the organic solvents and metallic residues) generated by an industrialized society. The control of these industrial or "special" wastes is discussed in this chapter.

Hazardous wastes are those wastes that could be harmful to the health of humans, other organisms, or the environment. More precise definitions, none of which has received wide acceptance, are used by various authorities. Ontario's Regulation 824/76 made under the Waste Management Act (and later placed under the Environmental Protection Act of 1975) provides a hazardous waste description that is typical of those used by many countries:

> Hazardous waste means waste that requires special precaution in its storage, collection, transportation, treatment or disposal to prevent damage to persons or property, and includes explosive, flammable, volatile, radioactive, toxic and pathological wastes.

An abbreviated definition of hazardous waste evolved by the United States Environmental Protection Agency (US EPA) reads as follows:

The term hazardous wastes means a solid waste or combination of solid wastes which because of the quantity, concentration, or physical, chemical, or infectious characteristics may

- cause, or significantly contribute to an increase in mortality or an increase in serious irreversible, or incapacitating reversible illness; or
- pose a substantial present or potential hazard to human health or the environment when improperly treated, stored, transported, or disposed of, or otherwise managed.

For the purpose of the foregoing definition, solid waste comprises a very broad spectrum of materials, including some liquids and gases.

The complete EPA treatise on hazardous waste is available in the May 19, 1980, issue of the *Federal Register*, Vol. 45, No. 98 (Book 2 of 3), pages 33119 to 33137. The lists of hazardous wastes in that document are quite extensive and include the following:

- Spent halogenated solvents used for degreasing, such as trichloroethylene, methylene chloride, and others.
- Spent nonhalogenated solvents such as xylene, acetone, ethyl benzene, ethyl ether, and others.
- Wastewater treatment sludges from electroplating operations.
- Dewatered air pollution control scrubber sludges from coke ovens and blast furnaces.
- Sludge generated during the production of various chromium compounds.
- API separator sludges from petroleum refineries.

Examples of wastes exempted from the hazardous waste lists are domestic sewage; irrigation return flows; mine tailings; animal manures; mining overburden; fly ash and bottom ash; drilling fluids; and wastes from crude oil, natural gas, or geothermal energy development. Also excluded from this listing are nuclear and other radioactive wastes, which, because of their special requirements, are controlled separately under the Atomic Energy Act of 1954 and amendments thereto.

The problem with every one of the definitions is the lack of classification of the degree of risk involved. Obviously, some wastes are more dangerous than others; for example, pesticides must receive greater attention than oily wastes. However, the definitions do not prioritize hazards, a deficiency that needs to be remedied in the future.

This chapter examines the management of three types of hazardous wastes from their source to ultimate disposal: (1) the radioactive materials, which are primarily the responsibility of federal authorities, (2) medical wastes, and (3) the nonradioactive liquid industrial wastes which are mainly under state or provincial jurisdiction.

Nuclear power is the root cause of a radioactive waste problem, initially from the mining and processing of uranium fuel and the resulting mine tailings and eventually from the spent fuel of nuclear power reactors. Problems in the safe management and disposal of these wastes and in the decommissioning of nuclear reactors are examined in Section 15.2. Medical wastes from hospitals and research centers and chemical wastes from industry are the other hazardous wastes considered. In complexity and magnitude, the industrial wastes present the greatest challenge, and the remainder of the chapter deals mainly with their management and disposal. The approaches to hazardous waste management in different countries are compared, and the options available for treatment and disposal of hazardous industrial wastes are explained in detail.

15.2 NUCLEAR WASTES

All nuclear reactors rely on uranium fuel. The uranium is extracted from ores and is refined in chemical processes. Uranium containing the isotopic composition found in nature of 0.72 percent U-235 and 99.2 percent U-238 is used in heavy-water reactors. However, for use in light-water reactors and advanced gas-cooled reactors, the U-235 content has to be increased four- to fivefold in an isotope enrichment plant. The uranium is fabricated into fuel rods which are inserted into the nuclear reactor, where they remain for up to two years. During this period they become intensely radioactive due to the production of fission products and the formation of transuranic elements such as plutonium, americium, and curium (elements 94, 95, and 96). Once the fuel rods are removed from the reactor, they must be cooled to remove the heat liberated in radioactive decay. This is most easily accomplished, at least for the first year or two, in a water-filled pool usually referred to as a spent-fuel storage bay. The fuel cycle is completed by separating the valuable residual plutonium from uranium and from fission products in a chemical reprocessing plant, or by disposing of

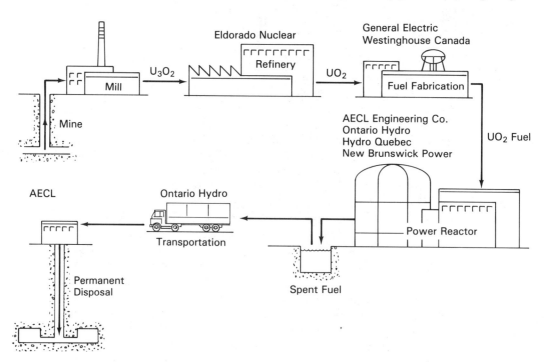

Figure 15–1 The once-through natural uranium fuel cycle in Canada

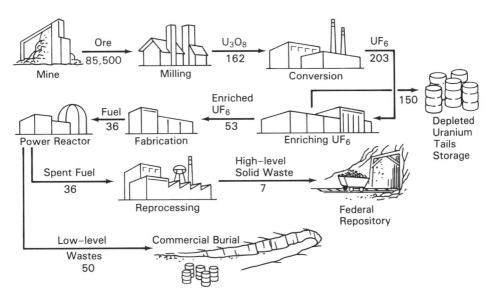

Figure 15–2 Annual quantities (tons) of materials required for routine (equilibrium) operation of a 1,000 MW light-water reactor

the fuel into an underground repository. If the former route is chosen, the fission products must be immobilized by incorporation in a glass or ceramic matrix before disposal in a similar repository. A flow diagram of the fuel cycle for the natural-uranium-fueled CANDU pressurized heavy water reactor is shown in Figure 15–1. The fuel cycle for light-water reactors is more complicated, because enriched fuel is required, as illustrated in Figure 15–2.

15.2.1 Health and Environmental Effects

The health and environmental effects from the operation of the nuclear fuel cycle can be separated into two categories, originating from the production of low- and high-level wastes. Most of the low-level wastes are generated by the uranium mining and refining operations which release small amounts of thorium-230, radium-226, radon-222, and lead-210, as well as nonradioactive ionic species to the environment. The damage to both people and their environment from such wastes usually occurs over a long period. For example, Ra-226 may be slowly eluted into streams, rivers, and lakes and could reach levels where ingestion of water over decades might result in blood and bone disorders. Similarly, some uranium miners in the past developed lung cancer from exposure to radon-222 after working underground for many years (Ham, 1976). This experience led the regulatory authorities to decrease exposure limits to Rn-222 by threefold in the early 1970s to the present level where miners can expect to work underground for 30 years without undue lung cancer risk from exposure to radon. The low-level waste problem is characterized, therefore, by large volumes of material and long exposure periods before serious health and environmental hazards emerge.

The high-level waste problem is quite the opposite. The waste is produced in a relatively small volume, the reactor fuel core, where in a short time the fresh fuel becomes intensely radioactive. If cooling of the fuel in the reactor should be interrupted, as was the case for the Three Mile Island (TMI) reactor, in Harrisburg, Pennsylvania, in March, 1979, the integrity of the fuel cladding can be destroyed and lethal amounts of fission products released inside the containment structure of the reactor building. Five years after the TMI accident, the radiation levels inside the building were still so high that personnel engaged in inspection and cleanup operations were allowed the equivalent of only a few minutes working time per month so that radiation exposure limits were not exceeded. Yet the amount of radioactive fission products that escaped from the TMI building to the surrounding area at the time of the accident was small. It has been estimated that the radiation received from these fission products may lead to less than a single excess fatal cancer among the two million persons living within 50 miles of the accident site over their remaining lifetimes (Fabrikant, 1981). Since this population is expected to develop a total of 541,000 cancers from other causes, it will be impossible to detect any increase due to the Three Mile Island accident (*ibid.*).

A more serious reactor accident occurred on April 26, 1986, when operators lost control of a water-cooled graphite-moderated reactor during a low-power test at Chernobyl in the Soviet Union. The core of the 1,000-MW electric reactor was destroyed, the 2,500-tonne graphite moderator caught fire, and the top of the light industrial-type building housing the unit was blown off. About 4–5 percent of the radioactive inventory escaped outside the building. Much settled on the ground nearby, but some was dispersed into the upper atmosphere to spread in a wide swath around the northern hemisphere. Many fires were started by ejected fuel and burning graphite on the roof of the turbine building, but these were extinguished within a few hours by firefighters working in high radiation fields.

All told, the number of reactor operators and firefighters who died as a result of the accident totalled 31 (IAEA, 1986). A total of some 200 additional people who were present at the accident scene received radiation injuries and required hospitalization. All of these individuals have since recovered. Estimates by Soviet experts suggest that up to 2,000 additional cancer cases could be produced by the released radioactivity over the next 50 years among the 75 million people living in the European part of the Soviet Union (IAEA, 1986). However, since the normal incidence of cancer will produce about 10 million cancers among this population, any extra ones resulting from the Chernobyl accident are so few as to be undetectable.

Although the world experienced perhaps the worst conceivable nuclear reactor accident, the consequences in terms of deaths and debilitating injuries were far less than had been predicted earlier by the world's reactor safety experts (US AEC, 1974). The Chernobyl experience suggests that reactor accidents are no more devastating than conventional accidents associated with large-scale production of energy.

In normally operating plants, once the nuclear fuel has reached the end of its irradiation period in the reactor after one to two years, it is removed so that fresh fuel can be inserted. The discharged or spent fuel is highly radioactive due to the presence of fission products and transuranic elements. The radioactive decay processes liberate thermal energy in the

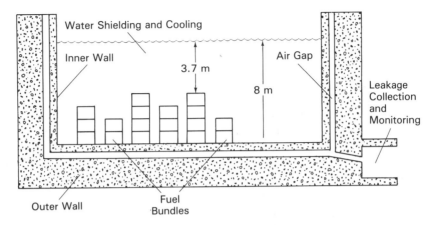

Figure 15–3   Irradiated fuel storage bay. Source: Ontario Hydro.

The main storage bays in the Pickering Generating Station are reinforced concrete structures approximately 46 meters long by 18 meters and 8 meters deep. The walls are about 2 meters deep. They are lined with fiberglass, a combination of fiberglass and stainless steel, or stainless steel. Corrosion-resistant zirconium fuel bundles contain the uranium pellets. Fuel bundles are stored in containers, stacked and covered with about 4 meters of water to provide cooling and shielding.

spent fuel, so it must be cooled to prevent overheating. Workers must be shielded from the radiation to avoid overexposure and consequent damage to human tissues. Both cooling and shielding can be achieved by immersing the spent fuel in a pool of water 5–7 m deep (Figure 15–3 and Photograph 15–1). The fuel is stored safely in this manner for several years before further steps are initiated to complete the fuel cycle as indicated earlier.

Photograph 15–1   Pickering nuclear generation station auxiliary fuel bay, where used fuel rods, stored under water, are visible but not harmful to the unprotected workers on the bridge. Source: Ontario Hydro.

Uranium, the universal fuel for the world's reactors, is a ubiquitous element which is present in the earth's crust in a concentration by weight of about 3 parts per million (ppm). Economic recovery from the host rock cannot be currently achieved, however, unless the uranium content is much higher, usually 300–3,000 ppm. The western world's production in 1986 was about 37,000 tonnes U (Runnalls, 1987). Seventy percent of the output came from Canada, the United States, South Africa, and Australia.

The normal recovery method involves crushing and grinding the ore to a finely divided state, followed by dissolution of the uranium-containing minerals in an acid or alkaline solution. The bulk of the ore treated in the uranium recovery operation is inert to the chemical treatment. Hence, a filtration step permits the solution to be separated from the solids; the solution can then be chemically treated further to recover the uranium. The large mass of residual solids can then be piped as a slurry to the tailings pond for disposal. For example, in the uranium mining area of Elliot Lake in Canada in 1986, 6 million tonnes of finely divided waste rock or tailings were generated in order to recover 5,000 tonnes U.

The tailings pose several problems because they contain mineral, chemical, and radioactive contaminants. It might be thought that they could simply be returned to the mine from which the ore was extracted. Unfortunately, however, this would interfere with mining of further ore, and in any event the volume of the tailings is usually 30–40 percent greater than that of the parent rock due to the comminution process. Thus, most uranium mining companies plan their operations so as to dispose of tailings in carefully engineered surface facilities.

Mineral contaminants vary according to the nature of the original ore. In the Elliot Lake conglomerate, for example, 5–8 percent is pyrite (FeS_2). Once the ore has been crushed and ground, the pyrite is more accessible to oxygen, water, and bacterial attack than in its original more massive state underground. A product of such attack is sulfuric acid, which can percolate slowly into nearby rivers and lakes, to their detriment.

Toxic metal contaminants such as arsenic, cadmium, iron, mercury, molybdenum, vanadium, and zinc are found in tailings. Other inorganic contaminants including ammonium, chloride, and sulfate ions, may also be present in such concentrations as to represent a threat to the nearby environment.

Radioactive contamination will be present as well in the residues from uranium mining. Most of the radioactivity originates from the uranium isotope U-238 and its daughters. Before the mining operation is conducted, all the radioactive daughters in the U-238 series are in equilibrium with the parent uranium. Once the uranium-containing crystals have been dissolved and the resulting solution processed, that equilibrium is upset. Up to 95 percent of the uranium is extracted and therefore separated from the decay products or daughters in the chain. About 85–90 percent of the radium-226 precipitates on the surface of small, solid particles which form part of the residual tailings. Much of the thorium is also deposited with the tailings, including Th-230 and Th-234 from the U-238 series, together with Th-232, a long-lived parent (half-life 1.4×10^{10} years) of another radioactive series. The main isotopes of environmental concern, however, are thorium-230, radium-226, and radon-222, together with the latter's radioactive daughters.

The primary concern over Th-230 is not related to its dispersion into the environment, because it has a relatively long half-life (7.7×10^4 years) and hence a relatively low specific activity. Rather, it is the parent of radium-226, an isotope with a much shorter half-life, 1,600 years, and higher chemical mobility. Thus, Th-230 represents a long-lived source of Ra-226 in tailings piles which will continue to be a low-level radioactive hazard for several hundred thousand years after disposal.

The Ra-226 found in freshly discharged tailings is deposited mainly on the surfaces of smaller particles, and in this position it is subject to slow dissolution and transport by waters percolating through the deposit. The principal environmental hazard from Ra-226, therefore, is that it could eventually move into surface waters and be ingested by humans. Radium is chemically similar to calcium and is transferred in humans into the bone structure. Long-term exposure to amounts as small as 1 microgram of Ra-226 retained in the body can produce serious bone damage.

The hazard posed by the daughter of Ra-226, Rn-222, is quite different, because it is an inert gas with a short half-life, only 3.8 days. Radon-222 is a member of the U-238 series and hence is to be found in association with uranium in the earth's crust. Much of the Rn-222 in the atmosphere has diffused through rock and soil from the top meter of the earth's land mass. The uranium content in this part of the earth's crust is roughly 1.5×10^9 tonnes. The uranium mined in 1986 in the world outside the Communist areas was more than four orders of magnitude less, viz., 3.7×10^4 tonnes (Runnalls, 1987). Thus, the amount of Rn-222 added to the atmospheric burden by uranium mining is insignificant compared to that diffusing from soil and rock.

Nevertheless, immediately above existing tailings piles, the radon concentration can be several hundred times higher than the normal background level. In Elliot Lake in Canada, for example, the Rn-222 release from tailings falls in the range of 100–700 picocuries* per square meter per second (pCi/m^2s). This radon is quickly diluted by the surrounding air mass, so that at distances beyond about 1 kilometer from the tailings deposits radiation levels above background are not normally detected.

There have been problems in past years resulting from the use of tailings as back-fill material in the construction of homes, schools, and industrial structures. Many millions of dollars have been spent since then in places such as Grand Junction, Colorado, and Port Hope, Ontario, in removing this material to reduce the Rn-222 hazard to building occupants. These costly lessons have pointed up the need to isolate tailings piles from the general public, to prevent their dispersion by water or wind, and to introduce natural barriers, reducing Ra-226 elution and Rn-222 release.

Although the foregoing radiation hazards from wastes have posed problems, by far the largest hazard results from the natural evolution of radon from the traces of uranium contained in soils, rocks, and building materials. As we have moved from caves to tents to buildings, and as those buildings have been better sealed to conserve energy, the radon content of the air we breathe has increased. Currently, the average North American spends 80–90 percent of the day inside, and the average exposure to atmospheric radon as a consequence is such as to produce an estimated 10,000 lung cancer fatalities per year in the United States (Cohen, 1980).

---

* 1 picocurie = 10^{-12} curie = $10^{-12} \times 3.7 \times 10^{10}$ disintegrations per second = 3.7×10^{-2} disintegrations per second = 3.7×10^{-2} becquerel.

15.2.2 Nuclear Wastes from Uranium Mining and Processing

Three of the largest uranium-producing countries in the world are Canada, the United States, and South Africa. Uranium mining operations in Canada are located within or near the granitic Precambrian Shield. The locations are characterized by the presence of large numbers of muskegs, rivers, and lakes, and precipitation rates which far exceed those of evaporation. The regulations in place, as well as those being considered, take account of the climatic environment. There is considerable emphasis, for example, on guarding against the leaching of Ra-226 into surface waters.

In contrast, most uranium mining in the United States is carried out in sandstone deposits. The associated milling operations are conducted in semi-arid regions in the midwest and southwest, where evaporation rates usually exceed the rate of precipitation. Hence, water seepage from the surface is not a significant problem. More serious is the dispersion of tailings by windstorms or flash floods. Thus, control regulations evolved by U.S. government agencies reflect the climatic conditions encountered. The major objective is to ensure that tailings are disposed of in such a way as to prevent erosion by wind or water.

In South Africa, most uranium is recovered as a by-product from gold mining operations. Usually, uranium concentrations are below 300 ppm. Hence, the radioactive hazard is considerably less than that from the richer North American ores.

The standards for radon emission from tailings piles have not yet been fully established. Under the United States Uranium Mill Tailings Radiation Control Act of 1978, the Environmental Protection Agency (EPA), is charged with evolving standards, the Nuclear Regulatory Commission (NRC) with licensing operating facilities, and the Department of Energy (DOE) with bringing the many inactive tailings sites into compliance with EPA standards. The standard proposed by EPA in January 1981 for the radon flux above inactive tailings piles was 2 pCi/m²s or less. This was equivalent to the natural background in many parts of the country. During the subsequent two years, that number was increased tenfold to a standard calling for a maximum of 20 pCi/m²s to be maintained for at least 200 years (*Nuclear Fuel*, 1982).

15.2.3 Nuclear Wastes from Power Reactors

More than 99 percent of the nuclear wastes produced in power reactors is in the form of fission products which are sealed within the spent-fuel bundles. Once such bundles have been discharged from the reactor, they are extremely radioactive due to the decay of fission products and transuranic isotopes and their daughters. Some of the decay energy is transmitted to the uranium matrix. Thus, the fuel must be cooled to prevent a rise in the cladding temperature to a point where it could react chemically in the atmosphere and lose its integrity. The cooling is accomplished by immersing the bundles in water in a spent-fuel storage bay. The water serves as a radiation shield as well. Thus, fuel handling can be carried out from above the surface of the pool without the need for any additional shielding.

A typical fuel bundle from a Canadian CANDU heavy-water reactor consists of zirconium-alloy-clad assemblies of sintered uranium dioxide pellets, each about 1 cm in diameter. In the Pickering Nuclear Generating Station near Toronto, Canada, 28 such rods

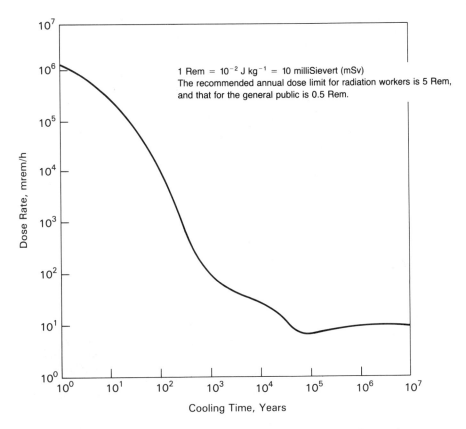

Figure 15–4 External gamma radiation dose rate from Pickering reference irradiated fuel bundle at a distance of 30 cm (average exit burnup of 7,500 MWd/tonne U). Source: Aikin, Harrison, and Hare, 1977. Reproduced with permission of the Minister of Supply and Services, Canada.

45 cm long are used in 11-cm-diameter assemblies weighing about 24 kg. The bundles are irradiated to an average burnup of 7,500 megawatt days thermal per tonne uranium. The gamma-radiation dose rate from one such spent-fuel bundle measured at 30 cm distance is plotted in Figure 15–4. As with light-water-reactor fuel, most of the fission products will have decayed after 600 years, and the residual activity will be due primarily to the transuranic isotopes and their decay products. In fact, the heat generation from the decay of transuranics and their daughters becomes dominant after about 100 years' storage (Figure 15–5). This is due to the predominance of α decay of the transuranic isotopes: the range of α particles is much shorter in the uranium matrix than either β or γ rays, so more energy is absorbed by the fuel.

Example 15.1

Using the data shown in Figure 15–4, how long could a member of the general public remain at 30 cm distance from an irradiated Pickering fuel bundle after it had been cooling for (a)

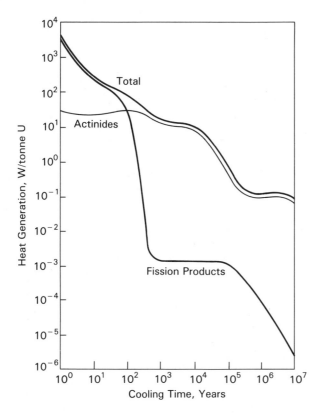

Figure 15–5 Decay heat generation: Pickering reference irradiated fuel (average exit burnup of 7,500 MWd/ tonne U). Source: Aikin, Harrison, and Hare, 1977. Reproduced with permission of the Minister of Supply and Services, Canada.

10 years, (b) 100 years, and (c) 1,000 years if that person were to receive the maximum recommended annual dose of gamma rays?

Solution   Dose rate after

(a)   10 years $= 2 \times 10^5$ mrem/h

(b)   100 years $= 1 \times 10^4$ mrem/h

(c) 1,000 years $= 1 \times 10^2$ mrem/h

Annual dose limit $= 0.5$ rem

Residence time at 30 cm distance after

(a)   10 years $= 9$ s

(b)   100 years $= 3$ min

(c) 1,000 years $= 5$ hr

Example 15.2

What is the heat output from one Pickering spent-fuel bundle after it has been cooling for (a) 1 year (b) 1,000 years?

Solution   From Figure 15–5, the heat output from irradiated Pickering fuel after 1 year is 4kW/t of uranium, and after 1,000 years it is 20W/t of uranium. Since one bundle contains 20 kg U, its heat output is 80 W after 1 year and 0.4 W after 1,000 years.

15.2.4 Management of Nuclear Wastes

Mine and mill tailings.   The storage and disposal of tailings are site-specific problems. Solutions will be very much dependent on local climate, topography, and geology, and on the nature of the mining operation, e.g., whether it is underground or open pit. The proximity of populated areas will also have an impact on the engineering design.

In the arid regions of the southwestern United States, emphasis is placed on disposal of tailings in a depression below ground level followed by covering the deposit with a protective thickness of soil, sand, gravel, or crushed rock. The covering layer may be 1–3 meters thick. The exact requirement will not be known until specific regulations relating to radon containment have been agreed upon by the various regulatory bodies.

Tailings management in areas where precipitation is higher, such as at Elliot Lake, Canada, is focused on designs which minimize dissolution of Ra-226, toxic metal ions, and other contaminants such as sulfates and chlorides.

Most of the technology is available to design and construct secure facilities for uranium mine and mill tailings disposal. Such areas would require few institutional controls and should provide environmentally safe storage for the long term. Intensive research and development efforts are continuing in uranium-producing countries to provide the remaining data required to ensure that uranium mine and mill tailings will not present a hazard to humans or the environment.

Refinery and fuel fabrication wastes.   Small amounts of uranium-bearing wastes are generated in refineries where uranium concentrates from the mines are purified. The uranium is so valuable, however, that the wastes are normally accepted back into the milling circuit at mine-mill sites at no charge, since the material, although of small quantity, is equivalent to high-grade ore from many mines. Another product of the refinery operation is ammonium nitrate. Its uranium content is reduced in the chemical processes involved to a concentration below that normally found in commercial fertilizers so that it can be marketed as an acceptable agricultural product.

In fuel fabrication operations, a small amount of uranium oxide waste is produced as a result of the pressing, sintering, and grinding operations. Once again, because the uranium is so valuable, it is simply recycled through the fabrication process by dissolution in nitric acid, precipitation as ammonium diuranate, and conversion to uranium dioxide powder. In effect, the circuit is a closed one, with no significant uranium wastes escaping to the environment.

Spent fuel.   The time that spent fuel should be stored under water is still open to question. Certainly, during the first one or two years after discharge from the reactor, the heat flux from decaying fission products is sufficiently high to make underwater storage the most desirable option. However, as time passes, another possibility emerges, viz., interim dry storage in concrete flasks. The specific route for interim storage will be determined by economics and the proposed ultimate disposition of the spent fuel. There appears to be a growing belief that the most economical method would be to store fuel underwater for perhaps the first 50 years. The high corrosion resistance of the zirconium alloy cladding material gives confidence that such a period should be attainable without significant damage due to corrosion. Some spent-fuel bundles from the Canadian NPD reactor, for example, have been stored in such a manner since about 1960.

There are others who hold the view that dry storage in concrete canisters may offer a more economical option for even longer storage times. Either route, however, offers the advantage of retaining the fuel bundles in readily recoverable form should reprocessing the fuel to separate and recycle the fissionable plutonium become desirable in the future.

When spent fuel is discharged from light-water reactors, it consists of a mixture of enriched uranium, plutonium, and fission products. The enriched uranium component has been reduced by irradiation from its original concentration of 3–3.5 percent U-235 to about 1.1 percent. The residual plutonium content is 0.8 percent. In the long term, the value of the enriched uranium alone should be high enough to underwrite the costs of reprocessing the fuel in order to recover the uranium and to separate the plutonium from the fission products. However, the price of newly-mined uranium today is so low and the cost of reprocessing so high, that many countries are discouraged from following the reprocessing route. Hence, serious consideration is being given in such countries to long-term storage of irradiated fuel bundles and perhaps even to the ultimate disposal of such bundles without reprocessing.

Spent fuel from CANDU reactors is less valuable, because its U-235 content has been decreased during irradiation to about 0.2 percent and its plutonium content is only 0.4 percent. Nonetheless, there is considerable long-term interest in the possibility of introducing advanced fuel cycles into CANDU reactors based on the utilization of plutonium and on the irradiation of thorium. An intermediate step in launching such a program would be to extract plutonium from natural uranium fuel in order to enrich initial charges of thorium. When thorium is irradiated, fissionable uranium-233 is produced. Fueling schemes based on the Th–U-233 cycle offer the prospect of much more efficient use of the available uranium and thorium resources. However, large-scale development of such advanced cycles may be several decades away.

In the meantime, a thorough research and development program is underway to outline a process for the safe disposal of unprocessed fuel bundles. The specific components of such a process are beginning to emerge, and there seem to be no technical barriers to its implementation early in the twenty-first century. Meanwhile, underwater storage facilities are adequate in Canada to store all the fuel that is likely to be produced by 2025. More underwater storage sites will need to be built in the United States, however, to accommodate the foreseen volume of discharged American reactor fuel.

The waste disposal process envisaged is one which takes advantage of multiple barriers to fission-product release to the environment. The disposal philosophy in Canada is that spent fuel should be placed in a geologically-stable granite repository 500–1,000 meters underground. There are many such granite formations in Canada, called **plutons,** which have not developed major new fracture systems due to any seismic, tectonic, or glacial activity for more than 10^9 years (Brown and McEwen, 1982).

A similar philosophy on disposal exists in the United States, except that emphasis there is currently being placed also on the use of underground salt domes. One further possibility being explored by an international team is that of disposal into the deep sea bed, where tectonic movement would slowly propel radioactive wastes downwards into the earth's crust.

The disposal route foreseen for CANDU fuel bundles is to load a number of these into a container of corrosion-resistant material, perhaps titanium, nickel-based alloys, or copper. The interstices between the fuel bundles and the container would be filled with a particulate support (such as glass beads) or by a lead-antimony alloy. A full-size prototype capable of holding 72 fuel bundles was fabricated for test purposes some years ago (Dixon and Rosinger, 1981).

The sealed containers would be placed in caverns drilled in the rock repository, and a buffer material placed around the container to separate it from the rock. One possible buffer is bentonite clay. It will expand when exposed to water, thus preventing further movement of water toward the container. After the filling of the repository, all rooms, boreholes, and shafts would be backfilled and plugged as shown in Figure 15–6 so that further supervision of the disposal site would not be required. Clandestine, unobserved access to the spent fuel in later years by a determined team of thieves or saboteurs is difficult to imagine.

The assumption being made in this plan is that the most likely transport mechanism

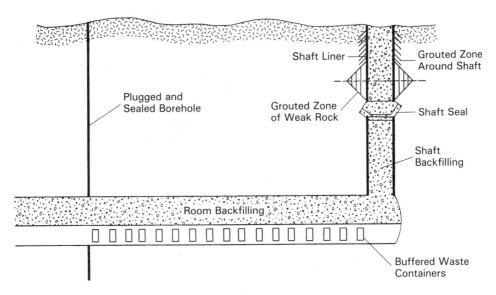

Figure 15–6   Vault sealing components (not to scale). Source: Dixon and Rosinger, 1981

for fission products to the surface would be in water. The radionuclides would have to pass several barriers before they would become a hazard to humans. First of all, many such fission products are chemically bound within the uranium oxide lattice. Then, the zirconium alloy sheath provides a corrosion-resistant container. The outer container is a further significant barrier that must be passed. The buffer is likely to be chosen so as to not only exclude water, but also to act as a chemical adsorber for fission product ions should they escape. The radionuclides migrating from the vault would move slowly in the subsurface waters through the surrounding rock formation to the surface. Only after arriving on the surface would they become incorporated into plants and surface waters to eventually reach the food chain.

With so many barriers to overcome, little radioactivity should escape to the environment. Scientists from the Whiteshell Nuclear Research Establishment of Atomic Energy of Canada Limited, in Manitoba, have established a comprehensive computer program into which a number of assumptions can be incorporated about corrosion rates, groundwater velocity, and chemical exchange effects. They examined the maximum annual radiation dose to humans from a nuclear waste repository which would contain all of Canada's fuel wastes (334,000 t) from the country's foreseen reactor program to 2055 (Rummery and Rosinger, 1984). For the assessment, 2086 simulations were performed. In 67 percent of these, there were no significant consequences for members of the exposed group in the first million years after disposal. (Actually, the consequences were zero, a radiation dose less than 10^{-10} mSv $\cdot$ a^{-1}, ten orders of magnitude below natural background radiation). Of the remaining 33 percent, about 32 percent showed consequences between zero and

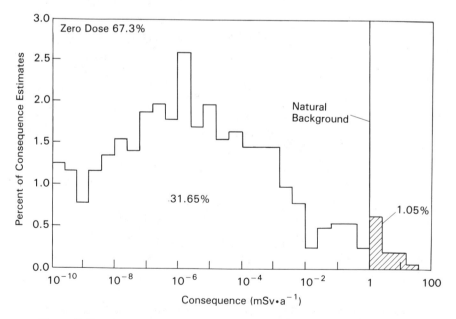

Figure 15–7   Histogram of frequency of consequence versus consequence from a radioactive waste repository. Source: Rummery and Rosinger, 1984

background, and 1 percent showed consequences greater than those from natural background radiation (see Figure 15–7).

Detailed examination of the results revealed that ingestion, primarily from well water, would be the predominant pathway to humans. The results further indicated that the geosphere is an excellent barrier and that only two radionuclides, iodine-129 and technetium-99, contribute significantly to dose.

The parameters used in these analyses were conservative, i.e., corrosion rates and groundwater velocity were chosen from the high end of the likely ranges, and chemical adsorption rates from the low end. Hence, the actual releases over these long periods of 10^6–10^7 years could be considerably less. The obvious conclusion, therefore, is that much of the technology is now available to design and construct underground repositories where high-level wastes from nuclear fuel can be safely disposed of in areas which should remain free of any disruptive seismic or tectonic activity for millions of years.

The technology also exists to permit separated fission products from chemical reprocessing operations to be incorporated into glass or ceramic blocks for safe disposal. From available information, chemical compositions and mineralogical structures can be chosen that should remain essentially inert to underground water dissolution for periods counted on a geological time scale.

15.2.5 Decommissioning of Nuclear Power Reactors

Power reactors will need to be decommissioned at the end of their useful life. Although many such units have been financed so as to amortize the initial capital investment over 30 years of operation, their actual lifetimes may be considerably longer.

Several possible routes are available in planning a decommissioning operation. A currently-favored one is to remove the main source of radioactivity, the nuclear fuel, and all useful nonradioactive or decontaminated equipment. The reactor building would then be sealed so that access was prevented for 30–50 years. During this time much of the residual radioactivity in the facility, due primarily to cobalt activation products, would have decayed away. Dismantling of the buildings and the residual equipment would then commence with the objective of returning the site to its original state. The relatively small volume of items containing residual radioactivity, about 5 percent of the total volume, would be transported in suitably shielded containers to a near-surface, secure disposal site.

The total costs of spent-fuel disposal plus reactor decommissioning are not expected to exceed a few percent of the cost of the electricity produced in that reactor. Several utilities have already established the practice of adding a charge for fuel disposal and decommissioning costs over the operating lifetime of a given reactor. The resultant funds can be utilized to meet the disposal and decommissioning costs as they arise in later years.

15.2.6 Concluding Remarks

Much of the general public appears to believe that there are serious unresolved environmental problems associated with the nuclear fuel cycle. This is a view which has been fostered largely by members of the media through the use of incomplete, exaggerated, or incorrect information. Although it must be admitted that there may be *perceived* problems, the facts clearly indicate that technical solutions are available for the safe, long-term management of uranium mine and mill tailings, and of power reactor wastes. The safe, economic decommissioning of reactors at the end of their useful life should also be achieved routinely in spite of the dire predictions of the ill-informed.

15.3 BIOMEDICAL WASTES

15.3.1 Types of Waste

Biomedical wastes from hospitals and other medical facilities such as clinics, research laboratories, and drug companies include the following pathological and infectious wastes:

- Pathological and surgical wastes
- Experimental animals and cadavers
- Drug and chemical residues and containers
- Discarded linens, clothing, and bandages
- Disposable needles, syringes, and surgical instruments
- Contaminated equipment, food, and other wastes.

Chemical and chemotherapy wastes, organic wastes (solvents), and radioactive wastes are not usually considered as biomedical wastes and are regulated under different legislation.

Solid wastes from hospitals have been estimated by the American Public Works Association (1970) to be about 5 kg (11 lb) per patient per day, with the hazardous wastes just noted accounting for two to four percent of this.

15.3.2 Control of Biomedical Wastes

Control of biomedical wastes is based on guidelines such as the US EPA's 1985 *Guidelines for Infectious Waste Management* and the Ontario Ministry of the Environment's 1986 *Guidelines for the Handling and Disposal of Biomedical Wastes*. Enactment and enforcement of regulations governing the collection, processing, transport, and disposal of biomedical wastes are generally the responsibility of the community in which the facilities are located. Arrangements for on-site handling of wastes and off-site disposal are left to the waste generators.

For hospitals, normal practice is on-site incineration of combustible solids in a specifically designed high-temperature incinerator with after-burners to heat the gases leaving the chamber to at least 700°C (1,300°F) for odor control. Ash disposal is to a sanitary landfill. Wastes from hospitals lacking incineration or sterilization facilities are segregated and packaged in special color-coded and labeled containers for transportation and treatment elsewhere prior to landfilling.

15.4 CHEMICAL WASTES

15.4.1 Need for Control

Modern science and technology have produced a host of new products which have greatly changed our lives from those of our ancestors. Television sets, heart pacers, earth satellites, aerosol cans, pesticides, and a spectrum of plastic materials are just a few examples of the broad range of products now available to us. Production of these goods, unfortunately, creates a multitude of industrial waste by-products, many of which are hazardous if mismanaged. Degreasing compounds, wood preservatives, pesticides, heavy metals, and other toxic contaminants discharged with liquid industrial wastes can have long-term effects on human health that do not become evident for years after their introduction into the environment. Under questioning before a New York Special Grand Jury looking into the problem of toxic wastes, a high-ranking official in the United States Environmental Protection Agency, testified that

> Over 90 percent of the hazardous waste generated in the United States today (1980) is handled improperly and may be or is causing detrimental effects to human health and the environment everyday.

Further testimony reported by Fine (1981) stated that

> Lax or non-existent government regulation has left the waste generator largely unhindered in the pursuit of the least costly means of disposing of hazardous waste.

The situation in Canada in the early 1980s was much the same. Both countries lagged several years behind Europe in the development of a plan for the management of hazardous wastes.

Concern over hazardous wastes is a recent phenomenon. It wasn't until the mid-seventies that legislation to control hazardous wastes evolved, not just in North America, but throughout Europe, Australia, Japan, and other developed countries. Incidents like New York State's now infamous Love Canal situation, which festered for 30 years before its dangers to humans were exposed, demonstrated the consequences of improper hazardous waste disposal. In fact, the few incidents reported may be only the tip of the iceberg. As new threats are revealed, additional technical, legal, and social measures will be implemented to control this insidious pollution. Most hazardous wastes cannot be handled by the conventional processes used in municipal wastewater treatment plants. Faced with increasing waste disposal costs, industries will try to recycle and reclaim more wastes and to minimize the quantity to be taken off-site. Large industries may be able to solve their own problems, but for smaller companies, an off-site facility for receiving hazardous wastes will be necessary. Methods for reducing hazardous wastes from various industries by changes in processes have been detailed by Campbell and Glenn (1982).

15.4.2 Environmental Effects

How serious is the hazard from improper management of industrial wastes? Is the public, spurred by media hyperbole, overreacting to the presence of toxic substances in the environment, or are we in fact inexorably poisoning our planet to the point where life will eventually cease? No doubt, the answer lies between these two extremes. The limited evidence available makes it difficult to evaluate the extent of the problem. It is clear, however, that it deserves our attention and that a better understanding of the behavior of toxic materials in the environment must precede our efforts.

All chemicals at some concentration are toxic to humans. What we need to know is the concentration at which various substances become toxic, in what form, by what environmental pathways, and with what persistence and biological concentration. Entry routes for hazardous chemicals into the environment are shown in Figure 15–8. A more detailed discussion of the environmental effects of toxic substances, including heavy metals and organic chemicals, is presented in Section 8.5.

Organic chemicals.   Organic chemicals of concern are those that persist (degrade slowly) in the environment and are fat soluble, since these can accumulate in the food chain. PCBs (polychlorinated biphenyls) and some pesticides are examples of organics that behave this way, causing problems ranging from immediate toxicity to long-term effects (carcinogenicity, mutagenicity, etc.). Many of these persistent pollutants are formed from the degradation of primary substances or from the burning of substances containing chlorine in some form. Fossil fuels, organic materials, and municipal solid wastes are common sources that release toxic chlorinated organics when incinerated. Adsorption of these organics on dust and fly ash permits wide atmospheric distribution of these pollutants.

Inorganic pollutants.   Many inorganic elements, such as mercury (Hg), lead (Pb), cadmium (Cd), and arsenic (As), are biological poisons at concentrations in the parts

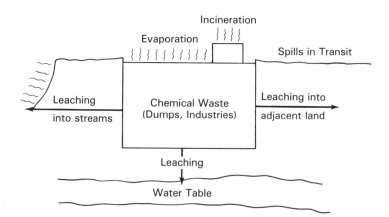

Figure 15–8 Entry routes for hazardous chemicals and wastes into the environment. Source: S. Safe, "Hazardous Organic Compounds," *Proceedings, Symposium on the Treatment and Disposal of Hazardous Wastes*, Toronto: University of Toronto, May, 1981

per billion (ppb) range. These and other toxic elements accumulate in organic matter in soil and sediments and are taken up by growing plants. Also, because they are poorly excreted by humans, they can build up in organs and tissues to toxic levels in the body. Toxic metals enter the atmosphere (from the burning of coal and the combustion of leaded gasoline), reach receiving waters (through atmospheric fallout anad leaching from mines and landfills), and contaminate land (as a result of sewage sludge application). Low pH caused by acid rain or the generation of CO_2 gas increases the transportability and hence availability of the metals by making them more soluble.

The pathways for the toxic contaminants from industrial sources into the environment and through it to humans are numerous and complex. Nevertheless, an understanding of these pathways is essential for controlling the environmental effects of chemical wastes.

15.5 IDENTIFYING A HAZARDOUS WASTE

15.5.1 Methods

As mentioned in the Introduction to this chapter, the definition of a hazardous waste is elusive. It is difficult to know when a material should be classified as hazardous. If you pour coffee in a fish bowl, the fish will die. Does that mean that coffee should be considered a hazardous substance? The simplest approach to identifying hazardous wastes is to consider them under general categories, such as radioactive, flammable, toxic, etc. The use of such a classification system enables fire departments to deal with urgent hazardous waste problems during fires, spills, and accidents in a safe manner.

Many countries—the United Kingdom, France, West Germany, and the Netherlands, among others—supplement a general classification system with detailed lists of substances, processes, industries, or wastes considered hazardous. Other countries expand on broad classifications by setting maximum concentrations for specific contaminants. In Japan, for example, four types of waste are considered toxic (sludges, slags, acid wastes, and alkaline wastes), as are wastes containing any of nine toxic materials (As, PCBs, CN, Cd, Cr^6, Pb, and Hg (total and alkyl) organic phosphates) in excess of allowable limits (Urata, 1980).

Algorithms that consider reactivity, toxicity, persistence, etc., may be used to provide preliminary screening for identifying hazardous wastes. Unfortunately, like all classifications systems, these models have shortcomings: the quantity of waste involved, its behavior in the environment, the degree of hazard, and its eventual effect on living creatures are seldom considered.

15.5.2 United States Practice

In the United States, the procedures for the identification of hazardous wastes are described in the *Federal Register* of May 19, 1980. These policies have been evolving since the introduction of the Resource Conservation and Recovery Act (RCRA) in 1976 and are still under constant review. In fact, amendments to the Hazardous Waste Regulations between

1980 and 1983 were more voluminous than the original standards promulgated in 1980. Five basic criteria are used by the US EPA to define and identify a hazardous waste:

- Is the material a solid waste as defined by RCRA?
- Has it been discarded?
- Is it specifically excluded by the regulations?
- Does it have toxic or hazardous characteristics?
- Is it listed as hazardous in the regulations?

Let us examine each of these in turn.

Solid waste.   **Solid waste** as defined under RCRA includes not just solids, but liquids, semiliquids, and contained gaseous material. This broad definition enables hazardous liquid wastes from industry to be covered by the solid wastes legislation and to be examined as potential hazards. This attempt to include hazardous liquids like solvents under the definition of solid wastes illustrates how legislation can neglect practical, technical considerations and as a result ensure confusion and future legal wrangling.

Discarded.   Has the waste been (legally) discarded after serving its original purpose? This stipulation includes wastes that are stored or treated prior to disposal, but materials to be recycled can be excluded.

Exclusions.   Is the waste specifically excluded by the regulations? For example, municipal solid wastes, agricultural waste, animal manures, and other wastes noted under the US EPA definitions of a hazardous waste in Section 15.1 are so excluded.

Characteristics.   Wastes designated as **acute hazardous wastes** possess characteristics that either

- Have the potential to increase mortality or illness, (i,e., may be toxic to humans) or
- Pose a substantial threat to human health or the environment because they are known to be flammable, corrosive, explosive, toxic, or hazardous.

In regard to human toxicity, in the absence of data on humans, a waste is considered hazardous if it has been shown to have (1) an oral LD50[1] toxicity of 50 mg/kg, or (2) a dermal LD50[1] toxicity of 200 mg/kg, or (3) an inhalation LC50[2] toxicity of 2 mg/kg.

---

[1] Lethal dose 50: dosage in mg/kg of body weight causing 50 percent mortality to test rats as a result of oral ingestion and to test rabbits as a result of dermal penetration.

[2] Lethal concentration 50: ambient concentration in mg/L of air causing 50 percent mortality to test rats during 4 h inhalation.

In regard to being hazardous to health and the environment, the following characteristics, detectable and measurable by standard tests, are the only ones considered:

- Ignitability, i.e., the substance causes or enhances fires
- Corrosivity, i.e., the substance destroys tissues or metals
- Reactivity, i.e., the substance reacts violently or causes explosions
- Toxicity, i.e., the substance is a threat to water supplies and health.

These four characteristics can be quantified by standard tests developed by the US EPA. In the case of toxicity, the major concern is the threat to groundwater. The 14 substances (8 metals, 4 pesticides, and 2 herbicides) listed in Table 15–1 for which limits for drinking water have been set serve as the basis for determining toxicity. The toxicity of the waste is determined from an "Extraction Procedure (EP) Toxicity Test" in which not less than 100 gm of the "solid" waste is agitated in 16 times its weight of deionized water at a controlled pH of 5 for 24 hours. The leachate is then analyzed for the 14 toxic substances in Table 15–1. When the concentration of these toxic substances exceeds 100 times the amount allowed in drinking water, the waste is designated as hazardous. Other tests, for radioactivity, infectiousness, phytotoxicity, and mutagenicity, are under development.

TABLE 15–1 CONTAMINANT CONCENTRATIONS[1] FOR IDENTIFYING HAZARDOUS WASTES (wastes containing any of these contaminants in excess of the limits noted are classified as hazardous wastes)

Contaminant[2]	Possible sources	mg/L[3]
Arsenic	Gold mine tailings, residues from arsenic pesticides	5.0
Barium	Printing industry, paints, lubricating oils, greases	100.0
Cadmium	Electroplating, metal alloys, sewage sludge, fertilizer	1.0
Chromium	Electroplating, tanning, industrial cooling water blowdown	5.0
Lead	Gasoline, paint, batteries, atmospheric fallout	5.0
Mercury	Chlor-alkali plants, paper mills, preservatives	0.2
Selenium	Manufacture of colored glass, electronic, photographic wastes	1.0
Silver	Silver plating, jewelry, photographic wastes	1.0
Endrin	Manufacture of pesticides, runoff, aerial spraying	0.02
Lindane	Manufacture of pesticides, runoff, aerial spraying	0.4
Methoxychlor	Manufacture of pesticides, runoff, aerial spraying	10.0
Toxaphene	Manufacture of pesticides, runoff, aerial spraying	0.5
2,4-D	Manufacture of herbicides, lawn clippings, runoff	10.0
2,4,5-TP	Sprayed foliage, brush piles, surface runoff	1.0

[1] As determined by the EPA "Extraction Procedure Toxicity Test."

[2] For additional information on some of these contaminants, see Section 8.5.

[3] These limiting concentrations are 100 times the concentrations allowed in drinking water. Regulations for drinking water are under review (1987) by the US EPA, and this may result in separate maximum concentration limits (MCL) for the different valences or forms of As, Cd, or Hg, higher values for Ba and Se, and a lower MCL for Pb.

TABLE 15–2 EXAMPLES OF TOXIC CHEMICALS FOUND IN INDUSTRIAL WASTES

Type (total number)	Example	Industrial use or source	Probable destination
1. Metals and inorganics (15)	Cyanides	Electroplating baths	Water
2. Pesticides (20)	Chlordane	Manufacture of pesticides	Sediment, Biota
3. Polychlorinated biphenyls (7)	PCB arochlors	Transformer coolant	Sediment, Biota
4. Halogenated aliphatics (24)	Dichloromethane (methylene chloride)	Solvent	Water
	Tetrachloromethane (carbon tetratchloride)	Solvent and degreaser	Water
	Chloroethene (vinyl chloride)	Manufacture of plastics	Water
5. Ethers (6)	2-chloroethyl (vinyl ether)	Pharmaceutical wastes	Water, Sediment
6. Monocyclic aromatics (12)	Ethylbenzene	Solvent	Sediment
	Toluene	Solvent	Sediment
7. Phenols and cresols (11)	Phenol	Refinery wastes	Water
	Pentachlorophenol	Wood preservation	Sediment, Biota
8. Phthalate esters (6)	Dimethyl phthalate	Manufacture of cellulose acetate	Sediment, Biota
9. Polyacrylic aromatics (18)	Naphthalene	Manufacture of dyes and synthetics	Sediment, Biota
	Phenanthrene	In coal tar	Sediment, Biota
10. Nitrosamines and others (7)	Acrilonitrile	Manufacture of plastics	Water, Sediment
Total 126			

Adapted from the US EPA list of 126 priority pollutants (Nov. 1980) as described by Chapman, M., Romberg, G. P., and Vigers, G. A., in "Design of Monitoring Studies for Priority Pollutants," *Journal of the Water Pollution Control Federation* 54 (1982): 292–297.

Wastes listed as hazardous.   Wastes known to be hazardous (carcinogenic, mutagenic, etc.) but not suited to EP toxicity tests are listed in three categories:

1. Nonspecific sources, which include spent solvents used in degreasing.
2. Specific sources, such as process waste from the preservation of wood, or the production of a variety of halogenated hydrocarbons.
3. Discarded products like toluene, mercury compounds, xylene, etc. (and their containers). From these listings, the US EPA has produced a list of 126 (originally 129) "priority pollutants." Examples of these toxic chemicals and their probable destinations in the environment are shown in Table 15–2.

Example 15.3

Grass clippings from a municipal golf course are composted each year for reuse as mulch on the fairways. By chance, several EP toxicity tests were conducted on the compost, with the following average results:

Cadmium	0.3 mg/L
Lead	1.0 mg/L
2,4-D	1.0 mg/L
2,4,5-TP	1.0 mg/L

The 2,4-D is used to control broadleaf weeds, and the 2,4,5-TP is used to control the growth of brush along road shoulders and fairway boundaries. Assess these results by considering:

(a) What contaminants make the compost a hazardous waste.
(b) How the 2,4,5-TP in the compost can be accounted for.
(c) How the presence of the two toxic metals might be explained.
(d) Whether the compost should continue to be used.
(e) How the problem could be avoided in the future.

Solution

(a) The 2,4,5-TP is the only contaminant in high enough concentration to indicate that the compost is a hazardous waste.
(b) Assuming that the 2,4,5-TP was sprayed only where intended, it appears that the brush along the roadways and fairways must have been collected and stockpiled with the grass clippings.
(c) Fertilizer (sewage sludge or chemical fertilizer) used on the lawns was the likely source of cadmium, and atmospheric fallout could probably account for the lead.
(d) Use of the existing compost should be stopped, and the material should be removed and taken to an approved disposal site.
(e) The use of 2,4,5-TP could either be discontinued in favor of cutting or, alternatively, measures to ensure that the sprayed foliage is not collected by the maintenance staff could be instituted.

15.6 HAZARDOUS WASTE MANAGEMENT

Elimination or reduction of hazardous wastes through processing changes or resource recovery is the first objective of a hazardous waste management plan. After this, it is essential that the remaining hazardous wastes be accounted for from their origin to ultimate disposal. This "cradle-to-grave" concept is dependent on extensive documentation by a "manifest" or "waybill" system for recording waste movements.

Federal regulations under RCRA in the United States recognize that the individual states are responsible for a hazardous waste plan with certain requirements, including

- The registration of all hazardous waste generators producing more than 100 kg of waste per month (reduced from 1,000 to 100 kg/month in 1984).
- The collection and transport of hazardous wastes only by licensed handlers operating under a manifest system.
- Ultimate disposal at an approved hazardous waste treatment or disposal facility.

Canada's practice, at least in theory, is similar, but the primary federal control of hazardous waste movement is through the Transportation of Dangerous Goods Act. Like other waste disposal problems in Canada, hazardous wastes are essentially a provincial concern.

15.6.1 Quantities of Hazardous Wastes Generated

The quantities of hazardous wastes produced by industry may vary from a negligible amount in countries with an agrarian economy to over 1 kg/capita per day in heavily industrialized societies. Table 15–3 contains rough estimates on the quantities of hazardous wastes produced in a few developed areas. The figures indicate that in many locations the quantity of hazardous waste handled by off-site disposal is about 45 kg/capita/year. Although the

TABLE 15–3 QUANTITIES OF HAZARDOUS WASTES PRODUCED IN INDUSTRIALIZED REGIONS

Location	Hazardous wastes (10^6 tonnes/yr)	Total population (millions)	kg per capita/yr (wet basis)	
			Total amount	Disposal off-site
United States	40.0	220	180	35
California	4.6	22	210	42
Ohio	5.0	12	420	85
Canada	4.0	25	160	—
Ontario	1.5	8	190	35
West Germany	3.5	62	56	45
Bavaria	0.5	11	45	35
Netherlands	1.2	13	90	70
Denmark	—	5	—	20

Source: Adapted from Henry, J. G., *Leachate from Hazardous Waste Landfills*, Toronto: University of Toronto, Solid and Hazardous Waste Management Publication No. WM82-02, 1982

amounts per capita for off-site disposal are comparable, the per capita quantities of waste generated in Europe are much less than those in North America. The greater emphasis in Europe on the recycling and reuse of wastes by industry could account for this difference.

The first step in ensuring that all hazardous wastes are accounted for is an inventory. Although information on the characteristics and amounts of wastes is scarce and difficult to obtain because of the reluctance of industries to release such data, it is essential. The amount of hazardous waste produced can vary from less than 5 to over 20 percent of all industrial wastes generated, depending to a large extent upon the basis used to define hazardous wastes. In Table 15–3, the U.S. value of 40 million tonnes/yr represents perhaps 5 percent of the industrial wastes generated. A national survey of the U.S. (US EPA, Office of Solid Waste, *Preliminary Highlights of Findings*, August 30, 1983) showed the total quantity of hazardous waste generated to be 150×10^6 t/yr, of which an estimated 5 percent (35 kg per capita/yr) was sent off-site for disposal. Later estimates (Bloom, 1986) placed hazardous waste production at 225–275 million tonnes per year. With such large discrepancies between estimates, the need for caution in using the data from the table and the importance of a site-specific hazardous waste inventory should be evident.

An inventory of the hazardous industrial wastes generated in the province of Ontario, Canada, in 1981, indicated the breakdown shown in Table 15–4. All these wastes, amounting

TABLE 15–4 HAZARDOUS INDUSTRIAL WASTES GENERATED (Typical Source) IN THE PROVINCE OF ONTARIO IN 1981

Inorganics	%*	Organics	%*	Sludges	%*
Acids (steel mill)	11.5	Spent solvents (electronics)	2.0	Metal sludges (electroplating)	4.0
Alkalis (steel mill)	5.1	Organic liquids (synthetics)	1.2	Lime sludges (coke plant)	3.5
Metal finishing (electroplating)	15.1	Polymeric wastes (plastics)	0.9	Cyanide sludges (electroplating)	1.8
Inorganic solutions (textile)	3.2	Waste oils (machinery)	3.1	Organic sludges (tannery)	4.5
Inorganic solids (dust, ash)	8.9	Oily wastes (refinery)	19.5	Other sludges (chemical)	0.5
Unused chemicals (photochemical)	8.9	Other organics (pesticides)	1.5	Lime dust (air filters)	3.7
Other inorganics (scrubber)	0.3			Miscellaneous waste (scrubber)	0.8
Total	53.0		28.2		18.8

Source: Ontario Waste Management Corp., *Ontario Waste Management Corporation Facilities Development Process: Phase I Report*, Toronto, Canada, Sept. 1982

* Percentages are percent by wet weight of the total of 1,490,300 tonnes per year generated in Ontario (approximate 1981 population, 8,000,000). 1 tonne ≈ 1 m³.

TABLE 15–5 HAZARDOUS WASTE SOURCES IN THE UNITED STATES

Industry	Basis	
	Dry	Wet
Primary metals	40%	29%
Inorganic chemicals	20%	12%
Organic chemicals	20%	24%
Electroplating	10%	18%
Other industries	10%	17%
Total	100%	100%

Source: Adapted from U.S. Environmental Protection Agency, *State Decision Makers' Guide for Hazardous Waste Management*, US EPA OSW SW 412, Washington, D.C., 1977

Note: The difference in the quantity of waste on a wet or dry basis is slight for such wastes as oils, paints, and pure solvents, but can be appreciable for wastes from tanneries, plating shops, and other industries where the contaminants are waterborne.

to about 190 kg per capita, may be considered hazardous, but there is a wide range in the degree of hazard involved between, for example, oily wastes and cyanide sludges.

Based on 14 individual waste studies, the US EPA (1977) estimated hazardous wastes in the United States to be about 14 percent of the total industrial wastes generated, distributed as indicated in Table 15–5. Wide variations in the amounts of hazardous waste from one region of the country to another can be expected, depending on the type of industry and the extent of industrialization found there.

Example 15.4

An industrialized area of the United States with 4 million residents is undertaking an inventory of its hazardous wastes. As a preliminary basis for this survey, estimate the annual amounts of:

(a) The total amount of liquid industrial waste (i.e., hazardous waste) generated by industry.

(b) The total amount of hazardous waste requiring off-site disposal.

(c) The quantity of the off-site wastes that could be incinerated.

Solution

(a) From the footnote to the 1981 Ontario Waste Inventory (Table 15–4), 1.5×10^6 tonnes of industrial waste were generated by a resident population of 8 million. Therefore, for this area, 7.5×10^5 tonnes/yr of liquid industrial wastes might be generated.

(b) If the amount hauled off-site represents 20% of the hazardous waste generated by industry (Table 15–3), then the total amount for off-site disposal would be

$$7.5 \times \frac{10^5 t}{yr} \times \frac{20}{100} = 150,000 \text{ t/yr } (150,000 \text{ m}^3/\text{yr})$$

(c) According to Table 15–4, 28.2% of the incoming organic wastes would be liquids and 4.5% would be sludges. Not all of these organics would be incinerated. Oily wastes (19.5%), for example, would likely undergo oil/water separation, and some waste oils and spent solvents might be recovered. Quantities recovered are small, and neglecting these, the organic wastes that could be incinerated would be

$$(28.2 + 4.5 - 19.5)\% \times 150,000 \text{ t/yr} = 13.2\% \times 150,000 \text{ m}^3/\text{yr} = 19,800 \text{ m}^3/\text{yr}$$

15.6.2 Components of a Hazardous Waste Plan

Once the inventory is completed, it permits hazardous waste sources and amounts to be identified so that the other necessary components of the system can be developed. These components provide for the storage, transport, spillage, and disposal of hazardous wastes:

Storage
: Industries need special on-site tanks or basins for storage of large quantities of hazardous wastes, or chemically resistant drums for holding smaller amounts of corrosive materials until these wastes can be removed.

Transport
: The stored wastes must be collected at regular intervals by licensed haulers and transported by tanker truck or rail car (for large volumes) or by flatbed truck (for drum-held waste) to disposal.

Spillage
: A well-publicized emergency plan should be in place for the protection of human health and the prevention of environmental damage in the event of spills or the release of contaminants. The recovery and safe disposal of spilled wastes, absorbents, and contaminated soil must also be considered.

Disposal
: Wastes are either hauled to a regional physical chemical treatment plant for processing and concentration, or taken directly to an approved hazardous waste treatment facility for final disposal.

15.6.3 Role of the Waste Exchange

On-site recycling or recovery of waste is one way of reducing the quantity of hazardous materials requiring disposal. Waste exchange is another possibility. The objective is to match the waste generated with those who could use this waste as raw material. Most of the 28 waste exchanges in the U.S. and Canada are nonprofit organizations supported by government or trade associations and act as clearinghouses for information. They allow the generator and user to negotiate their own arrangements. A few such exchanges, like the state-run California Waste Exchange, play an active role in the negotiations to transfer the waste. Others, like the American Chemical Exchange and the Ohio Resource Exchange (ORE), with in-house expertise, operate as material exchange brokers and collect a fee for their services.

Waste transfers are generally from continuous processes in larger companies to smaller firms able to reuse acids, alkalis, solvents, catalysts, and oils of low purity, or recover valuable metals and other materials from concentrated wastes. Experience in North America is that 10 to 15 percent of the number of wastes listed are matched, whereas in Europe, where exchanges have been in operation longer, users are found for 30 to 40 percent of the listings. The amount of wastes that can be reused is a small proportion, usually less than 1 percent of the total volume of hauled liquid waste.

Nonprofit waste exchanges face the prospect of a decreasing market and a decreasing number of participants as the number of readily usable short-haul wastes are claimed and the less desirable remote materials are left. Offsetting this trend is the increasing stringency of regulations governing the discharge of hazardous wastes, which will lead to a greater exchange of wastes. This should provide an opportunity for profit-making materials exchange specialists to implement a more efficient and aggressive approach to the reuse of wastes.

15.7 TREATMENT AND DISPOSAL OF CHEMICAL WASTES

15.7.1 Treatment and Disposal by Industry

Where recovery or reuse of industrial wastes is uneconomical, or no waste exchange is possible and no off-site facility for hazardous waste disposal is available, an industry must determine what pretreatment, if any, is necessary before the wastes can be safely disposed of on-site or discharged to receiving waters or municipal sewers. For some wastes, deep-well disposal may be a possibility. However, ideal underground conditions rarely occur and public opposition is usually so strong that this option is seldom pursued.

On-site treatment and disposal is widely used by industry in North America. In California, for example, 80 percent of the least hazardous wastes, such as oily wastes, are disposed of on-site, in lagoons, evaporation ponds, and landfills (Dallaire, 1981). The proportion handled on-site is believed to be similar throughout most of the US.

In Europe, the reverse situation seems to apply. In France, 73 percent of the hazardous wastes are taken to centralized hazardous waste treatment and disposal facilities. The corresponding proportion handled off-site in West Germany, Denmark, and the Netherlands is thought to be closer to 80% (Environment Canada, 1980).

When wastes are unsuitable for disposal, a method, or more frequently a combination of methods, of pretreating the waste for safe disposal must be selected. The following physical and chemical treatment (PCT) processes for use by industry could also be components in a regional PCT facility serving many industries.

Physical processes.   A number of physical processes are available for solid-liquid separation, including centrifugation, flotation, sedimentation, and filtration. Activated carbon is effective in extracting toxic organics. For the removal of specific components, one of the semipermeable membrane processes (reverse osmosis, dialysis, or electrodialysis) can be employed. Stripping and distillation are among additional useful physical processes for removing particular compounds.

Chemical processes.   Chemical treatment is an essential component of most hazardous waste treatment operations. These processes, together with some common examples of their application, are described by Calkins et al. (1979). They include the following:

- Oxidation (of cyanide to cyanates, by alkaline chlorination)
- Reduction (of Cr^6 to Cr^3, by SO_2)
- Precipitation (of Cd, Hg, etc., by sulfides)
- pH adjustment (of lime slurry, by spent pickle liquor)
- Ion exchange (removing dissolved metallic and nonmetallic inorganics)
- Chemical fixation (of inorganic sludges, by silicate-based agents).

Chemical fixation utilizes cement or lime-based agents to solidify or stabilize inorganic hazardous wastes into inert silicates or hydroxides. Chemfix, Liqua-Con, and the Stablex process are three of several successful silicate-based solidification processes. The technology of solidifying organic wastes is less advanced. Several extraction procedures to test for leaching after immersing the solid in various solutions for a fixed period have been proposed. No standard procedure has evolved. Until long-term stability of the end products has been fully established, a secure landfill (i.e., one that prevents leachate migration) is the best place for the solidified waste.

Biological processes.   Provided that the concentration of toxic substances is not excessive and the biological process can be acclimatized, some toxic organic materials (phenols, oils, and other refinery wastes, for example) can be successfully treated, and over 60 percent removal of many heavy metals (Cd, Pb, Cr^6, etc.) can be achieved. These results are most often accomplished in municipal wastewater treatment plants in combination with sanitary wastes which provide dilution and buffering capacity along with organics and nutrients for microbial growth. Aerobic biological processes with rapid microbial growth are less sensitive to toxic materials than anaerobic ones, which have slower microbial growth. Consequently, where wastewater treatment involves anaerobic sludge digestion, nonbiodegradable toxic materials accumulate in the sludge, and it is the digesters that determine the toxic limits for the plant.

Other factors that improve the ability of conventional biological plants to accept toxic wastes are a high active solids concentration, a long sludge retention time, and a high organic concentration (strong sewage). When sufficient sewage is unavailable for combining with the industrial waste, degradation of toxic organics or removal of metals by conventional processes may be impractical, but other biological methods may be possible. For example, land farming of oily wastes is proving to be an economical and effective treatment method provided that the waste is not applied too frequently and is spread in thin enough layers. A second example is biological leaching for metal recovery from low-grade ores, which has been practiced for many years. Today, according to McCready (1977), bacterial leaching accounts for 15 to 20 percent of the total annual copper production in the United States. In Canada, one uranium mine relies solely on bacterial leaching for its uranium production (*ibid.*). Utilization of bacteria to remove more concentrated metals from wastes is also a distinct possibility.

15.7.2 Off-Site Hazardous Waste Disposal

An industry that is unable to dispose of its hazardous waste on-site or into municipal sewers has three options for off-site disposal: deposition in a separate, secure landfill, codisposal with municipal refuse in a sanitary landfill, or delivery to a hazardous waste treatment facility.

Secure landfill.   Where a secure landfill is the only available option, it is intended that it accept and retain organic and inorganic hazardous wastes in as concentrated a form as possible for an indefinite period, perhaps in perpetuity. If necessary, leachate is removed for treatment and disposal.

Codisposal.   In the codisposal of hazardous waste with municipal refuse, it is intended that relatively small quantities of hazardous inorganic liquid wastes (and some organics) be absorbed by large quantities of refuse so that attenuation of the contaminants by the waste and the surrounding soil can occur.

Hazardous waste treatment facility.   In a hazardous waste treatment facility, organic wastes may be either incinerated or treated (physically and/or biologically) to produce an acceptable liquid effluent and a concentrated sludge to be landfilled. Inorganic wastes are detoxified, neutralized, and concentrated to produce an acceptable liquid waste and sludge which must be further concentrated and perhaps solidified for landfill disposal. No single technology is capable of solving all hazardous waste management problems. Rather, the treatment and/or disposal of industrial waste by-products requires a combination of several processes. The experience in industrialized societies has indicated a trend toward the use of the following facilities:

1. High-temperature (1,200°C or more) incineration to dispose of organic wastes such as oily sludges, PCBs, and banned pesticides.
2. A variety of physical chemical treatment processes to treat inorganic wastes (and some oil-water mixtures). Chemical fixation (solidification) of particularly hazardous inorganic wastes may also be desirable as an alternative or supplement to the more conventional physical chemical treatment processes.
3. A secure landfill to receive both the residues generated by the preceding treatment methods and a limited volume of wastes for which no satisfactory treatment processes exist (for example, residues containing such toxic materials as mercury, arsenic, and cadmium). Unless alternative repositories such as salt mines are available, a secure landfill is an essential component of any hazardous waste management plan.

Figure 15–9 is a schematic arrangement of the major components that might be included in a hazardous waste treatment facility.

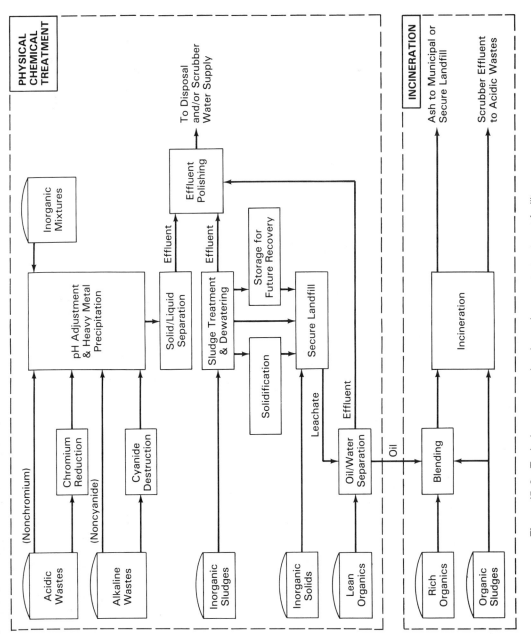

Figure 15—9 Typical components in a hazardous waste treatment facility. *Source:* Henry and Martini, 1982

15.7.3 Quantities to Be Landfilled

Estimating the volume of hazardous wastes residuals to be landfilled is extremely difficult because the densities and volumes vary widely with the characteristics of the wastes and the treatment processes used. Table 15–6 provides a rough guide for estimating sludge and residue volumes resulting from different concentration processes (see Example 15–5(b)). Densities of hazardous wastes can range from about 1 t/m^3 for liquids and fluid sludges to perhaps $1\frac{1}{2}$ t/m^3 for dry or dewatered inorganic solids. The values noted in the table are subject to wide fluctuation and should be used with discretion. Proper design of hazardous waste treatment facilities needs to be based on a detailed inventory of the wastes to be received and an analysis of their characteristics before and after various types of treatment.

15.7.4 Practices in Western Europe and the United Kingdom

The approach to hazardous waste treatment and disposal varies widely among developed countries, ranging from simple codisposal as practiced in the U.K. to the sophisticated treatment facilities employed in Europe. The government role also differs in various countries. In France, all 14 hazardous waste treatment facilities are privately owned and producers of the waste are responsible for its management and disposal. At the other extreme is Denmark, where the government treats and disposes of hazardous wastes in partnership with industry.

TABLE 15–6   SLUDGE AND RESIDUE VOLUMES FROM TREATMENT PROCESSES

Treatment process (example)	Feed volume	Sludge or residue volume
Solid/liquid separation (sedimentation)	100%	1–2%
Oil/water separation (flotation)	100%	1–2%
Solids dewatering (vacuum filtration, centrifugation)	100%	40–60%
Aerobic oxidation		
• Liquids (activated sludge)	100%	1–2%
• Sludges (aerobic digestion)	100%	60–70%
Anaerobic fermentation		
• Liquids (anaerobic treatment)	100%	0.05–0.1%
• Sludges (anaerobic digestion)	100%	60–70%
Incineration		
• Organic solvents[1,2]	100%	0.1–0.2%
• Organic solids/sludges[1,2]	100%	2–5%
• Municipal solid waste[3]	100%	5–15%

[1] Rotary kiln.

[2] Fluidized bed.

[3] Multiple hearth furnace.

Western Europe.   Information presented here on European methods of disposing of hazardous waste was gathered by a fact-finding mission on Western Europe (Environment Canada, 1980) and a site visit to West Germany (Henry, 1986). Although some European countries had started earlier, the European Common Market (ECM), which subscribes to the "polluter pay" principle got into hazardous waste management in 1974. Because of its earlier start, Europe is well ahead of North America in the control of hazardous waste. The policy in West Germany, which is similar to other continental European countries, is to detoxify and neutralize inorganic wastes by chemical treatment, thicken and dewater the resulting sludges by physical treatment, and treat the separated water by a combination of methods. Normally, this would be done at decentralized collection centers. Large centralized facilities would then provide for high-temperature (1,200°C or more) incineration of organic wastes and secure landfilling of the inorganic residues from the physical chemical treatment (PCT) processes and from the high-temperature incinerator. About 80 percent of the hazardous wastes in West Germany, where they are called special wastes, are handled in 15 hazardous waste treatment facilities (HWTFs). Codisposal with municipal refuse is not favored. The reason given is that better control can be exercised when the two wastes are kept separate.

The GSB (*Gesellschaft zur Beseitigung von Sondermüll*) facility at Ebenhausen in Bavaria was 70 percent publicly owned when it was commissioned in 1976. It is both a collection and treatment center and includes many of the standard components of an HWTF. Before the "special" wastes leave the generator's premises, they are analyzed by GSB personnel so that only physical spot checking is required upon arrival at the Ebenhausen plant. Incineration and physical chemical treatment (PCT) are the two principal means of treatment, both of which produce residues that go to an off-site secure landfill.

Incineration.   Wastes for incineration are stored in pits (solids and heavy sludges), in heated or mixed tanks (semisolids and liquids), or in a lined storage area (drummed organics). The solvent storage tanks are shown in Photograph 15–2(a). Two rotary kilns with a common afterburner handle 50,000 tonnes of organic wastes per year. Organic solids and sludges and drummed hazardous organics such as PCBs are incinerated, (including the drums when necessary) in the kilns. Liquid organics and nonhazardous drummed wastes (like waste oils) are emptied into the afterburner for destruction at 1,400°C. Empty drums are crushed for landfill disposal. Gases from incineration pass through a waste heat boiler, an electrostatic precipitator, and a scrubber. Scrubber effluent is treated for pH adjustment and sulfide precipitation of heavy metals. Power generated by incineration is used to run the plant and feeds into the local power grid.

Physical Chemical Treatment (PCT).   The PCT section receives 30,000 tonnes per year of oil-water mixtures for separation, inorganic solutions (acids and alkalis), and incinerator scrubber effluent and leachate from the hazardous waste landfill. Chemical treatment of these wastes consists of (1) Oxidation of cyanide to cyanate (2) Reduction of hexavalent chromium (3) Precipitation of heavy metals. The physical treatment processes employed are: (1) Solids-liquids separation by gravity clarifiers and Lamella separators (2) Oil-water separation by API-type separators followed by centrifuges and vacuum filters for sludge concentration (3) Dewatering of metallic hydroxides with plate and frame filter presses.

(a)

(b)

Photograph 15–2 (a) and (b) Hazardous waste treatment and disposal

The GSB hazardous waste treatment facility at Ebenhausen in Bavaria, West Germany, treats about 80,000 tons of hazardous wastes per year with the residues from treatment being hauled 55 km to a secure landfill at Gallenbach.

(a) Incoming organic wastes are stored in concrete pits, solvent storage tanks (as shown here) or in drums in a lined storage area, until burned in rotary kiln incinerators. Power from incineration runs the facility and feeds into the local power grid (photo courtesy J. G. Henry).

(b) One of the four secure landfill cells at Gallenbach in Bavaria is shown here after excavation and prior to placement of the high density polyethylene (HDPE) liner. Inorganic residues from physical-chemical treatment, ash from incineration of organics and inorganic solids are brought here for burial. (photo courtesy J. G. Henry)

The disposition of the PCT residues varies with their characteristics. Dewatered inorganic sludges are hauled to an off-site secure landfill, organic sludges and oily wastes are incinerated, and the treated effluent is discharged after storage to a stream or to a sanitary sewer depending upon the effluent quality.

Two features of the GSB facility are worth noting. First, no drummed wastes are taken to the landfill except those containing extremely toxic inorganics such as arsenic or mercury. In these cases the drums are encapsulated in concrete. The second important aspect is that no liquid wastes and no organic wastes go to the landfill.

Secure Landfill.   The secure landfill to which the GSB residues are hauled is located on a 15-ha (40-Ac) site in an agricultural area at Gallenbach, 55 km from Ebenhausen. Four cells with a total area of 9.5 ha and a capacity of 1.4 million m^3 are expected to last for about 20 years at their present rate of use. Photograph 15–2(b) shows one of the cells. The nonliquid, nonorganic residues from Ebenhausen are deposited along with small quantities of similar industrial wastes from Bavaria and other parts of West Germany. No municipal wastes and no liquid wastes are accepted, although some drummed wastes are allowed.

Factors contributing to the security of the landfill against leachate migration include the following:

- Low permeability of the surrounding clay soil ($K = 1 \times 10^{-7}$ cm/s).
- The existence of a clay layer and high-density polyethylene (HDPE) liner on the bottom of the landfill, between each 6-m lift and covering the completed cell.
- An elaborate leachate collection and removal system which transfers leachate to a holding pond, pending treatment at Ebenhausen.
- Diversion of surface runoff into holding ponds sized to accommodate a 100-year storm.
- Monitoring of groundwater quality downstream from the landfill.

The volume of leachate returned by truck to Ebenhausen for treatment amounts to about 60 percent of the annual precipitation falling on the cells (700 mm/yr).

Example 15.5

The inorganic wastes received annually at the GSB hazardous waste treatment and disposal facilities consist of solids for direct burial (90,000 t), sludges to be dewatered (10,000 t), and liquids for pH adjustment (10,000 t). The incoming organics include oily wastes (10,000 t), solvents and other rich organics (42,000 t) from which about 5% recovery is achieved, and organic sludges (10,000 t).

(a) Estimate the population served by this facility.

(b) Determine the volume to be landfilled and the length of time until the landfill capacity of 1.4 million m^3 is filled.

Solution

(a) Quantities of incoming waste:

INORGANICS

Solids (dusts, ash, piping, equipment, and their contaminated solid wastes, including drums to be encapsulated and sludges pretreated by industry)   = 90,000 t/yr

Sludges, to be dewatered   = 10,000 "

Liquids, for pH adjustment   = 10,000 "

ORGANIC

Lean, for oil-water separation   = 10,000 "

Rich, total before any recovery   = 42,000 "

Sludges, to be incinerated   = 10,000 "

Total   = 172,000 t/yr

POPULATION SERVED:

Assuming that 35 kg per year of hazardous waste per capita require off-site disposal (Table 15–3), we have

$$\frac{172,000 \text{ t/yr} \times 1,000 \text{ kg/t}}{35 \text{ kg/capita/yr}} = 4.9 \text{ million people capable of being served by the facility}$$

(b) Volume to be landfilled

The following table summarizes the quantities of waste coming into and going out of the facility:

Type of hazardous waste	Incoming waste			Residue volumes						Volume for landfill
	Quantity	Density	Volume	Separation		Dewatering		Incineration		
	t/yr	t/m³	m³/yr	%*	m³	%*	m³	%*	m³	m³/yr
Inorganics										
• Solids	90,000	1½	60,000	—	—	—	—	—	—	60,000
• Sludges	10,000	1	10,000	—	—	50	5,000	—	—	5,000
• Liquids	10,000	1	10,000	2	200	50	100	—	—	100
Organics										
• Lean	10,000	1	10,000	2	200	—	—	0.2	0.4	—
• Rich	40,000	1	40,000	—	—	—	—	0.2	80	80
• Sludges	10,000	1	10,000	—	—	—	—	5	500	500
Totals	170,000									65,680

* From Table 15–6.

The expected life of the landfill is given by

$$\frac{\text{Capacity}}{\text{Volume/yr}} = \frac{1,400,000 \text{ m}^3}{65,680 \text{ m}^3/\text{yr}} = 21.3 \text{ years}$$

Note that the volume of residue from physical chemical treatment and the incineration of organics is negligible, and the life of the landfill depends almost entirely upon the quantity of inorganic solids and dewatered sludges buried directly without treatment.

In other countries of the European Common Market, incineration and PCT are common, followed by stabilization of inorganic residues in a few cases. In France oil recovery is normal practice, and some codisposal is permitted. However, the main difference in continental Europe seems to be whether the responsibility for hazardous waste management is delegated to industry, as in France and the Netherlands, operated mainly by the government, as in Denmark, or shared, as in West Germany.

United Kingdom.   In the United Kingdom, a mixture of private and public ownership of facilities exists. There is some incineration, PCT, and solidification, but the main emphasis is on codisposal with municipal refuse. At the codisposal landfill at Pitsea, England, 70 km east of London, 43 million (US) gallons of liquid industrial wastes are trucked to the site each year and mixed with 400,000 tons of refuse, barged down the Thames from London. The 1,300-acre site, in operation since 1900, is reportedly one place in England where evaporation exceeds rainfall. It is believed that the relatively small ratio of industrial to municipal wastes (450 L/tonne) can be safely landfilled because of attenuation of contaminants by natural processes and natural dilution. Feates and Parker (1980, p. 56) mention that "the body has a natural ability to resist unwanted and dangerous materials if at low enough concentration and therefore they can be regarded as safe." The authors suggest that as long as we can ensure that wastes are dispersed to a concentration below the hazard level, this should be safe and acceptable. They indicate that landfilling of wastes has been practiced with few problems in Britain since the industrial revolution, over 200 years ago, and they recommend that it continue, with the quality (i.e., security) of the landfill site being matched to the degree of hazard of the waste, and that "fixation" of contaminants in landfills due to chemical precipitation and absorption be utilized.

In the opinion of Feates and Parker, the philosophy of fix, dilute, and disperse is gaining popularity in the U.K. Some justification for this philosophy stems from the results of a three-year research program conducted on 19 existing sanitary landfills in which all types of industrial wastes had been deposited. The behavior of pollutants, none of which was detected more than 300 m from the site, was as follows:

1. *Phenols* were greatly reduced, presumably by biodegradation, and not detected beyond 200 m.

2. Of 4,000 tonnes of *cyanide* deposited 10 years earlier, less than 3 percent remained in the landfill. The loss was attributed to oxidation to thiocyanate, conversion to gaseous hydrogen cyanide, and precipitation as metallic cyanides.

3. *Heavy metals* did not seem to get beyond the landfill unless low pH occurred. Fixation was the suggested explanation.

4. Investigations at three sites indicated no *mercury* outside the landfill in significant quantities. No organic mercury was formed, and it was postulated that the mercury precipitated as the sulfide, which has low solubility, and remained fixed.

5. No evidence of transport of *organics* was presented, but concern was expressed that organics in the landfill would be more likely to cause environmental problems than inorganic pollutants. It was noted that:

   • PCBs persist in the food chain. (Incineration of PCBs is now the normal practice.)

- Phenols are water soluble and slow to degrade, and their absorption by refuse could be reversible.

Additional studies of attenuation at laboratory and pilot scale have provided direction for long-term, full-scale research in England. Cells 50 m by 15 m with gas and leachate collection are receiving various combinations of municipal refuse and industrial wastes. Meaningful results from the elaborate monitoring system will not be available for some time.

15.7.5 Practices in North America

United States.   In the United States private enterprise is encouraged, and waste management firms like Waste Management, Inc., Browning Ferris Industries, Chem-Trol (a division of SCA), and others have appeared on the scene. Control of hazardous wastes is covered by the Resource Conservation and Recovery Act (RCRA) of 1976. As mentioned in Section 15.5.2, discarded wastes not specifically excluded by the regulations are classified as hazardous if they have certain characteristics or are listed as hazardous. The major concern with toxic materials is their effect on groundwater and surface water, which supply about one-fourth and three-fourths, respectively, of U.S. water requirements.

The U.S. EPA objective in hazardous waste management is to maximize reuse and recycling of wastes before their complete destruction or detoxification and prior to full containment of the residues. Full containment implies that the waste will be retained indefinitely, perhaps in perpetuity. No allowance is made for natural assimilation or dispersion, or for degradation in the environment to an acceptable level such as ''safe'' drinking water standards. The rationale for this approach, outlined by Dietrich (1980), is that:

- Modeling of attenuation and predictive tools are not available.
- The technology for rectifying environmental damage is not well advanced.
- The level of contaminant in drinking water that is safe is unknown.
- The public is unlikely to accept partial containment.

The introduction of RCRA was a significant step in the regulation of hazardous wastes, but the major obstacle to implementation of the program was left unresolved. What is lacking is a mechanism for approving sites deemed suitable and for dealing with the inevitable, legitimate objections that arise with any location selected. The single classification for hazardous wastes and for site requirements is another deficiency in RCRA. Economical management of hazardous wastes requires differentiation between different degrees of hazard. One of the early omissions of RCRA, namely, correction of the environmental problems created by hazardous waste sites, has been provided for in the Superfund legislation, passed in December 1980 (42 USC 9601), which authorizes funds ($1.6 billion) for remedial measures on an estimated 33,000 to 50,000 abandoned sites (Bloom, 1986), for emergency responses, and for long-term monitoring starting five years after closure. A tax on imported oil and on chemical- and crude-oil-producing industries covers $87\frac{1}{2}$ percent of the cost, with the balance raised by appropriation.

Until 1976, methods for the control of hazardous wastes varied considerably from state to state. Illinois allowed codisposal of industrial wastes if their heavy metal content was less than specified values and not more than 10 gallons of the wastes were added to each cubic yard of uncompacted refuse ($\approx$ 450 L/tonne). A simple classification system resulted in the most hazardous waste being directed to the best landfill ($K = 1 \times 10^{-8}$ cm/s). Ohio, one of the most industrialized states with 450 kg of hazardous wastes per capita per year, had no legislation before the passing of RCRA. The practice now is to prohibit landfills in areas of groundwater use. Despite California's dependence on its groundwater resources, hazardous waste disposal in the state, whether on- or off-site, was predominantly land based. Lagoons, evaporation ponds, landfills, soil farming, deep-well injection, and codisposal were employed, in preference to incineration because of concern over air pollution.

Although land disposal was considered safe when RCRA was originally passed in 1976, later amendments to the Act (proposed in the January 14, 1986, *Federal Register*) prohibited this method for untreated wastes, unless specifically approved by the U.S. EPA. The timetable for banning the most dangerous pollutants over a five-year period is as follows:

- Spent solvents, and wastes containing 23 specific hazardous constituents including dioxins and furans, are prohibited after November 8, 1986.
- Liquids containing chlorinated solvents, and eight metals including arsenic and chromium, are banned after July 8, 1987.
- About 50 specified process wastes and over 300 specific hazardous waste constituents are to be phased out between August 8, 1988, and May 8, 1990.

Under these 1986 amendments, the only wastes permitted in secure landfills after 1992 will be inorganic residues with organic constituents in trace amounts.

Canada.   Management of hazardous wastes in Canada, a provincial responsibility, is not as advanced as in Europe or the United States. The maritime provinces have no proper industrial waste disposal of any kind. Quebec has developed private disposal arrangements, and a facility for solidifying inorganic sludges (Stablex "Sealosafe") and incinerating organic ones has been built north of Montreal. Ontario's program has been on hold since 1982, pending approval of a preferred 135 ha (324 Ac) site in the Niagara Peninsula, west of Niagara Falls, for a complete 150,000-t/yr HWTF and secure landfill. Unsatisfactory handling of hazardous wastes will have to continue into the 1990s before the facility is available. Of the four western provinces, only Alberta has resolved the hazardous waste problem. Stabilized residues from the state-of-the-art (1988) HWTF at Swan Hills in northern Alberta are deposited in a secure landfill, set in almost impermeable glacial till, 50 m above the groundwater table. The facility is 40 percent government (Alberta) owned and 60 percent privately (Chem Security) owned. Charges range from $300/t for waste oils and sludges up to $800/t for more difficult material.

15.8 THE SECURE LANDFILL

15.8.1 Function

Although ocean disposal may be an option for some detoxified hazardous wastes under carefully controlled conditions (as is practiced in Japan), most hazardous waste systems must rely on land disposal at some stage. In the past, in North America, a supposedly secure landfill has been the usual repository for inadequately treated industrial wastes. In fact in New York, California, and Illinois, landfills were the primary means of disposal for untreated hazardous wastes. This was in contrast to the practice in Europe where secure landfills were used only for the residues from PCT, incineration, and solidification. Obviously, the European approach considerably reduced the waste and leachate volumes to be stored, treated, and monitored indefinitely. In Denmark 20 percent of hazardous wastes are landfilled, and even those are expected to be "mined" for heavy metal recovery in the future (Dallaire, 1981).

15.8.2 Acceptable Wastes

Before recoverable materials are extracted, and before deposition in a secure landfill, the solid or solidified wastes and well-dewatered sludges would have been detoxified and concentrated to minimize the hazard and the volume. Incompatible wastes would also be segregated within the landfill to reduce the possibility of interaction between the constituents. Table 15–7 is an abbreviated summary of a U.S. EPA list indicating the wastes in Group A which are incompatible with the corresponding wastes in Group B, and which should therefore be kept in separate cells. The potential consequences of improperly mixing the wastes in Group A with those noted opposite in Group B are also indicated. Incompatible wastes can of course be mixed so as to avoid risks—for example, by adding acid to water (not water to acid), by neutralizing strong acids with strong bases, and by confining flammable gases that are generated.

TABLE 15–7 INCOMPATIBLE WASTES

Group A	Group B	Risk
1. Alkaline liquids	Acidic wastes	Heat generation, violent reaction
2. Metals, metal hydroxides	1A, 1B	H_2 gas generation, fire, explosion
3. Alcohols, water	1A, 1B, 3A, water-reactive wastes	Fire, explosion, generation of heat and flammable or toxic gases
4. Alcohols, other hydrocarbons	1A, 1B, 2A	Fire, explosion, violent reaction
5. Cyanide or sulfur solutions	1B	Generation of toxic HCN or H_2S gas
6. Oxidizing agents	1A, 2A, 4A, organic acids, flammable wastes	Fire, explosion, violent reaction

Source: U.S. EPA, *Federal Register*, Vol. 45, No. 98, May 19, 1980, p. 33258

As noted earlier, RCRA requirements since 1976 have increasingly limited land disposal of liquids, so that by 1992 landfills are to receive only treated residues. Surface impoundment, landfilling, and deep-well injection of liquids will be prohibited.

Not all materials which are deemed hazardous need go to a secure landfill. For example, empty containers which held pesticides that degrade in a reasonable period of time should be acceptable in any well-designed municipal landfill. Commercial products containing some asbestos in a cement or similar matrix, such as broken asbestos-cement products, can similarly be directed to a municipal landfill.

Factors important in the development of an environmentally safe, socially acceptable, secure landfill include site selection criteria, design and construction details, and leachate collection and treatment.

15.8.3 Site Selection and Approval

Ideally, the land on which a secure landfill is to be located should have marginal agricultural potential and should be located in a rural area, at least 8 km from population centers and 750 m from the nearest neighbor (MacLaren, 1980). It should have a 1 to 5 percent surface slope overlying deep impermeable clay ($K = 1 \times 10^{-7}$ cm/s) and the maximum groundwater table should be at least 1.5 m below the bottom of the landfill.[1] Vehicle access, surface drainage, flooding potential, wildlife, and relation to major water resources and to community water supplies are among many factors which may become reasons for rejection of the site. Between the relatively minor technical difficulties and the major problems arising from public opposition, it can easily take five years, and it often takes more, to find and obtain approval for a suitable site.

For example, in the province of Ontario, Canada, it took about three years and 30 million dollars for the Ontario Waste Management Corporation, a Crown corporation formed in 1981, to select a preferred site for an industrial waste treatment and disposal facility. Then two more years were spent preparing a 7,000-page, 22-volume environmental assessment to explain why the project was needed, what alternatives were considered, and why the particular site was selected. Operation of the facility, what the impacts would be, and how these impacts would be managed were also described in the draft assessment. Review, modification, and finally acceptance of the assessment by the Ministry of the Environment and other government agencies will precede public hearings, which account for another one or two years before the two-year construction program can begin. In the end, ten years or more will have passed from the time the plan was first authorized until the system goes into service. This protracted approval process has been experienced in other jurisdictions for similar undertakings. In the Regional Municipality of Halton, Ontario, over 17 years was spent trying to get approval for a sanitary landfill receiving only municipal solid wastes from an existing and projected population of 210,000 and 400,000, respectively. It appears that gaining public acceptance of disposal sites has become the most formidable problem in solid and hazardous waste management.

[1] For an explanation of permeability and flow through uniform soil according to Darcy's Law, see Section 14.8.3.

15.8.4 Design and Construction

The secure landfill is the repository for all hazardous waste residuals, and it should be considered the last resort when all other efforts to eliminate or alleviate the problem have been evaluated. Protection of groundwater is the major concern in designing a secure landfill. Because so little is known about the migration, degradation, and synergistic effects of hazardous pollutants in the soil, the practice in North America is to design secure landfills for total containment. This is unlike the approach in the U.K., where attenuation by the soil is a consideration in the design.

Many designs are possible. A typical secure landfill would have an area of about 10 ha (24 Ac) and be located in heavy clay, partly above and partly below the original ground, to provide an overall depth of 10 m or more. The CECOS International site in Niagara Falls, N.Y., was constructed in this way. Three lifts were allowed for several cells, with each lift and cell being separated by a clay or absorbent barrier. After excavation and sloping of the clay bottom, an underdrain monitoring system was installed below the landfill for future surveillance. A synthetic membrane liner, sandwiched between protecting clay layers, was then placed over the sloped bottom. Sump pipes (900 mm in diameter) for leachate monitoring were installed vertically as the filling proceeded. Provision for gas venting was provided if burial of organic wastes was a possibility. Finally, when filling was completed, a sloping clay cap with a synthetic membrane (optional) was placed and finished with topsoil and a grass cover. Figure 15–10 illustrates some of the features of secure landfills constructed during the 1970s, when liquid wastes were not restricted and single liners were common.

Four types of linings are available to control seepage: admixtures, sealants, natural soils, and polymeric membranes. **Admixtures** of various cementing materials are mixed with the natural soil and have been in use for a long time. Most are asphalt based and therefore are attacked by solvents. **Sealants,** applied over the natural soil, are specifically for particular soils and do not seal all soils effectively. Latex has been found to be unsuitable,

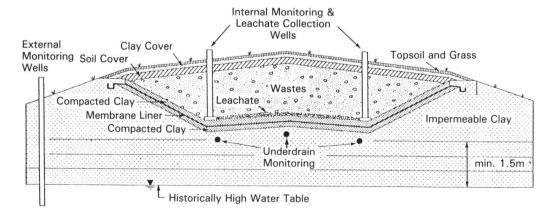

Figure 15–10 Typical landfill cell

for example, and soil cement is attacked by acid. **Natural Soils,** such as bentonite, provide adequate "waterproof" barriers for some wastes, but compacted clays are affected by acids. **Membranes** to cover the natural soil are made from many polymeric materials. Polyvinyl chloride (PVC) and Hyapalon have been the two most common. However, the new high-density polyethylene (HDPE) has become increasingly popular because of its extra thickness, good flexibility, wider width, and resistance to many different agents.

15.8.5 Problems

"Secure" landfills have been the dominant disposal method in New York State. They include many long-abandoned sites, like the Love Canal, as well as those recently closed. Of the sites active during the seventies, six were owned and operated by Chem-Trol Pollution Services (a division of SCA) and three by CECOS International, Inc. (successor of NEWCO Chemical Waste Services). Using these "secure" landfills as examples, Skinner (1981) described the problems that occurred after only three to eight years in operation. Subsidence, fires, explosions, organic migration, and erosion were continuing concerns which the proponents of each new application claimed to have resolved with improvements in design, materials, maintenance, contingency programs, etc. Skinner's observations could apply to the many other sites currently in use, as well as to the thousands of abandoned hazardous waste landfills in North America. Among these observations are the following:

1. A four-foot soil cap was found to be inadequate and, even when increased to eight to ten feet, the cap still cracked and leaked after five years.
2. Organic leachate, high in PCBs, proved to be too difficult to treat and had to be drummed for burial.
3. Plugging of the dewatering systems continued to be a common occurrence.
4. After closure of the site, excessive leachate quantities have required monthly pumping rates of 100,000–150,000 U.S. gallons per acre of landfill. (Average monthly rainfall is about 80,000 gallons per acre.) For new landfills, CECOS optimistically estimated monthly leachate quantities after closure at roughly 15,000 gallons/acre for the first four to six years, diminishing eventually to a constant annual infiltration rate of 2,000 gal/Ac of cap.
5. Degradation of organics in the landfill should not be expected since suitable conditions for microbial growth are not present.
6. Solvent-bearing wastes enhanced the migration of contaminants and degraded both soil and synthetic liners.
7. Leachate withdrawal improved consolidation, but resulted in cap settlement.
8. The characteristics and liquid content of drummed wastes could not be adequately checked, so frozen liquids, prohibited solvents in liquids, and other restricted wastes escaped detection.
9. The period required for maintenance and monitoring of the site by the landfill operator after closure and preceding perpetual care arrangements by the state was too short and should be at least ten years, as it is in Missouri.

10. A secure landfill as the sole means for waste disposal is unsatisfactory. It should be part of an overall system which first reduces waste volume and hazard by incineration, PCT, and solidification.

15.9 TREATMENT AND DISPOSAL OF LEACHATE

The characteristics of leachate are extremely waste and site specific and vary widely, depending upon (1) the type of waste and the pretreatment it received before deposition in the landfill, (2) the rate of evaporation and the net precipitation retained in the landfill, and (3) the amount of leachate which migrates into the surrounding soil.

15.9.1 Combined Treatment

Treatment of leachate may be accomplished by (a) combining it with municipal wastewater, where a municipal sewer is available, the leachate meets sewer-use bylaw requirements, and the municipal plant can handle the leachate; or (b) by combining it with hazardous industrial waste, where a hazardous waste treatment facility (HWTF) is available. Treatment with municipal wastes is seldom possible, so the second option is the more likely solution. Treating leachate is not a difficult problem when facilities are already available for processing the industrial wastes delivered to the site. The nature of these facilities, i.e., whether they include incineration, PCT, solidification, etc., establishes the type of material that goes into the secure landfill and, hence, the characteristics of the leachate produced.

15.9.2 Separate Treatment

When a secure landfill has been the repository for incoming hazardous wastes that have received little or no pretreatment, separate treatment of leachate can be difficult. Such a situation should not occur in the future with new sites. Unfortunately, it is typical of current sites in which treatment of leachate from abandoned hazardous waste landfills is necessary. This is an area in which there is little past experience. What little is known has been summarized by Shuckrow, Pajak and Touhill (US EPA 1982).

Where leachate is available, treatability studies can be used to evaluate potential treatment processes. This was the situation at Love Canal in Niagara Falls, N.Y. Treatability studies there indicated that biological treatment of the leachate was possible only with a 1:5 dilution of leachate to municipal wastewater if nutrients were added and the pH was controlled. The pilot studies also revealed that granular activated carbon (GAC) was the best means for removing priority pollutants (toxic organics). Based on these studies, a permanent plant to treat about 27 m³ (7,000 gal) of leachate per day was built at the Love Canal site (Haycock, 1981). Equalization, neutralization, sedimentation, filtration, and adsorption with GAC were the processes used. A limitation of 300 mg/L TOC (an easily monitored, but not particularly meaningful parameter) was required for the effluent to discharge to a city sewer for further treatment at the Niagara Falls PCT wastewater treatment plant. In the full-scale system, rapid exhaustion of the carbon occurred, and in 1985 biological treatment in a sequencing batch reactor (SBR) was added to reduce the load on the carbon beds.

At sites where leachate is not available for study, information on the type of waste to be deposited would be helpful in characterizing the resulting leachate and selecting potential treatment methods. Shuckrow, Pajak, and Touhill (US EPA 1982) analyzed the leachate from existing hazardous waste landfills (based on groundwater and surface-water analyses) and concluded that leachate would tend to be either primarily organic or predominantly inorganic at relatively high concentrations. That is, high concentrations of both organic and inorganic contaminants in the same landfill were uncommon. Their recommendations for dealing with leachate of unknown quality were as follows:

1. Consider interim storage of leachate until there is a sufficient quantity to analyze its characteristics.
2. Consider leachate recycling. Recycling has been investigated only for sanitary landfills and may have no effect on hazardous waste landfills. However, it could hasten stabilization of leachate characteristics by blending the constituents.
3. The use of mobile or temporary treatment facilities may avoid building the wrong type of plant or having to modify the process if the type of waste being removed changes.
4. Staging of facilities is advisable, starting with basic units and adding specialized components as required. Future options are thereby kept open.

15.10 FUTURE CHALLENGES

The out-of-sight, out-of-mind philosophy of hazardous waste disposal prevalent for many decades is no longer acceptable in industrialized societies. The insidious environmental consequences of improper disposal have become too apparent. If we intend that industrial production and the corresponding improvement in the quality of life continue, then we must accept industrial wastes as a by-product of this progress. Fire, water, and electricity have potentially harmful effects that we accept because of their offsetting benefits. Industrialization is no different: to enjoy its benefits, we must accept the need for safe disposal of the resulting hazardous wastes.

The legislative activity of the seventies has been followed by efforts to correct the improper disposal practices of the past and to develop new methods that will neither degrade our water resources nor threaten human health. Methods for dealing with hazardous wastes are still in the development stage. The pioneering procedures developed for the movement of radioactive waste could serve as a guide to the hauling of hazardous liquid industrial wastes. The approach of the nuclear industry to waste disposal, namely, complete isolation in geologically suitable locations, is also appropriate for hazardous industrial waste residues. Research by nuclear scientists indicates that it is technically feasible to immobilize nuclear wastes successfully in a form whose integrity can be assured for at least several thousand years. The same assurances cannot be given for industrial wastes at this time because of inadequate research.

The roles that should be played by incineration, solidification, codisposal, and secure landfilling are not well defined. For example, incineration of all organic wastes is common in Europe, but in California, because of concern over air pollution, incineration has not been widely used. During the seventies California landfilled most of its toxic organics, but pressure due to growing environmental concerns has modified this practice and burial of toxic organics is now considered only as a last resort. Solidification is established in the U.K., France, and Japan, but not elsewhere in Europe and only to a limited extent in North America. This means that in North America, wastes with high toxic metal concentrations (Hg, Pb, Cd) can be subjected to leaching from landfills, an undesirable situation that solidification could alleviate.

In the case of hazardous organics, it would seem that all such wastes, including PCBs and residues from solvent recovery operations, should be destroyed by incineration at temperatures above 1,200°C. Destruction in a cement kiln would be an ideal alternative solution for PCBs and a number of other halogenated organics. They could serve as supplementary fuel for the kiln, provide the chlorides needed in the manufacture of cement, and be completely destroyed in the process.

One promising area undergoing intensive study in the treatment of hazardous organic wastes is the adaptation of microorganisms that can detoxify or remove toxic anthropogenic (synthetic) organic substances. Already developed are microbes that can degrade oil, the pesticide parathion, 2,4-D, DDT, and PCBs (Sullivan, 1981). One scientist predicts that it will soon be possible to "lab-engineer" a bug that eats any toxic chemical or compound. With 111 organic compounds now designated by the U.S. EPA as priority pollutants, the application of this new concept of microbial removal to full-scale systems will provide an interesting challenge for scientists and engineers for many years to come.

More conventional options are available for dealing with hazardous inorganic wastes. Large industries can perhaps treat their wastes on-site. They would rely primarily on physical and chemical processes to neutralize and concentrate the wastes, so that in many cases the liquid effluent could be discharged to the municipal sewer system and the separated wet sludge could be shipped in bulk or drums to a hazardous waste facility. Smaller industries should be able to ship their untreated acids and bases in drums to an approved treatment facility or, where this is too distant, to a regional plant where the wastes could be neutralized and concentrated. Though common in Europe, hazardous waste processing and treatment facilities are scarce in the United States and Canada. Such limited capability will make it necessary to rank hazardous wastes, i.e., to assess the degree of hazard associated with different types of waste. This will permit more flexibility in treatment and disposal, enabling sites to be utilized for less hazardous wastes, such as oil-water mixtures that would be unsuitable for more toxic materials. The approach would require the classification of landfill sites into three or four categories based on their suitability for long-term storage of wastes.

A secure landfill will be an essential component of any hazardous waste management plan. It should serve, however, not as the primary receptacle for untreated hazardous wastes, but only as the ultimate resting place for the unrecyclable, detoxified, solidified, hazardous waste residues. Even so, a leachate is formed and should be collected and treated to specified requirements before disposal. Hazardous wastes are as variable as the processes that create them. As a result, leachate arising from the landfilling of their residues is equally variable.

It is this diversity in leachate composition, along with the differences in site characteristics and climate, that makes a standardized approach to leachate treatment difficult. What has been suitable in one instance or region may be inappropriate in another. Leachate treatment is at an early stage, and new techniques are continually being evaluated. Research at the University of Toronto using an upflow anaerobic filter to treat toxic organics looks promising. Anaerobic fluidized beds also seem to have potential for treating leachate.

However, technical problems will not be the major obstacle to the implementation of a hazardous waste management plan. Jurisdictional questions, sufficient motivation, financial arrangements, mechanisms, for site approval, provisions for perpetual care, public acceptance, and other nontechnical considerations will be far more crucial and, hence, deserving of first priority. Of these nontechnical problems, perhaps the two most troublesome are the siting of new hazardous waste treatment and disposal facilities and the environmental impact from old abandoned landfills. In the first case, there would seem to be a need for an impartial siting board or similar body to deal with the vigorous objections that inevitably arise from nearby property owners whenever a site is proposed. The "Not in My Backyard" (NIMBY) syndrome may be the most difficult of the many hazardous waste problems to be resolved. Regulatory controls that consign responsibility to those who created the problem or who now have title to the property may work in some cases. In others the only remedy may be action by the government, with rehabilitation costs borne by the taxpayer. In most cases, an acceptable solution will probably lie somewhere between these two extremes.

PROBLEMS

15.1. A 50-MW electric reactor discharges 1,120 MW of thermal energy to its condenser cooling water. The water flowrate through the condenser is 1.42×10^6 liters per minute. The specific heat of water is 4.2 J/g°C. How much does the cooling water increase in temperature?

15.2. The annual rate of uranium production from the Elliot Lake area in Canada is 5,000 tonnes. The water that drains from the area flows into Lake Huron via the Serpent River system. Assume that the average annual flowrate at the mouth of the Serpent River is 1,000 cubic feet per second and that the radium-226 concentration is 2 pCi/L. If all the radium came from the Elliot Lake ore where the original concentration was $4 \times 10^{-5}\%$ Ra in U, calculate the weight of radium that is being lost into Lake Huron each year. (1 g Ra-226 $\rightarrow$ 1 Ci). What percentage of the parent radium is this?

15.3. Assume that the waters of the Serpent River are uniformly mixed in Lake Huron so that all of the radium eventually is discharged into Lake Erie through the Detroit River at an average annual flowrate of 5,000 m³/s. How much water must a Detroit resident drink each day if the human body is to accumulate no more than the recommended limit of 0.1 μ Ci of Ra-226 after 50 years? Note that the body retains 1 part per 1,000 of that ingested.

15.4. In 1980, the total amount of uranium produced in the Western world was 44,000 tonnes. The three largest producers were the United States, with 16,800 t, Canada, with 6,820 t, and South Africa, with 6,146 t. Assuming that the average uranium recovery from the ore was 0.12, 0.09, and 0.03%, respectively, calculate the area in hectares that would be occupied in each country by the tailings produced in 1980 if they were stored in vertical-walled pits with a packing density of 2500 kg/m³.

15.5. Some of the important characteristics of each of the reactors at the Pickering Nuclear Generating Station are as follows:

Thermal output: 1,744 MW

Number of fuel channels: 390

Fuel bundles per channel: 12

Uranium content per fuel bundle: 19.8 kg

Total weight of fuel bundle: 23.7 kg

Length of fuel bundle: 49.5 cm

Outside diameter of fuel bundle: 10.2 cm

Average burnup of fuel: 7,500 MWd thermal/tonne U

Assume that each reactor achieves a capacity factor CF of 0.8, where

$$CF = \frac{\text{Actual energy produced}}{\text{Perfect production}}$$

Calculate:

(a) The number of new fuel bundles per reactor required for each day of operation

(b) The length of a rectangular container of cross section 2.6 m × 2.6 m required to accommodate one year's output of spent fuel from each reactor. Assume that each bundle fits in a rectangular box of outside dimensions 10.4 cm × 10.4 cm × 50 cm.

15.6. Concern about the environmental effects from the improper disposal of liquid industrial wastes is a recent phenomenon. Explain why these problems were not recognized earlier.

15.7. Drainage from an abandoned gold mine discharges into a river which serves as the water supply for several communities. Because of the large river flow, arsenic from drainage water never exceeds a concentration of 0.01 mg/L in the river, even during periods of drought flow. However, the average concentration of arsenic from EP toxicity tests on the mine tailings is 7 mg/L.

(a) Is the river water acceptable as a water supply?

(b) Would the mine tailings be designated as a hazardous waste, if tailings were not excluded from the RCRA regulations?

(c) What action, if any, should be taken to protect this water supply?

15.8. In the industrialized area of Example 15.4, it is proposed that codisposal of inorganic hazardous wastes with municipal refuse be allowed pending the completion of a physical chemical treatment facility.

(a) What percentage of the hazardous wastes requiring off-site disposal could be accommodated in this way? Is codisposal a good idea? Why?

(b) Assuming that all inorganic wastes could be handled by codisposal, prepare a diagram showing the possible components of a hazardous waste treatment facility that accepts only organic wastes and has no sewer outlet or receiving water available.

(c) Using Table 15–3 as a guide, show the likely disposition of the various organic wastes entering the plant if they are distributed in the same proportions as in the total hazardous waste generated. Estimate the amount of ash (t/yr) for disposal.

15.9. The following excerpt from an article entitled "Hazardous Waste Time Bomb Still Ticking" appeared in the *Toronto Globe and Mail* on February 20, 1982:

Long the standard means of industrial waste disposal, a landfill can be either a hole in the ground or, in more sophisticated situations, a quarry or specially excavated site resting on a bed of clay.

Clay is the key to the technology. Supporters of landfill disposal say a clay bed of adequate thickness is proof against poisonous leakage. But Pollution Probe (a public advocacy organization) remains unconvinced. "We're not supporters of landfill technology," Miss Wordsworth said. "Clay just slows down the leachate: it doesn't contain it. It's really a dinosaur technology."

(a) Comment on the views of landfill supporters and opponents as stated in the excerpt. Do you agree or disagree with either side, or neither, and what effect should these views have on the choice of methods for the disposal of hazardous industrial wastes? Explain.

(b) List several basic waste management alternatives (excluding deep-well injection and storage in salt caverns) that can be used by industry to lessen the amount of industrial waste residues being placed in a landfill. Note one advantage and one disadvantage for each alternative.

15.10. An industrialized region in North America with a resident population of 5 million is developing a plan for the disposal of its hazardous wastes.

(a) Estimate the approximate quantities of organic and inorganic wastes that could require off-site treatment and disposal.

(b) Prepare a flow diagram showing the components of a suitable hazardous waste treatment facility (HWTF) to handle these wastes, assuming that no incineration units are included and a municipal sanitary sewer is available.

(c) Estimate the volume of the secure landfill required to serve this HWTF for 20 years.

15.11. Suppose the off-site hazardous wastes from the industrial area in Problem 15.10 are treated in a hazardous waste treatment facility (HWTF) that provides physical chemical treatment (PCT) with sludge dewatering, incineration of all organics, and landfilling of concentrated residues.

(a) Prepare a flowdiagram showing the components of a suitable HWTF for the incoming wastes, and indicate the approximate proportions into which the inorganic and organic fractions might be further subdivided for subsequent treatment and disposal.

(b) Estimate the volume of the secure landfill required to serve this HWTF for 20 years.

REFERENCES

AIKIN, A. M., HARRISON, J. M., and HARE, F. K. *The Management of Canada's Nuclear Wastes*, EP77-6, Energy, Mines and Resources Report, Ottawa: August 31, 1977.

American Public Works Association. *Municipal Refuse Disposal*. Chicago: Public Administration Service, 1970.

BLOOM, G. F. "The Hidden Liability of Hazardous Waste Cleanup." *Technical Review*, February–March 1986, pp. 59–65.

BROWN, P. A., and McEWEN, J. "Plutons as Hosts." *GEOS* 11 No. 4 (1982): 12–15.

CALKINS, R. J., BISHOP, W. J., and BORGERDING, J. "Institutional Approach toward Combined Industrial Waste Treatment." *Journal Water Pollution Control Federation* 51 (1979): 612–618.

CAMPBELL, M. E., and GLENN, W. M. *Profit from Pollution Prevention*, Toronto: Pollution Probe Foundation, 1982.

COHEN, B. L. "Health Effects of Radon from Insulation of Buildings." *Health Physics* 39 (1980): 937–941.

CONNOLLY, R. "Hazardous Waste Legislation and Disposal in Australia." In *Toxic and Hazardous Waste Disposal*, vol. 3, edited by R. B. Pojasek, pp. 45–55. Ann Arbor, Mich.: Ann Arbor Science, 1980.

DALLAIRE, G. "Hazardous Waste Management in California: Lessons for the U.S." *Civil Engineering* 51, March 1981: 53–56.

DIETRICH, G. N. "Ultimate Disposal of Hazardous Wastes." In *Toxic and Hazardous Waste Disposal*, vol. 3, edited by R. B. Pojasek, pp. 1–11. Ann Arbor, Mich.: Ann Arbor Science, 1980.

DIXON, R. S., and ROSINGER, E. L. J. *Third Annual Report of the Canadian Nuclear Fuel Waste Management Program*. Pinawa Manitoba: Atomic Energy of Canada Limited, Report AECL-6821, December 1981.

Environment Canada. *Canadian Fact Finding Mission on Hazardous Waste Management in Europe*, Edited by G. Rivoche, Waste Management Branch, Environment Canada, 1980.

FABRIKANT, J. L. "Health Effects of the Nuclear Accident at Three Mile Island." *Health Physics* 40, (1981): 151–161.

FEATES, F. S., and PARKER, A. "Codes of Practice and Research Relating to Landfill Disposal of Hazardous Waste in the United Kingdom." In *Toxic and Hazardous Waste Disposal*, vol. 2, edited by R. B. Pojasek, pp. 51–66. Ann Arbor, Mich.: Ann Arbor Science, 1980.

FINE, J. C. "Toxic Waste Dangers." *Water Spectrum* 12, No. 1 (1981): 23–30.

HAM, J. M. *Report of the Royal Commission on the Health and Safety of Workers in Mines*. Toronto: Ministry of the Attorney General, Government of Ontario, 1976.

HAYCOCK, D. "Corrective Measures at the Love Canal." Paper presented at Seminar on Hazardous Wastes, Environment Canada, Wastewater Technology Centre, Burlington, Ontario, March 18–20, 1981.

HENRY, J. G. *Personal Site Visit to G.S.B. Facility at Ebenhausen and Secure Landfill Site at Gallenbach, June, 1986*.

HENRY, J. G., and MARTINI, O. V. *Hazardous Waste Treatment and Disposal*. Toronto: University of Toronto Solid and Hazardous Waste Management Publication No. WM 82-01, 1982.

IAEA (International Atomic Energy Agency). Experts' Meeting, "The Accident at the Chernobyl Nuclear Power Plant and Its Consequences." Vienna, August 25–29, 1986.

JOHNSON, R. A. "Secure Landfills for Chemical Waste Disposal." In *Toxic and Hazardous Waste Disposal*, vol. 4, edited by R. B. Pojasek, pp. 67–73. Ann Arbor, Mich.: Ann Arbor Science, 1980.

MACLAREN (MacLaren Engineers, Planners, Scientists, Inc.). *Annex (II) Site Selection*. Toronto: Technical Report to the Ministry of the Environment, 1980.

McCREADY, R. G. L. *Microbiology in the Mineral Industry*. Ottawa: Canada Centre for Mineral and Energy Technology, Report MRP/MSL 77-327J, October 1977.

MOFFA, R. P., and BROWN, D. P. "Hazardous Waste Management in Ohio." In *Toxic and Hazardous Waste Disposal*, vol. 3, edited by R. B. Pojasek, pp. 81–96. Ann Arbor, Mich.: Ann Arbor Science, 1980.

MILLER, M. L. "Special Waste Disposal-Illinois Style." In *Toxic and Hazardous Waste Disposal*, vol. 3, edited by R. B. Pojasek, pp. 67–79. Ann Arbor, Mich.: Ann Arbor Science, 1980.

Nuclear Fuel. "EPA Relaxes Standards." New York: McGraw Hill, Dec. 20, 1982.

RUMMERY, T. E., and ROSINGER, E. L. J. *The Canadian Nuclear Fuel Waste Management Program*. Pinawa, Manitoba: Atomic Energy of Canada Limited, Report AECL-8374, December, 1984.

RUNNALLS, O. J. C. "Prospects for the Canadian Uranium Industry." CIM Bulletin, Canadian Institute of Mining and Metallurgy, Montreal, Vol. 81, No. 911, March, 1988. pp. 137–142.

SKINNER, P. N. "Performance Difficulties of 'Secure' Landfills for Chemical Waste and Available Mitigation Systems." In *The Hazardous Waste Dilemma: Issues and Solutions*, edited by J. P. Collins and W. P. Saukin, pp. 32–55. New York: American Society of Civil Engineers, 1981.

SULLIVAN, N. "Poison Eater." OMNI 3 (June 1981): pp. 18 and 143.

URATA, J. "Treatment of Toxic Wastes in Japan." In *Toxic and Hazardous Waste Disposal*, vol. 3, edited by R. B. Pojasek, pp. 55–66. Ann Arbor, Mich.: Ann Arbor Science, 1980.

U.S. AEC (U.S. Atomic Energy Commission). *Reactor Safety Study: An Assessment of Accident Risks in U.S. Commercial Nuclear Power Plants*. Germantown, Maryland: WASH-1400, August 1974.

U. S. EPA (U.S. Environmental Protection Agency). "Hazardous Waste and Consolidated Permit Regulations." *Federal Register* Vol. 45, No. 98, May 19, 1980.

U.S. EPA (U.S. Environmental Protection Agency). *Management of Hazardous Waste Leachate*. Report by Shuckrow, A. J., Pajak, A. P., and Touhill, C. J. Contract #68-03 2766. Cincinnati: U.S. EPA SW 871, September 1982.

WILSON, R., and JONES, W. J. *Energy, Ecology and the Environment*. New York: Academic Press, 1974.

16

Environmental Management

R. Ted Munn, Gary W. Heinke and J. Glynn Henry

16.1 INTRODUCTION

Managing the environment wisely and equitably requires the balancing of a number of conflicting interests. An industry discharging untreated wastes to the atmosphere or to water bodies may argue that it is employing the cheapest disposal method resulting in decreased production cost, enabling it to sell its product cheaply. This produces benefits not only to the industry but to the public as well, in the form of an inexpensive product and increased employment. However, the same public is also paying indirect costs associated with industrial pollution. It must pay to treat its drinking water to a greater extent than might be necessary without industrial pollution, and it may pay in the form of increased sickness and decreased aesthetic quality of the environment.

If a large enough region and population are considered, it is not difficult to see that all of us are either directly or indirectly polluters and all of us share the disadvantages and cleanup costs of pollution-generating activities, although often not in an equal and equitable manner. It is not simply a case of the polluting industry versus the public. The underlying causes of environmental stress were presented in Part 1 of this book. It should be clear from reading those chapters that the blame for the change in environmental quality over the years cannot be laid on any one industry or group of industries. Wastes from residential and commercial sources are also partly responsible.

In this chapter, three major considerations in the implementation of an environmental policy are discussed and illustrated by case studies. The examples clarify the procedure for identifying and predicting environmental impacts, explain the strategies which governments use to control pollution, and point out the importance of ethics in environmental management.

16.2 ENVIRONMENTAL IMPACT ASSESSMENT

An **environmental impact assessment** (EIA) is an activity designed to identify and predict the impact on the biogeophysical environment and on human health and well-being, of legislative proposals, policies, programs, projects, and operational procedures, and to interpret and communicate information about the impacts. (Munn, 1979).

An **environmental impact statement** (EIS) is a public document written in a format specified by authorized national, state, and/or local agencies.

An **environmental inventory** is a description of the environment as it exists in an area where a particular proposed action is being considered.

16.2.1 Historical Perspective

When a new project or development is planned which might affect environmental quality, an environmental impact assessment (EIA) may be carried out. In most jurisdictions, an EIA is mandatory before permission is given to proceed with designated classes of engineering works. This is certainly the case for major developments such as power stations, flood-control systems, and smelters. However, smaller projects may also be included within EIA regulations. A list of development projects that may require an EIA is given in Table 16–1.

As early as the 1950s, EIAs of major developments were undertaken, particularly in North America, Europe, and Japan. The main objective was to ensure that public safety and health were adequately protected. Separate documents were submitted to each of the regulatory agencies involved (e.g., water authority, air pollution control branch, etc.), and no attempt was made to prepare a comprehensive overview. In the case of an EIA for a dam, calculations were undertaken to ensure that the structure could withstand major floods, based on return periods of precipitation and runoff extremes. Similarly, chimneys were designed on the basis of atmospheric diffusion models to ensure that ground-level concentrations of the pollutants would not exceed air quality standards. These analyses were reviewed by the appropriate regulatory bodies, although not by the public. In the case of the nuclear energy industry, detailed field investigations were also required before an operating permit would be authorized.

The environmental movement of the 1960s and early 1970s, pioneered by people such as Rachel Carson, Barbara Ward, and Barry Commoner, resulted in environmental

TABLE 16–1 DEVELOPMENT PROJECTS THAT MAY REQUIRE AN ENVIRONMENTAL IMPACT ASSESSMENT

Type of project	Example
1. Land use and transformation	Urban; industrial; agricultural; airport; transportation; transmission lines; offshore structures.
2. Resource extraction	Drilling; mining; blasting; lumbering; commercial fishing and hunting.
3. Resource renewal	Reforestation; wildlife management; fertilization; waste recycling; flood control.
4. Agricultural processes	Farming; ranching; dairying; feedlots; irrigation.
5. Industrial processes	Iron and steel mills; petrochemical industry; smelters; pulp and paper plants.
6. Transportation	Railways; aircraft; automobiles; trucks; shipping; pipelines.
7. Energy	Artificial lakes; dams; oil exploration, refining, and transmission; coal-fired and nuclear power stations.
8. Waste disposal and treatment	Ocean dumping; landfill; environmental contaminants and toxic substances; underground storage; biological emissions.
9. Chemical treatment	Insect control (insecticides); weed control (herbicides).
10. Recreation	Hunting areas; parks; resort development; all-terrain vehicles.

Source: Adapted from Munn, 1979 (SCOPE 5, *Environmental Impact Assessment*)

groups becoming politically active in many countries. As a partial response to these pressure groups, governments accepted the principle that citizens' organizations should have an opportunity to participate in the decision-making process of those major developments which could have significant environmental impacts.

The first comprehensive environmental legislation in the United States, the National Environmental Policy Act (NEPA), came into force on January 1, 1970. The act contained three main sections:

1. A declaration of national policy, and with it, a prescription of environmental goals which the federal government was to pursue
2. A specification that federal agencies must prepare EISs on those federal actions significantly affecting the quality of the human environment
3. The institutionalization of the EIA process in the Executive Office of the President through the establishment of the Council on Environmental Quality.

In Canada, the federal government established an Environmental Assessment and Review Process in 1973 to ensure that

1. Environmental effects would be taken into account early in the planning of new federal projects, programs, and activities.

2. An environmental assessment would be carried out before commitments or irrevocable decisions were made, for all projects which might have an adverse effect on the environment (projects with potentially significant environmental effects would be submitted to the Department of the Environment for review).

3. The results of the assessments would be used in planning, decision making, and implementation.

Federal projects include those initiated by federal departments and agencies, those for which federal funds are solicited, and those involving federal property. This definition covers projects that may originate outside the federal government but involve a particular federal department through funding or property considerations. In such cases, the federal department sponsoring the project is responsible for the EIA. All federal organizations are bound by the federal decision, except for proprietory Crown corporations and regulatory agencies, which are invited, rather than directed, to participate in the process (Duffy and Tait, 1979).

The need for carrying out an environmental impact assessment has been accepted in many other jurisdictions, including most U.S. states, Canadian provinces, Japan, Australia, and several European countries. Summaries of practices in several jurisdictions are given in Appendix 2 of Munn (1979).

The EIA process has been evolving since the early 1970s. At that time the emphasis was on measurable physical factors, particularly those for which there were standards and codes (e.g., air quality, water quality, solid waste disposal). After a few years, EIAs began to include biological and ecological factors, even though they were difficult to quantify. More recently, EIAs were broadened even further to include socioeconomic factors (employment opportunities, cultural impacts, recreational factors, etc.), so that trade-offs among socioeconomic and environmental factors could be evaluated. In some jurisdictions, the EIA process is also being used to evaluate class actions (e.g., to ban a pesticide and to regulate the lead content of gasoline).

The EIA system has been welcomed in principle by many scientists, engineers, citizens' groups and others. In practice, however, the process has left much to be desired. For example, EISs are often too long and too technical (for the citizens' groups), and do not deal with the environment in a holistic way. However, a learning process is taking place, which is improving the usefulness of EISs as decision-shaping documents.

16.2.2 Elements of the Environmental Impact Assessment Process

The first step in the EIA process is to determine whether a project falls within the relevant act or regulations, and whether the development is likely to create significant environmental disruption. If so, an assessment is undertaken, leading to the preparation of an EIS. In some jurisdictions, the EIS is open to public scrutiny and may be reviewed at public hearings. Eventually a decision is made at the political level as to whether to (1) accept the development, (2) accept an amended form of the proposed development, (3) accept an alternative proposal, or (4) reject the development.

Participants.   The various participants in the EIA process should be clearly identified. Following is a list of those that are normally involved:

Decision maker	Can be a head of state, a group of ministers, an elected body, or a single designated individual
Assessor	Is the person, agency, or company having responsibility for preparing the EIS
Proponent	Can be a government agency or a private firm wishing to initiate the project
Reviewer	Is the person, agency, or board with responsibility for reviewing the EIS and assuring compliance with published guidelines or regulations
Other agencies of government	Are agencies with a special interest in the project; they may be components of the national government services, or they may be associated with provinces, states, cities, or villages
Expert advisors	Are persons with the specialized knowledge required to evaluate the proposed action; they may come from within or outside the government service
Public at large	Includes citizens and the media
Special-interest groups	Includes environmental organizations, labor unions, professional societies, and local associations.

Contents.   The EIS should:

1. Describe a proposed action, as well as alternatives (including that of no action)
2. Estimate the nature and magnitude of the likely environmental effects of all alternatives
3. Identify the relevant human concerns
4. Define the criteria to be used in measuring the significance of environmental changes, including the relative weights to be assigned in comparing different kinds of changes
5. Estimate the significance of the predicted environmental changes (i.e., estimate the impacts of the proposed action)
6. Make recommendations for either acceptance of the project, remedial action, acceptance of one or more alternatives, or rejection of the project
7. Make recommendations for monitoring procedures to be followed during and after implementation of the action. In some jurisdictions, the EIA process stops short of making recommendations.

Often a baseline report must be prepared for submission in advance of or at the same time as the EIS. The baseline report contains an environmental inventory, i.e., a factual account of environmental conditions in the region at the time of the report, together with trends that may have recently occurred. Where data are deficient, the assessor may be required to undertake a field program.

The description of the proposed action should include a description of the construction phase of the project, the operating phase, and in some cases the shutdown phase as well. An important consideration is the selection of alternatives to the proposed action. The alternatives should include different ways of building and operating the project. The EIS may describe the project at only one specific site. However, it is easier to compare impacts at a variety of sites than it is to determine the absolute value of any impact at one site. Without a selection of alternative sites, there is a real difficulty, particularly at public hearings, with "not-in-my-backyard" problems, in which there is no disagreement with the project, as long as it is sited somewhere else. Electric transmission lines, highways, and waste disposal sites are typical examples of this type of problem.

The nature and magnitude of the environmental changes that are likely to occur and that must be estimated and included in the EIS fall into three main classes:

1. Physical, e.g., earthquake probabilities; water quality in groundwater, rivers, and lakes; soil and air quality
2. Biological, e.g., vegetation; wildlife; sport and commercial fish species; and endangered species
3. Socioeconomic, e.g., demographic; economic; and social values and attitudes.

Where socioeconomic impacts are not included in an EIS, they are usually on a "hidden agenda" of the participants at public hearings and at the political level.

Using good scientific practice, predictions of physical impacts are relatively easy to make; several methods have been described in earlier chapters. Predictions of biological and ecological impacts are much more uncertain. This is because living organisms and communities of organisms are subject to many natural stresses from droughts, floods, overgrazing, etc., as well as to stresses created by society. In addition, living organisms are adaptive and contain great genetic variability, so that their reactions to multiple environmental stresses are not always predictable. Early EISs included censuses of flora and fauna in the affected area, particularly with respect to endangered species. However, there is now a shift toward studies of life cycles, habitats, and food chains of representative species, with particular emphasis on processes thought to control the behavior of environmental components (Beanlands and Duinker, 1983).

Predictions of socioeconomic impacts are extremely uncertain even in a qualitative sense. Nevertheless, these impacts are of fundamental importance to the persons affected and may cause biological and/or ecological changes over and above those caused by the project itself. Note that socioeconomic impacts may vary across a community: some people may support a highway proposal, because it will reduce their commuting times; others will be opposed, because of noise, air pollution, or the loss of a scenic view.

Finally, cumulative impacts may be important in some cases, with many small impacts leading to a major crisis many years later (CEARC, 1986).

Determining which environmental changes are relevant is critical to the validity and credibility of the impact assessment process. Unfortunately, it is also very difficult. To ensure that an EIS will be a useful document, it is necessary at the very outset to keep the assessment bounded in time, space, and number of factors to be considered (Duinker and Baskerville, 1986).

Time frames.   In most cases, it is recommended that the EIA process cover the following time frames:

1. The present (baseline report or environmental inventory)
2. The construction phase
3. The time immediately after completion of the project
4. Several decades later. (What will happen to the environment when the development outlives its usefulness and is decommissioned?)

The disruptions occurring during construction are temporary; nevertheless, the impacts (noise, dust, influx of workers, etc.) may be severe. In some very large projects, such as the Alaska Highway pipeline, the impacts during this phase may bring about the most significant environmental and social changes. The distant future also should not be overlooked. The project may attract secondary industry to the region, or there may be a long-term buildup of solid or toxic wastes which will eventually have to be disposed of. For example, the future decommissioning or abandoning of old nuclear generating stations, exhausted mines, or gravel pits may pose severe problems.

Space frame.   The spatial reference frame should enclose an area larger than the area immediately encompassed by the proposed action. Downstream effects of effluents discharged to the air and water must be taken into account. Putting a boundary on socioeconomic impacts may be more difficult. The major priority should be given to those impacts which are directly related to the project. The establishment of a sanitary landfill would have fairly local boundaries. On the other hand, huge energy development projects, such as the Alaska Highway or MacKenzie Valley pipelines, may have economic impacts across the whole country. In any case, the area to be examined must be bounded in some way, even though arguments can often be made that there will be effects over a larger region.

Factors to be considered for inclusion.   The number of factors to be considered in the EIS should be determined very early in the assessment process. A desirable first step is prescreening (sometimes called scoping), in which the factors to be considered are reduced to a reasonable number. The resulting list should contain not only factors selected by specialists, but also factors considered to be important by citizens. In particular, low-probability, high-impact concerns should be considered in the EIS (Grima et al., 1986).

There are two extreme points of view on the scope of an EIS. On one side, there are traditionalists who believe that an EIS should concentrate on factors for which there are laws or regulations (e.g., emission standards, building codes, or city bylaws). This view often results in an EIS which is merely a collection of predictions conveniently brought together in one publication. Certainly, the EIS should have a wider scope than this. For example, if the air quality calculations have already been checked and approved by a control agency, why is there need for scrutiny by nonspecialists at public hearings? And how are alternative sites to be ranked if all of them meet the control regulations unless other, less tangible factors are considered?

At the other extreme are those who assert that an EIS should be all–inclusive. This often results in a document hundreds of pages long which is read by very few, if any. Schindler (1976, p. 509) refers to the great EIS ''boondoggle'' arising from the creation of a ''gray literature so diffuse, so voluminous, and so limited in distribution, that its conclusions and recommendations are never scrutinized by the scientific community at large.''

It is then essential to include relevant factors other than those presently covered by law and regulation, but it is also important to limit the scope of the assessment in time, space, and number of factors in order to make the EIS a useful document. The proper middle ground of the EIS will depend on the nature of each project and the wisdom of the participants in the EIA.

Weighting factors.   Once the factors to be included in an EIS have been selected, their future magnitudes must be predicted, even if only on a qualitative scale. A weighting system may then be devised to facilitate comparison of different kinds of impacts. The weights may be estimated by the Delphi method, in which each member of a group of people is asked to rank the importance of various factors or effects. Each person is then advised of the answers of other members and is invited to review and amend his or her own responses; group consensus is sometimes achieved by this process.

One of the simplest approaches to providing a visual assessment is the Leopold matrix (Leopold et al., 1971). The method, introduced in Chapter 2, is rather qualitative and should be used with much caution. Figure 16–1 outlines a procedure used for EIA. The matrix provides space for both the magnitude and importance of each effect. Provided that the matrix is not much larger than 10×10, this kind of display may be useful. In this connection, a flagging procedure may be used if some impacts (with respect to an historic site or a wildlife preserve, for example) are so important that they will produce a ''no-go'' situation.

16.2.3 Design of an Environmental Impact Assessment

A wide variety of approaches to EIA is available, and interested persons should consult at least the following references: Canter (1977), Holling (1978), Munn (1979), PADC (1983), Ortolano (1984), and Westman (1985). They should also examine EISs that have been prepared for similar kinds of projects. The following is a practical list of questions to consider in the design of an EIA.

PROJECT DESIGN AND CONSTRUCTION

- What type of project is being considered?
- What are the physical dimensions of the area under consideration?

Instructions

1. Identify all actions (located across the top of the matrix) that are part of the proposed project. List the relevant environmental characteristics or conditions down the side of the matrix.

2. Under each of the proposed actions, place a slash at the intersection with each item on the side of the matrix if an impact is possible.

3. Having completed the matrix, in the upper left-hand corner of each box with a slash, place a number from 1 to 10 which indicates the *magnitude* of the possible impact. The number 10 represents the greatest magnitude of impact, and 1 the least (there are no zeros). Before each number, place + if the impact would be beneficial. In the lower right-hand corner of the box, place a number from 1 to 10 which indicates the *importance* of the possible impact (e.g., regional vs. local); again, 10 represents the greatest importance, and 1 the least.

4. The text which accompanies the matrix should be a discussion of the significant impacts, of those columns and rows with large numbers of boxes marked, and of those individual boxes which have larger numbers.

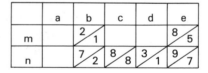

Figure 16–1 Instructions for using the Leopold matrix. Source: Leopold et al, 1971

• How much time will be required to implement the project?
• Is there an irretrievable commitment of land?
• Is the project a critical phase of a larger development?
• What are the longer-term plans of the proponent?
• Does the project make optimal use of local workers, renewable resources, etc.?
• Will there be serious environmental disruptions during construction?

PROJECT OPERATIONS

• How will hazardous wastes and waste products be handled?
• What provisions have been made for training employees in environmental protection?
• What contingency plans have been developed to cope with accidents?

- What plans have been made for environmental monitoring?
- Will safety equipment be checked regularly?

SITE CHARACTERISTICS

- Is the terrain complex, creating difficulties in predicting groundwater characteristics, air pollution transport, etc.?
- Is the site likely to be particularly susceptible to natural disasters, e.g., floods or earthquakes?
- Will many people be displaced by the project?
- Will historic sites or traditional thoroughfares be endangered?
- Will the project interfere with the movements of important migratory animal and fish populations?
- What are the main attributes of local flora and fauna? Conway and O'Connell (1978) suggest the following list of attributes: protein or calorie content, weed or pest status, domesticity, carnivorousness, rarity of species.
- Is the local environment unsuitable for the project to be a complete success?

INSTITUTIONAL AND SOCIOPOLITICAL FRAMEWORK

- What are the relevant governmental and intergovernmental regulations and procedures?
- What are the political factors to be considered?
- Are the participants in the EIA process clearly identified?
- What implementation difficulties can be expected during construction and operation of the project?

POSSIBLE IMPACTS

- For this class of project, what are the possible impacts on the environment? (During construction? After construction? Long term?)
- Who would be affected by these impacts?

SOCIOECONOMIC ANALYSIS

- Who will gain and who will lose by the project?
- What are the trade-offs?
- Will the project reduce inequalities between occupational, ethnic, sex, and age groups?
- Will it blend with or enhance valuable elements and patterns in the local, national, or regional culture?

ALTERNATIVES

- Could the project proposal be modified to reduce the environmental impacts?
- Is an alternative possible? (E.g., the same project at a different site? A different project at the same site?)

AVAILABILITY OF INFORMATION

- What are the relevant environmental standards, criteria, objectives, and bylaws?
- Is there information on the impacts of similar projects?
- What are the sources of relevant environmental data?
- What are the views of the general public and of specialist groups about the project proposal?

AVAILABILITY OF RESOURCES

- Are there local experts from whom advice can be sought with respect to specific impacts?
- Is there a possibility of seeking outside advice from a specialist?
- Are there publications (technical memoranda, guidelines, etc.) that would help in identifying possible impacts for the particular type of project?

After the magnitude and the significance of the impacts have been determined, the EIA is essentially complete. The final step of making recommendations as to whether the project should be accepted, rejected, or accepted with alterations is a political decision.

16.2.4 International EIAs

Some actions may cause environmental impacts in more than one jurisdiction or country or in international waters. Examples are the discharge of industrial and municipal wastes into rivers that border or pass through several countries, the long-range acid rain problem, integrated pest management, food import and export, the global CO_2 climate warming, international whaling, desertification, the environmental management of Antarctica, and the stratospheric ozone problem.

Many existing international EIAs are interesting treatises, but have had little impact on decision makers because of a lack of focus or a lack of understanding of the ways in which international consensus is achieved. In particular, the affected parties are rarely identified and the benefits and disadvantages are rarely compared. A special problem is that there frequently is no single decision maker. Also, often the benefits will accrue to one country and the disbenefits to another. Finally, environmental standards may not be uniform across the various jurisdictions involved.

The scientific problems are equally intractable, particularly on the global scale. Questions of risk need to be examined in special ways in an international context. Some of the outcomes (e.g., CO_2-induced climate warming and stratospheric ozone depletion) are very uncertain. Yet if no action is taken within a particular time period, it may be too late to prevent an irreversible trend. To complicate the assessment further, the trend may be harmful to some countries and/or economic sectors but beneficial to others. There is thus a need for frameworks for intergovernmental EIAs, including the development of guidelines for research priorities.

16.2.5 Conclusions

The main objective of an EIS is the reconciliation of environmental and socioeconomic considerations with respect to development and other proposed actions. Accordingly, the EIA system is a potentially useful component of good environmental management; however, as currently practiced, it is far from perfect. Some of the criticisms that have been leveled against it are as follows:

1. The EIA system delays projects, particularly if there are public hearings and court appeals. (This applies especially in jurisdictions where legislative bodies feel strongly about citizens' rights.)

2. The preparation of an EIS is costly. (Direct and indirect costs of an EIS may occasionally exceed one million dollars on large projects; however, this is often only about 0.1 percent of the capital cost of a project. By contrast, engineering and feasibility design studies may cost as much as 10 percent of capital costs.)

3. The predictions of EISs are too uncertain. (This is often the case with respect to biological and socioeconomic impacts. There is a real need for serious attempts to check the predictions of EISs once the proposed developments have taken place.)

4. The EIS is a glossy document written to impress or educate citizens' groups. (This is a fair criticism in some cases. If a public hearing is planned, the assessor should present the environmental case as objectively as possible, seeking to avoid jargon.)

5. The EIS is prepared too quickly and is not subject to peer review. (As the EIA concept develops, increasing numbers of EISs of improved quality should result.)

16.2.6 Case Study: The Atmospheric Component of an EIA for a Coal-Fired Power Station

To illustrate some of the ideas discussed here, suppose that a coal-fired power station is to be built in New England and that three possible sites are being examined: Site 1, on the Atlantic coast in flat, open countryside; Site 2, in a river valley 10 km wide with sides rising 300 m above the valley floor, and Site 3, beside a saltwater tidal flat within a built-up area. The following discussion outlines the approach to be taken in preparing the *atmospheric* component of an EIA for this development proposal. Of course, the detailed atmospheric assessment should be undertaken by specialists. However, the example will give an idea of the factors involved and should help the nonspecialist in deciding whether the atmospheric part of an EIA under review is reasonably complete.

Phase I: Information gathering.   Various kinds of information will be required in order to begin the assessment, including:

1. Engineering design characteristics (megawatt output; type of coal to be used, including sulfur content; chimney height, gas temperature, and exit velocity; water coolant system; life expectancy of the station)

2. Site characteristics (topography, proximity to bodies of water)
3. Federal and state regulations (emission standards, air quality standards, guidelines for dispersion calculations)
4. Inventories of meteorological data (from hourly weather-observing stations, daily climatological stations, twice-daily upper air radiosonde stations, and special data sources)
5. Inventories of air quality data (from industrial, urban, and rural sites; for air concentrations of SO_2, NO_x, and SPM (suspended particulate matter); and for concentrations of SO_4^{-2}, NO_3^{-2}, and H^+ in precipitation).

Phase II: Defining the issues.   The atmosphere-related impacts that need to be considered include the following:

1. Air pollution (impacts on human health, vegetation, and materials due to SO_2, NO_x, oxidants, and SPM)
2. Acidic deposition (impacts on the biosphere due to wet and dry deposition of acidic material)
3. Climatic impacts (not likely, except for fog and ice in the immediate vicinity of cooling towers and at warm-water discharge points)
4. Noise
5. Ecological effects of high-voltage transmission lines (not really an atmospheric problem, but may be a major public concern; check with other members of the EIA team to ensure that this question is considered by somebody)
6. Safety considerations (Is the structure capable of withstanding weather extremes, e.g., hurricanes, freezing rainstorms, etc.?)
7. Impairment of the natural beauty of the area.

Phase III: Defining the mesoclimate.   The mesoclimate refers to variations in climate that occur over distances of from 10 to 100 km due to the existence of valleys, coastlines, cities, and similar influences. Most textbook dispersion formulae assume that the underlying surface is homogenous. However, power stations are usually built in river valleys, along coasts, or in urban areas. The meteorology of the area needs careful attention, therefore, through studies of topographic maps and thorough inspections, not only of the proposed power station sites, but also of the locations of the meteorological observing stations identified in Phase I, item 4. One question to ask is, how representative are the historical data files? For example, if weather observations were made in an east-west valley, the wind climatology could not be used to predict dispersion in a north-south valley.

The meteorological conditions most likely to cause high ground-level concentrations downwind of a tall chimney in the given terrain should be identified. Among these might be

1. Limited mixing of air near ground level (because of a temperature inversion based a few hundred meters above the top of the stack)

2. Strong and steady winds

3. Fumigations due to the breakup of nighttime radiation inversions and of inversions associated with daytime coastal sea breezes

4. Light-wind strong daytime convective looping.

The assessor should try to estimate the frequency of occurrence of each of the problem conditions.

Where a representative upper air station is available, as in Portland or Caribou, Maine, the vertical structure of the atmosphere can be analyzed and frequency distributions obtained. If this is not possible, wind roses from a representative surface station may be obtained, separately by season and by day and night. (In the case of mesoscale wind fields, there will usually be a wind direction reversal near sunset and shortly after sunrise—e.g., land vs. sea breezes, and up-slope vs. down-slope valley winds; separate day and night wind roses will reveal these phenomena.)

Where current meteorological information is clearly deficient, field studies may be warranted during periods likely to be associated with relatively high ground-level pollution concentrations. These field programs might include balloon observations (minisondes) of temperature and surface wind measurements from 10-m towers.

Phase IV: Estimating the ground-level pollution concentration field.

Clearly, as a first step in this phase, it is necessary to estimate the background pollution concentration fields prior to construction of the new power station. This will lead to the loading to which the contributions of the new power station must be added. If existing information of air quality in the region is inadequate, two to three additional monitoring stations should be established as soon as possible.

The next step is to select the mathematical formulations to be used in the diffusion calculations of plume rise and downwind dispersion. This question has already been discussed in Chapter 13. Then, for the atmospheric situations mentioned in Phase III that are likely to cause high ground-level concentrations, dispersion calculations are undertaken and isopleths computed for an area that encompasses and extends well beyond the point of maximum ground-level concentrations. Included in the dispersion calculations should be estimates of their frequencies of occurrence, obtained from the meteorological analysis undertaken in Phase III.

Finally, the predicted concentrations (including the background component) are compared with the air quality standards mandated for the region. If the standards are exceeded, the planned emissions will need to be reduced, or the location will have to be dropped from the list of possible sites.

If mean annual air quality standards exist, it may be necessary to undertake dispersion calculations not only for episodes, but for all hours of a year. Usually, these calculations are based on five to ten years of weather observations from a representative station. (Sometimes the representativeness of the station may be hotly debated at a public hearing.)

Phase V: Analysis of other atmospheric impacts.   **Acidic deposition** is likely to be an issue, although a single power station contributes only a very small fraction

(one percent or less) of the wet deposition falling on the surrounding countryside. This is because the emissions contributing to wet deposition come from many sources extending over distances of many kilometers. Nevertheless, estimates of wet and dry deposition should be made, using appropriate formulae, based on a reasonable historical time series of hourly observations of wind and precipitation. A five-to ten-year period of record is usually appropriate.

Climatic impacts are not likely to be significant in the case of coal-fired power stations, except possibly if a cooling tower is proposed, in which case an adjacent highway could be affected. (Fog may impair visibility, and surface wetness or ice may cause slippery conditions.)

Noise will be an issue only during construction. However, the EIA should not overlook the problem.

Safety considerations are usually dealt with in the engineering design section of an EIA. However, the atmospheric section should contain the basic climatological statistics on extremes of temperature, wind, and precipitation.

Application of these ideas to the proposed power station.   Without attempting detailed dispersion calculations, it may be helpful to identify the air pollution situations of most interest at each of the proposed sites:

SITE 1 (Open coastal site)

1. Daytime sea-breeze fumigations in spring and summer
2. Light-wind daytime convective looping
3. Strong and persistently steady winds.

SITE 2 (In a valley)

1. Limited mixing of air at the valley floor
2. Morning fumigation
3. Cross-valley strong-wind downwash.

SITE 3 (Near a tidal flat within a built-up area)

1. Daytime sea-breeze fumigations in spring and summer
2. Aerodynamic downwash around tall apartment buildings
3. Strong and persistently steady winds.

16.3 POLLUTION CONTROL STRATEGIES

Improvement in the quality of our environment will come about only if action is taken by someone to correct an existing pollution problem or to prevent a new problem from occurring. Knowledge of the cause of the pollution problem and the means to correct it are, of course, necessary for a successful resolution, but unless the polluter takes action, voluntarily or through enforcement, nothing will change. There are many examples of pollution problems where the cause and the remedy are well known, but no action has been taken. Why is this so, and what can be done about it? Who should do something about it? These questions, as well as the economic and legal aspects involved in dealing with them, are considered next.

16.3.1 Economic Aspects

Homeowners wishing to improve the physical environment of their surroundings can do so by purchasing goods (trees, shrubs, furniture, roofing material, etc.) and labor (a gardener, carpenter, plumber, etc.) to accomplish this. The person who owns the land and building provides or borrows the money to make the improvements, enjoys the benefits of the improvements, and suffers the consequences of not making the improvements (e.g., water damage caused by a leaking roof). In extreme cases of neglect of the home or the yard, the local municipality may take action under a section of the Public Health Act or Weed Control Bylaw. Otherwise, there is no need for governmental interference. But what about the situation where the homeowner is concerned about the quality of the water in the creek at the rear of the property, or the quality of the surrounding air? What actions can be undertaken to correct these problems? The owner is powerless to do anything independently and will most likely complain to the local government, which may or may not be in a position to take action. The concept of ownership of resources is important in such a situation: the air and water are not owned by any individual, but are **public goods,** as compared to the lot, house, furniture, etc., which are **private goods.**

The economic system of capitalist countries, and to a lesser extent that of state-controlled socialist countries, has been remarkably efficient in producing consumer goods of all kinds, priced so that there is a balance between supply and demand. Neither of these economic systems has been as successful in providing public goods, such as clean air and water. Since no individual or identifiable group of individuals owns these resources, no one has taken the responsibility to maintain their quality. An individual or an industry which pollutes the air or water reduces the quality of that resource for everyone in the area. Everyone suffers the consequence for the activities or negligence of the polluter, although not always equally.

The costs of pollution have been generally external to the polluter. That is, the costs do not appear on the polluter's balance sheet, nor do they make up a part of the cost of production. This is true for a steel mill discharging untreated wastes or for a town discharging untreated sewage. These external costs are paid by society in two ways. One of these is the cost of pollution damage: the reduction in the value of available resources. For example, the value of a public beach is reduced if it is too polluted for swimming. And a fishing lodge operator loses business if the fish in the nearby stream or lake are killed by pollution.

The other component of external cost is the expense of repairing or reducing the damage caused by pollution. The cost of building and operating a municipal water treatment plant is an example: the water which has been "damaged" by pollution must be restored in order for it to be fit to drink.

Because the rights to public goods are not owned by anyone, they cannot be bought and sold, and therefore there is no marketplace in which values can be placed on them. Sometimes compensation for these external costs can be determined relatively easily. For example, suppose that the fish in a certain pond had been grown by a fishing lodge operator. If the effluent from an upstream pulp mill were to kill the fish, the lodge owner might have the chance to collect the value of the lost fishery operation from the pulp mill—often, however, not without a court case. To avoid this heavy toll, the pulp mill, by applying normal business management principles, would determine how much it needed to spend on pollution abatement to avoid killing the fish and paying the consequent damages. In this simple example, the external costs of pollution would be transferred to the polluter. Of course, in the long run, these costs would be passed on to the consumers of pulp and paper and, therefore, to society as a whole.

It is much more difficult to establish a value for the loss of enjoyment of the recreational and aesthetic pleasures of a river or lake by a group of cottagers once the river or lake is polluted by the same pulp mill. It is also more difficult for the cottagers, as individuals or as a group, to initiate a possibly expensive court action against the pulp mill. Accordingly, we look to our elected representatives in government, at the local, state, or national level, for protection of our public resources—for clean air and clean water. Governments are more likely to respond if they perceive that a cause is very important to the electorate. Many citizens' groups concerned about environmental protection have therefore made it their business to use every available means to pressure governments to enact legislation and to introduce incentives which will make the polluter implement pollution control measures. In one way or another, these methods ''internalize'' the external costs of pollution as much as possible. When it is not possible to achieve this, such external costs will have to be borne by society as a whole, as a social cost of pollution, just as we share other social costs, such as those created by unemployment, welfare, and occupational health requirements.

16.3.2 Legal Aspects and the Role of Government

Earlier in Part 3, legislation to control the quality of water and air and the disposal of solid and hazardous wastes was introduced, with particular emphasis on the United States and Canada. It is the responsibility of government, at the national, state, provincial, or local level, to enact and update environmental control legislation. This legislation is generally written to provide the broad goals and objectives for environmental quality. It does not provide the means and methods by which these goals are to be achieved. Nor need it provide the details which are necessary to monitor and control the performance of pollution control facilities. It is, therefore, necessary for governments at all levels to establish regulatory strategies, in order to implement the broadly stated objectives of general legislation.

The goal of environmental management strategies is to maintain or improve the quality of the ambient or surrounding environment. **Ambient standards** are determined for a number of different characteristics or pollutants within a medium such as air or water. These standards are designed to minimize risks to the health of humans, animals, or the environment in general. The components for which these ambient standards are set must be quantifiable and scientifically measurable. In water and air, criteria are set for allowable concentrations of a variety of pollutants. Furthermore, the pollutants for which ambient standards are set must be related to their sources. A regulatory agency can set ambient standards and monitor ambient conditions, but it cannot control or manage conditions except by controlling the sources of the pollutants which affect the ambient conditions. For example, in the air, it is desirable to maintain the concentration of particulates below a certain level. To do this, we must determine the possible sources of the particulates. Some of these sources may be identifiable, such as a smokestack or a burning garbage dump. But much of the particulate matter may come from unidentifiable or nonpoint sources, such as open fields, highways, or a forest fire many miles away. After the sources have been identified, it is necessary to relate the rate at which the pollutants are being released from the sources to the ambient concentrations. When this is done, it is possible to set allowable limits on the discharge of pollutants at the sources. This forms the basis for effluent standards.

Effluent standards are more useful for environmental management than ambient standards because they can be monitored and controlled in many cases. Even though the ambient quality is what we are interested in preserving, we normally try to achieve this by controlling effluent quality and quantity.

Three main instruments are available to government for environmental control: direct regulation, polluter subsidies, and polluter charges. They are all a means of controlling effluents or discharges of pollutants, and they all work to internalize pollution costs to the polluter. They can be applied independently, but are usually applied in combination. Each of these instruments appears in a variety of forms. We shall consider some of the more common forms in which they are applied, as well as other interesting possibilities for controlling pollution.

Direct regulation.   The government can use its legislative powers to regulate the actions of individuals, corporations, and lower levels of government. Therefore, through direct legislative action, the quantity, quality, and location of discharges of pollutants can be regulated. The main forms of direct regulation are zoning; prohibition, or zero discharge; and effluent standards.

Zoning.   Zoning regulations are one of the simplest and oldest forms of pollution control and are still a part of almost every pollution control strategy. The objective is to separate the polluter from the rest of society by either space or time. A result of the so-called sanitary awakening in mid-nineteenth century Britain was the realization that open garbage dumps had to be removed from areas of dense population and kept away from public water supplies. Local bylaws were enacted to ensure that this was done so that the benefits to public health were realized. The prohibition on the burning of coal in nineteenth-century London while parliament was in session is another example of this type of zoning. More recent examples of zoning to separate polluters from the public are the location of airports, the use of curfews on airport operations, and the construction of tall chimneys or long marine sewage outfalls.

Zoning results in variable ambient quality. In areas where pollutants are discharged, quality will be low. Government could of course divide the country into regions with different expected ambient standards and therefore different allowable effluent standards.

Polluters would then determine for themselves the trade-offs in abatement costs and location or relocation costs. People could also choose the level of environmental quality they wished to live in. However, the disadvantages of such a scheme are many. For example, industries might be forced to locate far from their markets, thus increasing the cost of goods. Also, during the implementation of such a scheme, there would be many "not-in-my-backyard" types of arguments. And natural resource industries, such as forestry and mining, cannot be moved from their supplies. Another disadvantage would be that industrial workers and workers in related service industries would be essentially doomed to live in poor-quality environments. In fact, industry, population density, and pollution have always gone together, and it would take more than simple legislation to separate them.

Prohibition, or Zero Discharge.   Another form of direct regulation of pollution is prohibition, also known as zero discharge. The advantages of such a concept are obvious. First and foremost, there would be no change in environmental quality. Moreover, all resources would have to be completely converted into useful products or stored indefinitely. And the legislation would appear to be equitable, since the same regulation would apply to everyone. Such a concept, however, is normally impossible to realize. A simple materials balance shows that any resource taken from the environment, including energy, must be returned in some form. Even if it were conceivable to recycle all wastes into new products, there would still be a large energy requirement to achieve this. For most activities, zero discharge would be expensive if not impossible to achieve. At present, producers of extremely hazardous wastes, for which no treatment is available, are the only ones subjected to zero discharge requirements. They must store their wastes until a means of safe disposal is found.

Effluent Standards.   Effluent discharge standards are the most common and the most useful form of direct regulation. They can be in the form of across-the-board standards which require that effluents of all polluters meet the same criteria, or they may be individually developed for each polluter. The advantages of an across-the-board type of approach are that it is easy to administer, it appears fair to all polluters, and it provides the most rigid control over environmental quality. The disadvantages are that it may be uneconomical, and therefore impractical, to insist that all polluters meet the same effluent standards. Some polluters may easily meet standards that others will be unable to meet at all, or only at a very high cost. The different assimilative capacities of the environment in different locations can be taken into account only on a case-by-case basis. For example, a large, fast-moving river can accept a much larger amount of organic pollution than a small creek, and therefore pollutant concentrations from point-source discharges could be much higher before river quality is seriously affected. Nevertheless, most jurisdictions prefer to set common effluent discharge guidelines, which must be met unless the contributor is specifically exempted.

Regulation through zoning and effluent standards has for many years been the basis for controlling the pollution we put into our environment. However, the legislation has not been sufficient to end pollution. For the controls to be successful, the polluter must be willing to respect the law. If the polluter considers the law unjust, economic pressure in the form of fines may force compliance with effluent requirements. However, fines have traditionally been very low, providing no economic incentive to polluters to comply with regulations. Other enforcement measures include the withholding of various permits or licenses without which the polluter cannot lawfully operate. Using a court injunction to force a polluter to cease operations is a drastic measure that is rarely used.

Enforcement is the major problem with any form of direct regulation. Those who have been discharging wastes freely resent and protest any curtailing of their "rights." Industries often threaten to close down if regulations are enforced. This would cause local unemployment and is politically unacceptable. Municipalities forced to upgrade their waste disposal systems often insist they simply have no money to do so. It is difficult for any level of government to force polluters into actions that may have politically unacceptable side effects.

Subsidies.   One method of encouraging polluters to comply with regulations is to provide money to help cover their costs. These subsidies may be in the form of direct payments or grants based on a percentage of the cost of pollution abatement or on a percentage reduction in effluent quantity or strength. They may also take the form of low-interest loans for the capital costs of improved treatment facilities. Alternatively, governments can reduce or defer taxes or relax other government requirements to encourage spending on pollution control.

The main advantage of subsidies is that they reduce the costs of pollution abatement to the polluter and limit the associated increase in production costs. Government grants can be used to cover capital costs, and tax incentives can be used to relieve operation and maintenance costs. Subsidies (the carrot) combined with regulations (the stick) can be used by government to reduce stress on the environment and at the same time encourage research and development by industry in pollution abatement technology. The main disadvantage of polluter subsidies is that the government will have to increase taxes or direct money from other programs in order to pay the subsidies. This is partially offset by decreased expenditures needed to correct the effects of damage due to pollution (i.e., expenditures on water treatment plants or public health care). However, these returns may be small compared to the costs involved. A general tax increase may seem fair when everyone benefits from an increase in environmental quality. In fact, however, people benefit to varying degrees, and some may balk at paying money for what appears to be someone else's problem.

Another serious drawback to the subsidy system is that it can be easily abused. The idea of paying someone to stop damaging the environment sounds suspiciously like a criminal protection racket. All potential polluters will want to be paid for not polluting. Companies may find that the subsidy available for waste reduction exceeds their actual costs of making the change. They may then increase their production above normal simply to receive a subsidy and go on to dump the extra goods at a lower price. In this situation, a polluting industry has been rewarded while its competitors who already treat their wastes adequately get no benefit.

Subsidies to municipal governments for the construction of wastewater collection and treatment facilities are common. The details of such subsidies vary depending on the government involved, but the purpose is to enable municipalities to meet effluent regulations at an affordable cost. Subsidies are generally in the form of grants or long-term, low, fixed-interest-rate loans on a portion of the capital cost of the project. A local municipal taxpayer is more likely to approve a spending bylaw if the municipality is going to receive a large grant from a higher level of government.

Service charges.   Service or user charges are similar to subsidies in that monetary means are used to encourage a polluter to comply with effluent requirements. Charges are the most direct way of internalizing the costs of pollution to a polluter. There are numerous types of service charges, but in general, money is paid to the local government or agency in proportion to the amount of pollution. The government or agency may then use the money to pay for and operate central pollution control facilities.

The obvious advantage of a service charge is that it is the polluter who pays for the costs of polluting. The system rewards those industries that are clean and efficiently run and penalizes those that are dirty and wasteful. Also, it does not encourage increases in polluting activity, as a subsidy system might. Finally, the administration of such a system is relatively easy, requiring only the monitoring of discharges.

The disadvantages are that production and operating costs for the industries connected may rise. If the service charges are nominal, industries may find it less expensive to simply continue polluting. If the charges are high enough to force an industry to stop or severely restrict its effluent discharges, the industry may close down. In any event, the charges will be passed on in the form of increased prices for the industry's products. Since each industry has different capabilities and costs related to controlling its wastes, a uniform service charge could upset the economic balance between competing industries. However, to customize effluent charges for each polluter would be an administrative burden and appear to be unfair.

We are all familiar with charges for municipal services. In urban areas, we pay through property taxes or special levies to have refuse and sewage removed from our homes. In the same way, industries may find it more convenient to pay to have their untreated wastes removed and disposed of at a central treatment facility. In some of the heavily industrialized areas of Europe, this has been found to be an attractive and efficient way to dispose of industrial wastes. In many cases, the extra cost of waste collection is offset by the economy of scale of large, specialized treatment plants.

For industries in urban areas, it may be possible to combine liquid wastes with municipal wastewater. Most municipalities with a separate or combined sewer system will have a sewer-use bylaw or, if industrial waste discharges are involved, an industrial waste bylaw. Typically, the bylaw in question would exclude wastes with temperatures exceeding 65°C, as well as those containing undesirable materials which might burn, explode, give off odors, plug sewers, or upset the plant. The bylaw would establish a maximum, and probably nominal, penalty per day for noncompliance. If the bylaw sets inflexible limits on BOD_5, suspended solids, and other constituents, with no provision for accepting and treating wastes stronger than normal, industry may be forced to pretreat its waste when in many cases such waste could be more economically treated in a municipal plant.

An industrial waste bylaw may have two serious drawbacks. First, an overly restrictive bylaw could put a municipality at a disadvantage in competing for industry with other cities having a less restrictive bylaw. Second, enforcement is left in the hands of the local municipality, which naturally is reluctant to deal harshly with a tax-paying industry that provides employment in the community. The task is even more difficult when that industry's competitors in another locality have no such restrictions imposed on them. The implementation of a national code, applying to all incorporated municipalities, would correct these problems. However, pressure to preserve local autonomy will undoubtedly limit future solutions to modifications of present pollution control methods within well-defined geographical (and political) areas.

In general, all wastes which do not harm the system or affect the operation of the treatment plant should be accepted without pretreatment. If the wastes are stronger than "normal" sewage, then a charge, or more correctly, a surcharge, should be assessed against the industry for the extra cost of sewage treatment. For this approach, a surcharge formula setting out the charges for accepting wastes stronger than normal would have to be included in the industrial waste bylaw. Ideally, charges for sewage treatment should be related to the cost of providing the facilities and the benefits received. The practical application of this method is difficult, however, and various methods of charging for industrial wastes have evolved. Three of these are outlined.

Quantity Formula.   The quantity method, in which the charge is determined from the volume of the waste, is the one used in most municipalities where a basis other than assessment is employed. It is practical where the industrial waste is about the same strength as domestic sewage. Often the sewage service charge is made equal to the water bill for industries with large volumes of waste. The method is simple to administer and places the financial burden on users.

Quantity-Quality Formula.   According to the quantity-quality formula, charges for disposal of industrial wastes are determined from both the volume and strength of the wastes. For example, in addition to the basic charge, depending on volume, as outlined in the preceding paragraph, a surcharge may be added to wastes with BOD_5 in excess of 300 mg/L and suspended solids greater than 400 mg/L. Many municipalities follow this practice.

Example 16.1

An industry produces 800 m³/day of organic waste with a BOD_5 of 635 mg/L. If wastes with BOD_5 less than 300 mg/L are covered by basic sewer charges, through taxes or as part of the water bill, what annual surcharge would this industry pay if the operating cost for removing BOD_5 is \$0.25/kg BOD_5 removed.

Solution

$$\text{Daily flow} = 800,000 \text{ Lpd}$$

$$\text{Daily } BOD_5 = \frac{800,000 \times 635}{10^6} = 508 \text{ kg}$$

$$BOD_5/\text{year} = 508 \times 365 = 185,420 \text{ kg}$$

$$BOD_5 \text{ allowed} = 300 \text{ mg/L}$$

$$BOD_5 \text{ allowed/yr} = \frac{800,000 \times 300}{10^6} \times 365 = \underline{87,600 \text{ kg}}$$

Thus, the excess BOD_5/year is 97,820 kg, and the annual surcharge is 978,820 × \$0.25 = \$24,455/yr.

Assessment/Surcharge Formula.   Another way of charging for sewer service is to allow all users the same amount of treatment per assessment dollar, after which a surcharge is applied. Municipalities sometimes allow industries up to twice the amount of treatment per assessment dollar without a surcharge; that is, wastes two times stronger than normal are accepted without extra cost. After this, a surcharge based on the actual cost of BOD_5 and suspended solids removal at the treatment plant is added.

Example 16.2

If the average residential assessment is $5,000 per house, determine the surcharge that would be applied to the industry in Example 16.1 if the industry were assessed at $4 million. Assume that users are entitled to the same amount of treatment per assessment dollar.

Solution   For the average private residence assessed at $5,000 and accommodating an average of 3.5 persons, the BOD_5 contribution per home, assuming 76 gm/capita/day from Table 12–3, would be

$$BOD_5 = 3.5 \times 76 = 266 \text{ gms/day}$$
$$= 0.266 \times 365 = 97 \text{ kg/year}$$

The normal assessment factor F is therefore

$$F = \frac{\text{lb BOD}_5/\text{year}}{\text{assessment}} \times \frac{97}{5000} = 0.02$$

(Note that F could also be found from total BOD_5 to plant/total assessment).

For $F = 0.02$, users are allowed 0.02 kg BOD_5 per year per assessment dollar before a surcharge is applied. Applying this to the industry in Example 16.1, the surcharge is calculated as follows:

BOD_5/year	= 185,420 kg
Assessment	= $4,000,000
Assessment factor	= 0.02
BOD_5 allowed	= 4,000,000 × 0.02 = 80,000 kg
Excess BOD_5/year	= 105,420 kg
Cost of treatment/kg BOD_5 removed	= $0.25
Surcharge	= 105,420 kg × $0.25 = $26,355.00/yr.

Comment   The assessment/surcharge formula avoids the problem of a surcharge based only on quality, whereby the surcharge might be avoided by simple dilution of the waste to less than the concentration permitted in the industrial waste bylaw.

Another method for determining polluter charges has been suggested by Dales (1968): the government would sell "rights" to any interested party to pollute the air or water for a period of one or more years. The government would then be able to limit the total amount of pollutants discharged by limiting the number of rights or discharge warrants for sale based on the assimilative capacity of the receiving body. These warrants would be bought and sold in an open market, where their prices would be determined by normal marketplace dynamics. Advantages of such a system are that it is equitable (in an economic sense), that the government can control total pollution loads (ambient quality) without concern for point-by-point discharge permits, that the costs are internalized to the polluters in a way that they can control, and that the government receives revenue. Each polluter then has the choice of determining how much to pay in polluter charges and how much for pollution abatement. In such circumstances, the real costs of pollution would become evident to polluters and the community at large since the prices of the discharge warrants would not be set arbitrarily by government, but rather, would be determined by the polluters themselves. Discharge warrants could also be purchased and held by environmental interest groups in order to prevent these rights from being used by polluters.

The scheme to allocate pollution rights is similar to the way rights to other public resources, such as trees or Crown land, are determined. The marketplace, and therefore the polluters themselves, determine how great the polluter charges should be and how they should be apportioned. Unfortunately, this very feature is also its major drawback, because the general public would not be comfortable with the seeming lack of control by the government and too much control by those with money. So far the scheme is only of academic interest.

Historically, the most common methods used by government for pollution control have been direct regulation (zoning and effluent standards) and subsidies (grants and tax incentives). Enacting laws and regulations gives the government the appearance of doing something about pollution. It also seems much simpler and more effective in the short run to require polluters to pay the costs of pollution abatement than to try to entice them by offering economic rewards and penalties. Regulatory controls will probably always be a part of any government strategy for pollution abatement, regardless of the success of economic controls.

In any event, regulatory and economic controls are not enough to correct pollution problems. Public pressure on those responsible for environmental protection is also important. This activist role by the public is a recent phenomenon and, when supported by an inquisitive press and frank disclosures of unsatisfactory conditions, could be the most effective means for protecting environmental quality.

16.3.3 Case Study: Toxic Chemical Wastes—The Niagara River Problem

The pollution of the Niagara River is an existing environmental problem. It is a very complex problem for at least two reasons. First, there are large gaps in scientific knowledge of the behavior of toxic chemicals making it difficult to monitor and control them and to predict their long-term effects on humans. And second, the Niagara River is a boundary water between two countries, the United States and Canada, and between New York State and the Province of Ontario. Thus, any proposed action requires the cooperation of four governments, which complicates the planning and decision-making processes.

A summary of the nature of the Niagara River pollution problem and the steps taken so far by the U.S. and Canadian federal governments and those of the state of New York and the province of Ontario is outlined. The process of preparing an environmental impact assessment is in this case a lengthy one, done in stages, with interim cleanup action carried out as the assessment continues.

Nature of the problem.   The Niagara River joins Lake Erie with Lake Ontario, the two most heavily polluted of the Great Lakes. The Great Lakes are the most precious freshwater resource in North America. They are also an international resource, and hence an international responsibility. This was recognized as long ago as 1909, when the Boundary Waters Treaty was signed by Canada and the United States, establishing the International Joint Commission. In 1972 the Canada–United States Water Quality Agreement was signed, which pledged governments in both countries to work together to restore and protect the Great Lakes. The agreement was reaffirmed in 1985.

The increasing concentration of toxic chemicals in the Great Lakes is the latest challenge to be faced. Through years of neglect, water quality has been allowed to deteriorate by the discharge of all kinds of effluents into the lakes (Figure 16–2). With the development of sophisticated monitoring equipment, we have discovered that the Great Lakes, Erie and Ontario in particular, are contaminated with hazardous substances. In many cases, it is

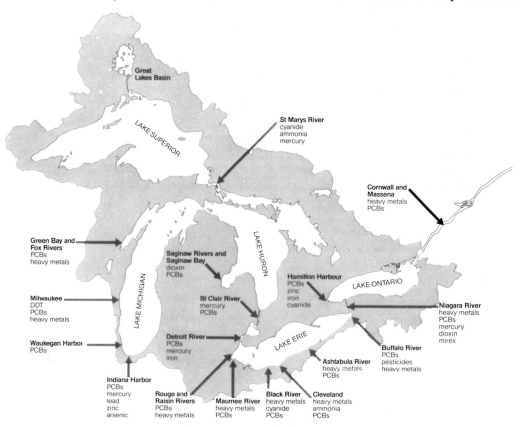

Figure 16–2   Toxic substances in the Great Lakes. Source: *Sweetwater Seas: The Legacy of the Great Lakes*, Environmental Canada, Ontario: Ministry of the Environment.

virtually impossible to say how much of a given chemical is too much. Some chemicals are so harmful that a trace amount could cause serious illness or even death to humans. The problem is compounded by the fact that many chemicals accumulate in the bodies of organisms that consume substances that have been contaminated. The higher a species is in the food chain, the more the effect is magnified. Since humans are at the top of the chain, the condition of the Niagara River, continues to be a matter of great concern to the 5 million residents of New York State and Ontario, who depend on the river or Lake Ontario as a source of drinking water.

The Niagara River is a continuing contributor of numerous trace metals and organic compounds to Lake Ontario because of the accumulation of these pollutants in Lake Erie

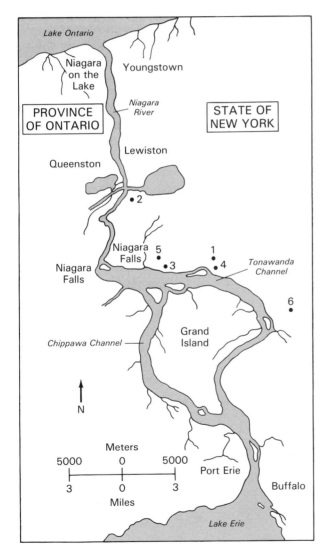

Figure 16–3 Location of major waste disposal sites on Niagara River. Source: Crabtree 1982.

(1) Love Canal   (4) 102nd Street
(2) Hyde Park   (5) Alphabet Sites
(3) S-Area   (6) Durez Property

and Niagara River sediments, and because of the discharges to the river from municipal and industrial sources and the leakage from chemical waste disposal sites. Figure 16–3 shows the location of the most important sources of wastes along the river. The Love Canal site achieved international notoriety during the 1970s, but it is by no means the only source of pollutants. In total, there are over 200 industrial and municipal point-source discharges on the New York side and 19 on the Ontario side. As examples of quantities involved, it is estimated that the daily loading of organic priority pollutants is about 400 kg from New York and 10 kg from Ontario. Total phenols are about 350 and 12 kg, respectively, and heavy metals comprise about 575 and 70 kg, respectively.

A preliminary assessment.   The adverse publicity on the waste disposal sites along the river resulted in ad-hoc cleanup measures and stricter controls for new facilities. Reductions in the levels of certain contaminants, such as phosphorus, DDT, Mirex, and PCBs were evident in the 1980s. At the same time, chemicals not previously identified in the river were detected in fish and sediments, and sometimes in the water itself. However, with the exception of the four chemicals stated above, it is too early to tell whether real progress is being made in cleaning up the river, because there are insufficient past data available on the various pollutants which can be compared with the more recent samples of water, sediments, and fish.

We can, however, assess whether the necessary framework for comprehensive action is in place by asking the following questions:

1. Is there a comprehensive, coordinated, and credible monitoring program in place to tell, now and in the future, what substances are in the river and where they come from?

2. Are there control programs in place to deal effectively with chemical substances that are being discharged in unacceptable quantities?

3. Do we know the effects of the chemicals that are present in the river on aquatic life and human health?

The status of these three issues can be summarized as follows.

Monitoring.   Canada has had a comprehensive monitoring program in place since the 1970s, progressively covering water, fish, plant, and sediment sampling. The first step to coordinate this work with U.S. efforts was taken early in 1981 when the Niagara River Toxics Committee (NRTC) was established, consisting of federal, state, and provincial scientists and engineers from both countries. The committee reported its findings in October 1984, including recommendations for research and a long-term monitoring program. There were difficulties in creating such a long-term program. (For example, the differences in detection limits used in the two countries and the very low levels at which many of the chemicals exist in the water make quantification difficult.)

From the intensive monitoring already done, the chemicals present in the river have been measured, priorities have been set on certain chemicals, and the gaps in sampling and analytical procedures have been identified. When the NRTC recommendations on monitoring are implemented, a monitoring system will exist that meets the criteria of being comprehensive, coordinated, and credible.

Discharge Permits.   For point-source discharges, there are control program frameworks in place on both sides of the border. In Ontario, there are Certificates of Approval and Control Orders; in New York, there are State Pollution Discharge Elimination System (SPDES) permits and Consent Agreements. As control systems, both can be effective provided the criteria selected are practical and enforced.

For landfill sites in operation today, control mechanisms are in place. Landfills are not the problem, however; it is the cleanup of existing inactive waste disposal sites that is the real challenge.

Effects of Chemicals.   From the monitoring work completed (and still continuing), a clearer picture has emerged of the extent of our knowledge of the effects of chemicals in the river and of the gaps in that knowledge as well. Since the first step in the process of improvement was the identification of gaps and the setting of priorities for filling them, progress has definitely been made. The Niagara River Toxics Committee completed this work as part of the task with which it was charged, and the Committee's report, issued in October 1984, contained this information. It is now up to the four governments involved to authorize the research needed to establish the effects of the chemicals and to develop the aquatic and health standards that do not presently exist. Progress is being made in both countries on the research identified in the report.

Factors influencing environmental decisions.   Decisions that relate to protecting or cleaning up the environment are not made on the basis of environmental factors alone. This is particularly true for the Niagara River. The international nature of the situation adds another element to the variety of influences normally present.

Economic Factors.   Protection of the river from current waste products of society is in itself expensive; correction of past errors and bad practices will add enormously to that cost. In facing this reality, we find points of view at two extremes. On the one hand, there is the argument that, because there are gaps in our knowledge of the health effects of many chemicals, their presence in the environment should be reduced to zero at any cost. On the other hand, there is the view that, because unemployment is high, priority must be given to jobs and investment in industry; any extra expenditure on the environment must wait until the economy improves.

Neither extreme is acceptable. The environmental risks are too great to wait for an indefinite time until economic conditions are better. The initial steps have been taken (as a result of public concern), but because of the length of lead time needed to put control and remedial measures in place, early commitments to specific programs are necessary. Of course, practical considerations (like costs and benefits) are important criteria in the development of these programs, whether they be for monitoring (the number, frequency, locations, and analytical requirements of samples must be examined) or corrective measures (a range of options must be investigated).

Movements of population have a significant impact on the economic strength of a region. During the 1970s, New York state experienced a reduction of 3.8 percent in its population, reflecting a general trend in the U.S. for people to move south to a warmer climate. Buffalo's population fell by a drastic 22.7 percent during this period. A number of industries in the area closed, with little chance that they will ever reopen. Inevitably, this situation affects environmental programs and stresses the need for a firm but balanced approach to river cleanup.

Social Factors.   The flood of information on environmental issues from government agencies, academic institutions, and special-interest groups has increased the awareness and knowledge of the general public about unsatisfactory waste disposal practices. As a result, there is steady pressure on governments and industries to correct these problems. This pressure, applied largely through the media without recourse to the law and its regulations, has been very effective, since neither the authorities nor the polluting industries want to be the subject of adverse publicity.

At the same time, there is a growing social conscience among the industries in the Niagara area, as elsewhere, which shows itself in a desire to be perceived as responsible corporate citizens. When infractions of discharge limitations are pointed out, many companies will voluntarily remedy the situation without the need for coercion by the authorities.

Political Factors.   Data published in both the United States and Canada indicate that hazardous contaminants are entering the Niagara River from the American side as well as from the Canadian side (about 90 and 10 percent, respectively). It is therefore necessary that there be not only political will on both sides of the border, but also close contact between the governments and agencies involved if cleanup of the river is to be effective.

For Ontario, the political route has two alternatives. One is direct contact between the Ontario Minister of the Environment and the Commissioner for the Department of Environmental Conservation of New York state. This route has been, and will continue to be, used to resolve problems that are within the jurisdiction of New York state and the province of Ontario, i.e., discharge permits or control orders and the majority of waste disposal sites.

The other alternative is to utilize the diplomatic channels through the Canadian Department of External Affairs and Environment Canada to the American State Department and the Environmental Protection Agency (EPA). This is the option for issues that are within EPA's jurisdiction, such as U.S. federal litigation on certain serious waste disposal sites, Superfund allocations, and the Clean Water Act. If appropriate, direct contact between the Minister of the Environment for Canada and the U.S. EPA administrator can be part of the process.

Diplomatic channels have also been used to try to develop arrangements for access to technical information on waste sites under litigation, and for each country to have an impact on the decisions regarding cleanup measures. In light of the confidentiality that necessarily envelops court cases, should these diplomatic means succeed, a major change in procedure will have been achieved.

Legal Factors.   There is a greater tendency in the United States than in Canada to resort to the courts for the resolution of environmental problems. The U.S. confrontational approach often results in delays lasting for years. Canada's view is that, because the Niagara

River is an international waterway subject to an international treaty, special measures should apply; the normal internal legislative procedures of either country should not be allowed to stand in the way of the quickest possible means to remove the sources of pollution.

Whatever approach is adopted, Canada and the United States will have to be familiar with each other's legal processes concerning environmental matters for meaningful interaction to occur.

Proposed action.   For long-term sustainable improvement to the Niagara River, the following activities are necessary:

1. Continuous reviews of SPDES permits and Certificates of Approval must be undertaken to ensure that water quality standards are met on both sides of the border. Upgrading of requirements should occur as detection and treatment technology improve, based on the need for more stringent water quality standards.

2. Regular reviews of the effectivenesss of waste site remedial programs must be held, and implementation of further work must be carried out if necessary.

3. A mass balance on the contaminants entering the river (from point and nonpoint sources and Lake Erie) and discharged into Lake Ontario must be performed.

4. A long-term monitoring program for the river must be initiated, with two objectives: to trace trends for specific chemicals that reflect changes in the overall quality of the river, and to keep a check on the most hazardous discharges from point and nonpoint sources.

5. Additional control programs must be set in motion based on the needs identified by the Niagara River Toxics Committee's work or by the long-term monitoring program.

6. An effective negotiation process between Canada and the United States must be developed which permits the flow of environmental information in both directions, as well as the opportunity to influence decisions on control programs and cleanup actions.

16.4 ENVIRONMENTAL ETHICS

A newspaper headline in the *Toronto Star* on August 12, 1987, read,

"58 New York Officials Charged with Corruption."

The article described a $2\frac{1}{2}$-year FBI undercover operation which resulted in charges ranging from conspiracy and mail fraud to perjury and bribery against officials, past and present, of more than 40 cities, towns, and villages in New York state.

Has morality changed since the days of the pilgrims?

16.4.1 Historical Perspective

Before we consider environmental ethics, which is the application of morality to one aspect of contemporary life, we should look at ethics in a general way. Ethics, from the Greek "ethika," meaning character, is the science concerned with the obligations of individuals or groups to one another or to society. It deals with principles of fairness, justice, morality, obligation, and duty. The study of western philosophical ethics started with the Greek philosophers, notably Pythagoras in the sixth century B.C., and was developed later by Socrates, Aristotle, and Plato. They believed that people would do the proper thing if they knew what it was. Education was viewed by these ancient philosophers as the key to moral behavior, although they applied it only to an elite and not to the general population. The appeal of Christianity was that it extended the concept of morality to all people, including slaves. Religion, i.e., a belief in God and adherence to the Ten Commandments and the Golden Rule ("So whatever you wish that men would do to you, do so to them": Matthew, 7:12) were the conditions necessary to achieve goodness. The influence of Judeo-Christian ethical beliefs and practices diminished during the Renaissance, and ethics began shifting from an individual responsibility to obedience to authority and tradition. This led to the development of modern secular ethics and eventually to legislation based on the political and civic duties of people as envisaged in ancient Roman law. Thomas Hobbes, an English philosopher, argued in 1651 that an organized society and political power were the most important instruments for regulating conduct. Since that time, other philosophers have theorized that a system of ethics could be formulated from various concepts, among which are the following:

- Human needs and interests should determine what is right and wrong.
- A rational system of ethics should be based on the laws of nature.
- That which provides the greatest happiness for the greatest number is morally right.
- Morality is the result of habits acquired by humans during evolution.
- Mutual aid and cooperation among people are natural and further the survival of the species.

16.4.2 Current Influences

What are the major influences on behavior today? Television, particularly with young viewers, may be having the single most serious effect. Education, religion, and legislation are still factors, but the emphasis seems to have shifted, with the school and church being less important in establishing moral attitudes and legislation taking on a larger role. In effect, individual responsibility (as emphasized by the school and church) has decreased and group obligations (as a result of government regulations) have increased.

Individually and collectively, we, as a people have obtained our ethical approach to life from the three traditional sources as well as from our families and other people. These influences and the 2,500 years of ethical philosophy behind us have provided guidelines for acceptable behavior among individuals, groups, corporations, governments, etc., when dealing with one another. However, where society's activities affect the environment, we have little historical precedent regarding the moral obligations of the various sectors of society to protect the environment. The "polluter pays principle" is probably the closest we have come to a philosophical approach.

Concern about our deteriorating environment is a recent phenomenon, triggered by the increasing urbanization and industrialization of the last 30 years. In the name of progress, we have allowed such gross insults to the environment as inadequate treatment and disposal of hazardous wastes, a reduction in volume of the ozone layer, and acid and toxic rain, with the resulting sterility of lakes and damage to forests and human health. Why has society tolerated and even encouraged unfettered industrialization and development that have brought about environmental degradation? The blame usually falls on religion, government, and technology. The arguments go something like this:

Religion. Judeo-Christian religions have taught that God gave man dominion over all the earth, and, the taming of nature and the exploitation of resources have been regarded as ethical pursuits by most of society (except for those native populations that revere and protect nature).

Government. In a capitalistic system, self-interest (and even greed) are regarded as virtues that encourage competition, which reduces the cost of goods and services and creates more production, which in turn means more jobs, which then means more buying power—and so the treadmill gains momentum. The economy really rolls (but at what cost to the environment?).

Technology. The advances in science have made our high level of development possible. (However, they have created wastes which the environment cannot assimilate.)

Can we really blame institutions like religion and government or hide behind words like "technology"? A full understanding of Judeo-Christian religions involves the concept of stewardship. This important tenet, properly understood and practiced, would prevent the spoiling of the environment. As for government, the condition of the environment in state-controlled economies, such as China, is far worse than in our free-enterprise society, so the "system" isn't to blame. And what about technology? Primitive societies in Africa have been guilty of stripping the land until it no longer sustained vegetation and so became desert. Thus, modern technology by itself is not the culprit.

In fact, it can be argued that religion, government, and technology are all positive factors in our efforts to protect the environment. It is *our* efforts that must be emphasized. The institution doesn't think, choose, or make decisions. The people in them do that, and in our democratic society we all have this opportunity. Our environment will be only as good as the people in it.

16.4.3 The Problem Is People

Early in this book we saw that environmental problems were due to people: their numbers and their activities. The irony is that these problems can be remedied only by the actions of these same people, actions that are directed by a deep appreciation of the benefits and limitations of our environment and a sense of responsibility for preserving it. Since we are dealing with a moral issue (i.e., lives are endangered), a strong moral stance is required of everyone if our global home and its inhabitants are to survive.

The urge to "beat the system" is a human characteristic that, to many, is the key to prosperity in life. If ethics inhibits the quest for success, morality becomes more flexible.

The following essay from an Ann Lander's column (December 27, 1986) illustrates the progression from minor transgressions to major misdemeanors.

Everybody Does It!

When Johnny was six years old, he was with his father when they were caught speeding. His father handed the officer a five-dollar bill with his driver's licence. "It's OK, son," his father said as he drove off. "Everybody does it."

When he was eight, he was permitted to sit in a family seminar, presided over by Uncle George, on how to shave points off an income tax return. "It's OK, kid," his uncle said. "Everybody does it."

When he was nine, his mother took him to his first theatre production. The box office man couldn't find any seats until his mother discovered an extra two dollars in her purse. "It's OK, son," she said. "Everybody does it."

When he was 12, he broke his glasses on the way to school. His Aunt Francine convinced the insurance company that they had been stolen and collected $27. "It's OK, kid," she said. "Everybody does it."

When he was 15, he made right guard on the high school football team. His coach showed him how to block and at the same time grab the opposing end by the shirt so the official couldn't see it. "It's OK, kid," the coach said. "Everybody does it."

When he was 16, he took his first summer job at the neighborhood supermarket. His assignment was to put overripe tomatoes in the bottom of the boxes and the good ones on top where they would show. "It's OK, kid," the manager said. "Everybody does it."

When he was 18, Johnny and a neighbor applied for a college scholarship. Johnny was a marginal student. His neighbor was in the upper three per cent of his class but he couldn't play right guard. Johnny got the scholarship. "It's OK," they told him. "Everybody does it."

When he was 19, he was approached by an upper class man who offered the test answers for three dollars. "It's OK, kid," he said. "Everybody does it."

Johnny was caught and sent home in disgrace. "How could you do this to your mother and me?" his father asked. "You never learned anything like this at home."

If there's anything the adult world can't stand, it's a kid who cheats.

Examples of unethical behavior in the corporate world are not hard to find, either. A few taken from newspaper clippings are briefly described.

Third-world exploitation. Central America has become the dumping ground for pesticides that North American and European chemical companies are banned from selling or, in many cases, even producing, in their own countries. One example is dibromochlorpropane (DBCP), which renders men sterile. After its manufacture and sale were prohibited in the U.S. in the late 1970s, millions of pounds were imported into Costa Rica for use as a wormicide on banana plantations. An estimated 2,000 Costa Rican workers were rendered sterile for life, some of whom have filed suits in U.S. courts against Dow Chemical and Shell Oil, the manufacturers of the pesticide. Evidence indicated that these manufacturers knew as early as the 1950s of the pesticide's dangerous properties. (Todd, 1987)

The sale of Nestle's baby formula in the 1970s to African mothers, who had no way of sterilizing water to mix with it, is another example of questionable corporate behavior since it caused the deaths of thousands of babies.

The exploitation of Third-World countries for raw materials produced by low-paid workers for whom no regulations exist covering working hours or health hazards is a continuing ethical problem.

Violation of environmental standards.   The senior management of Hooker Chemical and Plastics Corporation (now Occidental Chemical Corp.) knew in 1975 that their plant in Lathrop, California, was polluting underground water supplies with toxic pesticides. They also knew that their plant in White Springs, Florida, was seriously violating air pollution limits in smokestack emissions. The violations were never reported to any state or federal authorities, and it wasn't until pollution from another Hooker source, the Love Canal in Niagara Falls, New York, reached major proportions that the truth began to emerge. (Sagel, 1985)

There are 164 chemical waste dumps within 3 miles of the Niagara River and toxic contaminants from four of them are believed to have leaked directly into the river.

Health and safety vs. cost.   In 1973, Thomas Robertson, director of development at Firestone Tire in the United States, sent top management a memo. "We are making an inferior quality radial tire that will subject us to belt-edge separation at high mileage," it warned. That tire was the Firestone 500.

His advice was ignored. Instead, management kept producing the tire and sold more than 24 million in the next five years. Despite incidents of blowouts and separations, management kept issuing statements that the 500s were completely safe. By 1979, blowouts were said to have caused a number of deaths and serious injuries. The company replaced millions of the tires and settled a large number of injury cases. (Sagel, 1985)

Two other even more familiar examples of cost taking precedence over health and safety can be found in the automotive and asbestos industries. The automobile was the Ford Pinto, and its gas tank was located so close to the back of the car that a rear-end collision often resulted in fire and explosion. Senior officials of Ford apparently knew of this defect before the car went into production in 1971, but decided it would cost less to compensate the families of the killed and injured than to modify the car's design. (Sweet, 1986)

The scandal of asbestos is that the industry was aware of the harmful effects of asbestos in the 1930s and 1940s, long before the mid-1960s, when it claimed to have first learned of the serious health hazards. As a result of deaths and disabling diseases contracted by workers and others exposed to asbestos, more than 1,000 lawsuits have been filed against asbestos companies in the United States and Canada. (Sagel, 1985)

16.4.4 Finding a Solution

Even though it is difficult to legislate ethical behavior, we must keep trying. At the same time, we need to counteract the philosophy that self-interest is an admirable trait in all business dealings. In the early 1980s, after Ivan Boesky had amassed $100 million by inside trading on the stock market (and before his fall from grace), he spoke to the students at the Harvard Business School. He told them matter-of-factly that "Greed is healthy. You can be greedy and feel good about yourself." Vigorous applause was the response from the students.

It is not surprising that the attitudes of business are carried over into other aspects of life, including environmental protection. Projects involving public health and safety cannot be completely free of risk, so the question of how much should be spent to save a human life becomes a consideration. Researchers from Tufts and Harvard Universities and Oak Ridge National Laboratory in Tennessee found that agencies of the U.S. federal government seemed to use a figure of $2 million per life saved in enacting legislation regulating cancer-causing chemicals (Travis et al., 1987). If the cost per life saved was higher than this, the chemical was not regulated. But we can't always put a price on a human life or such intangibles as aesthetics, quality of life, and enjoyment of nature; ethics has to play a role. At the Fourth Conference on Environmental Engineering Education in 1981, Lord Eric Ashby noted that we would still have slavery if the decision had been based on a cost-benefit analysis. Instead, one man made the ethical decision that slavery should be abolished despite the inevitable consequences he knew would ensue.

How do people acquire moral behavior? As noted earlier, the decreasing role of the school and the church in influencing ethical behavior means that other influences—the home, friends, entertainment—will have a greater effect in establishing behavior patterns. Television and movies provide conflicting lessons about life. Morality, ethics, integrity, and other old-fashioned virtues are less prominent these days, and concealment, misinformation, and dishonesty have become more common, representing, no doubt, attitudes of the times. The distinction between good and evil, as well as between right and wrong, has become even less clear than it used to be. Business recruiters have reported finding a morally indifferent attitude in many of the graduates from today's schools. What has happened to the idealism and altruism of the 1960s?

The picture of ethical behavior that has been presented in this section is a disheartening one, but there are hopeful signs. In 1987, John Shad, a former chairman of the American Securities and Exchange Commission, gave Harvard Business School $20 million for the initiation and support of a program of ethics. Also in 1987, in Switzerland, more than 1,000 top European corporate managers attended lectures by the Greenpeace Foundation, an environmental advocacy group. Many engineering students, like those at the University of Toronto, now take a mandatory course on the philosophical concepts of right and wrong and the impact of their technology on the human environment.

Another development in the engineering profession that should curtail unethical behavior in the corporate world is the introduction of whistleblowing legislation. Whistleblowers are employees who report on an organization's illegal, dangerous, or unethical conduct, generally after an unsuccessful attempt at having the situation corrected internally. Usually, in this situation, the employee is fired. In 1981, Michigan became the first jurisdiction anywhere in North America to enact a "Whistleblowers Protection Act." The legislation arose from Michigan's PBB (polybrominated biphenyls) tragedy of the mid-1970s. In that incident, the Michigan Chemical Co. shipped a poisonous fire retardant to feed-grain cooperatives instead of the intended nutritional supplement. Thousands of cattle died, and the health of the residents was seriously damaged by eating PBB-contaminated food. Employees were warned by the company not to provide information to the investigators, or they would be fired. Because of Michigan's pioneering legislation, similar measures are under consideration by other jurisdictions and self-disciplinary professional bodies (Sagel, 1985).

16.4.5 Conclusions

Education, legislation, and corporate attitudes are important factors in shaping morality, but ethical behavior is still largely the responsibility of individuals. It is up to each of us, student, parent, environmental scientist, and engineer, as responsible citizens of the world, to recognize unethical conduct when we see it, and blow the whistle when necessary. An old saying applies here:

> All that is necessary for the triumph of evil is that good men do nothing.

According to philosophers, our concept of morality is developed early in life—before the age of 10, some claim. This means that the early years of childhood, while the family is still the major influence, are critical in establishing proper moral standards. We have an obligation to teach the young, by example where possible, about the ethical problems they will encounter in life and what responsibilities they must accept for their actions. They don't need to be ruthless, aggressive, competitive, profit-motivated individuals to succeed. In fact, mathematician Anatol Rapoport, director of the Institute for Advanced Studies in Vienna and formerly a professor at the University of Michigan, maintains that cooperation, sharing, and following the golden rule lead to success in life. The notion that life is a zero-sum game (for every winner, there must be a loser) is false, he says. Life is a mixed-motive game in which people's interests partly coincide and partly conflict. His concept is illustrated with a simple game called The Prisoner's Dilemma, in which the players, to get what they want, have to cooperate. They must trust each other consistently and be prepared to share the rewards. The principle, he believes, demonstrates an important moral lesson that has application to life in many areas, from parental training to nuclear disarmament, and presumably to environmental protection.

This chapter and book end with a Sioux prayer (*The United Church Observer*, vol. 50, No. 4, October, 1986, p. 48) in which the creation-centered values of native people contrast sharply with the modern world in which nature has been subdued, the land overcultivated, the cities overpopulated, and technology geared to minimizing costs:

Ho! Great Spirit, Grandfather, you have made everything and are in everything. You sustain everything, guide everything, provide everything, and protect everything, because everything belongs to you. I am weak, poor and lowly, nevertheless, help me to care, in appreciation and gratitude to you and for everything.

I love the stars, the sun and the moon, and I thank you for our beautiful mother, the

Earth, whose many breasts nourish the fish, the fowls and the animals, too. May I never deceive Mother Earth; may I never deceive my people, may I never deceive myself, and above all may I never deceive you.

—Sioux Prayer

PROBLEMS

16.1. As either a group or individual assignment, search the files of your local newspaper over the past year for articles related to environmental impact assessment. List the types of projects encountered.

16.2. Select one of the projects encountered in Problem 16.1, and prepare a 1,000-word summary statement on the project, the environmental impact assessment, and the actual results of the project. If available, obtain a copy of the EIS and other relevant material from the organizations involved.

16.3. Draw a flowchart showing the major role players and the main steps to be taken in carrying out the EIA process in your state or province.

16.4. The EIA process can be time consuming and costly to the proponent of the project. What are the possible benefits of having to carry out an EIA to the proponent?

16.5. In an article on pollution in an economics journal, the phrase "internalize the external costs" is used. What does this mean?

16.6. Costs for pollution abatement incurred by the public at large are sometimes referred to as social costs. List as many other social costs as you can think of.

16.7. The term "zero discharge" has been frequently used in regards to pollution in the recent past, particularly in the United States. Why is insistence on "zero discharge" not practical in most instances?

16.8. Why is direct government involvement necessary for the control of environmental quality?

16.9. A number of instruments for environmental control have been introduced in this chapter. Which of these methods, or combinations thereof, or schemes of your own (describe) would you choose for the control of the following pollution problems?

- Motor vehicle exhausts
- Pulp mill effluents and exhausts
- Noise and dust conditions from a quarry located close to a village
- Operation of a major airport close to a metropolitan area.

State the reasons for your choice(s), and briefly explain how the system might work.

16.10. Read the article by Schindler (1976), and comment on it briefly.

16.11. A tannery is intending to locate in a town of 10,000 people. The industry is asking the town council how much it would cost to have its industrial wastes treated at the local treatment plant. The industry proposes to construct an equalization tank on its land so that it can release wastes of uniform quantity and quality to the town sewer. The following information is available:

	Town	Tannery
Daily flow, m³/day	4,000	1,000
Organic strength, mg/L BOD₅	200	1,000
Cost of treatment operations per kg BOD₅ removed	$0.25	N.A.
Cost per m³ of town sewage treated	$0.06	N.A.

The town's treatment plant has some excess capacity, but will need to be extended. The industry has agreed to pay for the capital cost of expansion. Your job is to devise a formula for sharing operating costs and advise the industry of the likely annual costs.

16.12. You are the administrator of pollution control in your regional municipality. One of your field supervisors reports the following case to you and asks for instructions on how to proceed:

> Hijinks Industry, Inc., is located on the Sweetwater River. The industry has a permit to withdraw sufficient water from the river for its needs and to discharge its wastes to the river, provided that the concentration of pollutant X, the major contaminant in this industrial waste, does not exceed 10 mg/L, as required by your agency. The industry's waste treatment plant has had a history of exceeding this effluent standard. Recently, it has installed new pumping facilities and is pumping water from the river, mixing it with the plant effluent, thereby lowering the concentration to less than 10 mg/L. It claims it is thus meeting the agency's guidelines.

What will you tell your field supervisor?

16.13. Find, and properly reference, one quotation from any religious writing that sets out our responsibilities as stewards of the earth.

16.14. Discuss how one of religion, free enterprise, and technology can be positive factors in our efforts to protect the environment.

16.15. For a one-week period, find from your local newspaper, as many examples as you can of unethical human behavior. List these in order of decreasing severity, and indicate in each case what you think the proper response should have been.

16.16. There are those who contend that whistleblowing is incompatible with being a professional. As examples, they cite priests, lawyers, and doctors who protect the confidentiality of information obtained from their clients. Discuss how this view can be reconciled with whistleblowing, or, if you feel it cannot, then suggest which practice a professional should follow and why.

REFERENCES

BEANLANDS, G. E., and DUINKER, P. N. *An Ecological Framework for Environmental Impact Assessment in Canada.* Halifax, N.S.: Dalhousie University, Institute for Resource and Environmental Studies, 1983.

BURROWS, P. *The Economic Theory of Pollution Control.* Cambridge, Mass.: M.I.T. Press, 1980.

CANTER, L. W. *Environmental Impact Assessment.* New York: McGraw-Hill, 1977.

CEARC. *Cumulative Environmental Effects: A Binational Perspective.* Ottawa: Environment Canada, CEARC/FEARO, 1986.

CLARKE, R. O., and LIST, P. C. *Environmental Spectrum, Social and Economic Views on the Quality of Life,* New York: D. Van Nostrum Company, 1974.

CONWAY, G. R., and O'CONNELL, P. E. *The Analysis of the Environmental Component of Water Resource Management Projects*. London: University of London, Imperial College Centre for Environmental Technology, 1978.

CRABTREE, P. "Niagara River—An International Challenge." Paper presented at the Ontario Ministry of Environment Seminar, Toronto, November 3, 1982.

CRABTREE, P. "Niagara River—The Complexities of Clean-Up." Paper presented at the 30th Industrial Waste Conference at Toronto, June 14, 1983.

DALES, J. H. *Pollution, Property and Prices*, Toronto: University of Toronto Press, 1968.

DUFFY, P. J., and TAIT, W. S. "Canada's Policy on Environmental Assessment for Federal Activities." In *Environmental Impact Assessment*, edited by R. E. Munn, pp. 106–112. Toronto: John Wiley and Sons, 1979.

DUINKER, P. N., and BASKERVILLE, G. L. "A Systematic Approach to Forecasting in Environmental Impact Assessment." *Journal of Environmental Management* 23 (1986): 271–290.

Environment Canada, and Ontario Ministry of the Environment. *Sweetwater Seas, The Legacy of the Great Lakes*.

Environment Canada, Ontario Ministry of the Environment, U.S. Environmental Protection Agency, and New York State Dept. of Environmental Conservation. *Report of the Niagara River Toxics Committee (NRTC)*. Toronto: Niagara River Toxics Committee, October 1984.

GRIMA, A. P., TIMMERMAN, P., FOWLE, C. D., and BYER, P. *Risk Management and EIA: Research Needs and Opportunities*. Ottawa: Environment Canada, CEARC/FEARO, 1986.

HOLLING, C. S. (ed.), *Adaptive Environmental Assessment and Management*, Toronto: John Wiley and Sons, 1978.

LEOPOLD, L. B., CLARKE, F. E., HANSHAW, B. B., and BALSEY, J. R., *A Procedure for Evaluating Environmental Impact*, U.S. Geological Survey Circular 645, Government Printing Office, Washington, D.C., 1971.

MACNEILL, J. W., *Environmental Management*, Information Canada, Cat. No. CP 32-12/1971, Ottawa, 1971.

MUNN, R. E., ed. *Environmental Impact Assessment*, Scope 5, Toronto: Wiley, 1975, 1979.

Ontario Ministry of the Environment. *Water Management: Goals, Policies, Objectives and Implementation Procedures of the Ministry of the Environment*. Toronto: Ontario Ministry of the Environment, 1978.

ORTOLANO, L. *Environmental Planning and Decision Making*. New York and Chichester, England: John Wiley and Sons, 1984.

PADC (Project Appraisal for Development Control). *Environmental Impact Assessment*. Boston and The Hague: Martinus Nyhoff Publishers, 1983.

SAGEL, J. F. "Time for a Whistleblower Law." *Toronto Globe and Mail*, January 14, 1985.

SCHINDLER, W. "The Impact Statement Boondoggle." *Science* 192 No. 4239 (1976): 509.

SWEET, L. "Corporate Criminals Must Pay." *Toronto Daily Star*, April 14, 1986.

TODD, DAVE. "Central America Still Uses Banned Pesticides." *Toronto Daily Star*, July 8, 1987.

TRAVIS, C. C., RICHTER, S. A., CROUCH, E. A. C., WILSON, R., and KLEMA, E. D., "Cancer Risk Management." *Environmental Science & Technology* 21, No. 5, (1987): 415–420.

WESTMAN, WALTER. *Ecology, Impact Assessment and Environmental Planning*. New York: Wiley Interscience, 1985.

APPENDIX A

Symbols, Dimensions, and Units

A **symbol** is a specific designation, such as a letter of abbreviation, chosen to represent a certain item. For example, the letter **d** may be chosen to designate **diameter, A** to denote **area,** and **V** to indicate **volume.**

Dimension is a term given to the three basic units, mass [M], length [L], and time [t] in which all physical units may be expressed. In some cases, temperature [T] and amount of substance [mol] are also included as basic dimensions. For example, the dimensions of the items mentioned in the first paragraph are: d (diameter) $=$ [L], A (area) $=$ [L^2] and V (volume) $=$ [L^3]. The symbol $=$ [ ] means "has dimensions of." No matter which system of units is used—the SI, the American engineering system, the British system, or the metric system—the dimensions on each side of a complete equation must be the same. It is therefore very useful to indicate the dimensions of each term in engineering problems to ensure that each item has the correct dimensions.

Units are specific standards of measurements within a given system, such as the *Système International D'Unités* (*SI*) or the *American Engineering System* (*AES*). For example, the aforementioned items could have the following units:

Item	SI units	AES units
Diameter, d	Meter (m)	Feet (ft)
Area, A	Square meter (m^2)	Square feet (ft^2)
Volume, V	Cubic meter (m^3)	Cubic feet (ft^3)

It is very important to be consistent in the use of symbols and units. Even within a discipline, great inconsistencies may occur, causing unnecessary misunderstandings and confusion. In an applied field such as environmental engineering, which draws on several scientific and engineering disciplines, the problem is compounded. In this book, the symbols used consistently are those common in the environmental field.

APPENDIX A–1: INTERNATIONAL SYSTEM OF UNITS (SI)

The Système International D'Unités (SI) was established by the General Conference of Weights and Measures in 1960. It is popularly known as the metric system, although certain modifications of the metric system were made for the SI system. The metric system has been used for a long time in Europe and many parts of the world. The British system of measurement has been used in Britain and its former colonies throughout the world, including North America. In the United States, a variation of the British system referred to as the American Engineering System (AES) has been adopted. An example of a unit in the AES is the U.S. gallon, which is smaller than the Imperial gallon of the British system. The governments of many nations which have not used the metric system in the past have decided to switch to the SI system, because of the advantages of using a unified system in international trade and commerce. In many countries, this changeover is now occurring and causing considerable, but surmountable difficulties.

In science and in some technological fields, the metric system has been used for many years, even in North America. Therefore, this changeover is more difficult to make in some fields than in others. For example, the building industry will take much longer to adapt to the SI system than the electrical industry, because of past practice. For the education of scientists and engineers in North America, the SI system was introduced in the 1970s. In many disciplines the young graduate will have to be familiar with both systems for many years to come. Because there is so much valuable literature in the American Engineering System, the decision was made to use the SI system in this book, but to provide the AES equivalent wherever it was sensible. Illustrative examples and problems to be worked out by the student are given in one or the other system, to provide practice in each and to encourage conversion from one to the other. (Appendix A–3 provides information on conversions for units often used in environmental engineering.) There is a price to be paid for the decision to use both systems. One of the things you must learn from working through this book is a feel for realistic values for many items. For example, how much water do people require for their daily use? The answer is about 100 gallons, or 380 L. Obviously, with many new "facts" to learn, it is more difficult to do so in two systems, and therefore there is greater likelihood for error. Fortunately, for quite a few areas in the environmental

TABLE A–1.1 SI BASE AND SUPPLEMENTARY UNITS WITH EQUIVALENT UNITS
IN THE AES

Quantity	Symbol for quantity	SI		AES	
		Name of unit	Symbol for unit	Name of unit	Symbol for unit
Basic Units					
Length	L	Meter	m	Foot	ft
Mass	M	Kilogram	kg	Pound mass*	lb$_m$
Time	t	Second	s	Second	s
Electric current	I	Ampere	A	Ampere	A
Thermodynamic temperature	T	Kelvin	K	Rankin (Fahrenheit)	°R (°F)
Amount of substance	$\mathcal{M}$	Mole	mole	Mole	mol
Luminous intensity		Candela	cad	Foot candles	ft-candles
Supplementary units					
Plane angle		Radian	rad	Degrees	°
Solid angle		Steradian	sr	Degrees	°

* As distinct from the pound force (lb$_f$) as the unit of force.

field, the metric system has been used for a long time, so measurements in those areas are often only in the metric system. An example is the concentration of pollutants, which is usually reported in mg/L.

The SI system is based upon seven fundamental or base units and two supplementary units. These units, whose definitions and symbols have been internationally agreed upon, appear in Table A–1.1. The equivalent units in the AES are also shown. Selected SI derived units frequently used in this book are shown in Table A–1.2.

A further characteristic which typifies the building-block approach of the SI system is that a set of 16 prefixes has been adopted in order to form multiples and submultiples of any base or derived unit. These units may thus be changed from very small to very large measurements by the prefixes. The SI prefixes and their symbols appear in Table A–1.3. Some common examples are the following:

One hundredth of a meter (10^{-2} m) = 1 centimeter (cm)

One thousand meters (10^3 m) = 1 kilometer (km)

One thousandth of a gram (10^{-3} g) = 1 milligram (mg)

One millionth of a gram (10^{-6} g) = 1 microgram (μg)

One millionth of a meter (10^{-6} m) = 1 micrometer (μm) (formerly 1 micron (μ))

A number of units are not part of the SI system, but due to their widespread use, it would be impractical to discard them. Some of these appear in Table A–1.4.
Some additional comments on the units in the table may be helpful.

TABLE A-1.2 SELECTED SI DERIVED UNITS

Quantity			SI units		
Name	Symbol	Dimension	Name of unit	Symbol	Expression in terms of other SI units (where appropriate)
Area	A	L^2	Square meter	m^2	
Volume	V	L^3	Cubic meter	m^3	
Velocity	u	L/t	Meter per second	m/s	
Acceleration	a	L/t^2	Meter per second squared	m/s^2	
Force	F	M/t^2	Newton	N	$kg \cdot m/s^2$
Pressure	P	M/Lt^2	Pascal	Pa	$N/m^2 = kg/m \cdot s^2$
Work, Energy, Heat	—	$M\,L^2/t^2$	Joule	J	$N \cdot m = kg \cdot m^2/s^2$
Power	P	$M\,L^2/t^3$	Watt (joule per second)	W	$J/s = kg \cdot m^2/s^3$
Density	ρ	M/L^3	Kilogram per cubic meter	kg/m^3	
Specific volume	$v = 1/\rho$	L^3/M	Cubic meter per kilogram	m^3/kg	
Concentration (of amount of substance)	C	mol/L^3	Mole per cubic meter	mol/m^3	
Flowrate	Q	L^3/t	Cubic meter per second	m^3/s	
Dynamic viscosity	μ	M/Lt	Pascal second	$Pa \cdot s$	$kg/m \cdot s$
Kinematic viscosity	$\nu = \left(\dfrac{\mu}{\rho}\right)$	L^2/t	Meter squared per second	m^2/s	
Surface tension	—	M/t^2	Newton per meter	N/m	kg/s^2
Heat capacity	—	$M\,L^2/t^2\,T$	Joule per Kelvin	J/K	$kg \cdot m^2/s^2 \cdot K$
Specific heat capacity	—	$L^2t^2\,T$	Joule per kilogram Kelvin	$J/(kg \cdot K)$	$m^2/s^2 \cdot K$
Specific energy (enthalpy)	—	L^2/t^2	Joule per kilogram	J/kg	$m^2/s^2 \cdot kg$
Thermal conductivity	—	$ML/t^3\,T$	Watt per meter Kelvin	$W/m \cdot K$	$m/s^3 \cdot K$
Molar energy	—	$M\,L^2/t^2\,mol$	Joule per mole	J/mol	$kg \cdot m^2/s^2 \cdot mol$
Molar heat capacity	—	$M\,L^2/t^2\,mol\,T$	Joule per mole Kelvin	$J/mol \cdot K$	$kg \cdot m^2/s^2 \cdot mol \cdot K$

TABLE A–1.3 SI PREFIXES

Factor	Prefix	Symbol for prefix
10^{18}	Exa	E
10^{15}	Peta	P
10^{12}	Tera	T
10^9	Giga	G
10^6	Mega	M
10^3	Kilo	k
10^2	Hecto	h
10^1	Deka	da
10^{-1}	Deci	d
10^{-2}	Centi	c
10^{-3}	Milli	m
10^{-6}	Micro	μ
10^{-9}	Nano	n
10^{-12}	Pico	p
10^{-15}	Femto	f
10^{-18}	Atto	a

TABLE A–1.4 ADDITIONAL UNITS IN USE WITH SI

Unit name	Unit symbol	Value in SI units
Minute (time)	min	$1 \text{ min} = 60 \text{ s}$
Hour (time)	hr	$1 \text{ hr} = 60 \text{ min} = 3{,}600 \text{ s}$
Day (time)	d	$1 \text{ d} = 24 \text{ hr} = 86{,}400 \text{ s}$
Degree (angle)	°	$1° = \pi/180 \text{ rad}$
Minute (angle)	′	$1' = 1°/60 = \pi/10{,}800 \text{ rad}$
Second (angle)	″	$1'' = 1'/60 = \pi/648{,}000 \text{ rad}$
Liter	L*	$1 \text{ L} = 1 \text{ dm}^3 = 10^{-3} \text{ m}^3$
Metric ton (tonne)	t	$1 \text{ t} = 10^3 \text{ kg}$
Hectare	ha	$1 \text{ ha} = 10^4 \text{ m}^2$

* The symbol for liter is actually lowercase "l," which can easily be mistaken for the numeral "1." For this reason, the symbol "L" has been adopted for use in the United States and is used in this book.

Weight vs. Mass.   The term "mass" in the SI system is used to specify the quantity of matter *in* an object, whereas the term "weight" denotes the gravitational force acting *on* a material object. Therefore, the mass of an object is independent of its location, and conversely, weight (or force) can change with location.

The unit of mass is the kilogram (kg), and the unit in which weight or force is measured is the newton (N). One N equals $1 \text{ kg} \cdot \text{m/s}^2$, and quantitatively, the weight or force of an object is its mass (kg) times its gravitational acceleration g (m/s²). At sea level, $g = 9.80665 \text{ m/s}^2$.

The definitions of the units of mass and weight lead to the common SI quantities of density and unit weight. The *density* of an object is its mass per unit volume (kg/m³), and the unit weight of an object is its gravitational force per unit volume (N/m³).

The kilogram is the base unit of mass in the SI system and from Table A–1.3, 1,000 kilograms, would theoretically be 1 kilokilogram. However, double prefixes are not used in SI, and hence the correct expression for 1,000 kilograms is a megagram (1 million grams), or more simply, a metric ton (tonne).

Volume. The preferred unit of volume for engineering purposes is the cubic meter (m³), because quantities of concrete and earthwork were customarily expressed in cubic yards, which are of the same order of magnitude as the cubic meter. On the other hand, the liter is commonly used to measure liquid volumes, which are expressed in pints, quarts, or gallons. Multiples or submultiples of the liter, such as the megaliter (ML) and milliliter (mL) are acceptable measurements, replacing the quantities of million gallons (1 mgd $\approx$ 3.785 ML/d) and fluid ounces, (1 oz $\approx$ 29.5 mL) respectively.

Area. The basic unit of area is the square meter (m²). However, for larger measurements, the square hectometer called a hectare (ha) is commonly used. The hectare is of a similar order of magnitude as the acre (2.47 acres $\approx$ 1 ha). For larger area measurements, the square kilometer, which is about 250 acres, is used. The preferred unit for smaller measurements is the square millimeter (mm²) or square centimeter (cm²).

Temperature. The SI unit for thermodynamic temperature is the Kelvin (K). For practical use, temperatures are measured on the Celsius (°C) scale. One degree Kelvin is equal to one degree Celsius. The Celsius scale has 100 equal divisions between the freezing and boiling points of water, which are given the temperatures 0°C and 100°C, respectively. The simple conversion from Kelvin to degrees Celsius is given by

$$\text{Celsius temperature} = T - T_0$$

where T is the thermodynamic temperature in Kelvin and $T_0 = 273.15$ K.

APPENDIX A–2: CONVERSION PROBLEMS

For those unfamiliar with the SI system, a few examples, followed by a number of problems for converting measurements, are provided.

Example

Convert the following measurements from SI to the American Engineering System:

(a) Distance = 20 km

(b) Area = 10 ha

(c) Flowrate = 15 m³/s

(d) Temperature = 20°C

(e) Mass = 70 kg

(f) Force = 100 N

(g) Pressure = 100 kPa

(h) Density = 1,000 kg/m³

(i) Work = 200 J

(j) Power = 400 kW.

The quickest, but also the least informative, way of converting measurements is to use an SI-to-AES special calculator or to look in Appendix A or similar tables for the proper conversion factors. We shall do the problems by converting the basic units from SI to AES using a brief "bookkeeping" method. For routine calculations, the use of a conversion calculator or the appropriate tables in Appendix A–3 is sensible.

(a) Distance = 20 km = ? miles

The basic unit of length in SI is the meter (m), and in AES it is the foot (ft). The conversion is 1 ft = 0.3048 m. We thus have:

$$\frac{20.0 \text{ km}}{} \left| \frac{1000 \text{ m}}{\text{km}} \right| \frac{1 \text{ ft}}{0.3048 \text{ m}} \left| \frac{1 \text{ mile}}{5280 \text{ ft}} \right. = 12.4 \text{ miles}$$

(b) Area = 10 ha = ? acres, where 1 ha = 10,000 m² (100 m × 100 m)

$$\frac{10 \text{ ha}}{} \left| \frac{10,000 \text{ m}^2}{\text{ha}} \right| \frac{1 \text{ ft}^2}{(0.3048)^2 \text{ m}^2} \left| \frac{1 \text{ acre}}{43,560 \text{ ft}^2} \right. = 24.7 \text{ acres}$$

(c) Flowrate = 15 m³/s = ? ft³/s

$$15.0 \frac{\text{m}^3}{\text{s}} \left| \frac{1 \text{ ft}^3}{(0.3048)^3 \text{ m}^3} \right. = 530 \text{ ft}^3/\text{s}$$

(d) T = 20°C = ? °F (Note: Base for °C = 0; Base for °F = 32)

$$\frac{20°\text{C}}{} \left| \frac{9°\text{F}}{5°\text{C}} \right. + 32°\text{F} = 68°\text{F}$$

(e) Mass = 70 kg = ? lb$_m$

$$\frac{70 \text{ kg}}{} \left| \frac{1 \text{ lb}_m}{0.454 \text{ kg}} \right. = 154 \text{ lb}_m$$

*(f) Force = 100 N = ? lb$_f$, where 1 N = kg · m/s²

$$100 \frac{\text{kg·m}}{\text{s}^2} \left| \frac{1 \text{ lb}_m}{0.454 \text{ kg}} \right| \frac{1 \text{ ft}}{0.3048 \text{ m}} \left| \frac{1}{32.174 \frac{\text{lb}_m \cdot \text{ft}}{\text{lb}_f \text{ s}^2}} \right. = 22.5 \text{ lb}_f$$

*(g) Pressure = 100 kPa = ? lb$_f$/ft², where 1 Pa = 1 N/m²

$$100,000 \frac{\text{N}}{\text{m}^2} \left| \frac{1 \frac{\text{kg m}}{\text{s}^2}}{1 \text{ N}} \right| \frac{1 \text{ lb}_m}{0.454 \text{ kg}} \left| \frac{1 \text{ ft}}{0.3048 \text{ m}} \right| \frac{1}{32.174 \frac{\text{lb}_m \text{ ft}}{\text{lb}_f \text{ s}^2}} \left| \frac{(0.3048)^2 \text{ m}^2}{1 \text{ ft}^2} \right. = 2087 \frac{\text{lb}_f}{\text{ft}^2}$$

$$= 14.49 \frac{\text{lb}}{\text{in}^2}$$

(h) Density $= 1,000$ kg/m³ $= ?$ lb$_m$/ft³

$$1000 \frac{kg}{m^3} \left| \frac{1 \, lb_m}{0.454 \, kg} \right| \frac{(0.3048)^3 \, m^3}{1 \, ft^3} = 62.4 \, \frac{lb_m}{ft^3}.$$

*(i) Work $= 200$ J $= ?$ ft·lb$_f$, where 1 J $= 1$ N·m

$$\frac{200 \, N \, m \left| 1 \frac{kg \, m}{s^2} \right| \frac{1 \, lb_m}{0.454 \, kg} \left| \frac{1 \, ft}{0.3048 \, m} \right|}{1 \, N} \left| \frac{1 \, ft}{32.174 \, \frac{lb_m \, ft}{lb_f \, s^2}} \right| \frac{1 \, ft}{0.3048 \, m} = 147 \, \text{ft-lb}_f$$

*(j) Power $= 400$ kW $= ?$ ft·lb$_f$, where 1 W $= 1$ J/s $= 1$ N·m/s

$$400,00 \frac{N \, m}{s} \left| \frac{1 \frac{kg \, m}{s^2}}{1 \, N} \right| \frac{1 \, lb_m}{0.454 \, kg} \left| \frac{1 \, ft}{0.3048 \, m} \right| \frac{1 \, ft}{32.174 \, \frac{lb_m \, ft}{lb_f \, s^2}} \left| \frac{1 \, ft}{0.3048} \right| = 295,000 \, \frac{\text{ft-lb}_f}{s}$$

$$= \frac{295,000}{550} = 536 \, hp$$

* The conversion factor $g_c \left(32.174 \, \frac{lb_m \cdot ft}{lb_f \cdot s_2} \right)$ is required in the AES in order to have the numerical values of force and mass equal at the earth's surface.

PROBLEMS

The following problems provide practice in converting between SI and AES. Be sure to indicate the dimensions for each parameter involved.

1. The velocity of flow in a river is $u = 1.2$ ft/s. Express this in m/s.

2. The average daily water consumption per person in a small town is 250 L/d, and the maximum daily consumption is 500 L/d. The maximum hourly water consumption is 700 L/d. Calculate the equivalent values in AES.

3. The pressure in a watermain is 50 lb/in² (psi). What is the pressure in N/m² and in kPa?

4. One atmosphere is the equivalent of ____ psi and ____ kPa.

5. A buyer of a 100-Ac farm in Kansas pays $1,500 per acre. He wants to know how much this is in $/ha.

6. An oil tanker leaked 50,000 bbl of oil near a beach. If the thickness of the oil slick is 2 cm and the width about 400 m, what length of beach is affected?

7. The heat of combustion in an incinerator burning solid waste and sludge is about 1400°F. How much is this in °C and in K?

8. Requirements for irrigation water are given as 620 acre-ft/d. Express this in an appropriate SI unit.

9. Digested sewage sludge is a low-grade fuel with an energy content (fuel value) of 9,100 kJ/kg of dry solids. What is the fuel value in BTU/lb?

10. The rate of flow of water for fire fighting in a town is specified as 500 L/s for a minimum duration of 8 hr. Calculate the required rate of flow in gpm and the size of the storage reservoir in both m³ and million gallons.

11. The density and viscosity of an oil at 100°F are 55 lb/ft³ and 1.4×10^{-3} lb$_f$·s/ft², respectively. Calculate the equivalent SI values.

12. Equation 6.6 gives the terminal settling velocity of a spherical particle settling in a fluid as

$$u_t = \left[\frac{4}{3} \frac{g(\rho_p - \rho)}{C_D \rho} d_p \right]^{1/2}$$

What are the dimensions of C_D, the drag coefficient?

13. Papers submitted to most American journals must now be given in both SI and AES units, the latter in parentheses. Complete the missing information in the text below, from a recent journal submission.

> The Sarnia wastewater treatment plant has four rectangular, horizontal flow settling tanks 41.1 m long (   ), 9.2 m wide (   ), and average depth of 2.7 m (   ). The average flow rate is 35,000 m³/d (   ). Velocity measurements of the flow through the tank averaged 14 mm/s (   ) at an overflow rate of 40 m³/m² · d (   ) to 30 mm/s (   ) at an overflow rate of 115 m³/m² · d (   ).

14. The Darcy-Weisbach equation allows the calculation of friction loss for flow through circular pipes from

$$h_L = f \frac{L u^2}{2gd}$$

where h_L = frictional head loss
       f = friction factor
       L = length of pipe
       d = diameter of pipe
       u = average velocity of flow in pipe
       g = acceleration due to gravity

If the head loss has dimensions of length, what dimensions must f have for the equation to be dimensionally correct?

REFERENCES

For information regarding the International System of Units and its application, contact, in the U.S.A.,

> Office of Technical Publications
> National Bureau of Standards
> U.S. Department of Commerce
> Washington, DC 20234

and in Canada,

> Metric Commission
> 235 Queen Street
> Ottawa, Ontario,
> K1A 0H5

American Water Works Association Committee. *Final Report on Metric Units and Sizes. Journal of the American Water Works Association* 74 (1982): 27.

Canadian Standards Assn. CSA Standard Z234.1–76. *Metric Practice Guide.* Canadian Standards Association; Rexdale, Ontario, 1976.

International Standards Organization. *SI Units and Recommendations for Use of Their Multiples and of Certain Other Units, ISO 1000,* Geneva, Switzerland: International Standards Organization Central Secretariate, 1973.

National Bureau of Standards. *The Metric System of Measurement (SI).* U.S. Department of Commerce, Fed. Reg. December 10, 1976.

Units of Expression for Wastewater Treatment (With Emphasis on Metrication). WPCF Manual of Practice No. 6. October 1976.

APPENDIX A–3: FREQUENTLY USED CONVERSION FACTORS

In the following tables, the conversion from an initial or given set of units to the desired units is made by multiplying the original units by the factor given. The conversions are listed alphabetically within each category. Although most conversions from SI to AES are the same as those between SI and British units, exceptions are duly noted.

LENGTH

centimeter	× 0.3937	= in
	× 0.0328	= ft
foot	× 30.48	= cm
	× 0.3048	= m
inch	× 2.540	= cm
	× 0.0254	= m
kilometer	× 3281	= ft
	× 0.6214	= mile (statute)
meter	× 3.281	= ft
	× 39.37	= in
	× 1.094	= yd
yard	× 91.44	= cm
	× 0.9144	= m

AREA

acre	× 43,560	= sq ft
	× 4,046.85	= sq m
	× 0.404685	= ha
hectare	× 2.471	= Ac
	× 10,000	= sq m
square centimeter	× 0.155	= sq in
	× 10^2	= sq mm
	× 10^{-4}	= sq m

square foot	$\times$ 929.0304	= sq cm
	$\times$ 0.111111	= sq yd
	$\times$ 144.0	= sq in
	$\times$ 0.0929	= sq m
square inch	$\times$ 6.452	= sq cm
	$\times$ 645.2	= sq m
	$\times$ 0.006944	= sq ft
square meter	$\times$ 10.764	= sq ft
	$\times$ 1550	= sq in
	$\times$ 1.196	= sq yd
	$\times$ 10^4	= sq cm
	$\times$ 10^6	= sq mm
square yard	$\times$ 8,361.27	= sq cm
	$\times$ 1,296	= sq in
	$\times$ 0.836127	= sq m

VOLUME

acre foot	$\times$ 43,560	= cu ft
	$\times$ 1,233.49	= cu m
	$\times$ 1,613.3	= cu yd
	$\times$ 271,325.7	= gal (Imperial)
	$\times$ 325,851	= gal (U.S.)
barrel (petroleum)	$\times$ 42	= gal (U.S.)
	$\times$ 158.983	= L
cubic centimeter	$\times$ 0.061	= cu in
	$\times$ 1,000	= cu mm
	$\times$ 1	= mL
	$\times$ 10^{-6}	= cu m
cubic foot	$\times$ 28,316.85	= cu cm
	$\times$ 0.02832	= cu m
	$\times$ 0.03704	= cu yd
	$\times$ 6.2288	= gal (Imperial)
	$\times$ 7.4805	= gal (U.S.)
	$\times$ 28.32	= L
cubic inch	$\times$ 16.38700	= cu cm
cubic meter	$\times$ 0.000811	= Ac ft
	$\times$ 35.314667	= cu ft
	$\times$ 1.307950	= cu yd
	$\times$ 219.9694	= gal (Imperial)
	$\times$ 264.172	= gal (U.S.)
	$\times$ 1,000	= L

cubic yard	× 46,656	= cu in
	× 0.7646	= cu cm
	× 764.6	= L
gallon (Imperial)	× 4,546.087	= cu cm
	× 277.42	= cu in
	× 1.20095	= gal (U.S.)
	× 4.546	= L
gallon (U.S.)	× 3,785.412	= cu cm
	× 231	= cu in
	× 3.785	= L
liter	× 0.219975	= gal (Imperial)
	× 0.264179	= gal (U.S.)

MASS

gram	× 0.002204	= lb (avdp)
	× 10^3	= mg
	× 10^{-3}	= kg
kilogram	× 35.273962	= oz (avdp)
	× 2.205	= lb (avdp)
pound mass (avdp)	× 453.592	= g
	× 0.453592	= kg
ton (avdp), short	× 2,000	= lb (avdp)
	× 907.2	= kg
	× 0.9072	= tonne
tonne	× 1,000	= kg
	× 2,204.622	= lb (avdp)
	× 1.1023	= short tons (2,000 lb avdp)
	× 0.984206	= long ton

FLOWRATE

acre foot per day	× 0.50417	= cu ft per sec (cfs)
	× 1,233.5	= cu m per day
	× 0.014274	= cu m per sec
	× 188.4	= gal per min (Imp. gpm)
	× 226.3	= gal per min (U.S. gpm)
	× 0.3258	= million gal per day (U.S. mgd)
cubic foot per second	× 28.31605	= L per sec
	× 0.028317	= cu m per sec
cubic meter per second	× 35.315	= cu ft per sec (cfs)
	× 19.006	= million gal per day (Imp. mgd)
	× 22.825	= million gal per day (U.S. mgd)

cubic meter per minute	× 3.666	= gal per sec (Imp.)
	× 4.403	= gal per sec (U.S.)
gallon per day (Imp.)	× 0.05262	= cu cm per sec
gallon per day (U.S.)	× 0.04381	= cu cm per sec
liter per minute	× 0.035316	= cu ft per min
	× 0.000588	= cu ft per sec (cfs)
	× 0.219969	= gal per min (Imp. gpm)
	× 0.264178	= gal per min (U.S. gpm)
liter per second	× 13.1985	= gal per min (Imp. gpm)
	× 15.850	= gal per min (U.S. gpm)
	× 0.01905	= million gal per day (Imp. mgd)
	× 0.022825	= million gal per day (U.S. mgd)

CONCENTRATION

milligram per liter	× 1.0	= parts per million (ppm)
in water	× 0.001	= g per L
(specific gravity = 1.0)	× 0.058416	= grains per gal (U.S.)
	× 0.070155	= grains per gal (Imp.)
	× 1.0	= g per m^3
	× 8.34	= lb per million gal
milligram per kilogram	× 0.002	= lb per ton (short)

HYDRAULIC LOADING RATE

| cubic meter per square | × 24.54 | = gal (U.S.) per ft^2 per d |
| meter per day | × 20.43 | = gal (Imp.) per ft^2 per d |

DENSITY

gram per cubic centimeter	× 0.036127	= lb per cu in
	× 62.427961	= lb per cu ft
kilogram per cubic meter	× 1.68556	= lb per cu yd
	× 0.062428	= lb per cu ft

PRESSURE

atmosphere	× 76	= cm of Hg
	× 1,033.260	= cm of H$_2$O (4°C)
	× 33.8995	= ft of H$_2$O (32°F)
	× 29.921	= in of Hg (32°F)
	× 101.325	= kPa
	× 14.696	= lb per sq in (psi)
inch of mercury (32°F)	× 3,386.38	= Pa
(60°F)	× 3,376.85	= Pa

kilonewton per square meter	× 1	= kPa
	× 0.145	= lb_f per sq in
	× 20.884	= lb_f per sq ft
millimeter of mercury (0°C)	× 133.322	= Pa
(15°C)	× 0.535	= in of H_2O
millimeter of water	× 9.80638	= Pa
pound (force)*	× 4.448222	= N
pound force per square foot	× 0.047880	= kilonewton per sq m (kPa)
	× 47.880	= Pa

ENERGY

BTU	× 252	= cal
	× 1.055	= kJ
	× 0.2930	= Wh
BTU per second	× 1.414	= horsepower (hp)
	× 1.055	= kW
calorie	× 4.1855	= J
	× 4.1855	= Ws
foot pound force per second	× 1.356	= W = J per sec
horsepower	× 42.42	= BTU per min
	× 550	= ft lb_f per sec
	× 746	= W
	× 746	= J per sec
kilojoule	× 0.9478	= BTU
	× 238.8	= cal
	× 1	= kWs
megawatt-hour	× 3.6	= GJ
watt	× 3.412	= BTU per hr
	× 1	= J per sec

    * It is an inconvenient aspect of the AES and British systems that they use the same word, "pound," as a unit for both mass and force. (In contrast, the SI system represents mass in kilograms and force in newtons.) The pound mass is a measure of the amount of material contained in an object, i.e., the quantity of atomic and subatomic particles in the object, taken in total. This value is invariant. The pound force, on the other hand, is a measure of weight—the force that a mass exerts as a result of the force exerted upon it by a gravitational field. Thus, a one pound force is the force necessary to impart an acceleration of 32.2 ft/s² to one pound of mass. As a result of this relationship, the gravitational "constant" g is itself a mass-force conversion factor, i.e.,

$$1\ lb_m \times g = 1\ lb_f$$

$$1\ lb_m \times 32.174\ \frac{ft^2}{s^2} = 32.174\ lb_f\ \frac{ft}{s^2}$$

TEMPERATURE

°C
$$°C + 273.15 = K$$
$$°C \times (1.8) + 32 = °F$$
$$°C \times (1.8) + 491.58 = °R$$

°F
$$°F + 459.48 = °R$$
$$(°F - 32) \times 5/9 = °C$$
$$(°F - 32) \times 5/9 + 273.15 = K$$

K
$$K - 273.15 = °C$$
$$(K - 273.15) \times 1.8 + 32 = °F$$

°R
$$°R - 459.48 = °F$$
$$°R - 491.58 \times 5/9 = °C$$
$$(°R - 491.58) \times 5/9 + 273.15 = K$$

VISCOSITY, DYNAMIC (20°C)

centipoise	× 0.001	= N s per sq m = Pa s
newton-second	× 1000	= centipoise
per square meter	× 10	= poise
poise	× 0.1	= N s per sq m = Pa s
pound force-second per	× 47,880	= centipoise
square foot	× 47.880	= Pa s

VISCOSITY, KINEMATIC (20°C)

centistoke	× 1	= sq mm per sec
	× 0.038750	= sq ft per hr
square foot per hour	× 25.807	= centistokes
	× 0.09290	= sq m per hr
square meter per hour	× 10.764	= sq ft per hr

APPENDIX A–4: FREQUENTLY USED SYMBOLS AND UNITS IN ENVIRONMENTAL ENGINEERING

Parameter	SI Unit	SI Symbol	AES/British Unit	AES/British Symbol
Basic Units				
Mass	gram	g	ounce	oz
	kilogram	kg	pound-mass	lb_m
	tonne	t	ton	ton
Length	millimeter	mm	inch	in
	meter	m	foot	ft
	kilometer	km	mile	mi
Force	newton	N	pound-force	lb_f
Power	watt	W	British thermal unit	BTU
Pressure	pascal	Pa	pound per sq inch	psi
	kilopascal	kPa	pound per sq foot	psf
	megapascal	MPa		
Water Resources				
Aquifer thickness	meter	m	foot	ft
Catchment area	sq meter	m^2	square foot	sq ft
	hectare	ha	acres	Ac
	square kilometer	km^2	square miles	sq mi
Intake depth	meter	m	foot	ft
diameter	millimeter	mm	inch	in
velocity	meter/second	m/s	foot per second	fps
well volume	cubic meter	m^3	cubic foot	cu ft
Precipitation, rain	millimeter	mm	inch	in
snow	centimeter	cm	inch	in
rate	millimeter per day	mm/d	inch per day	in/day
River flow, velocity rate	meter per second	m/s	foot per second	ft/sec (fps)
	cubic meter per second	m^3/s	cubic foot per second	cu ft/sec (cfs)
Water stage	meter	m	foot	ft

Well drawdown	meter	m	foot	ft
level	meter	m	foot	ft
radius	millimeter	mm	inch	in

Water Treatment

Plant capacity	cubic meter per day	m³/d	gallon per day	gpd
	megalites per day	ML/d	million gallon per day	mgd
Raw water flow	cubic meter per second	m³/s	gallon/minute	gpm
	liter per second	L/s	cubic foot/minute	cu ft/min
Raw water temperature	degree Celsius	°C	degree Fahrenheit	°F
Raw water turbidity	turbidity units	TU	turbidity units	TU
Chemical dosage	milligram per liter	mg/L	part per million	ppm
Chemical feed rate	kilogram per day	kg/d	pound (mass) per day	lb_m/day
	liter per hour	L/h	gallon per hour	gph
Gas feeder				
supply pressure	kilopascals	k Pa	pounds per square inch	psi
differential pressure	kilopascals	k Pa	inches of water	in H_2O
vacuum pressure	kilopascals	k Pa	inches of mercury	in Hg
Displacement velocity	meter per second	m/s	foot per second	fps
Detention time	hour	h	hour	hr
Weir overflow rate	liter per meter per second	L/m · s	gallon/foot/day	gpd/ft
Filter head loss	meter	m	foot	ft
Filtration rate	meter per hour	m/h	gallon per minute per square foot	gpm/sq ft
	cubic meter per square meter per hour	m³/m² · h	gallon per minute per square foot	gpm/sq ft
Filter backwash rate	meter per hour	m/h	gallon per minute per square foot	gpm/sq ft
	cubic meter per square meter per hour	m³/m² · h	gallon per minute per square foot	gpm/sq ft

Water Distribution

Area	square kilometer	km²	square mile	sq mi
Elevation	meter	m	foot	ft
Head loss	meter	m	foot	ft

	SI		AES/British	
	Unit	Symbol	Unit	Symbol
Hydrant spacing	meter	m	foot	ft
Hydraulic gradient	meter per kilometer	m/km	foot per thousand feet	ft/1,000 ft
Level gauging	meter	m	foot	ft
Location of mains	meter	m	foot	ft
Pipe cross section	square millimeter	mm²	square inch	sq in
Diameter	millimeter (meter if large)	mm	inch	in
Flow velocity	meter per second	m/s	foot per second	fps
Length	meter	m	foot	ft
Pressure	kilopascal	k Pa	pound per square inch	psi
Pump capacity	cubic meter per second	m³/s	gallon per minute	gpm
	liter per second	L/s	cubic foot per minute	cfm
Residual chlorine	milligram per liter	mg/L	part per million	ppm
Storage				
elevation	meter	m	foot	ft
volume	cubic meter	m³	gallons	gal
volume	cubic meter	m³	cubic foot	cu ft
Water consumption	liter	L	gallon	gal
	cubic meter	m³	cubic foot	cu ft

Wastewater Treatment

	SI		AES/British	
Plant capacity	(see Water Treatment)			
Sewage flow per capita	liter per capita per day	Lpcd	gallons per capita per day	gpcd
Sewage velocity	meter per second	m/s	foot per second	fps
Concentration	milligram/liter	mg/L	part per million	ppm
Screening, and grit, volumes	cubic meter per cubic megameter	m³/Mm³	cubic foot per million gallon	cu ft/mil gal
Surface overflow rate	cubic meter per square meter per day	m³/m²-d	gallon per day per square foot	gpd/sq ft

Parameter	SI description	SI symbol	US description	US symbol
Air supply	cubic meter per meter per minute	m³/m-min	cubic foot per minute per foot	cfm/ft
Volume of sludge	cubic meter per cubic megameter	m³/Mm³	gallon per million gallon	gal/mil gal
Weight of dry sludge solids	gram per cubic meter	g/m³	pound (mass) per million gallon	lb$_m$/mil gal
Weir overflow rate	(see Water Treatment)			
Volumetric organic load (suspended growth system)	kilogram BOD per cubic meter per day	kg BOD/m³ · d	pound BOD per day per 1,000 cubic foot	lb BOD/day/1,000 cu ft
F/M ratio	kilogram BOD per day per kilogram MLVSS	kg BOD/d/ kg MLVSS	lb BOD per day per lb of MLVSS	lb BOD/day/lb MLVSS
Air requirements	cubic meter per kilogram BOD	m³/kg BOD	cubic foot per pound BOD	cu ft/lb BOD
Oxygen requirements	kilogram O₂ per kilogram BOD	kg O₂/kg BOD	pound O₂ per pound BOD	lb O₂/lb BOD
Air flowrate	cubic meter per second	m³/s	cubic foot per minute	cfm
Oxygen transfer rate	gram O₂ per cubic meter per hour	g O₂/m³ · h	lb O₂ per hour per 1,000 cubic feet	lb O₂/hr/1,000 cu ft
Diffused air mixing	cubic meter per 1000 cubic meter per minute	m³/1000 m³ · min	cubic foot per minute per 1,000 cubic feet	cfm/1,000 cu ft
Mechanical mixing	kilowatt per 1000 cubic meter	kW/1000 m³	horsepower per 1,000 cubic feet	hp/1,000 cu ft
Surface organic load (land-based treatment)	Kilogram BOD per hectare per day	kg BOD/ha · d	lb BOD per acre per day	lb BOD/Ac/day
Chlorine feed rate	kilogram per day	kg/d	pound per day	lb/day
Sludge Processing				
Digester loading (volatile solids)	kg per cubic meter per day	kg/m³ · d	pound per day per 1000 cubic foot	lb/day/1000 cu ft
Digester gas production (from volatile solids)	cubic meter per kilogram	m³/kg	cubic foot per pound	cu ft/lb
Filter yield (dry solids)	kilogram per square meter per hour	kg/m² · h	pound per square foot per hour	lb/sq ft/hr

	SI		AES/British	
	Unit	Symbol	Unit	Symbol
Hydraulic loading	cubic meter per square meter per hour	$m^3/m^2 \cdot h$	gallons per minute per square foot	gpm/sq ft
Sludge thickening (solids loading)	kilogram per square meter per day	$kg/m^2 \cdot d$	pound per square foot per day	lb/sq ft/day
Air Pollution				
Dustfall	gram per square meter kilogram per square kilometer	g/m^2 kg/km^2	gram per square foot pound per acre	g/sq ft lb/Ac
Airborne particulate matter	gram per cubic meter (various submultiples of grams permissible)	g/m^3	pound per 1,000 cubic feet	lb/1,000 cu ft
Atmospheric gases	microgram per cubic meter	$\mu g/m^3$	part per million by volume @ 20°C part per million by weight	ppm by volume ppm by weight
Particle count	number per cubic centimeter	n/cm^3	number per cubic foot	n/cu ft
Atmospheric pressure	millibar millimeter of mercury	mbar mm Hg	atmosphere inches of mercury	atm in Hg

APPENDIX B

Physical Properties and Constants

APPENDIX B–1: FREQUENTLY USED ATOMIC WEIGHTS

Element	Symbol	Atomic weight	Element	Symbol	Atomic weight
Aluminum	Al	26.98	Magnesium	Mg	24.31
Antimony	Sb	121.75	Manganese	Mn	54.94
Argon	Ar	39.95	Mercury	Hg	200.59
Arsenic	As	74.92	Molybdenum	Mo	95.94
Barium	Ba	137.34	Nickel	Ni	58.71
Beryllium	Be	9.01	Nitrogen	N	14.01
Bismuth	Bi	208.98	Oxygen	O	16.00
Boron	B	10.81	Phosphorus	P	30.97
Bromine	Br	79.90	Platinum	Pt	195.09
Cadmium	Cd	112.40	Potassium	K	39.10
Calcium	Ca	40.08	Selenium	Se	78.96
Carbon	C	12.01	Silicon	Si	28.09
Chlorine	Cl	35.45	Silver	Ag	107.87
Chromium	Cr	52.00	Sodium	Na	22.99
Cobalt	Co	58.93	Strontium	Sr	87.62
Copper	Cu	63.55	Sulfur	S	32.06
Fluorine	F	19.00	Tantalum	Ta	189.95
Gold	Au	196.97	Tin	Sn	118.69
Helium	He	4.00	Titanium	Ti	47.90
Hydrogen	H	1.01	Tungsten	W	183.85
Iodine	I	126.90	Uranium	U	238.03
Iron	Fe	55.85	Vanadium	V	50.94
Lead	Pb	207.19	Zinc	Zn	65.37
Lithium	Li	6.94	Zirconium	Zr	91.22

APPENDIX B–2: APPROXIMATE PHYSICAL PROPERTIES OF WATER AT ATMOSPHERIC PRESSURE

Temperature	Density, ρ	Specific weight, γ	Dynamic viscosity, μ	Kinematic viscosity, ν	Vapor pressure
	kg/m³	N/m³	N · s/m²	m²/s	N/m²
0°C	1,000	9,810	1.79×10^{-3}	1.79×10^{-6}	611
5°C	1,000	9,810	1.51×10^{-3}	1.51×10^{-6}	872
10°C	1,000	9,810	1.31×10^{-3}	1.31×10^{-6}	1,230
15°C	999	9,800	1.14×10^{-3}	1.14×10^{-6}	1,700
20°C	998	9,790	1.00×10^{-3}	1.00×10^{-6}	2,340
25°C	997	9,781	8.91×10^{-4}	8.94×10^{-7}	3,170
30°C	996	9,771	7.96×10^{-4}	7.99×10^{-7}	4,250
35°C	994	9,751	7.20×10^{-4}	7.24×10^{-7}	5,630
40°C	992	9,732	6.53×10^{-4}	6.58×10^{-7}	7,380
50°C	988	9,693	5.47×10^{-4}	5.54×10^{-7}	12,300
60°C	983	9,643	4.66×10^{-4}	4.74×10^{-7}	20,000
70°C	978	9,594	4.04×10^{-4}	4.13×10^{-7}	31,200
80°C	972	9,535	3.54×10^{-4}	3.64×10^{-7}	47,400
90°C	965	9,467	3.15×10^{-4}	3.26×10^{-7}	70,100
100°C	958	9,398	2.82×10^{-4}	2.94×10^{-7}	101,300
	lb$_m$/ft³	lb$_f$/ft³	lb$_f$-sec/ft²	ft²/sec	psia
40°F	1.94	62.43	3.23×10^{-5}	1.66×10^{-5}	0.122
50°F	1.94	62.40	2.73×10^{-5}	1.41×10^{-5}	0.178
60°F	1.94	62.37	2.36×10^{-5}	1.22×10^{-5}	0.256
70°F	1.94	62.30	2.05×10^{-5}	1.06×10^{-5}	0.363
80°F	1.93	62.22	1.80×10^{-5}	0.930×10^{-5}	0.506
100°F	1.93	62.00	1.42×10^{-5}	0.739×10^{-5}	0.949
120°F	1.92	61.72	1.17×10^{-5}	0.609×10^{-5}	1.69
140°F	1.91	61.38	0.981×10^{-5}	0.514×10^{-5}	2.89
160°F	1.90	61.00	0.838×10^{-5}	0.442×10^{-5}	4.74
180°F	1.88	60.58	0.726×10^{-5}	0.385×10^{-5}	7.51
200°F	1.87	60.12	0.637×10^{-5}	0.341×10^{-5}	11.53
212°F	1.86	59.83	0.593×10^{-5}	0.319×10^{-5}	14.70

Source: Roberson, J. A., and Crowe, C. T., *Engineering Fluid Mechanics*, 2d ed., Boston: Houghton Mifflin, 1980; data from Bolz, R. E., and Tuve, G. L., *Handbook of Tables for Applied Engineering Science*, Cleveland: CRC Press, 1973

APPENDIX B–3: MECHANICAL PROPERTIES OF AIR AT STANDARD ATMOSPHERIC PRESSURE

Temperature	Density, ρ	Specific weight, γ	Dynamic viscosity, μ	Kinematic viscosity, ν
	kg/m^3	N/m^3	$N \cdot s/m^2$	m^2/s
$-20°C$	1.40	13.7	1.61×10^{-5}	1.16×10^{-5}
$-10°C$	1.34	13.2	1.67×10^{-5}	1.24×10^{-5}
$0°C$	1.29	12.7	1.72×10^{-5}	1.33×10^{-5}
$10°C$	1.25	12.2	1.76×10^{-5}	1.41×10^{-5}
$20°C$	1.20	11.8	1.81×10^{-5}	1.51×10^{-5}
$30°C$	1.17	11.4	1.86×10^{-5}	1.60×10^{-5}
$40°C$	1.13	11.1	1.91×10^{-5}	1.69×10^{-5}
$50°C$	1.09	10.7	1.95×10^{-5}	1.79×10^{-5}
$60°C$	1.06	10.4	2.00×10^{-5}	1.89×10^{-5}
$70°C$	1.03	10.1	2.04×10^{-5}	1.99×10^{-5}
$80°C$	1.00	9.81	2.09×10^{-5}	2.09×10^{-5}
$90°C$	0.97	9.54	2.13×10^{-5}	2.19×10^{-5}
$100°C$	0.95	9.28	2.17×10^{-5}	2.29×10^{-5}
$120°C$	0.90	8.82	2.26×10^{-5}	2.51×10^{-5}
$140°C$	0.85	8.38	2.34×10^{-5}	2.74×10^{-5}
$160°C$	0.81	7.99	2.42×10^{-5}	2.97×10^{-5}
$180°C$	0.78	7.65	2.50×10^{-5}	3.20×10^{-5}
$200°C$	0.75	7.32	2.57×10^{-5}	3.44×10^{-5}
	lb_m/ft^3	lb_f/ft^3	$lb_f\text{-}sec/ft^2$	ft^2/sec
$0°F$	0.00269	0.0866	3.39×10^{-7}	1.26×10^{-4}
$20°F$	0.00257	0.0828	3.51×10^{-7}	1.37×10^{-4}
$40°F$	0.00247	0.0794	3.63×10^{-7}	1.47×10^{-4}
$60°F$	0.00237	0.0764	3.74×10^{-7}	1.58×10^{-4}
$80°F$	0.00228	0.0735	3.85×10^{-7}	1.69×10^{-4}
$100°F$	0.00220	0.0709	3.96×10^{-7}	1.80×10^{-4}
$120°F$	0.00213	0.0685	2.47×10^{-7}	1.16×10^{-4}
$150°F$	0.00202	0.0651	4.23×10^{-7}	2.09×10^{-4}
$200°F$	0.00187	0.0601	4.48×10^{-7}	2.40×10^{-4}
$300°F$	0.00162	0.0522	4.96×10^{-7}	3.05×10^{-4}
$400°F$	0.00143	0.0462	5.40×10^{-7}	3.77×10^{-4}

Source: Roberson, J. A., and Crowe, C. T., *Engineering Fluid Mechanics*, 2d ed., Boston: Houghton Mifflin, 1980; data from Bolz, R. E., and Tuve, G. L., *Handbook of Tables for Applied Engineering Science*, Cleveland: CRC Press, 1973

APPENDIX B-4: THE U.S. STANDARD ATMOSPHERE

			AES Units		
Altitude, ft	Temperature, °F	Absolute pressure, psia	Specific weight, γ, lb_f/ft^3	Density, ρ, lb_m/ft^3	Dynamic viscosity $\times 10^7$, μ, $lb_f \cdot s/ft^2$
0	59.00	14.696	0.07647	0.002377	3.737
5,000	41.17	12.243	0.06587	0.002048	3.637
10,000	23.36	10.108	0.05643	0.001756	3.534
15,000	5.55	8.297	0.04807	0.001496	3.430
20,000	−12.26	6.759	0.04069	0.001267	3.325
25,000	−30.05	5.461	0.03418	0.001066	3.217
30,000	−47.83	4.373	0.02857	0.000891	3.107
35,000	−65.61	3.468	0.02367	0.000738	2.995
40,000	−69.70	2.730	0.01882	0.000587	2.969
45,000	−69.70	2.149	0.01481	0.000462	2.969
50,000	−69.70	1.690	0.01165	0.000364	2.969
55,000	−69.70	1.331	0.00917	0.000287	2.969
60,000	−69.70	1.049	0.00722	0.000226	2.969
65,000	−69.70	0.826	0.00568	0.000178	2.969
70,000	−67.42	0.651	0.00445	0.000139	2.984
75,000	−64.70	0.514	0.00349	0.000109	3.001
80,000	−61.98	0.404	0.00263	0.000086	3.018
85,000	−59.26	0.322	0.00215	0.000067	3.035
90,000	−56.54	0.255	0.00170	0.000053	3.052
95,000	−53.82	0.203	0.00134	0.000042	3.070
100,000	−51.10	0.162	0.00106	0.000033	3.087

			SI Units		
Altitude, km	Temperature, °C	Absolute pressure, kPa	Specific weight, γ, N/m^3	Density, ρ, kg/m^3	Dynamic viscosity, $\times 10^5$, μ $N \cdot s/m^2$
0	15.00	101.33	12.01	1.225	1.789
2	2.00	79.50	9.86	1.007	1.726
4	−4.49	70.12	8.02	0.909	1.661
6	−23.96	47.22	6.46	0.660	1.595
8	−36.94	35.65	5.14	0.526	1.527
10	−49.90	26.50	4.04	0.414	1.458
12	−56.50	19.40	3.05	0.312	1.422
14	−56.50	14.17	2.22	0.228	1.422
16	−56.50	10.35	1.62	0.166	1.422
18	−56.50	7.57	1.19	0.122	1.422
20	−56.50	5.53	0.87	0.089	1.422
22	−54.58	4.05	0.63	0.065	1.432
24	−52.59	2.97	0.46	0.047	1.443
26	−50.61	2.19	0.33	0.034	1.454
28	−48.62	1.62	0.24	0.025	1.465
30	−46.64	1.20	0.18	0.018	1.475

Source: Vennard, J. K., and Street, R. L., *Elementary Fluid Mechanics*, 6th ed., New York: John Wiley & Sons, 1982; data from *U.S. Standard Atmosphere, 1962*, U.S. Government Printing Office, 1962. Data agree with ICAO standard atmosphere to 20 km and with ICAO proposed extension to 30 km

APPENDIX B–5: KINEMATIC VISCOSITIES OF VARIOUS FLUIDS

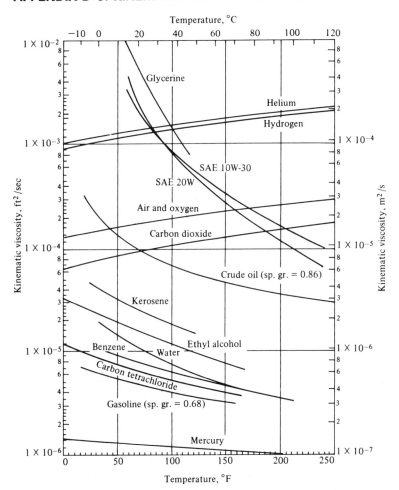

Source: Roberson, J. A., and Crowe, C. T., *Engineering Fluid Mechanics*, 2d ed., Boston: Houghton Mifflin, 1980; adapted from V. L. Streeter, *Fluid Mechanics*, 5th ed., New York: McGraw-Hill, 1971

APPENDIX B–6: HEAT VALUE (ENERGY CONTENT) OF VARIOUS FUELS IN SI AND AES/BRITISH UNITS

SI UNITS

Energy form	*Heat value (Terajoules)*
Petroleum	
(per thousand cubic meters)	
Crude oil	38.512
Liquefied petroleum gases	27.177
Motor gasoline	34.656
Aviation gasoline	33.518
Aviation turbo fuel	35.934
Kerosene	37.676
Diesel and light fuel oil	38.675
Heavy fuel oil and still gas	41.727
Petroleum coke	42.376
Natural gas	
(per million cubic meters)	37.229
Coal (per thousand tonnes)	
Anthracite	29.527
Imported bituminous	29.993
Canadian bituminous	29.295
Subbituminous	19.763
Lignite	15.345
Coke (per thousand tonnes)	28.830
Coke oven gas (per million cubic meters)	18.614
Electricity (per gigawatt hour)	
Primary energy calculations	10.500
Secondary energy calculations	3.600
(see Chapter 3, Table 3–3)	
Propane (per thousand cubic meters)	25.600
Ethanol (per thousand cubic meters)	23.200
Methanol (per thousand cubic meters)	18.100
Wood (oven-dried per thousand tonnes)	20.000
Wood (50% moisture per thousand tonnes)	10.000
Municipal solid waste (per thousand tonnes)	11.400

AES/BRITISH UNITS

Energy form	*Heat value (Million BTU)*
Petroleum (per barrel)	
Crude oil	5.8030
Liquefied petroleum gases	4.0950
Motor gasoline	5.2220
Aviation gasoline	5.0505
Aviation turbo fuel	5.4145
Kerosene	5.6770
Diesel & light fuel oil	5.8275
Heavy fuel oil & still gas	6.2874
Petroleum coke	6.3852
Natural gas (per thousand cubic feet)	1.0000
Coal (per short ton)	
Anthracite	25.4000
Imported bituminous	25.8000
Canadian bituminous	25.2000
Subbituminous	17.0000
Lignite	13.2000
Coke (per short ton)	24.8000
Coke oven gas (per thousand cubic feet)	0.5000
Electricity (per megawatt hour)	3.4120 (use)
	10.000 (output)

CONVERSION

Energy form	SI to AES/British	AES/British to SI
Crude oil	1 cubic meter (m³) (at 15°C) = 6.293 barrels (at 60°F) 1 barrel = 35 imperial gallons or 42 U.S. gallons 1 imperial gallon = 4.546 liters 1 metric tonne of crude oil = 7.428 barrels (crude Canadian average)	1 barrel (at 60°F) = 0.15891 cubic meters (m³) (at 15°C)
Natural gas	1 cubic meter (m³) (at 101.325 kPa and 15°C) = 35.301 cubic feet (at 14.73 psia and 60°F)	1 thousand cubic feet (at 14.73 psia and 60°F) = 28.32784 cubic meters (m³) (at 101.325 kPa and 15°C)
Coal	1 tonne (t) = 1.102 short tons	1 short ton = 0.907185 tonne (t)
Oil, gas, coal	1 joule (J) = 0.000948 BTU	1 BTU = 1.054615 kilojoules (kJ)

APPENDIX C

Abbreviations and Symbols

A	area, also van't Hoff Arrhenius coefficient
a	acceleration
BOD	biochemical oxygen demand
$\mathscr{C}$	fraction of tracer present (pulse input)
BTU	British thermal unit
C	concentration, also coefficient for fire flow
C_D	drag coefficient
COD	chemical oxygen demand
CFC_s	chlorofluorocarbons
CSTR	continuously stirred tank reactor
d	day, also distance and diameter
D	depth, also distance and diameter
DO	dissolved oxygen
DWF	dry weather flow
E_a	Arrhenius activity energy
E_c	electric field strength
EIA	environmental impact assessment
EIS	environmental impact statement
ESP	electrostatic precipitator
F	fraction of tracer present (continuous input),
F	force, fire flow, assessment factor and plume buoyancy factor
F/M	food to microorganism ratio

g	acceleration due to gravity, also gram
gpcd	gallons per capita per day
G	generation time, also normalized Gaussian curve
GSB	Gesellschaft zur Beseitigung von Sondermüll
GNP	Gross National Product
H	depth, also height of plume and distance
[H],[H$_3$O]	hydrogen ion concentration
HWTF	hazardous waste treatment facility
K	equilibrium constant, also coefficient of hydraulic conductivity
K_a, K_b, K_w	dissociation constants for acids, bases, and water, respectively
k	reaction rate constant
K_H	Henry's constant
L	liter, also dimensions of length and BOD remaining
LC50	lethal concentration causing 50% mortality
LD50	lethal dose causing 50% mortality
LDR	less developed regions
ln	natural logarithm
mgd	million gallons per day
mg/L	milligrams per liter
M	molarity, also mass
MDR	more developed regions
m	molality, also mass
mL	milliliter
MLSS	mixed liquor suspended solids
MLVSS	mixed liquor volatile suspended solids
MPN	most probable number
n	number of moles, also soil porosity and valency
N	number of moles of gas
N	normality
NO$_x$	nitrogen oxides
NTU	nephelometric turbidity unit
[OH]	hydroxyl ion concentration
p	pressure
pH	negative logarithm of the hydrogen ion concentration
ppb	parts per billion
ppm	parts per million
P, PE	population, and population equivalent
PCT	physical chemical treatment
PFTR	plug flow tubular reactor
PNA, PAH	polynuclear aromatic hydrocarbons
Q	volumetric flowrate, also source emission rate
q_p	charge on a particle
q	volumetric flow rate, also substrate conversion rate
R	universal gas constant

r	radius of curvature, also rate of reaction
R_e	Reynold's number
RBC	rotating biological contactor
RO	reverse osmosis
RDF	refuse derived fuel
S	hydraulic gradient, also substrate concentration and stability parameter
SS	suspended solids
SO_x	sulfur oxides
$\bar{t}$	detention time
t	time, also tonne
T	temperature, also time and theoretical detention time
TOC	total organic carbon
u, v	velocity
V	volume
VSS	volatile suspended solids
W	exit velocity of the plume
X	mole fraction, also mass of MLSS or MLVSS
$[X]$	concentration of MLSS or MLVSS
Y	yield
Z	distance above ground

Not included in the above List of Symbols are the special symbols used in Chapter 7 which are listed at the end of that chapter, or the Greek symbols used elsewhere which are listed below.

SYMBOLS USED

γ	actual lapse rate
Γ	adiabatic lapse rate
ΔH	difference in elevation
η	fractional removal
θ	temperature coefficient
Θ_c	mean cell residence time (sludge age)
μ	viscosity, also specific growth rate
ρ	density
σ	dispersion coefficient
ω	rotational speed

THE GREEK ALPHABET

A	α	alpha	=	a	N	ν	nu	=	n
B	β	bēta	=	b	Ξ	ξ	xi	=	x
Γ	γ	gamma	=	g	O	o	omīcron	=	o
Δ	δ	delta	=	d	Π	π	pi	=	p
E	ε	epsīlon	=	e	P	ρ	rho	=	rh, r
Z	ζ	zēta	=	z	Σ	σ ς	sigma	=	s
H	η	ēta	=	ē	T	τ	tau	=	t
Θ	θ ϑ	thēta	=	th	Υ	υ	upsīlon	=	ü
I	ι	iōta	=	i	Φ	φ	phi	=	ph
K	κ	kappa	=	k	X	χ	chi	=	kh
Λ	λ	lambda	=	l	Ψ	ψ	psi	=	ps
M	μ	mu	=	m	Ω	ω	ōmega	=	ō

Index

A

Abbreviations, 700–702
Abiotic, defined, 297
Absolute humidity, 232
Absorbed solar radiation, 221
Absorption processes for gas emission control, 511–12
Acceptable wastes for secure landfill, 622–23
Acceptance of natural hazards, 106
Acid(s)
  conjugate, 159, 160
  defined, 158
  examples of strong, 159
Acid-base reactions, 158–61
Acid dissociation constant, 159, 160, 161
Acidic deposition, 647–48
Acidic solution, 162
Acidification from coal usage, 74. *See also* Acid rain
Acidity, 166–69
Acid rain, 2, 124–39, 292, 476
  areas affected by, 130–31
  effects of, 129–37
    on aquatic systems, 129–31
    on groundwater, materials, and buildings, 135–37
    on terrestrial ecosystems, 131–35
  nature of problem, 124–25
  remedial and control measures, 137–39, 140
  sources and distribution of, 125–28

Activated carbon, 396–97
Acute hazardous wastes, 602–3
Adjustments to natural environmental hazards, 100–108
  classification of, 107–8
  industrial approach, 102–5
  postindustrial approach, 105–7
  preindustrial approach, 100–102
Admixtures, 624
Adsorption, 147
  processes for gas emission control, 512–13
Advection inversion, 525
Aedes aegypti, 286
Aerated lagoons, 439, 440
Aeration, 396
  extended aeration process, 445, 452
Aerobic/anoxic processes for wastewater treatment, 432–33
Aerobic bacteria, 259
Aerobic fungi, 265
Aeronomy, 213
Aerosol and particle load in atmosphere, 243
Aesthetic considerations in landfilling, 564
Agent Orange, 293
Age structure, 18, 19–20
Agnes (tropical storm, 1972), 95
Agrarian society waste cycle, 5
Agribusiness, 359, 366
Agricultural industries, 31